MOLECULAR BIOLOGY OF FIBRINOGEN AND FIBRIN

ANNALS OF THE NEW YORK ACADEMY OF SCIENCES

Volume 408

MOLECULAR BIOLOGY OF FIBRINOGEN AND FIBRIN

Edited by Michael W. Mosesson and Russell F. Doolittle

The New York Academy of Sciences
New York, New York
1983

Cover: The cover background shows a high magnification transmission electron micrograph of a fibrin clot. Many of the fibers display typical cross-striated banding with a characteristic periodicity of 220Å. A model of a single dimeric fibrinogen molecule is depicted in the inset and shows the approximate major dimensions (A), the arrangement of the three types of polypeptide chains (Aα, Bβ, γ), the disulfide bridges (-ss-), and the carbohydrate groups (○). Fibrinopeptides A and B are located at the NH_2-terminal ends of the Aα and Bβ chains, respectively, and are cleaved by thrombin from fibrinogen during the enzymatic conversion of fibrinogen to fibrin. As a result of this enzymatic process, the fibrin molecules polymerize to form a clot, such as is shown on the cover.

Library of Congress Cataloging in Publication Data
Main entry under title:
The Molecular biology of fibrinogen and fibrin.

(Annals of the New York Academy of Sciences; v. 408)
Papers presented at a conference held June 2–4, 1982
and sponsored by the New York Academy of Sciences.
Includes bibliographical references and index.
1. Fibrinogen—Congresses.   2. Fibrin—Congresses.
3. Molecular biology—Congresses.   I. Mosesson,
Michael W., 1934–   .   II. Doolittle, Russell F.
III. Series.   [DNLM: 1. Fibrinogen—Congresses.
2. Fibrin—Congresses. W1 AN626YL v. 408 /WH 310 M718
1982]

Q11.N5 vol. 408 [QP93.5]   500s [599'.0113]   83–13068
ISBN 0–89766–208–3
ISBN 0–89766–209–1 (pbk.)

PCP
Printed in the United States of America
ISBN 0–89766–208–3 (cloth)
ISBN 0–89766–209–1 (paper)

ANNALS OF THE NEW YORK ACADEMY OF SCIENCES

VOLUME 408

June 27, 1983

MOLECULAR BIOLOGY OF FIBRINOGEN AND FIBRIN *

Editors and Conference Organizers
Michael W. Mosesson and Russell F. Doolittle

———◆———

CONTENTS

* This series of papers is the result of a conference entitled Molecular Biology of Fibrinogen and Fibrin held June 2–4, 1982 by the New York Academy of Sciences.

Financial assistance was received from:

- AMERICAN HEART ASSOCIATION
- BAYER AG/CUTTER/MILES LABORATORIES
- GENERAL DIAGNOSTICS DIVISION OF WARNER-LAMBERT COMPANY
- NATIONAL HEART, LUNG AND BLOOD INSTITUTE—NATIONAL INSTITUTES OF HEALTH
- OFFICE OF NAVAL RESEARCH

OPENING REMARKS

Michael W. Mosesson

Department of Medicine
Mount Sinai Medical Center
Milwaukee, Wisconsin 53233

Russell F. Doolittle

Department of Chemistry
University of California–San Diego
La Jolla, California 92037

The idea to organize a conference about the molecular biology of fibrinogen and fibrin came from several considerations. First of all, it seemed most timely to propose convening a multidisciplinary scientific forum for summarizing and discussing the large body of knowledge that has accumulated on the subject. It was reasoned that such a forum would serve for (1) identifying more "developed" subject areas that had been intensely scrutinized over the years and about which there was now a general consensus; (2) outlining present areas of controversy or uncertainty; and (3) indicating the latest investigative thrusts that point the way to presently active and future areas of investigation.

Secondly, the envisioned conference could provide a productive setting that would serve to catalyze interactions among the participants, and these interactions would ultimately lead to a more coherent and comprehensive understanding of the biology of fibrinogen.

Thirdly, the published proceedings of such a conference would provide a comprehensive and cohesive written summary of knowledge on fibrinogen/fibrin in book form that could serve as a definitive source of information for all workers in this field.

In approaching these goals, this conference brings together active investigators representing a variety of scientific disciplines. We will discuss the structure and conformation of fibrinogen, its conversion to fibrin and its self-assembly into ordered forms, its interaction with other proteins and with cells, the biological activities of fibrinogen and fragments of fibrinogen, as well as aspects of the regulation of fibrinogen biosynthesis including genetic programming, synthesis, and secretion.

Several of the subjects reflect areas that have been completely, or nearly completely, elucidated (e.g., amino acid sequence, intramolecular covalent bridges), whereas others represent less "mature" subject areas of great interest and intense activity (e.g., fibrinogen biosynthesis). In any event, we hope that these topics and the formatting we selected for their presentation will serve to meet the goals outlined above, particularly in establishing a solid basis for approaching present and future areas of inquiry. Finally, we would particularly like to thank our conference subcommittee for its advice and support in planning this conference, and the NYAS staff for its aid in implementing the conference. We also are grateful to all of the conference speakers, discussants, and poster-presenters for their contributions and cooperation.

THE CONVERSION OF FIBRINOGEN TO FIBRIN: EVENTS AND RECOLLECTIONS FROM 1942 TO 1982 *

John D. Ferry

Department of Chemistry
University of Wisconsin
Madison, Wisconsin 53706

This account begins in 1942 because that is when fibrinogen first became available in ample quantities from large-scale plasma fractionation. It would be too much to say that this revolutionized the course of research on blood coagulation, since much of the best work has always been done by people who isolated and purified their own fibrinogen directly from whole blood, collected under their own controlled conditions. However, research was certainly stimulated by the ease of obtaining either human or bovine Fraction I from commercial or institutional sources. Many investigators used Fraction I from these sources either directly or as starting material for further purification, as one can see by examining the literature from that time forward.

It was also in 1942 that I was called back part time from another war assignment to the Harvard Medical School to examine possible practical applications of human Fraction I. This material was accumulating as a by-product of the plasma fractionation procedure developed by Edwin J. Cohn and his associates. The story of that wartime enterprise is well known. Plasma from the Red Cross was fractionated by use of ethanol at low temperatures, with precise specification of pH, ionic strength, and other conditions. The methods were worked out first at the Harvard Pilot Plant, with progressive modification and improvement, and were eventually employed at seven industrial firms to produce protein fractions on a commercial scale for use by the armed forces. The primary product was plasma albumin for treatment of shock. It was packaged as a sterile and stable 25% solution that could be injected immediately without the delay involved in reconstitution of dried plasma. By-products included Fraction I, which contained most of the fibrinogen; immunoglobulins; and others, including thrombin, in smaller amounts. Investigations of the components involved in blood clotting, including more highly purified fibrinogen obtained by subfractionation of Fraction I, antihemophilic factor, cold-insoluble globulin, and thrombin, were coordinated by John T. Edsall.

Peter R. Morrison and I were supplied with Fraction I, of which about 60% of the protein was clottable, in ample amounts for experiments to explore practical applications of this material. The first proposal was to use fibrin clots in the surface therapy of burns. We found that the adherence of a clot to a surface and its friability, toughness, and syneresis could be regulated by the conditions of its formation such as pH, ionic strength, and the presence of additives such as glycerol. Although the intended clinical use did not develop very far, we explored the effects of these variables on the physical properties of clots in detail, using mostly refractionated material with a higher

---

* After dinner speech.

0077–8923/83/0408–0001 $01.75/0 © 1983, NYAS

fibrinogen content than Fraction I. We measured opacity, elastic modulus, tensile strength, and rate of conversion to fibrin during clotting. There was a gradation of properties, from the fine, transparent clot for which we proposed a structure of fibrinogen units joined in chainlike fashion primarily end-to-end, to the coarse, opaque clot in which we pictured such chains as aggregated in large bundles.

Because of the easy syneresis of a coarse clot, it occurred to us that it could be compacted in one dimension to make a denser film with greater strength and stability. We turned out such films two feet long and eight inches wide, about a millimeter thick, with a fibrin content of about 30%. The saline solution remaining in the film could be replaced by glycerol to obtain a tough, translucent, rubbery sheet that could be stretched over 100% and re-covered most of its deformation when released from tension (FIG. 1).

We now had a product with clinical value. There were many head injuries in the War which left the brain exposed. No satisfactory material had been found to replace the dura membrane over the brain; rubber, other natural membranes, etc. caused subsequent growth of damaging adhesions. Fibrin film turned out to be a benign shield between the brain and adjacent tissue. It was gradually absorbed after surgery and replaced by a natural membrane of fibrous tissue with no adhesions.

Sterilization of this film for surgical use was a problem. Heating the film with its original content of saline solution produced the equivalent of a hard-boiled egg. The film could not be dried for heat treatment without irreversible change to a brittle solid. It was very difficult to attempt the whole preparation under aseptic conditions. We tried dipping the glycerol-plasticized film into hot glycerol, but this was messy and not suitable for large-scale production. Finally we found that water-plasticized film partially desiccated to a water content of 20% and treated with saturated steam at 121°C. gave an undamaged sterile product. Changes in several physical properties showed that some chemical cross-linking had taken place, and the absorption of this film *in vivo* took place more slowly, but it too was replaced by a natural membrane with no adhesions.

Fibrin film was also made as seamless tubing and used to join severed blood vessels in experimental animals, but this form was not used clinically.

At the same time, another clinical product, fibrin foam, was developed, primarily by Edgar Bering. A mixture of appropriate concentrations of fibrinogen and thrombin was agitated to a froth just before clotting, and subsequently dried from the frozen state. The spongy solid, whose properties could be varied by adjusting the conditions of preparation, was used as a hemostatic agent after it was soaked in a thrombin solution. It provided a surgical sponge that could be left in place and would gradually be absorbed with essentially no tissue reactions.

Although these products were manufactured by several industrial firms and widely used during the War, they are now obsolete. Human Fraction I was used for a while to treat certain afibrinogenemias; it was abandoned because of danger of transmission of hepatitis. (A subfraction containing antihemo-philic factor is still used, however.) Fibrin film has been replaced, mostly by poly(dimethyl siloxane), "silicone rubber." Fibrin foam has been replaced by similar spongy products prepared from gelatin.

At the time of these developments, there were a few physical chemical studies of fibrinogen and fibrin in the literature; E. Wöhlisch had published a

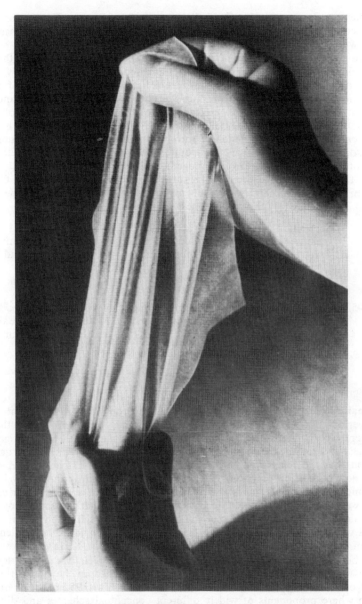

FIGURE 1. Fibrin film, plasticized with glycerol.

comprehensive review in 1940. Both Wöhlisch and R. Signer had concluded from flow birefringence that fibrinogen was an elongated molecule, and Wöhlisch had observed non-Newtonian viscosity, possibly due to aggregates of fibrinogen. Wide-angle x-ray scattering of fibrin had been studied by J. Katz, and electron microscopy by C. H. Wolpers and H. Ruska. Fibrin was often regarded as an insoluble form of fibrinogen and was therefore sometimes confused with precipitates of denatured fibrinogen.

We regarded the formation of fibrin as a polymerization rather than an insolubilization process, and we were impressed by the astonishingly small concentration of fibrin needed to form a coherent gel. The theory of three-dimensional polymerization to produce gel networks had just been formulated by P. J. Flory. However, we were ignorant of the chemistry of the polymerization, even as to whether the bonds involved were covalent or not. It had not yet been reported that the action of thrombin releases fibrinopeptides, nor that the properties of the fibrin clot are quite different depending on whether an additional enzyme besides thrombin together with calcium ion is present.

In the postwar years, there was a surge of research activity on the fibrinogen-fibrin conversion, and several major advances were made in a brief time period, as illustrated by these milestones:

1945	Separation of thrombin action from subsequent polymerization	K. Laki and W. F. H. M. Mommaerts
1947	Electron micrographs of fibrin, cross-striations	C. v. Z. Hawn and K. R. Porter
1948	Covalent bonds introduced by a new enzyme	K. Laki; L. Lorand
1950–52	Action of thrombin in releasing fibrino-peptides to form fibrin monomer	L. Lorand and W. R. Middlebrook; K. Bailey and F. R. Bettelheim
1950	Difference in isoelectric point between fibrinogen and fibrin monomer	E. Mihályi
1951	Similarity in hydrodynamic properties of fibrinogen and fibrin monomer	K. Laki

The names of several Hungarian scientists are prominent in the literature of this period—K. Laki, E. Mihályi, and L. Lorand, who worked together in A. Szent-Györgyi's institute in Budapest and then carried their investigations to Leeds and to Stockholm and eventually to the United States. Their demonstration that the release of fibrinopeptides by thrombin left an altered molecule (now called fibrin monomer), indistinguishable from fibrinogen by hydrodynamic measurements and stabe at pH 5 or in certain solvents such as 3M urea, was a striking advance. Moreover, they showed that the same molecular species could be obtained by dissolving a fibrin clot in such a solvent, provided that it was formed in the absence of calcium ion and/or an as yet unidentified "serum factor" now recognized as fibrinoligase (Factor XIIIa).

Other aspects of the blood coagulation process, especially the steps leading to the formation of thrombin, were intensively studied in the postwar years. Much of this work was reviewed annually from 1948 to 1952 at an interesting forum where proponents of widely different conceptual schemes and different vocabularies were brought together. This was the annual Conference on Blood Clotting and Allied Problems, sponsored by the Josiah Macy, Jr. Foundation and convened in New York each January. The chairman of these meetings was Irving S. Wright and the participants included K. M. Brinkhous, J. H. Ferguson, L. B. Jaques, K. P. Link, A. J. Quick, and W. H. Seegers. Experimental results of different protagonists that seemed to conflict were sometimes resolved when conditions were clearly defined, and progress was made toward clarifying the nomenclature, with a table of synonyms for the numerous factors that were identified only by their effects with as yet no chemical isolation. (One factor had six synonyms.) The fibrinogen-fibrin conversion was a minor part of the agenda of these conferences, but it was discussed at several of them by J. T. Edsall, K. Laki, D. F. Waugh, and myself.

I had moved to the University of Wisconsin and resumed work on following the fibrinogen-fibrin conversion by physical-chemical measurements. (Peter Morrison also came to Wisconsin, and we had joint research support for a while, but we pursued different lines of investigation.) My first student to undertake this problem was Sidney Shulman, who found that fibrin formation could be greatly retarded or prevented altogether by a wide variety of additives, especially dihydric alcohols. For example, a solution would remain unclotted for days but would clot quickly upon removal of the glycol by dialysis.

When Shulman in late 1949 ran such a solution in the ultracentrifuge, he obtained perhaps the most startling result ever observed in our laboratory (FIG. 2). There was one species with the sedimentation constant of fibrinogen, 8 S (extrapolated to zero concentration and reduced to water at 20°C.) and another with 23 S, which would be attributed to an intermediate polymer or oligomer. The proportion of monomer converted to polymer increased slowly with time in the presence of the inhibitor.

In 1945, Erwin Chargaff had written:

> The existence of a soluble precursor of fibrin, recently termed profibrin, which occupies a position intermediate between fibrinogen and fibrin, has been postulated by several investigators. It is quite conceivable that the chemical changes taking place along the fibrinogen molecule, which lead finally to its conversion to the insoluble fibrin, could, under proper circumstances, be arrested before the precipitation of fibrin takes place; but a decision regarding the existence and the properties of such an intermediary product will have to await further experimental work.

By 1951, the results of Laki, Lorand, and other identifying fibrin monomer, combined with ours, had shown the presence of two intermediates, as in the scheme

$$F \xrightarrow{T} f + P, \; nf \rightarrow f_n, \; f_n \rightarrow clot$$

where F is fibrinogen, T is thrombin, f is fibrin monomer; P symbolizes the fibrinopeptides, and f_n the intermediate polymer or polymers.

We pursued studies of the intermediate polymer, or polymers, for several years with Shulman, P. Ehrlich, S. Katz, K. Gutfreund, I. Tinoco, E. F. Casassa, and I. H. Billick. These oligomers could be identified also in solutions where clotting was inhibited by suitable concentrations of urea, lithium bromide, or sodium iodide; by adjusting the pH to 10.0, and, as first shown by Harold A. Scheraga and collaborators, even in solutions without inhibitor provided the thrombin concentration was low enough to allow for an ultracentrifuge run before clotting. Moreover, if a clot (unligated) was dissolved in 3.5 M urea to dissociate it to fibrin monomer and the urea concentration was subsequently lowered to 2.35 M, oligomers were formed. Thus the polymerization steps in the above scheme were reversible in suitable solvents.

There was no reason to think that the single oligomer peak in the ultracentrifuge meant a single oligomeric species, since if there was a distribution

FIGURE 2. Sedimentation diagram of clotting system inhibited by 1.0 M urea, pH 6.3, ionic strength 0.15, reaction time 25 hr.; photographs at 20 and 32 min., respectively, after attaining top speed (Ehrlich, Shulman, and Ferry (1952), reproduced with permission from the *Journal of the American Chemical Society*).

of rodlike oligomers with different lengths but the same cross-section area the sedimentation coefficient would be relatively insensitive to length, especially with boundary sharpening due to the Johnston-Ogston effect.

We could see that the oligomers were indeed rodlike, or at least very elongated in shape; this was evident from the changes in viscosity, dependence of viscosity on shear rate, and development of flow birefringence during the polymerization process. P. Ehrlich measured viscosities at different average shear rates by a tedious technique, following the rate of fall of barely visible glass spheres of the kind used in self-reflecting highway signs. For measurements of flow birefringence, Shulman and I carried our solutions to Ames, Iowa where Joseph F. Foster (who had measured flow birefringence of fibrinogen with John T. Edsall at Harvard) had the necessary apparatus. After the first experiment, Foster was obliged to modify his instrument completely to provide much lower velocity gradients for the intermediate polymers, but he was able to do this resourcefully in a few hours. It was clear from the dependence of extinction angle on velocity gradient that there was a wide range of polymer lengths, averaging roughly 4000Å. (Lorand had observed in 1948 flow birefringence during the dissolution of fibrin in concentrated urea, and Mihályi had reported in 1950 a large enhancement of viscosity in redissolved fibrin in certain ranges of urea concentration and pH; these results gave perhaps the earliest evidence of intermediate polymers.)

From light-scattering measurements with S. Katz and K. Gutfreund, we concluded that oligomers stabilized in hexamethylene glycol at pH 6.2 had a weight-average molecular weight corresponding to a degree of polymerization of about 15 and an average length corresponding to a mass/length ratio about twice that of fibrinogen. An assembly of two parallel chains of monomer units end-to-end with 15 to 20 units would give a sedimentation coefficient in agreement with that observed. This geometry led to the idea of assembly with staggered lateral overlapping, based partly on the teleological argument (never mentioned) that such a design was good biological engineering (FIG. 3).

In those light-scattering measurements the angular dependence was determined only down to 25° and the necessary extrapolation to zero angle was barely possible because of the extreme length of the polymer. Soon thereafter, E. F. Casassa joined us and made painstaking measurements at pH 9.5, both with and without hexamethylene glycol, where the degree of conversion to oligomers and the oligomer lengths attained are even greater than at lower pH. He showed that, although both the molecular weight and length were indeterminate, a different extrapolation method gave the mass/length ratio explicitly, and this was found to correspond to a polymer/monomer cross-section area ratio of 2.0 to 2.3. The staggered overlapping pattern has since been supported by electron micrographs and other evidence.

Our light-scattering studies also gave some information about the reversibility of the polymerization. When the protein concentration of a polymerized system in 0.5 M hexamethylene glycol was diluted from 4 to 0.1–0.2 g/l., the molecular weight and length reverted nearly to those of the fibrin monomer, within a period of an hour or two. Similarly, the flow birefringence at low velocity gradients diminished, indicating depolymerization. This dissociation of the oligomers was prevented by the presence of calcium ion. Although the nature of ligation was not yet understood, we knew that our fibrinogen contained the "serum factor" of Laki and Lorand and that its action was responsible for the difference.

Similar experiments were in progress in other laboratories. R. F. Steiner and Laki had, a year before our light-scattering results, studied the light scattering of a slowly polymerizing system with no inhibitor and concluded that the intermediate species had lengths several times that of fibrinogen but mass/length ratios only twice. Extensive studies on oligomers were made by H. A. Scheraga and his collaborators at Cornell, including measurements of flow double refraction, sedimentation, and light scattering. They focused attention on the reversibility of the polymerization and made many experiments starting

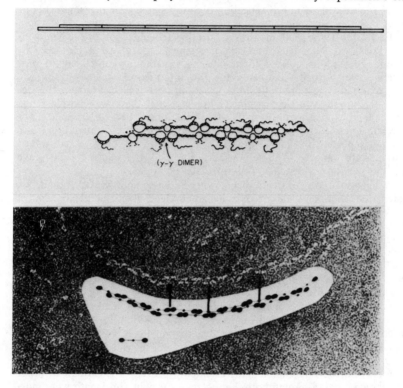

FIGURE 3. The half-staggered overlapping pattern for assembly of fibrin monomers. (a) Ferry, Katz, and Tinoco (1954), reproduced from the *Journal of Polymer Science* with permission; (b) Doolittle (1973), reproduced from *Advances in Protein Chemistry;* (c) electron micrograph by Fowler *et al.* (1981), reproduced from the *Proceedings of the National Academy of Sciences.*

with fibrin monomer obtained by dissolving fibrin clots in 1 M sodium bromide at pH 5.3. In this way they avoided the presence of unmodified fibrinogen. Results from the two laboratories were in gratifying agreement. We did have a disagreement about terminology; we had been calling the species f "activated fibrinogen" and the Scheraga group preferred the name "fibrin monomer." Later we adopted their term.

There was much speculation about the nature of the noncovalent bonding between fibrin monomers in the oligomeric species. There was a tendency to overemphasize electrostatic forces, because the fibrinopeptides carry off a substantial negative charge, and pH has a strong influence on degree and rate of polymerization. Now, of course, the role of specific interaction sites with particular amino acid sequences is recognized.

The unanswered questions of the postwar decade were greatly clarified by a new sequence of discoveries surging especially since 1966, including the following milestones:

1959	Electron micrographs of fibrinogen, trinodular rod	C. E. Hall and H. S. Slayter
1966–68	Chemical identification of ligation sites	L. Lorand; R. F. Doolittle; others
1972	Electron micrographs of oligomers	W. Krakow, H. A. Scheraga, and collaborators
1977–79	Complete amino acid sequencing of fibrinogen; identification of molecular domains	A. Henschen and collaborators; R. F. Doolittle and collaborators
1981	Electron micrographs of oligomers (negative staining)	R. C. Williams; W. E. Fowler, J. Hermans, and collaborators

The result has been a remarkably detailed understanding of the geometry of the fibrinogen molecule and of its assembly into oligomers by noncovalent association, as well as the location and detailed chemistry of the covalent bonds introduced by fibrinoligase (Factor XIIIa). But this is not yet ancient history and it belongs in the scientific sessions of this Conference.

Study of the fibrinogen-fibrin conversion in our own laboratory was resumed in 1971 after a lapse of fifteen years. Not only had the understanding of the polymerization, and especially of the ligation, of fibrin advanced considerably, but we had developed improved methods for measuring mechanical and rheological properties that could be applied to both fibrin clots and fibrin films. The nature of elasticity in these structures was still obscure; in spite of the superficial resemblance of fibrin film to a rubbery sheet, its molecular structure certainly had very little in common with the highly flexible and mobile molecular network of a rubbery polymer.

That the tensile strength of fibrin clots was affected by calcium had been noted by H. Wagreich and I. M. Tarlov in 1945, and confirmed by W. H. Seegers and collaborators in 1947. They thought the mechanism was ionic— and indeed later work shows that calcium ion alone has some effect on mechanical properties even without fibrinoligase. Early mechanical measurements of ours, with M. Miller and S. Shulman in 1951, showed a more striking effect. Without calcium, the elastic stress in a deformed clot gradually relaxed, indicating some kind of structural rearrangement. Calcium largely eliminated this relaxation and also increased the modulus of elasticity, indicating the presence of additional stronger bonds, which we attributed to action of the factor now known as fibrinoligase (FIG. 4). More recently, profiting from the advances meanwhile in understanding α-α and γ-γ ligation, G. W. Nelb and others in our laboratory have measured viscoelastic properties of unligated clots, ligated fine clots (in which essentially only γ-γ covalent bonds are formed), and ligated coarse clots (with both γ-γ and α-α bonds). Mechanical creep in the unligated clots shows that structural rearrangements take place

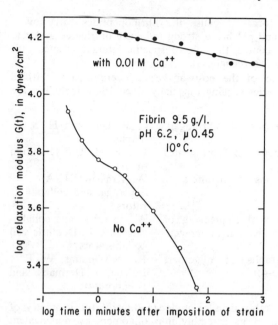

FIGURE 4. Stress relaxation modulus at small shear strains in fibrin clots with and without calcium—i.e., ligated and unligated (data of Ferry, Miller, and Shulman, 1951).

under stress, but without permanent damage as long as the deformations are small. The rearrangements are more prominent in coarse clots and probably reflect slippage of the protofibrils within the fiber bundles. They are eliminated by α-α ligation.

Very recent measurements by F. J. Roska on mechanical properties, birefringence, and small-angle x-ray scattering of fibrin film have thrown some light on the nature of its elasticity. Elastic deformation seems to take place primarily by reversible bending and orientation of the large fiber bundles, although there is also a reversible stretch attributable to some kind of internal transition within the fibrin monomer units. In stretched unligated films, there is slow relaxation of stress and permanent deformation, probably due to slippage within the fiber bundles.

We have also taken up again the problem of distribution of oligomer sizes in the early stages of polymerization. In experiments initiated by G. W. Nelb and continued by others, the changes in distribution of ligated oligomers with time have been followed by agarose gel electrophoresis as introduced by M. Moroi. Conclusions have been made about the kinetic sequence of fibrinopeptide release, polymerization, and ligation. But these experiments, too, are not yet history, and this is not the place to survey the rapid developments that are taking place in many laboratories to elucidate these processes.

I mentioned at the beginning that my first work on fibrinogen and fibrin was part-time, and I have never been more than a part-time coagulationist. But I have always had the conviction that the conversion of fibrinogen to fibrin is one of the most fascinating processes that nature has provided us to study. It is a problem of challenging intricacy, yet simple enough on the scale of biological phenomena so that complete understanding should soon be within our grasp.

INTRODUCTORY REMARKS

Sadaaki Iwanaga

Department of Biology, Faculty of Science
Kyushu University
Fukuoka-812, Japan

It is a great pleasure for me to have this opportunity to present these introductory remarks in the first session of this conference. Since it is more than ten years since I have been actively engaged in structural studies on fibrinogen, I cannot talk personally about the present status of fibrinogen structure; this will be done in detail by three distinguished investigators, Doctors Doolittle, Henschen, and Plow. I would, however, like to spend a little time sharing some memories of my early investigations of the primary structure of fibrinogen. These began about 18 years ago when I first joined Dr. Birger Blombäck at the Karolinska Institute in Stockholm. As possible subjects for investigation during my employment in his laboratory, he introduced several interesting projects. Among these were: (1) the determination of positions of tyrosine-*o*-sulfate and serine-*o*-phosphate in various fibrinopeptides, (2) investigation of the postulated abnormality of fibrinogen in hemo-

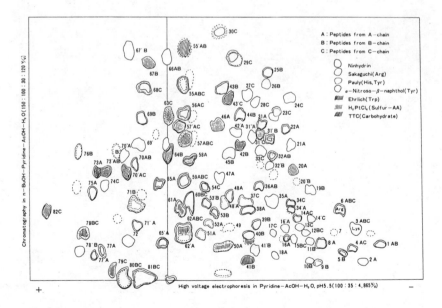

FIGURE 1. Two dimensional "fingerprint" of the tryptic peptides from human sulfitolyzed fibrinogen, Aα(A), Bβ(B) and γ(C) chains (From S. Iwanaga, A. Henschen, and B. Blombäck. 1966. Acta Chem. Scand. **20:** 1183–1185).

11

0077-8923/83/0408-0011 $01.75/0 © 1983, NYAS

philia, (3) determination of the structural difference between platelet fibrinogen and plasma fibrinogen, (4) studies on the primary structure of human fibrinogen, and (5) determination of the cross-linking site in stabilized fibrin. Because my knowledge of fibrinogen was very poor at that time, I could not decide which one should be selected, and I asked him which subject he considered to be most important in this field. He said immediately "I want to study the whole structure of fibrinogen because fibrinogen is my life."

I consequently began the studies on the primary structure of human fibrinogen. It's not necessary to say any more about Birger Blombäck's great contributions to the fibrinogen field, both past and present, since everybody here knows of his brilliant work. I would like to show a figure that is one reminder of many dear memories of fibrinogen. The figure shows a two-dimensional "fingerprint" of the tryptic peptides from human sulfitolyzed fibrinogen and its three chains, Aα, Bβ, and γ, which, at that time, we called the A, B and C chains. In all, 82 reproducible spots can be detected. Spot 75A is fibrinopeptide A and spot B is fibrinopeptide B. Spot 34′A is Gly-Pro-Arg, now known to be at one of the polymerization sites in the Aα chain. Spot 79C is the unique glycopeptide found in γ chain and spot 82C is the NH_2-terminal peptide derived from γ chain. Because amino acid sequence studies of these peptides are a major subject in this session, Dr. Doolittle, our first speaker, will now give us his story of the structure of fibrinogen.

THE STRUCTURE AND EVOLUTION OF VERTEBRATE FIBRINOGEN

Russell F. Doolittle

Department of Chemistry
University of California–San Diego
La Jolla, California 92093

The past three decades have given rise to extraordinary advances in our understanding of the fibrinogen-fibrin system. Thus, at the beginning of the 1950s it was discovered that a small amount of peptide material was proteolytically released from fibrinogen by thrombin, and that as a result the parent molecules spontaneously assembled into a fibrin gel.[1-4] In the intervening thirty years the general shape of the fibrinogen molecule has been determined,[5-8] its complete covalent structure established,[9-15] the mode of its polymerization described,[16, 17] the details of its covalent stabilization revealed,[18, 19] and very much more discovered, some of which is reviewed elsewhere,[20, 21] and many other aspects of which will be reviewed or reported at this meeting. The polymerization event itself is well enough understood that peptide antipolymerants have been designed on the basis of that understanding, synthesized, and shown to be effective.[22, 23] It has been a privilege merely to witness the astonishing progress during this period, much less to participate in it.

This is not to say that no mysteries about the fibrinogen-fibrin system remain. There are still many points that are poorly understood, including, for example, details of the lateral packing in fibrin fibers and the geometry of covalently cross-linked α chain multimers. Moreover, the exact contacts in the initial polymerization event are still unknown, although they may be tantalizingly close to revealment.[24, 25] We still suffer from the lack of a detailed x-ray structure, and the mechanics of the intracellular bioassembly of the molecule have yet to be unveiled.

The mystery I want to focus on in this article, however, is one that I have been stuck on for the last 25 years: how did the vertebrate fibrinogen molecule evolve? The problem is that the fibrinogen molecule seems just to "appear"— fully developed—with the dawning of vertebrate animals.[26] This is to say that fibrinogen of the type found in vertebrates has so far not been found in an invertebrate or a protochordate. Certain invertebrates are known to have coagulable proteins,[27, 28] but these molecules have no obvious common ancestry with the vertebrate type, nor are they activated by thrombin.

As far as we know, all vertebrates have fibrinogen molecules composed of three pairs of nonidentical chains ($\alpha_2\beta_2\gamma_2$), and all are converted to fibrin by thrombin-releasing fibrinopeptides from the amino-terminal segments of the α and β chains (TABLE 1). The selective advantage of a fibrinogen-fibrin system for vertebrates, what with their valuable pressurized and iron-laden blood, is obvious. What is not obvious is where the ingredients for the fibrinogen-fibrin system ever came from. Certainly a protein as complex as fibrinogen is not invented instantaneously as the result of a random assemblage of amino acids. In this article we will first consider the structure of the human fibrinogen molecule, making some inferences about fibrinogen evolution on the basis of

13

0077-8923/83/0408-0013 $01.75/0 © 1983, NYAS

that structure per se, and then compare its amino acid sequence with those of other mammals. At the same time we will digress a bit to consider the evolution of the hemoglobin molecule in order to establish a frame of reference for the changes we witness among the fibrinogen chains. We will resume our major theme with some observations about the structure of lamprey fibrinogen. Lampreys are the most primitive of the extant vertebrates, and any features that lamprey and human fibrinogens have in common will be presumed to have been in existence at the time lampreys and other vertebrates diverged more than 400 million years ago. In all this we will reflect on when and where those features common to lamprey and human fibrinogen may have originated. Particular emphasis will be paid to the evolutionary origins of the individual polypeptide chains.

TABLE 1

AMINO ACID SEQUENCES AT THE AMINO-TERMINI OF FIBRIN α- AND β-CHAINS FROM VARIOUS SPECIES *

Human α-chain	Gly-Pro-Arg-Val-Val-Glu-Arg. . .
Bovine α-chain	Gly-Pro-Arg-Leu-Val-Glu-Lys. . .
Dog α-chain	Gly-Pro-Arg-Ile-Val-Glu-Arg. . .
Chicken α-chain	Gly-Pro-Arg-Ile-Leu-Glu-Asn. . .
Lamprey α-chain	Gly-Pro-Arg-Leu- ? -Glx-Glx. . .
Human β-chain	Gly-His-Arg-Pro-Leu-Asp-Lys. . .
Bovine β-chain	Gly-His-Arg-Pro-Tyr-Asx-Lys. . .
Dog β-chain	Gly-His-Arg-Pro-Leu-Asp-Lys. . .
Chicken β-chain	Gly-His-Arg-Pro-Leu-Asp-Lys. . .
Lamprey β-chain	Gly-Val-Arg-Pro-Leu-Pro- ? . . .

* Compiled from references 36–39.

THE STRUCTURE OF HUMAN FIBRINOGEN

Like all other vertebrate fibrinogens, the human kind consists of three pairs of nonidentical chains bound together by a network of disulfide bonds. The arrangement of these disulfide linkages is such that the amino-terminals of all six chains are clustered in the central zone of the molecule, a critical feature in that the fibrinopeptide material released by thrombin consists of the amino-terminal segments of the α and β chains (FIGURE 1). Direct comparison of the amino acid sequences of the three polypeptide chains reveals that all three are homologous, which is to say, their sequences are similar enough that they must have evolved from a common ancestor. When they are optimally aligned (FIGURE 2), the majority of their cysteines are remarkably in register, indicating that the disulfide network is an ancient and essential structural feature. The alignment also indicates that the γ chain lacks the corresponding amino-terminal region where the thrombin-sensitive junctions and fibrinopeptides occur in the α and β chains.

How long ago did the chains diverge? What was the order of divergence? At first glance it would seem that the divergence point for the β and γ chains was the much more recent event, since the degree of sequence resemblance

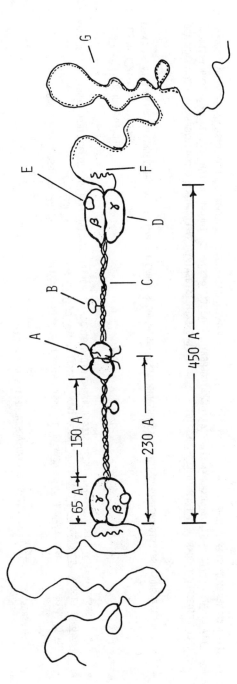

FIGURE 1. Schematic depiction of vertebrate fibrinogen molecule. A, central domain including fibrinopeptides; B, γ chain carbohydrate cluster; C, "coiled-coils" interdomainal connector; D, terminal domain consisting of homologous β and γ chain segments; E, β chain carbohydrate cluster; F, γ chain cross-linking site; and G, α chain carboxy-terminal extension.

```
α  ADSGEGDFLAEGGGVRGPRVVERHQSA                                  CKDSDWPFC  SDEDWNYKCPSGCRMKGLIDEVNQDFTNRINKLKNSLFEYQKNNK DSHSLTT
β  QGVNDNEEGFFSARGHRPLDKKREEAPSLRPAPPPISGGGYRARPAKAAATQKKVERKAPDAGGCLHADPDLGVLCPTGCQLQEALLQQERPIRNSVDELNNNVEAVSQTSSSSQFYMYL
γ                                                              YVATRDNCCILDERFGSYCPTTCGIADFLSTYQTKVDKDLQSLEDILHQVENKTS EVKQLIK

α  NIMEILRGDFSSANNRDNTYNRVSEDLRSRIEVLKRKVIQKVQHIQLLEKNVRAQLVDMKRLEVDIDIKIRSCRGSCSRALAREVDLKNYEDQQKQLEQVIAKDLLPSRDRQHLPLIKMK
β  LKDLWQKRQKQVKDNENVVNEYSSELEKHQLYIDETVNSNIPTNLRVLRSILENLRSKIQKLESDVSAQMEYCRTPCTVSCDIPVVSG   KECEEIIRKGGETSEMYLIQPDSSVK
γ  AIQLTYNPDESSKPNMIDAATLKSRKMLEEIMKYEASILTHDSSIRYLQEIYNSNNQKIVNLKEKVAQLEAQQEPCKDTVQIHDITG   KDCQDIANKGAKQSGLYFIKPLKANQ

α  PVPNLVPGNFKSQLQKVPPEWKALTDMPQMRM                ELERPGGNEITRGGSTSYGTGSETESPRNPSSAGSWNSGSSGPGSTGNRNPGSSGTGSGATW
β  PYRYVCDMNTENGGMTVIQNRQDGSVDFGRKWDPYKQGFGNV ATNTDGKBYCGLPGEYWLGBBKISELTRMGPTELLI EMEDWKGDKVKAHYGGFTVQNEANKYQISVNKYRGTAGNA
γ  QFLVYCEIDGSGNGMTVFQKRLDGSVDFKKNWIQYKEGFGHLSPTGTT  EFWLGNEKIHLISTQSAIPYALRVELEDWNGRTSTADYAMFKVGPEADKYRLTYAYFAGGDAGD

α  KPGSSGPGSTGSWNNSGSSGGTGSTGNQNPGSPRPGSTGTWNPGSSERGSAGHNTSESSVSGSTGQWHSESGSFRPDSPGSGNARPNDPNWGTFEEVSGNVSPGTRREYHTEKLVTSKGDKELR
β                                                                              LMDGASQL MGENRTMTIHNGMFFS
γ                                                                              AFDGFDFGDDPSDKFFTSHNGMQFS

α  TGKEKVTSGSTTTTRRSCSKTVTKTVIGPDGHKEVTKEVVTSEDGSDCPEAMDLGTLSGIGTLDGFRHRHPDEAAFFDTASTGKTFPGFFSPMLGEFVSETESRGSESGIFTNTKESSSHHP
β  TYDRDNDGWLTSDPRKQCSKEDGGGWWYBRCHAANPNGRYYWGGQYTWDMAKHGTBDGVVWMNWKGSWYSMRKMSM  KIRPFFPQQ
γ  TWDNDNDKFEGN   CAEQDGSGWWMNKCHAGHLNGVYYQGGTYSKASTPNGYDNGIIWATWKTRMYSMKKTTM  KIIPFNRLTIGEGQQHHLGGAKQAGDV

α  GIAEFPSRGKSSSYSKQFTSSTSYNRGDSTFESKSYKMADEAGSEADHEGTHSTKRGHAKSRPV
β
γ
```

FIGURE 2. Alignment of α, β, and γ chains of human fibrinogen (from reference 12).

for these two polypeptides is much greater than for either with the α chain (TABLE 2). How are we to know when that was? There are two different ways to estimate the times involved, both of which are ultimately based on the paleontological record. In the first instance we can assume, as a rough approximation, that the rate of change exhibited by the system on a chain-to-chain basis is the same as is observed for the individual chains on a species-to-species basis. The second approach is more direct: it merely depends on the existence of the individual gene products in species that can be related through the fossil record. For example, if birds and mammals both contain the α and β chains of a protein—hemoglobin, for example—then the gene duplication that gave rise to these two types must antedate the most recent common ancestor of birds and mammals.

In the case of the β and γ chains of fibrinogen, 32% of the amino acids are at identical positions when the chains are optimally aligned (TABLE 2). How does that compare with interspecific change for these two chains? At

TABLE 2

DEGREES OF HOMOLOGY OF α, β AND γ CHAINS OF HUMAN FIBRINOGEN *

	Hα	Hβ	Hγ	
Hα	–	6.1	5.6	
				Std. Dev.†
Hβ	18	–	30.1	
Hγ	17	32	–	
	Percent Identity			

* The comparison of the β and γ chains covers their entire sequences (461 and 411 residues, respectively), whereas the comparisons of the α chain with the β and γ chains only cover segments containing the first 239 residues from the amino terminals.

† The number of standard deviations (Std. Dev.) reflects the degree of confidence that these resemblances are not the result of chance, anything greater than 3.0 standard deviations being regarded as reflecting common ancestry.

this point we have sequence data from human and bovine for the β chains,[29] and from human and rat for the γ chains.[30] In both cases the nonhuman data are the result of DNA sequencing.

As it happens, the rat and human γ chains are 82% identical, whereas the bovine and human β chains are 84% identical over the corresponding lengths (the amino-terminal segment of the β chain, which is absent in γ chains, will not be included in these comparisons). By comparison, the hemoglobin α chains of mouse, bovine and human are 88% identical, one to another. Rodents, bovids and humans all shared a common ancestor about 80 million years ago. Can we set our clock with these data? In fact, hemoglobin sequences from many different vertebrates are available, and a good calibration curve can be constructed for this system.

In the course of this discussion it will be convenient to describe the "differentness" of two sequences by a "Difference Coefficient" whereby

$$DC = -\ln(1 - d/n)$$

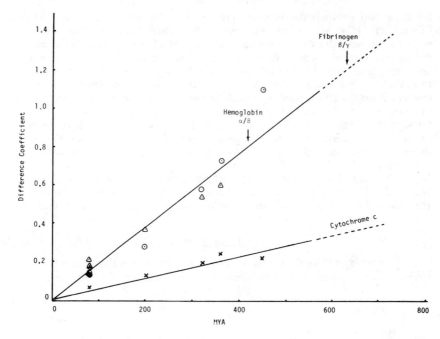

FIGURE 3. Plot of difference coefficients $[DC = -\ln(1 - d/h]$ against time since last common ancestor (MYA = millions of years ago). ⊙, hemoglobin α chains; ▲ hemoglobin β chains; X, cytochromes C; ◇, fibrinogen β chains, ◊ fibrinogen γ chain. The cytochrome comparisons (X) involve the same species pairs and are shown as an example of a slower changing protein.

where d is the number of differences and n is the number of residues being compared. The use of the negative expontial tends to compensate, to a degree, for back mutations and multiple changes at the same position. When the difference coefficients for various comparisons are plotted against the time since a pair of creatures last had a common ancestor, a reasonably straight line is observed (FIGURE 3).

If we take this simple approach for the hemoglobin chains, then we can estimate when the α and β chains diverged. In FIGURE 3 the difference co-efficients for hemoglobin chains from various pairs of hemoglobin comparisons are plotted, several species being compared with human in the cases of both the α and β chains. What we find is that the difference coefficient for a direct comparison of the α and β chains, regardless of species, averages about 0.83 (TABLE 3). This corresponds to the occurrence of a gene duplication about 400 million years ago. This value is in good agreement with the actual occurrence of the two kinds of chain. Thus, lampreys, which last had a common ancestor with other vertebrate types about 450 million years ago, have a single-chained (α-type) hemoglobin. In contrast, teleost fish such as the carp have tetrameric hemoglobins consisting of α and β chains. These creatures branched off from the rest of vertebrate stock about 360 million years ago (FIGURE 3). Clearly the invention of non-α chains occurred some-time during the interval between these two divergences (FIGURE 4).

We can make a similar reckoning for the β and γ chains of fibrinogen, although in this case our species comparisons are still limited to pairs of recently diverged mammals. Still, the data we have fall right amongst those for the hemoglobin chains, and if we assume a similar and constant rate of change for the β and γ chains since their initial divergence, the gene duplication we are focusing on must have occurred about 600 million years ago (FIGURE 3).

THE LAMPREY FIBRINOGEN MOLECULE

Lampreys and human last shared a common ancestor some 400–500 million years ago. The fact that lamprey fibrinogen has well-differentiated α, β and γ chains, each of which exhibits definite homology with its human counter-part,[31] indicates that the gene duplications that gave rise to the individual chains occurred well in advance of 450 million years ago. In this regard the lamprey fibrinogen situation is different from that of hemoglobin, the latter being a primitive single chain structure that has not yet enjoyed the luxury of differentiation that can come from "twinning." Not only does lamprey fibrino-

TABLE 3

DIFFERENCE COEFFICIENTS * FOR VARIOUS PAIRS OF SEQUENCES FROM ASSORTED FIBRINOGENS AND HEMOGLOBINS

Hemoglobin α Chain Species	DC	Divergence Time (Million Years)
Human/Mouse	0.136	80
Human/Bovine	0.129	80
Human/Chicken	0.285	200
Human/Newt	0.580	320
Human/Carp	0.729	360
Human/Lamprey	1.097	450
Hemoglobin β Chain		
Human/Rat	0.189	80
Human/Bovine	0.214	80
Human/Chicken	0.368	200
Human/Frog	0.541	320
Human/Carp	0.603	360
Hemoglobin α/β Chains (average)	0.826	(see FIGURE 3)
Fibrinogen β Chain		
Human/Bovine	0.177	80
Fibrinogen γ Chain		
Human/Rat	0.189	80
Human Fibrinogen β/γ	1.23	(see FIGURE 3)

* $DC = -\ln(1 - d/n)$.

gen have three pairs of nonidentical polypeptides, but those chains already had well-defined functions at the time of the last common ancestry with other vertebrate types. Thus, the amino-terminal segments of the α and β chains are fibrinopeptides removed by thrombin. The γ chain resembles the fibrinogen of all other vertebrates in that it lacks a fibrinopeptide region; it also becomes covalently cross-linked by factor XIII during the polymerization process. Moreover, the molecular weight and other properties of the lamprey γ chain are virtually identical to those of mammalian fibrinogen γ chains. These observations are in accord with the invention of fibrinogen α, β and γ chains occurring long before the divergence of lampreys and other vertebrates (FIGURE 4).

CONSERVED AND UNCONSERVED SEQUENCES IN THE SAME GENE PRODUCT

These comparisons based on percent identities have to be regarded with some caution. The fact is, although it is generally known that different proteins change at characteristic rates, it is less well appreciated that different parts of the same protein can change at radically different rates. In the case of vertebrate fibrinogen, there are several regions which change very rapidly and others that have not changed significantly since the age of fishes. The best known of the fast-changing regions are the fibrinopeptides, entities which in themselves have long been recognized as among the most mutable peptide structures known.[32] The reason lies in the rather general and undemanding nature of their role in masking potential polymerization sites. The carboxy-

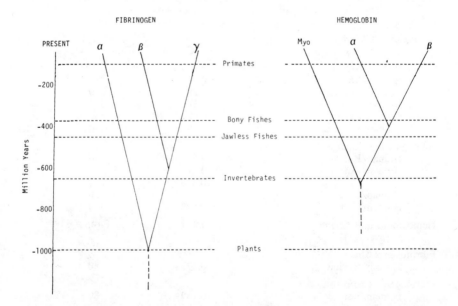

FIGURE 4. Divergence of fibrinogen α, β, and γ chains after gene duplications. Compared with divergence of myoglobin and hemoglobin α and β chains. Broken lines indicate divergence times for key organismic groups.

TABLE 4

HOMOLOGY OF AMINO-TERMINAL SEGMENTS OF α, β AND γ CHAINS OF HUMAN
FIBRINOGEN AND γ CHAIN OF RAT FIBRINOGEN

	Hα	Hβ	Hγ	Rγ	
Hα *	–	1.71	1.83		Diff.
Hβ	18.0	–	1.56		Coeff.†
Hγ	15.8	20.9	–	0.19	
Rγ	15.3	19.8	74.6	–	

Percent Identity

Hα * = Res. 29–209, Hβ = Res. 58–235, Hγ = Res. 1–173, and Rγ = Res. 2–174.
† Diff. Coefficient = − ln(1 − d/n). H = Human; R = Rat.

terminal halves of α chains are also well known for the variability.[16] Less
well appreciated is the fact that the interdomainal "coiled coils" are also given
to great variability.[33] It appears that a three-stranded rope does not have
stringent structural constraints; as long as polar residues replace polar residues
and nonpolars replace nonpolars, the integrity of the "coiled coils" can be
preserved.

The regional variability in evolutionary conservation of sequence is dramatically revealed in computer-derived differential plots of resemblance between the β and γ chains of human fibrinogen (FIGURE 5). Comparisons of
rat-human γ chains and bovine-human β chains, yield profiles that are qualitatively similar, indicating in both cases more rapid change in the "coiled-coil"
region and showing highly conserved sectors nearer the carboxy-termini.

These differences in degree of structure conservation are especially important to us in making inferences about the times of origin for the individual
chains. On the basis of the overall sequences, for example, the α chain is so
different from the other two that it would appear that its divergence time
must have been a very long time ago, at least antedating he divergence of
plants and animals, and perhaps extending back all the way to the prokaryotic
world. This may be misleading, however, for if we consider only the 180-
residue segments corresponding to the first 40% of the γ chain, a region that
includes the rapidly changing "coiled coils," then we see that α chains are not
much more different from β and γ chains than the latter two are from each
other (TABLE 4). This is not simply a consequence of the variability reaching some blurred limit, since the ratio of difference coefficients for these segments of human and rat γ chains is the same as is found when the entire γ
chain sequences are used. The same is true in the case of β chains from bovine
and human.

The implication is that some event has occurred during the evolution of
the fibrinogen molecule that has left the amino-terminal portions of all the
chains similar in structure, but allowed the carboxy-terminal two-thirds of the
α chain to escape the confines of the conservative natural selection which has
kept the β and γ chains so similar in these regions. The nature of this hypothetical event may become clearer when the DNA sequences of all three genes
are available. It is already known that the distribution of introns follows a
similar pattern in all three genes,[34] and the distinctiveness of the α chain may

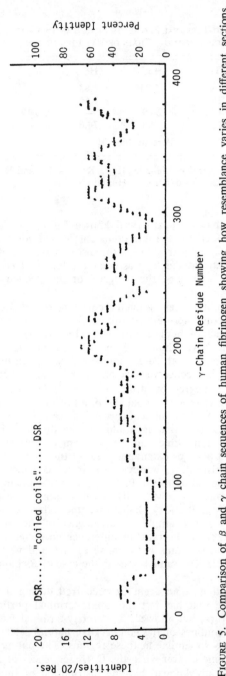

FIGURE 5. Comparison of β and γ chain sequences of human fibrinogen showing how resemblance varies in different sections. The most conserved parts are near the carboxy-terminus. Similar profiles are obtained when interspecies comparisons are made for either chain (from reference 35).

follow upon some simple event such as a frameshift and/or expression of a segment which is intronic in the β and γ chains.

The Paradox

But if the individual genes for the various fibrinogen polypeptide chains existed long ago—perhaps 600–1000 million years ago—where did they go? Why hasn't fibrinogen been found in protochordates or invertebrates? There are several possibilities. For one, nonvertebrates may indeed have fibrinogen, but it may be present inside blood cells (as is the horseshoe crab coagulogen) where it has eluded identification in previous studies. Or it may be present in much smaller amounts. Or it may have become obsolete as a functional entity.

Searching for fibrinogen-related proteins among these creatures has been a difficult and so-far unrewarding pastime. But recent developments have

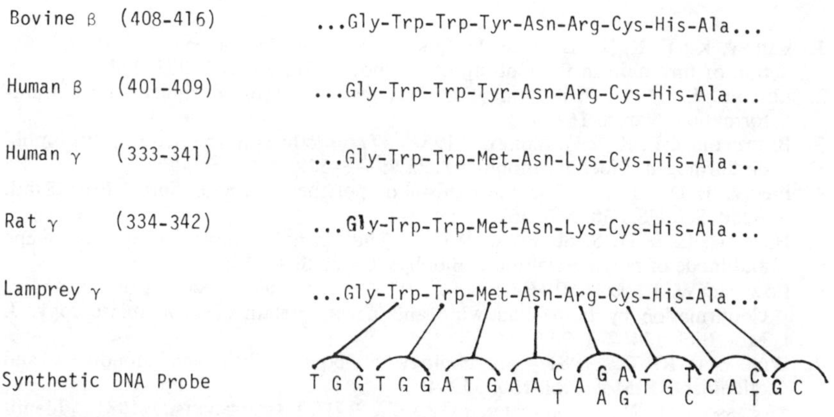

Bovine β (408-416)   ...Gly-Trp-Trp-Tyr-Asn-Arg-Cys-His-Ala...

Human β   (401-409)   ...Gly-Trp-Trp-Tyr-Asn-Arg-Cys-His-Ala...

Human γ   (333-341)   ...Gly-Trp-Trp-Met-Asn-Lys-Cys-His-Ala...

Rat γ   (334-342)   ...Gly-Trp-Trp-Met-Asn-Lys-Cys-His-Ala...

Lamprey γ   ...Gly-Trp-Trp-Met-Asn-Arg-Cys-His-Ala...

Synthetic DNA Probe   T GG T GG A T G A A T A A G A T G C C A T G C

FIGURE 6. A highly conserved portion of fibrinogen β and γ chains that can serve as a basis for a synthetic DNA mixed probe. The bovine β chain sequence is from Chung et al.,[29] the rat γ chain from Crabtree & Kant [30] and the lamprey sequence from Doolittle and Cottrell (unpublished).

provided a much more promising approach. Instead of looking for the gene products, why not look for the genes? In fact, even if the products are no longer made in these creatures, the vestigial genes may still be lingering in some genomes.

To this end we are currently undertaking a serious exploration among protochordates and echinoderms in search of antecedent vertebrate fibrinogen genes. Our strategy depends on identifying those sequences that are most conserved in vertebrate fibrinogen molecules, the presumption being that those features would also have been preserved in the more ancient proteins. For example, our studies on lamprey fibrinogen have revealed that its γ chain is more than 50% identical with its human equivalent over the last third of its length. Indeed, there is one particular stretch where the lamprey and mammalian β and γ chains have seven or eight of nine residues identically situated (FIG-URE 6). We are now synthesizing a DNA probe, based on these sequences, which ought to be sufficiently matchable to bind to all the genes corresponding to the amino acid sequences shown in FIGURE 6. As such, we refer to this as the "universal betagamma 23-mer." If conditions of appropriate stringency can be found whereby the probe behaves appropriately with the known vertebrate genes, then we will use it to search for similar structures in the DNA of protochordates and invertebrates. Using this molecular fishhook, perhaps we will be able to bootstrap our way back to the origins of the fibrinogen molecule.

ACKNOWLEDGMENT

I would like to thank Drs. Crabtree and Kant for showing me their manuscript (reference 30) prior to its publication.

REFERENCES

1. BAILEY, K., F. R. BETTELHEIM, L. LORAND & W. R. MIDDLEBROOK. 1951. Action of thrombin in the clotting of fibrinogen. Nature 167: 233–234.
2. LORAND, L. 1951. Fibrinopeptide. New aspects of the fibrinogen-fibrin transformation. Nature 167: 992.
3. BETTELHEIM, F. R. & K. BAILEY. 1952. The products of the action of thrombin on fibrinogen. Biochim. Biophys. Acta 9: 578–579.
4. FERRY, J. D. 1952. The mechanism of polymerization of fibrin. Proc. Natl. Acad. Sci. USA 38: 566–569.
5. HALL, C. E. & H. S. SLAYTER. 1959. The fibrinogen molecule: its size, shape and mode of polymerization. J. Biophys. Cytol. 5: 11–15.
6. FOWLER, W. E. & H. P. ERICKSON. 1979. Trinodular structure of fibrinogen. Confirmation by both shadowing and negative stain electron microscopy. J. Mol. Biol. 134: 241–249.
7. WILLIAMS, R. C. 1981. Morphology of bovine fibrinogen monomers and oligomers. J. Mol. Biol. 150: 399–408.
8. MOSESSON, M. W., J. HAINFELD, J. WALL & R. H. HASCHEMEYER. 1981. Identification and mass analysis of human fibrinogen molecules and their domains by scanning transmission electron microscopy. J. Mol. Biol. 153: 695–718.
9. HENSCHEN, A. & F. LOTTSPEICH. 1977. Amino acid sequence of human fibrin. Preliminary note on the γ chain sequence. Hoppe-Seyler's Z. Physiol. Chem. 358: 935–938.
10. HENSCHEN, A. & F. LOTTSPEICH, F. 1977. Amino acid sequence of human fibrin. Preliminary note on the completion of the β chain sequence. Hoppe-Seyler's Z. Physiol. Chem. 358: 1643–1646.
11. WATT, K. W. K., T. TAKAGI & R. F. DOOLITTLE. 1979. Amino acid sequence of the β chain of human fibrinogen. Biochemistry 18: 68–76.
12. DOOLITTLE, R. F., K. W. K. WATT, B. A. COTTRELL, D. D. STRONG & M. RILEY. 1979. The amino acid sequence of the α chain of human fibrinogen. Nature 280: 464–468.
13. BLOMBACK, B., B. HESSEL & D. HOGG. 1976. Disulfide bridges in NH_2-terminal part of human fibrinogen. Thromb. Res. 8: 639–658.
14. BOUMA, H., T. TAKAGI, R. F. DOOLITTLE. 1978. The arrangement of disulfide bonds in fragment D from human fibrinogen. Thromb. Res. 13: 557–562.
15. HENSCHEN, A. 1978. Disulfide bridges in the middle part of human fibrinogen. Hoppe-Seyler's Z. Physiol. Chem. 359: 1757–1770.
16. DOOLITTLE, R. F. 1973. Structural aspects of the fibrinogen-fibrin conversion. Advances Prot. Chem. 27: 1–109.

17. KUDRYK, B. J., D. COLLEN, K. R. WOODS & B. BLOMBACK. 1974. Evidence for localization of polymerization sites in fibrinogen. J. Biol. Chem. **249:** 3322–3325.

18. CHEN, R. & R. F. DOOLITTLE. 1970. Isolation, characterization and location of a donor-acceptor unit from cross-linked fibrin. Proc. Natl. Acad. Sci. USA **66:** 472–479.

19. MCKEE, P. A., P. MATTOCK & R. L. HILL. 1970. Subunit structure of human fibrinogen, soluble fibrin and cross-linked insoluble fibrin. Proc. Natl. Acad. Sci. USA **66:** 738–744.

20. DOOLITTLE, R. F. 1981. Fibrinogen and fibrin. In Haemostasis and Thrombosis. A. L. Bloom & D. P. Thomas, Eds.: 163–191. Churchill-Livingstone. London, England.

21. HERMANS, J. & J. MCDONAGH. 1982. Fibrin: Structure and interactions. Seminars Thromb. Haem. **8:** 11–24.

22. LAUDANO, A. P. & R. F. DOOLITTLE. 1978. Synthetic peptide derivatives which bind to fibrinogen and prevent the polymerization of fibrin monomers. Proc. Natl. Acad. Sci. USA **75:** 3085–3089.

23. LAUDANO, A. P. & R. F. DOOLITTLE. 1980. Studies on synthetic peptides that bind to fibrinogen and prevent fibrin polymerization. Structural requirements, numbers of binding sites and species differences. Biochemistry **19:** 1013–1019.

24. DOOLITTLE, R. F. & A. P. LAUDANO. 1980. Synthetic peptide probes and the location of fibrin polymerization sites. In Protides of Biological Fluids. 28th Colloquium. D. H. Peters, Ed.: 311–316. Pergamon Press.

25. OLEXA, S. A. & A. Z. BUDZYNSKI. 1981. Localization of a fibrin polymerization site. J. Biol. Chem. **258:** 3544–3549.

26. DOOLITTLE, R. F., J. L. ONCLEY & D. M. SURGENOR. 1962. Species differences in the interaction of thrombin and fibrinogen. J. Biol. Chem. **237:** 3123–3127.

27. FULLER, G. M. & R. F. DOOLITTLE. 1971. Studies of invertebrate fibrinogen. Purification and characterization of fibrinogen from the spiny lobster. Biochemistry **10:** 1305–1311.

28. NAKAMURA, S., T. TAKAGI, S. IWANAGA, M. NIWA & K. TAKAHASHI. 1976. A clottable protein (coagulogen) of horseshoe crab hemocytes. J. Biochem. **80:** 649–652.

29. CHUNG, D. W., M. W. RIXON, R. T. A. MACGILLIVRAY & E. W. DAVIE. 1981. Characterization of a cDNA clone coding for the β chain of bovine fibrinogen. Proc. Natl. Acad. Sci. USA **78:** 1466–1470.

30. CRABTREE, G. R. & J. A. KANT. 1982. Organization of the rat fibrinogen gene: alternate mRNA splice patterns produce the γA and γB (γ') chains of fibrinogen. Cell. In press.

31. DOOLITTLE, R. F., B. A. COTTRELL & M. RILEY. Amino acid compositions of the subunit chains of lamprey fibrinogen. Biochim. Biophys. Acta **453:** 439–452.

32. DOOLITTLE, R. F. & B. BLOMBACK. 1964. Amino acid sequence studies on fibrinopeptides from various mammals: evolutionary implications. Nature **202:** 147–152.

33. DOOLITTLE, R. F., D. M. GOLDBAUM & L. R. DOOLITTLE.. 1978. Designation of sequences involved in the "coiled coil" interdomainal connector in fibrinogen: construction of an atomic scale model. J. Mol. Biol. **120:** 311–325.

34. CRABTREE, G. R., & J. A. KANT. 1982. Structure and regulation rat fibrinogen genes. This volume.

35. DOOLITTLE, R. F. 1980. The evolution of vertebrate fibrinogen. In Protides of Biological Fluids. 28th Colloquium. H. Peeters, Ed.: 41–46. Pergamon Press.

36. IWANAGA, S., P. WALLEN, N. J. GRONDAHL, A. HENSCHEN & B. BLOMBACK. 1967. Isolation and characterization of N-terminal fragments obtained by plasmin digestion of human fibrinogen. Biochem. Biophys. Acta **147:** 606–609.

37. COTTRELL, B. A. & R. F. DOOLITTLE. 1976. Amino acid sequences of lamprey fibrinopeptides A and B and characterization of the junctions split by lamprey and mammalian thrombins. Biochim. Biophys. Acta **453**: 426–438.
38. BIRKEN, S., G. D. WILNER & R. E. CANFIELD. 1975. Studies of the structure of canine fibrinogen. Thromb. Res. **7**: 599–610.
39. MURANO, G., D. WALZ, L. WILLIAMS, J. PINDYCK & M. W. MOSESSON. 1977. Primary structure of the amino terminal regions of chicken fibrin chains. Thromb. Res. **11**: 1–10.

DISCUSSION OF THE PAPER

H. A. SCHERAGA (*Department of Chemistry, Cornell University, Ithaca, NY*): When you use the term "coiled-coil" for the intervening sections, do you imply that these are coiled α helices and if so, are the substitutions that you see between lamprey and human fibrinogens consistently substituting one helix favoring one residue for another? You mentioned polar for polar and nonpolar for nonpolar, but have you looked more deeply into the nature of the substituents?

R. F. DOOLITTLE (*Department of Chemistry, University of California–San Diego, La Jolla, CA*): Yes. I should say that coiled coils certainly do imply coiled α helices. This is based on the diffraction patterns that were originally seen by Astbury's group. The distance between the disulfide rings is 111 or 112 residues. Multiplied by 1.43—the number of angstroms of translation one would expect from the compressed α helices that exist in a coiled coil—this gives one a value of just a little bit more than 150 angstroms, the same distance predicted by the Slater model back in 1959. It might be just a coincidence. Still, I regard it as a remarkable outcome: using an amino acid sequence as a ruler.

Now about the substitutions, in the case of the lamprey, we do not have the sequence of that part of that structure completed; we have been working primarily at the C-terminal ends of the β and γ chains and the terminal domain. The substitutions that I referred to are based on the rat γ chain sequence compared with the human. I know that bovine β chain data are available also, but I tallied up the difference in rat and human γ chains. Just to give you an example, the overall comparison of the rat γ chain with human has about 80% identity. However, the coiled coil region the identity drops to about 60%, whereas the part where the polymerization sites are thought to be, is up close to 90% identical. In the coiled coil sector the bulk of those 40% substitutions fall into the realm of polar for polar, nonpolar for nonpolar.

A. BUDZYNSKI (*Temple University Health Sciences Center, Philadelphia, PA*): Yes, I think as you mentioned precursors of fibrinogen there are two animals in which clottable proteins are present: one is the horseshoe crab, *Limulus polyphemus,* and the second is the lobster. *Limulus* coagulogen is clottable by endotoxin, but it is a single chain protein; maybe you can tell us what is known as far as its relationship as a precursor of fibrinogen? And lobster protein is directly coagulable by a transglutaminase-like enzyme. Aren't these the precursors of fibrinogen in that term which you have mentioned?

DOOLITTLE: I started work on the lobster about the same time as on the lamprey, using Star Market lobsters back when I was a graduate student. In fact, I spent a long time on this. Later Gerry Fuller wrote his Ph.D thesis on the topic, and I think that both he and I would agree that this is an independently evolved creation in every way. Fuller's work was mainly on the physical chemistry, and although we do not have a sequence in this case yet, I would be very surprised if these molecules have any evolutionary common ancestry. However, time will show.

In the case of the horseshoe crab, Dr. Mosesson just told me that the full sequence of the *Limulus*, or its Japanese equivalent, coagulogen is going to be presented here at this symposium. Dr. Takagi sent me a 90% completed sequence, and although at one point some people thought there might be some relationship of that sequence and the human molecule, from the point of view of comparing amino acid sequences, there is no significant homology apparent. There may be a connection, but it is too early to tell that now. Only by getting a bunch of intermediates would it be possible to show that there has been a persistent structure. It may be, but there is no scientific basis for thinking so at this point, as far as I know.

COVALENT STRUCTURE OF FIBRINOGEN

Agnes Henschen, Friedrich Lottspeich, Maria Kehl, and
Christopher Southan

Max-Planck Institute for Biochemistry
D-8033 Martinsried/Munich, Federal Republic of Germany

INTRODUCTION

It is well established that the fibrinogen molecule is made up of two identical halves, each containing three different peptide chains. The overall structure may therefore be described as $(A\alpha, B\beta, \gamma)_2$. On thrombin-digestion the fibrino-peptides A and B are released and fibrin, with the structure $(\alpha, \beta, \gamma)_2$, is formed. The complete primary structure of human fibrinogen is known.[1,15] The sequence information may be used to study relationships between structure, function and evolution. In genetically determined abnormal fibrinogens the correlation between the structural error and the dysfunction of the molecule may reveal the functional importance of single amino acid residues.

AMINO ACID SEQUENCES

The complete amino acid sequences of the three human fibrinogen chains have recently been eluciated. The sequence analyses have all been conducted in one or more of three laboratories, i.e., that of Blombäck in Stockholm, that of Doolittle in La Jolla, and our own in München-Martinsried (detailed references to the results of the other research groups are contained in the references quoted below). Various strategies have been employed for the isolation of the fibrinogen fragments to be used in sequence analysis. Our own strategy has been to prepare first the three peptide chains and then fragments of each separate chain. The peptide chains were isolated by CM-cellulose chromatography of S-carboxymethylated fibrin(ogen).[2] The individual chains were cleaved by chemical and enzymatic methods into fragments of a size which could be handled by sequencing methods, cyanogen bromide and trypsin cleavage being primarily used.[3-5] The sequence of the γ chain, with 411 amino acid residues, was completed in 1977.[3] In the same year the 461 residues of the $B\beta$ chain were elucidated,[4] and two years later the 610 residues of the $A\alpha$ chain.[5,6] The sequences of all three chains are shown in FIGURE 1 and their compositions and molecular weights in TABLE 1.

DISULFIDE BRIDGES

The six peptide chains in fibrinogen are held together by 29 disulfide bonds.[7] The molecule contains no free sulfhydryl groups. From the amino acid sequences it is obvious that the half-cystine residues are concentrated in three clusters along each chain, i.e., one N-terminal, one intermediate and one C-terminal cluster. In order to find the unique partner of each half-cystine residue native fibrinogen may be cleaved by cyanogen bromide. Hereby five

28

0077–8923/83/0408–0028 $01.75/0 © 1983, NYAS

```
  1 A D S G E G D F L A E G G G V R↓G P R V V E R H Q S A C K D S D W P F C S D E D
 41 W N Y K C P S G C R M K G L I D E V N Q D F T N R I N K L K N S L F E Y Q K N N
 81 K D S H S L T T N I M E I L R G D F S S A N N R D N T Y N R V S E D L R S R I E
121 V L K R K V I E K V Q H I Q L L Q K N V R A Q L V D M K R L E V D I D I K I R S
161 C R G S C S R A L A R E V D L K D Y E D Q Q K Q L E Q V I A K D L L P R S D R Q
201 H L P L I K M K P V P D L V P G N F K S Q L Q K V P P E W K A L T D M P Q M R M
241 E L E R P G G N E I T R G G S T S Y G T G S E T E S P R N P S S A G S W N S G(G)
281 S G P G(G)T G N R N P G S S S G T G G T A T W K P G S S G P G S T G S W N S G S S
321 G T G S T G N Q N P G S P R P G S T G T W N P G S S E R G S A G H W T S E S S V
361 S G S T G Q W H S E S G S F R P D S P G S G N A R P N N P D W G T F E E V S G N
401 V S P G T R R E Y H T E K L V T S K G D K E L R T G K E K V T S G S T T T T R R
441 S C S K T V T K T V I G P D G H K E V T K E V V T S E D G S D C P E A M D L G T
481 L S G I G T L D G F R H R H P D E A A F F D T A S T G K T F P G F F S P M L G E
521 F V S E T E S R G S E S G I F T N T K E S S S H H P G I A E F P S R G K S S S Y
561 S K Q F T S S T S Y N R G D S T F E S K S Y K M A D E A G S E A D H E G T H S T
601 K R G H A K S R P V
```

```
  1 Z G V N D N E E G F F S A R↓G H R P L D K K R E E A P S L R P A P P P I S G G G
 41 Y R A R P A K A A A T Q K K V E R K A P D A G G C L H A D P D L G V L C P T G C
 81 Q L Q E A L L Q Q E R P I R N S V D E L N N N V E A V Q(S,T)S S S S Q F Y M Y L
121 L K D L W Q K R Q K Q V K D N E N V V N E Y S S E L E K H Q L Y I D E T V N S N
161 I P T N L R V L R S I L E N L R S K I Q K L E S D V S A Q M E Y C R T P C T V S
201 C N I P V V S G K E C E E I I R K G G E T S E M Y L I Q P D S S V K P Y R V Y C
241 D M N T E N G G W T V I Q N R Q D G S V D F G R K W D P Y K Q G F G N V A T N T
281 D G K N Y C G L P G E Y W L G N D K I S Q L T R M G P T E L L I E M E D W K G D
321 K V K A H Y G G F T V Q N E A N K Y Q I S V N K Y R G T A G N A L M D G A S Q L
361 M G E N⁺R T M T I H N G M F F S T Y D R D N D G W L T S D P R K Q C S K E D G G
401 G W W Y N R C H A A N P N G R Y Y W G G Q Y T W D M A K H G T D D G V V W M N W
441 K G S W Y S M R K M S M K I R P F F P Q Q
```

```
  1 Y V A T R D N C C I L D E R F G S Y C P T T C G I A D F L S T Y Q T K V D K D L
 41 Q S L E D I L H Q V E N⁺K T S E V K Q L I K A I Q L T Y N P D E S S K P N M I D
 81 A A T L K S R K M L E E I M K Y E A S I L T H D S S I R Y L Q E I Y N S N N Q K
121 I V N L K E K V A Q L E A Q C Q E P C K D T V Q I H D I T G K D C Q D I A N K G
161 A K Q S G L Y F I K P L K A N Q Q F L V Y C E I D G S G N G W T V F Q K R L D G
201 S V D F K K N W I Q Y K E G F G H L S P T G T T E F W L G N E K I H L I S T Q S
241 A I P Y A L R V E L E D W N G R T S T A D Y A M F K V G P E A D K Y R L T Y A Y
281 F A G G D A G D A F D G F D F G D D P S D K F F T S H N G M Q F S T W D N D N D
321 K F E G N C A E Q D G(S)G W(W)M N K C H A G H L N G V Y Y Q G G T Y S K A S T P
361 N G Y D N G I I(W)A T(W)K T R W Y S M K K T T M K I I P F N R L T I G E G Q Q H
401 H L G G A K Q A G D V
```

FIGURE 1. Amino acid sequences in human fibrinogen Aα-, Bβ- and γ-chains. The arrows indicate thrombin cleavage sites, the + carbohydrate side chains.

so-called disulfide knots, containing all disulfide bridges of the molecule, are formed.[8-12] Some properties of these knots, which may be donated FCB 1–5 (*f*ibrinogen-*c*yanogen-*b*romide fragment), are summarized in TABLE 2.

The largest disulfide knot, FCB 1, encompasses the N-terminal regions of all six chains.[8] Like fibrinogen itself it is dimeric, three of the 11 constituent sulfur bridges connecting the two halves of the fibrinogen molecule, one of these between the Aα chains and two between the γ chains,[9] as indicated in FIGURE 2. It is not yet known if the two disulfide bonds between the γ chains link the γ chain N-termini in parallel or anti-parallel. It is obvious that the configuration here is of decisive importance for the symmetry of the whole molecule. It seems, however, likely that the γ chains are linked in parallel.

The second-largest disulfide knot, FCB2, contains intermediate parts of all three chains.[10, 11] The three smaller disulfide knots, FCB 3-5, correspond to C-terminal loops, one from each chain.[10, 12] The positions of and the connections within all the knots are shown in FIGURE 2. In four regions of the fibrinogen molecule the three chains are held together by a characteristic triple-connection in a ring-like structure, i.e., once within each of the N-terminal clusters and once within each of the intermediate clusters.[11]

CARBOHYDRATE SIDE CHAINS

Two of the peptide chains, i.e., the Bβ and γ chains,[13, 14] contain N-glyco-sidically linked carbohydrate side chains, made up of sialic acid, galactose,

TABLE 1

AMINO ACID AND CARBOHYDRATE COMPOSITION OF HUMAN FIBRINOGEN AND ITS PEPTIDE CHAINS: RESIDUES PER MOL, AND CALCULATED MOLECULAR WEIGHTS

Aminoacids		Fibrinogen	Aα-Chain	Bβ-Chain	γ-Chain
Alanine	A	142	22	23	26
Arginine	R	154	40	27	10
Asparagine	N	170	30	32	23
Aspartic acid	D	188	34	28	32
Cysteine	C	58	8	11	10
Glutamine	Q	134	18	25	24
Glutamic acid	E	192	44	30	22
Glycine	G	296	71	42	35
Histidine	H	64	15	7	10
Isoleucine	I	114	16	16	25
Leucine	L	166	29	28	26
Lysine	K	208	39	31	34
Methionine	M	66	10	15	8
Phenylalanine	F	94	19	10	18
Proline	P	138	35	23	11
Pyroglutamic acid	Z	2	–	1	–
Serine	S	280	84	31	25
Threonine	T	196	48	22	28
Tryptophan	W	66	10	13	10
Tyrosine	Y	100	9	21	20
Valine	V	134	28	25	14
Sum		2964	610	461	411
Carbohydrates					
N-Acetylglucosamine	GlcNAc	16	–	4	4
Galactose	Gal	8	–	2	2
Mannose	Man	12	–	3	3
Sialic acid	Neu	4–8	–	1–2	1–2
Molecular weight		337 897	66 062	54 358	48 529

TABLE 2

DISULFIDE KNOTS IN HUMAN FIBRINOGEN

Designation	FCB 1 (NDSK)	FCB 2	FCB 3	FCB 4	FCB 5
Position	N-terminal	Intermediate	C-terminal	C-terminal	C-terminal
Number per molecule	1	2	2	2	2
Chain components	Aα, Bβ, γ	Aα, Bβ, γ	Aα	Bβ	γ
Number of components	2 × 3	5	1	1	2
Number of sulfur bridges	11	6	1	1	1
Number of residues, per component and total	2 × (51 + 118 + 78) = 494	(60 + 34 + 18 + 63 + 170) = 345	236	53	(43 + 26) = 69

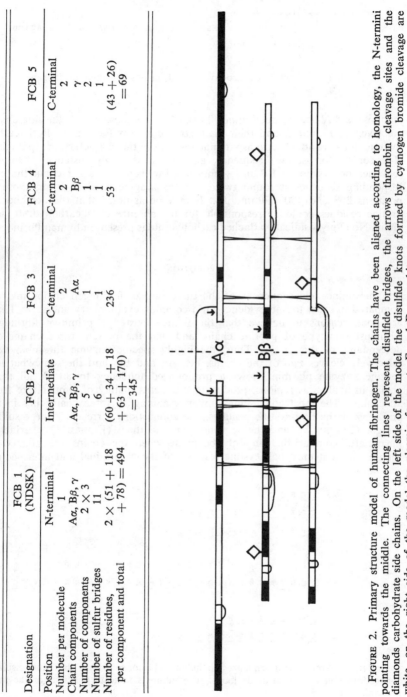

FIGURE 2. Primary structure model of human fibrinogen. The chains have been aligned according to homology, the N-termini pointing towards the middle. The connecting lines represent disulfide bridges, the arrows thrombin cleavage sites and the diamonds carbohydrate side chains. On the left side of the model the disulfide knots formed by cyanogen bromide cleavage are white, on the right side of the model the plasmic fragments E and D are white.

```
Neu → Gal → GlcNAc → Man
                          ↘
                           Man → GlcNAc → GlcNAc → Asparagine
                          ↗
(Neu)→ Gal → GlcNAc → Man
```

FIGURE 3. Structure of carbohydrate side chains in human fibrinogen.

mannose and N-acetylglucosamine (TABLE 1). The positions of the side chains are given in FIGURE 2 and their main structure [15] in FIGURE 3. Both carbohydrate units are attached to asparagine residues in the characteristic asparagine-X-threonine (or -serine) sequence, and this is the only instance when this sequence occurs in the Bβ and γ chains. However, in the Aα chain the corresponding sequence is found twice,[5] the asparagine residues being those in positions 269 and 400 (FIGURE 1). It may be assumed that the neighboring proline residues could be responsible for the absence of a carbohydrate side chain. No O-glycosidically attached carbohydrate is present in human fibrinogen.

EVOLUTION

The amino acid sequences of the three human fibrinogen chains may be compared in order to find evidence for a common evolutionary ancestor. There are good reasons to believe that there once existed a primitive fibrinogen with just one type of peptide chain, and that the present three chains are descendants of this chain. Thus, the highly specific enzyme thrombin very selectively cleaves two of the chains, i.e., Aα and Bβ, and the somewhat less specific enzyme plasmin shows a pronounced tendency to attack all three chains in analogous regions, producing fragments mainly of similar size from all chains. The most convincing similarity among all three chains lies in the positions of the half-cystine residues, the connections formed by the disulfide bridges (FIGURE 2), and the sequences around them (FIGURE 4). In fact, a meaningful comparison can only be made when the chains are aligned in such a way that most half-cystine residues of the N-terminal and intermediate

```
Aα    45  C P S G C     161  C R G S C

Bβ    76  C P T G C     193  C R T P C

γ     19  C P T T C     135  C Q E P C


Bβ   211  C E E I I R K G G E T S E M Y L I Q P D S S V K P Y R V Y C

γ    153  C Q D I A N K G A K Q S G L Y F I K P L K A N Q Q F L V Y C


Bβ   394  C S K E D G G G W W Y N R C

γ    326  C A E Q D G S G W W M N K C
```

FIGURE 4. Amino acid sequences in human fibrinogen within the triple-disulfide connections and certain disulfide loops. The numbers indicate positions of the most N-terminal residues.

cluster are brought into register. This type of alignment is indicated in FIGURE 2.

When the complete sequences of the chains are compared a remarkable homology is found for the pair Bβ-γ, i.e. a large number of positions are occupied by the identical amino acid.[16–18] The overall degree of identity is 33%, when the long N-terminal "extension" of the Bβ chain and the short C-terminal "extension" of the γ chain are disregarded. The identical positions are unevenly distributed along the chains, the C-terminal two thirds being much more similar. A detailed comparison between the sequences can be made by calculating the number of identities within a 20-residue segment, which is shifted along the chains.[17, 18] The resulting curve, which is shown in FIGURE 5 (lower, heavy line), contains pronounced maxima and minima. Maximal identity is found around positions 65-85, 200-220, 250-270, 320-340,

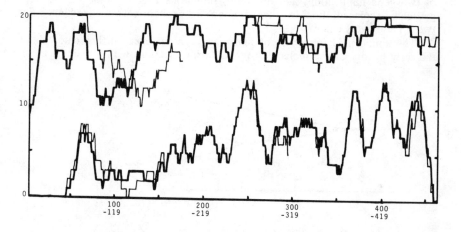

FIGURE 5. Number of identical residues in 20-residue segments of aligned human and bovine fibrinogen Bβ- and γ-chains. The lower heavy line corresponds to human Bβ- against γ-chain, the lower light line bovine Bβ- against γ-chain, the upper heavy line human against bovine Bβ-chain, and the upper light line human against bovine γ-chain.

370-390, 400-420, and 440-460. The first two peaks seem to correspond to the N-terminal and intermediate disulfide knot, respectively. The three last peaks might be related to the calcium-binding region (see below), the C-terminal disulfide loop and the C-terminal polymerization site, respectively. The fact that the Bβ-chain shows an N-terminal "extension," when the Bβ and γ chains are aligned, could serve to explain the absence of a fibrinopeptide C, the γ chain being "too short" at the N-terminus.

When the Aα-chain sequence is compared with either the Bβ or the γ chain sequence only about 12% of the positions show identity, even when regions with chain length differences are disregarded. In fact, the homology is often too week to allow a unique correct alignment. The agreement between the chains is better in the N-terminal part of the chains. Here, at least, the degree of homology is higher than would be expected from a comparison between randomized sequences.

The sequence similarities make it likely that all three chains have a common evolutionary ancestor, and that the Bβ and γ chain share a more recent ancestor, as illustrated by FIGURE 6. The γ and Bβ chain sequences must have diverged during evolution more than 450 million years ago, as even primitive vertebrates have fibrinogen molecules with three types of peptide chains. The Aα chain sequence could have diverged from the Bβ-γ chain-ancestral sequence 1500 million years ago, as the Aα chain is much less similar to the other chains. This would then have happened at about the same time as when thrombin and trypsin separated from each other.[19] In this way the highly specific enzyme thrombin and its substrate, the primitive Aα chain, would have been able to evolve together in a process of continuous mutual adaptation. It may be concluded that the original fibrinogen molecule should have been a hexamer of identical ancestral Aα chains, and that as the result of two gene duplications and numerous mutations four of these chains were gradually replaced, first by the common Bβ-γ chain-ancestor and then by two Bβ and two γ chains.

The amino acid sequences of fibrinogen contain even further clues about the early history of the molecule. When various sections within a single chain

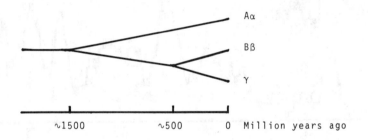

FIGURE 6. Evolution of the three fibrinogen chains.

are compared with each other numerous internal repeats are found, which may be the visible remnants of gene duplications of the type resulting in chain elongation. Thus, in all three chains the characteristic cysteine-X-Y-Z-cysteine sequence is found twice, once in the N-terminal and once in the intermediate half-cystine cluster. These sequences are shown in FIGURE 4. It is obvious that the similarity is higher within each cluster than within each chain, and this would mean that the gene duplication leading to chain elongation took place before the one leading to chain differentiation.

Additional internal repeats may be observed within the γ and the Aα chain. In the γ chain a similar sequence is found in the middle of the chain as at the C-terminus, the amino acid residues involved in the cross-linking of the C-terminus being present in the homologous sequence as well [20] (FIGURE 7). The internal repeat of the Aα chain is an at least ninefold, continuous repeat of a 13-residue segment in the middle part of the chain [5, 6] (FIGURE 8), over 60% of the positions being identical with an "average" sequence. Of the two last-mentioned internal repeats the one in the γ chain ought to be due to the gene duplication before chain differentiation, the one in the Aα chain, however, concerns only this chain and ought accordingly to have occurred later.

FIGURE 7. Internal sequence repeat in human fibrinogen γ-chain. The dots indicate the crosslinking sites at the C-terminus.

γ   148   I T G K D C Q D I A N K G A K Q S G L Y

γ   392   L T I G E G Q Q H H L G G A K Q A G D V

For a deeper understanding of the evolution of the fibrinogen molecule the amino acid sequences in different species should be compared. However, very limited information has been available about nonhuman sequences, except for those of the fibrinopeptides, the extreme variability of which has been well documented (see FIGURE 9). Recently, a considerable part of the bovine fibrinogen sequence has been elucidated. Thus, 85% of the γ chain [17, 21, 22] and, partly deduced from cDNA analysis, the total Bβ chain [17, 21-23] are know known. When these two sequences are aligned with each other and with the corresponding human sequences and again the number of identical residues are calculated the additional patterns in FIGURE 5 are obtained. It is striking how similar the human and bovine curves for the Bβ-γ alignment are (FIGURE 5, lower two lines). Comparisons between the Bβ or γ chain sequences from the two species give curves in which, surprisingly, most maxima and minima coincide with those of the Bβ-γ alignment curves (FIGURE 5, upper two lines). This would mean that those sections of the chains which were functionally important when the Bβ and γ chain diverged during evolution still are important and therefore conserved. Occasionally, a section from one chain is more conserved than the corresponding section of the other chain, and this would then indicate a specialized function of the more conserved section.

It is obvious that the evolutionary pattern which could be recognized by studying the human sequences is confirmed by results of the inter-species comparisons. The Bβ and γ chains change more slowly in their C-terminal two thirds. It should also be mentioned in this context that the Aα-chain is much more conserved in its N-terminal third, the C-terminal two thirds representing a hypervariable region of the fibrinogen molecule,[17] the degree of variability being similar to that of the fibrinopeptides.

The fibrinogen sequences may also be compared with other protein sequences in order to find evolutionary or function-dependent relationships. Here only one example will be given, demonstrating the structural similarity [18] between the calcium-binding region in calmodulin-related proteins [24] and cer-

Aα   253   G G S T S Y G - T G S E T

     265   E S P R N P S S A G S W N

     278   S G G S G P G G T G N R N

     291   P G S S G T G G T A T W K

     304   P G S S G P G S T G S W N

     317   S G S S G T G S T G N Q N

     330   P G S P R P G S T G T W N

     343   P G S S E R G S A G H W T

     356   S E S S V S G S T G Q W H

Aα-average (P) G S S (G P) G S T G - W N

FIGURE 8. Internal sequence repeats in human fibrinogen Aα-chain.

```
                       A
                       D
                       E
             G   A K   G             D
             P   E G P K             E
             S   K P Q Q             G
             T   N T T D S D E       H
     A D D V T V V E T S S   E S A       D G             I
     E T G - - - - T - - T S I T K   A T I ↓         L G   Q P   Q     E T G
Aα   - - - A D S G - E G D F L A E G G G V R↑G P R V V E R H Q S A Ċ K D S D


                       A             A
     A A A             D             G
     F F D             D   A   G   A H
     G I H             E   E   I   D L
     H L I   E       D G D G   K   L P L
     I P L   D   D E E Q E N V P K R R P D
     S T S D S   E T V S G Y R T P S T S G G             I
     Z Z Y T Y Y D H - - - - - - V V V V V I V ↓         Y   R R K       L
Bβ   - - - - - - - - Z G V N D N E E G F F S A R↑G H R P L D K K R E E A P S


     ⋎ T         E           Q                             Y
  γ  Y V A T R D N C C I L D E R F G S Y C P T T C G I A D F L S T Y Q T K V

Abn.       A D S G E G D F L A E G G G V R↑G P R V V E R H Q S A C K D S D
                         N           V       C       S
                                             H       N
```

FIGURE 9. N-Terminal amino acid sequence variations in mammalian fibrinogens and in human abnormal fibrinogens. The continuous line represents the normal human sequence; the arrows, the thrombin cleavage sites; the — indicates a deletion (i.e. shorter sequence); ⋎, a blocked tyrosine. Fibrinopeptide positions correspond to about 40 species, fibrin positions to only 5–9 species.

tain Bβ and γ chain segments from the calcium-binding region in fibrinogen.[25] The pertinent sequences are aligned in FIGURE 10, and the known calcium ligands in calmodulin as well as the suggested ligands in the γ and/or Bβ chain are indicated. Supporting evidence for the relevance of the hypothesis is given by the facts that in the γ chain the lysyl-phenylalanine bond on the N-terminal side of this sequence is protected against plasmin digestion and the disulfide bridge on the C-terminal side of the sequence is protected against mercaptolysis by calcium ions.[25] The Bβ chain might be independently protected against plasmin attack by the neighboring carbohydrate side chain.

```
                            x    y    z    -y  -x   -z
Calmodulin       + E X + + X X + + X D + O + O G - + I O + + E X + + X X + + X
γ 301     D K↑F F T S H N G M Q F S T W D N D N D K E E G N C A E Q D G S G W W M
Bβ364     N⁺R T M T I H N G M F F S T Y D R D N D G - W L T S - D P R K Q C S K E
```

FIGURE 10. Homology between calcium-binding sequence in calmodulin and suggested calcium-binding sequence in human fibrinogen γ- and Bβ- chain. The x — -z denote calcium ligands in calmodulin; X, hydrophobic residues; O, oxygen-containing ligand residues; —, deletion; +, arbitrary residue; and + over N denotes carbohydrate side chain. Underlined are residues in agreement with calmodulin "test sequence."

ABNORMAL FIBRINOGENS

Comparisons between the structure of normal fibrinogen and that of genetically abnormal fibrinogens give a detailed insight into the functional importance of single amino acid residues or regions of the molecule. Close to 100 families with defective blood coagulation due to a molecular dysfunction of fibrinogen have already been identified. However, until recently the corresponding structural error had only been exactly located in Fibrinogen Detroit, in which the arginine residue in position 19 of the Aα chain is replaced by serine [26] (see FIGURE 9). During the last two years 12 additional abnormal fibrinogens have been successfully analyzed, 10 of these by our group.[27-31]

The structural error could, in principle, be located in any of the 1482 different positions of the abnormal fibrinogen molecule. It seems, however, reasonable to assume that amino acid substitutions in certain regions of the molecule would give rise to more severe functional defects. Experience has shown that most dysfibrinogenemias are discovered because of the prolonged thrombin and/or reptilase clotting times, the delay being caused by defective fibrinopeptide release, defective fibrin polymerization, or both. It is therefore meaningful to start the search for the structural error by analyzing the fibrinopeptides released by thrombin. As a first attempt the number and the properties of the peptides should be determined. As a second attempt the kinetics of release for both fibrinopeptides should be estimated. Finally, when a fibrinopeptide release defect has been ruled out, the functional state of the polymerization domains should be tested.

For fibrinopeptide analysis the newly developed high-performance liquid chromatography procedure is the method of choice.[32, 33] The method allows the simultaneous, quantitative determination and micro-preparative isolation of all fibrinopeptides and their common degradation products. The chromatographic pattern obtained with normal peptides is shown in the upper part of FIGURE 11. In order to study the fibrinopeptides released, normal fibrinogen and 28 different genetically abnormal fibrinogens were incubated for about 20 hours with a high thrombin concentration, and the supernatant liquid after heat precipitation analyzed. Comparing the results with the different samples five characteristic release patterns could be discerned, as indicated in TABLE 3. The interpretation of these patterns led in many cases to the localization of the structural error.

The first pattern (TABLE 3) is characterized by no A-peptide release, only B-peptide release. This pattern (FIGURE 11, middle) was found only in a single case, Fibrinogen Metz.[28-30] Sequence analysis of the isolated Aα-chain gave the explanation why no A-peptide could be cleaved off. The arginine residue of the thrombin cleavage site, i.e., position 16 in the Aα chain, is homozygously replaced by cyst(e)ine. This abnormal fibrinogen is the first one to coagulate exclusively by B-peptide release. When the kinetics of B-peptide release was determined it was a surprise to find that Fibrinogen Metz will release the peptide at an even slightly higher rate than normal fibrinogen. It has earlier been postulated that thrombin cleavage at the A-peptide site and conformational change should precede a rapid cleavage at the B-peptide site. However, it is obvious that at least for Fibrinogen Metz this does not apply.

The second pattern is recognizable by the appearance of only one mole of A-peptide versus two moles of B-peptide in the chromatography. This pat-

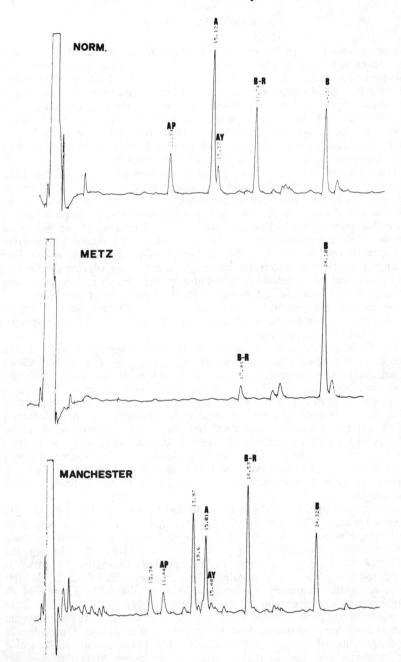

FIGURE 11. High-performance liquid chromatographies of the fibrinopeptides released by thrombin from normal human fibrinogen and from Fibrinogen Metz and Fibrinogen Manchester.

tern was obtained with eight fibrinogens. However, the structural error has so far been identified only in one case, i.e., Fibrinogen Zürich I. Here the amino acid substitution is the same one as in Fibrinogen Metz, though in the heterozygous form.[28, 30]

In the third pattern of fibrinopeptide release as analyzed by high-performance liquid chromatography only A-peptides with abnormal retention times are found in addition to the B-peptides. A single sample, that of Fibrinogen Bicêtre, belonged to this group. Determination of amino acid composition and sequence of the abnormal A-peptide proved that its C-terminal arginine residue is homozygously exchanged against histidine. The fact that thrombin may cleave a histidyl bond selectively and quantitatively, even though at a considerably decreased rate, has until recently been quite unexpected.

TABLE 3

FIBRINOPEPTIDES RELEASED BY THROMBIN IN NORMAL AND GENETICALLY ABNORMAL FIBRINOGENS, MOLES PER MOLE

A		B	
Abnormal	Normal		Fibrinogen
–	2	2	Normal
–	–	2	Metz
–	1	2	Amsterdam, Frankfurt II, Frankfurt III, London III, Marseille, Stony Brook, Schwarzach, Zürich I
2	–	2	Bicêtre
1	1	2	Louisville, Manchester, New Albany, Rouen, Sydney I, Sydney II
–	2	2	Copenhagen, Frankfurt I, London IV, München I, Münster I, Oslo I, Oslo II, Oslo IV, Philadelphia II, Pontoise, Temple I, Temple II

The fourth pattern showed the presence of one mole of A-peptide with abnormal retention time and one mole with normal retention time. In five out of the six fibrinogens from this group the abnormal A-peptides were identical, and furthermore, they were identical with the abnormal peptide from group three (above), i.e. in all these abnormal fibrinogens the arginine residue in position 16 of the Aα chain is substituted by histidine. The cases in this group are the heterozygous versions of the case in group three. The chromatography obtained with the peptides of Fibrinogen Manchester,[34] a representative of this group, is shown in the lower part of FIGURE 11. In the sixth case from this group, Fibrinogen Rouen, the abnormal A-peptide had a much longer retention time. Here the substitution was found to be glycine against valine in position 12 of the Aα chain.

In the remaining abnormal fibrinogens the normal amount and type of fibrinopeptide was found to be released, i.e., a fifth pattern (TABLE 3). However, in some instances a delayed A-peptide release was observed, indicating the presence of a structural error on the fibrin side of the thrombin cleavage site. Thus, in Fibrinogen München I sequence analysis of the Aα and α chain demonstrated that the arginine residue in position 19 is exchanged heterozygously against asparagine,[27, 28, 30] i.e., the same position is altered as in Fibrinogen Detroit.[26] In some other cases from this heterogeneous group elongated peptide chains had already been discovered, in still others a structural error in a polymerization domain seems to cause the functional defect. It is obvious that several different strategies will have to be applied in order to find the various primary structure defects.

The 13 so-far-identified amino acid substitutions in genetically abnormal fibrinogens have all been indicated in the lower part of FIGURE 9. Out of these 13 cases 10 have been described above, the remaining three being Fibrinogen Detroit, with the substitution in position 19 of the Aα chain,[26]

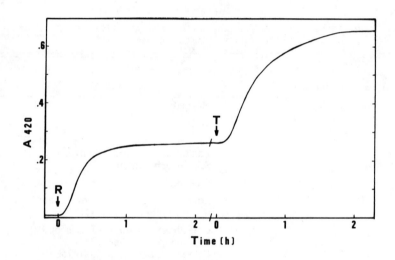

FIGURE 12. Optical density measurement on Fibrinogen Louisville during coagulation with first reptilase, R, and then thrombin, T.

Fibrinogen Lille, with the substitution in position 7 of aspartic acid against asparagine,[35] and Fibrinogen Petoskey, with the substitution in position 16 of arginine against histidine.[36] It might at first seem remarkable that amino acid exchanges have been found only in such a narrow section of the molecule. The explanation lies no doubt in the great functional importance of this region of the fibrinogen molecule and in the hitherto established methods to identify structural errors in fibrinogen. It is noteworthy that nine of the 13 cases imply a modification at the thrombin cleavage site, seven of these being replacements by histidine. A comparison between mammalian evolutionary variants, in the upper part of FIGURE 9, and human genetically determined abnormal variants, in the lower part of the FIGURE, illustrates the difference between "allowed" and "not allowed" structure in relation to appropriate and nonappropriate function. It is striking that all substitutions in the abnormal fibrinogens, except in Fibrinogen Lille, have occurred in evolutionary strictly nonvariant positions. The dysfunction of a molecule with a substitution in such a position is not a surprise.

With some of the abnormal fibrinogens in the heterozygous form it is possible to separate out fractions that differ in reptilase-inducible clottability. This fact would imply that different molecular species are present. When abnormal fibrinogens with the second or fourth peptide release pattern (TABLE 3), i.e. those which contain a single mole of normal A-peptide, are treated with reptilase only the normal A-peptide is cleaved off, causing exclusively the corresponding molecules to coagulate. The remaining molecules coagulate on treatment with thrombin, in fibrinogens with the second release pattern because B-peptide is cleaved off and in those with the fourth pattern because abnormal A-peptide and B-peptide are cleaved off. The two steps of coagulation with first reptilase and then thrombin are shown for Fibrinogen Louisville [37] in FIGURE 12. It was observed that the reptilase-clottable part of this fibrinogen released exclusively B-peptide and the reptilase-unclottable part exclusively abnormal A-peptide and B-peptide on thrombin treatment. The interpretation of this would be that only completely normal and completely abnormal molecules were present in this sample of abnormal fibrinogen, i.e., no hybrid molecules seem to exist. The finding that in heterozygous cases of genetically abnormal fibrinogen no hybrids between normal and abnormal molecules appear will be of great importance for the understanding of peptide chain assembly during fibrinogen biosynthesis.

REFERENCES

1. HENSCHEN, A. 1981. Fibrinogen—Blutgerinnungsfaktor I—Biochemische Aspekte. Hämostaseologie 1: 30–40.
2. HENSCHEN, A. & P. EDMAN. 1972. Large scale preparation of S-carboxymethylated chains of human fibrinogen and fibrin and the occurrence of γ chain variants. Biochim. Biophys. Acta 263: 351–367.
3. LOTTSPEICH, F. & A. HENSCHEN. 1977. Amino acid sequence of human fibrin. Preliminary note on the completion of the γ-chain sequence. Hoppe-Seyler's Z. Physiol. Chem. 358: 935–938.
4. HENSCHEN, A. & F. LOTTSPEICH. 1977. Amino acid sequence of human fibrin. Preliminary note on the completion of the β-chain sequence. Hoppe-Seyler's Z. Physiol. Chem. 358: 1643–1646.
5. HENSCHEN, A., F. LOTTSPEICH & B. HESSEL. 1979. Amino acid sequence of human fibrin. Preliminary note on the completion of the intermediate part of the α-chain sequence. Hoppe-Seyler's Z. Physiol. Chem. 360: 1951–1956.
6. DOOLITTLE, R. F., K. W. K. WATT, B. A. COTTRELL, D. D. STRONG & M. RILEY. 1979. The amino acid sequence of the α-chain of human fibrinogen. Nature 280: 464–468.
7. HENSCHEN, A. 1964. Number and reactivity of disulfide bonds in fibrinogen and fibrin. Arkiv Kemi. 22: 355–373.
8. BLOMBÄCK, B., M. BLOMBÄCK, A. HENSCHEN, B. HESSEL, S. IWANAGA & K. R. WOODS. 1968. The N-terminal disulphide knot of human fibrinogen. Nature 218: 130–134.
9. BLOMBÄCK, B., B. HESSEL & D. HOGG. 1976. Disulfide bridges in NH$_2$-terminal part of human fibrinogen. Thromb. Res. 8: 639–658.
10. GÅRDLUND, B., B. HESSEL, G. MARGUERIE, G. MURANO & B. BLOMBÄCK. 1977. Primary structure of human fibrinogen. Eur. J. Biochem. 77: 595–610.

11. HENSCHEN, A. 1978. Disulfide bridges in the middle part of human fibrinogen. Hoppe-Seyler's Z. Physiol. Chem. **359:** 1757–1770.

12. HENSCHEN, A., F. LOTTSPEICH & B. HESSEL. 1978. Amino acid sequence of human fibrin. Preliminary note on a disulfide-containing internal peptide of the α-chain, obtained by plasmic digestion of fibrinogen. Hoppe-Seyler's Z. Physiol. Chem. **359:** 1607–1610.

13. TÖPFER-PETERSEN, E., F. LOTTSPEICH & A. HENSCHEN. 1976. Carbohydrate linkage site in the β-chain of human fibrin. Hoppe-Seyler's Z. Physiol. Chem. **357:** 1509–1513.

14. BLOMBÄCK, B., N. J. GRÖNDAHL, B. HESSEL, S. IWANAGA & P. WALLEN. 1973. Primary structure of human fibrinogen and fibrin. J. Biol. Chem. **248:** 5806–5820.

15. TÖPFER-PETERSEN, E. 1980. Kohlehydratseitenketten im Fibrinogen. *In* Fibrinogen, fibrin and fibrin glue. K. Schimpf, Ed.: 43–45. Schattauer, Stuttgart.

16. HENSCHEN, A. & F. LOTTSPEICH. 1977. Sequence homology between γ-chain and β-chain in human fibrin. Thromb. Res. **11:** 869–880.

17. HENSCHEN, A., F. LOTTSPEICH, E. TÖPFER-PETERSEN, M. KEHL & R. TIMPL. 1980. Fibrinogen evolution: intra- and interspecies comparisons. Protides Biol. Fluids **28:** 47–50.

18. HENSCHEN, A., F. LOTTSPEICH, M. KEHL, C. SOUTHAN & J. LUCAS. 1982. Structure-function-evolution relationship in fibrinogen. *In* Fibrinogen–Recent Biochemical and Medical Aspects. A. Henschen, H. Graeff & F. Lottspeich, Eds.: 67–89. de Gruyter. Berlin.

19. DAYHOFF, M. O. 1972. Atlas of protein sequence and structure. 5. National Biomedical Research Foundation. Washington, D.C.

20. LOTTSPEICH, F. & A. HENSCHEN. 1977. Amino acid sequence of human fibrin. Preliminary note on an internal peptide obtained by cleaving the γ-chain at the arginyl bonds and showing sequence homology with the C-terminus. Hoppe-Seyler's Z. Physiol. Chem. **358:** 703–707.

21. TÖPFER-PETERSEN, E. & A. HENSCHEN. 1977. Carbohydrate carrying peptides isolated from bovine β- and γ-chain. Thromb. Res. **11:** 881–892.

22. TIMPL, R., P. P. FIETZEK, V. VAN DELDEN & G. LANDRATH. 1978. Disulfide-linked cyanogen bromide peptides of bovine fibrinogen. III. Isolation and identification by sequence analysis of the constituents in peptides F-CB2, F-CB4 and F-CB5. Hoppe-Seyler's Z. Physiol. Chem. **359:** 1553–1560.

23. CHUNG, D. W., M. W. RIXON, R. T. A. MACGILLIVRAY & E. W. DAVIE. 1981. Characterization of a cDNA clone coding for the β-chain of bovine fibrinogen. Proc. Natl. Acad. Sci. USA **78:** 1466–1470.

24. TUFTY, R. M. & R. H. KRETSINGER. 1975. Troponin and parvalbumin calcium binding regions predicted in myosin light chain and T4 lysozyme. Science **187:** 167–169.

25. HAVERKATE, F. & G. TIMAN. 1977. Protective effect of calcium in the plasmin degradation of fibrinogen and fibrin fragments D. Thromb. Res. **10:** 803–812.

26. BLOMBÄCK, M., B. BLOMBÄCK, E. F. MAMMEN & A. S. PRASAD. 1968. Fibrinogen Detroit–a molecular defect in the N-terminal disulphide knot of human fibrinogen? Nature **218:** 134–137.

27. HENSCHEN, A. & C. SOUTHAN. 1980. Methode zur Isolierung abnormer Fibrinogene aus Plasma. *In* Fibrinolyse, Thrombose, Hämostase. E. Deutsch & K. Lechner. Eds.: 290–293. Schattauer. Stuttgart.

28. HENSCHEN, A., C. SOUTHAN, M. KEHL & F. LOTTSPEICH. 1981. The structural error and its relation to the malfunction in some abnormal fibrinogens. Throm. Haemostas. **46:** 181.

29. SORIA, J., C. SORIA, M. SAMAMA, A. HENSCHEN & C. SOUTHAN. 1982. Detection of fibrinogen abnormality in dysfibrinogenemia: Special report on Fibrinogen

Metz characterised by an amino acid substitution located at the peptide bond cleaved by thrombin. *In* Fibrinogen–Recent Biochemical and Medical Aspects. A. Henschen, H. Graeff & F. Lottspeich. Eds.: 129–143. de Gruyter, Berlin.

30. SOUTHAN, C., A. HENSCHEN & F. LOTTSPEICH. 1982. The search for molecular defects in abnormal fibrinogens. *In* Fibrinogen–Recent Biochemical and Medical Aspects. A. Henschen, H. Graeff & F. Lottspeich. Eds.: 153–166. de Gruyter, Berlin.

31. HENSCHEN, A., M. KEHL, C. SOUTHAN & F. LOTTSPEICH. 1983. Genetically abnormal fibrinogens—an overview. *In* Fibrinogen–Structure, Functional Aspects and Metabolism. F. Haverkate, A. Henschen, W. Nieuwenhuizen & P. W. Straub. Eds.: 125–144. de Gruyter. Berlin.

32. KEHL, M., F. LOTTSPEICH & A. HENSCHEN. 1981. Analysis of human fibrinopeptides by high-performance liquid chromatography. Hoppe-Seyler's Z. Physiol. Chem. **362:** 1661–1664.

33. KEHL, M., F. LOTTSPEICH & A. HENSCHEN. 1982. Analysis of human fibrinopeptides by high-performance liquid chromatography: Quantitative determination and kinetics of release. *In* Fibrinogen–Recent Biochemical and Medical Aspects. A. Henschen, H. Graeff & F. Lottspeich. Eds.: 217–226. de Gruyter. Berlin.

34. SOUTHAN, C., M. KEHL, A HENSCHEN & D. A. LANE. 1983. Fibrinogen Manchester, Characterisation of an abnormal fibrinopeptide A, released by thrombin and with an arginine → histidine substitution in position 16 of the Aα-chain. Brit. J. Haematol. In press.

35. MORRIS, S., M. H. DENNINGER, J. S. FINLAYSON & D. MENACHE. 1981. Fibrinogen Lille: Aα 7 Asp → Asn. Thromb. Haemost. **46:** 104.

36. HIGGINS, D. L. & J. A. SHAFER. 1981. Fibrinogen Petoskey, a dysfibrinogenemia characterized by replacement of Arg-Aα16 by a histidyl residue. J. Biol. Chem. **256:** 12013–12017.

37. GALANAKIS, D. K., A. HENSCHEN, M. KEELING, M. KEHL, R. DISMORE & E. I. PEERSCHKE. 1983. Fibrinogen Louisville: An Aα16 Arg → His defect which forms no hybrid molecules in heterozygous individuals and inhibits aggregation of normal fibrin monomers. Ann. N.Y. Acad. Sci. This volume.

DISCUSSION OF THE PAPER

L. LORAND (*Northwestern University, Evanston, IL*): How conservative is the mid-portion of the α chain where the crosslinking sites are?

A. HENSCHEN: It seems to be very variable; the C-terminals of the α chain, which have the same kind of substitution rate as fibrinopeptides, are especially variable.

IMMUNOCHEMICAL ANALYSIS OF THE CONFORMATION OF FIBRINOGEN *

E. F. Plow and T. S. Edgington

Department of Molecular Immunology
Research Institute of Scripps Clinic
La Jolla, California 92037

C. S. Cierniewski

Department of Biophysics
Institute of Physiology and Biochemistry
Medical School of Lodz
Lodz, Poland

INTRODUCTION

Fibrinogen is a multi-functional molecule possessing diverse biological activities within the coagulation and hemostatic systems and broadly participating in extravascular inflammatory responses. The dramatic, visual nature of the fibrin transition has served to focus the greatest attention upon the transition of fibrinogen to fibrin and the polymerization of the fibrin derivative. The capacity of fibrinogen to mediate platelet aggregation induced by a wide variety of physiologic and pharmacologic stimuli indicates the existence of a fibrinogen-dependent pathway of platelet aggregation [1] and establishes a role for the molecule in the primary phase of hemostasis, a role of potential significance equal to that of the fibrin transition. These two hemostatic effector functions exemplify biological activities of fibrinogen, which are critically dependent upon the native structure of the molecule as exemplified by the marked reduction or modulation of these activities by limited proteolysis.[2, 3] In contrast, other physiologic functions of fibrinogen require degradation to generate effector fragments. The capacity to regulate fibrinogen synthesis,[4] the expression of vasoactive properties,[5] and the suppression of the immune response [6] represent biological activities which are contingent upon proteolytic degradation of fibrinogen. Although the primary sequence of fibrinogen contains the structural elements requisite for all these activities, the tertiary folding and overall conformation of the molecule dictate their expression. Therefore, a clear perspective of the conformation of fibrinogen is required to establish precise relationships between structure and functions.

During the past three decades, the conformation of fibrinogen has been a topic of extensive interest. A variety of biophysical approaches have stressed the anomalous behavior of fibrinogen; but the contributions of asymmetry versus hydration have not been fully differentiated.[7] The direct visualization of fibrinogen by electron microscopy is a particularly appealing approach to the conformational analysis of the molecule. Such studies, however, have disclosed a variety of conformational formats for the molecule. These have ranged in

* This work was supported by National Institutes of Health Grant POI HL–16411.

0077–8923/83/0408–0044 $01.75/0 © 1983, NYAS

extremes from globular to extended molecular profiles.[8, 9] Other approaches to conformational analysis of fibrinogen have provided pertinent information and have established constraints which must ultimately be resolved and integrated into a comprehensive model. As an independent approach, immunochemical analyses of fibrinogen are particularly relevant as they permit examination of the conformation of the molecule as it exists in solution and under relatively physiologic conditions.

The underlying premise of conformational analyses with antibodies can be stated at the most elementary level, as follows. In order for a discrete intramolecular locus, an epitope, to be recognized by a specific antibody (the analytical probe), the epitope must be expressed at the hydrated surface of the molecule in a conformation accessible and complementary to the binding site of the antibody probe. By characterizing a number of epitopes of fibrinogen in sufficient detail so as to classify their status as "exposed" or "unexpressed" sites, a putative topographical map of the molecule may be developed. Changes in epitope expression induced by the fibrin transition or by plasmic proteolysis identify local changes in conformation associated with known physiologic modifications of the molecule.

An antigenic epitope exists minimally of a six amino acid stretch; and epitope expression is established by: (1) the local secondary conformation imposed by the amino acid sequence; and, (2) the tertiary effects of contiguous amino and carboxy terminal sequences and even by distant sequences which are brought into close proximity of the epitope by the overall folding of the molecule. Although the characterization of individual epitopes provides a more precise perspective of conformation than the examination of single amino acid residues, the size of individual epitopes is quite limited relative to the more than 1,500 residues per half fibrinogen molecule. Thus, immunochemical data are most effectively interpreted within the context of a hypothesized molecular format. The immunochemical analyses may indicate precise architectural details within the overall shape of the molecule. The lack of consensus as to the number of possible conformations of fibrinogen in a hydrate free or monomeric state (a circumstance that may be distinct from formats selected by surface association as protein:protein interactions) has indeed handicapped the full interpretation of molecular details provided by immunochemical studies. Recently, a number of independent laboratories have consistently visualized fibrinogen by electron microscopy in an extended trinodular format.[10-15] The coiled coils, predicted from amino acid sequence, can be readily envisioned as the interdomainal connections between nodules.[16] This model seems most reasonably to represent one of the molecular formats and one within which we will interpret the conformational details of fibrinogen identified by immunochemical studies. We have recently reviewed the epitopes of fibrinogen in detail,[17] and we shall attempt to avoid a recapitulation of this review. Rather, this analysis will focus on the agreements, the constraints and the complexities of the trinodular format of fibrinogen imposed by immunochemical analyses.

THE TRINODULAR MODEL

In 1959, Hall and Slater [8] produced electron micrographs of fibrinogen in which the molecule appeared in an extended trinodular format. The essential features of this model are depicted in FIGURE 1. The overall length of the molecule was 475 ± 25A, with two larger peripheral nodules (65A) connected to a smaller central nodule (50A) by thin interconnecting threads of 8 to 15A diameter. Despite the capacity of this model to attractively accommodate an increasing number of biochemical and biophysical observations, particularly the polymerization of fibrin, the early failure of other laboratories to reproduce these structures was troublesome. During the past four years, the trinodular format, or subtle variations thereof, has been visualized in several independent laboratories.[10-15] Particularly convincing has been the scanning transmission electromicrographs of Estis and Haschmeyer [14] and more recently by Mosesson et al.[15] who were able to demonstrate trinodular structure (among other forms) in nonstained preparations. Electron microscopy using antibodies to the domains of fibrinogen has clearly established that the two peripheral nodules correspond to the D domains and the central nodule to the E domain.[11, 13, 18] The thread-like structures interconnecting the D and E domains could represent the postulated coiled coils deduced from predictive analysis of the primary amino acid sequence by Doolittle et al.[16] The coiled coils would

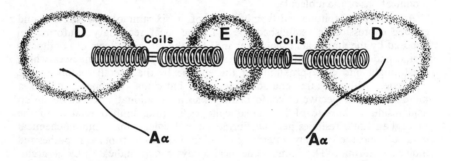

FIGURE 1. Schematic representation of the trinodular model of fibrinogen. The domainal identifications, the coiled-coils and the carboxy terminal of the Aα chain have been superimposed on the model originally proposed by Hall and Slayter.[8]

be encompassed within residues 50–160 of the Aα chain, 81–192 of the Bβ chain, and 24–134 of the γ chain. These sequences are highly helix permissive; and, when brought into registry between the sets of disulfide collars, they can be intertwined into a three-stranded super-helical structure. The coiled coils are initiated within the E domain, relax centrally at the protease sensitive interdomainal cleavage sites, recoil and terminate within the D domains. The carboxy terminal two-thirds of the Aα chains was not accounted for in the original trinodular model, and these may be considered as a pair of independent domains. The predominance of hydrophilic amino acids and the lack of helical content may permit this region to be a "free-swimming appendage" of the molecule.[7] Immunolocalization has visualized this region as being in close proximity but extending from the pair of D domains of the molecule.[11] Within this general format, the hydrodynamic properties, the domainal organization, the plasmin susceptibility, crystallographic data and fibrin formation can be readily accommodated. Thus, a growing acceptance of the trinodular format of the molecule as "the" true conformation of fibrinogen predominates in the current literature.

IMPLICATIONS OF THE TRINODULAR MODEL

Inherent in the extended trinodular model of fibrinogen as outlined above are a number of specific implications. First, the D and E domains would be conformationally independent of one another as they are separated by the coiled-coils. Second, with the initial sites of plasmic cleavage localized to the carboxy terminal of the Aα chain and to the interdomainal regions, proteolysis should result in only limited perturbations of domainal conformation. Indeed, electron micrographs of fragments D and E indicate that they retain the dimensions of the D and E domains in the trinodular display of the intact molecule. Third, the predicated coiled-coils should present a systematic and ordered structure. Fourth, the carboxy-terminal region of the Aα chain should be surface-oriented and exist in a relatively unordered structure. Immunochemical analyses relevant to these implications are addressed.

THE INDEPENDENCE OF THE D AND E DOMAINS

Fragments D and E exhibit classical immunochemical nonidentity by a variety of methods.[19] Quantitative immunochemical analyses permit more incisive assessment of the antigenic independence of the fragments, as illustrated in FIGURE 2. Using antibody to the intact fibrinogen molecule and specifying epitope recognition to fragment D by using ^{125}I-fgD as the ligand in competitive inhibition analysis, fragment E is at least 1,000-fold less reactive than fragment D (<0.1% cross-reactivity). Similarly, with ^{125}I-fgE as the ligand, fragment D is at least 500-fold less reactive than fragment E (<0.2% cross-reactivity). Such analyses demonstrate that D and E *fragments* do not share epitopes in common but do not exclude the possibility that the D and E *domains* might be interactive or share epitopes when integrated into native fibrinogen.

The immunochemical analysis shown in FIGURE 3 bears more directly on the interaction of the D and E domains. In this analysis, native fibrinogen epitopes are examined by utilizing ^{125}I-fibrinogen as the ligand and anti-fibrinogen antibodies. The preeminent conclusion from this analysis is that the late stage plasmic digest, consisting exclusively of fragments D and E as the high molecular weight polypeptide fragments, exhibits markedly enhanced epitope expression relative to the sum of the isolated fragments D and E. Dialysis of the terminal digest was without effect, thus dismissing a contribution of small peptides. These observations are readily interpretable in terms of an interaction between fragments D and E in the terminal digest with conformational effects sufficient to impart greater expression of epitopes characteristic of native fibrinogen but expressed poorly by the separated D and E domains.[20] Since this interaction recapitulates the native conformation of fibrinogen, this argues against the contribution of random or nonspecific associations in contributing to this phenomenon.[20, 21]

More supportive evidence for the conformational interaction between the D and E domains has been obtained from analysis of a single epitope set, the D:E dependent site (DEDS). DEDS represents one of the few γ chain epitopes which is expressed by native fibrinogen. It is identified in a system consisting of ^{125}I-fibrinogen ligand and anti-γ chain serum.[22] The terminal plasmic digest presents the DEDS less well than intact fibrinogen, but isolated

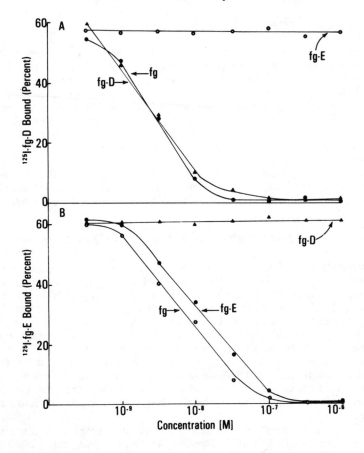

FIGURE 2. The antigenic independence of fragments D and E. ^{125}I-fragment D and ^{125}I-fragment E are the ligands in Panels A and B, respectively, and the antiserum is anti-fibrinogen. In this configuration fibrinogen epitopes expressed by fragment D (Panel A) on fragment E (Panel B) are specified, and no evidence of epitopes shared by the fragments is observed.

D and E fragments are entirely devoid of expression (FIGURE 4). Thus, these results corroborate those presented in FIGURE 3 but are independently demonstrated with a very restricted epitope set. When fragment D is reduced and alkylated and its γ chain isolated, this 24,000 molecular weight species contains the DEDS locus. Thus, the Dγ chain expresses an epitope which is shared by the "D:E complex" and by native fibrinogen but not by the isolated D and E fragments. Since the Dγ chain is separated from the carboxy-terminus of the Eγ chain by at least 50 amino acids, the interdomainal interactions which maintain the DEDS site must result from conformational interplay between the domains rather than from contiguity of primary sequence. The reexpression of DEDS by the isolated Dγ chain corroborates this conclusion. In summary, the immunochemical analyses are more consistent with a model permitting frequent conformational interaction and communication between domains when in a hydrated state under relatively physiologic conditions rather than a model based on spatially distant and totally independent domains. This, however, does not exclude an extended format as one of a number of conformations occurring at a finite frequency or in equilibrium with a compact format.

EFFECTS OF PLASMIC CLEAVAGE

The generation of the major plasmic cleavage products of fibrinogen can be attributed to the proteolysis of a limited number of peptide bonds. These cleavage sites reside primarily in the carboxy-terminal of the Aα chain and within the interdomainal connectors. The cleavage sites within the D domain and the sequence of their plasmic cleavage result in a progressive shortening of the major residual Dα and the Dγ chains. In the latter case, calcium ions protect the carboxy-terminal aspects of the γ chain from proteolysis.[23] In the E domain, the amino-terminal aspect of the Bβ chain is a favored site of plasmic cleavage in which the peptide bond at Bβ 42–43 is first attacked.[24, 25] The amino-terminal region of the Aα chain and the carboxy-terminal region

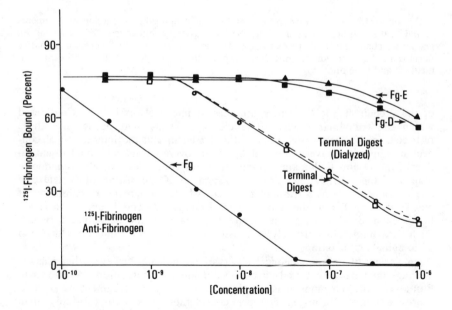

FIGURE 3. Expression of native fibrinogen epitopes by fibrinogen and its plasmic degradation products. The radioimmunoassay system consists of ^{125}I-fibrinogen and anti-fibrinogen, specifying the recognition of native fibrinogen epitopes. The terminal digest, containing only fragments D and E as the high molecular weight derivatives, expresses these epitopes although less well than intact fibrinogen. The individual D and E fragments, isolated from the same digest by DEAE-cellulose chromatography, exhibit minimal expression of these epitopes.

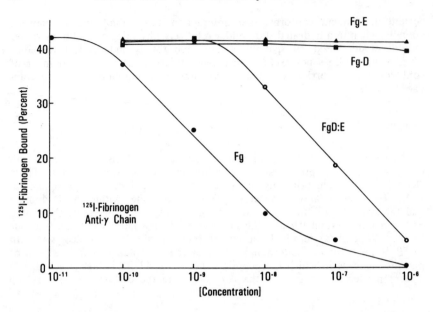

FIGURE 4. Identification of epitope set dependent upon the interaction of fragments D and E, the D:E dependent site (DEDS). The assay system consists of ^{125}I-fibrinogen and anti-γ chain and limits recognition to the few γ chain epitopes expressed by native fibrinogen. The terminal plasmic digest express these epitopes but its constituent fragments D and E do not.

of the γ chain in the E domain are sites of late proteolysis.[24, 26] Considered within the trinodular model, these cleavage events, and particularly those restricted to the carboxy-terminal of the Aα chain and the interdomainal connectors, appear to have limited effect on the overall conformational organization of the D and E domains. Supporting this conclusion are electron micrographs of the degradation products. Fragment X, consisting of two D and one E domain, retains a trinodular format; whereas fragment Y, consisting of one D and one E domain, is binodular. Fragment D and E appear to retain the essential dimensions of the D and E domains in the trinodular format.[27] Such data imply that proteolysis is associated with only limited perturbation of domainal organization.

A variety of immunochemical observations indicate, however, that plasmic cleavage does have profound effects on domainal conformation. As shown in FIGURE 3, the terminal plasmic digest consisting of fragment D and E expresses native fibrinogen epitopes considerably less well (50-fold) than intact fibrinogen indicating the perturbation of these epitopes. This represents conformational perturbations since most of the epitopes are expressed but have a lower affinity for their specific antibodies. The generation of a cleavage associated neoantigen in each domain, fgDneo and fgEneo, indicates that proteolysis also alters domainal conformation sufficiently to engender a new epitope set in each domain.[28, 29] Systematic immunochemical analysis of the γ chain have also indicated that the amino-terminal two-thirds of this chain undergoes progressive reconformation and/or exposure during proteolysis.[30, 31] In summary, these observations attest to the dynamic changes in the conformation of both the D and E domain in association with plasmaic proteolysis.

Of particular import in considering the conformational organization of fibrinogen is the apparent capacity of a proteolytic event restricted to one domain to influence an independent domain. This is a highly significant event. The cleavage associated neoantigen in the D domain, fgDneo, serves to illustrate this point. Antiserum to fragment D is rendered specific for the fg-Dneo neoantigen by absorption with excess fibrinogen or fresh plasma.[28, 32] Analysis uses ^{125}I-fgD as the ligand with the specific absorbed antiserum (anti-fgDneo). As defined, fgDneo is not expressed by intact fibrinogen (<1 mole fgDneo per 10^5 moles fibrinogen) but is fully expressed on fragment D (1 mole fg-Dneo per fragment D) [FIGURE 5]. Loss of the carboxy-terminal region of the Aα chain such as for high solubility fibrinogen fraction I-9 does not result in expression of fgDneo. The intermediate plasmic degradation product fragment X, however, fully expresses the neoantigen. The single structural difference between I-9 and fragment X is in the deletion of Bβ 1–42 from the E domain of fibrinogen. Indeed, we have verified that the generation of amino-terminal alanine at Bβ 43 and the expression of fgDneo are coincident events. We have recently localized fgDneo to γ 160–197. This indicates that cleavage of Bβ 1–42 in the E domain results in exposure of γ160–197 in the D domain. This clearly documents the dynamic effect of plasmic cleavage on the conformation of the molecule and requires either direct energetic interplay

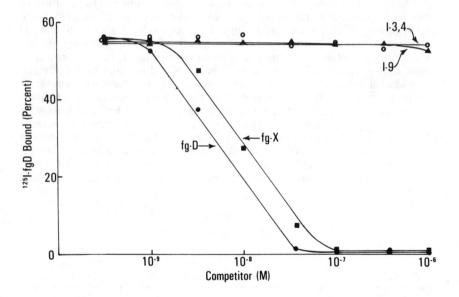

FIGURE 5. The cleavage associated neoantigen of the D domain, fg-Dneo. The ligand is ^{125}I-fragment D, and the antiserum, anti-fgDneo, is prepared by extensive absorbance of anti-fragment D with plasma fibrinogen. The recognized epitope, fg-Dneo, is not expressed by fibrinogen I-3,4 or I-9 but is fully expressed by fragments X and D.

between the D and E domains or the relatively long distance transmission of forces from one domain to the other so as to impart reconformation.

THE COILED COILS

Regions of fibrinogen residing between the cysteine residues 49 and 161 of the Aα chain, 80 and 192 of the Bβ chains and 23 and 135 of the γ chain are α-helix permissive by predictive analysis from their amino acid sequences.[16] Furthermore, alignment of these helices between the disulfides permits them to intertwine into a super-helical arrangement. If present in this format, such a coiled-coil arrangement should markedly influence the ability of simple epitopes (amino acid sequences determined) to interact with antibody and thus serve as effective antigenic epitopes. Immunochemical analyses of epitopes within this region, although in some cases incomplete, are consistent with this prediction. As summarized in TABLE 1, antibodies have been elicited to at least one epitope set within each constituent chain of the putative coiled-coil region. In each case, the epitope set is not expressed by intact fibrinogen indicating that their accessibility to antibody is severely restricted. Such effects on antibody accessibility are clearly consistent with a complex ordering of structure which would be imparted by the supercoiling. In collagen, for example, epitopes of the consistent chains are not expressed when integrated into the triple helical arrangement, and the triple helix in itself is poorly immunogenic.[33]

Of the epitopes in the coiled-coil region, the cleavage-associated antigen in the E domain, fgEneo, has been the most completely characterized. FgEneo has been localized to the γ36–53 sequence and thus resides in the amino-terminal aspects of the putative coiled-coils.[34] As shown in FIGURE 6, plasmic cleavage is associated with a progressive exposure of fgEneo. Although neither low nor high solubility fibrinogen expresses the neoantigen, the sequential generation of fragment X followed by fragment Y and followed by the terminal degradation products D and E is accompanied by a progressive increase in fgEneo expression. Within the context of the coiled-coil structures, these observations can be attributed to a sequential unravelling of the super-helix with progressive plasmic proteolysis. However, unravelling at its extreme as induced by reduction and alkylation of fragment E results in a 100-fold diminution in FgEneo expression.[34] Conformational changes induced by plasmic cleavage, e.g., unravelling, must differ significantly from the reorganization of this region permitted by disulfide bond disruption. Thus, the immunochemical observations

TABLE 1

EPITOPES OF THE COILED-COILS

Coiled-coil Region	Number of Epitopes Analyzed	Expression by Fibrinogen *
Aα 50–160	> 1	None
Bβ 81–191	> 1	None
γ 24–134	3	None

* A 1000-fold molar excess of fibrinogen relative to radiolabelled peptides containing the recognized epitopes produced no competitive inhibition.

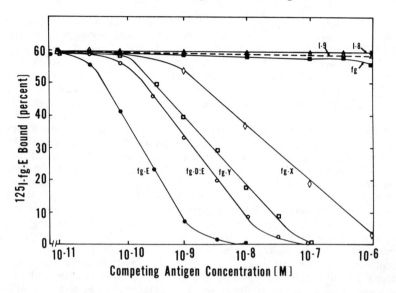

FIGURE 6. The cleavage associated neoantigen of the E domain, fg-Eneo. The ligand is ^{125}I-fragment E, and the antiserum, anti-fgEneo, is prepared by extensive absorbance of anti-fragment E with plasma fibrinogen. The progressive "exposure" of the recognized epitope, fgEneo, which resides in the γ chain segment of the coiled-coils, with sequential plasmic proteolysis is indicated. Reprinted from Plow & Edgington [34] with permission from *J. Biol. Chem.*

are consistent with a systematic ordering of this region in all permitted conformations and as might be imposed by a super-helical structure. Further, systematic changes in conformation are induced by plasmic cleavage. These systematic changes arise from peptide bond cleavages within the region itself and also from proteolytic events at sites peripheral to the region (Bβ 42–43).

THE CARBOXY-TERMINAL REGION OF THE Aα CHAIN

The hydrophilic amino acid composition and the absence of α helical content by predictive and experimental analyses suggests that the carboxy-terminal two-thirds of the Aα chain is surface-oriented and organized into a random coil.[35, 36] The ready susceptibility of this region to proteolytic degradation strongly supports this interpretation. For immunochemical analyses of this region, we have isolated and characterized the two largest cyanogen bromide cleavage fragments of the carboxy-terminals of the Aα chain, Aα 239–476 and Aα 508–584. Together, these two fragments make up 80% of the mass of this region. In these studies, we have utilized antibodies elicited to the intact Aα chain of fibrinogen to exclude antibodies recognizing modifications associated with cyanogen bromide cleavage *per se*. Specificity for the epitopes within these regions has been established by using each peptide as a radiolabelled ligand. The data summarized in TABLE 2 verifies the surface orientation and accessibility of all or part of these regions to antibody probes in native hydrated fibrinogen. With ^{125}I-Aα 239–476 as the ligand, the recognized epitopes are readily accessible in Aα 239–476, the Aα chain and fibrinogen. Similarly, with ^{125}I-Aα 518–584 as the ligand, the homologous peptide, the Aα chain and fibrinogen also express the recognized epitopes in a surface-accessible format. There is no evidence for epitopes being shared by the two peptides.

Two lines of evidence indicate that, despite the surface orinetation, these regions are ordering when integrated into fibrinogen. First, as shown in FIGURE 7, brief plasmic digestion exhibits markedly enhanced epitope expression, approximately 100-fold, relative to intact fibrinogen. This indicates that limited plasmic proteolysis leads to changes which enhance the availability and/or local conformation of the epitopes of this region. The second line of evidence is derived from experiments in which the antibodies to the Aα chain have been fractionated by affinity chromatography on fibrinogen-Sepharose. Although the vast majority of the antibody was bound to fibrinogen, the unbound fraction devoid of affinity for fibrinogen was of sufficient concentration and affinity for the respective epitopes to use in the analyses summarized in TABLE 3. With either peptide as the radiolabelled ligand, fibrinogen was noncompetitive verifying the complete depletion of antibodies recognizing the parent molecule. In both systems, however, the homologous peptide and the Aα chain exhibit full expression of the recognized epitopes. In addition, a prolonged plasmic digest of fibrinogen also exhibits full epitope expression. Taken together, the results of these analyses indicate that plasmic cleavage significantly enhances the expression of certain epitopes within the carboxy-terminal aspects of the Aα chain. This implies that the conformation of this region must be specifically ordered as contrasted to a random conformation so as to restrict epitope expression. Plasmic cleavage appears to relax the conformational constraints on this region so as to render the epitopes more accessible to antibodies.

SUMMATION

In this analysis, we have attempted to interpret immunochemical data relevant to the conformation of fibrinogen within the context of the trinodular model of the molecule. Four major implications of the trinodular model have been considered: (1) the independence of the D and E domain; (2) the effect

TABLE 2

EXPRESSION OF CARBOXY-TERMINAL Aα CHAIN EPITOPES BY FIBRINOGEN
AND DERIVATIVES *

Ligand	Concentrations (nM) Required for 50% Competitive Inhibition			
	Fibrinogen	Aα chain	Aα 239–476	Aα 508–584
^{125}I-Aα 239–476	8.7	14.8	19.4	> 1000
^{125}I-Aα 508–584	2.1	7.9	> 1000	9.7

* The radioimmunoassays consisted of the radiolabelled peptides as ligands and antiserum to the Aα chain.

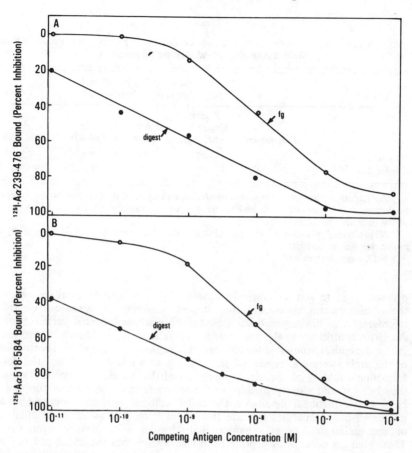

FIGURE 7. The effect of plasmic cleavage on epitope expression in the carboxy-terminal region of the Aα chain. The ligands in Panels A and B are ^{125}I-Aα 239—476 and ^{125}I-Aα 508–584, respectively, and the antiserum has been raised to the Aα chain. The competing antigens are intact fibrinogen or a short-term plasmic digest of fibrinogen [1:250 (w/w) ratio of plasmin to fibrinogen for 20 min at 37°C]. An enhanced expression of epitopes within these peptide regions with plasmic proteolysis is observed.

of plasmic cleavage; (3) the coiled-coil structure interconnecting the D and E domains; and; (4) the existence of the carboxy-terminal aspects of the Aα chain as a pair of independent surface-oriented domains. The latter two implications of the trinodular model are readily supported by immunochemical analyses. The surface orientation of the carboxy-terminal of the Aα chain is substantiated by the accessibility of this region to antibody probes. Juxtaposed, the inaccessibility of the intradomainal region to antibodies is consistent with the systematic organization of these regions into an ordered structure. The immunochemical studies of both these regions have also provided detailed information on the effects of plasmic cleavage on their organization. In native fibrinogen, the carboxy-terminal aspects of the Aα chain must be sufficiently

TABLE 3

IDENTIFICATION OF EPITOPES IN THE CARBOXY-TERMINAL REGION OF THE Aα CHAIN
WHICH ARE NOT EXPRESSED BY FIBRINOGEN

Ligand	Concentrations (nM) Required for 50% Competitive Inhibition *				
	Fibrinogen	Plasmic Digest † of Fibrinogen	Aα chain	Aα 239–476	Aα 508–584
^{125}I-Aα 239–476	> 1000	46	49	53	N.D.‡
^{125}I-Aα 518–584	> 1000	78	73	N.D.‡	60

* Antiserum to the Aα chain was subjected to affinity chromatography on fibrinogen-Sepharose. The unbound antibodies were used in radioimmunoassays with each of the indicated peptides as the ligands.

† A prolonged plasmic digest containing only fragments D and E as the high molecular weight derivatives.

‡ N.D.—not determined.

ordered so as to restrict antibody accessibility, and plasmic cleavage relaxes these conformational constraints. Recent scanning transmission electron micrographs of fibrinogen have indicated the carboxy-terminal region of the Aα chain contributes to the mass integration ratio of the E domain.[15] Previous immunochemical studies indicated that this region influences the accessibility of the amino terminal region of the γ chain to antibodies.[30] Thus, the conformational folding of the carboxy-terminal of the Aα chain is systematic and may bring it into close proximity to the amino terminal aspects of the molecule. Conformational details on the coiled coils of fibrinogen derived from immunochemical analyses indicate that at least the gamma chain component of the region undergoes organizational changes with plasmic proteolysis. These changes occur with peptide bond cleavages within the coiled-coil region itself as well as beyond the coils and are not mimicked by a "simple" unravelling of the coiled coils.

Analysis of multiple epitopes indicate that plasmic cleavage has profound and long range effects on the conformation of fibrinogen and its derivatives. The unfolding of the molecule so as to expose significant portions of the γ chain in both the D and E domain appears to be a general phenomenon. The effect of the single plasmic cleavage at Bβ 42–43 of the E domain on conformation is particularly dramatic as this event results in the initial exposure of both fgEneo in the proximal coiled coil and fgDneo in the D domain. Both of these neoantigens reside in the γ chain of the molecule. Thus, deletion of the Bβ 1–42 segment culminates in effects that appear to reverberate throughout the molecule. The apparent conformational liability of fibrinogen is not in itself inconsistent with the trinodular format of the molecule. The transmission of allosteric changes as originally documented with the binding of effector ligands is a well-recognized phenomenon in protein biology. Indeed, the coiled coils may provide excellent conduits for the transmission of such allosteric changes from one domain to another.

The immunochemical data having the greatest impact on the trinodular model of fibrinogen are the observations suggesting conformational interaction between the D and E domains. Such interaction would not be predicted for spatially distant domains as in an extended format of the molecule. Considerable caution must be given to the interpretation of the immunochemical data suggesting intradomainal interactions. These observations are based primarily upon the characterization of plasmic digests of fibrinogen and not upon the native molecule per se. The apparent association of fragments D and E in plasmic digests, however, appears to recapitulate the native conformation of fibrinogen to a much greater extent than its isolated constituted fragments. This heavily favors the D and E association as being a native conformational element of fibrinogen rather than arising from intermolecular interactions contingent upon proteolysis.

The possibility of considerable molecular flexibility would accommodate both interdomainal interactions and an extended format for fibrinogen. Accordingly, overlapping versus not interacting domains would represent the extreme in formats existing in equilibrium. Perturbations of this equilibrium by the method of analysis could favor the selection of a particular format. Surface interactions may favor the extended format whereas reaction with antibody could drive the equilibrium toward the more compacted arrangement. Recent fluorescent polarization studies [37] and scanning transmission electron micrographs [14, 15] have indeed been interpreted in terms of the flexibility of the fibrinogen molecule. The constraints imparted by immunochemical analysis reinforce the need to carefully evaluate the contributions of molecular flexibility on the overall format(s) of fibrinogen.

ACKNOWLEDGMENTS

The authors wish to thank Ellen Schmeding for the preparation of the manuscript.

REFERENCES

1. MARGUERIE, G. A. & E. F. PLOW. 1983. Ann. N.Y. Acad. Sci. **408:**. This volume.
2. HERMANS, J. & J. McDONAGH. 1982. Sem. in Thromb. Hemost. **8:** 11–24.
3. NIEWIAROWSKI, S., A. Z. BUDZYNSKI & B. LIPINSKI. 1977. Blood **49:** 635–644.
4. KESSLER, C. M. & W. K. BELL. 1979. J. Lab. Clin. Med. **93:** 758–764.
5. BELEW, M., B. GERDIN, J. PORATH & T. SALDEEN. 1978. Thromb. Res. **13:** 983–994.
6. GIRMANN, G., H. PEES, G. SCHWARZE & P. G. SCHEURLEN. 1976. Nature **259:** 399–401.
7. DOOLITTLE, R. F. 1973. Adv. Protein Chem. **27:** 1–109.
8. HALL, C. E. & H. S. SLAYTER. 1959. J. Biophys. Biochem. Cytol. **5:** 11–15.
9. KOPPEL, G. 1966. Nature **212:** 1608–1609.
10. FOWLER, W. E. & H. P. ERICKSON. 1979. J. Mol. Biol. **134:** 241–249.
11. PRICE, T. M., D. D. STRONG, M. L. RUDEE & R. F. DOOLITTLE. 1981. Proc. Natl. Acad. Sci. USA **78:** 200–204.
12. WEISEL, J. W., G. N. PHILLIPS & C. COHEN. 1981. Nature **289:** 263–267.
13. TELFORD, J. N., J. A. NAGY, P. A. HATCHER & H. A. SCHERAGA. 1980. Proc. Natl. Acad. Sci. USA **77:** 2372–2376.
14. ESTIS, L. F. & R. H. HASCHEMEYER. 1980. Proc. Natl. Acad. Sci. USA **77:** 3139–3143.

15. MOSESSON, M. W., J. HAINFELD, J. WALL & R. H. HASCHEMEYER. 1981. J. Mol. Biol. **153:** 695–718.
16. DOOLITTLE, R. F., D. M. GOLDBAUM & L. R. DOOLITTLE. 1978. J. Mol. Biol. **120:** 311–325.
17. PLOW, E. F. & T. S. EDGINGTON. 1981. Sem. Thromb. Hemost. **8:** 36–56.
18. NORTON, P. A. & H. S. SLAYTER. 1981. Proc. Natl. Acad. Sci. USA **78:** 1661–1665.
19. NUSSENZWEIG, V., M. SELIGMANN & P. GRABAR. 1961. Ann. Inst. Pasteur **100:** 490–508.
20. PLOW, E. F. & T. S. EDGINGTON. 1972. Thromb. Res. **1:** 605–618.
21. PLOW, E. F., C. CIERNIEWSKI & T. S. EDGINGTON. 1977. Thromb. Res. **10:** 175–181.
22. CIERNIEWSKI, C., E. F. PLOW & T. S. EDGINGTON. 1977. J. Biol. Chem. **252:** 8917–8923.
23. HAVERKATE, F. & G. TIMAN. 1977. Thromb. Res. **10:** 803–812.
24. BUDZYNSKI, A. Z., V. J. MARDER & J. R. SHAINOFF. 1974. J. Biol. Chem. **249:** 2294–2302.
25. NOSSEL, H. L. 1981. Nature **291:** 165–168.
26. WALLEN, P. 1971. Scand. J. Haemat. Supple. **13:** 3–14.
27. FOWLER, W. E., L. J. FRETTO, H. D. ERICKSON & P. A. MCKEE. 1980. J. Clin. Invest. **66:** 50–56.
28. PLOW, E. F. & T. S. EDGINGTON. 1973. J. Clin. Invest. **52:** 273–282.
29. PLOW, E. F. & T. S. EDGINGTON. 1975. J. Biol. Chem. **250:** 3386–3392.
30. CIERNIEWSKI, C. S. & T. S. EDGINGTON. 1979. Biochim. Biophys. Acta **580:** 32–43.
31. FAIR, D. S., T. S. EDGINGTON & E. F. PLOW. 1981. J. Biol. Chem. **256:** 8018–8023.
32. PLOW, E. F., C. HOUGIE & T. S. EDGINGTON. 1971. J. Immunol. **5:** 1496–1500.
33. TIMPL, R. 1978. Thromb. Haemost. Suppl. **63:** 163–170.
34. PLOW, E. F. & T. S. EDGINGTON. 1979. J. Biol. Chem. **254:** 672–678.
35. DOOLITTLE, R. F., K. W. K. WATT & B. A. COTTRELL. 1979. Nature **280:** 464–468.
36. HUSEBY, R. M., M. W. MOSESSON & M. MURRAY. 1970. Phys. Chem. Phys. **2:** 374–384.
37. HANTGAN, R. 1982. Biochemistry **21:** 1821–1829.

DISCUSSION OF THE PAPER

R. F. DOOLITTLE (*University of California–San Diego, La Jolla*): I think there are other interpretations of data that would accommodate your first two points, but which of them may apply will depend on your answer to this question. If you take your fragments D and E and put them back together again, do they behave the way your so-called D-E complex behaves?

E. PLOW: That really depends on what epitope we are looking at.

DOOLITTLE: Well let's take the two you showed. One was for fibrinogen and one was for the γ chain.

PLOW: Right, in one case the native fibrinogen epitopes do become recapitulated when you mix D and E, and in the other case they do not.

DOOLITTLE: In the case where they do not, what's the difference between them? Why don't they?

PLOW: I do not have an answer for that.

DOOLITTLE: With regard to D and E getting together, I do not know what kind of Ds and Es you are using, but they must represent a distribution. I realize that you are over two orders of magnitude from the native case, but I would think you ought to refer to this as an unfractionated digest and not presume it's a D-E complex; that's presumptuous. If you, in fact, have some Es in which the terminal glycines are still there, the fibrinopeptide A is off, and you have Ds that have latent complementary sites, there is not a doubt some of that the cleaved products could get together the way one gets "D_2–E" complexes. It would be a mistake to presume that this had anything to do with the original virginal molecule back in the beginning, however. I think we are going in a wrong direction here to think that just because one sees things in the EMs that are balled up or something that the terminal domains can be interacting with one central domain in the native molecule. I think more emphasis ought to be given to all the alternative interpretations of these data.

PLOW: We are saying that these interactions express the epitopes of native fibrinogen to a much greater extent than the isolated constituents. If this were merely an association between fragments D and E in a random format, or even in a highly ordered format, one would not expect that. There is no reason for such interactions to behave more like native fibrinogen than the terminal digest; indeed they should not. They should interiorize antigenic determinants. Furthermore, take a look at the sequence of fragments from X to Y to D and E. In terms of the γ chain epitope, the X and the Y [X and the Y] and the D-E complex behave identically, again emphasizing that this reflects something is occurring within each fibrinogen molecule as opposed to an interaction that occurs among fibrinogen molecules.

As to the status of the D and E, in most cases we used late fragment D. The γ chain begins at about residue 109, so that it is separated from the E by the γ chain of fragment E by at least 50 residues in primary sequence.

KINETICS AND MOLECULAR MECHANISM OF THE PROTEOLYTIC FRAGMENTATION OF FIBRINOGEN

Elemer Mihalyi

Hematology Service
Clinical Pathology Department
National Institutes of Health
Bethesda, Maryland 20205

The events of the fragmentation of fibrinogen by proteolytic enzymes have to be related to the primary process, that is, the cleavage of peptide bonds. This gives a frame of reference independent of the kinetic variables of the digestion. Various aspects of this facet of the reaction will be discussed in the first section of this paper. The discussion of the fragmentation of the molecule will follow, with the second section dealing with the large fragments that have a compact, protein-like secondary structure, and the last section with the peptide fragments of various sizes.

Cleavage of Peptide Bonds

An easy and reasonably accurate method for estimating cleavage of peptide bonds relies on the pH stat. What one measures here is the amount of hydrogen ions liberated by the cleavage process; therefore, this method is totally nonspecific with regard to the residues forming the scissile bond. The number of bonds cleaved can be calculated from the equivalents of hydrogen ions liberated, knowing the pH of the reaction and the pK of the generated α-amino groups or by performing the reaction at successively increasing pH values and estimating the plateau value of the liberated hydrogen ions.

Kinetically, the cleavage of peptide bonds during digestion of fibrinogen by either trypsin or plasmin, appears to be composed of three simultaneous reactions: fast, slow and very slow.[1-3] There are roughly ten times more bonds cleaved in the slow reaction than in the fast one and the rates also differ by about one order of magnitude. The third reaction is two or three orders of magnitude slower than the slow one and with trypsin comprises probably all of the remaining susceptible bonds, leading to the complete disintegration of the molecule into small peptides.[3] The number of bonds cleaved in each of the first two classes for the four systems studied is given in TABLE 1 and the rates in TABLE 5.

A further method of estimating proteolytic degradation is the quantitative determination of the small peptides formed. This is performed usually by removing the undigested protein by a precipitating agent, most frequently trichloroacetic acid (TCA), and estimation of the soluble peptides in the supernatants by optical density or total nitrogen determination. This is also a nonspecific method up to this point. However, it can be made more specific by isolating and analyzing the peptides. In most of the studies reported on the chemical identification of the cleavage points,[4-7] heat coagulation of the undigested protein was used.

0077–8923/83/0408–0060 $01.75/0 © 1983, NYAS

TABLE 1

NUMBER OF BONDS CLEAVED BY TRYPSIN OR PLASMIN IN BOVINE
OR HUMAN FIBRINOGEN AS DETERMINED IN THE pH-STAT

Digestion Mixture	Bonds Cleaved per Mole of Fibrinogen	
	Fast Reaction	Slow Reaction
Bovine fibrinogen-trypsin	13.2	91.3
Bovine fibrinogen-plasmin	3.1	54.3
Human fibrinogen-trypsin	25.7	79.1
Human fibrinogen-plasmin	5.2	50.1

With the digestion of fibrinogen, the kinetics of the formation of TCA soluble optical density follows a first order reaction course within experimental error. Bovine and human fibrinogen digested with trypsin have an equal rate that is comparable to the rate of the slow reaction in the pH stat. Furthermore, when TCA soluble optical density is plotted against the number of bonds cleaved as monitored by the pH stat, a straight line relationship is obtained. This means that in the reaction aspect corresponding to these measurements cleavage of each peptide bond results in the same average optical density increase in the supernatant. Thus, these peptides must originate from a part of the molecule that is rapidly chopped up into small peptides and cannot be connected with the more elaborate fragmentation process.

Further information can be extracted from the amino acid composition of the TCA supernatants and from their peptide maps. In supernatants of tryptic digests it may be assumed that the digestion proceeded to cleave all susceptible bonds present there. This is reasonable because refractory bonds, for example Lys or Arg, followed by Pro are relatively rare. Thus, with this reservation, the total Lys + Arg content of the supernatant should be equal to the number of peptide bonds cleaved which resulted in liberation of small peptides if these cleavages were from the N-terminus of the chain. If the cleavages were from the C-terminus, the Lys + Arg content is one less than the number of peptide bonds cleaved. If there were bonds cleaved that resulted in large fragments precipitated by TCA or the molecule was only nicked by the enzyme, the Lys + Arg content in the supernatant will be below the number of peptide bonds cleaved estimated by the pH stat. Finally, peptide mapping, if each susceptible peptide bond was quantitatively cleaved, will give a number of spots equal to the number of bonds cleaved. There are a few complicating factors in the latter reckoning: repeating sequences will give identical spots, thus diminishing the number of spots observed. Further, incomplete cleavage at a susceptible bond will work the other way, increasing the number by adding extra spots for the large peptides that were not cleaved. Fibrinogen is a symmetrical molecule built of two identical halves. Therefore, each spot on the peptide map will actually represent two peptides and the number of bonds cleaved to produce these peptides will be double the number of spots observed.

All the estimates mentioned above were performed on the same digests of bovine and human fibrinogen with trypsin. The digestions were arrested at the point where the fragmentation of the molecule was just completed. The results are summarized in TABLE 2. Comparison of the number of bonds cleaved estimated with the pH stat with the Lys + Arg content of the TCA supernatants reveals that the former is larger by 6 to 10 bonds per molecule of fibrinogen. Although the errors are large, the figures being differences of fairly large numbers, these data suggest that some of the peptide bonds cleaved do not produce small peptides. These are the ones that are involved in the large fragment formation and as will become apparent in the next section, their number estimated above is in agreement with the one expected from the reaction scheme of the fragmentation. Twice the number of peptide spots in the maps is only slightly larger than the figure estimated from the pH stat. Although some compensation may have occurred, this suggests that the reaction must have gone near to completion, with all the susceptible bonds involved in this phase of the reaction being completely cleaved. Further, the agreement of all these numbers indicates that internal cleavages, so-called nicks, do not occur in this reaction. This is in agreement with the observations of the molecular weights of the chain segments in the isolated fragments.

The data with plasmin digests clearly show that this enzyme is more selective in cleaving the lysyl-arginyl peptide bonds than trypsin. The number of bonds cleaved, given in TABLE 1, is only about ½ to ⅔ of those seen with trypsin and the number of peptide spots in the TCA supernatants also corresponds to this magnitude. On the other hand, the Lys + Arg content of the TCA supernatants is approximately the same as with trypsin. Thus, apparently the same segment of the molecule is removed by plasmin as with trypsin, but it is cleaved into fewer peptides.

Neither of the above data specify the site of the cleavages in the molecule. This can be obtained by determinations of the new N- and C-terminal sequences resulting from the cleavages. These should include both the small peptide fractions and the large fragments. Most of these studies were performed in Doolittle's [4-7] and Blombäck's [8-10] laboratory on plasmic digests of human fibrinogen. The results are assembled in TABLE 3. The number of identified cleavage sites is 42, below the expected number of 55 from the pH stat studies. This is because the cleavages in the latter phase of the reaction yielding the small peptides from the C-terminus of the α chain were not identified. In the chain segment α 493–610 there are 13 susceptible sites, thus accounting for the above difference. The cleavage points in the table help delineate the

TABLE 2

CORRELATION OF THE NUMBER OF BONDS CLEAVED WITH FORMATION
OF SMALL PEPTIDES IN TRYPTIC DIGESTS

	Bovine Fibrinogen	Human Fibrinogen
Number of bonds cleaved estimated with the pH stat (moles per mole of fibrinogen)	84.1	97.7
Lys + Arg content of the TCA supernatant (moles per mole of fibrinogen)	77.6	87.1
Twice the number of spots in peptide maps of the TCA supernatants	92.0	84.0

TABLE 3

PLASMIC CLEAVAGE POINTS OF HUMAN FIBRINOGEN

α Chain Midportion	Connecting Segment	N-Terminal Segment (fibrinopeptides)	γ Chain C-Terminal Segment
α Arg 199–Gln 200	α Lys 78–Asn 79	α Arg 16–Gly 17	γ Lys 405–Gln 406
α Lys 206–Met 207	α Lys 81–Asp 82	α Arg 19–Val 20	
α Lys 219–Ser 220	α Arg 101–Asp 102	β Arg 14–Gly 15	
α Lys 230–Ala 231	α Arg 110–Val 111	β Arg 42–Ala 43	
α Arg 239–Met 240	β Lys 122–Asp 123	β Lys 53–Lys 54	
	β Lys 133–Asp 134		
	γ Lys 53–Thr 54		
	γ Lys 58–Gln 59		
	γ Lys 62–Ala 63		
	γ Lys 85–Ser 86		

segments of the chains comprised in the large fragments.[11] The molecular weights estimated with these are in good agreement with those found experimentally for the chains of isolated fragments. There are no data on the cleavage points with trypsin with either bovine or human fibrinogen or with plasmin with the bovine protein.

FORMATION OF THE LARGE FRAGMENTS

We have undertaken to characterize these reactions by estimating the relative proportion of each one of these fragments along the reaction path.[2] A useful method for this is gel-filtration chromatography on a Sephadex G-200 column, a technique introduced by Marder et al.[12] However, even with a 100 cm long column, fragments D and E were not separated. A more convenient method is SDS-polyacrylamide gel electrophoresis, performed in quartz tubes on 5% gels. The components are quantitated by direct scanning in the tubes at 280 nm.[2] With this method, however, native fibrinogen and fragment X are not separated. Therefore, in some instances one method has to be complemented with the other one to obtain complete resolution of the fragments. The two methods were checked against each other and the relative concentration of the components was found to agree to better than 2%. Such a comparison of two digests is shown in FIGURE 1. The peptide fractions are better resolved in a 10% gel system after removal of the large fragments by heat coagulation. Furthermore, the small peptide peak in either gel filtration or gel electrophoresis is equivalent to the peptides in the TCA supernatants. All three kinds of values were combined to obtain the complete analysis of the studied digestion systems.

The relative areas of the peaks obtained as described above can be used directly for kinetic calculations. However, these represent fractions of optical density in each component at 280 nm. To convert these values into relative weight concentrations, the specific optical density of each fragment had to be determined. The relative optical density of each component divided by its specific optical density is proportional to the weight concentration of the respective component. The sum of these quotients for all the components

SEPHADEX G200 CHROMATOGRAPHY

SDS-GEL ELECTROPHORESIS

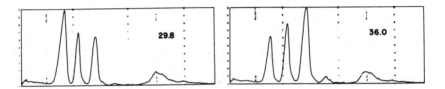

FIGURE 1. Comparison of the resolution of two digests into the fragmentation products by gel filtration chromatography on Sephadex G-200 column and by SDS-polyacrylamide gel electrophoresis. The numbers indicate the extent of digestion by trypsin, given by the moles of base per mole of bovine fibrinogen consumed in the pH-stat. In upper patterns, filled symbols are experimental, empty symbols computed points. The drawn curves are the individual Gaussians into which the overall curve was resolved. Peaks from left to right are X,Y,D,P1 and P2, with E apparent between D and P1 only in the gel electrophoresis patterns.

TABLE 4

RELATIVE WEIGHT CONCENTRATION OF THE FRAGMENTS IN TRYPTIC DIGESTS OF BOVINE FIBRINOGEN

Reaction Time (min)	Fragments							
	Fibg.*	X	Y	D	E	P1	P2	Sum
0	67.66	—	—	—	—	—	—	67.66
0.7	32.31	28.40	—	0.90	—	1.02	2.17	64.80
2.05	17.65	35.46	0.68	0.77	—	4.74	4.43	63.73
4.35	7.60	37.30	4.05	2.07	—	5.97	7.20	64.19
8.40	5.06	27.43	10.43	6.00	0.92	6.37	8.06	64.24
13.25	1.85	11.80	15.33	15.75	2.32	6.39	13.39	66.83
20.00	1.07	4.29	10.82	23.88	5.94	6.57	15.02	67.59
27.05	1.06	1.44	4.78	30.20	7.89	6.45	15.02	66.84
42.90	1.49	0.92	2.20	31.20	9.25	6.78	15.71	67.55
95.00	0.63	—	—	33.16	10.37	7.03	17.34	68.53

* Fibg. = fibrinogen; in calculating these numbers, the following specific optical densities (E_{280}/mg/ml) were used: Fibg. 1.478, X 1.662, Y 1.561, D 2.004, E 0.897, P1 2.173, P2 0.460.

should be constant because of the law of conservation of mass at each stage of the reaction. This relationship was satisfied with all the systems investigated. TABLE 4 shows as an example the data obtained with tryptic digests of bovine fibrinogen. The sum, at the initial phase of the reaction where fragment X is abundant, is slightly below the expected value. This is because fragment X is a mixture of species of gradually decreasing sizes and the specific optical density was estimated for the lowest molecular weight last fragment of this kind. The specific optical density of the mixture of fragment X species increases gradually during digestion from close to that of fibrinogen to this final value. Correction for this effect brought up the sums to their correct value.

The kinetic analysis of the data obtained was based on the scheme proposed by Marder et al.[12] If the probability of cleavage of a certain bond is proportional to its concentration, i.e., the reaction is first order with respect to this bond, then the fraction of bonds cleaved, α, at any time t is given by the equation:

$$a = 1 - e^{-kt}$$

where k is the rate constant of a first order reaction and the other terms have been defined previously. If a fragment is formed by severing n bonds, then the relative concentration of this fragment will be given by the product of the probabilities of cleavage at each one of these n bonds, i.e., by the product of n terms of $(1 - e^{-k_i t})$ form, where k_i denotes the n individual rate constants.

In the following equations the molar fractions of each component are represented by the capital letters used to denote these fragments. Formation of fragment X was not resolved in these experiments into its subreactions and its overall formation was fast compared to the subsequent reactions. Therefore, it could be approximately with a single reaction with rate constant k_X: $X = 1 - e^{-k_x t}$. The connecting segments between the domains may be formed by a single polypeptide chain, or any number of these. It was assumed that each chain has to be cleaved at a single point to effect separation of the fragments. To each cleavage an individual rate constant was assigned. Best fit to the experimental points was obtained with a model containing three chains in each connecting segment. Further, no better fit was obtained when the rate constants were allowed to vary independently than with all six of them constrained to a single value (k_D). Thus, neither asymmetry nor cooperativity was evident in these rates. If the D fragments derived from the right side of the molecule were denoted by subscript R and those from the left side by suscript L, the following equations may be written for the fragmentation process:

$$D = D_R + D_L = 2(1 - e^{-k_D t})^3$$

$$E = (1 - e^{-k_D t})^6 = \tfrac{1}{4} D^2$$

$$Y = D - 2E$$

$$X = 1 - (Y + E)$$

The expressions simply reflect that a fragment D is formed whenever all three connecting chains are severed on one side, and a fragment E is formed when all six connecting chains are severed, i.e., three on either side. Concentration of Y is equal to the concentration of those fragments D which came from unilateral cleavage of the molecule. Thus, from the total number of D fragments two for each E fragment should be subtracted. Finally, fragment X

decays by forming Y, and each decayed Y forms an E. Thus, the total of Y formed and decayed is $Y + E$ and the remaining X is $1 - (Y + E)$. All of these quantities are normalized concentrations (dimensionless) that vary from 0 to 1, except for D that varies from 0 to 2. As mentioned above, formation of X is fast compared to the subsequent reactions and the above treatment is simplified by assuming that the reaction starts with X fully formed. The whole model of the reaction is extremely simple, relying only on two rate constants. Nevertheless, the fit to the experimental data was excellent, as shown on the examples given in FIGURE 2. Naturally, kinetic schemes can be complicated to any extent. Numerous other schemes were tried and gave acceptable fit to the data. For the details of these the original publication should be consulted.[13]

TABLE 5

RATES OF THE VARIOUS PROCESSES OF THE FRAGMENTATION OF HUMAN
AND BOVINE FIBRINOGEN WITH EITHER TRYPSIN OR PLASMIN *

	pH Stat		Fragmentation Into			
Digestion Mixture	Fast Reaction	Slow Reaction	X	D+E	P1	P2
Bovine fibrinogen-trypsin	0.69	0.055	0.270	0.113	0.48	0.11
Bovine fibrinogen-plasmin	0.61	0.024	1.58	0.072	1.03	0.04
Human fibrinogen-trypsin	0.39	0.052	0.398	0.063	0.48	0.11
Human fibrinogen-plasmin	0.99	0.048	0.99	0.090	0.33	0.03
			21 K		10 K+8 K	
Bovine plasmic P1-trypsin	0.91	0.066	0.41		0.089	
Human plasmic P1-trypsin	1.93	0.023	—		—	

* Rate constants were calculated with natural logarithms and time in minutes. They were converted to 10 U/ml trypsin and 1 CTA U/ml plasmin, assuming linear relationship between rate and enzyme concentration. One unit of trypsin is defined as the amount that cleaves 1μ mole of tosyl-arginine-L methyl ester per min at pH 8.00 at 25°C.

All the rate constants of the various facets of the fragmentation reaction are assembled in TABLE 5.

FORMATION OF THE SMALL FRAGMENTS (PEPTIDES)

These can be divided into two subgroups, denoted P1 and P2. From these, P1 contains large peptides from 10,000 to 35,000 molecular weight, whereas P2 is made up of small peptides of less than 10,000 molecular weight. FIGURE 3 shows these fractions in Sephadex G-200 chromatography. The digestion of these samples was conducted to the point where fragments X and Y disappeared and the formation of fragments D and E was just completed. As it is apparent, the P2 peak seems to be nearly identical with all four digests. The P1 peak, on the other hand, has the same size but it is situated in a region indicating much higher molecular weight for the plasmic than for the tryptic

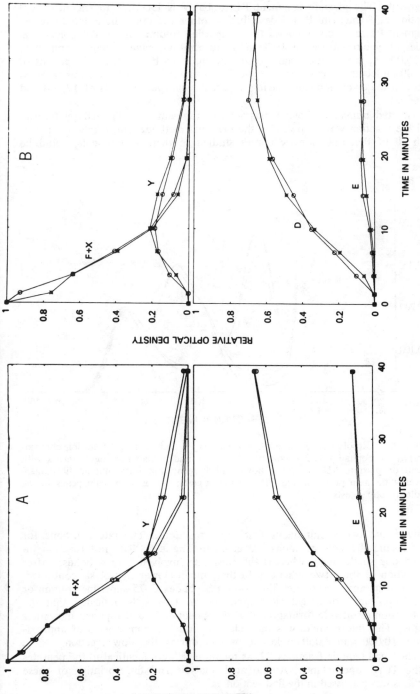

FIGURE 2. Kinetic fitting of the fragmentation reaction of bovine fibrinogen based on the equations given in the text: **A**. digestion with trypsin, and **B**. with plasmin. The empty symbols represent experimental findings, the filled symbols computed values.

fragment. The plasmic fragment P1 probably is identical to fragment A,[14] fraction H,[15] fragment P23 [16] described by other authors. The isolated plasmic fragment P1 rerun on Sephadex G-75 superfine column, or on SDS-polyacrylamide gel electrophoresis at 10% gel concentration, gave a major component of 26,000 apparent molecular weight accompanied by a minor component of 21,000 molecular weight. The tryptic P1 fragment in similar runs was resolved into two major components having apparent molecular weights of 17,600 and 14,600.

Isolated human or bovine plasmic P1 fragment was digested by trypsin, and the reaction was followed by the pH shift method. The reaction curve as with all of the other kinetic curves studied in these experiments, could be

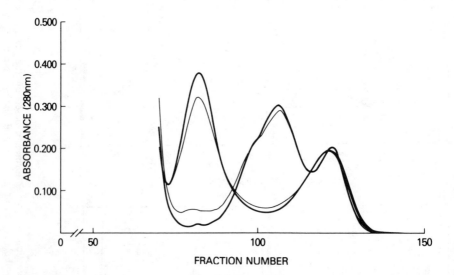

FIGURE 3. Resolution of the peptide fractions P1 and P2 in gel filtration chromatography on Sephadex G-200 columns of bovine and human fibrinogen digests with trypsin or plasmin. Heavy lines, bovine fibrinogen; thin lines, human fibrinogen. Curves with main peaks on the left, plasmic digests; curves with main peaks in the middle, tryptic digests.

resolved into two simultaneous first order reactions. The rate constants for the fast and the slow reactions are given in TABLE 5. The fast reaction involved one single bond, whereas the slow one involved 2 to 4 bonds. After completion of these two reactions further, much slower hydrolysis continued. Serial digests analyzed by gel filtration on Sephadex G-75 superfine column or by SDS-polyacrylamide gel electrophoresis on 10% gels indicated that the fast reaction resulted in formation of a component of 21,000 apparent molecular weight. This intermediate was then cleaved into two components of approximately 10,000 and 8,000 molecular weight during the slow reaction.

The small peptide fractions P2, as mentioned, were identical to the peptides in the TCA supernatants. Amino acid analysis and fingerprinting of these fractions was discussed in the first section.

DISCUSSION

All of the various facets of the fragmentation of fibrinogen by trypsin and plasmin studied in these experiments had one common feature: the reaction could be resolved into two simultaneous first order reactions. The rate constants in the two classes differed approximately by a factor of 10. It is reasonable to assume that this division reflects the structural characteristics of the portion of the molecule being attacked. The bonds that are cleaved fast are probably exposed close to the availability of free peptides. Those in the slow class are in some restricted conformation. It is interesting that the susceptible bonds in the connecting segment between the D and E domains and in the C-terminal tail segment of the α chain, all belong to this category and they are cleaved with the same approximate rate. This, in itself, indicates some structure for the α chain appendage. Further indication of even tighter structure is furnished by the resistance of the P1 segment to plasmin and its restricted cleavage by trypsin. Digestion by plasmin removes the whole appendage in a fast reaction. This is reflected in the fast reacting two-three bonds seen in the pH stat. With trypsin there are up to 13 bonds in this class with bovine fibrinogen. The additional highly susceptible bonds to this enzyme are supplied by the cleavage points of the fibrinopeptides and the fast cleaving bonds forming tryptic fragments of the plasmic P1 fragment. It is possible that the fragmentation at the connection between the D and E domains also conains a fast reacting bond.[17] Severance of the domains is probably the result of a single cleavage through each chain. There are multiple sites in this region, as shown in TABLE 3, but it is more likely that these are cleaved sequentially. If they were equivalent and cleavage at any point would result in fragmentation, the reaction would be faster, equal to the number of susceptible bonds times the average rate constant. This is not evident in the experimental data.

It is remarkable how well all of these various data can be assembled into one logical model, governed by only two rate constants. All of the information on the molecular weights of the intermediates and final fragments is also in complete agreement with the assumed structure.[2]

REFERENCES

1. MIHALYI, E. & J. E. GODFREY. 1963. Biochim. Biophys. Acta **67:** 73–89.
2. MIHALYI, E., R. M. WEINBERG, D. W. TOWNE & M. E. FRIEDMAN. 1976. Biochemistry **15:** 5372–5381.
3. EISELE, J. W. & E. MIHALYI. 1975. Thromb. Res. **6:** 511–522.
4. TAKAGI, T. & R. F. DOOLITTLE. 1975. Biochemistry **14:** 940–946.
5. TAKAGI, T. & R. F. DOOLITTLE. 1975. Biochemistry **14:** 5149–5156.
6. COTTRELL, B. A. & R. F. DOOLITTLE. 1976. Biochem. Biophys. Res. Commun. **71:** 754–761.
7. DOOLITTLE, R. F., K. G. CASSMAN, B. A. COTTRELL, S. J. FRIEZNER & T. TAKAGI. 1977. Biochemistry **16:** 1710–1715.
8. KOWALSKA-LOTH, B., B. GÅRDLUND, N. EGBERG & B. BLOMBÄCK. 1973. Thromb. Res. **2:** 423–450.
9. COLLEN, D., B. KUDRYK, B. HESSEL & B. BLOMBÄCK. 1975. J. Biol. Chem. **250:** 5808–5817.
10. GÅRDLUND, B. 1977. Thromb. Res. **10:** 689–702.
11. MIHALYI, E. 1980. Section I. Hematology. *In* CRC Handbook Series in Clinical Laboratory Science. **3:** 51–60. CRC Press, Boca Raton, FL.

12. MARDER, V. J., N. R. SHULMAN & W. R. CARROLL. 1969. J. Biol. Chem. **244:** 2111–2119.
13. SHRAGER, R. I., E. MIHALYI & D. W. TOWNE. 1976. Biochemistry **15:** 5382–5386.
14. NUSSENZWEIG, V., M. SELIGMANN, J. PELMONT & P. GRABAR. 1961. Ann. Inst. Pasteur. **100:** 377–389.
15. HARFENIST, E. J. & R. E. CANFIELD. 1975. Biochemistry **14:** 4110–4117.
16. FRETTO, L. J., E. W. FERGUSON, H. M. STEINMAN & P. A. MCKEE. 1978. J. Biol. Chem. **253:** 2184–2195.
17. FURLAN, M., T. SEELICH & E. A. BECK. 1975. Biochim. Biophys. Acta **400:** 112–120.

THE RELATIONSHIP OF FIBRINOGEN STRUCTURE TO PLASMINOGEN ACTIVATION AND PLASMIN ACTIVITY DURING FIBRINOLYSIS *

Marsha Adams Lucas, Larry J. Fretto,† and Patrick A. McKee ‡

Howard Hughes Medical Institute Laboratories
Department of Medicine and Department of Biochemistry
Duke University Medical Center
Durham, North Carolina 27710

INTRODUCTION

Over the past several years, investigators have used many different approaches to the study of the fibrinogen molecule. Most of these have contributed to an expanding knowledge about the structural details of fibrinogen, particularly as related to functional properties such as the conversion of fibrinogen to fibrin or the digestion of fibrinogen or fibrin by plasmin. We used three approaches that have yielded information about these interactions. First, we analyzed the digestion of fibrinogen to determine if its native structure could be characterized by the accessibility of various regions to plasmin attack.[1-3] Our findings were consonant with the proposed asymmetric cleavage model of Marder *et al.*[4] in which fibrinogen is cleaved progressively to Fragment X (symmetrical), then one Fragment D and Fragment Y (asymmetrical), and finally the later is digested to another Fragment D and one Fragment E. This interpretation received challenge from Mosesson and colleagues [5] who proposed that fibrinogen was cleaved symmetrically to yield terminal fragments consisting of only one Fragment D and one Fragment E. This latter model depended upon the existence of intramolecular disulfide bond(s) between the two Fragment D portions of the fibrinogen molecule. As will be shown, our work addressed this issue and resolved it in favor of the asymmetric model.[6] Our second major approach to the study of fibrinogen structure was the electron microscopic analysis of Fragments X, Y, D and E.[7] These results documented the structural fidelity of the major fibrinogen domains and defined the relationship of each plasmic cleavage fragment to the trinodular model of Hall and Slayter.[8] In a third set of experiments we used the plasmic fragments of fibrinogen to examine the subtleties of the interactions of plasminogen with fibrinogen or fibrin. In these studies we identified the regions of the fibrinogen molecule that facilitate the activation of plasminogen to plasmin and its subsequent binding to fibrin. These latter mechanisms help explain how maximal plasmin activity becomes localized on a fibrin clot.

STRUCTURAL ANALYSIS OF PLASMIC FRAGMENTS FROM FIBRINOGEN AND FIBRIN

The characterization of the subunits of the nonreduced and reduced products generated when fibrinogen, noncrosslinked fibrin, or crosslinked fibrin

* This work was supported in part by National Heart, Lung, and Blood Institute Grant HL 15615.
† Senior associate, Howard Hughes Medical Institute.
‡ Investigator, Howard Hughes Medical Institute.

0077-8923/83/0408-0071 $01.75/0 © 1983, NYAS

are digested by plasmin not only provided insights about the processes of fibrinogen and fibrin digestion, but also about the structure of fibrinogen and the mechanism of fibrin formation.[1-3, 6, 9-12] As alluded to earlier, the asymmetric cleavage model was deduced by Marder et al.[4] from sedimentation velocity ultracentrifugation studies and immunological findings. By correlating the order of appearance of these fragments with their subunit structures we were able to develop a structural model for the digestion of fibrinogen and fibrin.[1, 2, 6] In essence, we found that the Aα chains, which contain no detectable carbohydrate, were the first subunits to be degraded by plasmin. Later the Bβ chains, which do contain carbohydrate, were degraded. The initial, but transient degradation product, Fragment X, proved structurally heterogeneous, having forms that varied in the extent to which their Aα and Bβ chains were degraded.[13] However, the γ chains of fibrinogen, which also contain carbohydrate, were much more resistant to plasmic attack. Fragment X is subsequently cleaved by plasmin to produce one Fragment D and Fragment Y, the latter being a very transient species which after further proteolysis, yields another Fragment D and the amino-terminal core region of fibrinogen, designated Fragment E. The structure of this latter product is closely similar to that of the amino-terminal disulfide knot produced by cyanogen bromide cleavage of fibrinogen.[14] Each of the major plasmic fragments contains proteolytic derivatives of the Aα, Bβ and γ chains since, as shown in FIGURE 1, plasmin cleaves all the subunit polypeptide chains of the intact fibrinogen molecule. There appear to be at least three major species of Fragment D that can be formed in vitro, each differing primarily in the extent of carboxyl-terminal degradation of the γ chain derivatives. The largest species, Fragment D_1, possesses the γ chain crosslink donor and acceptor sites,[6] an ability to inhibit fibrin polymerization,[15] a calcium binding site [16] and a binding site for the platelet receptor for fibrinogen [17, 18]—all of which are lacking in Fragments D_2 and D_3, although they all contain the same subunit derivatives from the Aα and Bβ chains (FIGURE 1). By summation of subunit molecular weights from sodium dodecyl sulfate polyacrylamide gel electrophoresis, the three Fragment D species had molecular weights of 92,000, 86,000 and 82,000. By sedimentation equilibrium ultracentrifugation, the corresponding apparent molecular weights were 94,000, 88,000 and 76,000, depending on the values used for the partial specific volumes.[6] It is of interest that calcium concentrations comparable to those occurring in blood protect Fragment D_1 from further plasmic cleavage.[19]

As part of our evidence that two Fragment D molecules were derived from one fibrinogen molecule, we deduced that the formation of γ dimers between different fibrin monomers should yield a terminal digestion product consisting of a dimeric Fragment D (i.e., two Fragment D molecules crosslinked through their γ chains) when fully crosslinked fibrin was completely digested by plasmin. In contrast, a Fragment D polymer should occur if indeed the two Fragment D regions per fibrinogen molecule were linked by disulfide bonds and if each Fragment D region became intermolecularly crosslinked to another Fragment D. After subjecting fully crosslinked fibrin to plasmic cleavage, each major product was isolated in pure form and identified by molecular weight and amino acid composition; the largest terminal digestion product was shown conclusively to be a dimeric structure composed of two Fragment D_1 species.[6] The molecular weight of fibrin Fragment D was 184,000 by summation of subunit molecular weights and between 190,000 to 175,000

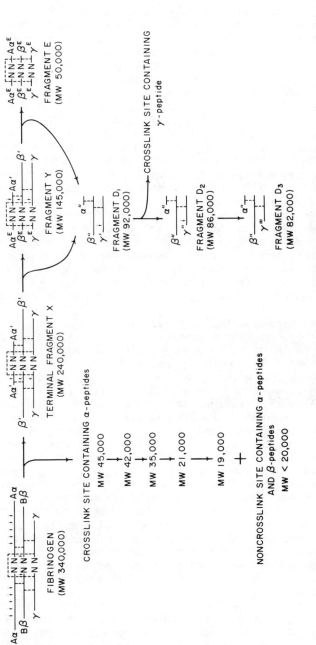

FIGURE 1. Schematic of fibrinogen digestion by plasmin. The small vertical arrows represent plasmin cleavages; the dashed lines indicate the minimal number of disulfide bonds present in each fragment.

by sedimentation equilibrium.[6] The formation of crosslinks between two γ chains or the incorporation of the site-specific fluorescent label, monodansyl cadaverine, into the γ chain crosslink acceptor site, makes the carboxyl-terminal portion of these subunits resistant to degradation by plasmin; hence, cross-linked Fragment D, otherwise known as fibrin Fragment D is not cleaved to domains similar to the monomeric Fragment D species from fibrinogen. From these results we concluded that the plasmic digestion of one fibrinogen molecule clearly results in the production of two Fragment D molecules.

Fragment E derived from fibrinogen is essentially identical to the Fragment E derived from fully crosslinked fibrin. Two major fibrinogen Fragment E species are generated by plasmin with time, the first of which is lacking fibrinopeptide B but not fibrinopeptide A.[13] The second form of Fragment E from fibrinogen is thought to be the most terminal form and lacks both fibrinopeptides. Fibrinogen Fragment E and fibrin Fragment E each contain three pairs of peptides which are connected by disulfide bonds and range in molecular weight from 7,000 to 12,000 daltons.[1, 9] The sedimentation equilibrium molecular weight for intact fibrinogen Fragment E or fibrin Fragment E is about 50,000 daltons. The amino acid compositions and the amino-terminal sequences of the final fibrinogen Fragment E species are essentially identical to those obtained for Fragment E from fully crosslinked fibrin.[9] Slight differences in the tryptic peptide maps between fibrinogen Fragment E and fibrin Fragment E suggest minor variations in the extent of plasmic cleavage. While Fragment E plays a role in fibrin polymerization,[20] it is not involved in crosslink formation.

Next, we attempted to localize the residues of the Aα chain that were involved in the formation of α-polymers as the last step in fibrin crosslinking.[10, 11] To do this, we used monodansyl cadaverine, which had been shown by Lorand et al.[21] to be a site-specific label for the glutaminyl acceptor group of the γ-glutaminyl(ϵ-lysyl)isopeptide crosslink. As a consequence, only the α and γ chains, and not the β chains, of fibrin monomer become fluorescently labeled by monodansyl cadaverine. We had previously shown that, unlike the γ chain, the incorporation of monodansyl cadaverine into the crosslink acceptor sites of the α chain did not alter its plasmic digestion pattern.[6] Knowing that the Aα chain is degraded very early in the course of plasmic degradation of fibrinogen, we first labeled monomeric fibrin with monodansyl cadaverine and then subjected it to digestion by plasmin. As shown by the schematic in FIGURE 1, several large peptides are cleaved from the α chain very early. The five largest peptides, 40,000, 37,000, 32,000, 21,000 and 19,000 daltons, were released almost simultaneously and each contained at least one crosslink acceptor site as demonstrated by fluorescence. With time, the large peptides were progressively degraded to the 19,000-dalton fluorescent peptide. Non-fluorescent peptides larger than about 15,000 daltons were not seen. The same digestion sequence was observed for unlabeled fibrinogen. The sodium dodecyl sulfate gel electrophoretic mobilities of the fluorescent α-peptides were not significantly altered by reduction of disulfide bonds with β-mercaptoethanol, thus indicating the absence of a subunit structure or large disulfide loops. With the generation of the 19,000-dalton fluorescent peptide, a weakly fluorescent peptide of 2,500 daltons became apparent, suggesting that all α chain crosslink acceptor sites were contained within a segment of 21,000 daltons.[6] As a consequence of its early appearance during fibrin cleavage by plasmin, it could be deduced that the 21,000-dalton peptide was in the carboxyl-terminal half of the Aα chain and was not bound into the rest of the fibrinogen molecule by disulfide bonds. Functional studies that had been carried out by us and others were then interpretable as these observations were made. For example, we had observed that fibrinogen began losing its clottability during the earliest stage of plasmin digestion, namely Aα chain cleavage. The partially degraded fibrinogen (primarily Fragment X species) could still form fibrin, albeit friable. In the presence of fibrin-stabilizing factor, this fibrin became crosslinked by virtue of γ-dimerization, but no α-polymer formation was evident. When the results from the plasmic cleavage of the Aα chain are interpreted in concert with the cyanogen bromide cleavage and ancrod cleavage of monodansyl cadaverine-labeled fibrin, three acceptor sites were positioned at approximately residues 255, 310 and 385 of the Aα chain.[10] Sequence analysis [22] now shows these glutaminyl residues to be at positions 237, 328 and 366. Subsequently by plasmic and cyanogen bromide cleavage of highly crosslinked fibrin (containing four ε(γ-glutaminyl)lysyl crosslinks per monomeric unit) we produced a 19,000-dalton peptide of α chain origin crosslinked to a smaller peptide from a neighboring α chain. Cyanogen bromide cleavage of highly crosslinked fibrin formed *in vitro,* or fibrin from an arterial embolus spontaneously formed *in vivo,* yielded three species of crosslinked α chain peptides that ranged in molecular weight from 30,000 to 40,000 daltons.[11] A 24,000-dalton segment (CN29) that contained two acceptor sites for monodansyl cadaverine was present within each of these peptides. The donor peptide could be localized to a region of the α chain very near the carboxy terminus as judged by difference amino acid compositions and amino terminal analyses.[11, 23-25] Minor amounts of crosslinked fragments with molecular weights of 68,000–70,000 daltons also appeared to be derived from crosslinked α chains.[11] Integration of all these data eventually allowed the construction of a model for the α-polymer structure as shown in FIGURE 2. In short, the acceptor-site containing region of the α chain is located in its midportion, whereas the donor peptide region is at its carboxyl-terminal end. Interestingly, plasmin in its initial attack on fibrinogen or fibrin monomer cleaves and releases the α chain region that contains both the donor and acceptor residues, so that α chain polymerization can not occur during fibrin formation.

As stated earlier, the two largest terminal fragments generated by plasmic digestion of fully crosslinked fibrin are dimeric Fragment D and Fragment E. Knowing that these two fragments have a high affinity for each other [20] and that fibrin monomers polymerize in a half-staggered overlap pattern,[26] the various permutations of the very large fragments produced in the earlier stages of crosslinked fibrin digestion by plasmin can be predicted. Francis and Marder [27] have established that these are indeed cleaved and released with time and have characterized them in detail.

ELECTRON MICROSCOPIC STUDIES OF PLASMIC CLEAVAGE OF FIBRINOGEN

The early physicochemical studies on the plasmic degradation products from fibrinogen [1-4, 28-32] produced models that were consistent with its then generally accepted trinodular structure deduced by Hall and Slayter electron microscopic studies using heavy metal shadowing.[8] Subsequ ports on negatively stained [33, 34] and even on shadowed preparations fibrinogen, however, showed a variety of different structures. Thes

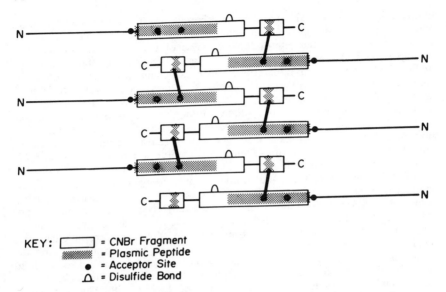

KEY: ⬜ = CNBr Fragment
▨ = Plasmic Peptide
● = Acceptor Site
⌂ = Disulfide Bond

FIGURE 2. Model for an α-polymer structure which utilizes the acceptor site and donor site peptides that appeared to be preferred in highly crosslinked fibrin formed under physiologic conditions. The heavy lines between the schematic α-chains represent the ε(γ-glutaminyl)lysine crosslinks. The largest CNBr fragment depicted in the α chains is 24,000 daltons beginning at Glu 241 and the larger of the plasmic peptides is about 19,000 daltons and starts at Met 240.

suggested that one or both of the methods of specimen preparation did not show the true molecular structure of fibrinogen. This discrepancy has been resolved recently by Fowler and Erickson,[39] who observed that virtually all of the fibrinogen molecules in negatively stained and in shadowed preparations indeed have the extended trinodular structure first described by Hall and Slayter.[8] The overall length is about 45 nm with elongated outer nodules of 7 by 11 nm and a smaller central nodule of 4 by 6 nm. Until recently the identification of the plasmic fragments in terms of the trinodular model was based upon the demonstration of the asymmetric cleavage of fibrinogen to give a fragment D and fragment Y, which is then cleaved to a second fragment D and fragment E, as discussed above. With the development of electron microscopic techniques that reliably show the trinodular structure of the intact molecule,[39] we were able to examine each of the purified Fragments X, Y, D and E and establish their relationship to the trinodular structure of fibrinogen.

FIGURE 3 shows the sodium dodecyl sulfate gel electrophoretic analyses of fibrinogen digestion mixtures that were terminated at various times. Four major bands are observed in the 45-min digest: the doublet migrating just below the position of intact fibrinogen corresponds to early and late species of Fragment X; the next two bands, which are about equal in intensity, correspond to Fragments Y and D. A minor band near the top of the gel appears resistant to limited digestion by plasmin and is attributed to fibronectin which is present in trace amounts. The minor band in the bottom half of the gel, which becomes more prominent at later digestion times, corresponds to Fragment E. The relative molar quantities of Fragments X, Y, D and E in the 45-min digest were 33, 21, 38 and 8%, respectively, as estimated from densitometry of the gel and the molecular weights. In the presence of 5 mM calcium, Fragment D is resistant to further digestion by plasmin.[19] Thus the species of Fragment D that is present in the 45-min digest is the largest Fragment D and appears at the same position in a gel of the 10-hr digest.

An ion-exchange chromatography method for separating the terminal plasmic products of fibrinogen [40] was used to fractionate limited digests that contained, in addition to Fragments D and E, the intermediate degradation products, Fragments X and Y. The first and third peaks from the QAE-Sephadex step were each gel filtered once to obtain Fragment D and Fragment E, respectively, with purity >96%. Fragment X and Fragment Y were only partially resolved when the major peak from the ion-exchange column was chromatographed on Sephacryl S-200. However, when fractions from the leading and trailing halves of this peak were rechromatographed separately on the same gel filtration column, these fragments were each obtained ~96% pure. A recent alternate purification scheme using lysine-agarose, which gives good yields of Fragments D_1, D_3, E and X + Y, may prove to be more efficient.[41]

FIGURE 3 shows sodium dodecyl sulfate gel electrophoretic analyses of the purified fragments, which were subsequently examined by electron microscopy. Each appears as a single band by this method of analysis and clearly represents the major species of each fragment present in the whole digest. The smaller of the two forms of Fragment X observed in the 45-min digest was isolated with an overall yield of 35%. Fragment Y was isolated at about 20% yield and was the only form of Fragment Y observed at any time of digestion. The Fragment D and E species were also unique and were obtained with 50% and 40% yields, respectively. Sodium dodecyl sulfate gel electrophoretic analyses of these isolated fragments after reduction with β-mercaptoethanol provided accurate assessment of their subunit molecular weights.[6] As indicated earlier, these could be summed to give the following total molecular weights: Fragment X = 240,000; Fragment Y = 150,000; Fragment D = 93,000; Fragment E = 50,000.

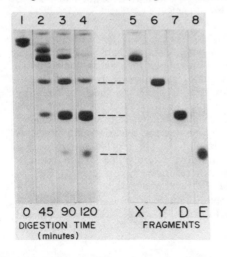

FIGURE 3. SDS-5% polyacrylamide gel electrophoresis of nonreduced samples of fibrinogen (gel 1), whole plasmic digests of fibrinogen (gels 2–4), and the final preparations of the purified fibrinogen fragments (gels 5–8) that were examined by electron microscopy. The digestion was performed in the presence of 5 mM calcium.

0 45 90 120   X Y D E
DIGESTION TIME   FRAGMENTS
(minutes)

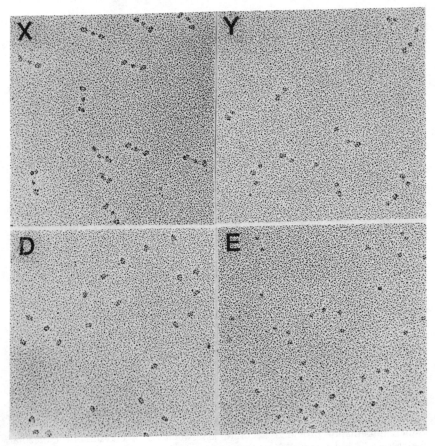

FIGURE 4. Electron micrographs of rotary shadowed preparations of purified Fragments X, Y, D and E sprayed onto mica at a final concentration of 20 to 30 µg/ml. Note the apparent elongation of the Fragment D molecules; ×150,000.

These purified fragments are all clearly distinguishable in the rotary shadowed preparations shown in FIGURE 4. Although their resolution is limited by the size of the platinum grains, the major structural features such as the number, size and shape of the globular regions present in each of the fragments are sufficiently distinct to identify them with respect to the trinodular model. The specimens in FIGURE 4 were all prepared and shadowed simultaneously, thereby obviating the need to account for a variable thickness of platinum when comparing the relative sizes of the different fragments. However, the absolute molecular dimensions are obviously affected by the platinum thickness and the values in TABLE 1 have therefore been corrected for a 2-nm platinum shell, which completely surrounds each molecule.[39]

As in fibrinogen, the outer nodules of Fragment X appear to be equal in size and larger than the central nodule (TABLE 1). In general, the corrected average length and nodule dimension of Fragment X are somewhat less than those of fibrinogen, but these differences are not significant given the standard deviations of the measurements and the possible error of the correction factor (0.6 nm per 2 nm thickness).[39] Thus, the carboxyl-terminal 70% of the Aα-chains and residues Bβ 1–42, which are cleaved from fibrinogen to form Fragment X, do not contribute significantly to the trinodular structure observed for intact fibrinogen by electron microscopy. However, as noted by Erickson and Fowler [42] there is a fourth nodule in intact fibrinogen that is somewhat smaller than the central E nodule and lies about 10 nm away on one side of the molecular axis. The absence of this structure in shadowed specimens of Fragment X supports its identification as the carboxyl-terminal halves of the α chains. FIGURE 4 also shows that a few of the outer nodules of the Fragment X molecules appear to be split lengthwise into two domains. The subdivision of this outer nodule into two or more domains has also been suggested by recent thermodynamic [43] and other electron microscopic studies [44] and may represent a separation of the regions of the β and γ chains that are present in both Fragment X and fibrinogen.

Fragment Y, the next largest intermediate, consists of two linked nodules as shown in FIGURE 4. As with fibrinogen, the actual linkage between nodules is not well resolved in the rotary shadowed specimen. Most of the molecules have one nodule clearly larger than the other. The sizes of these nodules and the distance between them are very similar to those obtained by measuring one outer nodule and the central nodule in the intact fibrinogen molecule (TABLE 1).

Both terminal plasmic digestion products, Fragments D and E, consist of single nodules (FIGURE 4). Fragment D is almost exactly the same size as the outer nodules of intact fibrinogen (TABLE 1) and is elongated, which is a characteristic reported for the outer nodules of the fibrinogen molecule.[39, 45] Fragment E is clearly smaller than Fragment D and, as shown in TABLE 1, is comparable in size to the central nodule of intact fibrinogen. The orienta-

TABLE 1

COMPARISON OF THE NODULE DIMENSION OF THE PLASMIC FRAGMENTS
WITH THOSE OF INTACT FIBRINOGEN *

		Outer Nodule		Central Nodule	
Overall Length		Length	Width	Length	Width
		nm		nm	
Fbg	43.8 ± 1.7	11.3 ± 1.3	6.8 ± 1.3	6.1 ± 0.6	4.0 ± 0.6
X	41.9 ± 2.6	10.3 ± 1.3	7.0 ± 1.3	5.1 ± 1.0	4.0 ± 0.8
Y	22.7 ± 2.5	9.3 ± 1.3	6.4 ± 0.8	4.6 ± 1.0	3.5 ± 0.8
D		11.3 ± 1.1	6.1 ± 0.8		
E				diameter	
				5.3 ± 0.8	

* Values are ± SD and were obtained by subtracting 4 nm from the actual measured value. This correction reflects the contribution of two thicknesses of a thick platinum shell surrounding each molecule.[39] At least 25 molecules preparations that were rotary shadowed simultaneously were measured in t termination of each value.

tion of Fragment E molecules could not be determined; therefore, the only dimension given for Fragment E in TABLE 1 is its diameter, which is midway between the values measured for length and width of the central nodule of intact fibrinogen. Although never fully substantiated, it was frequently proposed on the basis of biochemical data that Fragments D and E correspond to the middle and outer nodules of the trinodular fibrinogen molecule, respectively.

Our electron microscopic findings established definitively the identification of the plasmic fragments within the trinodular structure of fibrinogen and also fully validated asymmetric plasmic cleavage of fibrinogen. Subsequent electron microscopic studies of antibody-labeled fibrinogen have also supported this assignment for the nodules.[46-48]

EFFECT OF FIBRINOGEN/FIBRIN PLASMIC CLEAVAGE FRAGMENTS ON THE CONVERSION OF PLASMINOGEN TO PLASMIN

Fibrin is believed to provide a surface on which the major reactions of fibrinolysis occur, i.e., the conversion of plasminogen to plasmin, the cleavage of fibrin by plasmin and the inhibition of plasmin by α_2-antiplasmin and other inactivators.[49, 50] These reactions are localized on the fibrin surface probably because of binding sites on fibrin that are complementary to the lysine-binding sites on plasminogen[51] as well as the recently sequenced homologous sites on tissue activator and urokinase-like activators.[52] Studies examining the relative affinities of the plasminogen molecule for various ligands indicate that the optimal complementary structure consists of an amino nitrogen atom and a carboxyl carbon separated by 0.68 nm.[53] Although it is generally accepted that fibrin does bind plasminogen through such interactions with the lysine-binding sites on plasminogen,[54] these interactions have not been fully characterized and therefore we undertook such studies. To examine the binding between plasminogen and fibrinogen, native human Glu-plasminogen, which has an amino-terminal glutamic acid, was isolated from fresh-frozen plasma by lysine-agarose affinity chromatography.[55] Lysyl amino-terminal plasminogen was prepared by treating Glu-plasminogen with urokinase-free plasmin which was then removed by insolubilized bovine pancreatic trypsin inhibitor. This Lys-plasminogen, which is activated to plasmin more rapidly than is Glu-plasminogen, is said to have a greater affinity for fibrin. By sucrose density ultracentrifugation experiments, fibrinogen or its fragments could not be demonstrated to bind to Glu-plasminogen. However, as shown in FIGURE 5, they did bind to Lys-plasminogen. Panel A shows that when fibrinogen was centrifuged with ^{125}I-Lys-plasminogen, some of the radioactivity shifted to form a new peak, coincident with the fibrinogen absorbance peak indicating that binding had occurred. As shown in Panel B, when ^{125}I-Fragment E, the central nodule of fibrinogen, was centrifuged with Lys-plasminogen, binding between these two molecules was again evident by the development of a new radioactive peak. In contrast to Fragment E, the isolated Fragment D domain of fibrinogen did not bind to Lys-plasminogen.

To localize the region in Lys-plasminogen that was binding to fibrinogen, three plasminogen domains were isolated after digestion by elastase as described by Sottrup-Jensen et al.[56] Plasminogen contains five triple-disulfide loop structures, called kringles, that have a high degree of sequence homology with each other and are believed to correspond to lysine-binding sites. Elastase cleaves between kringles 3 and 4 and between 4 and 5, producing three major fragments called K123, K4 and Val_{442}-plasminogen, which is a fully activatable zymogen. Fibrinogen, Fragment D or Fragment E was then incubated with each of these plasminogen domains and analyzed by ultracentrifugation. In Panel A of FIGURE 6, fibrinogen bound to K123, but to a much lesser extent than to Lys-plasminogen. However, in Panel B, Fragment E showed a higher affinity for K123 than for Lys-plasminogen as indicated by the shift of the entire radioactive peak. Fibrinogen or its fragments did not bind either K4 or Val_{442}-plasminogen.

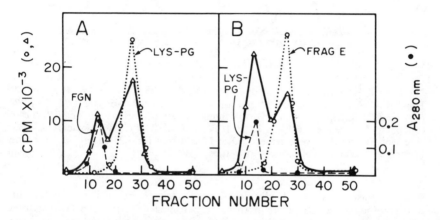

FIGURE 5. Sucrose density gradient ultracentrifugation of Lys-plasminogen with fibrinogen or Fragment E. The dashed and dotted lines indicate absorbance and radioactivity, respectively, of the proteins when centrifuged alone. The solid lines are the radioactive profiles of the mixtures. *Panel A*: Centrifugation of tubes containing fibrinogen (●), ^{125}I-Lys-plasminogen (○), or ^{125}I-Lys-plasminogen (△) with a two-fold molar excess of fibrinogen at 65,000 rpm for 2.5 hr at 4°C. *Panel B*: Simultaneous centrifugation of Lys-plasminogen alone (●), ^{125}I-Fragment E alone (○), and ^{125}I-Fragment E (△) with a twofold molar excess of Lys-plasminogen at 65,000 rpm for 3.5 hr at 4°C. Fraction 1 represents the bottom of the centrifuge tube.

Since fibrin may have a greater affinity for plasminogen than does fibrinogen, the binding of both Glu- and Lys-plasminogen to fibrin was also studied. Recently, several investigators have demonstrated that fibrin does indeed adsorb plasminogen,[57-59] but their quantitations of the affinity have differed by 1000-fold.[60, 61] In those studies, purified fibrinogen was converted to fibrin in the presence of radiolabeled plasminogen, the fibrin washed free of unbound plasminogen, and then binding constants were obtained from the amount of radiolabeled protein remaining. No correction was made for nonspecific binding despite the fact that radiolabeled human serum albumin can be nonspecifically incorporated into crosslinked fibrin to the same extent as Lys-plasminogen.[62] In addition, the equilibria conditions must change during the extensive washing of the fibrin clot and may cause artificially low affinities. Both of these problems were overcome by a technique using sonicated suspen-

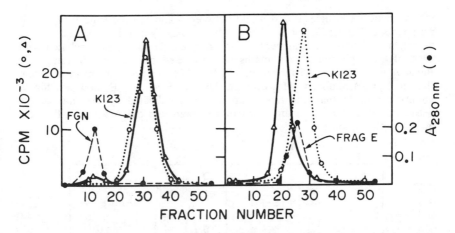

FIGURE 6. Density gradient analysis of ^{125}I-K123 with fibrinogen or Fragment E. *Panel A*: Centrifugation at 65,000 rpm for 3.25 hr at 4°C of fibrinogen (●), ^{125}I-K123 (○), and ^{125}I-K123 (△) with a 24-fold molar excess of fibrinogen. *Panel B*: Tubes containing Fragment E (●), ^{125}I-K123 (○) or ^{125}I-K123 (△) with a tenfold molar excess of Fragment E were centrifuged at 65,000 rpm for 4.25 hr at 4°C. Fraction 1 represents the bottom of the centrifuge tube.

sions of plasminogen-free, fully crosslinked fibrin.[63] Following reduction by β-mercaptoethanol, the fibrin in these suspensions gave the expected sodium dodecyl sulfate electrophoretic pattern;[11] no extra bands were present. The homogeneously dispersed fibrin suspensions proved useful in binding studies by allowing the rapid separation of free from bound radiolabeled plasminogen, thereby ensuring that the measured amount of binding reflected the equilibrium conditions immediately prior to the separation step. Centrifugation of a radiolabeled plasminogen-fibrin suspension at 6900 × g for 10 minutes pelleted more than 96% of the radioactivity. Nonspecific incorporation by the crosslinked fibrin suspension of various proteins such as albumin and human IgG was negligible. As shown in the left panel of FIGURE 7, when the fibrin suspensions were incubated with increasing concentrations of ^{125}I-Lys-plasminogen, the total binding of ^{125}I-Lys-plasminogen increased. When nonspecific binding, determined from the amount of radiolabeled Lys-plasminogen not displaced by excess unlabeled Lys-plasminogen, was subtracted from the total binding, the resultant specific binding was found to be saturable. As shown by the lower lines in the left panel, nonspecific trapping of ^{125}I-labeled human γ-globulin, Fragment E or human serum albumin by the crosslinked fibrin suspension was minimal. To determine the binding parameters for the Lys-plasminogen-fibrin interaction, saturation binding curves were subjected to Scatchard analysis, which gave an apparent dissociation constant of 0.68 μM and about two plasminogen sites per fibrin monomer as shown in FIGURE 7. The binding to fibrin of Glu-plasminogen, K123, K4 and Val$_{442}$-plasminogen was examined by the same technique. As shown in TABLE 2, the affinity of Glu-plasminogen for fibrin was about 55-fold weaker than that of Lys-plasminogen. K123 had an affinity similar to that of Glu-plasminogen, while those of K4 and Val$_{442}$-plasminogen were about 3-fold weaker. Despite these differences, fibrin had about the same number of binding sites for each type of plasminogen or plasminogen fragment.

Ligands such as lysine and its analogues, ε-aminocaproic acid (EACA) and trans-4-(aminomethyl)cyclohexanecarboxylic acid (t-AMCHA) bind to plasminogen and inhibit fibrinolysis.[64-66] Markus *et al.*[51, 67, 68] have shown that plasminogen has one high affinity binding site for these ligands and four or five weaker ones. To demonstrate that the sites on fibrin with which plasminogen or its fragments interact are indeed structurally similar to lysine or its analogues, EACA was used to compete with fibrin for binding to ^{125}I-Lys-plasminogen. As shown by the curve designated with closed circles in FIGURE 8, 1.0 mM EACA completely inhibited the interaction between these two proteins; this could not be accomplished with high salt concentration or unrelated amino acids, such as glycine. Nonlinear least squares computer analysis of this competition curve indicated that dissociation of Lys-plasminogen from fibrin occurred with a K_d of 18 μM. This value is similar to the K_d of 35 μM established by ultrafiltration for the high affinity binding site of Lys-plasminogen.[68] Recent studies have shown that the first kringle, or K1, of the K123 region corresponds to the Lys-plasminogen high affinity site, having a K_d of 18 μM for EACA.[69, 70] In experiments not shown, EACA inhibited the binding of K123 to fibrin with a K_d of 16 μM, which is close to the value found by ultrafiltration. These

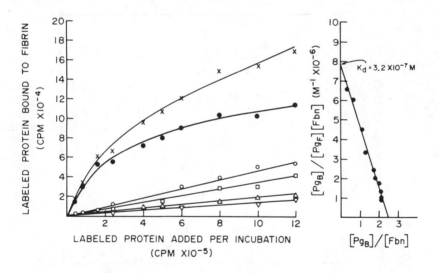

FIGURE 7. Binding of ^{125}I-Lys-plasminogen to fibrin suspensions. In the left panel, specific binding (●) was calculated by subtracting nonspecifically bound ^{125}I-Lys-plasminogen (○) from the total amount bound (×). Trapping of various proteins in the centrifuged fibrin pellet was low: ^{125}I-human γ-globulin (□); ^{125}I-Fragment E (△); ^{125}I-human serum albumin (▽). Incubations were done at 20°C for 10 min in a 200 μl volume with a fibrin concentration of 8.5×10^{-8} M and labeled protein having a specific activity of 2.37×10^{15} cpm per mole. Scatchard analysis of the specific binding is shown in the right panel. [Pg$_B$], molar concentration of bound ^{125}I-Lys-plasminogen; [PgF], concentration of free ^{125}I-Lys-plasminogen; [Fbn], concentration of fibrin monomer in suspension.

TABLE 2

DISSOCIATION CONSTANTS FOR THE INTERACTIONS OF PLASMINOGEN OR ITS
FRAGMENTS WITH FIBRIN

Plasminogen Component	K_d (μM)	n *
Lys-plasminogen	0.68	2.6
Glu-plasminogen	38	2.0
K123	51	1.8
K4	164	2.5
Val$_{442}$-plasminogen	170	2.5

* Number of plasminogen binding sites per fibrin monomer.

results implicate the K1 region of Lys-plasminogen in binding to a lysine-like
charge array in fibrin.

When the inhibition of the binding of Glu-plasminogen to fibrin by EACA
was studied in a similar manner, computer modeling of the competition curve
indicated that two binding sites were involved in this interaction in contrast
to the one site found with Lys-plasminogen. As shown by the curve designated

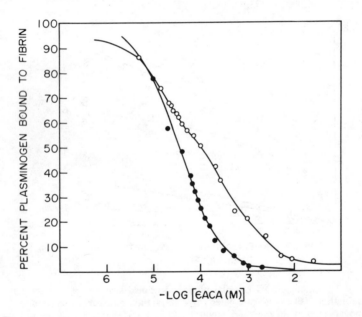

FIGURE 8. Inhibition of specific ^{125}I-plasminogen binding to fibrin by EACA. For
^{125}I-Lys-plasminogen (●), the line represents the computer-fit of the data points,
each the average of three determinations, to a model describing the interaction of a
ligand with a single type of binding site. The interaction of EACA with the site on
Lys-plasminogen, which inhibits binding to fibrin, has a K_d of 18 μM. With ^{125}I-Glu-
plasminogen (○), the best computer-fit was to a model for two types of binding
sites, one with a K_d of 10 μM and the other with a K_d of 440 μM.

with open circles in FIGURE 8, one site had a K_d of 10 μM while the second site had a 40-fold lower affinity, with a K_d of 440 μM. These results compare with data from ultrafiltration experiments in which Glu-plasminogen had one EACA binding site with a K_d of 9 μM and four or five sites with a K_d of 5mM.[68] Although the values for the strong site found by ultrafiltration and by our competition studies are essentially identical, the competition studies indicate the second site has an affinity about 10 times stronger than that of the four or five low affinity sites on Glu-plasminogen. This finding suggests that, on binding to fibrin, a change in the conformation of Glu-plasminogen occurs that is analogous to the events that take place when Glu-plasminogen is proteolytically modified to Lys-plasminogen. As a consequence of this conversion, the molecule concomitantly undergoes a conformational change in which one of the low affinity sites develops an intermediate strength affinity, with a K_d of 300 μM.[68] This value is close to the affinity of the second site on Glu-plasminogen that was interacting with fibrin. Hence, fibrin may induce a similar conformational change in Glu-plasminogen when they bind so that the strength of one of the four or five low affinity sites on Glu-plasminogen is increased tenfold. It is known that the binding of lysine analogues at one or more of the weak lysine-binding sites of Glu-plasminogen is responsible for a major conformational change and an increased rate of activation.[51, 68] Since recent studies have indicated an enhancement effect by fibrinogen on Glu-plasminogen activation [71-75] and since our ultracentrifugation experiments suggested that the high affinity site is not involved in fibrinogen binding of Glu-plasminogen, we blocked the high affinity binding site by EACA and then examined the effect of fibrinogen on the activation of Glu-plasminogen. We found that plasminogen activation was enhanced by fibrinogen whether or not the high affinity site was blocked; this suggests that an interaction indeed occurs through one or more of the lower affinity sites. As in the interaction of Glu-plasminogen with fibrin, a conformational change similar to that caused by lysine analogues may be induced by fibrinogen. Yet unlike the binding of Glu-plasminogen to fibrin, which occurs through both the high and low affinity sites, the fibrinogen-Glu-plasminogen interaction occurs only through the low affinity site. This binding at the weak site appears to induce a conformational change that enhances the activation rate in a manner analogous to the effect of lysine analogues on Glu-plasminogen.

To examine in more detail the potentiating effect of fibrinogen on plasminogen activation, the major degradation products of fibrinogen digestion, Fragment D, Fragment E and P21 (a 19,000-dalton peptide from the middle of the α chain), were each assessed for their ability to enhance this reaction when urokinase was used as the activator. Studies of the effect of Fragments D and E with streptokinase had demonstrated that the steady-state kinetic parameters were relatively unchanged, but that both fibrinogen and Fragment D increased the rate of formation of the streptokinase-plasminogen activator complex for all plasminogen forms.[76] With urokinase, however, fibrinogen, Fragment D, and Fragment E, but not P21, increased the steady-state second order rate constant only when Glu-plasminogen was the substrate. While they did not have an effect on the initial rate of activation of Lys-plasminogen, there was a significant enhancement and prolongation of Lys-plasmin activity, which suggests a reduction in the rate of plasmin autolysis. To determine whether this diminished autolysis involved interaction of fibrinogen, Fragment D or Fragment E with the lysine-binding sites of plasmin, the effect of each on the activation of Val$_{442}$-plasminogen, which lacks four sites including the high affinity site, was examined. As with Lys-plasminogen there was no effect on the initial activation rate, but in addition, there was also no effect on the rate of loss of activity. Thus, one or more of the lysine-binding sites that Val$_{442}$-plasminogen lacks must interact with fibrinogen or its fragments to protect plasmin from autolysis. At physiologic ratios with Lys-plasmin, each of the fragments significantly reduced the second order rate constant for activity loss.

In Vivo *Implications*

FIGURE 9 summarizes the interactions of fibrin with plasminogen and the consequent effects on fibrinolysis. Following the removal of the fibrinopeptides from fibrinogen, fibrin monomers become polymerized and crosslinked through Fragment D domains and then interact with Glu-plasminogen with a moderately weak K_d but more strongly than does fibrinogen. The binding occurs through two interactions: one is between the high affinity lysine-binding site of Glu-plasminogen and probably the Fragment E domain of fibrin; the other is between one of the weaker lysine-binding sites and the fibrin Fragment D domain. By virtue of the latter interaction, the conformation of Glu-plasminogen is induced to become more like that of Lys-plasminogen and Glu-plas-

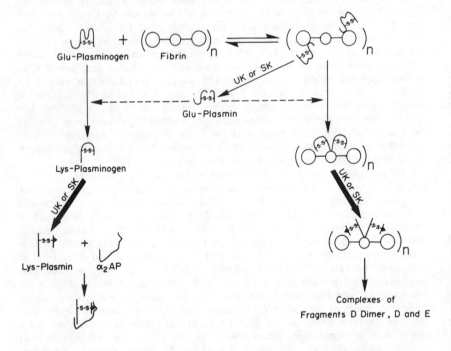

FIGURE 9. The proposed interactions of plasminogen with fibrin domains during fibrinolysis.

minogen activation to Glu-plasmin is enhanced. Glu-plasmin then converts free
or fibrin-bound Glu-plasminogen to Lys-plasminogen. The lysine-binding sites
on unbound Lys-plasminogen or free Lys-plasmin are not protected by interaction with fibrin and therefore become rapidly involved in the initial interaction
with α_2-antiplasmin, the primary inhibitor of plasmin. In contrast, when Lys-
plasminogen is tightly bound to the Fragment E domain, it is rapidly activated
by urokinase. Through this latter interaction, plasmin activity is protected
from its circulating inhibitor, α_2-antiplasmin, as well as from autolysis. Efficient
fibrinolysis subsequently occurs, with the production of various complexes
composed of D dimer, Fragment D and Fragment E.

REFERENCES

1. PIZZO, S. V., M. L. SCHWARTZ, R. L. HILL & P. A. McKEE. 1972. The effect
   of plasmin on the subunit structure of human fibrinogen. J. Biol. Chem. **247:**
   636–645.
2. PIZZO, S. V., M. L. SCHWARTZ, R. L. HILL & P. A. McKEE. 1973. The effect
   of plasmin on the subunit structure of human fibrin. J. Biol. Chem. **248:**
   4574–4583.
3. PIZZO, S. V., M. L. SCHWARTZ, R. L. HILL & P. A. McKEE. 1973. Subunit
   structure of Fragment D from fibrinogen and cross-linked fibrin. J. Biol.
   Chem. **248:** 4584–4590.
4. MARDER, V. J., N. R. SHULMAN & W. R. CARROLL. 1969. High molecular
   weight derivatives of human fibrinogen produced by plasmin. J. Biol. Chem.
   244: 2111–2119.
5. MOSESSON, M. W., J. S. FINLAYSON & D. K. GALANAKIS. 1973. The essential
   covalent structure of human fibrinogen evinced by analysis of derivatives
   formed during plasmic hydrolysis. J. Biol. Chem. **248:** 7913–7929.
6. FERGUSON, E. W., L. J. FRETTO & P. A. McKEE. 1975. A re-examination of
   the cleavage of fibrinogen and fibrin by plasmin. J. Biol. Chem. **250:** 7210–
   7218.
7. FOWLER, W. E., L. J. FRETTO, H. P. ERICKSON & P. A. McKEE. 1980. Electron microscopy of plasmic fragments of human fibrinogen as related to
   trinodular structure of the intact molecule. J. Clin. Invest. **66:** 50–56.
8. HALL, C. & H. SLAYTER. 1959. The fibrinogen molecule: its size, shape, and
   mode of polymerization. J. Biophys. Biochem. Cytol. **5:** 11–15.
9. SLADE, C. L., S. V. PIZZO, L. M. TAYLOR, JR., H. M. STEINMAN & P. A. McKEE.
   1976. Characterization of fragment E from fibrinogen and cross-linked
   fibrin. J. Biol. Chem. **251:** 1591–1596.
10. FRETTO, L. J., E. W. FERGUSON, H. M. STEINMAN & P. A. McKEE. 1978.
    Localization of the α-chain cross-link acceptor sites of human fibrin. J. Biol.
    Chem. **253:** 2184–2195.
11. FRETTO, L. J. & P. A. McKEE. 1978. Structure of α-polymer from *in vitro*
    and *in vivo* highly cross-linked human fibrin. J. Biol. Chem. **253:** 6614–6622.
12. MIHALYI, E. 1980. Proteolytic fragmentation of fibrinogen. *In* Handbook
    Series in Clinical Laboratory Science. R. M. Schmidt, Ed. Section I: Hematol-
    ogy Vol. III: 51–60. CRC Press. Boca Raton, FL.
13. BUDZYNSKI, A. Z., V. J. MARDER & J. R. SHAINOFF. 1974. Structure of plasmic
    degradation products of human fibrinogen. J. Biol. Chem. **249:** 2294–2302.
14. KOWALSKA-LOTH, B., B. GARDLUND, N. EGBERG & B. BLOMBACK. 1973. Plas-
    mic degradation products of human fibrinogen. II. Chemical and immuno-
    logical relation between fragment E and N-DSK. Thromb. Res. **2:** 423–450.

88   Annals New York Academy of Sciences

15. HAVERKATE, F., G. TIMAN & W. NIEUWENHUIZEN. 1979. Anticlotting properties of fragment D from human fibrinogen and fibrin. Eur. J. Clin. Invest. 9: 253–255.
16. NIEWENHUIZEN, W., A. VERMOND, W. J. NOOIJEN & F. HAVERKATE. 1979. Calcium-binding properties of human fibrin(ogen) and degradation products. FEBS Letters 98: 257–259.
17. KLOCZEWIAK, M., S. TIMMONS & J. HAWIGER. 1982. Localization of a site interacting with human platelet receptor on carboxy-terminal segment of human fibrinogen γ chain. Biochem. Biophys. Res. Commun. 107: 181–187.
18. MARGUERIE, G. A., N. ARDAILLOU, G. CHEREL & E. F. PLOW. 1982. The binding of fibrinogen to its platelet receptor. J. Biol. Chem. 257: 11872–11875.
19. HAVERKATE, F. & G. TIMAN. 1977. Protective effect of calcium in the plasmin degradation of fibrinogen and fibrin fragments D. Thromb. Res. 10: 803–812.
20. OLEXA, S. A. & A. Z. BUDZYNSKI. 1979. Binding phenomena of isolated unique plasmic degradation products of human cross-linked fibrin. J. Biol. Chem. 254: 4925–4932.
21. LORAND, L., D. CHENOWETH & A. GRAY. 1972. Titration of the acceptor cross-linking sites in fibrin. Ann. NY Acad. Sci. 202: 155–171.
22. WATT, K. W. K., B. A. COTTRELL, D. D. STRONG & R. F. DOOLITTLE. 1979. Amino acid sequence studies on the α chain of human fibrinogen. Overlapping sequences providing the complete sequence. Biochemistry 18: 5410–5416.
23. CORCORAN, D. H., E. W. FERGUSON, L. J. FRETTO & P. A. McKEE. 1980. Localization of a cross-link donor site in the α-chain of human fibrin. Thromb. Res. 19: 883–888.
24. DOOLITTLE, R. F., K. G. CASSMAN, B. A. COTTRELL & S. J. FRIEZNER. 1977. Amino acid sequence studies on the α chain of human fibrinogen. Isolation and characterization of two linked α-chain cyanogen bromide fragments from fully cross-linked fibrin. Biochemistry 16: 1715–1719.
25. SOBEL, J. H., J. A. KOEHN, R. FRIEDMAN & R. E. CANFIELD. 1982. Alpha chain crosslinking of human fibrin: Purification and radioimmunoassay development for two Aα chain regions involved in crosslinking. Thromb. Res. 26: 411–424.
26. FOWLER, W. E., R. R. HANTGAN, J. HERMANS & H. P. ERICKSON. 1981. Structure of the fibrin protofibril. Proc. Natl. Acad. Sci. USA 78: 4872–4876.
27. FRANCIS, C. W. & V. J. MARDER. 1982. A molecular model of plasmic degradation of crosslinked fibrin. Semin. Thromb. Haemostas. 8: 25–35.
28. FURLAN, M. & E. A. BECK. 1972. Plasmic degradation of human fibrinogen. I. Structural characterization of degradation products. Biochim. Biophys. Acta 263: 631–644.
29. MILLS, D. A. 1972. A molecular model for the proteolysis of human fibrinogen by plasmin. Biochim. Biophys. Acta 263: 619–630.
30. NUSSENZWEIG, V., M. SELIGMANN, J. PELMONT & P. GRABAR. 1961. Les produits de degradation du fibrinogene humain par la plasmine. 1. Separation et proprietes physico-chimiques. Ann. Inst. Pasteur (Paris) 100: 377–389.
31. BUDZYNSKI, A. Z., V. J. MARDER & J. R. SHAINOFF. 1974. Structure of plasmic degradation products of human fibrinogen. J. Biol. Chem. 249: 2294–2302.
32. DONOVAN, J. W. & E. MIHALYI. 1974. Conformation of fibrinogen: calorimetric evidence for a three-nodular structure. Proc. Natl. Acad. Sci. USA 71: 4125–4128.
33. KOPPEL, G. 1967. Elektronenmikrokopische Untersuchungen zur Gestalt und zum Makromolekularen bau des Fibrinogenmolekuls und der Fibrinfasern. Z. Zellforsch. Mikrosk. Anat. 77: 443–517.
34. POUIT, L., G. MARCILLE, M. SUSCILLON & D. HOLLARD. 1972. Etude en microscopie electronique de differentes stapes de la fibrinoformation. Thromb. Diath. Haemorrh. 27: 559–572.

35. BACHMANN, L., W. SCHMITT-FUMIAN, R. HAMMEL & K. LEDERER. 1975. Size and shape of fibrinogen, I. Electron microscopy of the hydrated molecule. Die Makromolekulare Chemie **176:** 2603–2618.

36. MOSESSON, M. W., J. ESCAIG & G. FELDMANN. 1979. Electron microscopy of metal-shadowed fibrinogen molecules deposited on carbon films at different concentrations. Thromb. Haemostasis. **42:** 88.

37. STEWART, G. 1971. Comments and discussion. Scand. J. Haematol. Suppl. **13:** 63–67.

38. BLAKEY, P., M. GROOM & R. TURNER. 1977. The conformation of fibrinogen and fibrin: An electron microscope study. Br. J. Haematol. **35:** 437–440.

39. FOWLER, W. E. & H. P. ERICKSON. 1979. The trinodular structure of fibrinogen: confirmation by both shadowing and negative stain electron microscopy. J. Mol. Biol. **134:** 241–249.

40. CHEN, J. P., H. M. SHURLEY & M. F. VICKROY. 1974. A facile separation of fragments D and E from the fibrinogen/fibrin degradation products of three mammalian species. Biochem. Biophys. Res. Commun. **61:** 66–71.

41. RUPP, C., R. SIEVI & M. FURLAN. 1982. Fraction of plasmic fibrinogen digest on lysine-agarose. Isolation of two fragments D, fragment E and simultaneous removal of plasmin. Thromb. Res. **27:** 117–121.

42. ERICKSON, H. P. & W. E. FOWLER. 1983. Electron microscopy of fibrinogen, its plasmic fragments and small polymers. Ann. NY Acad. Sci. **408:**. This volume.

43. PRIVALOV, P. L. & L. V. MEDVED. 1982. Domains in the fibrinogen molecule. J. Mol. Biol. **159:** 665–683.

44. WEISEL, J. W., G. N. PHILLIPS, JR. & C. COHEN. 1981. A model from electron microscopy for the molecular structure of fibrinogen and fibrin. Nature **289:** 263–267.

45. SLAYTER, H. 1976. High-resolution metal replication of macromolecules. Ultramicroscopy. **1:** 341–357.

46. TELFORD, J. N., J. A. NAGY, P. A. HATCHER & H. A. SCHERAGA. 1980. Location of peptide fragments in the fibrinogen molecule by immunoelectron microscopy. Proc. Natl. Acad. Sci. USA **77:** 2372–2376.

47. PRICE, T. M., D. D. STRONG, M. L. RUDEE & R. F. DOOLITTLE. 1981. Shadow-cast electron microscopy of fibrinogen with antibody fragments bound to specific regions. Proc. Natl. Acad. Sci. USA **78:** 200–204.

48. NORTON, P. A. & H. S. SLAYTER. 1981. Immune labeling of the D and E regions of human fibrinogen by electron microscopy. Proc. Natl. Acad. Sci. USA **78:** 1661–1665.

49. WIMAN, B. & D. COLLEN. 1978. Molecular mechanism of physiological fibrinolysis. Nature **272:** 549–550.

50. LIJNEN, H. R. & D. COLLEN. 1982. Interaction of plasminogen activators and inhibitors with plasminogen and fibrin. Semin. Thromb. Haemostas. **8:** 2–10.

51. MARKUS, G., J. L. DEPASQUALE & F. C. WISSLER. 1978. Quantitative determination of the binding of ε-aminocaproic acid to native plasminogen. J. Biol. Chem. **253:** 727–732.

52. GUNZLER, W. A., G. J. STEFFENS, F. OTTING, S.-A. A. KIM, E. FRANKUS & L. FLOHE. 1982. The primary structure of high molecular mass urokinase from human urine. Hoppe-Seyler's Z. Physiol. Chem. **363:** 1155–1165.

53. WINN, E. S., S.-P. HU, S. M. HOCHSCHWENDER & R. A. LAURSEN. 1980. Studies on the lysine-binding sites of human plasminogen: The effect of ligand structure on the binding of lysine analogs to plasminogen. Eur. J. Biochem. **104:** 579–586.

54. WIMAN, B. & P. WALLEN. 1977. A physiological role of the lysine binding site in plasminogen. Thromb. Res. **10:** 213–222.

55. DEUTSCH, D. G. & E. T. MERTZ. 1970. Plasminogen: Purification from human plasma by affinity chromatography. Science **170:** 1095–1096.

56. SOTTRUP-JENSEN, L., H. CLAEYS, M. ZAJDEL, T. E. PETERSEN & S. MAGNUSSON. 1977. The primary structure of human plasminogen: Isolation of two lysine-binding fragments and one "mini-"plasminogen (MW, 38,000) by elastase-catalyzed-specific limited proteolysis. In Progress in Chemical Fibrinolysis and Thrombolysis. J. F. Davidson, R. M. Rowan, M. Samama & P. C. Desnoyers, Eds. 3: 191–209. Raven Press. New York.

57. RAKOCZI, I., B. WIMAN & D. COLLEN. 1978. On the biological significance of the specific interaction between fibrin, plasminogen and antiplasmin. Biochim. Biophys. Acta 540: 295–300.

58. THORSEN, S. 1975. Differences in the binding to fibrin of native plasminogen and plasminogen modified by proteolytic degradation. Influence of ω-aminocarboxylic acids. Biochim. Biophys. Acta 393: 55–65.

59. CEDERHOLM-WILLIAMS, S. A. 1977. The binding of plasminogen (mol. wt. 84,000) and plasmin to fibrin. Thromb. Res. 11: 421–423.

60. CEDERHOLM-WILLIAMS, S. A. 1977. The binding of plasminogen, plasmin and streptokinase-plasminogen activator to fibrin. Biochem. Soc. Trans. 5: 1441–1443.

61. SUENSON, E. & S. THORSEN. 1981. Secondary-site binding of Glu-plasmin, Lys-plasmin and miniplasmin to fibrin. Biochem. J. 197: 619–628.

62. SUMMARIA, L., I. G. BOREISHA, L. ARZADON & K. C. ROBBINS. 1977. The dissolution of human cross-linked plasma fibrin clots by the equimolar human plasmin-derived light (B) chain-streptokinase complex. The acceleration of clot lysis by pretreatment of fibrin clots with the light (B) chain. Thromb. Res. 11: 377–389.

63. LUCAS, M. A., L. J. FRETTO & P. A. MCKEE. 1983. The binding of human plasminogen to fibrin and fibrinogen. J. Biol. Chem. In press.

64. OKAMOTO, S., S. OSHIBA, H. MIHARA & U. OKAMOTO. 1968. Synthetic inhibitors of fibrinolysis: In vitro and in vivo mode of action. Ann. NY Acad. Sci. 146: 414–429.

65. AMBRUS, C. M., J. L. AMBRUS, H. B. LASSMAN & I. B. MINK. 1968. Studies on the mechanism of action of inhibitors of the fibrinolysis system. Ann. N.Y. Acad. Sci. 146: 430–447.

66. SKOZA, L., A. O. TSE, M. SEMAR & A. J. JOHNSON. 1968. Comparative activities of amino acid and polypeptide inhibitors on natural and synthetic substrates. Ann. NY Acad. Sci. 146: 659–672.

67. MARKUS, G., J. L. EVERS & G. H. HOBIKA. 1978. Comparison of some properties of native (Glu) and modified (Lys) human plasminogen. J. Biol. Chem. 253: 733–739.

68. MARKUS, G., R. L. PRIORE & F. C. WISSLER. 1979. The binding of tranexamic acid to native (Glu) and modified (Lys) human plasminogen and its effect on conformation. J. Biol. Chem. 254: 1211–1216.

69. VALI, Z. & L. PATTHY. 1982. Location of the intermediate and high affinity ω-aminocarboxylic acid-binding sites in human plasminogen. J. Biol. Chem. 257: 2104–2110.

70. LERCH, P. G., E. E. RICKLI, W. LERGIER & D. GILLESSEN. 1980. Localization of individual lysine-binding regions in human plasminogen and investigations on their complex-forming properties. Eur. J. Biochem. 107: 7–13.

71. TAKADA, A., K. MOCHIZUKI & Y. TAKADA. 1980. Further characterization of SK-potentiators of plasminogen. Thromb. Res. 19: 485–492.

72. TAKADA, A. & Y. TAKADA. 1982. Potentiation of the activation of Glu-plasminogen by streptokinase and urokinase in the presence of fibrinogen degradation products. Thromb. Res. 25: 229–235.

73. SAMAMA, M., M. CASTEL, O. MATSUO, M. HOYLAERTS & H. R. LIJNEN. 1982. Comparative study of the activity of high and low molecular weight urokinase in the presence of fibrin. Thromb. Haemostas. 47: 36–40.

74. HOYLAERTS, M., D. C. RIJKEN, H. R. LIJNEN & D. COLLEN. 1982. Kinetics of the activation of plasminogen by human tissue plasminogen activator. J. Biol. Chem. **257:** 2912–2919.
75. HISHIKAWA-ITOH, Y., I. SUGIE, H. KATO & S. IWANAGA. 1982. Streptokinase-dependent potentiating factor (SK-Potentiator) for plasminogen activation from human plasma: Its identification as a fibrinogen degradation product. J. Biochem. **92:** 1129–1140.
76. STRICKLAND, D. K., J. P. MORRIS & F. J. CASTELLINO. 1982. Enhancement of the streptokinase-catalyzed activation of human plasminogen by human fibrinogen and its plasminolysis products. Biochemistry **21:** 721–728.

DISCUSSION OF THE PAPER

E. F. PLOW (*Research Institute of Scripps Clinic, La Jolla, CA*): Why do you think that Lys-plasminogen binds so much more avidly than Glu-plasminogen? Physiologically, you would think that you would need to be able to bind Glu-plasminogen.

P. A. MC KEE: There is a huge conformational difference between those two molecules, and I think you must look at the binding site on Lys-plasminogen as being much more available to latch on to fibrinogen.

CALCIUM-BINDING REGIONS IN FIBRINOGEN *

Willem Nieuwenhuizen and Frits Haverkate

Gaubius Institute
Health Research Division TNO
2313 AD Leiden, the Netherlands

The effect of calcium ions on the clotting of fibrinogen is well known.[1-3] Calcium also increases the heat stability of fibrinogen [4] and appears to act as calcium-regulating factor in the factor XIIIa-catalyzed cross-linking reaction.[5] Furthermore, calcium influences the structure of late plasmic degradation products D.[6-8] On the basis of these data it is conceivable that fibrinogen is a calcium-binding protein. Indeed, it has been shown in the past five years that fibrinogens from man,[7, 9, 10] rat,[9, 11] and cow [12] have three calcium-binding sites, and most of the effects of calcium might be explained by saturation of one or more of these sites. For reasons that will become clear below, we wish to divide the three sites into D-related calcium-binding sites and the third calcium-binding site.

D-RELATED CALCIUM-BINDING SITES

The D-fragments in a late plasmic digest or fibrinogen are usually described as a heterogeneous group of fragments ranging in molecular weight from about 100K to 80K. However, we discovered [6, 13] that this heterogeneity disappears when the digestion is carried out under strictly controlled conditions i.e., either in the presence of calcium ions or of EGTA.[14, 15] Under those two different incubation conditions homogeneous D-fragments are formed, the molecular weight of which depends on whether calcium or EGTA was present during their formation. In the presence of calcium ions (2–10 mM) a D fragment with molecular weight 93K is formed whereas in the presence of EGTA the molecular weight is 80K. These D fragments were designated by us as Dcate and D EGTA, respectively.[6, 13] Purified Dcate [13] is resistant against plasmin degradation in the presence of calcium, even with 2 M urea in the medium,[14, 15] but is rapidly degraded in their absence.

To explain these findings we looked for calcium-binding sites in Dcate and D EGTA and found that Dcate has 0.8 binding sites ($K_d = 1.7 \times 10^{-5}$ M), whereas D EGTA has none.[9-11] We concluded from this that apparently the binding of calcium makes Dcate stable against plasmin attack. The first indications as to which part of the structure is essential for the calcium-binding site in Dcate came from comparison of the chains of Dcate and D EGTA. We found that Dcate consists of remnants of the Aα, Bβ and γ chains of fibrinogen with molecular weights of 12K, 43K and 38K, respectively [6, 13] (FIGURE 1, left). Only the γ chain remnant differs in D EGTA and has a molecular weight of 25K (FIGURE 1, right).

The amino-terminal amino acids are identical in Dcate and D EGTA i.e., Val, Asp and Ser for their Aα, Bβ and γ chain remnants, respectively.[13] On

* This work was supported by NATO, grant number RG 80.017.

the basis of these results we concluded [6, 10-13] that a 13K stretch at the carboxyl-terminal end of the γ chain is a prerequisite for the D-related calcium-binding sites. Purves *et al.*[7] and Lawrie and Kemp [8] have arrived at similar conclusions.

Recently [16] we have prepared a D-fragment of a size intermediate between that of Dcate and D EGTA. We designate this as fragment D intermediate

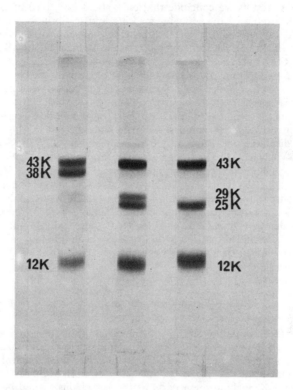

FIGURE 1. SDS-polyacrylamide gel electrophoresis under reducing conditions of a typical D int preparation (center) in comparison with Dcate (left) and D EGTA (right). D int was prepared as follows: Dcate was dissolved in 0.05 M Tris-HCl buffer pH 7.5, containing 0.15 M NaCl and 10 mM EGTA to give a final concentration of 0.8 mg/ml. Plasmin (5 μg active plasmin[19] per ml Dcate solution) was added to the solution and the mixture incubated for four hours at 37°C. Plasmin was inactivated with 10 μl 0.1 M DFP in dry isopropanol per ml incubation mixture. The digest was concentrated by freeze-drying and desalted on a Sephadex G-50 column.

(D int). A typical D int preparation is depicted in FIGURE 1, center. From the banding pattern on the gels in FIGURE 1 it can be concluded that the preparation consists of a mixture of D int and D EGTA and that the Aα, Bβ and γ chain remnants of D int have molecular weights of 12K, 43K and 29K, respectively. D int thus has a γ chain only about 4K longer at its carboxyl-terminal end than that of D EGTA and, obviously, 9K shorter than that of Dcate. Using a D int preparation consisting of 70% D int and 30%

D EGTA (as judged by the densitometric scanning of the 25K and 29K bands) we have carried out calcium binding studies as in reference 11. We found (after correction for the 30% contaminating D EGTA, which has no calcium-binding sites [9-11]) that D int has 0.7 binding sites with $K_d = 4.5 \times 10^{-5}$ M. This is comparable with our results obtained with Dcate i.e. 0.8 sites with $K_d = 1.7 \times 10^{-5}$ M.[9]

From these results we conclude that only about one third of the carboxyl-terminal 13K stretch of the γ chain in Dcate and in fibrinogen is involved in the D-related calcium-binding sites. Since the γ chain in fragment D EGTA ends at position γ303 [17] and since the γ chain in D int is about 4K longer than that of D EGTA, cleavages of the γ chain at positions 321 or 356 [17, 18] in the transition of Dcate to D int are the most likely candidates. On gels (not shown) the molecular weight of nonreduced D int appears lower than that of Dcate. A cleavage at position γ338 would not have caused a lower molecular weight of D int as compared with Dcate, because the carboxyl-terminal stretch starting from residue γ339 would stay linked with the rest of the molecule via the internal disulfide loop formed by cysteine residues γ327 and γ339. Studies to define the cleavage site (γ321 or γ356) are in progress.

It is conceivable that calcium ions exert their protective effect against plasmic attack in the region of residue γ303 by inducing a conformational change. From velocity sedimentation experiments it was concluded that no gross overall but rather local conformational changes occur in D-cate upon binding of calcium (Dr A.C.M. van der Drift, personal communication). In summary, it is concluded that the two D-related calcium-binding sites in fibrinogen are dependent on stretches of the γ chains, about 4K long, which start at residues γ303.

The Third Calcium-Binding Site

An odd calcium-binding site in a symmetrical molecule such as fibrinogen is *a priori* likely to be localized in some symmetrical entity. Marguerie suggested [4] that such a structure would be induced by calcium ions on the carboxyl-terminal ends of the two Aα chains of fibrinogen. If this were the case, fragments X, lacking these parts of the Aα chains, would be expected to have lost the third site. However, recently we have prepared, purified and characterized fragments X [19] and found that they possess 2.4 calcium-binding sites with $K_d = 0.9 \times 10^{-5}$ M.

This is in contrast with recent results by Marguerie [20] who finds only two binding sites with $K_d = 9 \times 10^{-6}$ M in an X-like product, isolated from Cohn fraction I. If characterized for chain composition, comparison of this product with our X-fragments may give clues as to the localization of the third site.

A symmetrical moiety of the fibrinogen molecule is the E-domain. However, we have not been able to find a calcium-binding site in late plasmic fragments E [9-11] allowing the conclusion that the third site exists in fragments X but is lost during later stages of plasmic digestion. As a logical step to identify this stage we recently prepared [21] fragments Y fulfilling the criteria listed by Marder and Budzynski.[22] Fragments Y appeared to have two calcium-binding sites with $K_d = 1.9 \times 10^{-5}$ M. Since fragments Y comprise one D-domain (with one site) and the central symmetrical domain it may be concluded that the third site is still intact in fragments Y and that it may be localized in the central domain as it exists in fragments Y (or be related to the parts of the chains which connect D with the central domain).

As stated above late plasmic fragments E do not bind calcium. This may be because they have lost structures that are essential for calcium binding during their formation. An alternative to study the central domain with a somewhat different structure is to prepare the N-terminal disulfide knot (NDSK) by CNBr digestion of fibrinogen and of fibrin, according to Blombäck et al.[23] Recently, we performed such studies [24] and found that NDSK-fibrinogen has 0.8 binding sites with $K_d = 1.8 \times 10^{-4}$ M and that NDSK-fibrin has 0.9 sites with $K_d = 3 \times 10^{-4}$ M. This, again, strongly suggests that the third site is localized in the central, symmetrical part of the fibrin(ogen) molecule.

The observed dissociation constant for the site in NDSK is a factor of ten higher than that of the sites in intact fibrinogen. This may be due to the harsh conditions applied to prepare NDSK i.e., 70% formic acid for 24 hours at room temperature. Under those conditions disulfide interchanges and deamidation reactions might occur leading to a structure with a diminished affinity for calcium. Another possibility is that the pH-optimum for the binding of calcium in the central domain has shifted as a result of the cleavage of NDSK from the rest of the firbinogen molecule. Marguerie has shown [12] that one of the three calcium-binding sites in bovine fibrinogen is relatively pH-sensitive and is lost at pH 5.5.

Fibrinopeptides A and B do not seem to be involved in the third calcium-binding site, since NDSK-fibrinogen and NDSK-fibrin have the same calcium-binding properties. NDSK-fibrin has the general formula (Aα 17–51, Bβ 15–118, γ 1–78)$_2$. The exact structure of our late fragments E, which do not bind calcium, is not known at present. However, it is conceivable that they are identical with fragments E$_3$,[25] since they are prepared by long plasmic digestion. The formula of fragment E$_3$ has recently been published [26] and is (Aα 20–78, Bβ 54–120, γ 1–53)$_2$. Comparison of NDSK-fibrin with fragment E$_3$ would suggest a direct involvement of residues Aα 17–19, Bβ 15–53 and/or γ 54–78 in the calcium binding site of NDSK and as a consequence in the third site of fibrinogen.

Residues Aα 17–19 and Bβ 15–17 are involved in fibrin polymerization.[26] This, combined with the possibility that they are also involved in calcium binding may be important in understanding the accelerating effect of calcium on fibrin formation. In summary, evidence has been found for the localization of the third calcium-binding site in the central domain of the fibrinogen molecule.

ACKNOWLEDGMENTS

Thanks are due to Dr. A.C.M. van der Drift, Medical Biological Laboratory TNO, Rijswijk, the Netherlands, for stimulating discussions and for his information on the sedimentation coefficients of Dcate, prior to publication.

REFERENCES

1. DOOLITTLE, R. F.   1973.   Adv. Prot. Chem. 27: 1–109.
2. ENDRES, G. F. & M. A. SCHERAGA.   1972.   Arch. Biochem. Biophys. 153: 266–278.

3. LY, B. & H. C. GODAL. 1973. Haemostasis 1: 204–209.
4. MARGUERIE, G. 1977. Biochim. Biophys. Acta 494: 172–181.
5. CREDO, R. B., C. G. CURIS & LT. LORAND. 1978. Proc. Natl. Acad. Sci. USA 75: 4234–4237.
6. HAVERKATE, F. & G. TIMAN. 1977. Thromb. Res. 10: 803–812.
7. PURVES, L. R., G. G. LINDSEY & J. J. FRANKS. 1978. S. Afr. J. Sci. 74: 202–209.
8. LAWRIE, J. S. & G. KEMP. 1978. Biochem. Soc. Trans. 6: 769–771.
9. NIEUWENHUIZEN, W., I. A. M. VAN RUIJVEN-VERMEER, W. J. NOOIJEN, A. VERMOND, F. HAVERKATE & J. HERMANS. 1981. Thromb. Res. 22: 653–657.
10. NIEUWENHUIZEN, W., A. VERMOND, W. J. NOOIJEN & F. HAVERKATE. 1979. FEBS Letters 98: 257–259.
11. VAN RUIJVEN-VERMEER, I. A. M., W. NIEUWENHUIZEN & W. J. NOOIJEN. 1978. FEBS Letters 93: 177–180.
12. MARGUERIE, G. 1977. Biochim. Biophys. Acta 490: 94–103.
13. VAN RUIJVEN-VERMEER, I. A. M., W. NIEUWENHUIZEN, F. HAVERKATE & T. TIMAN. 1979. Hoppe-Seyler's Z. Physiol. Chem. 360: 633–637.
14. NIEUWENHUIZEN, W., A. VERMOND & F. HAVERKATE. 1981. Biochim. Biophys. Acta 667: 321–327.
15. NIEUWENHUIZEN, W., A. VERMOND & F. HAVERKATE. 1982. In Fibrinogen, Recent Biochemical and Medical Aspects. A. Henschen, H. Graeff & F. Lottspeich, Eds. Walter de Gruyter. Berlin. In press.
16. NIEUWENHUIZEN, W., M. VOSKUILEN, A. VERMOND, F. HAVERKATE & J. HERMANS. 1982. Biochim. Biophys. Acta 707: 190–192.
17. HENSCHEN, A. 1981. Haemostaseologie 1: 30–40.
18. HENSCHEN, A., F. LOTTSPEICH, E. TOPFER-PETERSEN & R. WARBINEK. 1979. Thrombos. Haemostas. 41: 662–669.
19. NIEUWENHUIZEN, W. & M. GRAVESEN. 1981. Biochim. Biophys. Acta 668: 81–88.
20. MARGUERIE, G. & N. ARDAILLOU. 1982. Biochim. Biophys. Acta 701: 410–412.
21. NIEUWENHUIZEN, W., M. VOSKUILEN & J. HERMANS. 1982. Biochim. Biophys. Acta 708: 313–316.
22. MARDER, V. J. & A. Z. BUDZYNSKI. 1975. Thromb. Diath. Haemorrh. 33: 199–207.
23. BLOMBÄCK, B., M. BLOMBÄCK, A. HENSCHEN, B. HESSEL, S. IWANAGA & K. R. WOODS. 1968. Nature (London) 218: 130–134.
24. NIEUWENHUIZEN, W., A. VERMOND & J. HERMANS. Manuscript submitted for publication.
25. OLEXA, S. A., A. Z. BUDZYNSKI, R. F. DOOLITTLE, B. A. COTTRELL & F. C. GREENE. 1981. Biochemistry 21: 6139–6145.
26. LAUDANO, A. P. & R. F. DOOLITTLE. 1981. Science 212: 457–459.

FIBRINOGEN HETEROGENEITY *

Michael W. Mosesson

Department of Medicine
Mount Sinai Medical Center
Milwaukee, Wisconsin 53233

INTRODUCTION

It has been widely appreciated for many years [1] that certain protein preparations display microheterogeneity as judged by one or more analytical criteria. This type of behavior, leaving preparation impurities aside, was taken to indicate that such proteins existed as families of closely related, but not necessarily identical, molecules. Among the proposed explanations for this occurrence were differences in molecular size, charge density, configuration, amino acid composition, and/or biologic activity.

Subsequent investigations over the ensuing decades have provided both more examples of protein microheterogeneity plus greater molecular details concerning those proteins manifesting molecular heterogeneity. Fibrinogen was not listed among the proteins exhibiting heterogeneity in 1954,[1] but is now known to exemplify several aspects of the microheterogeneity alluded to above. Furthermore, fibrinogen heterogeneity can now largely be understood in specific molecular terms and it is my purpose in this review to present a summary in those terms concerning the major structural forms of this family of thrombin-coagulable proteins. Most data to be presented involve human fibrinogen, although other species' fibrinogen will be considered whenever appropriate.

Aα CHAIN HETEROGENEITY

Fibrinogen is among the least soluble of the plasma proteins, and is the major protein component first precipitated from plasma by a variety of standard fractionation procedures.[2-5] Investigations of plasma fibrinogen subfractions of higher solubility than those found in Cohn fraction I [8] (FIGURES 1 and 2) led to the recognition [6, 8] that a major proportion of circulating fibrinogen is composed of coagulable species that have undergone varying degrees of proteolytic catabolic attack resulting in the cleavage and release of COOH-terminal Aα chain peptides.[9-11] In general, the smaller the residual Aα chain peptides remaining with a given molecule, the greater is the solubility of the subfraction with which it precipitates. Similarly, a concomitant functional consequence of the loss of peptide material containing the COOH-terminus, is a delay in the fibrin aggregation rate (i.e., thrombin clotting time) of the molecules comprising that fraction (FIGURE 3). It also follows that the thrombin time of a given subfraction is directly related to its solubility (i.e., the higher the solubility, the longer the clotting time).

In addition to classifying plasma fibrinogen subfractions according to their solubility, it is also useful to classify them in terms of their size, according to

* This work was supported by National Institutes of Health Grant HL–28444.

97

their electrophoretic migration rate in NaDodSO$_4$-containing polyacrylamide gels (FIGURE 4).[12-14] This system is particularly valuable for relating the Aα chain structure of the various subfractions to the major fibrinogen components found in unfractionated normal [13, 15-18] and pathological plasmas.[19-22]

Upon NaDodSO$_4$ gel electrophoresis, the bands obtained from unreduced plasma subfractions can be assigned to five more-or-less well-defined regions,[12, 13] designated "band I" through "band V", in order of increasing anodal migration rate (i.e., decreasing size). Band I fibrinogen contains the largest molecules (FIGURE 5), and can readily be prepared in virtually pure form.[13, 16, 23] There appears to be general agreement on the Aα chain popu-

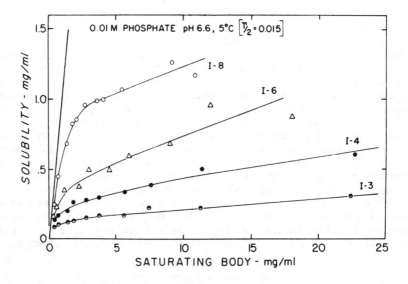

FIGURE 1. Demonstration of solubility differences among human plasma fibrinogen subfractions, which are designated: I–3, I–4, etc. in order of increasing solubility. The concentration of fibrinogen remaining soluble (ordinate) after dialysis at 5°C under the solvent conditions shown in the diagram is plotted against the initial protein concentration ("saturating body"). The solid line through the origin represents the theoretical solubility curve of a completely soluble preparation. (From Mosesson and Sherry.[6] By permission of *Biochemistry*.)

lation of this material; band I molecules contain intact Aα chains or slightly smaller Aα-chains, termed Aα/2.

In contrast, there have been sharp differences in opinion regarding the Aα-chain subunit composition of fibrinogen comprising bands II and III,[13, 24-26] particularly with regard to the structural contribution of an Aα chain remnant, termed Aα/4, which is only slightly smaller than a Bβ chain. The size similarity makes it difficult to appreciate the presence of Aα/4 by gel electrophoresis of reduced samples when Bβ chains are abundant. Recent detailed analyses,[14] as summarized below, should serve to resolve the two divergent interpretations, which evidently have arisen from differences in methodological approaches and specimen inhomogeneities. A modified fibrinogen subfraction, I–6, which con-

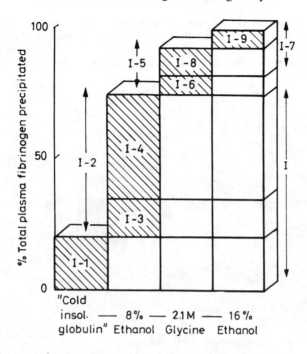

FIGURE 2. Schematic distribution of subfractions of human plasma fibrinogen. General conditions favoring precipitation increase from left to right. Designation and approximate distribution of subfractions are indicated on the ordinate. (From Mosesson.[7] By permission of *Thrombosis et Diathesis Haemorrhagica*.)

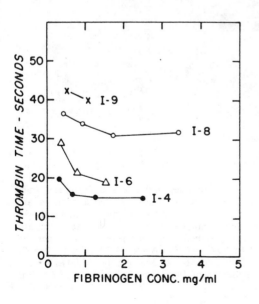

FIGURE 3. Thrombin clotting times of fibrinogen subfractions of different solubilities. (From Mosesson and Sherry.[6] By permission of *Biochemistry*.)

sisted almost exclusively of band II material,[14] was prepared and analyzed. Quantitative analysis indicated that $A\alpha/4$ chains in this fraction accounted for more than 70% of the total $A\alpha$ chain population.[13, 14] One may conclude from this that molecules with two such chains characterize the structure of

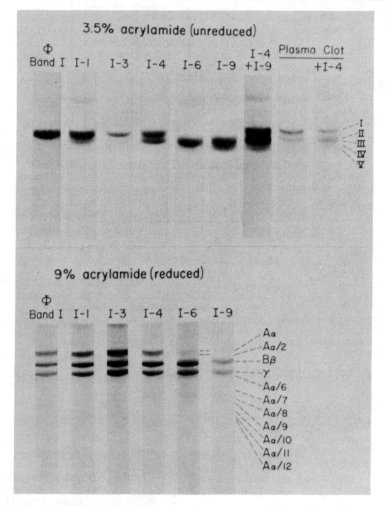

FIGURE 4. $NaDodSO_4$ gel electrophoretic analyses of unreduced fibrinogen fractions, a noncrosslinked plasma clot, or mixtures of these. Reduced samples are shown in the lower portion of the figure. The molecular weights of chains shown to be of $A\alpha$ origin are as follows: chains smaller than the intact $A\alpha$-chain are designated $A\alpha/n$ [12] where n = the number assigned to a given peptide, the larger the number the smaller the size; $A\alpha$, 70,900; $A\alpha/2$, 67,300; $A\alpha/4$, 57,700; $A\alpha/6$, 46,500; $A\alpha/7$, 37,600; $A\alpha/8$, 33,900; $A\alpha/9$, 31,800; $A\alpha/10$, 29,200; $A\alpha/11$, 25,000; $A\alpha/12$, 22,600. (From Galanakis et al.[13] By permission of Journal of Laboratory and Clinical Medicine.)

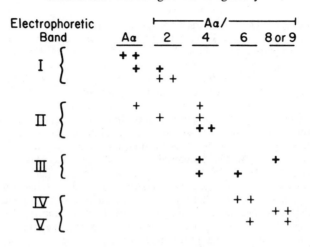

FIGURE 5. Schematic distribution of chains of Aα origin in plasma fibrinogen species migrating in electrophoretic bands I through V. Each horizontal pair of crosses represents the components of dimeric molecules comprising the species within these bands. Major species are indicated by thickly drawn crosses. For simplification, minor Aα remnants Aα/7, Aα/10, Aα/11, Aα/12 were omitted from the diagram. (From Galanakis *et al.*[13] By permission of *Journal of Laboratory and Clinical Medicine.*)

band II fibrinogen. Other minor but significant components in band II fibrinogen include molecules with an Aα/4 chain plus a larger chain (i.e., Aα or Aα/2). Band II fibrinogen is always very prominent when normal plasma is analyzed (FIGURE 4),[13, 14, 16] a finding that is completely consistent with data suggesting that Aα/4, which characterizes band II, is the most abundant Aα-chain derivative in plasma fibrinogen [13, 14, 27, 28] (FIGURE 6).

Band III fibrinogen is found predominately in fractions I–8 and I–9.[12, 13] Since these fractions are relatively minor components of plasma fibrinogen (FIGURE 2), it follows that this type of fibrinogen is also a relatively minor component of plasma. Mixtures of bands II and III fibrinogen are incompletely resolved by electrophoresis[12, 13] although they are distinguishable from one another by their characteristic solubilities [6] and by their Aα-chain composition.[12-14] Thus, typical band III molecules consist of an Aα/4 chain plus a smaller Aα chain derivative (FIGURE 5). Given the minor size differences between the smallest components in band II and the largest in band III, it is not at all surprising that these bands overlap each other.

Bands IV and V are very minor components of plasma fibrinogen [12, 13, 15] and are concentrated in fraction I–9. They have not been analyzed in as great detail as the other electrophoretic bands, but it is clear that they must consist of combinations of Aα chains which are smaller than Aα/4.

Isoelectric focusing has been very useful for characterizing charge heterogeneity of fibrinogen chains (FIGURE 7).[13, 24, 28, 31-34] Chains of Aα origin are the most cathodal and exhibit the greatest polydispersity (in our hands at least 23 bands were distinguishable between pI 6.2 and 7.5, reference 13). There was little overlap, if any, between bands of Aα origin and those of Bβ or γ chain origin. The isoelectric band patterns of Aα-chain fractions that had first been separated by DEAE-cellulose gradient chromatography were consistent with their chromatographic elution behavior (cf. FIGURE 7, gels 5 to 7).

There are several facors that can account for the marked polydispersity within the Aα chain population: (1) Three major types of Aα-chains have been identified that differ with respect to their fibrinopeptide A composition.[35] One type (AP) contains a phosphorylatcd serine residue whereas another does not (A). The fetal form of fibrinogen contains almost twice as much phosphorous as does adult fibrinogen,[36] but this has no apparent effect on the kinetics of the fibrinogen-fibrin conversion.[37] A third type of fibrinopeptide (Y) lacks the NH$_2$-terminal acid (Ala) characterizing A and AP. (2) There are several types of catabolic Aα-chain remnants, as summarized earlier in this section. (3) Some of the microheterogeneity could conceivably be a reflection of genetic individuality, inasmuch as Gaffney[31] reported finding less isoelectric heterogeneity in single donor samples compared with pooled samples. On the other hand, Kuyas et al.[34] have failed to detect differences among single donor specimens using the same analytical technique.

The origin of Aα chain heterogeneity remains incompletely understood. To be sure, the demonstration of in vivo conversion of low to high solubility forms of fibrinogen[38] indicates a precursor-product relationship reflecting post-secretory fibrinogen catabolism. The physiologic importance of circulating fibrinogen catabolites, in terms of their functional effect on fibrin aggregation rates, as well as in terms of their lacking certain functional binding regions that are present in COOH-terminal regions of intact Aα-chains (e.g., binding regions for plasma fibronectin, reference 23), remains to be established. Furthermore, the proteolytic processes that mediate their formation are incompletely understood. In vitro plasmin attack results in formation of deriva-

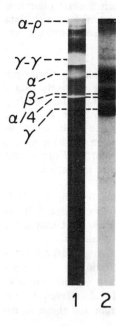

FIGURE 6. NaDodSO$_4$ gel electrophoresis[29] in 9% polyacrylamide gels of reduced fibrin samples from freshly pooled (n = 6) citrated plasma. Gel 1, UV fluorescence of a reduced clot that had been partially crosslinked in the presence of factor XIIIa and dansylcadaverine. Fluorescence is due to incorporation of dansylcadaverine into crosslinking acceptor sites (α-polymer, αP; γ-dimers, γ-γ). Since fibrin β chains do not incorporate dansylcadaverine in the presence of factor XIIIa,[30] the fluorescence in the β chain position is due solely to material of Aα origin, namely α/4. Gel 2, noncrosslinked reduced fibrin from the same pool stained with Coomassie blue. (From Galanakis and Mosesson[14])

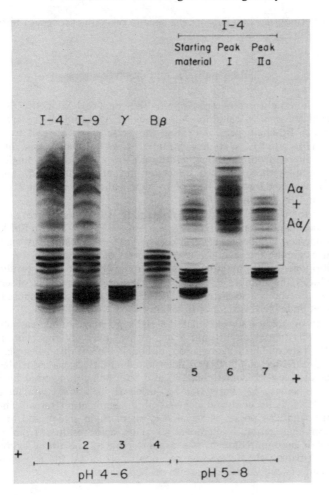

FIGURE 7. Acrylamide gel isoelectric focusing of reduced, S-carboxymethylated fibrinogen samples stained with Coomassie blue. Ampholytes yielding a pH gradient of 4 to 6 were best for resolution of Bβ or γ chains (gels 1 to 4), whereas those yielding a pH gradient of 5 to 8 were best suited for resolution of Aα chains (gels 5 to 7). The region occupied by chains of Aα origin (pH 5 to 8) is indicated, whereas the dashed lines delineate the positions of Bβ and γ chains, respectively. (From Galanakis et al.[13] By permission of Journal of Laboratory and Clinical Medicine.)

tives that are similar to but not identical with these found in vivo,[11, 12, 15] suggesting that other proteolytic enzymes play an important role in their formation. For one thing, it is known that leukocytes can secrete fibrinogenolytic proteases other than plasmin.[39] Further, the demonstration of plasmin-independent fibrinolytic activity in plasma [40] coupled with the recent demonstration that a vitamin K-dependent protein, termed "protein C," induces plasmin-independent fibrinolysis,[41] underscore the need to relate the process of fibrinogen catabolism to these systems.

Bβ CHAIN HETEROGENEITY

Gradient elution chromatography of S-sulfo [42] or S-carboxymethyl [43] derivatives of fibrinogen on CM-cellulose [42, 43] or on DEAE-cellulose [13, 44] results in a heterogeneous Bβ-chain peak indicating the presence of at least two major components (FIGURE 8). Gati and Straub [43] have evaluated this difference using CM-cellulose chromatography, and have shown that in both single donor and pooled samples there are two Bβ chain variants (termed $B\beta_L$ and $B\beta_R$, respectively), which differ with respect to their sialic acid content (two sialic acid residues in $B\beta_L$ and one in $B\beta_R$).

Isoelectric focusing of preparations containing Bβ chains has revealed four to five distinct bands representing Bβ variants [13, 24,28, 31, 33, 34] (FIGURE 7). NaDodSO$_4$ polyacrylamide gel electrophoresis has not revealed any significant size differences among the several isoelectric forms of Bβ chains.[24, 28, 34] Treatment of fibrinogen with neuraminidase to remove sialic acid residues reportedly reduced the number of isoelectric bands to two.[34] This observation suggests that the sialic acid heterogeneity ($B\beta_L/B\beta_R$) is superimposed upon another, as yet undefined, variant of Bβ chains.

A number of studies suggest that the sialic acid-containing carbohydrate groups that are situated on Bβ and γ chains [45, 46] play an important role in regulating the rate of fibrin aggregation, although the specific roles played by each of the groups has not yet been determined. Fibrinogens associated with fibrotic liver disease,[47–49] hematoma,[50, 51] and with some congenital fibrinogen abnormalities (Nancy, 52; Paris II, 53; Zurich II, 43) are characterized by elevated levels of sialic acid and delayed fibrin aggregation rates. The fetal form of fibrinogen also possesses elevated levels of sialic acid [54] and manifests delayed fibrin aggregation.[55] Removal of sialic acid residues from normal fibrinogen with neuraminidase causes a shortening of the fibrin aggregation rate,[56] and normalizes the delayed aggregation rate of fetal fibrin [54] as well as that from subjects with liver disease [48] or hepatoma.[51]

γ CHAIN HETEROGENEITY

Human fibrinogen can be resolved by DEAE-cellulose gradient elution chromatography into two major peaks—"peak 1" and "peak 2" fibrinogen.[57] Peak 2 fibrinogen accounts for about 15% of the total [11, 57] and differs from peak 1 fibrinogen by the presence of a unique class of γ chain variant, termed γ'.[42] The γ' chains, which are more negatively charged than γ chains, are distinguishable by gel electrophoresis at pH 8.6 or 2.7,[42] by isoelectric focusing,[44] or by DEAE-cellulose ion exchange chromatography (FIGURE 8). They account for approximately half of all γ chains in the peak 2 fraction,[42] indicating that the general formula for molecules found in that peak is $(A\alpha)_2(B\beta)_2\gamma\gamma'$. It should be pointed out that the existence of γ/γ' heterodimer molecules (e.g., peak 2 fibrinogen) indicates that individual hepatocytes can synthesize fibrinogen molecules containing both types of chain, evidence

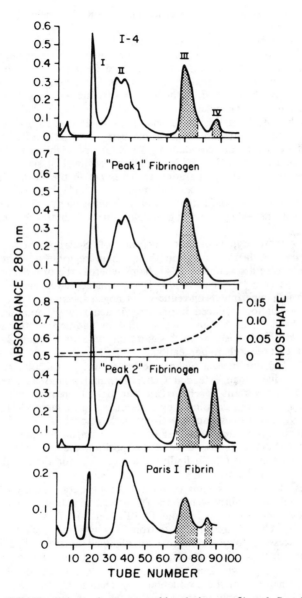

FIGURE 8. DEAE-cellulose chromatographic elution profile of S-carboxymethyl derivatives of: fraction I-4; peak 1 fibrinogen; peak 2-rich fibrinogen; a fraction from cross-linked fibrinogen Paris I fibrin that had been separated from γ dimers by gel sieving chromatography. Pooling of peak fractions is indicated by the shaded areas. The theoretical gradient is shown in the middle portion of the figure. Peak II contains the Bβ chains, peak III the γ chains, and peak IV and γ' chains. (From Stathakis and Mosesson.[44] By permission of *Thrombosis Research*.)

that is consistent with the "single gene" mechanism (see below) that evidently regulates their production.

A similar chromatographic heterogeneity has been found for several other animal fibrinogens [58-60]; in the case of bovine fibrinogen, molecules containing two γ' chains have been demonstrated in a third, later-eluting, chromatographic peak.[59] A minor third fibrinogen peak, presumably having the same $\gamma'\gamma'$ composition, has also been observed in human specimens [11, 57] but has not been characterized in the same detail.

Studies by Martinez [61] showed that peak 1 and peak 2 fibrinogens are cleared from the circulation at the same rate and that neither type is converted to the other in the circulation. There are no significant differences in their sialic acid contents.[57, 61, 62] Furthermore, removal of sialic acid does not alter the behavior of the preparation with respect to the peak 1/peak 2 elution profile.[44] In this regard, another type of γ chain heterogeneity, termed γ-1/γ-2,[42, 63] is superimposed upon the γ/γ' heterogeneity. Evidence has been presented that the γ-1/γ-2 heterogeneity is due to differences in sialic acid content.[43]

In our laboratory, isoelectric focusing of S-carboxymethyl fibrinogen yielded at least 8 distinct bands in the γ chain region.[13, 44] The four most anodal bands were associated with γ' chains whereas the other four were associated with γ chains.[44] Kuyas et al.[34] have resolved fewer bands in their isoelectric focusing experiments on reduced fibrinogen specimens (three γ, two γ'). Following desialation, reduced fibrinogen yielded two γ bands and only one distinct γ' band. This finding suggests that in addition to the γ-1/γ-2 and γ/γ' variants, there exists a third significant type of γ chain heterogeneity involving at least the γ chain population.

Human γ' chains have a higher molecular weight ($\sim$2000 daltons) than γ chains [64, 65] due to an extended COOH-terminal amino acid sequence.[64,66] (FIGURE 9) The variant γ' chains are functionally normal with respect to their ability to undergo crosslinking in the presence of factor XIIIa, and they participate in this process nonselectively with respect to γ and γ' chains.[64] The γ' sequence is the same as that of the γ chain up to four residues from the γ chain COOH-terminus. Beyond this position the γ' sequence is unique and is rich in Asp and Glu, a finding that accounts for their higher negative charge relative to γ chains.[42, 44]

Analysis of mutant γ chains (γ Paris I) from a congenitally dysfunctional fibrinogen molecule (fibrinogen Paris I) showed that both γ and γ' features were expressed in the γ Paris I chains (44, FIGURE 8). This occurrence implies that γ and γ' chains are produced from a single gene, since otherwise it would be necessary to postulate that an identical mutation had occurred in two separate structural γ chain genes, a seemingly less likely possibility.

Assuming the "single gene" hypothesis to be correct, there are several possible explanations for the production of γ' chains. On the basis of sequence data alone, it seems highly improbable that the γ' chain represents a naturally occurring "readthrough" protein [67] or an unprocessed "precursor" of γ chains, since in both cases the terminal sequence of the γ chain would first be completed and then followed by additional amino acids. Instead, as suggested in our recent report [66] and from more direct evidence summarized below, it appears that the different mRNAs for γ and γ' chains, respectively, are transcribed from a single gene and produced by differential RNA splicing.[68] Support for this explanation has recently been obtained by Crabtree and Kant,[69] who have shown in rats that a single gene gives rise to two mRNAs coding for rat γ and γ' chains. This mechanism of regulation of gene transcription has also been proposed to account for the existence of membrane-bound and secreted forms of immunoglobulin mu chains [70-72] and gamma chains,[73] which differ in each case by their COOH-terminal sequences.

The γ' chains may prove to have a unique affinity for certain components of the factor VIII complex, based upon recently reported studies on the heparin precipitable fraction of plasma (HPF).[74] These investigations demonstrated that factor VIII-related antigen (VIIIR:Ag), a major amount of which is usually precipitated in the HPF from normal plasma, is not precipitated when the HPF is prepared from afibrinogenemic plasma. Addition of fibrino-

FIBRIN MOLECULE A

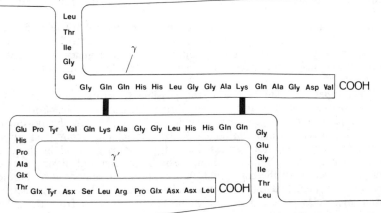

FIBRIN MOLECULE B

FIGURE 9. Amino acid sequence of the COOH-terminal region of the γ' and γ chains. For ease of comparison, the sequence data are depicted in terms of a factor XIIIa-crosslinked γγ' dimer. The reciprocal ε-(γ-glutamyl)lysine crosslinks are represented by solid bars bridging the anti-parallel chains.

gen fraction I–4 tended to correct this situation.[74] Preliminary follow up studies of this phenomenon (D. Amrani and M. Mosesson, unpublished observations) indicate that the precipitation of VIIIR:Ag is considerably augmented when the added fibrinogen molecules contain γ' chains (e.g., fraction I–4 fibrinogen or peak 2 fibrinogen) as compared with the amount of VIIIR:Ag that precipitates when the added fibrinogen contains only γ chains (e.g., peak 1 fibrinogen). Since fibrinogen itself becomes incorporated in the HPF because of the affinity of its Aα chains for plasma fibronectin (CIg),[23] it would appear that VIIIR:Ag becomes incorporated in this precipitate because of a binding preference for γ'-containing fibrinogen molecules. Further detailed studies will be necessary to properly interpret these observations.

PLATELET FIBRINOGEN

The foregoing discussion has focused on the structural differences that exist in plasma fibrinogen. However, since fibrinogen is also found within platelets and seems to have certain features that set it apart from the plasma form, it would be appropriate to summarize these features in the present context of fibrinogen heterogeneity. Fibrinogen is a major protein in platelets (5 to 15% of total cellular proteins) and is present at the level of ~100 to 200 $\mu g/10^9$ cells.[75-78] Most data indicate that this fibrinogen is located exclusively within α-secretory granules [79-82] and that the platelet pool is separate from the plasma fibrinogen pool.[75, 83-89] Despite several studies which indicate that the molecular weight of platelet fibrinogen is the same as that from plasma [76, 90-92] and that the fibrinopeptides are the same,[90] significant physicochemical,[92] biochemical [92, 93] and immunochemical [94] differences have been recorded.

Ganguly and coworkers [95-97] contended that "native" platelet fibrinogen has a lower molecular weight than plasma fibrinogen, and offered this difference as the rationale for proposing different structural genes. As mentioned above, these size differences have not been confirmed by other investigators.[90-92, 98]

Other types of evidence bearing on possible structural (and genetic) differences between plasma and platelet fibrinogen are available. Soria et al.[93] analyzed platelet fibrinogen from a subject with a congenitally dysfunctional plasma fibrinogen termed fibrinogen Metz. Aα chains from their subject's platelet fibrinogen did not express the charge anomaly found in the plasma fibrinogen Aα chains. Similarly, comparison of platelet and plasma fibrinogen from the subject with the congenital plasma fibrinogen abnormality, fibrinogen Paris I,[98] indicated that the mutant γ chains (termed γ Paris I) characterizing plasma molecules were not present in fibrinogen obtained from the subject's platelets. It is interesting to record here that preliminary electrophoretic analyses of crosslinked platelet fibrin from normal subjects have failed to demonstrate γ dimers comprised of plasma fibrinogen γ' chain variants (M. W. Mosesson and D. L. Amrani, unpublished experiments). The absence of mutant chains from the Metz and Paris I subjects' platelet fibrinogen molecules and the failure to find γ' chains in platelet fibrinogen support the hypothesis that there are differences in the structural genes or in the programming and/or regulation [99] of platelet and plasma fibrinogen production, at least as far as Aα and γ chains are concerned.

CONCLUDING REMARKS

In this report I have summarized existing knowledge on the molecular heterogeneity of human plasma fibrinogen (TABLE 1) and have attempted to relate these findings to known functional characteristics of the molecule.

1. Solubility differences reflect size heterogeneity due to catabolic proteolytic cleavage and release from parent molecules of portions of the Aα chains containing its COOH-terminal region. The fibrinogen catabolites evidently are formed after secretion from hepatocytes. Another structural heterogeneity involving the NH_2-terminal fibrinopeptide A (A,AP,Y) is superimposed upon the size heterogeneity.

2. Two main types of Bβ chains have been described which differ with respect to their sialic acid content (Bβ_L/Bβ_R); this heterogeneity may be superimposed upon another, as yet undefined, heterogeneity. A fetal form of fibrinogen, as well as functionally abnormal fibrinogens (delayed fibrin aggregation) associated with fibrotic liver disease, hepatomas, and certain congenital fibrinogen abnormalities are characterized by elevated sialic acid levels. The

TABLE 1

SUMMARY OF KNOWN MAJOR VARIANTS OF HUMAN PLASMA FIBRINOGEN CHAINS

Terminology	Variation	Essential Structural Difference	Comment
A	Fibrinopeptide A structure	—	
AP		phosphorylated serine	unknown functional difference
Y		lacks NH$_2$-Ala	
(Aα/2, Aα/4, (Aα/6 . . . etc.)	Aα/chain size	lack of peptides containing COOH-terminal regions due to proteolysis and release from parent molecules	loss of such peptides results in delay in fibrin aggregation rate
Bβ_L	Bβ chain carbohydrate content (sialic acid)	2 sialic acid residues/ chain	
Bβ_R		1 sialic acid residue/ chain	desialation enhances fibrin aggregation rate *
γ–1	γ chain carbohydrate content (sialic acid)	2 sialic acid residues/chain	
γ–2		1 sialic acid residue/chain	
γ	γ chain amino acid sequence (chain size)	COOH-terminal sequences differ	unknown functional difference
γ'		(γ' larger than γ)	

* It is not known specifically which group(s) are functionally important.

delay in aggregation can be normalized by removal of sialic acid, but it is not known what specific relationship this functional modification has to the respective sialic acid-containing groups on Bβ and γ chains.

3. A sialic acid heterogeneity involving γ chains (γ-1/γ-2) is superimposed upon another type of heterogeneity (γ/γ') that reflects differences in COOH-terminal amino acid sequences. The γ/γ' heterogeneity appears to result from

differential mRNA splicing, resulting in production of two gene products from a single gene.

4. Platelet fibrinogen from a subject with a plasma fibrinogen Aα chain abnormality (fibrinogen Metz) and that from a subject with a plasma fibrinogen γ chain abnormality (fibrinogen Paris I) did not express the molecular defect that characterizes the respective plasma molecule. These findings plus the apparent reduction or lack of γ′ chains in platelet fibrinogen are consistent with the hypothesis that there are differences in the structural genes or in the programming and/or regulation of production and assembly of platelet and hepatic (plasma) fibrinogen molecules.

ACKNOWLEDGMENTS

I wish to acknowledge the contributions of my coworkers with whom I have carried out several of the studies cited in this paper, to thank David Amrani and Dennis Galanakis for helpful criticisms of this manuscript, and to thank Julie Erickson for her help in its preparation.

REFERENCES

1. COLVIN, J. R., D. B. SMITH & W. H. COOK. 1954. Chem. Rev. **54:** 687–711.
2. COHN, E. J., W. L. STRONG, W. L. HUGHES, JR., D. J. MULFORD, J. N. ASHWORTH, M. MELIN & H. L. TAYLOR. 1946. J. Amer. Chem. Soc. **68:** 459–475.
3. KEKWICK, R. A., M. E. MACKAY, M. H. NANCE & B. R. RECORD. 1955. Biochem. J. **60:** 671–683.
4. BLOMBÄCK, B. & M. BLOMBÄCK. 1956. Ark. Kemi. **10:** 415–443.
5. KAZAL, L. A., S. AMSEL, O. P. MILLER & L. M. TOCANTINS. 1963. Proc. Soc. Exp. Biol. Med. **113:** 989–994.
6. MOSESSON, M. W. & S. SHERRY. 1966. Biochemistry **5:** 2829–2835.
7. MOSESSON, M. W. 1970. Thromb. Diath. Haemorrh. (Suppl. 39): 63–70.
8. MOSESSON, M. W., N. ALKJAERSIG, B. SWEET & S. SHERRY. 1967. Biochemistry **6:** 3279–3287.
9. SHERMAN, L. A., M. W. MOSESSON & S. SHERRY. 1969. Biochemistry **8:** 1515–1523.
10. MILLS, D. & S. KARPATKIN. 1970. Biochem. Biophys. Res. Commun. **40:** 206–211.
11. MOSESSON, M. W., J. S. FINLAYSON, R. A. UMFLEET & D. GALANAKIS. 1972. J. Biol. Chem. **247:** 5210–5219.
12. MOSESSON, M. W., D. K. GALANAKIS & J. S. FINLAYSON. 1974. J. Biol. Chem. **249:** 4656–4664.
13. GALANAKIS, D. K., M. W. MOSESSON & N. E. STATHAKIS. 1978. J. Lab. Clin. Med. **92:** 376–386.
14. GALANAKIS, D. K. & M. W. MOSESSON. Human fibrinogen heterogeneities: determination of the major Aα chain derivatives in blood. Submitted for publication.
15. MILLS, D. & S. KARPATKIN. 1971. Biochim. Biophys. Acta **251:** 121–125.
16. LIPINSKA, I., B. LIPINSKI & V. GUREWICH. 1974. J. Lab. Clin. Med. **84:** 509–516.
17. LIPINSKA, B., A. NOWAK & V. GUREWICH. 1974. Experientia **30:** 84–85.
18. JILG, W. & H. HÖRMANN. 1974. Hoppe Seyler's Z. Physiol. Chem. **355:** 1316–1324.

19. LIPINSKA, I., B. LIPINSKI, V. GUREWICH & K. D. HOFFMAN. 1976. Am. J. Clin. Pathol. **66:** 958–966.
20. LIPINSKI, B., I. LIPINSKA, A. NOWAK & V. GUREWICH. 1977. J. Lab. Clin. Med. **90:** 187–194.
21. TSIANOS, E. B. & N. E. STATHAKIS. 1981. Thromb. Haemostas. (Stuttgart) **44:** 130–134.
22. GUREWICH, V. & I. LIPINSKA. 1981. Thromb. Res. **22:** 535–541.
23. STATHAKIS, N. E. & M. W. MOSESSON. 1977. J. Clin. Invest. **60:** 855–865.
24. WEINSTEIN, M. J. & D. DEYKIN. 1978. Thromb. Res. **13:** 361–377.
25. GALANAKIS, D. K. & M. W. MOSESSON. 1979. Thromb. Res. **15:** 287–289.
26. WEINSTEIN, M. J. & D. DEYKIN. 1980. Thromb. Res. **17:** 281–283.
27. SEMERARO, N., D. COLLEN & M. VERSTRAETE. 1977. Biochim. Biophys. Acta **492:** 204–214.
28. FERGUSON, E. 1980. J. Lab. Clin. Med. **96:** 710–721.
29. WEBER, K. & M. OSBORN. 1969. J. Biol. Chem. **244:** 4406–4412.
30. LORAND, L. 1972. Ann. N.Y. Acad. Sci. **202:** 155–171.
31. GAFFNEY, P. J. 1971. Nature (New Biol.) **230:** 54–56.
32. AGUERCIF, M., N. GIACOMETTI, O. M. NIGG, G. LACOURT & C. A. BOUVIER. 1973. Paediatrie **28:** 381–399.
33. KUYAS, C., A. HAEBERLI & P. W. STRAUB. 1982. J. Biol. Chem. **257:** 1107–1109.
34. KUYAS, C., A. HAEBERLI & P. W. STRAUB. 1982. Thromb. Haemostas. (Stuttgart) **47:** 19–21.
35. BLOMBÄCK, B., M. BLOMBÄCK, P. EDMAN & B. HESSEL. 1966. Biochim. Biophys. Acta **115:** 371–396.
36. WITT, I. & H. MÜLLER. 1970. Biochim. Biophys. Acta **221:** 402–404.
37. WITT, I. & K. HASLER. 1972. Biochim. Biophys. Acta **271:** 357–362.
38. SHERMAN, L. A., A. P. FLETCHER & S. SHERRY. 1969. J. Lab. Clin. Med. **73:** 574–583.
39. PLOW, E. F. & T. S. EDGINGTON. 1975. J. Clin. Invest. **56:** 30–38.
40. MOROZ, L. A. & N. J. GILMORE. 1976. Blood **48:** 531–545.
41. COMP, P. C. & C. T. ESMON. 1981. J. Clin. Invest. **68:** 1221–1228.
42. MOSESSON, M. W., J. S. FINLAYSON & R. A. UMFLEET. 1972. J. Biol. Chem. **247:** 5223–5227.
43. GATI, W. P. & P. W. STRAUB. 1978. J. Biol. Chem. **253:** 1315–1321.
44. STATHAKIS, N. E. & M. W. MOSESSON. 1978. Thromb. Res. **13:** 467–475.
45. GAFFNEY, P. J. 1972. Biochim. Biophys. Acta **263:** 453–458.
46. TÖPFER-PETERSEN, E., F. LOTTSPEICH, A. HENSCHEN. 1979. Thromb. Haemostas. (Stuttgart) **41:** 671–676.
47. PALASCAK, J. E. & J. MARTINEZ. 1977. J. Clin. Invest. **60:** 89–95.
48. MARTINEZ, J., J. E. PALASCAK & D. KWASNIAK. 1978. J. Clin. Invest. **61:** 535–538.
49. SORIA, J., C. SORIA, J. J. RYCKEWAERT, M. SAMAMA, J. M. THOMSON & L. POLLER. 1980. Thromb. Res. **19:** 29–41.
50. VON FELTON, A., P. W. STRAUB & P. G. FRICK. 1969. N. Eng. J. Med. **280:** 405–409.
51. GRALNICK, H. R., GIVELBER, H. & E. ABRAMS. 1978. N. Eng. J. Med. **299:** 221–226.
52. STREIFF, F., P. ALEXANDRE, C. VIGNERON, J. SORIA, C. SORIA & L. MESTER. 1971. Thromb. Diath. Haemorrh. **26:** 565–576.
53. MESTER, L. & L. SZABADOS. 1968. Bull. Soc. Chim. Biol. **50:** 2561–2566.
54. GALANAKIS, D. K. & M. W. MOSESSON. 1979. Thromb. Haemostas. (Stuttgart) **42:** 79 (abstr.)
55. GALANAKIS, D. K. & M. W. MOSESSON. 1976. Blood **48:** 109–117.
56. MARTINEZ, J., J. PALASCAK & C. PETERS. 1977. J. Lab. Clin. Med. **89:** 367–377.

57. FINLAYSON, J. S. & M. W. MOSESSON. 1963. Biochemistry **2:** 42–46.
58. FINLAYSON, J. S. & M. W. MOSESSON. 1964. Biochim. Biophys. Acta **82:** 415–417.
59. MOSHER, D. F. & E. R. BLOUT. 1973. J. Biol. Chem. **248:** 6896–6903.
60. LEGRELE, C. D., C. WOLFENSTEIN-TODEL, Y. HURBOURG & M. W. MOSESSON. 1982. Biochem. Biophys. Res. Commun. **105:** 521–529.
61. MARTINEZ, J. 1980. Blood **56:** 417–420.
62. MOSESSON, M. W. & J. S. FINLAYSON. 1963. J. Lab. Clin. Med. **62:** 663–674.
63. GERBECK, C. M., T. YOSHIKAWA & R. MONTGOMERY. 1969. Arch. Biochem. Biophys. **134:** 67–75.
64. WOLFENSTEIN-TODEL, C. & M. W. MOSESSON. 1980. Proc. Natl. Acad. Sci. USA **77:** 5069–5073.
65. FRANCIS, C. W., V. J. MARDER & S. E. MARTIN. 1980. J. Biol. Chem. **255:** 5599–5604.
66. WOLFENSTEIN-TODEL, C. & M. W. MOSESSON. 1981. Biochemistry **20:** 6146–6149.
67. KORNER, A. M., S. I. FEINSTEIN & S. ALTMAN. 1979. *In* Transfer RNA. S. Altman, Ed.: 105–135. MIT Press. Cambridge, MA.
68. DARNELL, J. E., JR. 1979. Prog. Nucleic Acid Res. Mol. Biol. **22:** 327–353.
69. CRABTREE, G. & J. A. KANT. 1981. Blood 58 (Suppl. 1):213a.
70. SINGER, P. A., H. H. SINGER & A. R. WILLIAMSON. 1980. Nature (Lond) **285:** 294–300.
71. EARLY, R., J. ROGERS, M. DAVID, K. CALAME, M. BOND, R. WALL & L. HOOD. 1980. Cell **20:** 313–319.
72. ALT, F. W., A. L. M. BOTHWELL, M. KNAPP, E. SIDEN, E. MATHER, M. KOSHLAND & D. BALTIMORE. 1980. Cell **20:** 293–301.
73. ROGERS, J., E. CHOI, L. SOUZA, C. CARTER, C. WORD, M. KUEHL, D. EISENBERG & R. WALL. 1981. Cell **26:** 19–27.
74. AMRANI, D. L., L. HOYER & M. W. MOSESSON. 1982. Blood **59:** 657–663.
75. NACHMAN, R. L. & A. J. MARCUS. 1968. Br. J. Haematol. **15:** 181–189.
76. SOLUM, N. O. & S. LAPACIUK. 1969. Thromb. Diath. Haemorrh. **21:** 419–427.
77. KARACA, M., I. M. NILSSON & U. HEDNER. 1971. J. Lab. Clin. Med. **77:** 485–489.
78. WEISS, H. J., L. D. WITTE, K. L. KAPLAN, B. A. LAGES, A. CHERNOFF, H. L. NOSSEL, D. S. GOODMAN & H. R. BAUMGARTNER. 1979. Blood **54:** 1296–1319.
79. BROEKMAN, M. J., N. P. WESTMORELAND & P. COHEN. 1974. J. Cell Biol. **60:** 507–519.
80. BROEKMAN, M. J., R. I. HANDIN & P. COHEN. 1975. Br. J. Haematol. **31:** 51–56.
81. KAPLAN, K. L., M. J. BROEKMAN, A. CERNOFF, G. R. LESZNIK & M. DRILLINGS. 1979. Blood **53:** 604–618.
82. GERARD, J. M., D. R. PHILLIPS, G. H. R. RAO, E. F. PLOW, D. A. WALZ, R. ROSS, L. A. HARKER & J. G. WHITE. 1981. J. Clin. Invest. **66:** 102–109.
83. WARE, A. E., J. L. FAHEY & W. H. SEEGERS. 1948. Am. J. Physiol. **154:** 140–147.
84. SALMON, J. & Y. BOUNAMEUX. 1958. Thromb. Diath. Haemorrh. **2:** 93–110.
85. GRETTE, K. 1962. Acta Physiol. Scand. (Suppl.) **195:** 1–93.
86. GOKCEN, M. & E. YUNIS. 1963. Nature **200:** 590–591.
87. CASTALDI, P. A. & J. CAEN. 1965. J. Clin. Pathol. **18:** 579–585.
88. NACHMAN, R. L. 1965. Blood **25:** 703–711.
89. NACHMAN, R. L., A. J. MARCUS & D. ZUCKER-FRANKLIN. 1967. J. Lab. Clin. Med. **69:** 651–658.
90. DOOLITTLE, R. F., T. TAKAGI & B. A. COTTRELL. 1974. Science **185:** 368–369.

91. DAVEY, M. D. & E. F. LUSCHER. 1967. *In* Biochemistry of Blood Platelets. E. Kowalski, & S. Niewiarowski, Eds.: 9–20. Academic Press. New York, N.Y.
92. SOLUM, N. O. & S. LOPACIUK. 1969. Thromb. Diath. Haemorrh. **21:** 428–440.
93. SORIA, J., C. SORIA, M. SAMAMA, E. POIROT & C. KLING. 1976. Pathol. Biol. (Suppl.) **24:** 15–17.
94. PLOW, E. F. & T. S. EDGINGTON. 1975. Thromb. Res. **7:** 729–742.
95. JAMES, H., P. GANGULY & C. W. JACKSON. 1977. Thromb. Haemostas. (Stuttgart) **38:** 939–954.
96. GANGULY, P. 1972. J. Biol. Chem. **247:** 1809–1816.
97. JAMES, H. L. & P. GANGULY. 1975. Biochem. Biophys. Res. Commun. **63:** 659–662.
98. JANDROT-PERRUS, M., M. W. MOSESSON, M.-H. DENNINGER & D. MÉNACHÉ. 1979. Blood **54:** 1109–1116.
99. LEE, D. C., G. S. McKNIGHT & R. D. PALMITER. 1978. J. Biol. Chem. **253:** 3494–3503.

THE PLASMA FIBRINOGEN FRACTION WITH ELEVATED SIALIC ACID CONTENT AND ELONGATED γ CHAINS *

DISCUSSION PAPER

P. W. Straub, C. Kuyas, and A. Herberli

Thrombosis Research Laboratory
University Department of Medicine
Inselspital
Bern, Switzerland

The long-known heterogeneity of human fibrinogen [1] is reflected by a heterogeneity of its constituent polypeptide chains.[2-7] Several years ago we demonstrated that the two main fractions of both the γ and Bβ chains, separated on CM-cellulose, are heterogeneous with respect to the sialic acid content,[7] one variant containing approximately two, the other one sialic acid per chain. The sialic acid-rich variant was found to be increased in fibrinogen Zürich II,[8] explaining the increased overall sialic acid content of this abnormal fibrinogen which shows a prolonged clotting time upon thrombin addition. As in some other abnormal fibrinogens, the thrombin clotting time is normalized after neuraminidase treatment, thus raising the question of whether changes in sialic acid alone might explain the heterogeneity found in both normal and abnormal fibrinogens. Recently, Wolfenstein and Mosesson [9] have found a normal γ chain variant with an extended carboxy-terminal sequence, indicating that at least in part peptides themselves may explain the heterogeneity.

We have therefore compared the polypeptide chains derived from normal fibrinogen and those derived from asialofibrinogen,[10] in an attempt to assign the individual chains to subfraction of fibrinogen as separated on DEAE-cellulose.[11] Upon isoelectric focusing of reduced fibrinogen in 7 M urea (FIGURE 1), three main γ and three main Bβ chains were found in addition to several Aα-chains. Furthermore, the elongated γ' chains described by Mosesson were also heterogeneous. Removal of sialic acid led to a shift of the apparent isoelectric point of the γ and Bβ variants but not of the Aα variants, the Aα chain being known to contain no carbohydrate.

In addition, neuraminidase treatment reduced the number of the Bβ and the γ chains from 3 to 2. Two-dimensional separation was achieved according to O'Farrell [12] using isoelectric focusing in the first dimension and SDS polyacrylamide electrophoresis in the second dimension. Aside from at least six Aα chain variants, all of the same molecular weight and thus excluding proteolytic Aα chain degradation, we found three isoelectrically different Bβ, three γ, and two γ' chains. Again, neuraminidase treatment lead to a reduction of the number of Bβ and γ bands. This confirms that the heterogeneity of the Bβ and γ chains can only in part be attributed to the differences in sialic acid content of the polypeptide chain variants. Further, the finding of groups of chain variants, each with the same molecular weight, suggested that the

* Supported by the Swiss National Science Foundation, Grant 3.915.0.80, and by the Emil Barrel Foundation (F. Hoffmann-La Roche, AG).

parent fibrinogen molecule must exist in several isoelectric forms. We are then confronted with the question of whether the sialic acid-dependent heterogeneity is of intramolecular nature—asymmetry of the molecule—or of intermolecular nature—different populations of symmetric molecules.

Using DEAE cellulose chromatography, Martinez [13] has found two fibrinogen fractions with identical carbohydrate content. We have used stepwise elution from DEAE cellulose and could achieve a better separation of both pooled and several single donor fibrinogens into 3 peaks (FIGURE 2), each with different sialic acid content. The most acidic subfraction 3 represents 22% of the protein and contains 8.2 moles of sialic acid as determined using the thiobarbituric assay, according to Aminoff [14] versus 6.2 moles in bulk fibrinogen, 5.2 in subfraction 1 and 5.7 in subfraction 2. After reduction, isoelectric

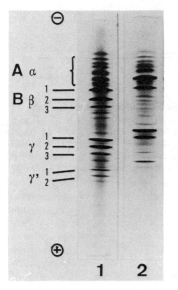

FIGURE 1. Isoelectric focusing on polyacrylamide gels of reduced fibrinogen (gel nr 1) and reduced asialofibrinogen (gel nr 2) in the presence of 7 M urea and 1% Triton B-1000. pH gradient 4–9. The bands are designated with arabic numbers according to the IUPAC-IUB Commission of Biochemical Nomenclature.

focusing (FIGURE 3) showed a preferential accumulation of the elongated γ' chains in this sialic acid-rich and most acidic subfraction. Similarly in two-dimensional electrophoresis (FIGURE 4) the faintly stained acidic bands found on the bottom of the gel of pooled fibrinogen are concentrated in the third most acidic fibrinogen fraction. Normal human fibrinogen, both pooled and from single donors, therefore contains a distinct subfraction with the following features:

(1) It contains approximately half normal and half elongated γ-chains and (2) its sialic acid content is significantly higher than that of the bulk fibrinogen. It is so far not clear whether the extra-polypeptide tail of γ chain carries an additional glycosylation site. Part of this tail has recently been sequenced,[15] but no possible site has been found.

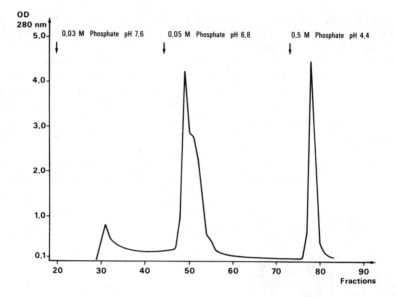

FIGURE 2. Stepwise (DE 52 ion exchange) DEAE-cellulose chromatography of human pool fibrinogen. 100 mg of protein were applied to the column (1.6 × 20 cm). Fibrinogen was eluted stepwise with three different buffers: buffer 1, 0.03 M phosphate/0.06 M Tris, pH 7.6; buffer 2, 0.05 M phosphate/0.08 M Tris, pH 6.8; and buffer 3, 0.5 M phosphate/0.5 M Tris, pH 4.4. The arrows indicate the start of the new buffer. The flow rate was 28 ml/hr and fractions of 3 ml were collected.

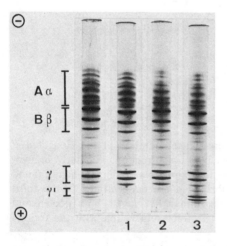

FIGURE 3. Isoelectric focusing on polyacrylamide gel in the presence of 7 M urea and 1% Triton X-100, pH gradient 4–9. 75 mg of protein/gel were loaded. The gel concentration was T = 7%/C = 25% (DATD). Left: reduced pool fibrinogen, 1: reduced fibrinogen subfraction 1, 2: reduced fibrinogen subfraction 2, and 3: reduced fibrinogen subfraction 3.

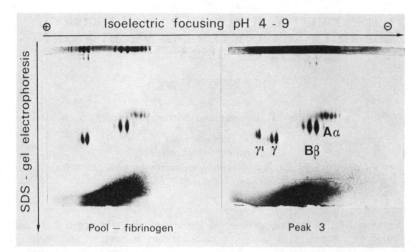

FIGURE 4. Two-dimensional separation of reduced pool fibrinogen (*left*) and reduced fibrinogen subfraction 3 (*right*). First dimension, isoelectric focusing was performed in gel rods (0,25 × 7 cm) (T = 7%, C = 25% DATD) in the presence of 8 M urea and 1% Triton X-100, the concentration of the pharmalyte (Pharmacia, Sweden) (pH 4–9) being 2,7%. The focusing time was 12 hr with 12 V/gel at 10°C. Immediately after isoelectric focusing, the gel rod was placed on a SDS-polyacrylamide slab gel (8 × 6,5 × 0,3 cm) (T = 10%, C = 4%, the cross-linking agent being N,N'-methylene-bisacrylamide). The electrophoresis buffer was 0.05 M Tris, 0.35 M glycine, 0.1% SDS, pH 8.5. The electrophoresis was carried out at 10°C with 20 mA/slab.

REFERENCES

1. FINLAYSON, J. S. & M. W. MOSESSON. 1963. Biochemistry **2**: 42–46.
2. MOSESSON, M. W., J. S. FINLAYSON & R. A. UMFLEET. 1972. J. Biol. Chem. **247**: 5223–5227.
3. GAFFNEY, P. J. 1972. Biochim. Biophys. Acta **263**: 453–458.
4. GALANAKIS, D. K., M. W. MOSESSON & N. E. STATHAKIS. 1978. J. Lab. Clin. Med. **92**: 376–386.
5. WEINSTEIN, M. J. & D. DEYKIN. 1978. Thrombos. Res. **13**: 361–377.
6. STATHAKIS, N. E., M. W. MOSESSON, D. K. GALANAKIS & D. MÉNACHÉ. 1978. Thrombos. Res. **13**: 467–475.
7. GATI, W. P. & P. W. STRAUB. 1978. J. Biol. Chem. **253**: 1315–1321.
8. STRAUB, P. W. & W. P. GATI. 1979. Thrombos. Haemostas. **41**: 714–717.
9. WOLFENSTEIN-TODEL, C. & M. W. MOSESSON. 1980. Proc. Natl. Acad. Sci. USA **77**: 5069–5073.
10. KUYAS, C., A. HAEBERLI & P. W. STRAUB. 1982. Thrombos. Haemostas. **47**: 19–21.
11. KUYAS, C., A. HAEBERLI & P. W. STRAUB. 1982. J. Biol. Chem. **257**: 1107–1109.
12. O'FARRELL, P. H. 1975. J. Biol. Chem. **250**: 4007–4021.
13. MARTINEZ, J. 1980. Blood. **56**: 417–420.
14. AMINOFF, D. 1961. Biochem. J. **81**: 384–392.
15. WOLFENSTEIN-TODEL, C. & M. W. MOSESSON. 1981. Biochemistry. **20**: 6146–6149.

HETEROGENEITY OF NORMAL HUMAN FIBRINOGEN DUE TO TWO HIGH MOLECULAR WEIGHT VARIANT γ CHAINS

DISCUSSION PAPER

Charles W. Francis and Victor J. Marder

Hematology Unit
Department of Medicine
University of Rochester School of Medicine and Dentistry
Rochester, New York 14642

Variants of the γ chain of normal human plasma fibrinogen have been attributed to differences in charge,[1-3] carbohydrate content,[4,5] and molecular size.[6-10] In 1979 we described a variant of higher molecular weight than the major form [7] and proposed that it might represent the γ' variant previously described by Mosesson and colleagues.[2] Subsequent studies have confirmed this finding and localized the site of variability to the carboxy-terminal portion of the polypeptide chain.[9,10]

We have now employed ion exchange chromatography to separate fibrinogen into three populations of molecules which differ in γ chain composition. Normal fibrinogen was applied to a column of DEAE Sephacel and eluted with a combined pH and ionic strength gradient. As we have shown,[11] three protein peaks emerged with 85% of protein in the first eluting peak, 6% in the second, and 9% in the third. The same elution pattern was found with commercial fibrinogen, fibrinogen from normal pooled plasma and that purified from plasma of normal individuals.

Protein eluting in the three peaks was electrophoresed on SDS-polyacrylamide gels before and after crosslinking, demonstrating a unique γ chain composition of each fibrinogen fraction. As shown in TABLE 1, fibrinogen in peak A contained only γ chains of molecular weight 50,000 which formed dimers of 100,000. Peak B fibrinogen demonstrated two γ monomers of molecular weights 50,000 and 55,000 in nearly equal amounts. These appeared to

TABLE 1

APPARENT MOLECULAR WEIGHTS OF γ CHAIN MONOMERS AND DIMERS

Peak	Monomer	Dimer
A	50,000	100,000
B	50,000	100,000
	55,000	105,000
		111,000
C	50,000	100,000
	57,500	108,000
		115,000

TABLE 2

EVIDENCE FOR γ CHAIN IDENTITY OF ADDITIONAL BANDS

1. Dimer formation by Factor XIII$_a$
2. Incorporation of dansyl cadaverine into monomers and dimers
3. PAS staining characteristics
4. Pattern of plasmic degradation
5. Identification of tyrosine as N-terminal residue of all dimers

crosslink randomly to form dimers with molecular weights of 100,000, 105,000 and 111,000. Peak C also contained γ chains of 50,000 and, in addition, had γ chains of molecular weight 57,500 which crosslinked to form dimers of 100,000, 108,000 and 115,000.

The larger γ chains in fibrinogen which eluted in peaks B and C demonstrated characteristic properties of γ chains as shown in TABLE 2. All demonstrated dimer formation and incorporated the lysine analogue dansyl cadaverine with Factor XIII$_a$. They stained positively with the periodic acid-Schiff reagent and had a pattern of plasmic degradation characteristic of γ chains. In addition, tyrosine was identified as the N-terminal amino acid residue in all crosslinked dimers following purification by preparative electrophoresis.

TABLE 3 shows the probable polypeptide chain composition of fibrinogen in each of the three peaks and their relative proportions in normal fibrinogen. Fibrinogen in the first eluting peak constitutes 89% of the total and contains two γ chains of molecular weight 50,000. Peak B fibrinogen is the fraction present in smallest amount and contains one γ chain of 50,000 and one of 55,000. Fibrinogen which elutes in peak C represents 10% of the total and contains a γ chain of molecular weight of 57,500 and one of 50,000.

As shown in TABLE 4, the starting material for chromatography contained 98% fibrinogen and 2% fibronectin. Peaks A and C contained no fibronectin while peak B contained 41% fibrinogen and 59% fibronectin. Considering the relative molecular weights of fibrinogen and fibronectin, the molar ratio of these two components in peak B is close to 1. This coelution may indicate an association of fibrinogen containing γ_{55} chains with fibronectin present in the preparation.

Our use of a different anion exchange resin has resulted in a distinct separation of three fibrinogen fractions rather than two previously reported by

TABLE 3

PROBABLE POLYPEPTIDE CHAIN STRUCTURE OF FIBRINOGEN
IN CHROMATOGRAPHIC PEAKS

Peak	Composition	% of Total Fibrinogen
A	$(A\alpha)_2 (B\beta)_2 (\gamma_{50})_2$	87
B	$(A\alpha)_2 (B\beta)_2 (\gamma_{50}\gamma_{55})$	3
C	$(A\alpha)_2 (B\beta)_2 (\gamma_{50}\gamma_{57.5})$	10

Mosesson and colleagues.[2] It is likely that the variant termed γ'^2 represents the $\gamma_{57.5}$ chain, which is found in the third chromatographic peak (FIGURE 1), and also corresponds to the larger γ variant termed γ_B which we previously reported.[7, 8] The electrophoretic studies of Kuyas et al.[5] demonstrate charge heterogeneity superimposed on γ chain size heterogeneity, but the method employed did not resolve the γ chain of 55,000 molecular weight because of its low proportion in unfractionated fibrinogen. The chromatographic purification we have now achieved has permitted the delineation of two, rather than one, high molecular weight γ chain variants.

TABLE 4

PROPORTION (%) OF FIBRINOGEN AND FIBRONECTIN IN THE STARTING MATERIAL AND IN EACH DEAE SEPHACEL PEAK

		Elution Peak		
	Starting Material	A	B	C
Fibrinogen	98	100	41	100
Fibronectin	2	0	59	0

REFERENCES

1. GAFFNEY, P. J. 1971. Nature New Biol. 230: 54–56.
2. MOSESSON, M. W., J. S. FINLAYSON & R. A. UMFLEET. 1972. J. Biol. Chem. 247: 5223–5227.
3. GALANAKIS, D. K., M. W. MOSESSON & N. E. STATHAKIS. 1978. J. Lab. Clin. Med. 92: 376–386.
4. GATI, W. P. & P. W. STRAUB. 1978. J. Biol. Chem. 253: 1315–1321.
5. KUYAS, C., A. HAEBERLI & P. W. STRAUB. 1982. Thromb. Haemostas. 47: 19–21.
6. HENSCHEN, A. & P. EDMAN. 1972. Biochim. Biophys. Acta 263: 351–367.
7. FRANCIS, C. W., V. J. MARDER & S. E. MARTIN. 1979. Blood 54: 279a.
8. FRANCIS, C. W., V. J. MARDER & S. E. MARTIN. 1980. J. Biol. Chem. 255: 5599–5604.
9. WOLFENSTEIN-TODAL, C. & M. W. MOSESSON. 1980. Proc. Natl. Acad. Sci. USA 77: 5069–5073.
10. WOLFENSTEIN-TODAL, C. & M. W. MOSESSON. 1981. Biochemistry 20: 6146–6149.
11. FRANCIS, C. W., D. H. KRAUS & V. J. MARDER. Biochim. Biophys. Acta. In press.

DISCUSSION OF THE PAPER

M. W. MOSESSON (Mt. Sinai Medical Center, Milwaukee, WI): I would like to ask three questions. Have you been able to show that the A, B, and C peaks are present after desialation? Have you any C-terminal sequence information that would indicate that these chains differ by their C-terminal? Have you done any studies to differentiate coelution of fibrinectin and fibrinogen from a true binding affinity between the two? My feeling is that they happen to coelute in the same positions.

C. W. FRANCIS: We have not desialated them. As for the second question, we have not done the C-terminal analysis you spoke of. Regarding the coelution question, as of now all I can say is that these two coelute at a molar ratio of one to one and studies are now in progress to examine that further.

CONGENITAL FIBRINOGEN ABNORMALITIES

D. Ménaché

Plasma and Plasma Derivatives Services
American Red Cross Blood Services Laboratories
Bethesda, Maryland 20814

INTRODUCTION

Congenital fibrinogen abnormalities are rare disorders and include afibrinogenemia, dysfibrinogenemia, and hypofibrinogenemia. Afibrinogenemia and dysfibrinogenemia are two relatively well-defined entities, the former being a quantitative defect characterized by the lack of circulating fibrinogen, whereas the latter is characterized by the presence of a qualitatively abnormal and functionally defective fibrinogen molecule. Following the suggestion of Beck,[1] a nomenclature similar to that formerly used for abnormal hemoglobins has been adopted and abnormal fibrinogens are designated according to the city of origin of the patient. Hypofibrinogenemia is less well defined. It may occur in both parents of a patient with afibrinogenemia where it clearly represents the heterozygous state. Low levels of fibrinogen determined by immunologic methods may also be found in some patients with an abnormal fibrinogen molecule (hypodysfibrinogenemia), e.g., Bethesda II,[2] Bethesda III,[3] Chapel Hill,[4] Giessen II,[5] Leuven,[6] New York,[7] Parma,[8] Philadelphia [9] and Valencia.[10] However, it is still not clear whether some patients with hypofibrinogenemia represent a totally separate condition.

INCIDENCE

As mentioned above, fibrinogen defects occur rarely. Although the precise incidence is not known, dysfibrinogenemia would appear to be the most frequent. Since the first documented congenital and inherited abnormal fibrinogen, Paris I, reported in 1963,[11] at least 69 additional families have been identified. Afibrinogenemia, which was first observed in 1920,[12] has been identified in about the same number of families. Hypofibrinogenemia, first recognized in 1935,[13] has a much lower incidence.

TRANSMISSION

Fibrinogen disorders occur in both sexes with equal frequency. The mode of inheritance for afibrinogenemia is autosomal recessive, as substantiated by the lack of clinical symptoms in both parents and a high frequency of both consanguinity and affected siblings. In some instances, the fibrinogen levels measured in the parents have been found to be low in the father and the mother (heterozygous state). In other cases, however, only one parent is affected, while in still other cases, both parents have been reported to have normal levels of fibrinogen. Interpretation of these data is difficult. The considerable range within which fibrinogen is considered to be at a "normal level"

0077-8923/83/0408-0121 $01.75/0 © 1983, NYAS

and the variety of techniques used to measure fibrinogen concentration, might explain these inconsistencies. With two exceptions, Parma [8] and Valencia,[10] autosomal dominant transmission has been found in all the dysfibrinogenemic families, including one homozygotic patient (Metz).[14]

In hypofibrinogenemia, the fact that siblings are affected in several families [15-25] favors a genetic transmission. It would appear, however, that hypofibrinogenemia includes a heterogeneous population of patients. Two types of autosomal transmission have been observed. In some families, absence of consanguinity between the parents,[26-29] hemorrhagic symptoms in one of the parents,[19, 20, 22, 24, 27] or low fibrinogen levels in only one of the parents,[27] or some offspring,[13, 23] favor dominant transmission. In other families, consanguinity between parents,[24, 30-35] low fibrinogen levels [21, 34] and lack of clinical symptoms in both parents,[28, 30, 32-37] favor recessive transmission.

CLINICAL SYMPTOMS AND REPLACEMENT THERAPY

Patients with afibrinogenemia suffer from a hemorrhagic tendency which is particularly severe during the neonatal period and childhood. In 66 percent of the reported cases a hemorrhagic episode occurs within the first few days of life, umbilical bleeding being the most frequent manifestation (72 cases out of 108), and the cause of death in thirteen patients.[38-50] It is interesting to note that the same high incidence of umbilical bleeding has been reported in patients with a Factor XIII deficiency. Internal bleeding in the central nervous system is also frequent and has caused the death of seven patients.[40, 50-54] Other manifestations such as hematemesis and melena have been reported, and gingival hemorrhage and epistaxis are frequent. Subcutaneous hematomas are particularly frequent and cephalohematomas are often reported.[44, 46, 55-62] Hemarthrosis is rare,[38, 54, 63-67] occurs in both sexes and, in contrast to hemophilia, usually leaves no sequelae. This plethora of hemorrhagic symptoms seems to decrease with age, and young adults often experience long intervals between bleeding episodes. However, because of the early onset and the severity of the hemorrhages, only a few patients are known to have reached the age of 20. One third of the patients described died following hemorrhage. No full-term pregnancy has been reported to date.

The prognosis is dependent upon both the rapidity with which replacement therapy is instituted and the tolerance to therapy. In the absence of replacement therapy, hemorrhage following surgery or dental extraction is the rule. Effective replacement therapy has been achieved using whole blood, plasma, fibrinogen and, more recently, cryoprecipitate. However, adverse reactions to transfusion, i.e., fever, nausea, headache, asthma, have been reported in several instances.[54, 56, 67, 68] In addition, two patients developed antibodies to fibrinogen,[51, 69] rendering replacement therapy ineffective and resulting in severe intolerance reactions and death.

In contrast to afibrinogenemia, there is no uniform clinical pattern which characterizes congenital dysfibrinogenemia. Sixty percent of the reported families are asymptomatic and the disease is found by chance through routine laboratory coagulation tests. Some patients suffer from hemorrhagic episodes (Bethesda, I,[70] Bethesda III,[3] Buenos Aires,[71] Caracas I,[72] Chapel Hill,[4] Cleveland II,[73] Giessen II,[5] Metz,[14] Montreal II,[74] Philadelphia,[9] New Orleans,[75] and Vancouver [76, 77]) but, with the exception of Detroit [78] and Bethesda III,[3] these are rarely severe. An association between a bleeding syndrome and thrombotic episodes has been observed (Baltimore,[79] Marburg [80] and Wiesbaden [81]), as have isolated thrombotic states (Chicago,[82] Copenhagen,[83] Haifa,[84] New York,[7] Oslo I,[85] and Paris II [86]). Poor wound healing has been reported in four instances (Buenos Aires,[71] Caracas I,[72] Cleveland I,[87] and Paris I [11, 88]), and has been the cause of death of one patient (Paris I,[11, 88]).

Patients with hypofibrinogenemia suffer from hemorrhages that occur with high frequency in the neonatal period, manifested in 15 percent of the cases by umbilical bleeding. In later life, the hemorrhagic symptoms are similar to those reported for afibrinogenemia. One fifth of the observed patients have died from hemorrhage, including two patients from cerebral hemorrhage,[16, 50] and four patients from umbilical bleeding.[15, 16] Of interest is the occurrence of multiple thromboses and pulmonary emboli in four patients,[22, 28, 89, 90] two of whom died.[89, 90]

LABORATORY FINDINGS

In afibrinogenemia, various techniques such as gravimetry, heat or salt precipitation and electrophoresis, all fail to detect fibrinogen. In some instances, trace amounts are detectable when using immunologic techniques.[46, 48, 49, 61, 64, 65, 67, 91-99] Owing to the lack of fibrinogen, all coagulation tests which depend for an end point on the formation of a clot are abnormal. These are corrected by the addition of fibrinogen. All the other clotting factors are within normal range. Prolonged bleeding time, unrelated to thrombocytopenia, has been reported in more than 50 percent of cases and platelet aggregation is also defective.

The diagnosis of dysfibrinogenemia is made when coagulation tests which are dependent only on the fibrinogen to fibrin conversion are impaired, together with the demonstration that the degree of abnormality is not related to either the level of circulating fibrinogen or to the presence of an abnormal circulating protein (other than fibrinogen) e.g., in gamma chain myeloma. With the exception of Oslo I,[85] in which conversion is accelerated, fibrinogen to fibrin conversion (as assessed by the relevant clotting tests) is delayed, the common finding being a prolonged thrombin time. Fibrinogen levels are usually method-dependent and found to be low when determined by clotting tests because they involve the velocity with which fibrinogen coagulates, whereas levels are found to be normal when determined by immunological methods. However, hypo-fibrinogenemia (50–180 mg/dl) is found in 12 percent of the reported cases regardless of the method used. On the rare occasions when the half-life of autologous fibrinogen has been measured, survival has been either normal (Bethesda II,[2] Chapel Hill [4]), or decreased (Bethesda III,[3] Philadelphia [9]).

Analysis of the functional properties of abnormal fibrinogens have provided evidence in some instances for delayed fibrinopeptide release and/or abnormal fibrin monomer aggregation, as well as for abnormal fibrin polymer stabilization by Factor XIIIa and calcium. The structural defect has been characterized for several abnormal fibrinogens. A single amino acid substitution in the Aα chain has been found to occur in position 7 (Asp → Asn) for Lille [100]; in position 12 (Gly → Val) for Rouen [101]; in position 16 (Arg → His) for both Petoskey [102] and Manchester,[103] while Arg is replaced by Cys for Metz [104] and Zurich II [105]; in position 19 (Arg → Ser) for Detroit,[106] while Arg is replaced by Asn for München.[107] All these fibrinogens exhibit delayed release of fibrinopeptide A which may or may not be associated with a delay in fibrin monomer aggregation. No single amino acid substitution has been reported so far in the Bβ or the γ chains. Rather than an amino acid substitution, a segmental deletion in the carboxy-terminal end of the Aα chain has been evoked to elicit the abnormal fibrin polymerization for Chapel Hill.[4] Defective γ chains have been reported for Bern I,[108] Haifa [84] and Paris I.[109] The heavier than normal molecular weight of the γ Paris I mutant chain favors an elongation of the chain.[110] These three abnormal fibrinogens exhibit normal release of fibrinopeptides and delayed fibrin monomer aggregation.

In hypofibrinogenemia, the level of fibrinogen is usually below 75 mg/dl and in routine coagulation tests the degree of abnormality is directly dependent upon the amount of fibrinogen present. In only a few cases [20, 27, 35, 89, 90, 111] has the low level of fibrinogen been confirmed by electrophoresis or immunologic techniques. This observation is important inasmuch as many dysfibrinogenemia patients may be incorrectly identified as being hypofibrinogenemic when a method other than immunologic is used for evaluation. A striking example of this is the family, initially diagnosed as hypofibrinogenemic,[76] which was later found to exhibit dysfibrinogenemia (Vancouver).[77]

CONCLUSIONS

Congenital abnormalities of the fibrinogen molecule appear to reflect a decreased or absent protein synthesis or the production of an abnormal molecule. Abnormal production may be associated with normal or decreased synthesis. These various abnormalities are referred to as afibrinogenemia, hypofibrinogenemia, dysfibrinogenemia and hypo-dysfibrinogenemia. It is not surprising that patients with afibrinogenemia suffer from a severe bleeding disorder following birth since hemostasis cannot be achieved. Why these patients often have long intervals between bleeding episodes after they reach young adulthood is not understood. It is also not surprising that patients with hypofibrinogenemia suffer from bleeding episodes. There is no good explanation for the thrombotic episodes seen in some patients in the absence of replacement therapy. Patients with dysfibrinogenemia exhibit various pathologic manifestations that are certainly related to the structural defect of the molecule, but at this time too little information is available to shed any light on the structural and clinical correlations.

The relatively recent characterization of patients with hypo-dysfibrinogenemia raises the question as to whether families that have been reported as "hypofibrinogenemic" with autosomal dominant transmission are not, in fact, families with dysfibrinogenemia. Whenever possible, reappraisal of patients with low functional or immunologic levels of fibrinogen should be undertaken in order to determine whether they are, in fact, hypofibrinogenemic, dysfibrinogenemic or hypo-dysfibrinogenemic.

REFERENCES

1. BECK, E. A. 1964. Abnormal fibrinogen (fibrinogen "Baltimore") as a cause of a familial hemorrhagic disorder. Blood 24: 853–854.
2. GRALNICK, H. R., H. M. GIVELBER & J. S. FINLAYSON. 1973. A new congenital abnormality of human fibrinogen, Fibrinogen Bethesda II. Thromb. Diath. Haemorrh. 28: 562–571.

3. GRALNICK, H. R., B. COLLER, J. C. FRATANTONI & J. MARTINEZ. 1979. Fibrinogen Bethesda III—A hypodysfibrinogenemia. Blood **53:** 28–46.
4. MCDONAGH, R. P., N. A. CARRELL, H. R. ROBERTS, P. M. BLATT & J. MC- DONAGH. 1980. Fibrinogen Chapel Hill: Hypodysfibrinogenemia with a tertiary polymerization defect. Amer. J. Hematol. **9:** 23–38.
5. KRAUSE, W. H., K. HUTH, D. L. HEENE & H. G. LASCH. 1975. Hypodysfibrinogenämie : Fibrinogen Giessen II. Klin. Wochenschr. **53:** 781–782.
6. VERHAEGHE, R., M. VERSTRAETE, J. VERMYLEN & C. VERMYLEN. 1974. Fibrinogen Leuven, another genetic variant. Br. J. Haematol. **26:** 421–433.
7. AL-MONDHIRY, H. A. B., S. B. BILEZKIAN & H. L. NOSSEL. 1975. Fibrinogen "New York"—An abnormal fibrinogen associated with thromboembolism: functional evaluation. Blood **45:** 607–619.
8. IMPERATO, DI C. & A. G. DETTORI. 1958. Ipofibrinogenemia congenita con fibrinoastenia. Helv. Paediatr. Acta **4:** 380–398.
9. MARTINEZ, J., R. R. HOLBURN, S. S. SHAPIRO & A. J. ERSLEV. 1974. Fibrinogen Philadelphia. A hereditary hypodysfibrinogenemia characterized by fibrinogen hypercatabolism. J. Clin. Invest. **53:** 600–611.
10. AZNAR, J., A. FERNANDEZ-PAVON, E. REGANON, V. VILA & F. ORELLANA. 1974. Fibrinogen Valencia. A new case of congenital dysfibrinogenemia. Thromb. Diath. Haemorrh. **32:** 564–577.
11. MÉNACHÉ, D. 1963. Dysfibrinogenemie constitutionnelle et familiale. Proc. 9th Congr. Europ. Soc. Haemat., Lisbon, 1963: 1255–1259. S. Karger. Basel/New York, N.Y.
12. RABE, F. & E. SALOMON. 1920. Über Faserstoffmangel im Blute bei einem Falle von Hämophilie. Dtsch. Arch. Klin. Med. **132:** 240–244.
13. RISAK, E. 1935. Die Fibrinopenie. Ztschr. Klin. Med. **128:** 605–629.
14. SORIA, J., C. SORIA, M. SAMAMA, E. POIROT & C. KLING. 1972. Fibrinogen Troyes-Fibrinogen Metz. Two new cases of congenital dysfibrinogenemia. Thromb. Diath. Haemorrh. **27:** 619–633.
15. WOLF, J. 1936. Verblutungslod wegen Fibrinmangels. Jahrb. Kinderheilkd. **148:** 33–37.
16. SCHONHOLZER, G. 1939. Die hereditäre Fibrinogenopenie. Dtsch. Arch. Klin. Med. **184:** 496–510.
17. HEINILD, S. 1944. On familial constitutional fibrinopenia. Some observations with regard to the simultaneous appearance of fibrinopenia, thrombopenia and hypoprothrombinemia. Acta Med. Scand. **118:** 479–488.
18. SEVERINO, A. 1937. Su due casi di diatesi emorragica a lipo emofilofibrinopenico (osservazioni e ricerche). Pediatria (Napoli) **45:** 627–653.
19. REVOL, L., F. BOREL MILHET & N. PERRIN. 1951. Grande hypofibrinemie familiale. Considerations genetiques et therapeutiques. Sang **22:** 747–752.
20. D'ABLAING, G. & R. L. SEARCY. 1959. Congenital hypofibrinogenemia. Laboratory evaluation of a case. J. Am. Osteopath. Assoc. **58:** 492–494.
21. HENSEN, D. A., P. G. HOORWEG & M. A. E. DEKKER. 1960. Een geval van congenitale hypofibrinogenemie. Ned. Tijdschr. Geneesk. **104:** 510–512.
22. CAEN, J., Y. FAUR, S. INCEMAN, J. CHASSIGNEUX, M. SELIGMANN, T. ANAGNOSTOPOULOS & J. BERNARD. 1964. Necrose ischemique bilaterale dans un cas de grande hypofibrinogenemie congenitale. Nouv. Rev. Fr. d'Hematol. **4:** 321–324.
23. HATTERSLEY, P. G. & M. L. DIMICK. 1969. Cryoprecipitates in treatment of congenital fibrinogen deficiency. Case report. Transfusion **9:** 261–264.
24. AZNAR, J., J. A. FERNANDEZ-REGO, M. A. FERNANDEZ-PLA & J. A. AZNAR. 1970. Congenital hypofibrinogenemia in three of a family. Coagulation **3:** 279–284.
25. HAHN, L. & P. A. LUNDBERG. 1978. Congenital hypofibrinogenaemia and recurrent abortion. Case report. Br. J. Obstet. Gynaecol. **85:** 790–793.

26. MINNIS, J. F. & E. H. GRIFFIN. 1961. Congenital heart disease associated with congenital hypofibrinogenemia. Report of a case subjected to corrective surgery. Am. J. Cardiol. **7:** 432–435.
27. PHILLIPS, L. L., V. SKRODELIS & J. A. WOLFF. 1963. Normal fibrinolytic system in two cases of familial hypofibrinogenemia. Acta Haemat. **30:** 244–252.
28. MARCHAL, G., G. DUHAMEL, M. SAMAMA & G. FLANDRIN. 1964. Thrombose massive des vaisseaux d'un membre au cours d'une hypofibrinemie congenitale. Hemostase **4:** 81–90.
29. PFLUGER, N. & D. GEHRIG. 1977. Untersuchungen bei einer Familie mit vererbter Hypofibrinogenämie. Schweiz. Med. Wochenschr. **107:** 1454–1455.
30. BIDDAU, I. & L. AMMANATI. 1946. Contributo allo studio della afibrinogenemia. Arch. Ital. Pediatr. Pueric. **11:** 374–407.
31. LIDA, E. & R. BANFI. 1953. Un caso de fibrinogenopenia congenita. Rev. Asociac. Bioquim. Argent. **18:** 93–95.
32. PENATI, F. & G. P. GAIDANO. 1953. Osservazione di un caso di fibrinogenopenia essenziale. Minerva Med. **44:** 518–522.
33. PAVLOVSKY, A. & L. J. BERGNA. 1962. Fibrinopenie congenitale. Commentaires sur l'evolution d'un cas suivi pendant 20 ans. Hemostase **2:** 239–242.
34. MANIOS, S. G., W. SCHENCK & W. KÜNZER. 1968. Congenital fibrinogen deficiency. Acta Paediatr. Scand. **57:** 145–150.
35. KRUSE, H., K. SCHULZ & H.-J. BLAU. 1969. Klinik und Krankheitsverlauf bei einem Fall von kongenitaler Hypofibrinogenämie. Kinderaerztl. Prax. **37:** 361–372.
36. PALIARD, F., N. JEUNE & L. REVOL. 1938. L'hemophilie chez la femme. J. Med. Lyon **19:** 679–687.
37. LUCKHAUS, M. 1949. Ein Fall von Fibrinogenopenie. Dtsch. Med. Wochenschr. **75:** 1475–1476.
38. MACFARLANE, R. G. 1938. A boy with no fibrinogen. Lancet. 5 February: 309–312.
39. HENDERSON, J. L., G. M. M. DONALDSON & H. SCARBOROUGH. 1945. Congenital afibrinogenaemia. Report of a case with a review of the literature. Q.J.M. New Series No. 54: 101–111.
40. BUCEK, A. VON. 1951. Ein Fall der angeborenen familiären Afibrinogenämie. Ann. Paediatr. **177:** 111–115.
41. LAWSON, H. A. 1953. Congenital afibrinogenemia. Report of a case. N. Engl. J. Med. **248:** 552–554.
42. ALBEGGIANI, A. & A. LA GRUTTA. 1954. Contributo alla conoscenza della afibrinogenemia primitiva. Haematologica **38:** 1169–1196.
43. CAUSSADE, L., N. NEIMANN, M. PIERSON & M. MANCIAUX. 1954. L'afibrinogenemie congenitale et familiale (a propos de trois observations). La Presse Medicale **62:** 1040–1042.
44. BLOCH, M. & G. SANCHO. 1956. Fibrinogenopenia congenita. Sangre **1:** 431–434.
45. EDLUND, S. 1958. Case of congenital afibrinogenemia. Acta Paediatr. (Upsala) **48:** 101–102.
46. NIEWIAROWSKI, S., J. KOZLOWSKA, A. GULMANTOWICZ & E. PELCZARSKA-KASPERKA. 1962. Afibrinogenemie congenitale. Etude biologique de deu cas. Hemostase **2:** 191–202.
47. BOMMER, W., W. KUNZER & H. SCHROER. 1963. Kongenitale Afibrinogenämie. Ann. Paediatr. **200:** 46–59.
48. ARAUJO, A. R. & M. C. MARQUES. 1969. Afibrinogenemia congenita. O Medico **52:** 54–67.
49. MONTGOMERY, R. & S. E. NATELSON. 1977. Afibrinogenemia with intracerebral hematoma. Am. J. Dis. Child. **131:** 555–556.

50. FRIED, K. & S. KAUFMAN. 1980. Congenital afibrinogenemia in 10 offspring of uncle-niece marriages. Clin. Genet. **17:** 223–227.
51. BRÖNNIMAN, R. VON 1954. Kongenitale Afibrinogenämie. Mitteilung eines Falles mit multiplen Knochencysten und Bildung eines spezifischen Antikörpers (Antifibrinogen) nach Bluttransfusion. Acta Haematol. **11:** 40–51.
52. FERNANDO, P. B. & B. D. DHARMASENA. 1957. A case of congenital afibrinogenemia. Blood **12:** 474–479.
53. GROSSMAN, B. J. & R. E. CARTER. 1957. Congenital afibrinogenemia. J. Pediatr. **50:** 708–713.
54. MACKINNON, H. H. & J. F. FEKETE. 1971. Congenital afibrinogenemia : Vascular changes and multiple thromboses induced by fibrinogen infusions and contraceptive medication. Can. Med. Assoc. J. **104:** 597–599.
55. VAN NUFFEL, E. & M. VERSTRAETE. 1953. Un syndrome hemorragique rare : l'afibrinogenie. Presentation clinique d'un cas. Acta Paediat. Belgica **7:** 185–191.
56. CAMELIN, A., L. REVOL, J. FAVRE-GILLY, J. VAILHE, R. ARDRY & R. MIERAL. 1955. Afibrinogenemie congenitale. A propos de deux nouveaux cas et de leur traitement par une solution de fibrinogene. Bull. Mem. Soc. Med. Hop. Paris Nos. 5 and 6: 124–132.
57. VAN CREVELD, S. & K. H. LIEM. 1958. Congenital afibrinogenemia. Etud. Neo-Natales **7:** 89–100.
58. GARBIN, S. & E. GAROFALO. 1960. Afibrinogenemia congenita con particolare rapporto di consanguineita negli ascendenti. Hematologica **45:** 405–412.
59. GUIMBRETIERE, J. & H. HAROUSSEAU. 1962. A propos de deux freres atteints d'afibrinogenemie congenitale. Hemostase **2:** 203–215.
60. SUAREZ, M., G. DIAZ DE IARAOLA, F. JIMENEZ-DIEZ & E. MARTIN. 1964. Afibrinogenemia congenita. Estudio de dos casos. Rev. Esp. Pediat. **20:** 631–644.
61. BENTEGEAT, J., P. VERGER, J. MOULINIER, F. MESNIER, J. Y. BEAUVIEUX, N. BOISSEAU & C. DE JOIGNY. 1965. Un cas d'afibrinogenemie congenitale. Proc. 10th Congr. Europ. Soc. Hemat. **2:** 977–987. Strasbourg, France.
62. MEHTA, B. C., P. C. SHAH, M. P. BHAGAT, S. S. WAGLE & J. C. PATEL. 1970. Congenital afibrinogenemia. Report of two cases. Indian J. Med. Sci. **24:** 208–211.
63. ALEXANDER, B., R. GOLDSTEIN, L. RICH, A. G. LE BOLLOC'H, K. L. DIAMOND & W. BORGES. 1954. Congenital afibrinogenemia. A study of some basic aspects of coagulation. Blood **9:** 843–865.
64. HARDISTY, R. M. & J. L. PINNIGER. 1956. Congenital afibrinogenaemia : Further observations on the blood coagulation mechanism. Br. J. Haematol. **2:** 139–152.
65. WERDER, E. 1963. Kongenitale Afibrinogenämie. Helv. Paediatr. Acta **18:** 208–229.
66. DELAGE, J. M. & A. BARRY. 1964. Afibrinogenemie congenitale : presentation de quatre cas. Can. Med. Assoc. J. **90:** 1270–1273.
67. EGBRING, R., K. ANDRASSY, H. EGLI & J. MEYER-LINDENBERG. 1971. Diagnostische und therapeutische Probleme bei congenitaler Afibrinogenämie. Blut **22:** 175–201.
68. MAUPIN, B., J. VIGNE, H. PERROT, M. E. LEROUX, C. RABY, J. STORCK & F. LACASSIE. 1962. Observation clinique et biologique de trois cas d'afibrinogenemie congenitale dans une meme fratrie. Hemostase **2:** 217–228.
69. DE VRIES, A., T. ROSENBERG, S. KOCHWA & J. H. BOSS. 1961. Precipitating antifibrinogen antibody appearing after fibrinogen infusions in a patient with congenital afibrinogenemia. Am. J. Med. **30:** 486–494.
70. GRALNICK, H. R., H. M. GIVELBER, J. R. SHAINOFF & J. S. FINLAYSON. 1971. Fibrinogen Bethesda : A congenital dysfibrinogenemia with delayed fibrinopeptide release. J. Clin. Invest. **50:** 1819–1830.

71. BURASCHI, J. A., E. S. SACK, E. QUIROGA & H. HENDLER. 1975. A new fibrinogen anomaly : Fibrinogen Buenos Aires. 5th Congr. ISTH. Abstr. 224: 244. Paris, France.

72. BOSCH, N. B. DE, C. L. AROCHA-PINANGO, J. SORIA, C. SORIA, A. RODRIGUEZ & S. RODRIGUEZ. 1977. An abnormal fibrinogen in a Venezuelan family. Thromb. Res. 10: 253–265.

73. CRUM, E. D., J. R. SHAINOFF, R. C. GRAHAM & O. D. RATNOFF. 1974. Fibrinogen Cleveland II. An abnormal fibrinogen with defective release of fibrinopeptide A. J. Clin. Invest. 53: 1308–1319.

74. D'ANGELO, G., M. LACOMBE, J. LEMAY, R. LAVALLEE, Y. BONNY & J. BOILEAU. 1975. A new congenital dysfibrinogenemia with hemorrhagic diathesis (Fibrinogen Montreal II). 5th Congr. ISTH, Abstr. 223: 244. Paris, France.

75. ANDES, W. A., S. I. CHAVIN, G. BELTRAN & W. J. STUCKEY. 1982. Fibrinogen New Orleans: Hereditary dysfibrinogenemia with an Aα chain abnormality. Thromb. Res. 25: 41–50.

76. HASSELBACK, R., R. B. MARION & J. W. THOMAS. 1963. Congenital hypofibrinogenemia in five members of a family. Can. Med. Assoc. J. 88: 19–22.

77. THOMAS, J. W. & E. A. BECK. Unpublished data. Quoted by E. A. Beck, 1968. Congenital variants of human fibrinogen. In Fibrinogen. K. Laki, Ed.: 269–275. Marcel Dekker. New York.

78. MAMMEN, E. F., A. S. PRASAD, M. I. BARNHART & C. C. AU. 1969. Congenital dysfibrinogenemia: Fibrinogen Detroit. J. Clin. Invest. 48: 235–249.

79. BECK, E. A., P. CHARACHE & D. P. JACKSON. 1965. A new inherited coagulation disorder caused by an abnormal fibrinogen: fibrinogen "Baltimore." Nature 208: 143–145.

80. FUCHS, G., R. EGBRING & K. HAVEMANN. 1977. Fibrinogen Marburg—New genetic variant of fibrinogen. Blut 34: 107–118.

81. WINCKELMANN, G., R. AUGUSTIN & K. BANDILLA. 1971. Congenital dysfibrinogenemia. Report of a new family (Fibrinogen Wiesbaden). ISTH II Congr. Oslo, 1971. Abstract volume: 64.

82. PAPP, A. C., R. M. SNOPKO, E. R. COLE, R. J. SASSETTI & K. K. WU. 1981. Recurrent venous thrombosis related to a hereditary dysfibrinogen with abnormal crossed immunoelectrophoretic pattern. Thromb. Haemostas. 46: 360A.

83. HANSEN, S. M., I. CLEMMENSEN & D. WINTHER. 1980. Fibrinogen Copenhagen; an abnormal fibrinogen with defective polymerization and release of fibrinopeptide A, but normal adsorption of plasminogen. Scand. J. Clin. Lab. Invest. 40: 221–226.

84. SORIA, J., C. SORIA, S. TAVORI, M. SAMAMA, A. RIMON & I. TATARSKY. 1981. A new fibrinogen variant with abnormal gamma chain: Fibrinogen Haifa. Thromb. Haemostas. 46: 359A.

85. EGEBERG, O. 1967. Inherited fibrinogen abnormality causing thrombophilia. Thromb. Diath. Haemorrh. 17: 176–187.

86. SAMAMA, M., J. SORIA, C. SORIA & J. BOUSSER. 1969. Dysfibrinogenemie congenitale et familiale sans tendance hemorragique. Nouv. Rev. Franc. d'Hemat. 9: 817–832.

87. FORMAN, W. B., O. D. RATNOFF & M. H. BOYER. 1968. An inherited qualitative abnormality in plasma fibrinogen: Fibrinogen Cleveland. J. Lab. Clin. Med. 72: 455–472.

88. MÉNACHÉ, D. 1964. Constitutional and familial abnormal fibrinogen. Trans. Int. Comm. on Blood Clotting Factors. Gleneagles, Scotland. Thromb. Diathes. Haemorrh. Suppl. 13: 173–185.

89. INGRAM, G. I. C., D. J. McBRIEN & H. SPENCER. 1966. Fatal pulmonary embolus in congenital fibrinopenia. Acta Haemat. 35: 56–62.

90. NILSSON, I. M., J.-E. NILEHN, S. CRONBERG & G. NORDEN. 1966. Hypofibrinogenaemia and massive thrombosis. Acta Med. Scand. 180: 65–76.

91. GITLIN, D. & W. H. BORGES. 1953. Studies on the metabolism of fibrinogen in two patients with congenital afibrinogenemia. Blood. **8:** 679–686.
92. PRICHARD, R. W. & R. L. VANN. 1954. Congenital afibrinogenemia. Report on a child without fibrinogen and review of the literature. Am. J. Dis. Child. **88:** 703–710.
93. OSEID, S. & H. M. SVENDSEN. 1963. Congenital afibrinogenemia. Acta Paediat. **52:** 129–132.
94. GROSS, R., G. SCHWICK, N. LANG, D. NIES, B. RAHN, M. BECKER & H. HENGSTMANN. 1963. Untersuchungen an einer angeborenen Afibrinogenämie. (Zur rolle der Blutgerrinnung bei der Blutstillung). Klin. Wochenschr. **41:** 695–706.
95. GUGLER, E., H. STILLHART, N. BURGER & R. BUTLER. 1964. Die kongenital Afibrinogenämie. Schweiz. Med. Wochenschr. **94:** 1469–1475.
96. WEISS, J. H. & J. ROGERS. 1971. Fibrinogen and platelets in the primary arrest of bleeding. Studies in two patients with congenital afibrinogenemia. N. Engl. J. Med. **285:** 369–374.
97. GIROLAMI, A., M. LAZZARIN & G. CELLA. 1971. Congenital afibrinogenemia. A case report. Coagulation **4:** 181–187.
98. GIROLAMI, A., G. ZACCHELLO & R. D'ELIA. 1971. Congenital afibrinogenemia. A case report with some considerations on the hereditary transmission of this disorder. Thromb. Diath. Haemorrh. **25:** 460–468.
99. GIROLAMI, A., R. VENTURELLI & G. BAREGGI. 1972. A report on a case of congenital afibrinogenemia., Blut **24:** 23–31.
100. MORRIS, S., M.-H. DENNINGER, J. S. FINLAYSON & D. MÉNACHÉ. 1981. Fibrinogen Lille: $A\alpha^{7 ASP \rightarrow ASN}$. 1981. Thromb. Haemostas. **46:** 104A.
101. LOTTSPEICH, F. 1981. Personal communication.
102. HIGGINS, D. L. & J. A. SCHAFER. 1981. Fibrinogen Petoskey, a dysfibrinogenemia characterized by replacement of Arg-Aα16 by a histidyl residue. J. Biol. Chem. **256:** 12013–12017.
103. HENSCHEN, A. 1981. Personal communication.
104. HENSCHEN, A., C. SOUTHAN, J. SORIA, C. SORIA & M. SAMAMA. 1981. Structure abnormality of Fibrinogen Metz and its relationship to the clotting defect. Thromb. Haemostas. **46:** 103A.
105. HENSCHEN, A., C. SOUTHAN, M. KEHL & F. LOTTSPEICH. 1981. The structural error and its relation to the malfunction in some abnormal fibrinogens. Thromb. Haemostas. **46:** 181A.
106. BLOMBÄCK, M., B. BLOMBÄCK, E. F. MAMMEN & A. S. PRASAD. 1968. Fibrinogen Detroit—A molecular defect in the N-terminal disulphide knot of human fibrinogen? Nature **218:** 134–137.
107. HENSCHEN, A., F. LOTTSPEICH, S. SOUTHAN & E. TÖPFER-PETERSEN. 1980. Human fibrinogen: Sequence, sulfur bridges, glycosylation and some structural variants. *In* Protides of the Biological Fluids. H. Peeters, Ed.: 51–56. Pergamon Press. New York, N.Y.
108. RUPP, C., C. KUYAS, A. HÄBERLIE, M. FURLAN & E. A. BECK. 1981. Fibrinogen Bern I: A hereditary fibrinogen variant with defective conformational stabilization by calcium ions. Thromb. Haemostas. **46:** 104A.
109. BUDZYNSKI, A. Z., V. J. MARDER, D. MÉNACHÉ & M. C. GUILLIN. 1974. Defect in the gamma polypeptide chain of a congenital abnormal fibrinogen (Paris I). Nature **252:** 66–68.
110. MOSESSON, M. W., D. L. AMRANI & D. MÉNACHÉ. 1976. Studies on the structural abnormality of Fibrinogen Paris I. J. Clin. Invest. **57:** 782–791.
111. KRIZ, K. 1956. Fortschreitende ischämische nekrose des Hypophysenvorderlappens (Scheehansches Syndrom) bei kongenitaler Hypofibrinogenämie. Zentralbl. Gynäkol. **34:** 1347–1353.

Discussion of the Paper

D. K. Galanakis (*State University of New York, Stony Brook*): These comments are intended to summarize the current status of information on fetal fibrinogen. Several laboratories have independently established that the prolongation of the clotting time in fetal blood is attributable to delayed fibrin aggregation. Fetal fibrin characteristically forms a more transparent clot than adult fibrin. Under electronmicroscopic conditions the fibrin fibrils are relatively thinner and shorter, suggesting incomplete lateral association during fibrin assembly. These distinguishing functional characteristics of fetal fibrin are demonstrated more easily at a high ionic strength. Moreover, in mixtures of fetal and adult fibrin the aggregation pattern is similar to that of fetal, indicating that fetal fibrin inhibits the aggregation of adult fibrin.

The only known structural characteristics of fetal fibrinogen are high phosphorus and sialic acid content. The functional characteristics relate to the high sialic acid content, since as far as we know, removal of the excess sialic acid results in correction of the delay in fibrin assembly and fibrinopeptide release. This suggests that all fetal fibrinogen characteristics are of post-translational origin. The evidence strongly implies that at full term there is a variable fetal and adult fibrinogen mixture.

ELECTRON MICROSCOPIC STUDIES OF FIBRINOGEN STRUCTURE: HISTORICAL PERSPECTIVES AND RECENT EXPERIMENTS *

Henry S. Slayter

Sidney Farber Cancer Institute
and
Department of Physiology and Biophysics
Harvard Medical School
Boston, Massachusetts 02115

Early attempts to visualize macromolecules in the electron microscope were fraught with difficulties of several types. Thirty years ago, instrumental resolving powers were still quite restricted; although 20 Å resolution was being approached, 50 Å was more typical during routine operation. The convenience of modern instruments was as yet undreamed of. Technologies for specimen preparation were inadequate. Although the shadow casting method had been introduced during the 1940s,[51, 52] making it possible to enhance the contrast of small particles, the technique remained rudimentary. In particular, irregularities of supporting films tended to be as large, and therefore as subject to contrast enhancement, as details of the specimens themselves. The negative contrast, or "negative staining" method, on the other hand, had yet to be proposed. Finally, many electronmicroscopists at that time were physicists, with little interest or training in the methods of preparative biochemistry.

In the face of these difficulties, prospects for visualizing particles in the molecular weight range from 30,000 to 500,000 d seemed problematic at best. Fibrinogen, however, was readily available at reasonably high purity and, as a subject of intense interest with respect to its role in the clotting process, was well characterized in physicochemical terms. It was to become one of the first subjects of macromolecular electron microscopy.

The first electron micrographs of fibrinogen were published in 1949 by Hall,[41, 42] who suggested that the structure of the molecule resembled a string of beads. Such was the level of background "noise," however, that neither the number of subunits nor the length of the molecule could be determined. Shortly afterwards, Porter and Hawn [43] suggested quite a different model: that of an oblate ellipsoid. Subsequently, Mitchel [64] found spherical particles in fibrinogen preparations, while Siegel, Mernan and Scheraga [45] returned to Hall's original string-of-beads suggestion.

In 1956 an important advance in the technology of specimen preparation appeared: the development, by Hall, of the so-called "mica method." Until that time, shadowing techniques had inevitably contrasted irregularities in the substrate film along with the surface contours of particles to be studied. Even the most carefully prepared plastic or carbon film substrates were rough enough

* This work was supported by Research Grants GM 14237 from the National Institute of General Medical Sciences and FR 05526 from the Division of Research Facilities and Resources, National Institutes of Health.

to obscure molecular detail, while potentially smoother substrates failed on the grounds of lacking electron transparency and/or mechanical stability. Hall, however, devised a new approach. He applied macromolecules to a surface which was smooth to molecular dimensions, coated with contrasting metal, and then stripped away the metallic layer, supporting it for viewing on a carbon film. Freshly cleaved surfaces of mica proved to be adequately smooth, readily available, and of a hydrophilic character that allowed relatively routine stripping of intact shadowing-metal coatings. Immediately, individual macromolecules that had previously been discerned with great difficulty presented themselves with delightful clarity.

At about the same time, Hall [54] also noted that tomato bushy stunt virus particles, which he was attempting to stain with various electron-dense compounds, sometimes were surrounded by pools of staining solution that did not penetrate the virus structure. Subsequently, this method of negatively contrasting stain-resistant particles was exploited by many workers, first for observation of bacteriophages and later for other viruses. Generally tightly packed and resistant to stain penetration, these particles provided ideal specimens for preparation by negative contrast. Gradually, however, that approach was applied to other, more loosely structured biological aggregates, and finally to individual macromolecules.

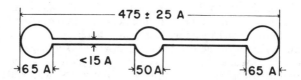

FIGURE 1. Trinodular model for fibrinogen proposed by Hall and Slayter,[13] based upon micrographs of unidirectionally shadowed platinum replicas and sedimentation equilibrium molecular weight measurements.

In 1959, on the basis of methodological improvements, Hall and Slayter [13] proposed the "triad" model of fibrinogen structure. We reported at that time that the macromolecular unit is 475 ± 25 Å long, consisting of "a linear array of three nodules held together by a very thin thread" which was estimated to have a diameter of 8–15 Å (FIGURE 1). Measurements of calibrated shadow lengths led to the assignment of 50 Å and 70 Å as the diameters of the medial (E) and distal (D) nodules, respectively. By adopting the relative dimensions from electron microscopy and the molecular volume from solution studies, and armed with the confidence of having determined by the semiequilibrium method, in duplicate, that the weight average of the molecular weight of the fibrinogen preparations from which micrographs were made was in fact 350,000, we calculated diameters for the E and D nodules. The values obtained, 50 Å and 65 Å, respectively, fitted remarkably well with the dimensions obtained by electron microscopy. At the same time, in an attempt to exclude the possibility that triads were simply aggregates of smaller units, a rigorous tabulation was made of *all* particle types observed in micrographs. Gratifyingly, the triad configuration was far more prevalent in preparations made from purified (bovine) fibrinogen than in those made from the relatively crude "Fraction I." In retrospect, however, it seems probable that the fibrinogen used at that time contained, or had been degraded to some extent by, contaminating proteins including plasmin. Nevertheless, when these results were published, there was no doubt in our minds that the triad model was indeed an accurate approximation to fibrinogen structure. It was a model on which further studies, whether in solution or by electron microscopy, could be based even if, as we acknowledged at the time, it was "almost certain that the nodules represented in outline contain incompletely resolved substructure."

The triad model accounts for the results of plasmin degradation studies, and for the presence of a disulfide "knot" which ties together, at the central nodule, all six amino terminal ends of the component polypeptide chains. The model is also consistent with preferred schemes for polymerization, and with much of the hydrodynamic data now available.[7, 8] The three pairs of polypeptide chains (Aα, Bβ, and γ) have now been sequenced,[53] revealing a primary structure that is consistent with a model of three globules joined by predominantly helical regions, the lengths of which are very close to those originally proposed by Hall and Slayter.[13]

The triad molecule was not supported by all subsequent electron microscopic studies. The variety of alternate shapes reported has ranged from sausages and spheres to discs and dodecahedra.[1, 2, 15, 20, 24, 28, 47, 48] All of these now appear to be incorrect, while other microscopists have repeatedly confirmed the details of the triad structure.[5, 6, 10, 12, 16, 33, 36] Among the first of these were Gorman et al. in 1971 [12] and Krakow et al. in 1972.[16] In 1979, also, Fowler and Erickson [10] compared metal-coated and negatively contrasted micrographs of fibrinogen, showing that both methods reveal the typical trinodular structure.

FIGURE 2 is a micrograph of bovine fibrinogen made in 1960,[50] using the original mica technique. Shadowing, with platinum, is unidirectional, at a mass thickness of 1.5×10^{-6} gm/cm.^{2} This type of image revealed the trinodular structure with its connecting links, and even suggested that the D region is elongated. Micrographs prepared by the same technique also revealed that bovine fibrinogen tends to collapse in the vicinity of the isoelectric point (pH 5.4), shortening considerably with respect to specimens prepared at pH values on either side. Due to some sort of structural collapse which occurs during drying, and which was attributed to loosening of the globular region, measured lengths at the isoelectric point were often less than 250 Å [14, 50] (FIGURE 3, from 1962). FIGURE 4, also from 1962, is a "spattered" preparation in which aggregated bovine fibrinogen has spread in strings upon impinging on the mica substrate at high velocity.[50, 55] Although individual trinodular particles are present, thin strings appear to unwind from the terminal regions, an effect which has also been reported recently by others.[35] The extended string-like projections may be related to α chain extensions.[9] We also found fibrin monomer to be virtually identical, morphologically, to fibrinogen.[14, 50]

At that time, however, it became clear that a major limitation to further detailing of the structure of fibrinogen (or of other macromolecules) was again background noise, now derived from the coarseness of the shadowing metal grain itself. Our first attempts to correct this situation, dating back to 1971, utilized lighter metal coatings and substituted tungsten for platinum. Micrographs prepared in this way, still by unidirectional shadowing, revealed a rather more elongated D region and a clearly defined connecting link.[26, 27, 49, 55] Subsequently, it was found that rotary coating can provide more information concerning macromolecular structure than does unidirectional shadowing, since no one side of any particle need be obscured by accumulating metal. The method was successfully applied to both myosin and fibrinogen.[29] More sophisticated metal coating procedures have also resorted to the use of very thin replicas

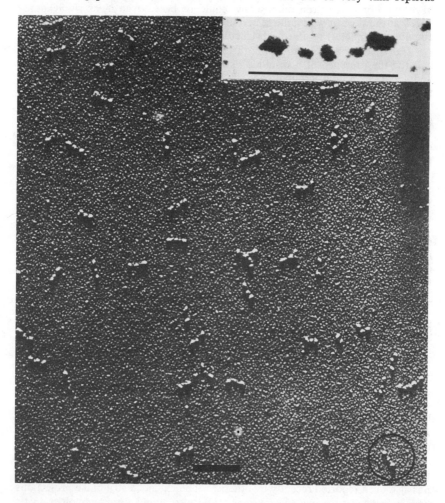

FIGURE 2. Bovine fibrinogen molecules unidirectionally shadowed with platinum. As would be expected, lightly unidirectionally shadowed tungsten replicas appear more filamentous than their platinum counterparts. This micrograph was taken in 1960. Unidirectionally shadowed particles lend themselves to height measurements through a shadow-to-height ratio obtained from an internal standard, e.g., polystyrene latex spheres. Rotary-shadowed particles do not, and observed dimensions can only be corrected empirically. Measured length is not significantly affected by the replica correction, but the effect on measured widths is critical and corrections must be applied to obtain meaningful small dimensions.[27] Scale bar 0.1 μm, (Inset, scale bar 0.05 μm). (From Slayter.[50])

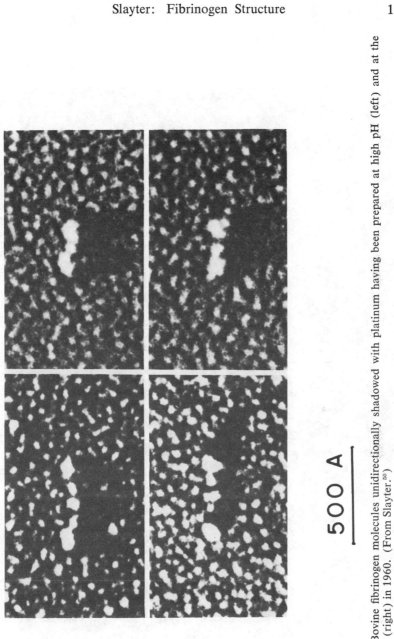

FIGURE 3. Bovine fibrinogen molecules unidirectionally shadowed with platinum having been prepared at high pH (left) and at the isoelectric point (right) in 1960. (From Slayter.[50])

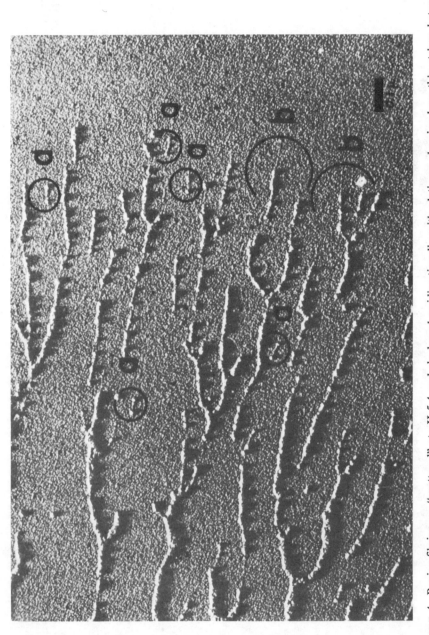

FIGURE 4. Bovine fibrinogen "spattered" at pH 5.4 and shadowed unidirectionally with platinum showing long thin strings between molecular units and also extending from terminal nodules (b), from 1960. (From Slayter.[50])

(mass thickness 4×10^{-7} gm/cm^2), refractory metals, low substrate temperatures, and inclusion of 50% glycerol in the macromolecular solution being prepared.[27, 49] These techniques for preparation of marginally contrasted replicas, together with the electron optics now available which permit high resolution imaging of low contrast specimens, have yielded micrographs of substantially improved quality (FIGURE 5).

When purified human fibrinogen is prepared for electron microscopy by fine-grain metal coating methods, not only are virtually all particles triads, but the D regions now are very frequently resolved into subdomains. (FIGURE 5) These subdomains appear whether or not Ca^{2+} is present in the solution from which the preparation is made, as suggested by Williams.[36] FIGURE 6 shows groups of selected particles from such preparations. Two subdomains are more frequently found only on one end of a given molecule than on both. Whether this is a statistical phenomenon or simply due to some twist in the structure that favors visualization of only one end at a time is not clear. The molecules are generally straight, although most are not perfectly so. The variety of slight bends in the long axis of the particles is great, but among the relatively extended particles no particular form is outstanding. Altogether, the images suggest a significant molecular flexibility. Occasionally, sigmoid forms such as those reported elsewhere [36] have been found—but one suspects not, given the range of other configurations seen in this structure, with a greater frequency than might be expected at random.

Incidental to an immune electron microscopic determination of the loci of the D- and E-fragments,[22] we reported that in negatively contrasted preparations, D-regions appeared to be subdivided into at least two domains. These were essentially proximal and distal to the E-region, forming an asymmetric array in which the mini-axis of the subdomains was often canted a little off the major axis of the triad. Williams,[36] working with bovine fraction I, compared rotary tungsten-shadowed and negatively contrasted molecules, again confirming the connecting link. He also noticed slightly canted and elongated D-regions, suggesting that most of the long axes of the molecules seemed to have a gently sigmoidal configuration of like handedness. Upon treatment with Ca^{2+}, the E-region of negatively contrasted preparations became more prominent, and the D-regions appeared to be divided.

An obvious difference between negatively stained and shadow-cast preparations is that the latter have usually been made from solutions dried at neutral pH, whereas negative contrast media, such as uranyl formate or acetate at a concentration of 1%, are generally reputed to be used between pH 3.5 and 4.5. This pH would be expected to produce a net positive charge on the specimen molecules and, thus, quite a different charge profile than occurs at physiological pH. Possibly, therefore, conformational changes may be induced. In our current work with negatively contrasted fibrinogen, the positions of subdomains of the D-regions appear to be distributed somewhat more laterally than longitudinally; that is, canted with respect to the long axis of the molecule. Similar effects have been noted in micrographs of negatively contrasted individual fibrinogen molecules published by several groups.[22, 34, 36] Our work and that of others has sometimes also suggested extreme variability of details of the molecular profile, albeit with several salient features (FIGURE 7). The triad structure is evident, with larger D-regions and smaller E-regions, interconnecting links, and frequent evidence for split D-regions, usually with subdomains lying near the long axis, but often with the distal subdomain lying

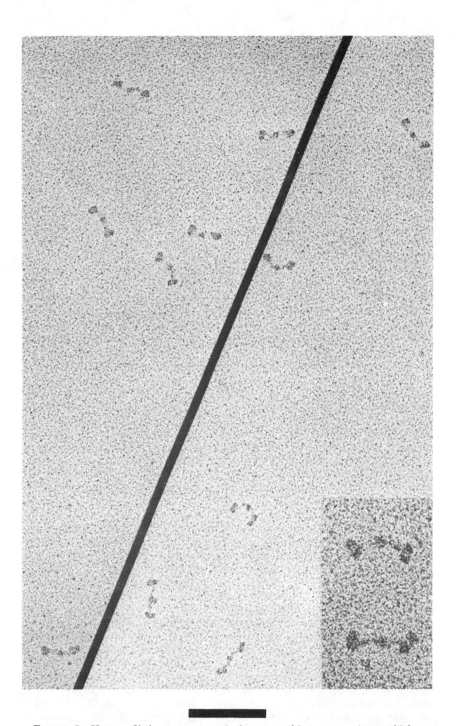

FIGURE 5. Human fibrinogen rotary shadow-cast with tungsten (mass thickness $= 4 \times 10^{-7}$ gm/cm^2) prepared by the mica replica method from a solution of 50 μgm/ml fibrinogen in 0.05 M NH$_4$OAC, containing 50% redistilled glycerol, stabilized with carbon and the coating thickness measured by means of a piezoelectric thickness monitor as described previously.[27, 67] Scale is 0.1 μm. (Inset width is 74 nm.) Human fibrinogen was prepared as in reference 22.

also slightly lateral to the medial subdomain (FIGURE 8, top). It is not difficult to reconcile these profiles obtained from negative contrast with that obtained by light metal coating (FIGURES 5, 6, 8. What seems clear, however, is that the probability of being able to observe each and every molecule in a field is far

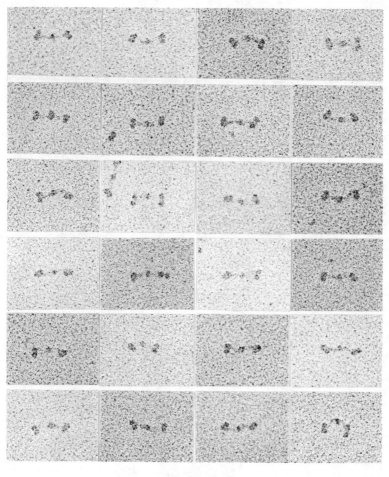

FIGURE 6. Human fibrinogen rotary shadow-cast with tungsten, prepared as in FIGURE 5. Individual particles show various small differences in the disposition of the subdomains of the D-region, as well as the limited apparent flexibility in the course of the axial spine. Clearly, length measurements would be affected by both of these parameters. Scale bar is 0.1 μm. α-chain extensions can occasionally be seen at the ends of the molecules.

less in the case of negative contrast, apparently because of the spurious way in which fibrinogen dries down in uranyl salts near or below its isoelectric point.

In our experience, the profile of fibrinogen determined by the negative contrast method ranges from a very thin string to a structure somewhat more expanded than that produced by metal coating. Overall, the range could be said to vary from stringy to fat. At the stringy end of the spectrum, apparent

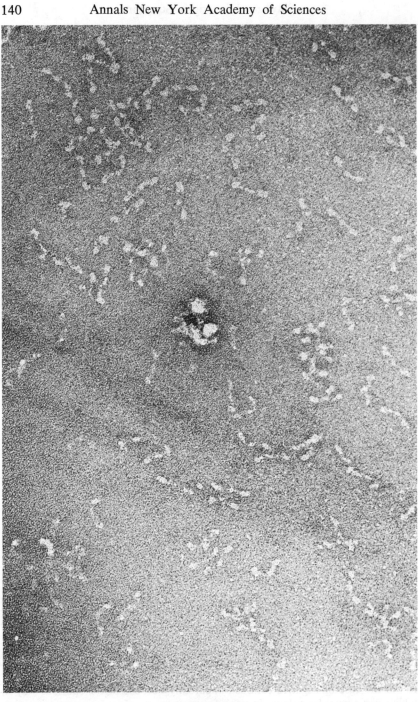

FIGURE 7. Human fibrinogen negatively contrasted with uranyl acetate. Scale bar is 0.1 μm. Prepared as in reference 22.

molecular volume is too small to represent all of the mass in one molecular unit. Thus, it seems likely that in the case of thinner images, and perhaps in general, a substantial portion of the molecule is so loosely packed that negative stain is not effectively excluded. Consequently, some individual polypeptide segments would be essentially uncontrasted and thus invisible.

Estes and Haschemeyer [33] have suggested that the variability of observed fibrinogen profiles in negative stained preparations may be related to interactions of the protein with the specimen substrate surface. Certainly, most microscopists will agree that the use of substrates freshly cleaned by glow discharge facilitates production of what we judge to be good specimens by various modifications of the diffusion method. It is possible that the very condition which facilitates sticking of protein also produces some form of surface denaturation? Such a denaturation might account for highly variables results, not only from molecule to molecule, but also from specimen to specimen and investigator to investigator.

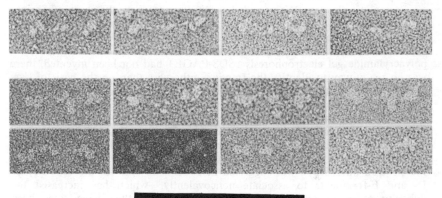

FIGURE 8. Selected human fibrinogen molecules negatively contrasted with uranyl acetate, top row; rows two and three are complexes of human fibrinogen with anti-E F_{ab}. Scale bar is 0.1 μm.

These considerations suggest, in brief, that any conclusions concerning the structure of fibrinogen (or any other macromolecule) should be based on information obtained by as many methods as possible. Recently, for example, an investigation of uncontrasted fibrinogen by scanning transmission electron microscopy (STEM) methods confirmed the trinodular structure, but negative contrast was required to deduce further details [34] similar to those reported by application of other methods.[22, 36]

The assignment of precise dimensions to the substructure of the E- and D-domains of the triad, from electron microscope measurements of single macromolecules, is probably not as relevant as it was 23 years ago because we now know almost everything about the detailed chemical structure of the subregions of the molecule. Nevertheless, to summarize present understanding: the D-region of the structure, as compared with the E-region, appears somewhat more variable in overall size in the range up to 55 Å. Yet, possibly due to variable diffuseness of this structure, the D-region may also appear more compact. Each subdomain of the region appears roughly spheroidal and, when presented at its most asymmetric aspect, measures about 45 $\times$ 90 Å. When least asymmetric, the D-region itself measures 65 Å. The width of the connecting link is about 10 Å.

Some additional information concerning the structure of fibrinogen has been derived from crystallographic approaches. Tooney and Cohen [37] demonstrated that the packing unit in crystalline states of modified fibrinogen is a nonpolar unit which is 450 Å long and therefore consistent with the triad model. Weisel et al.[30, 38, 39] produced electron micrographs of negatively contrasted microcrystals of fibrinogen modified by proteolytic degradation. These images were analyzed by optical and computer image-processing methods, with results that led the authors to conclude that D-regions are subdivided into one very small and two larger domains. Measured dimensions of the domains, they suggested, must be corrected upward to account for effects of dehydration of the microcrystals. However, the extent to which proteolytic removal of a significant proportion of molecular mass modifies the tertiary structure remains unclear. It may well be that no further definition of fibrinogen structures will be achieved until x-ray crystallographic study of native crystals can provide a three-dimensional plot of the structure at high resolution.

Proteolytic cleavage has provided another approach to the study of fibrinogen structure. In 1962, we fragmented purified bovine fibrinogen, isolated fractions by ion exchange chromatography, and characterized these fractions by means of electron microscopy.[50, 56] Since at that time, sodium dodecylsulfate polyacrylamide gel electrophoresis (SDS-PAGE) had not been invented, there was no ready way to identify the fractions otherwise than on the basis of size. In fact, however, dimensions were consistent with those of the D- and E-regions.

In 1969, Marder et al.[18] proposed a scheme, which has been widely accepted, for the proteolysis of fibrinogen to form core fragments "D" and "E" via high molecular weight intermediates "X" and "Y." Their results indicated that two D-fragments and one E-fragment are generated from each fibrinogen molecule; correspondence of the fragments with the three globular domains observed in the electron microscope was suggested in 1976.[25] In spite of the tendency of D- and E-fragments to associate noncovalently,[3] which has increased the difficulty of completely separating these fragments, they have since been demonstrated to be immunologically distinct [23] and to differ in molecular weight and other physicochemical characteristics.[11] Recently, Fowler et al.[32] have characterized Y-, D- and E-fragments by means of metal coating techniques. On the basis of proteolytic fragment size, the outer nodules of the triad were identified as the fragment-D-containing regions.

Preparation of specific antibodies affords an additional means of testing, by means of immune-labelling electron microscopy, the hypothesis that the D- and E-fragments represent the cleaved outer and central fibrinogen nodules, respectively.[17] Thus, F_{ab} fragments of immunoglobulin directed against either D- or E- fragment were complexed with native human fibrinogen and the complexes isolated by gel filtration. Electron micrographs of metal replicas and of negatively contrasted specimens of the complexes (FIGURE 9) were then scored for binding of F_{ab} to specific nodules.[22] Preparations of fibrinogen complexed with anti-D F_{ab} were found to include 91% of end-nodular complexes, and those with anti-E F_{ab} 82% of middle-nodular complexes.[24] The results of this analysis confirm the contention that the D- and E-fragments have their origin in the ends and center, respectively, of the fibrinogen macromolecule. Similar conclusions have also been deduced by other workers.[31]

The triad model of fibrinogen was proposed as a somewhat more detailed approximation to fibrinogen structure than a rod or ellipsoid. It was hoped that this might assist in the design of further experiments, and also permit more precise calculations from solution measurements. In this purpose, the model has apparently succeeded. Nevertheless, it would seem that, in terms of flexibility, the triad has been interpreted rather too rigidly.

Our early data, which led us to believe that fibrinogen could bend or even collapse, formed the basis for our suggestion of a possible alternate mode of polymerization by means of folding. A recent report by Plow and Edgington [58] suggests that flexed and linear forms of fibrinogen may coexist at equilibrium. Micrographs suggests, at least, that bending or rotation is possible, with the

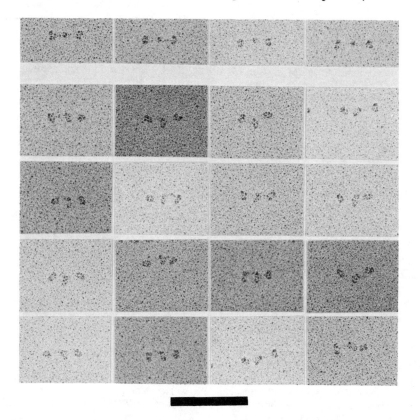

FIGURE 9. Selected human fibrinogen molecules complexed with anti-E F_{ab} rotary shadow cast with a very thin coat of tungsten as described in the caption to FIGURE 5. Scale bar is 0.1 μm.

lowest energy state depending on the environment. Furthermore, consistent evidence that a complement of free polypeptide exists in association with fibrinogen might now be interpreted in terms of α chain extensions. Finally, although the D- and E-fragments seem firmly associated with terminal and medial nodules, respectively, recent work has revealed subdomains within the D-region. The disposition of the latter seems also to be flexible. Future experiments will no doubt be directed toward understanding of the role of flexible fibrinogen substructure in critical intramolecular interactions.

ACKNOWLEDGMENTS

The assistance of Helen Brinkerhoff, Donald Gantz, and Christine Harnett, and the cooperation of Dr. Elizabeth Slayter in preparation of the manuscript, is gratefully acknowledged.

REFERENCES

1. BACHMAN, L., W. W. SCHMITT-FUMIAN, R. HAMMEL & K. LEDERER. 1975. Die Makromol. Chem. **176:** 2603–2618.
2. BLAKELY, P. R., M. J. GROOM & R. L. TURNER. 1977. Br. J. Haematol. **35:** 437–440.
3. BUDZYNSKI, A. Z., M. STAHL, M. KOPEC, Z. S. LATATTO, Z. WEGRZYNOWICZ & E. KOWALSKI. 1967. Biochim. Biophys. Acta **147:** 313–323.
4. COHEN, C., H. S. SLAYTER, L. G. GOLDSTEIN, J. KUCERA & C. E. HALL. 1966. J. Mol. Biol. **22:** 385–388.
5. CONIO, G., G. DONDERO, C. TROGLIA, V. TREFLETTI & E. PATRONE. 1975. Biopolymers **14:** 2363–2372.
6. DONOVAN, J. W. & E. MIHALYI. 1974. Proc. Natl. Acad. Sci. USA **71:** 4125–4128.
7. DOOLITTLE, R. F. 1973. *In* Advances in Protein Chemistry. C. B. Anfinsen, Ed. **27:** 1–109. Academic Press. New York.
8. DOOLITTLE, R. F. 1975. *In* The Plasma Proteins. F. W. Putnam, Ed. **2:** 109–161. Academic Press. New York.
9. DOOLITTLE, R. F., D. M. GOLDBAUM & L. R. DOOLITTLE. 1978. J. Mol. Biol. **120:** 311–325.
10. FOWLER, W. E. & H. P. ERICKSON. 1979. J. Mol. Biol. **134:** 241–249.
11. GAFFNEY, P. J. 1977. *In* Hemostasis: Biochemistry, Physiology and Pathology. D. Ogsten and B. Bennett, Eds.: 105–169. John Wiley. New York.
12. GORMAN, R. R., G. E. STONER & A. CATLIN. 1971. J. Phys. Chem. **75:** 2103–2107.
13. HALL, C. E. & H. S. SLAYTER. 1959b. J. Biophys. Biochem. Cytol. **5:** 11–17.
14. HALL, C. E. & H. S. SLAYTER. 1962. *In* Fifth Int. Cong. for Electron Microscopy. S. S. Breese, Ed. **1:** 31–32. Academic Press. New York.
15. KOPPEL, G. 1966. Nature (London) **212:** 1608–1609.
16. KRAKOW, W., G. F. ENDRES, B. M. SIEGEL & H. A. SHERAGA. 1972. J. Mol. Biol. **71:** 95–103.
17. LAKE, J. A. 1979. *In* Methods in Enzymology. C. H. Hirs and S. N. Timasheff, Eds. **61:** 250–257. Academic Press. New York.
18. MARDER, V. J., N. R. SHULMAN, W. R. CARROLL. 1969. J. Biol. Chem. **244:** 2111–2119.
19. MILLS, D. A. & D. C. TRIANTOPHYLLOPOULOS. 1969. Arch. Biochem. Biophys. **135:** 28–35.
20. MOSESSON, M. W. 1976. Thromb. Res. **8:** 737–744.
21. MOSESSON, M. W., N. ALKJAERSIG, B. SWEET & S. SHERRY. 1967. Biochemistry **6:** 3279–3287.
22. NORTON, P. & H. S. SLAYTER. 1981. Proc. Natl. Acad. Sci. USA **78**(3): 1661–1665.
23. NUSSENZWEIG, V., M. SELIGMANN & P. GRABAR. 1961. Ann. Inst. Past. **100:** 490–508.
24. POTUIT, L., G. MARCILLE, H. SUSCILLON & D. HOLLARD. 1972. Thromb. Diath. Haemorrhag. **27:** 559–572.
25. SHRAGER, R. I., E. MIHALYI & D. W. TOWNE. 1976. Biochemistry **15:** 5382–5386.

26. SLAYTER, H. S. 1971. *In* Proceedings, Electron Microscopy Society of America, 29th Annual Meeting. C. J. Arcenaux, Ed. 424–425. Claitor's. Baton Rouge, LA.
27. SLAYTER, H. S. 1976. Ultramicroscopy **1:** 341–357.
28. STEWART, G. 1971. Scand. J. Hematol. **13:** 63–67.
29. TOONEY, N. M. & C. COHEN. 1977. J. Mol. Biol. **110:** 363–385.
30. WEISEL, J. W., G. N. PHILLIPS & C. COHEN. 1981. Nature (London) **289:** 263–267.
31. PRICE, T. M., D. D. STRONG, M. L. RUDEE & R. F. DOOLITTLE. 1981. Proc. Natl. Acad. Sci. USA **78:** 200–204.
32. FOWLER, W. E., L. J. FRETTO, H. P. ERICKSON & P. A. MCKEE. 1980. J. Clin. Invest. **66:** 50–56.
33. ESTIS, L. F. & R. M. HASCHEMEYER. 1980. Proc. Natl. Acad. Sci. USA **77:** 3139–3143.
34. MOSESSON, M. W., J. HAINFELD, J. WALL & R. M. HASCHEMEYER. 1981. J. Mol. Biol. **153:** 695–718.
35. RUDEE, M. L. & T. M. PRICE. 1981. Ultramicroscopy **7:** 193–196.
36. WILLIAMS, R. C. 1981. J. Mol. Biol. **150:** 399–408.
37. TOONEY, N. M. & C. COHEN. 1972. Nature (London) **237:** 23–25.
38. WEISEL, J. W., S. G. WARREN & C. COHEN. 1978. J. Mol. Biol. **126:** 159–183.
39. WEISEL, J. W., N. M. TOONEY, I. KAPLAN, D. AMRANI & C. COHEN. 1980. J. Mol. Biol. **143:** 329–334.
40. SCHERAGA, H. A. & M. LASKOWSKI, JR. 1957. Adv. Prot. Chem. **12:** 1–131.
41. HALL, C. E. 1949. J. Am. Chem. Soc. **71:** 1138–1139.
42. HALL, C. E. 1949. J. Biol. Chem. **179:** 857–864.
43. PORTER, K. F. & C. V. Z. HAWN. 1949. J. Exp. Med. **90:** 225–232.
44. MITCHEL, R. F. 1952. Biochim. Biophys. Acta **9:** 430–442.
45. SIEGEL, B. M., J. P. MERNAN & H. A. SCHERAGA. 1953. Biochim. Biophys. Acta **11:** 329–336.
46. HALL, C. E. 1956. Proc. Natl. Acad. Sci. USA **42:** 801–806.
47. MOSESSON, M. W., J. ESCAIG & G. FELDMAN. 1979. Thromb. Hemostasis. **42:** 88.
48. KOPPEL, G..1967. Z. Zellforsch. Microsk. Anat. **77:** 443–517.
49. SLAYTER, H. S. 1978. *In* Principles and Techniques of Electron Microscopy: Biological Applications. M. A. Hayat, Ed. **9:** 175–245.
50. SLAYTER, H. S. 1962. Ph.D. Thesis, Mass. Inst. of Tech. 1–267.
51. WILLIAMS, R. & R. W. G. WYKOFF. 1944. J. Appl. Phys. **15:** 712–716.
52. MULLER, H. O. 1942. Kolloid-Z. **99:** 6–28.
53. DOOLITTLE, R. F. 1981. Scientific American **245**(6)**:** 126–135.
54. HALL, C. E. 1955. J. Biophys. Biochem. Cytol. **1:** 1–12.
55. SLAYTER, H. S. 1981. *In* Enzymes and Soluble Proteins. J. R. Harris, Ed. **1:** 197–254. Academic Press. London, England.
56. SLAYTER, H. S. & C. E. HALL. 1964. J. Mol. Biol. **8:** 593–601.
57. SLAYTER, H. S. 1980. *In* SEM Inc. O. Johari, Ed. **1:** 171–182.
58. PLOW, E. F. & T. S. EDGINGTON. 1982. Seminars in Thrombosis and Hemostasis **8:** 36–56.

ELECTRON MICROSCOPY OF FIBRINOGEN, ITS PLASMIC FRAGMENTS AND SMALL POLYMERS *

Harold P. Erickson and Walter E. Fowler †

Department of Anatomy
Duke University Medical Center
Durham, North Carolina 27710

THE TRINODULAR STRUCTURE VISUALIZED BY SHADOWING AND NEGATIVE STAIN

When we began our project on the electron microscopy of fibrinogen four years ago, the most widely accepted structural model was the trinodular rod proposed by Hall and Slayter in 1959.[1] They imaged individual fibrinogen molecules that had been sprayed on mica in a volatile buffer, dried in vacuum, and shadowed with heavy metal. There was some controversy over these results because other labs had difficulty reproducing the specimens. Nevertheless, most of the data from biochemistry and physical chemistry fit the trinodular model quite nicely. A more serious controversy arose from laboratories attempting to visualize the molecules by negative staining, which is generally recognized as a higher resolution technique than shadowing. Large globular particles were the predominant species found in these specimens.[2,3] Although strikingly different in shape from the trinodular rod, models based on these images of spherical particles could be made to accommodate most of the hydrodynamic and solution scattering data. We wanted to find which structure was correct and to demonstrate the same structure by these different specimen preparation techniques.

Our first goal was to obtain improved shadowed specimens that showed large fields of clean, unaggregated molecules. We achieved this goal with a crucial innovation in the shadowing technique: inclusion of glycerol, at a concentration of 20% to 50%, in the buffer in which the protein was sprayed onto the mica.[4] The glycerol seems to affect the spreading of the drops on the mica and the deposition of the protein molecules as the drop recedes during vacuum drying. Glycerol is not always required, in particular for fibrinogen, which has yielded good specimens from a wide range of solution conditions. In our experience, however, the use of glycerol results in a marked increase in both the frequency and the quality of good specimen areas. In the absence of glycerol the protein is often deposited in aggregates or the molecular structure is distorted by drying; in the presence of glycerol most of the droplets have clean areas of unaggregated molecules. Typical fields of unidirectional- and rotary-shadowed specimens of fibrinogen dried from glycerol are shown in FIGURES 1A and 1B.

It may be useful to recount briefly the history of the glycerol innovation. When we were beginning our project we were particularly impressed by images of shadowed myosin presented by Elliott and Offer.[5] Their specimen preparation involved freeze-drying a sample that had been sprayed on mica;

* This work was supported by NIH Research Grant HL-23454.

† Current address: Department of Cell Biology and Anatomy, Johns Hopkins Medical School, Baltimore, Maryland 21205.

0077–8923/83/0408–0146 $01.75/0 © 1983, NYAS

as a precaution against formation of ice crystals they added glycerol to the solution before spraying. We decided to apply their freeze-drying technique to fibrinogen, but we were also influenced by an unpublished protocol of J. Pullman, later described in his Ph.D. thesis (University of Chicago, 1979). A copy of this protocol, dated January, 1976, which had made its way to our lab, suggested that "glycerol, at a concentration between 10–30%, appears to aid in spreading and in preserving structure at the 20 Å level."

While we were setting up the freeze-drying procedure we decided to try preparing specimens at room temperature, keeping all other conditions, including the glycerol, the same. This turned out to be a much easier technique than the freeze drying and we immediately obtained specimens like that shown in FIGURE 1A. The technique is described in our paper [4] and also by Shotton et al. and Tyler and Branton,[6, 7] who developed essentially identical procedures for the study of spectrin and other proteins. Subsequent experiments in our lab and in others have shown that the important component of the technique of Elliott and Offer was the presence of glycerol. For imaging single protein molecules virtually identical results are obtained with either room temperature drying or freeze drying. We have recently found, however, that for large protein aggregates freeze drying may preserve a crystalline lattice that is not seen in specimens dried at room temperature (W. Fowler, unpublished results).

Over the past few years we have continued to experiment with specimen preparation conditions, including salt and glycerol concentrations, the angle of shadowing and the amount of metal deposited. Optimal conditions for one protein are not necessarily useful for another, and we have found it necessary to explore a wide range of conditions in each new study. With fibrinogen, we have found that increasing the amount of metal has given higher contrast specimens (FIGURE 1B), which show certain high resolution features more clearly than in our original study. The splitting of the outer nodules and the extra domain near the central nodule will be discussed in more detail in the next section.

The trinodular structure was so clear and reproducible in our shadowed specimens that we were convinced it was correct. Turning our attention to negative stain, we found that the standard carbon films were very hydrophobic and the stain would not spread unless the fibrinogen solution was applied at fairly high protein concentrations, on the order of 1 mg/ml. We were concerned that at this protein concentration the carbon film might be covered by a dense mat of protein, and that any structures seen would be random aggregates or holes rather than individual molecules. In order to get the negative stain to spread and stick to the carbon film at low protein concentrations it was necessary to use a surface of the carbon film that was hydrophilic. We originally used the flotation technique as described and referenced in our paper [4] to prepare specimens at very dilute protein concentrations. More recently we have used "glow discharge," in which the surface of the carbon film is cleaned and rendered hydrophilic by exposure to ion bombardment in a partial vacuum. We now prefer this technique because it gives larger areas of well-spread stain. We have found no difference in the structures seen by the two techniques.

With the hydrophilic carbon films we were able to prepare specimens at concentrations of 1 to 5 micrograms per ml, 1000 times less than those used in previous studies.[2, 3] These specimens showed fields of well-separated, rod-like

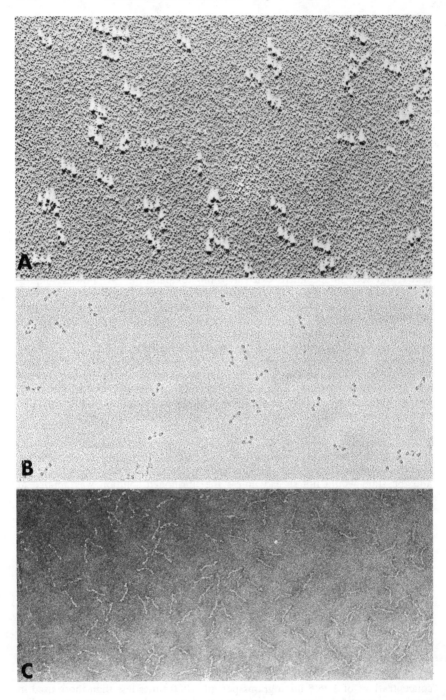

FIGURE 1. Three different techniques for visualizing the trinodular structure of fibrinogen. **A)** Human fibrinogen (Kabi) in 50 mM ammonium formate, 33% glycerol sprayed on mica, dried in vacuum and shadowed unidirectionally with Pt-C at an angle of ~ 10 degrees. (From Fowler and Erickson.[4] By permission of *Journal of Molecular Biology,* copyright: Academic Press, London). **B)** Human fibrinogen (Kabi) sprayed on mica from a buffer containing 0.2 M ammonium formate, 30% glycerol, dried in vacuum and rotary shadowed with Pt-C at an angle of 5–7 degrees. Several molecules in this recent specimen show the higher resolution features discussed later. **C)** Bovine fibrinogen (Sigma) applied to hydrophilic carbon films (glow discharged) and negatively stained with uranyl acetate. Human fibrinogen prepared by the same technique was identical in appearance. Magnification × 100,000.

molecules with a trinodular structure very similar to that seen in the shadowed specimens (FIGURE 1c). It may be noted that the negatively stained molecules appear much thinner and the nodules are less prominent than in the rotary-shadowed specimens. This is largely due to the ~2 nm shell of metal in the rotary-shadowed specimens, which exaggerates the thickness and the size of the globular domains.

In our earlier images of negatively stained fibrinogen we noted that the outer nodules were elongated. We frequently saw substructure within the outer nodules, but the images were quite variable. At that time we were most interested in the fact that we could demonstrate the same trinodular shape seen in the shadowed specimens. Subsequently Norton and Slayter [8] called attention to a division of the outer nodule into two domains. Williams has presented some of the most detailed images of negatively stained fibrinogen,[9] which show the elongation and the splitting of the end nodules. In addition, he noted that many of the molecules had a characteristic sigmoid or S shape. Remarkably, all of the S-shaped molecules in his preparations had the same handedness, implying that they were all attached to the carbon film by the same molecular face.

A selection of images from our recent study of negatively stained fibrinogen is shown in FIGURE 2. The splitting of the outer nodules and the S shape are clearly seen in some of the images and we believe these are real structural features of the molecule as viewed from certain directions. We still find a much

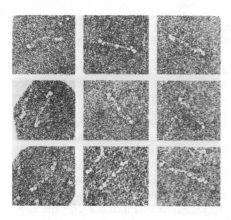

FIGURE 2. Selected images of negatively stained bovine fibrinogen enlarged and contrast enhanced by a computer image processing system, × 200,000. The image enhancement was done in the laboratory of Dr. U. Aebi at Johns Hopkins Medical School using programs for scanning and computer processing written by Mr. E. L. Buhle and by Dr. P. R. Smith.

greater variety of structures than is reported by Williams (reference 9, and this symposium), perhaps because the molecules in our preparations are more randomly oriented on the carbon film.

ELECTRON MICROSCOPY OF FRAGMENTS X, Y, D, AND E

Probably the strongest biochemical evidence supporting the trinodular model is the sequence of fragments produced by plasmin digestion. The interpretation was presented many years ago (reference 10, and McKee *et al.,* this symposium) that fragments D and E corresponded to the outer and middle nodules, respectively, Y was binodular, comprising one D and one E, and X was a trinodular form missing the carboxyl-terminal segment of the α chain. In collaboration with Larry Fretto and Patrick McKee, we prepared specimens of the purified fragments and confirmed that they did correspond to the anticipated nodular forms.[11] More recent micrographs of the fragments, rotary shadowed to give a higher contrast than our original images, are presented by McKee *et al.* in this symposium. The imaging of these purified fragments is probably the most direct mapping of the peptide sequences into the trinodular structure. A number of other laboratories have demonstrated the same mapping by electron microscopy of antibody-labeled fibrinogen.[8, 12, 13] For convenience in the remainder of this paper, we will refer to the outer and central nodules as the D-nodule and E-nodule.

In our original investigation of the fragments we noted that Fragment X was trinodular and could not be distinguished from native fibrinogen. The carboxyl-terminal segment of the α chain that is missing in Fragment X is really quite large, 20,000 to 40,000 daltons depending on the extent of digestion. We were concerned, therefore, that we had not been able to visualize any feature of the intact molecule that was missing on Fragment X. We concluded that this segment might be hidden if it were closely attached to the larger domains, or if, as it is depicted in many models, it were extended in solution as a strand that was too thin to see in the shadowed specimens. We now believe that the α chain tails are actually visible as a separate domain or nodule near the central nodule. This domain was occasionally seen in our earlier images but we have recently been able to visualize it more clearly.

IMPROVED MICROGRAPHS OF INTACT FIBRINOGEN AND FRAGMENT X SHOW NEW DETAILS OF MOLECULAR STRUCTURE: THE α CHAIN TAIL

In the past year we have obtained images of rotary-shadowed molecules that show structural details beyond the trinodular shape. These details are seen most clearly in more heavily shadowed specimens, especially in certain clean, high-contrast areas of the specimen. One feature, already observed in negative stain but not previously resolved in shadowed specimens, is a cleft that appears to split the outer nodules into two domains. This splitting is seen on a fraction of the molecules, probably corresponding to a certain orientation, and is observed in both intact fibrinogen (FIGURE 3) and Fragment X (FIGURE 4). In addition, certain views of the molecules show an S-shaped conformation very similar to that reported by Williams[9] in negative stain (see also our FIGURE 2). This view is clearly presented in FIGURE 3 (middle row, molecule on left side) and FIGURE 4 (middle row, molecule on right side; bottom row, two molecules on left side). Since these images have apparent two-fold symmetry, we suggest that this may be a view of the molecule down the molecular two-fold axis.

One of the most striking new features seen on intact fibrinogen is an extra nodule near the central nodule. A selection of images showing these

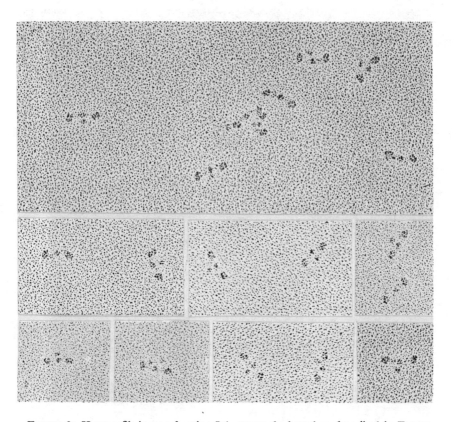

FIGURE 3. Human fibrinogen fraction I-4, rotary shadowed as described in FIGURE 1B, × 200,000. One large field and several isolated examples were selected to show the α chain nodule, centrally positioned about 10 nm away from the central nodule. Several molecules show a splitting of the outer nodules, and the left-most molecule in the middle row has a symmetrically curved S shape but no extra domain near the central nodule.

extra central nodules is given in FIGURE 3. Characteristically, the extra nodule is somewhat smaller than the E-nodule (5 nm in diameter) and lies about 10 nm away on one side of the molecular axis. Usually the extra nodule projects away from the E-nodule perpendicular to the axis, and is thus equidistant from the two outer nodules. Sometimes it is closer to one of the outer nodules, and occasionally it lies some distance from the molecular axis, apparently attached to one outer nodule by a (rarely visible) thin filament (FIG-URE 1B, center).

This extra nodule is seen most clearly in our recent, heavily shadowed specimens, although once one is aware of it it can be found in most images (e.g., FIGURE 2A of Fowler and Erickson, reference 4). Most of the speci-

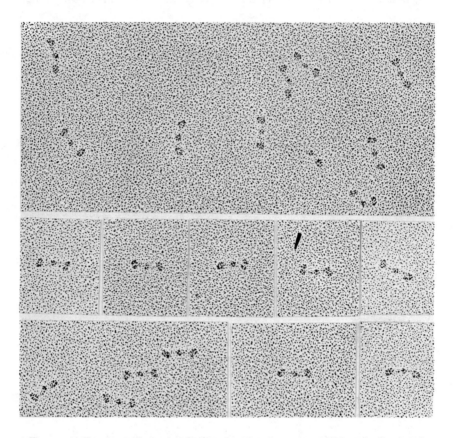

FIGURE 4. Fragment X, rotary shadowed under the same conditions as the fraction I-4 shown in FIGURES 1B and 3, ×200,000. None of the molecules show the extra (α chain) nodule found in intact fibrinogen, but the other high resolution details seen in FIGURE 3, the splitting of the outer nodules and the S shaped molecules, are also seen here. The Fragment X, with extensively digested α chains, and the I-4 fibrinogen, were prepared by Dr. Nadia Carrell.

mens used in our recent study were of fraction I-4 human fibrinogen, but commercial preparations of human and bovine fibrinogen, from Kabi and Sigma, also showed the extra nodule. In a good specimen area it is seen on about 20% to 40% of the molecules. This probably corresponds to the fraction of molecules in which the extra nodule projects to one side, rather than lying on top of the E-nodule.

The most important observation for the identification of this nodule is that it is always missing from Fragment X. Recent images of Fragment X, from specimens prepared under identical conditions to the I-4 fibrinogen, are shown in FIGURE 4. This preparation was a late Fragment X, missing a 40,000 dalton segment from each α chain. We conclude that the carboxyl-terminal segments of the α chains constitute the extra nodule, and will call this the α chain nodule.

Mosesson et al. (reference 14, and J. Wall in this symposium) have recently proposed that the carboxyl-terminal segments of the α chains are located at or near the central nodule. Their conclusion was based on measurements of the mass of the individual nodules by scanning transmission electron microscopy. They found that I-9 fibrinogen (an early Fragment X) was lighter than I-4 fibrinogen by an amount equivalent to the missing segments of the alpha chains, and that the mass was lost entirely from the region of the central nodule. Our findings are in complete agreement with their proposal, but our micrographs show the structural arrangement to be rather different from what they had imagined.

The α chain nodule is most often visualized lying a short distance from the E-nodule on one side of the molecule. If the molecule possess a twofold axis, this axis must be in the plane of such an image, passing through both the E-nodule and the α chain nodule. It is important to note that the views showing the S-shaped molecules (presumably views down the twofold axis) do not show the extra nodule. If that nodule does lie on the twofold axis it would be superimposed on the central nodule in this view, and therefore would not be seen separately. Thus the S-shaped molecules and those showing the extra nodule may be perpendicular views of the molecule.

Rudee and Price [15] have recently suggested that the α chain tail extends from the outer domain as a thin strand, which they claim to have resolved using a new type of specimen support film. The field of molecules shown in their Figure 1 appears, however, to be at the edge of a drop, where we have found aggregation to be a problem. It is quite possible that the long strands they see are aggregates of several molecules rather than extensions from single molecules. A much more extensive survey and a more critical testing of the new specimen preparation would be required to make their interpretation convincing.

In summary, our interpretation is that the α chain carboxyl tails originate from the two outer nodules as a thin strand. These two strands come together on one side of the molecule to form a compact domain, which lies next to the central nodule along the twofold axis. The arrangement is sketched in our final model, FIGURE 11.

ATTACHMENT OF FIBRONECTIN TO FIBRINOGEN

FIGURE 5 presents micrographs of another plasma protein, fibronectin, as well as some interesting molecules and complexes that we found in these preparations. This study was done in collaboration with Nadia Carrell and Jan McDonagh.[16] It was known from biochemical analysis that the purified fibronectin always contained a small amount of fibrinogen as a contaminant or a copurifying species. We did not expect the small amount of fibrinogen to present any problems for the microscopy because the trinodular structure is so easily recognized, and it was actually useful as an internal reference. Indeed, the structure of fibronectin is a striking contrast to that of fibrinogen: the fibronectin molecule is a long slender strand, about 140 nm long and only 2 nm diameter, with no evidence for large globular domains (FIGURE 5, top row.)

The main interest for the present discussion is the contaminating or co-purifying molecular species. More than half of the fibrinogen molecules were typical trinodular molecules, but there were a large number of complexes that

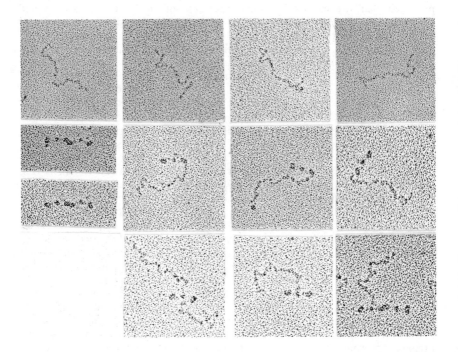

FIGURE 5. Rotary shadowed molecules and complexes from a purified fibrinonectin preparation; ×150,000. (From Erickson et al.[16] By permission of Journal of Cell Biology.) Top row: individual fibronectin molecules; middle row: left panel shows end-to-end fibrinogen dimers found in the fibronectin preparations; the other three panels shows complexes of fibronectin strands attached to the end nodule of a fibrinogen molecule, and bottom row: fibrinogen-fibronectin complexes in which the fibronectin appears to be attached to the middle nodule, or to the adjacent α-chain nodule.

appeared to be end-to-end fibrinogen dimers (FIGURE 5, middle row). These were identical in appearance to dimers that were prepared in vitro by cross-linking with Factor XIIIa at the γ chain cross-link site,[17] which will be discussed in the next section. On the basis of this similarity, and consistent with reports from other labs that such fibrinogen dimers are found in plasma,[18] it is a reasonable speculation that these dimers are cross-linked at the γ chain site.

In addition to fibrinogen dimers we found many examples of fibrinogen-fibronectin complexes. It is well established that Factor XIIIa will cross-link fibronectin to fibrin or fibrinogen.[19] The cross-link acceptor site has been identified as a glutamine three residues from the amino terminus of the fibronectin chain. Fibronectin is a dimeric molecule with carboxyl termini linked in the middle by a disulfide bond and an amino terminus at each end of the strand.[16] One important point is that the complexes always involve the very end of the fibronectin strand, exactly where we expect its cross-link acceptor site to be. In most cases (FIGURE 5, middle row) the strand was attached to an end nodule of the fibrinogen molecule, where we had identified the γ chain cross-link site (discussed below). Since Factor XIIIa cross-linking of fibrinogen occurs most rapidly at the γ chain site, and since we had apparent γ chain cross-linked fibrinogen dimers in the same preparation, it seemed reasonable to suggest that the fibronectin attachment to the end nodule involved the γ chain donor site of fibrinogen.

A number of examples were found, especially in certain preparations, where the fibronectin was apparently attached to the middle nodule of the fibrinogen (FIGURE 5, bottom row). Attachment to the central nodule was hard to understand while we were thinking of the α chain cross-link site as being on or extended from the outer nodule. These difficulties are eliminated, however, by the hypothesis presented in the previous section that the carboxyl tails of the α chains fold back to form a separate domain beside the central nodule. It is then reasonable to speculate that the apparent attachment of fibronectin to the central nodule is really to the α chain cross-link site. In some of the images the fibronectin actually seems to attach to a separate domain, slightly removed from the E-nodule itself.

Until we can demonstrate biochemically which fibrinogen chains are involved in the complexes these conclusions remain speculative. It should be noted that Mosher has demonstrated fibronectin attachment to the α chain of fibrin but not to the γ chain.[19] In that study, however, the γ chain sites were quickly used up in fibrin-fibrin cross-linking, and were probably unavailable for fibronectin attachment. It is reasonable to expect that fibronectin can also be cross-linked to the γ chain site if it is available. Our observation of the two sites for fibronectin attachment, exactly where we have identified the γ chain site (see below) and the α chain site (the α chain nodule proposed above), adds important circumstantial evidence in support of our model.

END-TO-END FIBRINOGEN DIMERS AND TRIMER COMPLEXES

We will now discuss the structure of end-to-end fibrinogen dimers that were prepared *in vitro* by cross-linking with Factor XIIIa.[17] This project and the work on trimer complexes and protofibrils was done in collaboration with Roy Hantgan and Jan Hermans, who prepared the material and characterized it biochemically. Gel electrophoresis demonstrated that the fibrinogen dimers were cross-linked almost exclusively at the γ chain site.[17] These dimers (FIGURES 6 and 7) are essentially identical in structure to the end-to-end dimers that we found subsequently in the fibronectin preparations (FIGURE 5). The important features are that the molecules are joined at the distal tips of the end nodules and they are usually colinear, forming a fairly straight strand.

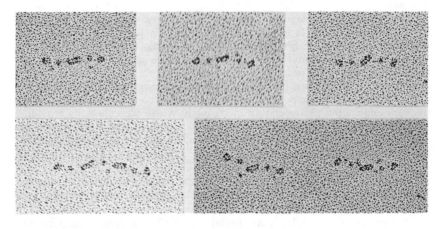

FIGURE 6. Rotary-shadowed end-to-end fibrinogen dimers (and one trimer) prepared *in vitro* by cross-linking with Factor XIIIa,[16] ×200,000.

We concluded that the end-to-end contact is one of the normal contacts in polymerized fibrin because fibrinogen dimers would form fibrin fibers of normal structure when they were activated by thrombin. Since this is a longitudinal contact between the outer or D-nodules, we have called it a

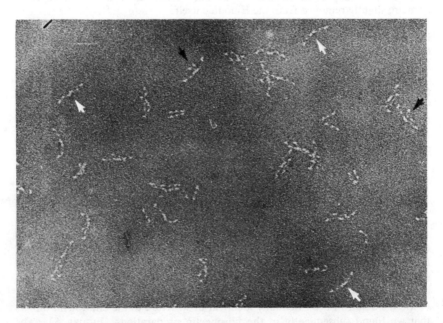

FIGURE 7. A field from a negatively stained specimen of trimer complexes, ×100,000. Some of the end-to-end dimers are indicated by white arrows, trimers by black arrows. (From Fowler *et al.*[21])

DD-long contact. It must, however, be very weak as a noncovalent bond, because it does not produce any significant association of fibrinogen, and it apparently plays little role in the polymerization kinetics.[20] We have been able to visualize this DD-long contact only by the artifice of stabilizing it by the covalent cross-link, a process that normally occurs after polymerization. In summary, we draw two conclusions from these dimers: that the linear, end-to-end contact is a normal structural arrangement in fibrin polymers, and that this DD-long contact is the site of the γ chain cross-link.

The fibrinogen dimers gave us the material to prepare a more interesting complex, which shows a second intermolecular contact. Trimer complexes were prepared by diluting activated fibrin monomers into an excess of fibrinogen dimers.[21] These dimers were not activated by thrombin so they would not polymerize themselves. In fact they inhibited the polymerization of the fibrin by tying up the activated monomers in stable trimer complexes. A low magnification view of a negatively stained specimen is shown in FIGURE 7. Fibrinogen or fibrin monomers, end-to-end dimers, and several trimer complexes may be seen.

Selected complexes are shown in FIGURE 8 with interpretative drawings above them. There are several points to note from the negatively stained complexes in FIGURE 8 and from the shadowed dimers in FIGURE 6. First, in the dimers the two end nodules are usually not seen separately but appear to be fused into one elongated mass. The curvature or slewing of the end nodules and the split into two domains is not seen with any regularity. The α chain nodules may be seen in some of the shadowed dimers, FIGURE 6. In the trimers the central nodule of the fibrin monomer is usually fused into the D dimer complex, presumably attached to each of the joined D-nodules.

The D-nodules of the fibrin monomer are frequently bent away from the E-nodules of the dimer. The central nodules of the dimer have not been activated by thrombin, so we assume that this interaction is blocked and may actually be repelled by the charge of the fibrinopeptides. The most important point is that the trimer complex clearly demonstrates a half-molecule stagger. It also shows the contact between the E-nodule of the activated fibrin monomer and the fused D-nodules of the dimer. We call this the DE-stag contact to indicate the domains involved and the half-molecule stagger between them.

THE STRUCTURE OF THE TWO-STRANDED FIBRIN PROTOFIBRIL

Three decades ago Ferry identified a two-stranded polymer of fibrin, which he called a protofibril, and proposed that this was an intermediate polymer in fibrin polymerization.[22] Twenty years later Krakow et al.[23] were able to visualize long thin polymers in shadowed specimens of partially clotted fibrin. Although the resolution of their shadowed specimens was somewhat limited, they interpreted the thinnest polymers to be two-stranded protofibrils with a half molecule stagger between strands. That was the strucure originally proposed by Ferry and the one that we have been able to confirm and show in some more detail.

We particularly wanted to visualize the protofibrils by negative staining, which can give higher resolution and better definition of details in complicated structures. We were fortunate to have the collaboration of Hantgan and Hermans, who were studying the polymerization reaction by light scattering.

In their kinetic experiments [20] they could identify two phases of fibrin polymerization. In the first phase the light scattering measurements indicated that the polymers were almost exclusively two-stranded protofibrils. In the second phase these protofibrils associated laterally to form thicker fibrin fibers. We were able to confirm this reaction scheme by making electron microscope specimens at different time points, as the reaction was being followed simultaneously by light scattering (reference 24, and Hantgan *et al.*, these proceedings). Specimens prepared during the first phase of polymerization showed an abundance of protofibrils.

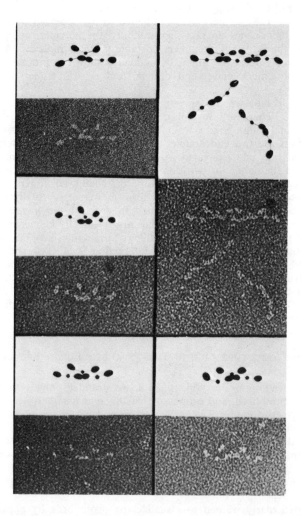

FIGURE 8. Selected images from the specimen of trimer complexes, with schematic drawings to show our interpretation of how the trinodular molecules are arranged, × 200,000. (From Fowler *et al.*[21])

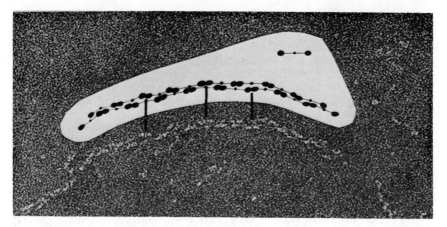

FIGURE 9. An image of a negatively stained two-stranded protofibril, with an interpretative drawing, ×200,000. The vertical lines indicate the register of the dashes in the micrograph with the joined D-nodules in the drawing. A single trinodular molecule may be seen to the right of the drawing (From Fowler et al.[21])

The main interest for the present discussion is the structure of these protofibrils. Our model for the protofibril structure [21] is based on our electron micrographs, but depends very much on considerations of symmetry that had not been spelled out previously. The main point of the model is to identify and locate the intermolecular contacts involved in protofibril formation and subsequent lateral growth.

A micrograph of a protofibril is shown in FIGURE 9, along with an interpretative drawing. The image was difficult to interpret at first because one can not identify individual trinodular molecules in the protofibril, even though single molecules can be clearly resolved in the same field. What the protofibril shows is a sequence of elongated dashes, which seem to alternate from side to side. The structural interpretation was greatly clarified by our images of the dimers and trimer complexes. As seen in FIGURES 6–8, the nodules joined at a DD-long contact are not resolved as separate structures but are fused into a single elongated structure. These are identical to the dashes we see in the protofibril.

With the dashes identified as the fused D-nodules the interpretation of the image was straightforward. The drawing shows the location of the molecules, and the vertical lines indicate the register of some of the D-D pairs with the dashes in the image. It might be noted that a faint density or nodule is sometimes seen between the dashes. In the interpretative drawing these are shown as unattached E-nodules. Since we expect that all of the E-nodules should be bound to the D-D nodules in the opposite strand, it is tempting to suggest that the faint density between the dashes might be the α chain nodules. A final point of information that we draw from the micrographs is that the two strands are seen to be separate and parallel over fairly long stretches. We conclude that the two strands are, in fact, parallel with little or no intrinsic helical twist.

In the interpretative drawing in FIGURE 9, as in many previous schematics, the fibrinogen molecules are represented with featureless spherical nodules. Although the structure or "face" of the nodules is not yet resolved in our electron micrographs, it is important to realize that they have a complex structure, including a polarity. When we construct a model it is essential to show the molecules facing in some specific direction. We will first present the model and then discuss the reasons for our particular choice of which direction the molecules face.

The model drawn in FIGURE 10 incorporates several important structural features: (i) The protofibril consists of two parallel strands of molecules. (ii) The strands are staggered axially by one half a molecular length. (iii) All the molecules in one strand face the same direction. (iv) The two strands face each other. These features may be summed up more concisely by saying that the molecules in the protofibril are related (at least approximately) by a twofold screw axis between the two strands. The intramolecular twofold axes then lie in the plane of the protofibril. A second set of twofold axes runs perpendicular to this plane and intersects the protofibril axis midway between the first set.

The symmetry of this model has important consequences for the arrangement of intermolecular bonds, and consequently for the pathway of fibrin polymerization. The most important feature is that all of the interfaces involved in the DE-stag contacts face inward, toward the protofibril axis, and all of the DE-stag contacts are saturated within the protofibril. The protofibril is therefore "closed" with respect to lateral growth, and this DE-stag contact is not available for adding a third strand to the polymer.

If the DE-stag contacts are all saturated within the protofibril, we must postulate a third type of contact to generate the second stage of fibrin polymerization, the lateral association of protofilaments. In contrast to the DE-stag and the DD-long contacts, which we have visualized in the small complexes and in the protofibrils, the nature of this third contact is still largely speculative. We suggested in our previous publication [21] that the lateral association of protofibrils could be effected by a single type of contact between the D-nodules, which we called DD-lat. In this model all of the protofibrils are in register, with the half stagger occurring only within the protofibrils. As an alternative the adjacent protofibrils could also be staggered, in which case the lateral association would probably involve additional contacts between D- and E-nodules. The former model has the aesthetic appeal of simplicity, but there are some reasons to think that adjacent protofibrils may also be staggered. In particular, if the α chain cross-link sites are near the central nodules, large α chain

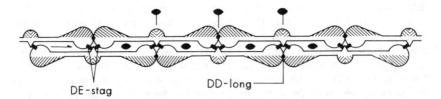

FIGURE 10. (a) Model of the two-stranded protofibril with the twofold screw axis between the strands and the intra- and intermolecular twofold axes indicated. The two intermolecular contact sites involved in protofibril formation are identified as DD-long and DE-stag. (From Fowler et al.[21])

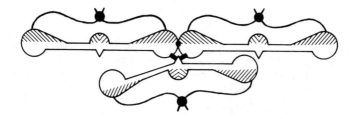

FIGURE 11. A diagram of a trimer complex (a short length of protofibril) showing the proposed location of the α chain nodules. These extra nodules are shown attached to the end nodules by a thin strand of the α chain, lying in the plane of the protofibril (on the molecular twofold axis) and projecting toward the outside.

polymers could be generated if adjacent protofibrils are staggered, but only a limited number of α chains could be cross-linked together if the protofibrils are in register.

Finally, we would like to draw some conclusions about the bond strengths of the different contacts and their role in polymerization. Thrombin activation involves removal of fibrinopeptides from the E-nodule, which we believe affects primarily the DE-stag bond. Before removal of the fibrinopeptide this bond is apparently quite weak, since it does not produce significant association of fibrinogen at the relatively high concentrations found in plasma. When the fibrinopeptides are removed the DE-stag contact increases enormously in bond strength and is the main force driving the polymerization of protofibrils. The DD-long contact serves as the site for eventual γ chain cross-linking, but as stated above it is a weak bond and probably plays little role in the polymerization. The DD-lat bond (or other bonds determining lateral association of protofibrils) is also weak, but it is subject to an amplification mechanism based on cooperativity. As the protofibrils grow longer, their lateral association involves the formation of more and more of these bonds. The cooperative formation of two or more bonds can be orders of magnitude more favorable than the formation of single bonds,[25] so beyond a certain length the lateral association of protofibrils becomes highly favored.

In FIGURE 11 we have drawn the α chain nodule as we described it in the previous section, showing its arrangement relative to the protofibril. The intramolecular twofold axis is in the plane of the protofibril, so we have drawn the α chain nodule in this plane projecting toward the outside, away from the DE-stag contact. In this configuration it is well situated to contact the α chains from adjacent protofibrils and participate in α chain cross-linking.

ACKNOWLEDGMENTS

We would like to thank Dr. Nadia Carrell for preparing the I-4 fibrinogen and the Fragment X used in these studies, and for critical comments and discussions during the preparation of this manuscript.

162   Annals New York Academy of Sciences

REFERENCES

1. HALL, C., & H. SLAYTER. 1959. The fibrinogen molecule: its size, shape and mode of polymerization. J. Biophys. Biochem. Cytology 5: 11–16.
2. KÖPPEL, G. 1967. Elektronenmikroskopische Untersuchungen zur Gestalt und zum Makromolekularen bau des Fibrinogenmoleküls und der Fibrinfasern. Z. Zellforschung 77: 443–517.
3. POUIT, L., G. MARCILLE, M. SUSCILLON & D. HOLLARD. 1972. Etude en microscopie électronique de differentes étapes de la fibrinoformation. Thromb. Diath. Haemorrh. 27: 559–572.
4. FOWLER, W., & H. ERICKSON. 1979. Trinodular structure of fibrinogen—confirmation by both shadowing and negative stain electron microscopy. J. Mol. Biol. 134: 241–249.
5. ELLIOTT, A., & G. OFFER. 1978. Shape and flexibility of the myosin molecule. J. Mol. Biol. 123: 505–519.
6. SHOTTON, D., B. BURKE & D. BRANTON. 1979. The molecular structure of human erythrocyte spectrin. J. Mol. Biol. 131: 303–329.
7. TYLER, J., & D. BRANTON. 1980. Rotary shadowing of extended molecules dried from glycerol. J. Ultrastr. Res. 71: 95–102.
8. NORTON, P., & H. SLAYTER. 1981. Immune labeling of the D and E regions of human fibrinogen by electron microscopy. Proc. Natl. Acad. Sci. USA 78: 1661–1665.
9. WILLIAMS, R. C. 1981. Morphology of bovine fibrinogen monomers and fibrin oligomers. J. Mol. Biol. 150: 399–408.
10. MARDER, V. J. 1970. Physicochemical studies of intermediate and final products of plasmin digestion products of human fibrinogen. Thromb. Diath. Haemorrh. 39 (Suppl): 187–195.
11. FOWLER, W., L. FRETTO, H. ERICKSON & P. McKEE. 1980. Electron microscopy of plasmic fragments of human fibrinogen as related to the trinodular structure of the intact molecule. J. Clin. Invest. 66: 50–56.
12. TELFORD, J., J. NAGY, P. HATCHER & H. SCHERAGA. 1980. Location of peptide fragments in the fibrinogen molecule by immunoelectron microscopy. Proc. Natl. Acad. Sci. USA 77: 2372–2376.
13. PRICE, T., D. STRONG, M. RUDEE & R. DOOLITTLE. 1981. Shadow-cast electron microscopy of fibrinogen with antibody fragments bound to specific regions. Proc. Natl. Acad. Sci. USA 78: 200–204.
14. MOSESSON, M., J. HAINFELD, J. WALL & R. HASCHEMEYER. 1981. Identification and mass analysis of human fibrinogen molecules and their domains by scanning transmission electron microscopy. J. Mol. Biol. 153: 695–781.
15. RUDEE, M., & T. PRICE. 1981. Observation of the alpha-chain extensions of fibrinogen through a new electron microscope specimen preparation technique. Ultramicroscopy 7: 193–196.
16. ERICKSON, H., N. CARRELL & J. McDONAGH. 1981. Fibronectin molecule visualized in electron microscopy: a long, thin, flexible strand. J. Cell Biol. 91: 673–678.
17. FOWLER, W., H. ERICKSON, R. HANTGAN, J. McDONAGH & J. HERMANS. 1981. Cross-linked fibrinogen dimers demonstrate a feature of the molecular packing in fibrin fibers. Science 211: 287–289.
18. KANAIDE, H., & J. SHAINOFF. 1975. Crosslinking of fibrinogen and fibrin by fibrin-stabilizing factor (factor XIIIa). J. Lab. Clin. Med. 85: 574–597.
19. MOSHER, D. 1975. Crosslinking of cold-insoluble globulin by fibrin-stabilizing factor. J. Biol. Chem. 250: 6614–6621.
20. HANTGAN, R. & J. HERMANS. 1979. Assembly of fibrin, a light scattering study. J. Biol. Chem. 254: 11272–11281.
21. FOWLER, W., R. HANTGAN, J. HERMANS & H. ERICKSON. 1981. Structure of the fibrin protofibril. Proc. Natl. Acad. Sci. USA 78: 4872–4876.

22. FERRY, J. D. 1952. The mechanism of polymerization of fibrinogen. Proc. Natl. Acad. Sci. USA **38**: 566–569.
23. KRAKOW, W., G. ENDRES, B. SIEGEL & H. SCHERAGA. 1972. An electron microscopic investigation of the polymerization of bovine fibrin monomer. J. Mol. Biol. **71**: 95–103.
24. HANTGAN, R., W. FOWLER, H. ERICKSON & J. HERMANS. 1980. Fibrin assembly: a comparison of electron microscopic and light scattering results. Thrombosis and Haemostasis **44**: 119–124.
25. ERICKSON, H., & D. PANTALONI. 1981. The role of subunit entropy in cooperative assembly. Biophysical J. **34**: 293–309.

DISCUSSION OF THE PAPER

G. A. MARGUERIE (*Hôpital de Bicêtre, Le Kremlin-Bicêtre, France*): Can you tell us whether the existence of this fourth nodule is dependent upon the presence of calcium or not?

H. P. ERICKSON: Usually I do not add calcium or a chelator. I assume that tightly bound calcium is still present, and the free calcium is probably on the order of 10 micromolar. Actually, I did do one set of experiments looking at molecules in the presence of 2 millimolar calcium and I found the extra nodule in those cases.

R. F. EBERT (*Johns Hopkins Hospital, Baltimore, MD*): Dr. Erickson, glycerol, aside from its effects on the viscosity of the medium, might also be considered a weak chaotropic agent. Have you tried other chaotropic agents? Have you ever seen these α nodules in preparations that lacked glycerol?

ERICKSON: We have made some specimens with one molar sodium bromide in addition to glycerol. The images are similar to those presented. The glycerol probably doesn't affect the shape of the molecule too much because we have done sedimentation studies in the presence of glycerol and we obtained the same sedimentation coefficient of about 8 S, as in other studies without glycerol. We have not been successful in making reasonable specimens without glycerol. Dr. Slayter's lab, for example, obtained nice specimens in the absence of glycerol, but for me glycerol is a magic bullet that makes clean reproducible specimens.

V. MARDER (*University of Rochester School of Medicine and Dentistry, Rochester, NY*): Has anyone studied polymerizing fibrin with regard to where the α chain appendage would be? Would it still be in the same place or does it flop out freely after the fibrinopeptides are liberated?

ERICKSON: I did some preliminary sedimentation studies of monomer fibrin and fibrinogen in acetic acid. The sedimentation coefficients were similar, which indicates that the α chain did not pop out drastically upon activation.

ANALYSIS OF HUMAN FIBRINOGEN BY SCANNING TRANSMISSION ELECTRON MICROSCOPY *

J. Wall and J. Hainfeld

Biology Department
Brookhaven National Laboratory
Upton, New York 11973

R. H. Haschemeyer

Department of Biochemistry
Cornell University Medical College
New York, New York 10021

M. W. Mosesson

Department of Medicine
Mount Sinai Medical Center
Milwaukee, Wisconsin 53201

INTRODUCTION

Electron microscopic studies by Hall and Slayter [1] led to a trinodular model of fibrinogen that appears to be compatible with much of the available structural and biochemical data. This model has received additional support from recent studies on fibrinogen,[2-5] on core fragments of fibrinogen,[6] and on assembled crystalline forms.[7] On the other hand, several studies have reported observations of more spheroidal shapes,[8-13] and even in those studies where elongated forms predominated, a significant portion of objects had more compact shapes.[1, 14, 15, 16, 17] Two explanations for the observed pleomorphism have been advanced: (1) the linear form arises from unfolding of a more compact spheroidal molecule on the specimen support or (2) the linear forms represent aggregates of fibrinogen molecules (see reference 5 for discussion).

In the trinodular model, the outer domains have been identified with the biochemically isolated D fragment and the inner domain with the E fragment through various studies. Fowler et al.[6] observed metal shadowed specimens of fibrinogen fragment Y which lacks one of the two D domains but retains the E domain. These molecules appeared dinodular with a length about ⅔ that of native molecules. Price et al.[18] have localized D and E domains in the outer and middle regions of the molecule, respectively, using metal shadowed specimens of fibrinogen reacted with anti-D or anti-E Fab' fragments. Telford, et al.[20] have observed metal shadowed specimens of complexes formed by reacting fibrinogen with the IgG fraction of an antibody directed against the NH_2-terminal domain, thought to be in the central region. Their images were consistent with the interpretation of trinodular molecules connected by IgG molecules bridging central domains. Price et al.[18] and Rudee and Price [21]

* This work was supported by grants from the National Heart, Lung and Blood Institute (HL-17419), the United States Department of Energy and the National Institutes of Health Biotechnology Resource Branch (RR–715).

164

report observations of shadowed specimens prepared by reacting fibrinogen with the Fab fragment of an antibody prepared against the central region of the Aα chain. They interpret their results as demonstrating that the Aα chain extends outward from the D fragment.

The scanning transmission electron microscope (STEM) offers several advantages in studying a molecule such as fibrinogen. (1) Molecules and their domains observed unstained in dark field can be identified directly by their mass. Thus, it is possible to correlate the STEM image directly with biochemical models. (2) Lack of phase contrast when using the large angle annular detector makes image intensity directly intrepretable in terms of mass thickness (no oscillations in the contrast transfer function).[22] (3) Certain specimen preparation artifacts such as denatured protein films are detected easily in the STEM and can be avoided through suitable specimen preparation techniques.[3]

The STEM mass measurement technique is a refinement of that originally developed by Zietler and Bahr [23] differing in that it takes advantage of superior electron detectors available in the STEM as well as serial readout of image information into an on-line computer. The technique has been applied in several laboratories [24-26] on a wide variety of problems. The theoretical and practical limits of accuracy are well characterized.[24, 27] Recent studies using STEM mass measurements have been carried out on nucleosomes,[28] ribosomes,[29] RNA polymerase bound to DNA,[30] intermediate filaments,[31] dynein,[32] low density lipoprotein [33] and glutamine synthetase.[34] For a single molecule the size of fibrinogen, the expected mass error is ~10% at a dose of 10el/$Å^2$ compared to the tobacco mosaic virus (TMV) used as an internal standard which gives a standard deviation of ~0.5%. Thus it should be possible to rule out the possibility that trinodular objects are actually three fibrinogen molecules (340 VS 1020kD), a distinction more difficult to make by dimensional measurements on shadowed specimens. Furthermore, it is possible to address the question of the identity of compact forms: do they also have a mass of 340 kD? By the same methods, the interpretation of antibody labeled specimens should be much more quantitative.

A substantial difficulty in obtaining reproducible specimens of fibrinogen has been its tendency to form denatured protein films (on air-water interfaces) which interfere with specimen attachment and interpretation. The flotation technique of Estes and Haschemeyer [3] largely overcomes this difficulty. Affected areas can be recognized in the STEM as regions of increased substrate thickness, often having a mottled appearance, whereas the presence of such areas is very difficult to appreciate in shadowed or negatively stained specimens. Thus it should be possible to recognize areas of clean background and interpret all objects situated in such areas as to identity and conformation.

MATERIALS AND METHODS

Proteins

Preparation of fibrinogen and various subfragments has been described previously.[5] Crosslinking with 0.1% glutaraldehyde was carried out in 0.05 M sodium phosphate buffer, pH 7, at a protein concentration of 3 to 10 μg/ml for 45 to 90 min at room temperature. Degree of crosslinking that had taken place was characterized by SDS gel electrophoreses and showed the following:

(1) production of some intermolecular covalent bonds giving rise to high molecular weight bands corresponding to crosslinked molecules, (2) retention of material in the band position corresponding to monomeric fibrinogen, and (3) failure to produce any electrophoretic bands corresponding bands normally observed, following reduction with DTT. This finding indicated that intramolecular crosslinking of chains had occurred.

Preparation of Fab Fragments

Affinity purified Fab fragments directed against fragments D or E were prepared from a rabbit antiserum to human fibrinogen in the following way. An ammonium sulfate fraction (precipitated at 50% saturation) prepared from the antiserum and solubilized in 0.15M NaCl, was passed through a fragment D-Sepharose or a fragment E-Sepharose column that had been prepared by coupling the respective fragment to CNBr-activated Sepharose 4B.[35] Adsorbed IgG molecules were desorbed with a 0.2M glycine buffer, pH 2.5,[36] dialyzed against 0.15M NaCl and freeze-dried. This material was subsequently processed by digestion with papain, followed by CM-cellulose chromatography to isolate the Fab fragments.[37] As assessed by immunodiffusion analysis, the anti-E and anti-D IgG fractions reacted only with fibrinogen and fragment E or D, respectively; the Fab fragments prepared from these materials gave no antigen-antibody precipitin reactions themselves but when mixed with the parent IgG fraction, inhibited the formation of the IgG-induced precipitin reaction. Upon electrophoresis in Na Dod SO_4-containing gels, these preparations yielded a single band migrating somewhat more anodally than fragment E_2.

Specimen Preparation

Carbon foil substrates 15–25Å thick were prepared by vacuum evaporation from a carbon arc onto freshly cleaved single crystal rocksalt. The carbon foil was floated on distilled water and picked up with 2.3 mm dia. titanium grids which had previously been coated with a holey carbon film. Grids were glow discharged in O_2 immediately prior to specimen preparation.

Specimens were mounted on substrates by injecting 2.5 μl of sample solution (2–10 μg/ml) into a 2.5 μl drop of buffer previously placed on the film. After 2–10 min. the grids were washed two times with water or ammonium acetate buffer and once with TMV solution (30 μg/ml). For negative staining, samples were washed once more with 2% uranyl sulfate, wicked, and air dried.

Samples for freeze drying were wicked from the edge with filter paper (leaving a water layer several microns thick), plunged into liquid nitrogen and transferred to an ion pumped freeze drying system. Following rapid evacuation with a sorption pump, the specimens were warmed slowly from −196°C using a feedback circuit which controlled the warming rate so as to maintain the system pressure below 10^{-7} Torr. Typical freeze dry times were 4–8 hr with the majority of that time spent at the sublimation temperature of ice.

Specimens were transferred to the STEM under vacuum in order to prevent rehydration. Specimen transfer was carried out at room temperature, although cold transfers are also possible with this system. Additional details of specimen preparation are given in reference 5.

Scanning Transmission Electron Microscope Imaging

Specimens were imaged at the Brookhaven STEM Biotechnology Resource using a 40KV probe focused to a 2.5Å diameter spot. During observation, specimens were maintained at −110 to −140°C and at 10^{-9} Torr. No specimen contamination or condensation of residual gases was observed under these conditions.

Electrons illuminated the specimen with a cone having a convergence half-angle 0.012 radian. Transmitted electrons struck one of three detectors subtending annular regions of 0.0 to 0.015 radian (bright field), 0.015 to 0.040 radian (small angle, dark field) and 0.040 to 0.2 radian (large angle dark field). Detectors consisted of CaF_2 scintillators coupled to photomultiplier tubes.

Detector signals were stored digitally as 8 bit numbers on a magnetic disc memory (512×512 points/image) and displayed on a TV screen. For recording low dose first scan images, the beam was focused on an area adjacent to that of interest, moved to the area of interest and scanned under computer control. When not focusing or acquiring data, the beam was blanked off. Images suitable for analysis were stored digitally on magnetic tape.

Mass Measurement

Mass measurements were performed off line using either the Brookhaven CDC 7600 computer or a DEC LSI 11/23 analysis computer as described by Hainfeld *et al.*[38] Fibrinogen and TMV particles were identified and the regions around them removed from the image for background computation. Since the fibrinogen molecules give a signal only about 10% above the intensity of the background carbon film, accurate background determination was essential.

Particles of TMV were used as an internal standard for mass measurements in order to eliminate ~5% differences observed in mass calibration from specimen to specimen. Regions containing TMV were identified and summed (after background was subtracted from each picture element). The known molecular weight of TMV, 39×10^6D, was divided by the summed data to give the mass per unit intensity.

Unknown particles were analyzed by enclosing them with a rectangular or circular boundary and computing the sum of intensity minus background for all points within the boundary and multiplying by the calibration factor. In order to minimize errors for trinodular images, a hybrid scheme was used as illustrated in FIGURE 6 with circles having their overlap regions removed. Further details are given in Mosesson *et al.*[5]

Results

Conformation of 340 kD Objects

Injection into buffer and freeze drying resulted in reproducible objects on a clean background as shown in FIGURE 1. The range of possible conformations under various pH conditions was investigated as shown in FIGURE 2. Specimens were prepared by dilution in buffer and injection into pH 2.7, pH 7.0 or pH 9.3 buffer, followed by freeze drying. The most common form of objects having mass near 340 kD was an elongated symmetrical structure 460 ± 20 Å in contour length and ~80 Å in greatest width. There were essentially no differences in the length, width and degree of bending in the forms observed under the three different pH conditions.

Most elongated forms were tridomainal and in some cases appeared S shaped (circled molecules, FIGURE 2). The outer domains were somewhat wider and more dense than the central domain. Some objects appeared to

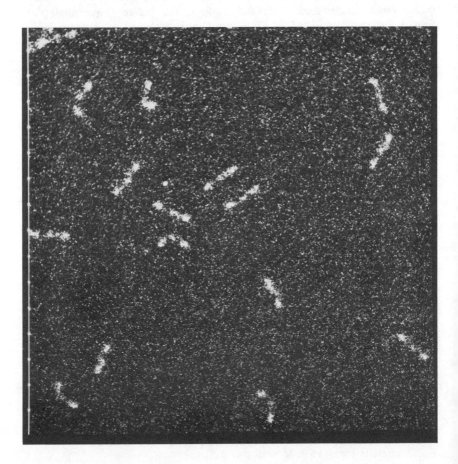

FIGURE 1. Unstained I-4 fibrinogen. 0.52 μm scan width.

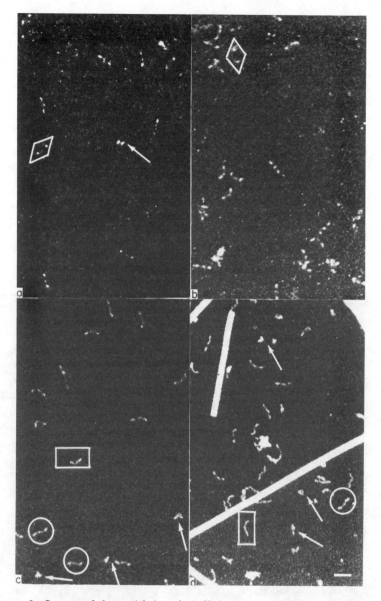

FIGURE 2. Images of freeze-dried peak 1 fibrinogen that had been deposited on carbon films under the following buffer conditions: (a) 1% acetic acid (pH 2.7); (b) 0.05 M-sodium phosphate buffer (pH 7); (c) 0.1% ammonium acetate buffer (pH 9.3); (d) glutaraldehyde-treated material in 0.1% ammonium acetate buffer (pH 8.4). Scale bar represents 50 nm. The material shown in (c) was freeze-dried directly after application, whereas other specimens were washed with water before freeze-drying. TMV was added to the specimen shown in (d) with the final water wash. TMV appear as cylinders $\sim$ 160 Å in diameter. The radiation dose for the images shown ranged from 7 to 10 electrons/Å². See the text for a description of the images indicated by arrows, circles, rectangles, and diamonds. (Reprinted from Mosesson *et al.*[5] with permission.)

have a more uniform mass distribution along their length (rectangular boxes, FIGURE 2). Some fields contained "dinodular" structures which appeared to lack a middle domain (diamonds, FIGURE 2)

In addition to linear and bent forms, compact objects, sometimes dinodular, comprised about 10% of the objects in the 340 kD size range (arrows, FIGURE 2). In an attempt to determine which conformation was more common in solution, glutaraldehyde fixed specimens were examined. The same general forms were observed (FIGURE 2d) with compact forms somewhat more common (~25%); however, the observed molecular weight was shifted systematically higher (see below).

Molecules lacking the COOH terminal of the Aα chain (I–9) were compared to intact molecules (I–4) and revealed no striking differences detectable by visual inspection (FIGURE 3). Retrospective inspection suggests less mass in the thread connecting the three domains.

Results of STEM mass analysis on these specimens are presented in TABLE 1. A radius of integration of 120 Å (per domain) was selected to include more than 95% of the particle mass while providing an acceptable noise level. The accuracy of background computations was tested by measuring

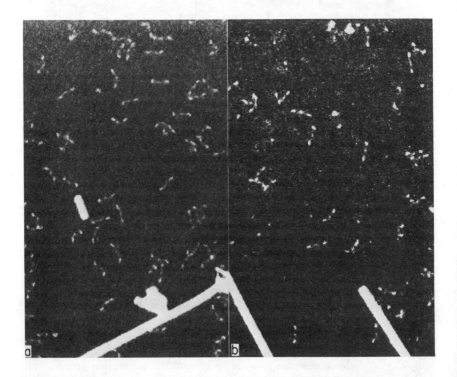

FIGURE 3. Images of freeze-dried I-4 (a) and I-9 (b) fibrinogen. The specimens were deposited in 0.15 M NaCl, 0.01 M Tris HCl buffer pH 7.4 (a) and 0.05 M sodium phosphate buffer (pH 7). The scale bar is 50 nm and the radiation doses are 21 and 8 el/Å in a and b, respectively. (Reprinted from Mosesson et al.[5] with permission.)

TABLE 1

MASS ANALYSIS OF FIBRINOGEN AND FRAGMENTS D_1 AND E_2

Preparation or object analyzed	Established mol. wt. $\times 10^{-3}$ from physical and other measurements	Mass by STEM analysis			
		Mol. wt. $\times 10^{-3}$	S.D.* $\times 10^{-3}$	S.E.* $\times 10^{-3}$	n*
I-4 tridomainal form	340±15	327 †	32,56	9	12,14
I-4 compact form	340±15	359	79	17	20
I-9 tridomainal form	300±15	297 †	41,55,56	11	12,20,9
I-4 outer domain	—	111 †	17,23	4	12,14
I-9 outer domain	—	112 †	13,22,13	3	12,20,9
I-4 middle domain	—	101 †	17,20	4	12,14
I-9 middle domain	—	76 †	27,36,37	5	12,20,9
Fragment D_1	100±5				
Fragment E_2	45±5				
Glutaraldehyde-fixed:					
I-4 tridomainal form ‡	—	435	42	10	17
I-4 compact form ‡	—	409	36	14	7
Blank film §	—	7	30	6	31

Table from Mosesson *et al.*[5]

* n = number of objects analyzed; S.D. = standard deviation; S.E. = standard error of the mean.

† Mass values for tridomainal molecules and their domains from two fields of fraction I-4 and three fields of fraction I-9 were combined according to Bevington[47] to obtain weighted mean molecular weights and standard errors.

‡ Fibrinogen contains ~ 185 lysine residues/molecule.[48] Assuming that monomolecular forms of fibrinogen bind three glutaraldehyde residues/lysine amino group,[49, 50] the expected size of the molecule would be ~ 400 KD.

§ Same measuring parameters as for tridomainal forms.

blank carbon film with the same parameters used for molecules. Mass loss of <5% due to imaging dose <10 el/Å² was compensated for by use of TMV as an internal standard which loses mass at nearly the same rate as fibrinogen.

Mass analysis of tridomainal structures from fraction I–4 (whole molecules) gave a mean of 327 kD while similar objects from fraction I–9 gave 297 kD. Mass analysis of outer domains of the two forms gave very similar values (111 VS 112 kD) whereas the mass within the central domain was about 25% less for I–9 as compared to I–4 molecules (76 vs. 101 kD).

Spherical forms in these images of I–4 preparations gave a mass of 359 ± 79 kD, after discarding two particles of 868 and 919 kD, presumed to be aggregates. The mean weight of folded or compact forms from glutaraldehyde-fixed preparations was 409 ± 36 kD, in comparison to the value of 435 ± 42 kD found for tridomainal objects in the same preparation.

Immunoelectron Microscopy

Information on the identity of the domains observed in the electron microscope in comparison with those isolated biochemically has been attained from immunoelectron microscopy and morphology of negatively stained specimens.

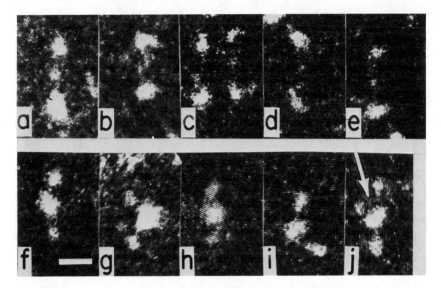

FIGURE 4. Images of freeze dried fibrinogen molecules (fraction I-4) that had been reacted in 0.15 M NaCl, 0.01 M Tris-HCl buffer, Ph 7.4, with Fab fragments directed against the D domain (panels a-e) or the E domain (panels f-j). The arrow in j indicates a free Fab fragment. Scale bar—20 nm.

Fab fragments mixed with fibrinogen in a 6:1 molar excess yielded two types of images which differed significantly from the appearance of fibrinogen itself (FIGURE 4). The first type, discrete, randomly distributed particles ~50 Å in diameter (arrow FIGURE 4j), was attributed to uncomplexed Fab fragments. The second type depended on the specificity of the Fab fragments. In the case of anti-D Fab fragments, the outer domains of tridomainal molecules showed increased density (FIGURE 4a–e). In the case of anti-E Fab most molecules showed increased density in the central domain. Mass measurement with a radius of 80 Å about the central nodule for the anti-E Fab preparation showed ~40% of molecules with a mass increase of 61 ± 21 kD and ~60% with mass increase of 115 ± 23 kD relative to unreacted molecules.

Negatively Stained Specimens

Images of negatively stained fibrinogen show significantly more structural detail than unstained images but their interpretation in terms of mass distribution is not as quantitative. The predominant form obtained in negatively stained fibrinogen preparations was also an elongated structure. Flexibility was evident in the wide variety of bent forms. Intensity distribution within the central region appeared more uniform, but tridomainal forms were common. Compact forms were observed occasionally.

In images of negatively stained fibrinogen the outer nodules were often resolved into two spheroidal regions forming an oblong unit 90–95 Å long and

~40 Å wide canted at an angle of 120–150° to the long axis (diamonds, FIGURE 5). Images of negatively stained D fragments resembled the outer domain of intact molecules, being resolved into two discrete subdomains as described above (arrows FIGURE 5c). The central domain in images of negatively stained fibrinogen was less sharply defined than the outer domains but almost always possessed a central dense area. Considerable variability was observed in this region, especially for intact molecules, as described below. Images of isolated E fragments appeared as roughly spherical or oblong shapes.

Images of negatively stained I-4 differed from those of I-9 preparations in suggesting the presence of a thread-like structure originating from the outer domain and extending toward the central domain (diamond, FIGURE 5a). In

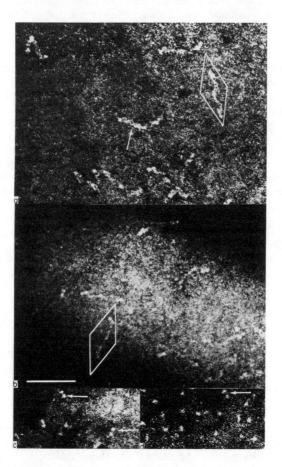

FIGURE 5. Images of negatively stained fibrinogen and fragments D and E. (a) Fraction I-4; (b) fraction I-9; (c) fragment D_1; (d) fragment E_2. The contrast was inverted electronically from signals obtained from the large angle detector. Scale bar represents 50 nm. The radiation dose (electrons/Å²) was 200 (a), 100 (b), 50 (c), and 35 (d), respectively. See the text for a description of the images indicated by arrows. (Reprinted from Mosesson *et al.*[5] with permission.)

some cases, the thread-like structure appeared to wind around the central axis (arrow, FIGURE 5a). The thread like structure was not observed in I–9 preparations (FIGURE 5b).

DISCUSSION

Conformation of 340 kD Objects

One of the advantages of STEM mass analysis is that measured mass is independent of molecular conformation, thus permitting quantitative comparison of objects observed in the STEM. The results presented above demonstrate that both trinodular and compact objects can have molecular weights expected for fibrinogen. The hypothesis that trinodular forms contain three fibrinogen monomers can be ruled out on the basis of mass measurements. These data suggest that the conformation of individual fibrinogen molecules may be flexible, either unfolded or relatively compact. The increased number of compact forms found in glutaraldehyde-fixed preparations suggests that this variability may exist in solution also. There does not appear to be any covalent bond to prevent the unfolding of the arms.

The accuracy of STEM mass measurements of fibrinogen under the conditions used is consistent with theoretical predictions of random errors (34 kD for mass $= 340$ kD, area of integration $= 1.4 \times 10^5 Å^2$, substrate thickness $= 20$ Å and dose $= 7$ electrons/$Å^2$).[5] The relatively large error arises because of the extended conformation of trinodular molecules. An additional source of variability may be heterogeneity in the preparations themselves. Approximately 30% of I–4 molecules are 325 kD and 70% are 350 kD, resulting in an intrinsic standard deviation of ~14 kD. For I–9, a continuum of sizes from 270–310 kD is present, resulting in an intrinsic standard deviation of 18 kD. Freeze drying, vacuum transfer to the STEM and observation at low temperature were used to minimize or eliminate errors due to salt artifacts, contamination, and mass loss. In some cases it was not possible to distinguish clumps of salt (which sometimes form during freeze drying) from compact forms of fibrinogen, which may lead to increased errors in that case. Use of TMV as an internal calibration standard removed microscope variations (typically $\sim5\%$ from specimen to specimen) as a source of error. The increased mass of glutaraldehyde-fixed molecules may be due to binding of glutaraldehyde to the ~185 lysines per fibrinogen or crosslinking of small protein fragments to whole molecules.

Identification of Domains

Evidence for identification of domains comes from immunoelectron microscopy, comparison of morphology of isolated domains with that of whole molecules and changes in appearance of whole molecules following selective proteolytic cleavage. Based on results presented above, we confirm the identification of D and E domains, confirm the observation of a new substructure within the D domain and present a new hypothesis concerning localization of the COOH terminal of the Aα chain.

Images of molecules that had been reacted with Fab fragments directed against either D or E domains provide evidence verifying the molecular domainal assignments (D, outer domain; E, central domain) suggested previously on the basis of biochemical [39-42] or electron microscopic analyses.[6, 20, 42, 18] With STEM mass measurements it is possible to state that no mass is added to the domain not targeted by the Fab fragments and to quantitate the number of Fab fragments bound to the targeted domain. In the case of the anti E Fab fragments, it appears that 40% of the fibrinogens observed had one added Fab and 60% had two added Fab fragments.

Additional evidence for the assignment of the D domain comes from the comparison of images of negatively stained molecules and isolated subdomains. The isolated doublets observed in the D fragment preparation bear a striking resemblance to the outer domain in the whole molecules. Images of E fragments do not show any characteristic features which would permit their assignment to a domain of the whole molecules. Williams [4] has reported resolving the outer domains of bovine fibrinogen that had been deposited in the presence of 1 to 2 mM $Ca,^{2+}$ into two spheroidal regions. The angle of attachment of this oblong structure to the long axis was more acute than we have observed. The orientation is reversed also, presumably due to the way images are printed. STEM images are presented as if viewed from below, with the specimen on top of the carbon film. Weisel *et al.*[44] suggest three globular domains linearly arranged in the D region, on the basis of analysis of assembled crystalline forms of the molecule. Establishing the relationship between these observations of the D region should prove most interesting.

In view of this assignment of D and E fragments, the lack of a significant mass difference between outer and central domains of unstained intact molecules raises an interesting question. The D and E fragments are known to be 100 ± 5 and 45 ± 5 kD, respectively. The observed value for the outer domain, $111 \pm$ kD, is in reasonable agreement with this value. However, the observed mass of the central domain, 101 ± 17 kD, is much larger than that expected for the E fragment. This similarity in masses between outer and central domains makes it impossible to arrive at an assignment of D and E domains by mass measurements of whole molecules only. However, the issue is clarified by examination of I–9 fibrinogen which has the COOH terminal of the Aα chain cleaved off. I–9 molecules have essentially the same mass in the outer domains as intact I–4 molecules, but 25% less mass in the central domain. Thus it appears that the COOH terminal of the Aα chain is located within the central region of the molecule, even though it is known to exit from the structure comprising the D fragment. (Note that this chain is cleaved off in preparation of plasmic fraction D1, reviewed by Mosesson and Finlayson.[45]) These observations could be reconciled if it is hypothesized that this portion of the Aα chain exists as a long, thin structure emanating from the outer (D) domain and extends centrally (perhaps winding around the axis of the molecule). This view is supported by the observation of just such a thread-like structure in negatively stained I–4 fibrinogen preparations but not in I–9 fibrinogen preparations, the latter of which have the COOH terminal of the Aα chain removed.

This view of the location of the Aα chains conflicts with that proposed by Price *et al* [18] on the basis of metal-shadowed molecules that had been reacted with Fab fragments directed against a segment from the middle portion of the Aα chain . They interpreted their images as showing this portion of the Aα chain as well as the contiguous COOH terminal to be situated beyond the outer domains of the molecule at a considerable distance (>225 Å) from the central domain. Our results are consistent with the possibility that a portion of the Aα chain could be situated beyond the outer domain. However, our results indicate that the COOH terminal segment is located close to the central domain (within 120Å) as depicted in FIGURE 6. These observations can be reconciled by noting that the segment of the Aα chain in question is probably very flexible and has no covalent attachments to other chains. Attachment of a molecule with a mass greater than that of the chain itself may favor a more open conformation, especially if shearing forces are present during specimen attachment to the substrate, as in the case reported by Rudee and Price.[21]

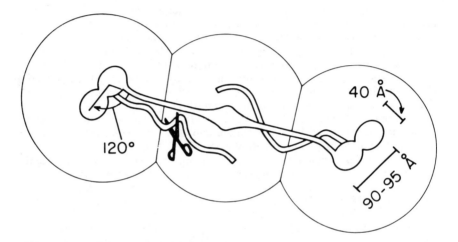

FIGURE 6. Schematic model of the fibrinogen molecule based upon data obtained in this investigation. The area of integration for mass determination of each domain is shown by three overlapping circles having a radius of 122Å. The scissors indicate a site on the Aα chain, cleavage of which results in molecules characteristic of fraction I-9. (Reprinted from Mosesson *et al.*[5] with permission.)

CONCLUSIONS

In this study we have demonstrated the capability of the STEM to establish a firm link between biochemistry and electron microscopy through quantitative image analysis (mass measurement). Although some of the results presented confirm previous observations of others, we feel that in several respects they offer a significantly higher confidence level. For example, it is not possible to interpret trinodular images of unstained, freeze-dried fibrinogen as being trimers in light of their measured mass of 330 kD. By the same token, the controversy between proponents suggesting a globular form as opposed to a trinodular form can be resolved with points in favor of both possibilities (although some previous observations of globular particles at high concentration were almost certainly due to the carpeting effect reported by Estes and Haschemeyer[3]). The flexibility of the fibrinogen is very evident in STEM dark field images where essentially every object in a field is detected and can be analyzed according to mass and conformation (see TABLE 1). Even denatured protein is detected as a signal above background, so it is difficult to overlook artifacts, in comparison to shadowing or negative staining which tend to obscure such artifacts.

Interpretation of antibody or Fab-labeled specimens is less ambiguous than with the conventional microscope due to one's ability to demonstrate that mass is added only to the region in question and only in the expected increments. However, for interpretation of the differences between I–4 and I–9 fibrinogen the STEM offers clear advantages. The ability to observe the molecule directly using only its own signal, rather than that of large attached labels such as antibodies, increases confidence in the structural integrity of the molecules. Finally, lack of phase contrast when using the large angle annular detector and the low dose, low temperature conditions used for observation of negatively stained specimens, allow interpretation of features as thin as the "thread-like structures" as being part of the Aα chain.

Future studies of fibrinogen with the STEM can be designed to take advantage of heavy atom cluster compounds [46] to label specific sites such as SH groups or sugar residues in order to locate them within the framework proposed in FIGURE 6. Extension of these studies to aggregates of fibrin should help to define the mode of interaction.

ACKNOWLEDGMENTS

We thank Mike Reedy for helpful discussions on glutaraldehyde fixation, Kristin Chung, Ed Desmond, Frank Kito, and George Latham for technical assistance, and Nancy Siemon for skillful secretarial assistance.

REFERENCES

1. HALL, C. E. & H. S. SLAYTER. 1959. J. Biophys. Biochem. Cytol. **5:** 11–16.
2. FOWLER, W. E. & H. P. ERICKSON. 1979. J. Mol. Biol. **134:** 241–249.
3. ESTIS, L. F. & R. H. HASCHEMEYER. 1980. Proc. Natl. Acad. Sci. USA **77:** 3139–3143.
4. WILLIAMS, R. C. 1981. J. Mol. Biol. **150:** 399.
5. MOSESSON, M. W., J. HAINFELD, R. H. HASCHEMEYER, & J. WALL. 1981. J. Mol. Biol. **153:** 695–718.
6. FOWLER, W. E., L. J. FRETTO, H. P. ERICKSON & P. A. MCKEE. 1980. J. Clin. Invest. **66:** 50–56.
7. WEISEL, J. W., N. M. TOONEY, I. KAPLAN, D. AMRANI & C. COHEN. 1980. J. Mol. Biol. **143:** 329–334.
8. MITCHELL, R. F. 1952. Biochim. Biophys. Acta **9:** 430–442.
9. KOPPEL, G. 1966. Nature (London) **212:** 1608–1609.
10. POUIT, L., G. MARCILLE, M. SUSCILLON & D. HOLLARD. 1972. Thromb. Diath. Haemorrh. **17:** 559–572.
11. BELITSER, V. A., T. V. VARETSKA & V. PH. MANJAKOV. 1973. Thromb. Res. **2:** 567–568.
12. BLAKEY, P. R., M. J. GROOM & R. L. TURNER. 1977. Br. J. Haematol. **35:** 437–440.
13. MOSESSON, M. W., J. ESCAIG & G. FELDMANN. 1979. Br. J. Haematol. **43:** 469–477.
14. BANG, N. U. 1967. *In* Blood Clotting Enzymology. W. H. Seegers, Ed.: 487–549. Academic Press. New York.

15. GORMAN, R. R., G. E. STONER & A. CATLIN. 1971. J. Phys. Chem. **75:** 2103–2107.
16. KRAKOW, K., G. F. ENDRES, B. M. SIEGEL & H. A. SCHERAGA. 1972. J. Mol. Biol. **71:** 95–103.
17. BACHMANN, L., W. W. SCHMIT-FUMIAN, R. HAMMEL & K. LEDERER. 1975. Makromol. Chem. **176:** 2603–2618.
18. PRICE, T. M., D. D. STRONG, M. L. RUDEE & R. F. DOOLITTLE. 1981. Proc. Natl. Acad. Sci. USA **78:** 200–204.
19. NORTON, P. A. & H. S. SLAYTER. 1981. Proc. Natl. Acad. Sci. USA **78:** 1661–1665.
20. TELFORD, J. N., J. A. NAGY, P. A. HATCHER & H. A. SCHERAGA. 1980. Proc. Natl. Acad. Sci. USA **77:** 2372–2376.
21. RUDEE, M. L. & T. M. PRICE. 1981. Ultramicroscopy **7:** 193–196.
22. FERTIG, J. & H. ROSE. 1977. Ultramicroscopy **2:** 269–279.
23. ZEITLER, E. & G. F. BAHR. 1962. J. Appl. Phys. **33:** 847–53.
24. WALL, J. 1979. *In* Introduction to Analytical Electron Microscopy. J. J. Hren, J. J. Goldstein & D. C. Joy, Eds.: 333–342. Plenum Pub. New York.
25. ENGEL, A. & J. MEYER. 1980. J. Ultrastruct. Res. **72:** 212–222.
26. FREEMAN R. & K. R. LEONARD. 1981. J. Micros. **122:** 275–286.
27. WALL, J. S. & J. F. HAINFELD. Introduction to Analytical Electron Microscopy. J. J. Hren, J. J. Goldstein & D. C. Joy, Eds. 2nd Edit. In press.
28. WOODCOCK, C. L. F., L.-L. Y. FRADO & J. S. WALL. 1980. Proc. Natl. Acad. Sci. USA **77:** 4818–4822.
29. BOUBLIK, M., N. ROBAKIS, W. HELLMAN & J. WALL. 1982. Eur. J. Cell Biol. **27:** 177–184.
30. HOUGH, P. V. C., I. MASTRANGELO, J. WALL, J. HAINFELD, M. SIMON & J. MANLEY. 1983. J. Mol. Biol. **160:** 375–386.
31. STEVEN, A. C., J. WALL, J. HAINFELD & P. STEINERT. 1982. Proc. Natl. Acad. Sci. USA **79:** 3101–3105.
32. JOHNSON, K., J. HAINFELD & J. WALL. J. Cell Biol. **96:** 669–678.
33. HASCHEMEYER, R. Personal communication.
34. HASCHEMEYER, R. H., J. WALL, J. HAINFELD & M. MAURIZI. 1982. J. Biol. Chem. **257:** 7252–7253.
35. AXEN, R., J. PORATH & S. ERNBACK. 1967. Nature **214:** 1302–1304.
36. AVRAMEAS, S. & T. TERNYCK. 1969. Immunochemistry **6:** 53–66.
37. PORTER, K. R. 1959. Biochem. J. **73:** 119–126.
38. HAINFELD, J. F., J. S. WALL & E. J. DESMOND. 1982. Ultramicroscopy **8:** 263–270.
39. MARDER, V. J., N. R. SHULMAN & W. R. CARROLL. 1969. J. Biol. Chem. **244:** 2111–2119.
40. PIZZO, S. V., M. L. SCHWARTZ, R. L. HILL & P. A. MCKEE. 1972. J. Biol. Chem. **247:** 636–645.
41. FERGUSON, E. W., L. J. FRETTO & P. A. MCKEE. 1975. J. Biol. Chem. **250:** 7210–7218.
42. COLLEN, D., B. KUDRYK, B. HESSEL & B. BLOMBACK. 1975. J. Biol. Chem. **250:** 5808–5817.
43. FOWLER, W. E., H. P. ERICKSON, R. R. HANTGAN, J. MCDONAGH & J. HERMANS. 1981. Science **211:** 287–289.
44. WEISEL, J. W., G. N. PHILLIP & C. COHEN. 1981. Nature (London) **289:** 263–266.
45. MOSESSON, M. W. & J. S. FINLAYSON. 1976. *In* Progress in Hemostasis and Thrombosis. T. H. Spaet, Ed. **3:** 61–107. Grune and Stratton. New York.

46. SAFER, D., J. S. WALL, J. F. HAINFELD & J. RIORDAN. 1982. Science **218:** 290–291.
47. BEVINGTON, P. R. 1969. *In* Data Reduction and Error Analysis for the Physical Sciences: 73. McGraw Hill. New York.
48. HENSCHEN, A. & B. BLOMBACK. 1964. Ark. Kemi. **22:** 347–353.
49. BOWES, J. H. & C. W. CATER. 1968. Biochim. Biophys. Acta **168:** 341–352.
50. MOREL, F. M. M., R. F. BAKER & H. WAYLAND. 1971. J. Cell Biol. **48:** 91–100.

MORPHOLOGY OF FIBRINOGEN MONOMERS AND OF FIBRIN PROTOFIBRILS *

Robley C. Williams

Department of Molecular Biology and Virus Laboratory
University of California
Berkeley, California 94720

INTRODUCTION

Most of our impressions of the morphology of the fibrinogen molecule have come from measurements of its hydrodynamic behavior in solution and from observations of its size and shape when examined in the electron microscope. Both kinds of study were initiated at about the same time, some thirty years ago, but whereas the classical hydrodynamic approach is now little undertaken, electron microscopic observations of the molecule and its assembly states continue under active pursuit. Electron microscopy is immensely attractive for the study of molecular morphology because it permits direct visualization of individual particles. Its pictorial results are usually convincing, at least at a scale of size that is large compared to the resolution limit of the instrument used. Thus, for example, even in the days before the introduction of methods of shadowing and of staining it showed unequivocally that some viruses, like tomato bushy stunt, were isometric, tobacco mosaic virus was a rigid rod, while particles of a T-even bacteriophage were tadpole shaped. Details of molecular morphology far smaller than the salient features of whole viruses are readily detectable nowadays, even down to dimensions of 1 nm or less. The modern success of the electron microscope in making the finest details visible to the eye, however, comes at a considerable cost. The methodology is one of sampling, with consequent destruction of the specimen, and with no possibility of directly monitoring the average behavior of a solution of molecules, as may be done with most hydrodynamic methods. More seriously, molecules initially in solution that are to be electron microscopically examined must undergo at least two operations that are likely to change their in-solution morphology: desiccation, and what may be called contrast enhancement. The observation of fibrinogen molecules, for example, requires that they be dried, either from a volatile fluid prior to a shadowing operation or from a solution of heavy-metal salt that forms a dry matrix around them (negative staining). The former procedure results in collapse and consequent disturbance of molecular shape due to forces of surface tension, to which deformation are added changes of shape and size owing to the condensation of shadowing metal.

Fortunately, molecules that are dried while in a heavy-metal salt (such as negative staining with uranyl acetate) appear to experience less collapse and distortion than do those dried from a volatile liquid, as demonstrated, for example, by the preservation of well-arranged capsomers on the capsid surface of λ phage.[1] But in observations of negatively stained material what is imaged in the electron micrograph is a two-dimensional projection of every-

* This investigation was supported by Research Grant PCM80–10650 from the National Science Foundation.

180

thing in the line of (electron) sight that scatters electrons. A molecule as tenuous in physical structure as fibrinogen, even if possessed of a distinctive three-dimensional morphology in solution, becomes flattened into something much more closely two-dimensional when dried in negative stain. The electron micrographic image of a negatively stained molecule has intramural "contrast" insofar as it exhibits varying degrees of dark and light over the image region taken to represent the molecule. The contrast is primarily determined by the degree to which stain is excluded from various parts of the molecule and from regions above and below it in the line of sight. Those parts that remain thick upon drying, and also impermeable to stain, will appear relatively light on the viewing screen (or in finished micrographs, as these are normally reproduced). Parts that become thin upon drying, or ones that remain thick but are stain permeable, will appear dark. Unfortunately, little of quantitative nature can be ascertained about the relative masses of molecular regions when seen in negative stain, although it is usually safe to infer that the lighter regions of a molecule contain more material in the line of sight than do darker regions.

As a consequence of the way in which image brightness is dependent upon stain thickness, portions of a slender molecule, such as fibrinogen, will exhibit apparent diameters or breadths that are severely influenced by the average thickness of the stain. They may even disappear into the background density if the stain above the molecule is quite thick. If the stain is very thin, some parts of the molecule may suffer surface-tension flattening upon drying and appear artifactually spread out. There are really no certain guides to acceptance of an image as one that "really" represents the size and shape of a moelcule as dried in negative stain. Even a further complication, of unknown magnitude, arises from the possibility that the negative stain employed is not passive in its effect on molecular shape, particularly inasmuch as its concentration is greatly increased during drying. (An example of nonpassive effect of stain is the disintegration of microtubules to particles too small to be detectable when they are immersed for even a few seconds in a 1–3% aqueous solution of Na- or K-phosphotungstate at neutral pH.) Inasmuch as certain stains are unstable except at pH 4.5, or less, the risk of stain-induced structural alteration is real. A similar uncertainty in interpreting images of negatively stained molecules comes from the possibility that the relative permeability of protein structures to stain is not proportional to their permeabiliy to water. Thus, two portions of a molecule may be equally thick and equally porous to water but they may appear unequally bright (or dark) in negative stain if they are unequally permeable to the stain. It should appear from the above that a study by the electron microscope of the fine-scale morphological features of a molecule such as fibrinogen is primarily a study of the way in which stain is distributed as the molecule becomes embedded within it.

BACKGROUND

The Fibrinogen Monomer

The earliest electron microscopic observations of fibrinogen molecules were those of Hall [2] on unidirectionally shadowed specimens. Objects described as "nodose" were seen. Ten years later Hall and Slayter [3] produced electron micrographs, also of unidirectionally shadowed specimens, that showed the molecule to be 45 nm long and characterized by a linear arrangement of three, apparently disconnected, spheres. Their conclusion was that the fibrinogen molecule had an elongated shape, with a spherical knob of 6.5 nm diameter at each end, a smaller one about 5 nm in diameter in the center, and some kind of connecting fiber too slender to cast a visibly detectable shadow. Although some subsequent reports [4,5] on the morphology of shadowed molecules have shown divergences from the conclusions of Hall and Slayter,[3] their "trinodular" characterization has been generally supported, particularly in the report of Fowler and Erickson.[6]

The visualization of the fibrinogen molecule by methods of negative staining has been characterized by conflicting reports about its morphology. Although the techniques of negative staining have a higher potential for structural discrimination than have those of shadowing, they suffer from the many ways in which variations in preparative procedures can produce widely differing electron microscopic appearances. Kay and Cuddigan [7] reported the molecule to be a linear array of small nodules, with an overall length of about 70 nm. Fowler and Erickson [6] examined fibrinogen molecules negatively stained with uranyl acetate in experiments that also included rotary shadowing, and obtained electron micrographs showing slender, elongated objects of 45 nm length, with some curvature, and, in many cases, having an enlargement or knob at each end and, much less distinctly visible, a small knob in the center. They interpreted this morphology to be the negative-stain equivalent of the three-nodule forms that they had seen after either unidirectional or rotary shadowing [6]—the classical Hall-Slayter [3] form. Estis and Haschemeyer [8] employed both sodium phosphotungstate and uranyl acetate as negative stains, and, using a scanning transmission electron microscope, also examined fibrinogen molecules that were unstained. The stained particles appeared slender and elongated, varying greatly in type and degree of curvature and showing beading, or nodulation, along their lengths but in no consistent pattern. Some of the molecules in the unstained material had fairly straight rod-like forms with definite signs of nodulation. Using the same scanning instrument Mosesson et al.[9] analyzed the mass distribution in fibrinogen molecules and found elongated, frequently curved structures that had a pronounced nodule at each end, symmetrically disposed about a lesser, central nodule. The outer domains were occasionally resolved into two contiguous subdomains. Somewhat earlier Williams [10] had shown that negatively stained molecules had slender, sigmoidal contours, and in addition to the minor central nodule, possessed at each end a protuberance that was distinctly elongated. After treatment with 1–2 mM Ca^{2+} this feature was found to be composed of two contiguous spheres.

A different approach to determining the morphology of fibrinogen by electron microscopy has been reported by C. Cohen and collaborators.[11] Rather than observing individual molecules, these investigators obtained electron micrographs of negatively stained, thin crystals of a "modified" fibrinogen, one that had been proteolytically cleaved to about 90% of its native mass. After optical processing to remove noise, the images of the crystalline arrays were matched against simulated patterns arising from computer-generated assemblies of "molecules" that were assigned various hypothetical mass distributions. The best-fitting model for the fibrinogen molecule, one that accounted closely for the observed crystal patterns and also for the band pattern observed in thrombin-induced fibrin fibers, was an extension of the Hall-Slayter model: [3] a thin, slightly bent spine along which were arrayed seven knobs (instead of three).

One of these was at the center, and on either side, symmetrically disposed, were three knobs of which the two distal ones were the larger and the proximal one quite small. In essence, the large, terminally situated knobs of the Hall-Slayter model were replaced by two smaller, nearly contiguous ones, and a new, small knob added about half-way toward the molecule's center.

Protofibrillar Assemblies

Negative-stain electron microscopy has recently been used [12] in studying the arrangement of fibrin monomers when they assemble into primitive bundles, "protofibrils," [13] in the earliest stages of formation of fibrin fibers. They characteristically have been seen to consist of two closely apposed, parallel chains of fibrin monomers seemingly bonded end to end. Williams [10] found, at least with short protofibrils, that the elements in one chain appeared to be attached to those of the other chain with a half-molecule staggered overlap and that, in a given protofibril, neither chain was ever longer than the other by more than one monomer unit. Fowler et al.[12] exhibited spectacular linear arrays of protofibrils in which, in their seemingly well-preserved regions, the prominent places of contact between the terminal knobs of contiguous monomers in one chain appeared to lie midway between those of the laterally adjacent chain; thus, a half-molecule staggered overlap was evident.

The present report deals with recent attempts to improve electron microscopic imagery of fibrinogen molecules and short protofibrils, primarily by use of negative stains, and to arrive at reasonable conjectures about the molecules' morphology when in solution.

EXPERIMENTAL

Materials

Examination was made of bovine and human fibrinogen, the former designated as Sigma lyophilized Fraction I, and the latter as Grade L, Kabi, Stockholm. Thrombin was from bovine plasma, Sigma, Grade 1, lyophilized.

Fibrinogen stock solutions were usually at 3–5 mg/ml, obtained by dissolving small aliquots of lyophilized material (20–30 mg) in the chosen solvent, with gentle stirring at room temperature. Several different solvents were employed at one time or another, all at pH 7.0–7.8, but no dependence of molecular morphology on solvent conditions was seen. Solvents were: ammonium formate, 50–200 mM, pH-adjusted with ammonium hydroxide; NaCl, 100–300 mM, buffered with 20 mM-50 mM concentrations of phosphate, or Tris-HCl, or MOPS (morpholinopropanesulfonic acid). The stock solutions were routinely discarded after one week, or earlier, although only those with ammonium formate as solvent were other than water-clear throughout the storage interval. With ammonium formate a visible flocculence formed during several hours at 4°C; this was routinely sedimented and discarded. Some stock solutions were dialyzed extensively (18 hr) against the pure solvents, but no effect of such treatment on the molecular morphology was ever noted. Dilution of both the fibrinogen and the thrombin to concentrations appropriate for electron microscopy was done in the solvent used for the stock solution.

The negative stains employed were water solutions of uranyl formate (Eastman) and of uranyl acetate (Mallinckrodt), the latter being used only in an occasional trial. A highly desirable characteristic of a negative stain is an ability to wet, or spread upon, a hydrophilic surface such as a glowed carbon film. Neither of the stains mentioned, as commonly prepared, will remain evenly spread as they dry to a thin film; rather, upon drying they will "bead," leaving much of the underlying carbon surface devoid of dried stain and a small portion of it covered with an unusually thick stain deposit. Phospho-tungstate stain, on the other hand, will spread as well as does pure water upon, for instance, a freshly cleaved mica surface or a carbon film that has been recently charged in a partial-vacuum plasma discharge. But only the uranyl salts seem to be satisfactory for the finest delineation of fibrinogen morphology. It was found that Eastman uranyl formate in powdered form could be used to make a solution of stain having wetting characteristics much like water if certain care was taken in its preparation.[10] The powdered material evidently has components of quite different solubility, with the most readily soluble portion having the best spreading properties. Consequently, the stain was prepared by gently adding water to a beaker containing enough powder to make a 3–4% solution if it were all dissolved. Subsequent to the addition the mixture was disturbed as little as possible.

From time to time samples were withdrawn by pipet from a level a few mm above the level of the dissolving powder. When the A_{450nm} value of $0.09–0.12^{cm-1}$ was reached, corresponding to a concentration of about 0.6%, about half the volume of the liquid initially added was removed by pipet and saved. It was then passed through a 0.022 μm millipore filter. The storage life of such stain was variable (being terminated by formation of a precipitate and loss of wettability characteristics), but if refrigeration in the dark is used it was usually a few weeks. It was interesting to note that at the end of stain preparation the residual, undissolved uranyl formate on the bottom of the beaker, at least 80% of the starting material, had formed a hard, relatively insoluble cake. While it has excellent physical properties in its habit of drying to a uniform, "glassy" film over large areas of the specimen, uranyl formate is less than ideal in having a very low pH, between 3.5 and 4.0. It is liable to precipitate if its pH is raised appreciably above 4.0.

Preparation of Specimens

Some specimens were prepared for electron microscopy by shadowing, both unidirectional and rotary. The shadowing source was tungsten wire [14] 0.5 mm in diameter and clamped horizontally between copper electrodes 3 cm apart. It was heated by a current of 32.5 amps until, after about 90 sec, it burned out. During this time about 10 mg of tungsten were evaporated. The shadow angle for rotary deposition was usually tan^{-1} 0.12, with the specimen grids laid on a 60 rpm rotating table whose center was 11.5–12.0 cm from the tungsten filament.

The specimen films 3HG 400 (Pelco) were filmed with Formvar spread on a water surface and, after drying, were coated *in vacuo* with a carbon film. They were made hydrophilic, or "charged," by a 10-sec plasma discharge at 5,000 volts in residual air at 0.1 torr and used for specimen deposition within a few minutes. The handles, or tabs, of the grids were kept covered during the glow discharge so as to avoid any creep of liquid toward the tips of the forceps in which they were subsequently held.

A chosen stock solution of fibrinogen monomers was diluted to 5–10 μg/ml and a 5-μl sample pipetted upon the horizontal surface of a forceps-clamped grid. Several seconds later the grid was rinsed by inverting it and skimming it several times over the surface of a 100-mM solution of ammonium acetate (or ammonium formate) followed by skimming over a 10-mM solution. If rotary shadowing was to follow, the grid was held face down for ~10 sec upon a 10-μl drop of 2%–4% ammonium acetate between the first and second rinses. Drying of the grid after the second rinse was achieved by suction application to its edge of the tip of a glass pipet, flame-drawn to ~¼ mm bore. Observation of the withdrawal of fluid was made through a ×10 binocular; any failure of the liquid to recede slowly and dry as a uniform, flat film was sure to portend a specimen in which the fibrinogen molecules were badly distributed. If negative stain was to be applied, there was no application of uranyl acetate between rinses. The fluid remaining on the grid after the second rinse was withdrawn to a thin film by suction pipet, but before drying took place the grid was inverted and touched to a 5–10 μl drop of negative stain previously placed upon a freshly scraped Teflon surface. After about 5 sec the grid was lifted along with its adherent drop of stain. The drop was almost entirely removed by aid of the suction pipet and drying allowed to ensue. It was observed that the region of the grid near the point touched by the pipet tip was the last to dry and hence was covered with thicker stain. The *average* thickness of dried stain could be crudely controlled by varying the time of contact between the stain drop and the end of the suction pipet. A drying time of 10–15 sec after application and removal of the pipet, at a relative humidity of 30–40%, was usually an indication that a satisfactory film of stain was being deposited.

On a few occasions a fixation step with glutaraldehyde was performed prior to the rinsing steps. The 5-μl drop of initial specimen solution was mostly removed by suction (without rinsing) and a 5–10 μl drop of 0.3–1.0% (w/v) of glutaraldehyde (Polysciences) applied for 5–10 min. In no case was any effect of the fixation treatment detectable, with specimens of either fibrinogen monomers or protofibrils.

The preparations of protofibrils were usually made by mixing equal volumes of thrombin at 0.4–2.0 NIH unit/ml and fibrinogen at 20–40 μg/ml, in 0.1 M NaCl, 30 mM Tris-HCl, pH 7.4, and incubating for 1–3 min at room temperature. Protofibrils were adsorbed to the grids most consistently when a 100-μl drop of mixed fibrinogen-thrombin was placed immediately upon a Teflon block and, after incubation, the inverted grid was touched to its surface.

Electron microscopy was performed with a JEOL-100 B instrument, usually at a primary magnification of ×45,000 as determined by calibration with particles of tobacco mosaic virus. All micrographs were obtained under conditions of minimal beam exposure [15] adapted to the JEOL instrument. [After visual estimation at low magnification and low illumination that a desirable specimen field lies within the grid square that is being examined, the electron beam is condensed and translated to the edge of the viewing screen nearest the observer. At ×45,000, when beam spot size "No. 3" is employed, the diffuse edge of the image of the condensed beam should come no closer than 2 cm to the near edge of the portion of the specimen image to be photographed.

The binocular viewer is tipped down and clamped to allow focusing in the area illuminated. The specimen is then translated so that the image moves toward the observer by at least 10 cm. Focus is then promptly achieved and four operations are performed in appropriate, but rapid, sequence: the "plate advance" button is depressed, the electron beam is decondensed and enlarged to a size predetermined to be appropriate for photography, the enlarged beam is centered, and the exposure is manually initiated in the usual way.] When micrographs are obtained in this manner, incidentally, there is no need for activation of the anti-contamination cold trap. An important consequence of the technique described, in addition to the primary effect of minimal damage to specimen objects, is that the negative stain retains a highly desirable glassy character. As can be easily noted by visual inspection of the viewing screen, even a brief exposure of a negative stain to an intense electron beam usually creates detectable granulation.

RESULTS AND DISCUSSION

Shadowed Fibrinogen Monomers

FIGURE 1 shows a field of bovine fibrinogen molecules that were unidirectionally shadowed. The anticipated features were seen: enlargements at the ends of the long molecule and a much less prominent knob in the center. A notable departure from previous published micrographs [3, 4, 6] was the prominence of the spine of the molecule, lending an appearance somewhat like that seen in freeze-etched specimens.[5] One difference in technique between the present and previous work that might explain the electron micrographic differences was the application of uranyl acetate to the specimen prior to the

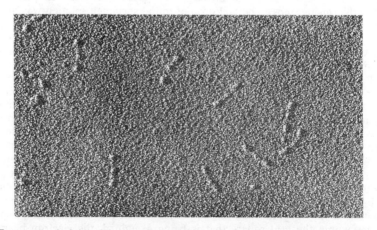

FIGURE 1. A field of bovine fibrinogen molecules shadowed with tungsten at an angle $\tan^{-1}$ 0.25. Material dissolved in 50 mM ammonium formate (pH 7.4) and applied at 5 μg/ml to specimen film. After 30s adsorption time the grid was rinsed in 100 mM ammonium acetate (pH 7.4), followed by 15s exposure to 2% uranyl acetate, followed by rinse in 10 mM ammonium acetate, $\times$200,000.

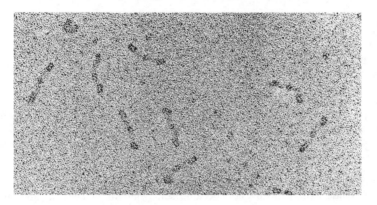

FIGURE 2. Bovine fibrinogen molecules, prepared as for FIGURE 1, but rotary shadowed with tungsten at an angle $\tan^{-1} 0.12$, $\times 220,000$.

final rinse. This step may have effected something in the way of "fixation" and prevented full-scale collapse of the material in the spine. Rotary-shadowed particles are shown in FIGURE 2. Not all molecules exhibited the same morphology, but there was a recognizably typical size and shape. Pertinent dimensions were found to be: length, 45 nm; length and breadth of terminal enlargements, 10 nm and 6 nm; central knob, estimated 4–5 nm; width of spine, ~3 nm. In consideration of the distortions that inevitably occur when a molecule of this shape and complexity dries from a volatile solvent, only general conclusions may be drawn. The molecule is distinctly elongated but not in the form of a stiff rod; it has a slight bulge at its center, and has one larger, elongated enlargement at each end. Inasmuch as the amount of condensed tungsten was very small (estimated to average 0.5 nm over the molecule), it is likely that the molecule was not significantly distorted during the actual shadowing. The primary merit of the rotary-shadowing procedure lies in its relative simplicity. The morphologies it discloses, while severely affected by surface-tension forces, are free of potentially larger distortions brought about by the alternative procedure of negative staining. As mentioned above for unidirectionally shadowed specimens, the molecular spine was distinctly more prominent than had been previously reported.[6, 16]

Negatively Stained Fibrinogen Monomers

Specimens for electron microscopy that were stained with uranyl formate by the procedures mentioned above usually demonstrated a variation of stain thickness over the area of the grid, the thickest region being near the place where the suction pipet was applied. A stain that was on the verge of unusability usually exhibited a blotchy distribution upon drying. At very low magnification (~$\times 4,000$) such a stain could be seen to vary in thickness from almost nothing to a quite heavy deposit within even a single opening of a 400-mesh grid; it was particularly thick near the corners of the opening. A stain that had shown good wetting properties while drying produced a dried deposit that appeared completely uniform over any chosen grid opening and seemed nearly uniform in thickness over many contiguous openings.

Variation in thickness of stain produced appreciable variation in the appearance of the fibrinogen molecules. At one extreme was very thin stain, in which the molecules were contrasted very poorly indeed, even with some signs of positive contrast, and the background film was noticeably granular. Somewhat thicker stain yielded negative contrast, but still with granular background, in which the spines of the molecules appeared relatively wide and with markedly jagged edges. Still thicker stain produced an exquisitely fine-grained background and showed molecular forms that had a spine of measurable thickness, an almost unnoticeable central knob, and a knob at each end that was usually single and elongated [10] but that sometimes vaguely appeared like two contiguous spheres. As the stain increased in thickness the spine appeared more slender, the central knob became somewhat more evident, and the terminal protrusions were clearly seen as two spheres (FIGURE 3). These electron microscopic morphologies were identical in bovine and human fibrinogen. The effect of the addition of millimolar amounts of Ca^{2+} was to cause the molecules to display the terminal spheres at stain thickness judged to be too small to show them in the absence of Ca^{2+}, and to render the central knob more visible.[10]

A striking feature of the morphology of the fibrinogen monomer as seen embedded in a film of uranyl formate (or acetate) was its consistently sinuous, or gently sigmoidal, shape (FIGURE 3). It has been noted before [6, 9] that the molecule had the overall appearance of a flexible rod, but with no sign of a typical conformation. Not all particles had a sigmoidal form, but the great majority of them did. Whenever the sigmoidal form was plainly evident it

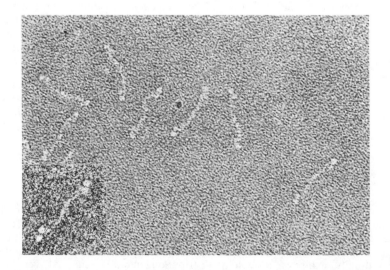

FIGURE 3. Molecules of human fibrinogen negatively stained with uranyl formate. For specimen preparation see EXPERIMENTAL. Note the generally evident sigmoidal shape and the two contiguous spheres at each terminus, $\times 350,000$. Inset: bovine fibrinogen, $\times 450,000$.

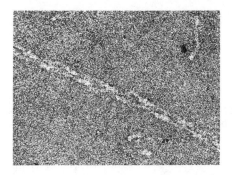

FIGURE 4. Segment of a protofibril of human fibrinogen stained with uranyl formate. Fibrinogen at 10 μg/ml in 0.1 M NaCl, 20 mM Tris-HCl (pH 7.4); thrombin at 0.2 units/ml. Mixture sampled after 2 min, × 225,000.

was always of the same handedness: that of a bacterial growth curve as it is ordinarily drawn with *time* as the abscissa. [Photographic processing was done so as to show the molecules as they would appear to an observer looking at the specimen grid with the film side uppermost.] The long axis of each terminal knob (when seen in thin stain), or the line connecting the centers of the terminal knobs (thicker stain), is tilted about 45° to the axis of the spine near its ends. Thus, the molecule showed symmetry of form about a line through its center, and drawn perpendicular to the specimen film.

The invariance of handedness of the sigmoidal shape of the dried fibrinogen molecule must be a reflection of an important aspect of its shape when in solution. It can be readily accounted for by postulating [10] that the molecule in solution has a right-handed helical form and that its center is the place where the initial binding to the charged support film takes place. As the stain dries, the ends of the molecule are forced down upon the support film, resulting in a two-dimensional form that is sigmoidal and of invariant handedness.

FIGURE 4 shows a proposed model of the morphology of the bovine (or human) fibrinogen molecule based primarily on micrographs like that shown in FIGURE 3. The extent of the presumptive helical segment is taken as one turn. Dimensions derived from the electron micrographs are: length of molecule, tip-to-tip, 45 nm; diameter of terminal spheres, 4.0–4.5 nm; width of spine, 2.0–2.5 nm. The central knob was too small to be measured with any reliability. The spine width, as estimated here, is less by some 0.5–1.0 nm than that given previously by Williams.[10] The spine must represent the coiled-coil interconnector region of the fibrinogen molecule.[17] Its estimated diameter agrees reasonably with that of a triple, polypeptide coiled coil, and its half-length, 15 nm, compares satisfactorily with the 16 nm calculated from the primary sequence data.[17]

The model shown in FIGURE 4 is not out of accord with the micrographs shown previously of fibrinogen negatively stained with uranyl acetate,[6, 10] nor, for that matter, fibrinogen molecules that had been rotary shadowed.[6] The new features incorporated within it are its explicit helical form and the dual spheres at the termini. These are most probably the carboxy-terminal ends of the β and γ chains of the molecule, the two together included in the D fragment found in plasmin digests.[18] The diameters of the spheres are consistent with molecular weights in the range of 30–40 kilodaltons, a figure in reasonable accord with that of a tightly coiled polypeptide of about 300 amino acid residues.[19] The morphological fate of the carboxy-terminal end of the negatively stained α chain of fibrinogen is a complete mystery. If it, in fact, retains a random coil configuration after drying in stain its invisibility is understandable.

The model recently proposed by Weisel et al.[11] for an enzymatically modified fibrinogen, based upon reconstruction analysis of electron micrographs of thin crystals and other ordered assemblies, differs from the model shown in FIGURE 4. The former presents the molecule as a straight rod with a slight bend in the center and possessing a total of seven nodules, one in the center and three near each end. The two most distal are nearly in contact and have diameters close to those reported here for the two terminal spheres. The third, more proximal, nodule shown in the model as located in each half-molecule is distinctly smaller than either of the distal two and slightly smaller than the central one. A difficulty encountered in placing a nodule in this position is that there seems to be nothing in the data for the primary sequence of the human fibrinogen molecule that would suggest its presence. Nor has there been any hint of such a nodule in any reported electron micrograph of fibrinogen molecules either stained or shadowed. It should be noted that a model derived from the micrographs of ordered assemblies may not be directly compared with ones (like FIGURE 3) that result from visual contemplation of electron micrographs of individual molecules. A primary constraint upon the former is that its morphology must be compatible for the banding pattern seen in fibrin. On the other hand, the morphology of the fibrinogen molecule as proposed here and elsewhere has not been conditioned by any requirement that it can be converted to fibrin.

Morphology of Protofibrils

The most primitive form of a protofibril, as the word was used by Ferry [13] when he introduced it, is a dimer of fibrin molecules consisting of two parallel, apposing monomers that are out of longitudinal register by one-half unit. Dimers that are end-to-end assemblies of fibrinogen molecules can be created, and stabilized by the action of Factor XIIIa,[20] but they are not considered to be protofibrils because of their single-chain character. Stable trimers were readily made by Fowler et al.[12] by mixing purified fibrin monomers with a molar excess of covalently linked, end-to-end dimers of fibrinogen molecules. Such an assembly is a pseudoprotofibril. Its fibrin unit bonds in its E domain with the two contiguous D domains of the fibrinogen dimer but, inasmuch as the E regions of these two molecules have retained their fibrinopeptides, no further interstrand binding takes place. This kind of assembly is dead-end to further growth. A true protofibrillar trimer would be one in which the fibrinopeptide A of all three units has been removed, with or without covalent linkage of D domains, leaving an unsatisfied bonding site at the E domain of each member of the end-to-end dimers. Such a trimer would be expected to grow as a two-chain assembly, a protofibril, so long as there are fibrin monomers, or fibrinogen and thrombin, in the solution. It has been reported,[10] at least in the case of short protofibrils, that the number of monomers in one chain does not differ by more than one unit from the number in the other chain. Evidently, to hold protofibrils together (and in the absence of Factor XIIIa) it is necessary to have bonding greater than would be afforded by the D-D longitudinal contacts alone.

Fowler *et al.*[12] observed spectacularly long stretches of two-chain proto-fibrils in the relative absence of other assembly species; this observation strengthened previous notions [13, 21] that the protofibril is a distinct intermediate stage in fibrin fiber assembly. They were unable to discern any morphological detail of contact between fibrillar elements except to observe the half-molecule stagger between the adjacent elements that were laterally connected.

The finding reported in the present paper that, with or without added Ca^{2+}, the termini of fibrinogen molecules (and, presumably, fibrin monomers) exhibited two contiguous spheres has made it seem possible that some morphological detail of protofilament assembly could be found. The search is confounded by the likelihood that even the simplest protofibril assembly is more convoluted, in solution, than is the fibrin monomer and, consequently, that it will suffer more distortion and some degree of disarray upon drying in negative stain. Contacts that exist in solution may frequently be broken and, occasionally, a putative contact that is spurious may appear.

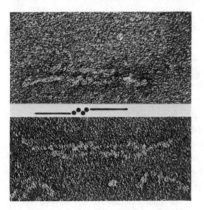

FIGURE 5. Two short protofibrils, a trimer (above) and a tetramer (below), showing detail of the D-D contact region wherein four contiguous spheres follow a zigzag path. Interpretive sketch between the two micrographs, × 390,000.

A likely place to look for detail of contact morphology is at the D-D longitudinal attachment sites. From the appearance of individual molecules, shown in FIGURE 3, it seems inevitable that four spheres will be positioned near each D-D contact. But which spheres actually touch and what is the orientation of their lines-of-center?

FIGURE 4 is a micrograph of a segment of a well-preserved protofibril in which can be seen periodic thickenings along each of the two chains; presumably these are the regions of D-D contact. Their separation is about 45 nm and their position indicates a half-molecule longitudinal stagger between the chains. But in this micrograph the existence of spheres in the D-D regions is only marginally evident. FIGURE 5 shows, or attempts to show, that the D-D contact region contains two pairs of spheres and that the spheres retain the orientation they had in the monomer (FIGURE 3). As a consequence the four-sphere contact region has a zigzag appearance, with the three lines connecting the centers of the four contacting spheres not colinear.

FIGURE 6. Three-dimensional model of a fibrinogen molecule. The extent of the right-handed helix is one full turn, with the two D domains shown as the four large spheres and the E domain as the small one.

Only regions of fairly thick stain, wherein the terminal spheres are evident, will yield the sorts of display seen, although marginally, in FIGURE 5. If the D-D junction is actually arrayed in the manner suggested here, the identification of the two terminal spheres seen in micrographs of the monomer is very likely established. The more distal sphere must represent the carboxyl end of the γ polypeptide chain, inasmuch as the covalent bond established in the transglutaminase action of factor XIIIa is known [22, 19] to be in this portion of the molecule.

A speculative model of the manner of assembly of a four-member protofibril is shown in FIGURE 7. Each of its "fibrin" elements has the same shape as that shown in FIGURE 6, the whole assembly forming a segment of a right-handed, paranemic helix. The view of FIGURE 7 is one in which the two sets of contacts between the E region (small sphere) and the two γ domains of the larger D spheres are alternately wholly visible and partially eclipsed. The zigzag presumed to exist in the D-D contact region is incorporated into the model. The two primary features of this model are the helical shape of the monomeric elements and the paranemic, helical form of the protofibril, and the placing of the D-D contact near the distal end of the outer of the two spherical termini. While it is easy to see that the "protofibril" can extend indefinitely as two parallel chains, the model makes no prediction about geometry of fibrin packing in three dimensions. Its helical form, however, does not seem to render it incompatible with proposed protofibrillar geometry underlying recent proposals of models of fibrin fibers.[23]

FIGURE 7. Model of a four-member protofibril. One of the D_2E contact regions is shown in full, while in the other (to the left) the E domain is eclipsed.

REFERENCES

1. WILLIAMS, R. C. & K. E. RICHARDS. 1974. J. Mol. Biol. **88:** 547–550.
2. HALL, C. E. 1949. J. Biol. Chem. **179:** 857–864.
3. HALL, C. E. & H. S. SLAYTER. 1959. J. Biophys. Biochem. Cytol. **5:** 11–16.
4. KRAKOW, W., G. F. ENDRES, B. M. SIEGEL & H. A. SCHERAGA. 1972. J. Mol. Biol. **71:** 95–103.
5. BACHMANN, L., W. SCHMITT-FUMIAN, R. HAMMEL & K. LEDERER. 1975. Die Makromolekulare Chemie **176:** 2603–2618.
6. FOWLER, W. E. & H. P. ERICKSON. 1979. J. Mol. Biol. **134:** 241–249.
7. KAY, D. & B. CUDDIGAN. 1967. Br. J. Haematol. **13:** 341–347.
8. ESTIS, L. F. & R. H. HASCHEMEYER. 1980. Proc. Natl. Acad. Sci. USA **77:** 3139–3143.
9. MOSESSON, M. W., J. HAINFELD, J. WALL & R. H. HASCHEMEYER. 1981. J. Mol. Biol. **153:** 695–719.
10. WILLIAMS, R. C. 1981. J. Mol. Biol. **150:** 399–408.
11. WEISEL, J. W., G. N. PHILLIPS, JR. & C. COHEN. 1981. Nature **289:** 263–267.
12. FOWLER, W. E., R. R. HANTGAN, J. HERMANS & H. P. ERICKSON. 1981. Proc. Natl. Acad. Sci. USA **78:** 4872–4876.
13. FERRY, J. D. 1952. Proc. Natl. Acad. Sci. USA **38:** 566–569.
14. WILLIAMS, R. C. 1977. Proc. Natl. Acad. Sci. USA **74:** 2311–2315.
15. WILLIAMS, R. C. & H. W. FISHER. 1970. J. Mol. Biol. **52:** 121–123.
16. SLAYTER, H. S. 1976. Ultramicroscopy **1:** 341–357.
17. DOOLITTLE, R. F., D. M. GOLDBAUM & L. R. DOOLITTLE. 1978. J. Mol. Biol. **120:** 311–325.
18. MARDER, V. J., N. R. SHULMAN & W. R. CARROLL. 1969. J. Biol. Chem. **244:** 2111–2119.
19. DOOLITTLE, R. F. 1981. Sci. Amer. **245**(6)**:** 126–135.
20. FOWLER, W. E., H. P. ERICKSON, R. R. HANTGAN, J. MCDONAGH & J. HERMANS. 1981. Science **211:** 287–289.
21. HANTGAN, R. R. & J. HERMANS. 1979. J. Biol. Chem. **254:** 11272–11281.
22. PIZZO, S. V., L. M. TAYLOR, JR., M. L. SCHWARTZ, R. L. HILL & P. A. MCKEE. 1973. J. Biol. Chem. **248:** 4584–4590.
23. HERMANS, J. 1979. Proc. Natl. Acad. Sci. USA **76:** 1189–1193.

DISCUSSION OF THE PAPER

H. P. ERICKSON (*Duke University Medical Center, Durham, NC*): If you construct the fibrinogen molecule in three dimensions as being a helix, I think you have to give up the intramolecular twofold axis, which, of course, is not established but is a very attractive idea.

R. C. WILLIAMS: No, it's established on a two-dimensional piece of paper.

ERICKSON: Right, but for a molecule that is a dimer, it is attractive to think that it might have a twofold axis. For me, one of the most exciting features of your images, and one of the hardest to explain, is why do all the molecules have the same handedness. I propose there is an extra nodule that lies along the twofold axis. It might be difficult for the fibrinogen to lie on the grid with this nodule on the bottom. If the α chain nodule always faced away from carbon film that would explain why the twofold views always have the same handedness.

THE STRUCTURE OF FIBRINOGEN AND FIBRIN: I. ELECTRON MICROSCOPY AND X-RAY CRYSTALLOGRAPHY OF FIBRINOGEN *

Carolyn Cohen, John W. Weisel †, George N. Phillips, Jr., Cynthia V. Stauffacher, James P. Fillers, and Elisabeth Daub

Rosenstiel Basic Medical Sciences Research Center
Brandeis University
Waltham, Massachusetts 02254

and

† Department of Anatomy
University of Pennsylvania
Philadelphia, Pennsylvania 19104

INTRODUCTION

One of the most basic and long-standing problems in the field of blood clotting has been the visualization of the three-dimensional structure of fibrinogen and its interactions to form fibrin. The very large size and complexity of the molecule have presented difficulties for structure determination. In the two decades since the Hall-Slayter model was first proposed, however, our understanding of protein conformation has advanced greatly, due largely to results from X-ray crystallography. At present, about 200 proteins have been solved to atomic resolution by this technique. One of the major findings to emerge is that there are a limited number of stable designs for polypeptide chain folding. During this period our understanding of fibrinogen has begun to reflect these new insights into protein architecture. For example, the establishment of the complete amino acid sequence of human fibrinogen, together with empirical rules derived from protein crystallography, have enabled alignment of the three pairs of chains so that the α-helical coiled-coil regions identified in images of the molecule [1] have been located in the sequence.[2, 3] Thus, we can now "see" fibrinogen—at least in our mind's eye—more clearly than was possible before.

But detailed information that reveals the precise polypeptide chain folding and arrangement of the various domains in the hydrated molecule can come only from X-ray crystallography. This large dynamic molecule poised to associate in a noncrystalline fibrous form—and present in a variety of heterogeneous states in the plasma—has not been easy to crystallize. A variety of crystals and microcrystals of modified fibrinogen are now available, however, and detailed structure determination is possible. In this paper we review our

* This work was supported by Grant AM17346–09 from the NIH to C.C. J.W.W. is an Established Investigator of the American Heart Association. G.N.P. was a fellow of The Medical Foundation, Inc. of Boston, Mass. J.P.F. is a fellow of the Muscular Dystrophy Association.

0077-8923/83/0408-0194 $01.75/0 © 1983, NYAS

general strategy for solving this structure and describe present results from electron microscopy and X-ray crystallography. We intend to show that the microcrystals and crystals provide the clearest pictures we now have of the molecule, and, as described in a later session,[4] its assembly as well into the fibrin clot.

ORDERED ARRAYS: MICROCRYSTALS

Our approach to this problem springs from fibrinogen's close relation to other fibrous proteins. Structural proteins are designed to self-assemble by weak specific bonds into the native fibrous structures. Solubilizing the protein and changing the conditions of precipitation can lead to the production *in vitro* of ordered assemblies. Very few of these giant molecules have in fact been crystallized, but paracrystals can often be obtained. These ordered forms can give precise information about molecular length, mass distribution and specific intermolecular interactions. Most determinations of the length of fibrous proteins, for example, have been made by analysis of the axial repeats of aggregates. Moreover, certain of these interactions may be very similar to those of the native fibers. The use of polymorphic *in vitro* forms and the recognition of this intrinsic self-assembly property allowed major advances to be made in our understanding of the structure and assembly of collagen, as well as proteins more closely related in folding to fibrinogen, such as the muscle proteins myosin, paramyosin and tropomyosin.[5] Additional advantages of these ordered arrays are that the molecules are stabilized by interactions so that they are less susceptible to specimen damage from both preparation and electron irradiation. Moreover, the regular repeats in the lattice of microcrystals allow the use of image processing techniques such as filtering or averaging to reduce noise. Thus a more detailed and reliable image can be obtained than is possible from studies of single molecules. The arrays seen in the electron microscope are two-dimensional projections of three-dimensional structures. Nevertheless, as we will describe, much information can be extracted from a close "reading" of the stain distribution of such images.

The unusual feature of fibrinogen in contrast to other fibrous proteins is that it occurs in two states: it is designed to be soluble in the plasma and to self-assemble into an insoluble form in fibrin. The strategy we have employed has been to focus on assemblies produced by fibrinogen, and to consider fibrin as but one other ordered state of the molecule. The justification for this approach is straightforward. First, the early wide-angle X-ray studies of Astbury, Bailey and Rudall showed that the α-helical domains in both molecules were similar, supporting the notion that "fibrin is no other than an insoluble form of fibrinogen."[6] More convincing evidence came from small-angle X-ray diffraction studies in the early sixties which showed that at high ionic strength both fibrinogen and fibrin give X-ray patterns with the same axial spacing of 225 Å and a similar intensity distribution.[7] This result demonstrated that the dimensions of the two molecules are the same, and that under these ionic conditions, where the charges were shielded, native fibrinogen can pack into an assembly very like that produced by fibrinogen upon loss of fibrinopeptides. Moreover, the dimensions available at that time from the Hall-Slayter model [8] indicated that this packing was likely to be a half-staggered arrangement of molecules 450 Å in length.[7] Somewhat later this finding was confirmed by electron microscopy: when native fibrinogen is precipitated at high ionic strength (above .1μ), the band pattern of the fibers formed is very similar to that of fibrin [9, 10] (FIGURE 1).

Thus the first ordered form of native fibrinogen to be discovered was a fibrin-like aggregate. A key point, however, is that although a simple half-staggered arrangement accounts for basic features of the fibrin assembly, neither the detailed structure of the molecule, nor its packing, can readily be deciphered from a study of fibrin-like forms. As Weisel et al.[4] describe, one must in fact analyze other highly ordered forms of fibrinogen to account for the complex organization of fibrin.

FIGURE 1. Electron micrograph of native bovine fibrinogen precipitated at high ionic strength (0.3M KF at neutral pH) negatively stained with uranyl acetate. The band pattern is very similar to that of fibrin. The axial repeat is 225 Å.

A happy accident led to the production of such forms. Rod-shaped aggregates were noted in samples of bovine fibrinogen precipitates that had been stored under nonsterile conditions at low ionic strength. These aggregates showed a good deal of order in the electron microscope; they were, in fact, microcrystals. Dr. Nancy Tooney, a postdoctoral fellow at the time, succeeded in growing the contaminating Pseudomonas bacteria and obtained a crude preparation of protease. This enzyme reacted with fibrinogen and produced a modified molecule that formed a wide variety of ordered arrays [10] (FIGURE 2). This finding that proteolytically modified fibrinogen could form highly organized structures was a basic breakthrough in our efforts to visualize the molecule.

Our studies have consisted of two distinct aspects: the first is the production of these ordered forms; the second is the analysis of the images to extract information about molecular structure and packing. We wish to emphasize that answers are often not accessible by simple inspection. What is required is

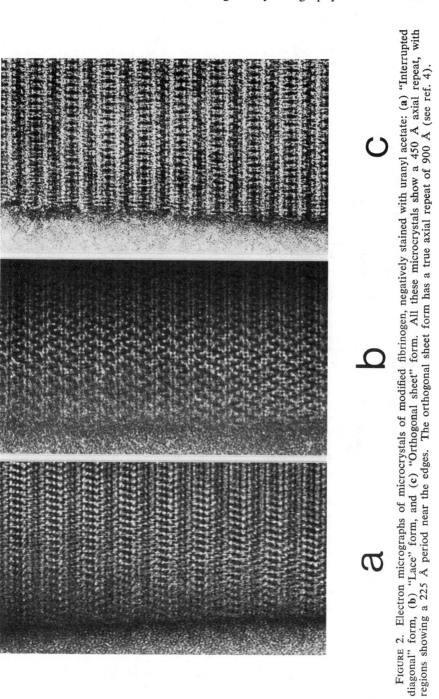

c

b

a

FIGURE 2. Electron micrographs of microcrystals of modified fibrinogen, negatively stained with uranyl acetate: (**a**) "Interrupted diagonal" form, (**b**) "Lace" form, and (**c**) "Orthogonal sheet" form. All these microcrystals show a 450 Å axial repeat, with regions showing a 225 Å period near the edges. The orthogonal sheet form has a true axial repeat of 900 Å (see ref. 4).

close scrutiny and image processing, and, basic to our strategy, the *coordinated examination* of as many ordered forms as possible.

We summarize here the essential facts about the molecule revealed by our initial review of these different microcrystalline forms. A basic finding is that, although a wide variety of band patterns have been observed, all of these forms of fibrinogen display a 450 Å axial repeat—or half this length (as in fibrin-like arrays or the so-called "simple axial form").[11] The interpretation of electron microscope images of simple rod-like molecules that are negatively contrasted allows a determination of the length and polarity of the molecules. In the case of both collagen and paramyosin, for example, the molecules pack with "gaps" and "overlaps" in the assemblies so that the axial period is considerably less than the molecular length.[5] Tropomyosin, however, bonds essentially end to end, and the period in paracrystals is close to the molecular length.[12] In contrast to these highly helical molecules, fibrinogen's complex shape makes interpretation of the nonuniform stain pattern of microcrystal images more difficult. Analysis of this problem shows, however, that the only consistent axial length for the molecules in all these forms is 450 Å. This finding indicates also that the fibrinogen molecules bond end to end to form filaments that build these different arrays.[11]

The arguments on molecular length are related as well to the problem of the symmetry of fibrinogen. The chemical evidence certainly indicates that fibrinogen is a dimer. Accordingly, it is a striking fact that all the microcrystalline forms and fibrin itself, have what we call "nonpolar" band patterns. "Polar" band patterns, in which the top and bottom of the repeating pattern are different, have not been observed in any aggregates. This observation is in marked contrast to results on proteins such as collagen, paramyosin or tropomyosin. All of these molecules are polar—that is, have a top and a bottom—and they form both polar and nonpolar arrays by specific parallel or antiparallel packing of molecules. The lack of polarity in all the arrays of fibrinogen so far observed implies that there is a twofold axis perpendicular to the long axis of the molecule.[11]

A final point that should be stressed here is that analysis of all these different microcrystalline forms, including fibrin, indicates that they may be accounted for by the superposition of a single basic array. Early in these studies a highly ordered rectangular array was discovered which showed packing units with detailed substructure. This so-called "orthogonal sheet" form (FIGURE 2c) could account for the band pattern of fibrin by a simple half staggering.[10, 11] Moreover, analysis of the images of the other forms indicated a close relationship of the orthogonal sheet form to virtually all these arrays. So the first information derived from a study of the paracrystalline arrays was that microcrystals of fibrinogen are built of molecular filaments making specific lateral contacts. And the key notion was advanced as well of a basic array related to the orthogonal sheet form, whose packing generated the great variety of structures, including fibrin, that were observed.[11]

CRYSTALS: PACKING MODEL FOR P2$_1$ FORM

The next advance came when it was found that, with further digestion, large single crystals, sometimes more than a millimeter in length, were produced[11, 13] (FIGURE 3). Thus, proteolysis by the crude enzyme preparation from *Pseudomonas* yielded a modified molecule that forms crystals. The crystals diffracted X-rays generally to 6 Å (and in certain directions to spacings as small as about 3 Å) so that the structure could be considered well ordered to that resolution (FIGURE 4). The molecular weight of the digested fibrinogen in the crystals was found to be about 320,000 daltons from a determination of the symmetry of the structure, the unit cell dimensions, and the crystal density. The modified molecule is therefore only about five to ten percent smaller than the native. Moreover, these molecules forming the crystals clot, and the fibrin so produced has a band pattern very similar to that of native fibrin.

We note here that proteolytic cleavage is now recognized as a useful way to produce modified protein molecules that crystallize. There are a number of

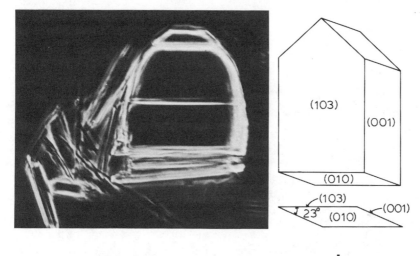

a   b

FIGURE 3. Crystals of modified bovine fibrinogen: (a) Phase contrast micrograph of P2₁ crystal. The crystals are thin plates; upon crushing, fragments break off along cleavage planes perpendicular to the long axis. The crystal width here is 0.25mm. (b) Diagram of typical crystal morphology. The (103) face is usually most prominent. The lower diagram shows an end-on view of the crystal.

examples: ribonuclease, elongation factor, chromatin. And, of course, the classic example is that of immunoglobulin; here, a variety of proteolytic products have been crucial in its structure determination. What proteolysis does in many cases is to excise wandering or randomly moving domains so that a more uniform and perhaps more static structure results, which can then crystallize.

In the case of bovine fibrinogen, limited cleavage which mainly affects the Aα chain produces highly ordered microcrystals. Further cleavage that may also affect the Bβ chain results in a molecule from which large three-dimensional crystals can be grown.[11, 13] Microcrystalline and crystalline forms of other species of fibrinogen, including rabbit, have also been produced in this

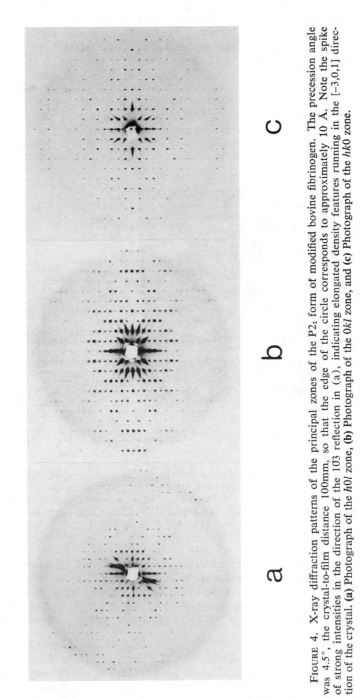

FIGURE 4. X-ray diffraction patterns of the principal zones of the P2₁ form of modified bovine fibrinogen. The precession angle was 4.5°, the crystal-to-film distance 100mm, so that the edge of the circle corresponds to approximately 10 Å. Note the spike of strong intensities in the direction of the 103 reflection in (a), indicating elongated density features running in the [−3,0,1] direction of the crystal. (a) Photograph of the h0l zone, (b) Photograph of the 0kl zone, and (c) Photograph of the hk0 zone.

fashion.[14] Although these forms provide considerable structural information, they are less suitable for single crystal structure determination than the bovine crystals. It is worth noting here, as well, that digestion with a number of other well-characterized enzymes has been tried with several species of fibrinogen, and that the products also yield both microcrystalline and macroscopic crystalline forms.[14] The molecule appears to be so designed that removal of some critical portions is necessary for crystal formation.

The proteolytically modified bovine fibrinogen crystals grown in low ionic strength buffer near the isoelectric point (pH 6.2) generally appear as thin birefringent plates. The crystal space group is monoclinic $P2_1$, with unit cell dimensions a = 135.3 Å, b = 98.1 Å, c = 176.0 Å, β = 92°. The crystal axes can be simply related to the morphology as shown in FIGURE 3. Density measurements give one molecule of fibrinogen per asymmetric unit and a solvent content of about 66.4% by volume. The relatively high solvent content accounts for the fragility of the crystals and suggests also that the crystal may comprise a somewhat open meshwork of molecules.

In general, one cannot deduce much information about a protein's structure or packing at this stage of the analysis. For fibrous proteins, however, some inferences can be made, although the crystallographic data do not provide strong constraints. The crystals readily cleave along the plane perpendicular to the long axis (FIGURE 3), indicating that the forces are weak between arrays of molecules related in this direction. When the X-ray beam is directed down this view of the crystal, the pattern shown in FIGURE 4a is obtained. One clearly sees a prominent spike of intensity. This diffraction feature arises from parallel arrays of the rod-like molecules aligned perpendicular to the spike. (An optical analog of an array of rods and its diffraction pattern is shown in FIGURE 5.) The direction perpendicular to the spike indicates the general run of these rod-shaped molecules which corresponds to the [−3,0,1] diagonal direction in the lattice, and the fall-off of intensity shows that the molecules are spaced about 40 Å apart in the crystal lattice.[14]

Ordinarily one cannot deduce a molecular dimension from unit cell parameters alone, because molecules can run across unit cell boundaries. Placing the axes of the molecules along the direction indicated by the spike of intensity, however, gives a distance of 448 Å between lattice points, so that molecules bonded end to end in this direction would have this length. On the basis of this argument, therefore, one can advance a plausible packing model for the fibrinogen molecule which yields a molecular length of 448 ± 6 Å (FIGURE 6). The presence of α-helical coiled-coil domains within fibrinogen also provides a marker for determining molecular orientation within the unit cell. The characteristic reflections from the turns of the α-helix would be expected to indicate the direction of the long axis of the molecule, and a crystal somewhat disordered (by glutaraldehyde fixation), did, in fact, show the α-fiber diagram perpendicular to the spike of intensity.[14] This packing also corresponds well with the crystal morphology (FIGURE 3).

An additional finding from these studies is that the critical diagonal dimension associated with the molecular length is preserved in a variety of states of the crystal. An interesting but sometimes troublesome feature of this crystal form is the tendency of the crystals to collapse, resulting in a large decrease in the c axis dimension accompanied by a change in β of up to 10°. (There seem to be two relatively stable points in this collapse, one with a change in unit cell parameters to c = 141 Å, β = 98° and another with c = 125 Å,

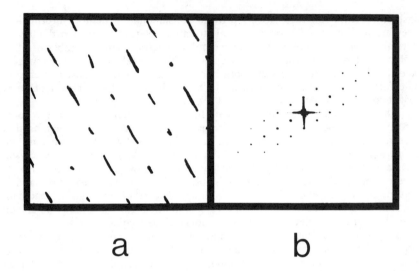

a   b

FIGURE 5. (a) An array of aligned rods located on a centered square net. (b) The optical diffraction pattern from (a) (From Berger *et al.*[18] By permission of Academic Press.)

$\beta = 100°$.) In all these cases, however, the length of the $[-3,0,1]$ lattice vector remains constant (within 2%), despite these unit cell changes. These findings extend observations made in the earlier stages of this work where fixation and staining of the crystals for electron microscopy resulted in very large unit cell changes [14]; here, too, the critical diagonal dimension associated with the molecular length is preserved throughout all stages of shrinkage. The invariance of this dimension is additional strong independent evidence that the molecules are 450 Å in length and bonded end to end. This strong linkage between molecular ends indicates that the fibrinogen crystal—like that of tropomyosin [12]—may be considered a lattice made up of molecular filaments.

Before discussing the problem of the shape or distribution of mass in the molecule, it is worthwhile to point out a basic constraint on molecular models that can now be advanced. As described in previous sections, we have shown that various aggregates of fibrinogen, including microcrystals or crystals—and fibrin as well—are made up of 450 Å long molecules bonded end to end. An important generalization springs from these observations: in all cases where a 450 Å axial repeat (or half this length) is maintained, the overall molecular shape and end-to-end contacts to form filaments may be quite similar. What appears to vary in all these different forms then is the lateral staggering of filaments.

The molecular shape can be determined in principle by these X-ray crystallographic studies. We are at present pursuing two levels of analysis: the first aims at an image of the hydrated molecule to a resolution of about 30 Å, where the general arrangement of domains can be determined, and some of the intermolecular contacts in the crystal identified. This study will also provide phase information for the higher resolution analysis. Data are being collected that will allow the molecule to be visualized to a resolution of about

6 Å. At this level, some details of the secondary structure (particularly coiled-coil regions) may be visualized. A brief progress report on this work will be given in a later section.

MOLECULAR MODELS FROM ELECTRON MICROSCOPY

One approach to solving the crystal diffraction pattern at low resolution involves refining a model for the molecule against the diffraction data. Although, in principle, one might in fact be able to phase the X-ray data using

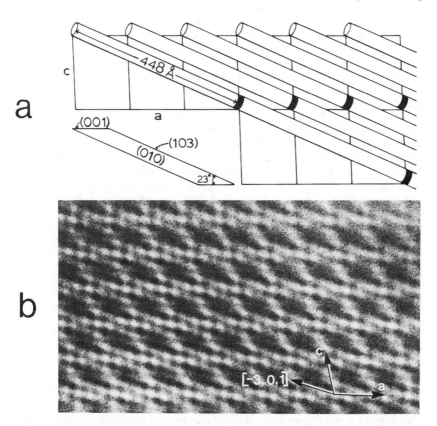

FIGURE 6. Diagram of the packing of fibrinogen molecules in P2₁ crystals of modified bovine fibrinogen. The molecules are represented schematically by a cylinder approximately 450 Å in length and 30 Å in diameter. (a) View of the P2₁ unit cell along the twofold screw (unique) axis. Only one molecule per unit cell is shown. The molecules run in the [−3,0,1] diagonal direction. Overall external morphology of the P2₁ crystal relative to the molecular axes (lower left). The dark bands represent end-to-end intermolecular contacts. (b) Optically filtered image of an electron micrograph of a net from a negatively stained, crushed crystal. Prominent features of macromolecular size, with a period of 450 Å, run in the [−3,0,1] diagonal direction (designated by the arrow), which is the molecular orientation proposed in the model of (a) above. Note that the unit cell is different from that of the untreated crystal in (a).

heavy atom derivatives alone, in practice—as in the case of tropomyosin, or virus structures—information about the general shape of the molecule reduces the magnitude of the problem. The most useful source of information for this mass distribution comes from electron microscopy. Since the resolution of this first stage of the crystal structure analysis is about the same as that of electron microscope images, a combination of the two methods is quite powerful.

Our strategy for determination of the model from electron microscopy is somewhat unusual: it depends on a coordinated analysis of virtually all the microcrystalline as well as macroscopic crystal forms and fibrin. We have argued above that the molecule (and certain of its contacts) is conserved in these different forms. Therefore, in the evaluation of a particular model, we consider that a persuasive test is internal coherence: that is to say, the ability of one model to account for all these various arrays.

We have described elsewhere,[15] and Weisel et al.[4] discuss in a later session, our approach to the improvement of the Hall-Slayter model. We emphasize here that there was no doubt early on that the shadowing technique, while providing direct images of individual molecules, could not resolve all the domains present. It has also been plain for some time that the Hall-Slayter model could not account for the details of the fibrin band pattern seen by negative staining or, in fact, for the X-ray intensities from hydrated gels of fibrin. Nor could this model account for the electron microscope images of the negatively stained $P2_1$ crystals or other microcrystalline forms.

There are a number of classes of images obtained from crushed $P2_1$ crystals, and by examination with "an informed eye" we were able to identify these views. Mesh images were found to be views along the unique axis of the crystal, that is to say, perpendicular to the (010) face, (FIGURE 6b) with differences in appearance depending on the degree of shrinkage or collapse of the crystal.[14, 15] In this projection, as expected from the molecular packing model, prominent features or strands of macromolecular size run in the $[-3,0,1]$ direction. These strands in the mesh are seen to be pairs of symmetry related molecular filaments in a cell distorted by dehydration. We cannot establish where the molecules begin or end, since this is a projection. Moreover, using a simple Hall-Slayter model, it was not possible to delineate individual molecules or to define the lateral packing. A striking fact, however, is that the size of the globular units seen is generally not greater than about 40–50 Å.

A form found useful in interpreting the crystal images was a "segment aggregate" obtained by Dr. Nancy Tooney from human fibrinogen modified by bacterial protease (FIGURE 7).[16] These segments displayed a strikingly simple band pattern in negative staining that reveals where the molecules begin and end. Analysis of this structure showed that the molecules are in register in layers which are tilted alternately with respect to the long axis. Here again the molecules were found to be about 450 Å long. Without describing the arguments in detail, the fine structure of the staining indicated that the Hall-Slayter model was valid at low resolution but that both the division of the outer domain into two units, and the addition of another smaller globular unit was necessary to account for the image. A new trial model—essentially a revision of the Hall-Slayter model—was thus formulated for fibrinogen. The molecule was represented as a heptad made up of seven globular regions connected by rod-like sections.[15]

Simulation of Electron Microscope Images of P2₁ Crystal

Using modifications of this model, computer simulations were carried out to develop a single molecular model whose substructure and packing could account for the negatively stained images of a variety of arrays seen in the electron microscope. The methods used have been described elsewhere,[15] and in another paper in this meeting we discuss further modeling for fibrin-like arrays.[4] The electron micrographs were filtered optically or by computer processing of a digitized image. Subjective criteria were used to alter parameters in comparing simulations to filtered electron microscope images. A range of molecular models was examined involving differences in the sizes of the globular units or changes in bending of the molecule. The first satisfactory fit was obtained with a slight bend in the center of the molecule.[15] Subsequent attempts to fit the X-ray data with this model indicated that some bending near the molecular ends was required. An improvement in the simulations could then be obtained with slight bends near the molecular ends (FIGURE 8). A basic criterion for goodness of fit was the agreement of all images using the same trial model.

Several images from P2₁ crystal fragments could be simulated. The net described previously and identified as the view down the unique axis of the crystal is characterized by strands having features of macromolecular size running in the horizontal direction (FIGURE 9a). The motif along each strand repeats every 450 Å and corresponds to the projection of two symmetry-related layers of molecular filaments in the unit cell. With a shift of about one-third the molecular length between filaments in the two layers the simulation shows good agreement with the filtered image of the electron micrograph (FIGURE 9b). Another commonly observed form has a mesh with a different appearance (FIGURE 9c). The motif is rather similar to that described above, although the intensities and positions of the bright units differ somewhat. The rod portion of the molecule can be visualized as well in some micrographs of the mesh.

FIGURE 7. Electron micrograph of "segment" aggregate of protease modified human fibrinogen negatively stained with uranyl acetate. The period is about 770 Å, and consists of alternately tilted layers of fibrinogen molecules 450 Å in length bonded end to end.

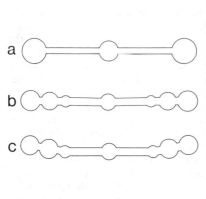

FIGURE 8. Models for the fibrinogen molecule: (a) Trinodular Hall-Slayter model, scaled to a length of 450 Å. The end domains are 65 Å in diameter and the central region is about 50 Å. (b) Heptad fibrinogen model derived from simulations of electron micrographs of crystals and microcrystals.[15] The molecule has a bend of a few degrees at the center. (c) "Bent" heptad model. The relative sizes of the domains of the model were determined from simulations of electron micrographs; the absolute scaling was based on the volume of the molecule in the crystal (3.8×10^5 Å³).[14] This model is characterized by an 11° bend near the ends. The diameters of the globular regions (starting from the end) are 48, 48, 34 and 45 Å. The diameter of the rod has been taken to be 22 Å.

Keeping the parameters of the molecular model used in the previous simulation, only a slight change in the relative stagger of the molecular filaments within each layer is required for a satisfactory fit (FIGURE 9d). A different view of the crystal could also be identified. Some preparations contained very thin crystals that tended to lie with the large face (103) on the grid. This image showed a distinctive pattern of alternating rows of bright and weak zig-zags (FIGURE 9e). The spacings indicated that this was the view along one of the crystal axes perpendicular to the previous view and nearly perpendicular to the large face of the crystal (see FIGURE 3). Using the same arrangement of molecules as before, and simply taking the appropriate projection, a satisfactory simulation can be achieved (FIGURE 9f). A similar analysis can be carried out for a variety of other microcrystalline forms and fibrin as well.[4]

In summary, simulations of electron microscope images of negatively stained crystals indicate that fibrinogen can be represented at low resolution as a molecule 450 Å long consisting of seven globular domains linked by rod-like regions. The present results indicate that a departure from colinearity of about 11° at each end gives a good fit with the data (FIGURE 10). This preliminary picture should be regarded as a working model to be used with the X-ray results for establishing the molecular boundaries.

MOLECULAR MODEL FROM X-RAY CRYSTALLOGRAPHY

The crystal diffraction data at low resolution can reveal the molecule in a hydrated unstained state and provide a more detailed image than models derived from simulations of electron microscope images. As indicated above, one approach to the solution of the low resolution structure involves the modeling of these diffraction data. Given the length of the molecule, the general packing in the crystal and a model for the mass distribution from the electron micrograph simulations, diffraction patterns for various models can be calculated and compared to the observed X-ray intensities. Adjustments to the model (for example, the size and locations of the globular regions or the packing positions for the molecular filaments) can be made to match the observed X-ray data as closely as possible. This procedure can be followed with a technical measure of goodness of fit (known as an "R factor") which decreases as the fit improves. Without describing the results in detail, we may summarize the development of a model for the 30 Å X-ray data.

It would be convenient if one could describe the X-ray pattern of the P2₁ fibrinogen crystal in relation to various physical features of a model. This is, in general, not possible for X-ray crystallographic data since all parts of the structure contribute to every reflection and the overall intensity distribution is "sampled" by the lattice. Nevertheless, fibrous protein molecules such as fibrinogen can produce certain characteristic intensity features in the X-ray pattern. As indicated earlier, a striking feature of the pattern is the spike of intensity (shown in FIGURE 4a) which reflects the run of the molecular filaments of fibrinogen along the [−3,0,1] direction of the crystal lattice. Although this spike clearly results from parallel packing of rod-like fibrinogen molecules, the distribution of intensity among the reflections is very sensitive to details of molecular shape and position along this diagonal direction. A similar study can be made of other regions of the diffraction pattern from the crystal, but it is difficult to extract definite features of the model by simple inspection. In the comparison with trial models, all these intensity data are taken into account.

For illustrative purposes, it may be useful to demonstrate the diffraction from a series of molecular models that have been advanced historically to account for fibrinogen, including a cylinder, the Hall-Slayter model, the simple linear heptad model and a heptad model with a slight bend at the ends. In building these models, constraints described previously have been applied: these include a 450 Å projected length for the molecule, a twofold symmetry axis in the molecule, a one-third stagger of the molecular filaments as predicted from electron microscope images, and a limit of 35% for the volume of the crystal occupied by the molecules. In FIGURE 11, diffraction from these models is compared with the crystal diffraction in the h0l projection, where the strong intensity spike is seen.

A simple cylindrical model for fibrinogen filaments (FIGURE 11a) produces diffraction intensities which account only for the basic molecular packing. As FIGURE 11a demonstrates, a relatively featureless pattern results from this model. All of the intensity is concentrated in a few strong reflections which indicate the stacking of the molecular filaments in the crystal. The diffraction from this model falls off sharply at 40 Å, reflecting the side-to-side packing of these rods. Modeling fibrinogen as a simple trinodular Hall-Slayter molecule (FIGURE 11b) significantly improves the fit to the X-ray pattern, although the intensity relations are clearly not correct. A simple linear heptad model further improves the fit, concentrating intensity in major strong reflections (FIGURE 11c). At the same time, however, the fit to other regions of the three-dimensional data is somewhat less good. A distinct improvement is obtained with a slight bend near the molecular ends. We may note here that the bend introduced is limited to a range of about 0–20° in order to preserve end-to-end bonding of the 450 Å long molecules. Moreover, the simulations of electron micrographs, described earlier, suggest that both ends may bend in the same direction; and that the degree of bending may be about 11°. Preliminary results from the X-ray modeling show that this model gives a pattern

a

b

c

d

e

f

FIGURE 9. Simulations of electron microscope images of the P2₁ crystal.

(a,c,e) Electron micrographs of negatively contrasted crystal fragments that have been filtered optically or by computer processing of a digitized image.[15] (a) and (c) are two crystal meshes with different amounts of shrinkage, which are viewed along the unique axis. In (a) and (c) the molecules run in a horizontal direction. The repeating unit along the strands is a group of seven stain-excluding regions of varying sizes (none larger than about 50 Å), followed by a dark-staining gap. The image in (e) corresponds to the same crystal form as in (c), but viewed in a perpendicular direction.

(b,d,f) Video display using bent heptad model of fibrinogen to simulate electron microscope images. The relative brightness of the different units of the motif depends on the size of the globular domains and their degree of overlap. In each simulation, the same molecular filaments are used with a slightly different lateral stagger in the two mesh images.

with an improved fit to the observed diffraction (FIGURE 11d). We are continuing to modify this model by combined studies of both electron microscope and X-ray results. Once a molecular shape is obtained which provides reliable initial phases, successive cycles of difference Fourier refinement can be used to improve the agreement between observed and calculated intensities until the molecular boundaries are established.

A quite different (and more objective) approach to solving the structure of fibrinogen is to phase the reflections by the use of isomorphous replacement, that is to say, to go directly from the X-ray pattern to the molecular image. (This method may be useful at low resolution, but is required for the data at 6 Å.) In order to use this technique, fibrinogen must be labeled with heavy atoms. This addition of electron density to the molecule causes measurable changes in the intensities of the X-ray reflections from which phases can be extracted. With these independently determined phases the molecular structure can then be calculated without any assumption of a model. The very high molecular weight of fibrinogen requires a very large addition of electron density, either by attaching a number of heavy atoms at enough sites to add up to significant intensity changes, or by use of a super heavy atom complex that

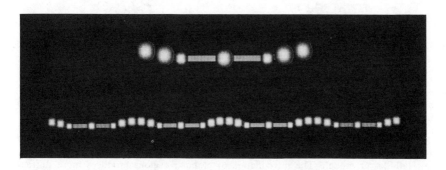

FIGURE 10. Computer drawing of bent heptad model, and molecular filaments. These are staggered by about one-third the molecular length in the crystals, and by one-half in fibrin.

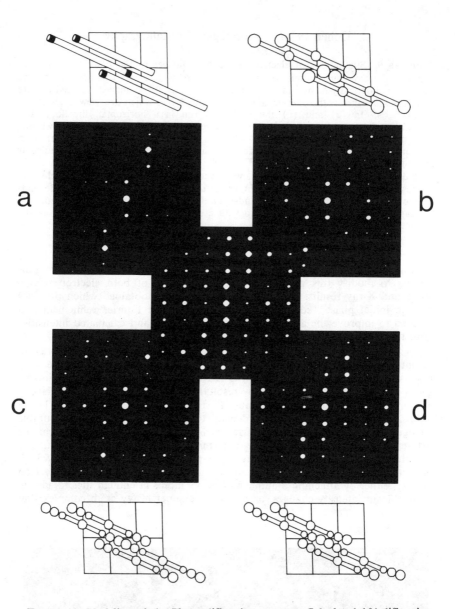

FIGURE 11. Modeling of the X-ray diffraction pattern. Calculated $h0l$ diffraction patterns from four fibrinogen models of increasing complexity are shown with the observed diffraction pattern in the center for comparison. The line drawings indicate the nature of the end-to-end bonding and the relative stagger of the molecules in the packing. The R-factor represents the fit of the three-dimensional X-ray data set to that generated for each model. (a) Cylindrical model (with varying density). Only a spike of intensity perpendicular to the molecular axis is generated (R = 91.4%). (b) Hall-Slayter model. The introduction of globular regions produces more diffraction, but the overall distribution of intensity does not match the observed pattern (R = 86.1%). (c) Linear heptad model. The division of the end globular regions into the domains of this model improves the agreement between the observed and calculated patterns (R = 78.3%). (d) Bent heptad model. Movement of density away from the [–3,0,1] direction of the end globular regions by a rotation of about 11° produces an improved fit between observed and calculated patterns (R = 59.6%).

might bind to a few regions. Both approaches are somewhat difficult technically but the problem is feasible.

At present we have surveyed specific derivatives which should be bound at only a few known sites on the molecule, and some of these may be of biological interest. Calcium ions in the three strong binding sites on fibrinogen, for example, can be substituted by the lanthanide chlorides or uranyl fluoride. These derivatives have thus far, however, resulted in a disruption of the crystals. A super heavy osmium complex that should bind the carbohydrate regions of the molecule appears more promising, as well as the standard gold and platinum chlorides which, though smaller in size, bind quite extensively to fibrinogen and produce intensity changes in the X-ray patterns.

Identification of good heavy atom derivatives will also be crucial for the high resolution study now under way. Data to about 3–6 Å are being collected by oscillation photography and the complete 6 Å native data set is now being processed (FIGURE 11). At this resolution, the "rotation function" [17] may also be useful. This analytical treatment of the data may allow the localization of an intramolecular twofold axis in fibrinogen. This information can then be combined with the heavy atom derivative data to improve the phase information. The results can provide an image of the hydrated structure that will clarify a number of puzzling features about the molecular shape. Moreover, at this resolution, the run of the polypeptide chains in fibrinogen may be visualized so that a correlation with sequence studies will finally be available.

PERSPECTIVE

The models described here represent the first stage in our approach to the detailed three-dimensional structure of fibrinogen. They were derived from a coordinated application of electron microscopy and X-ray crystallography to microcrystals and crystals of modified fibrinogen. We should emphasize, however, that the boundaries of the molecule can only be considered as tentative until the low resolution crystallographic study is completed. In another paper,[4] we discuss our related approach to the structure of fibrin. Our results indicate that the molecular filaments revealed by a study of the crystals and microcrystals of modified fibrinogen can account as well for the architecture of fibrin. In that paper the structural and biological significance of the different domains of the present molecular model will also be described.

The crystallographic analysis of fibrinogen is the most powerful and, indeed, the most objective method to determine the precise structure of this molecule. Such detailed information is essential for any real understanding of function. A variety of crystalline forms of the molecule are now becoming available in a number of laboratories, and a solution to this stubborn problem will certainly be forthcoming. When the detailed structure is finally revealed, we may expect that the simplified images of fibrinogen that we now contemplate will be surpassed by far in beauty and complexity by the actual design of this formidable molecule.

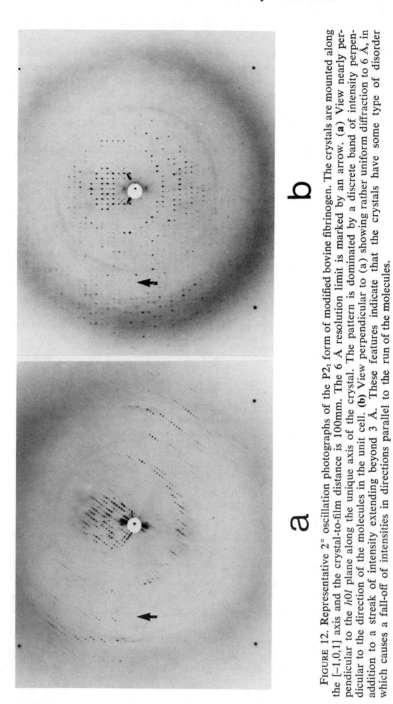

FIGURE 12. Representative 2° oscillation photographs of the P2₁ form of modified bovine fibrinogen. The crystals are mounted along the [-1,0,1] axis and the crystal-to-film distance is 100mm. The 6 Å resolution limit is marked by an arrow. (a) View nearly perpendicular to the *h0l* plane along the unique axis of the molecules in the unit cell. (b) View perpendicular to (a) showing rather uniform diffraction to 6 Å, in addition to a streak of intensity extending beyond 3 Å. These features indicate that the crystals have some type of disorder which causes a fall-off of intensities in directions parallel to the run of the molecules.

ACKNOWLEDGMENTS

We thank Paul Norton for electron microscopy, and William Saunders and Vicki Ragan for photography.

[**Addendum:** One of the memorable aspects of this Conference has been the presentation of remarkable new results from electron microscopy of individual fibrinogen molecules. There are now images from a number of laboratories showing more detailed structural features than had been visualized previously. Despite some differences in appearance, there is a good measure of agreement on overall molecular shape. It is important to recognize, however, that interpretations of our own model as well as these various images are all based on two-dimensional projections of a three-dimensional structure. Apparent dissimilarities between these pictures may be due, therefore, not only to differences in preparative techniques, but also to the fact that these models may represent different views of similar structures.]

REFERENCES

1. COHEN, C. 1961. J. Polym. Sci. **49:** 144–145.
2. PARRY, D. A. D. 1978. J. Mol. Biol. **120:** 545–551.
3. DOOLITTLE, R. F., D. M. GOLDBAUM & L. R. DOOLITTLE. 1978. J. Mol. Biol. **120:** 311–325.
4. WEISEL, J. W., G. N. PHILLIPS, JR. & C. COHEN. 1983. Ann. N.Y. Acad. Sci. **408:**. This volume.
5. COHEN, C., A. G. SZENT-GYORGYI & C. KENDRICK-JONES. 1971. J. Mol. Biol. **56:** 223–237.
6. BAILEY, K., W. T. ASTBURY & K. M. RUDALL. 1943. Nature **151:** 716–717.
7. STRYER, L., C. COHEN & R. LANGRIDGE. 1963. Nature **197:** 793–794.
8. HALL, C. E. & H. S. SLAYTER. 1959. J. Biophys. Biochem. Cytol. **5:** 11–16.
9. COHEN, C., H. SLAYTER, L. GOLDSTEIN, J. KUCERA & C. HALL. 1966. J. Mol. Biol. **22:** 385–388.
10. TOONEY, N. M. & C. COHEN. 1972. Nature **237:** 23–25.
11. TOONEY, N. M. & C. COHEN. 1977. J. Mol. Biol. **110:** 363–385.
12. COHEN, C. 1975. Sci. Amer. **233**(5): 36–45.
13. COHEN, C. & N. M. TOONEY. 1974. Nature **251:** 659–660.
14. WEISEL, J. W., S. G. WARREN & C. COHEN. 1978. J. Mol. Biol. **126:** 159–183.
15. WEISEL, J. W., G. N. PHILLIPS, JR. & C. COHEN. 1981. Nature **289:** 263–267.
16. WEISEL, J. W., N. M. TOONEY, I. KAPLAN, D. AMRANI & C. COHEN. 1980. J. Mol. Biol. **143:** 329–334.
17. ROSSMANN, M. G. & D. M. BLOW. 1962. Acta Crystallogr. **15:** 24–31.
18. BERGER, J. E., C. A. TAYLOR, D. SHECHTMAN & H. LIPSON. 1972. *In* Optical Transforms. H. Lipson, Ed.: 401–422. Academic Press. New York.

LASER DIFFRACTION OF ORIENTED FIBRINOGEN MOLECULES *

Rotraut Gollwitzer, Wolfram Bode, Hans-Joseph Schramm,
Dieter Typke, and Reinhard Guckenberger

Max Planck Institute for Biochemistry
D-8033 Martinsried/Munich, Federal Republic of Germany

INTRODUCTION

Fibrinogen is an extended protein, organized in a few characteristic domains, which possesses various sites for precise and strong interactions with enzymes, metals and neighboring fibrin(ogen) molecules. The overwhelming information accumulated in recent years about the chemical and immunological characteristics of the molecule and the elucidation of the complete sequence of human fibrinogen provided a more detailed understanding of the fibrinogen-fibrin conversion and the accommodation of the different domains in the aggregation process. However, the overall structure of fibrinogen is still a controversial subject and several contradictory models have been discussed. The three-nodular structure proposed by Hall and Slayter in 1959 [1] and confirmed in recent publications [2,3] reflects best our knowledge of fragments obtained by enzymatic or chemical cleavage and of the refined primary structure.

X-ray crystallographic analysis is unquestionably the method of choice to obtain a detailed three-dimensional structure. Cohen and coworkers [4,5] succeeded in producing crystals of partially digested bovine and rabbit fibrinogen. An X-ray structure analysis of these crystals has not yet been published. Electron microscopy has been shown to be a reasonable approach to gain information about the gross structure and especially the alignment of the single molecules in fibrinogen aggregates and in fibrin fibers.[6-8]

In this paper we present electron microscopic studies on needle-like (para)-crystals of native human fibrinogen and describe the evaluation and interpretation of these electron micrographs by laser diffraction and by autocorrelation.

MATERIALS AND METHODS

All reagents used were p.a. grade; all buffers and solutions were made with Ca^{2+}-free bidistilled water. Human fibrinogen was purified from fresh citrated plasma not older than one day. The first precipitation was carried out at 4° C by slowly adding glycine up to a concentration of 2.1 M during gentle stirring. The suspension was slowly stirred for at least two hours and centrifuged ($2000 \times g$, 4° C, 30 minutes). The precipitate was dissolved up to a concentration of 1.5% w/v in 0.055 M sodium citrate buffer, pH 6.35, at room temperature and dialyzed at 4°C against at least three changes of the same

* This work was supported by the Deutsche Forschungsgemeinschaft and the SFB 51.

0077–8923/83/0408–0214 $01.75/0 © 1983, NYAS

buffer overnight. This material was considered equal to Blombäck's fraction I-0, and further purification was performed according to Blombäck and Blombäck.[9] After centrifugation, the precipitated Blombäck fraction I-4 was immediately dissolved in the buffers used for crystallization and rotary shadowing of fibrin formation. These solutions were dialyzed at 4°C against at least three changes of buffer. None of the material used for crystallization or electron microscopy was lyophilized or concentrated mechanically.

Preparation of fibrin monomers was carried out by incubating 2 ml of a 3–5 mg/ml fibrinogen solution (fraction I-4) in 0.055 M citrate buffer, pH 7.5, with 10 units (NIH) of bovine thrombin (Thrombinum purum (R), Behringwerke) for two hours. Subsequently the resulting clots were synerized, washed extensively with 0.15 M sodium chloride solutions, and dissolved in 2% acidic acid. These solutions were immediately dialyzed against at least three changes of 0.05 M ammonium formate pH 5.7 at 4°C and diluted for rotary shadowing. The fibrin(ogen) concentration was determined spectroscopically according to Hörmann and Gollwitzer.[10]

Crystallization was performed by vapor diffusion, as described by Fehlhammer and Bode.[11] Crystalline needles were obtained at 20°C from fibrinogen solutions (15–25 mg/ml) either in citrate buffer (pH 7 to 8), starting at 0.05 M and concentrating to 0.15 M up to 0.3 M, or in 0.01 M phosphate buffered sodium chloride (pH 7 to 8), starting at 0.3 M and concentrating to 1.0 M up to 1.75 M. All solutions were centrifuged at low speed before crystallization. Furthermore, 0.02% sodium azide was added to prevent microbial growth.

For electron microscopy, the crystalline needles were picked up on copper grids with carbon film and negatively stained with 2% ammonium molybdate, pH 7.0, containing 0.5% glucose. Electron micrographs of the needles were taken in an electron microscope (Siemens Elmiskop 102) at 100 kV accelerating voltage and 53,000 electron-optical magnification (photographic material: Electron Image Plates from Kodak).

Laser diffraction was carried out on a light-optical bench with a helium-neon laser. Diffraction patterns were taken using several apertures of different diameter. The largest aperture, for which a diffraction spot in the direction perpendicular to the fibers occurred, was 5 mm. For the autocorrelation of cut-outs of the electron micrographs, an incoherent light-optical correlation device, developed by R. Guckenberger,[12] was used. This light-optical analog computer creates the two-dimensional correlation function of two (for autocorrelation identical) images with high speed.

For the preparation of the single molecules, the rotary shadowing technique from Shotton et al.[13] was adapted. Protein samples were dissolved in 0.05 M ammonium formate (20–30 μg/ml), pH 7.5, and after addition of an equal volume of 70% glycerol sprayed onto freshly cleaved mica discs from a distance of 30 cm. The samples were then immediately brought into a vacuum chamber of an Edwards vacuum coater model 306 and evacuated to 5×10^{-6} torr. An Edwards electron bombarded source was used for shadowing the proteins with platinum at an angle of 9°, followed by carbon coating at 90°. A platinum wire (length 5 cm, diameter 0.2 mm) was coiled around a tungsten rod (diameter 2 mm) and completely evaporated at 4 kV and an emission current of 50 mA. The distance to the mica discs mounted on a rotating table (120 rev/min) was 15 cm. Carbon was evaporated at 4 kV and an emission current between 50 and 100 mA for 10 sec. The replicas were cut into 2 to 3 mm diameter fragments, floated onto distilled water and picked up on 400 mesh copper grids. Specimens were examined in an electron microscope (Siemens Elmiskop 102). Electron micrographs were taken at magnifications between 20,000 and 50,000. The magnification was calibrated by photography of T4 phage tails (length 950 nm), negatively stained with uranyl acetate and catalase crystals (periodicity 8.75 nm). The absolute error in the magnification factor is estimated to be less than ±5%.

RESULTS

Small birefringent, needle-like crystals (FIGURE 1) were obtained from native human fibrinogen by vapor diffusion as described above. Optimal conditions were 20°C, protein concentration 15–18 mg/ml, and concentrating against 0.28 H citrate buffer at pH 7.5. The first needle-like crystals appeared after about three days and grew to a maximal length of 0.5 mm and a maximal diameter of ca. 0.01 mm. Whereas the single needles did not grow any further either in length or breadth, the number of needles increased rapidly. Finally the needles became interconnected by forming a loose matrix. Neither variation of protein- or salt-concentration or temperature, nor the addition of various metals led to an improvement of crystal size. Analogous experiments with highly purified bovine fibrinogen did not yield any crystals.

Isolation, washing and manipulation of the needles for mounting, e.g., in X-ray capillaries or on copper nets, was difficult due to their very soft property. Handling was much easier when the needles had been crosslinked with 0.1% glutaraldehyde in advance.

These needles consisted of native fibrinogen as proven by coagulation and gel electrophoresis. Clottability of the washed and dissolved crystals amounted to 88–95%. Sodium dodecylsulfate-gel electrophoresis of the nonreduced (not shown) and reduced samples (FIGURE 2) clearly showed that the needles consist solely of native fibrinogen, as the α-chain of the crystalline material is not detectably shortened. In comparison spherulites obtained from human fibrinogen digested by *Pseudomonas* proteases or by α-chymotrypsin under conditions described by Cohen et al.,[6] possess no band in α-position and a shortened β-chain.

At an early stage of their growth the crystals are suitable for electron microscopy. When negatively stained with ammonium molybdate at pH 7, they reveal a characteristic, periodic cross striation pattern, showing no polarity (FIGURE 3). One prominent continuous light band is adjacent to two less continuous and less distinct bands per period. The bands seem to consist of spherical, unstained globules of an approximate diameter of 3–5 nm. Each triad is separated from the next by a larger interval. The identity period along the horizontal axis of the micrographs (corresponding to the long axis of the needles) is 22.5 nm. There is only a weak periodicity in the direction perpendicular to the main axis of the needles indicating relatively poor order in this direction. Although this partial disorder shown by the electron micrographs could be a preparation artifact, we prefer in the absence of clear cut proof, to denominate the needles in the following as paracrystals.

To gain more information about the periodicity in the perpendicular direction, the micrographs were studied by laser-diffraction and by correlation with an incoherent light optical correlation device.[12] When diffraction was

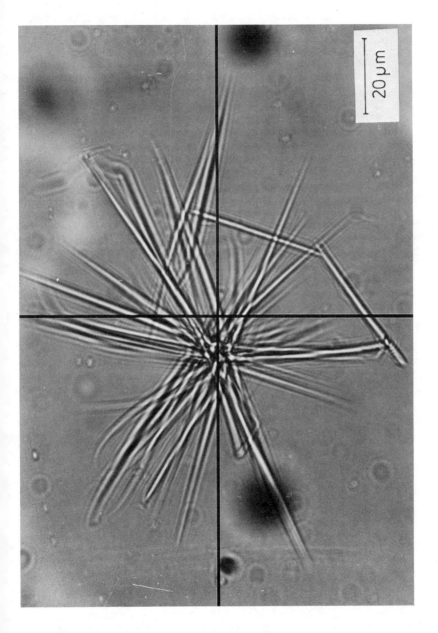

FIGURE 1. Crystalline needles of native human fibrinogen crosslinked with 0.1% glutaraldehyde.

carried out with a plane wave of laser light, it was necessary to reduce the diffracting area of the electron micrograph to ca. 5 mm in diameter (corresponding to 0.4 nm in the object) in order to obtain clear diffraction spots in the perpendicular direction (insert in FIGURE 4). In the horizontal direction, the spots corresponding to the third and fourth order of the 22.5 nm periodicity are easily observed. However, even with this small aperture, clear diffraction spots in the perpendicular direction were obtained for only a few regions of the image, although the whole area of the paracrystal in the original electron

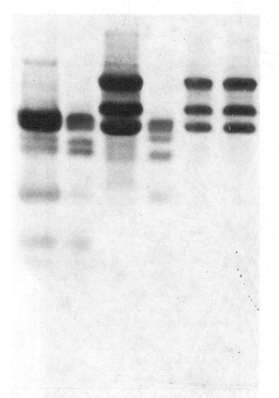

FIGURE 2. SDS-gel-electrophoresis (5.5% gel) of reduced samples: (A) native human fibrinogen; (B) paracrystals from human fibrinogen; (C) spherulites from human fibrinogen digested by α-chymotrypsin; (D) spherulites from human fibrinogen digested by *Ps. aeruginosa* proteinases.

micrograph (ca. 20 × 80 mm, corresponding to ca. 400 × 1600 nm of the paracrystal) was scanned. The observed optical diffraction maxima, obtained from different regions of the electron micrograph, correspond to vertical periodicities in the range of 6–8 nm, with periodicities from 7 to 7.5 nm predominating.

With the correlation device, the two-dimensional autocorrelation function of an image region or the cross-correlation function of different image regions can be determined. The autocorrelation function is in principle the Fourier

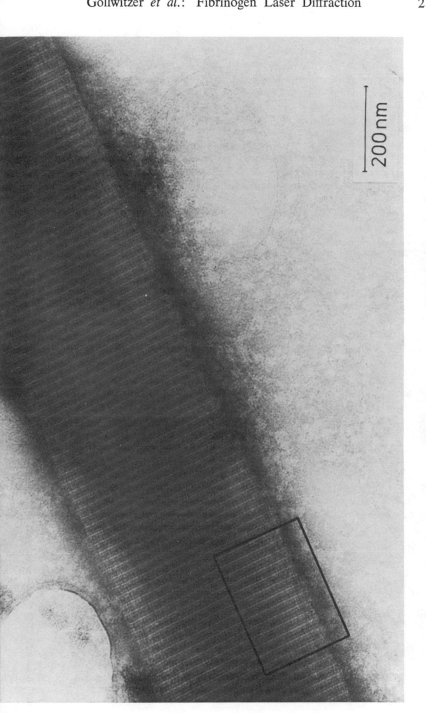

FIGURE 3. Electron micrograph of a crystalline needle from native, human fibrinogen, negatively stained with 2% ammonium molybdate, pH 7.0, containing 0.5% glucose. Original electronoptical magnification, × 53,000; reduced to 83% of original size. The framed section is reproduced in FIGURE 4.

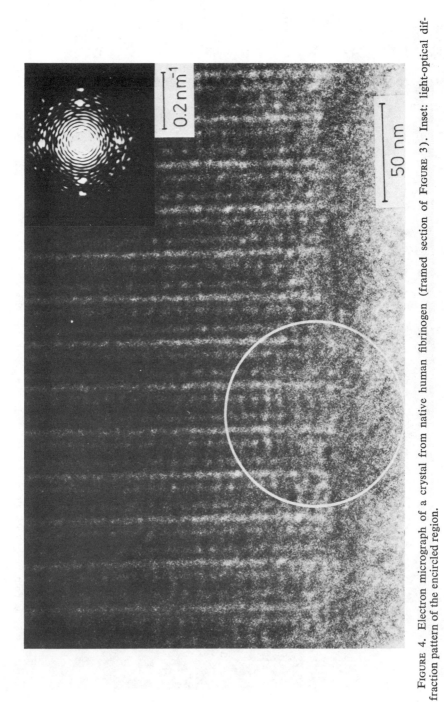

FIGURE 4. Electron micrograph of a crystal from native human fibrinogen (framed section of FIGURE 3). Inset: light-optical diffraction pattern of the encircled region.

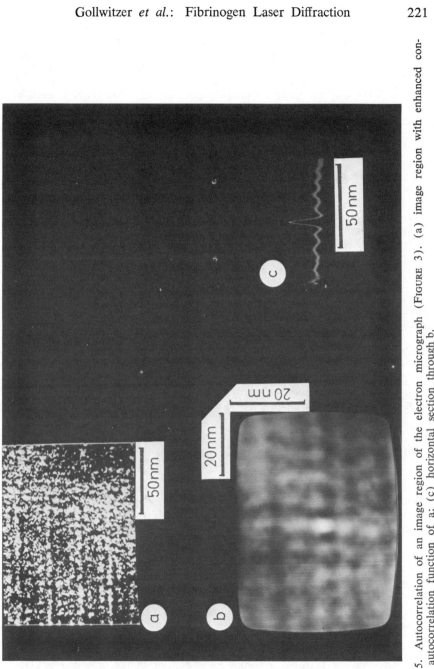

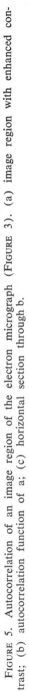

FIGURE 5. Autocorrelation of an image region of the electron micrograph (FIGURE 3). (a) image region with enhanced contrast; (b) autocorrelation function of a; (c) horizontal section through b.

transform of the diffraction pattern. It allows comparisons of different sub-regions within a studied region and can be used to determine periodicities or characteristic distances. An advantage of our correlation device is that pictures can be manipulated by electronic means for contrast enhancement. In FIGURE 5 the contrast-enhanced image region and its autocorrelation function are displayed on TV screens. A horizontal section through the autocorrelation function containing the highest peak is displayed on an oscilloscope. The Y direction of this trace corresponds to the light intensities, and the X direction to the distances along the line of intersection between the section and the auto-correlation function. A typical separation of 7.5 nm over at least four periods was observed in the direction perpendicular to the main axis of the para-crystal.

FIGURE 6 shows elctron micrographs of human fibrinogen and soluble fibrin after rotary shadowing. The overall length of straight molecules of fibrinogen and fibrin amounts to 42–45 nm. The three-nodular structure of the extended isolated molecules, reminding of the early model of Hall and Slayter,[1] is clearly to be seen. In many molecules the internodular connections are visible and seem to allow a flexible arrangement of the nodules. In several of the molecules a sigmoid, propeller-like shape of the string connecting the domains can be observed. If this feature is real, it would indicate that the molecule should have a twofold rotation axis perpendicular to the preparation plane, possibly suggesting an antiparallel linkage of both identical halves of the molecule. In most of the triade structures the outer nodules are essentially bigger than the central domain and have a somewhat ellipsoidal shape. An interesting feature are one or sometimes two satellite nodule(s) detectable in the neighborhood of the middle domain, especially in fibrinogen molecules. It is tempting to hypothesize that this could be part of the supposedly flexible C-terminal half of the α-chain.

DISCUSSION

A tentative alignment of the single molecules in the paracrystals was made (FIGURE 7). Owing to the twofold symmetry of fibrinogen, the molecular centers should be placed on the broad unstained bands that are apparent mirror lines in the electron micrograph. Because of the dimensions of single molecules (FIGURE 6) the most probable positions of both outer domains of each molecule are the first faint bands of the two triads on either side.

A model must allow for the feature that the apparent globular structures within the thin and the broad unstained bands seem to be arranged in lines perpendicular to the striation. Consequently, the central globular structures can in principle belong to two terminal globuli of the same line or of an adjacent line, thus leading either to a straight molecule (no model shown) or a bent molecule (as shown in FIGURE 7). Between the central and the terminal globuli is a broad stained band of about 13 nm width, which could accommodate the invisible, rod-like coiled-coil structures proposed by Doolittle [14] to connect the three globuli. The C-terminal part of the α-chain consisting of about 400 residues is still present in the paracrystals. It must be accommodated somewhere in the structure and may, possibly by means of many-fold contacts, be organized in a more compact form in the paracrystal

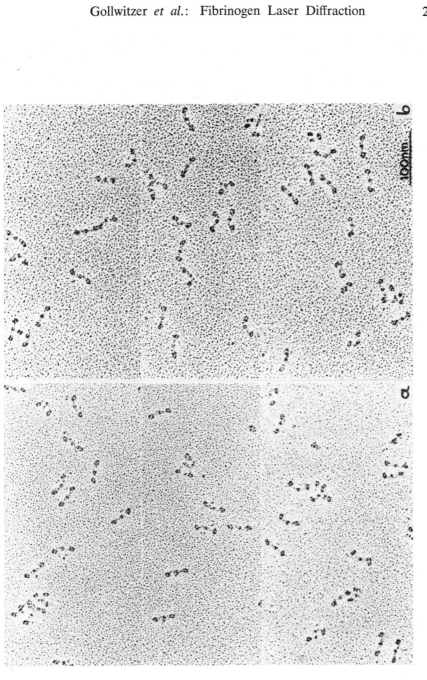

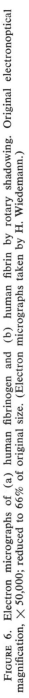

FIGURE 6. Electron micrographs of (a) human fibrinogen and (b) human fibrin by rotary shadowing. Original electronoptical magnification, × 50,000; reduced to 66% of original size. (Electron micrographs taken by H. Wiedemann.)

than in the isolated molecule (FIGURE 6). The corresponding C-terminal moieties could extend beyond the distal D domains and be positioned between the central domains of adjoining fibrinogen molecules.

Thus individual molecules can extend according to the electron micrographs from 38 nm to about 45 nm. Neighboring molecules overlap for about one third of their length. Surprisingly, this proposed arrangement of the fibrinogen molecules within the paracrystals is similar to that in fibrin, in which, due to the creation of new affinity centers, one E and two D domains of two different molecules are packed closely together.

However, we wish to point out that this tentative alignment of the molecules within the paracrystals does not necessarily need to resemble that of fibrin molecules within their respective fibers. In general, there are only a few inter-molecular contacts in protein crystals that stabilize the almost undistorted

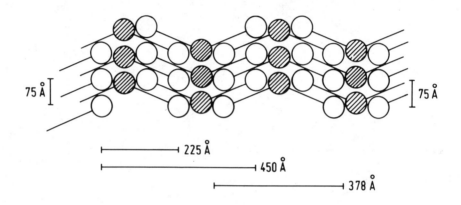

FIGURE 7. Tentative alignment of fibrinogen molecules in the paracrystal.

protein conformation of individual molecules in contrast to three-dimensional protein aggregates, in which protein molecules are mostly connected via large contact areas (see e.g., the contact area in the trypsin-trypsin inhibitor complex).[15]

Our model corresponds roughly to the fibrinogen models proposed by Cohen et al.,[16] based on their crystals and aggregates, which contain modified bovine fibrinogen. However, our electron micrographs do not display additional globular subunits. These different features could be due to the reduced length of the fibrinogen chains of their partially digested material.

ACKNOWLEDGMENTS

We thank Miss I. Mayr, Mrs. E. Weller, Mrs. Z. Cejka and Mrs. H. Wiedemann for their expert technical assistance.

REFERENCES

1. HALL, C. E. & H. S. SLAYTER. 1959. J. Biophys. Biochem. Cytol. **5:** 11–16.
2. KRAKOW, W., G. ENDRES, B. SIEGEL & W. SCHERAGA. 1972. J. Mol. Biol. **71:** 95–103.
3. FOWLER, W. E. & H. P. ERICKSON. 1979. J. Mol. Biol. **134:** 241–249.
4. COHEN, C., H. S. SLAYTER, L. GOLDSTEIN, J. KUCERA & C. E. HALL. 1966. J. Mol. Biol. **22:** 385–388.
5. TOONEY, N. M. & C. COHEN. 1977. J. Mol. Biol. **110:** 363–385.
6. WEISEL, J. W., S. E. WARREN & C. COHEN. 1979. J. Mol. Biol. **126:** 159–183.
7. WEISEL, J. W., C. COHEN, N. M. TOONEY, J. KAPLAN & D. ARMANI. 1980. J. Mol. Biol. **143:** 329–334.
8. GOLLWITZER, R., H. E. KARGES, H. HÖRMANN & K. KÜHN. 1970. Biochim. Biophys. Acta **207:** 445–455.
9. BLOMBÄCK, B. & M. BLOMBÄCK. 1956. Ark. Chem. **10:** 415–443.
10. HÖRMANN, H. & R. GOLLWITZER. 1966. Z. Physiol. Chem. **346:** 21–41.
11. FEHLHAMMER, H. & W. BODE. 1975. J. Mol. Biol. **98:** 683–692.
12. GUCKENBERGER, R. & W. HOPPE. 1980. Proceedings, Seventh European Congress of Electron Microscopy. **2:** 696. The Hague, the Netherlands.
13. SHOTTON, D. M., B. BURKE & D. BRANTON. 1979. J. Mol. Biol. **131:** 303–329.
14. DOOLITTLE, R. F. 1977. Horizons Biochem. Biophys. **3:** 164–191.
15. HUBER, R. & W. BODE. 1978. Acc. Chem. Res. **11:** 114–122.
16. WEISEL, J. W., G. N. PHILLIPS & C. COHEN. 1981. Nature **289:** 263–267.

NEW APPROACHES TO OLD PROBLEMS IN THE CLOTTING OF FIBRINOGEN *

L. Lorand

Department of Biochemistry, Molecular and Cell Biology
Northwestern University
Evanston, Illinois 50201

Research during the past thirty years enabled us to reconstruct with reasonable certainty the sequence of events surrounding the clotting of fibrinogen in normal plasma. The main features of regulation impinging on the temporal apex of the coagulation cascade in blood, though usually written at the bottom on charts of clotting schemes, are summarized in FIGURE 1. (For a recent review, see Lorand *et al.*[1]).

Thrombin, known since 1951 as a proteolytic enzyme of considerable specificity, controls the rate of fibrin formation as well as the velocity of generating a modified form of the fibrin stabilizing factor or Factor XIII zymogen by means of limited proteolysis. The XIII′ molecular ensemble $(a_2'b_2)$ serves as the immediate precursor of fibrinoligase, $XIII_a = a_2^*$, which endows the fibrin clot with added mechanical strength by catalyzing the formation of intermolecular γ-glutamyl-ϵ-lysine peptide bonds.

Some of the molecular transformations in this scheme are understood to a certain extent already; others still await elucidation. In this presentation, I shall try to focus attention on a few unresolved issues.

SOME KINETIC QUESTIONS

It has been known for some time from work with synthetic substrates that thrombin-catalyzed reactions proceeded by an acylation–deacylation pathway.[2-4] Yet, the relevant intermediates were omitted from FIGURE 1 not only for the sake of simplicity but also because they have not been characterized as yet. After the formation of appropriate Michaelis complexes, when the various fibrin molecules $[(\alpha_2\beta_2\gamma_2)AB_2, (\alpha_2\beta_2\gamma_2)B_2, (\alpha_2\beta_2\gamma_2)B$ and $(\alpha_2\beta_2\gamma_2)]$ and the hydrolytically modified Factor XIII′ species $[(aa'b_2)$ and $a_2'b_2)]$ arise, at least four fibrinopeptidyl-thrombin (two with fibrinopeptides A and two with B) and two acyl-thrombin intermediates with the activation fragments from each of the *a* subunits of Factor XIII must have also been produced.†

By measuring the rate of cleavage of the N-terminal activation fragment from Factor XIII by thrombin, we were recently able to show [4a] that fibrinogen exerted a marked acceleration on the reaction. This, in conjunction with the effect of fibrinogen on the hydrolytically modified zymogen species (i.e., XIII′) which modulates the Ca^{2+}-requirement for XIII_a generation, serves to har-

* This work was aided by U.S.P.H.S. Research Career Award 5KO6 HL–03512 and by Grants HL 02212 and 16346 from the National Institutes of Health.

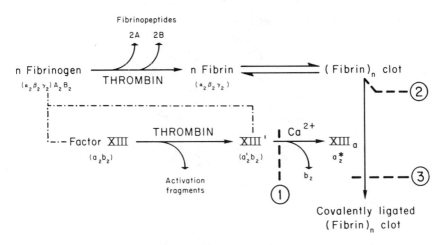

FIGURE 1. Outline of the clotting of fibrinogen in normal plasma, as reconstructed from studying individual reaction steps. Confirmation of this sequence of events deduced from *in vitro* experiments comes from a variety of clinical observations. In addition to afibrinogenemia and the hereditary deficiency of fibrin stabilizing factor (Factor XIII), in recent years three other hemorrhagic conditions (marked 1–3) due to the presence of acquired circulating inhibitors have been recognized.[1]

monize the scheme in FIGURE 1 and ensures that the formation of fibrin is coordinated with that of the cross-linking enzyme.

The latter enzyme is the only one of the blood coagulation cascade which functions with a cysteine rather than with a serine active center mechanism.[5, 6] As deduced again from studies with synthetic substrates,[7–9] Factor XIII_a-catalyzed reactions also proceed by an acylation-deacylation pathway sequence, as illustrated in FIGURE 2. If the second substrate is a fibrin molecule, fusion (ligation) and cross-linking ensues. If it is a synthetic amine, such as dansyl-cadaverine, incorporation of the amine into the reactive glutaminyl residues of the first fibrin molecule takes place in a site specific manner.[10–12] While we know that the γ chains of fibrin are the primary targets for amine incorporation by Factor XIII_a, but *not* by transglutaminase!) [12] and that the filling of the α chain sites occurs much slower, we still have to complete a detailed kinetic appraisal of these reactions. Nor do we know the identities of the fibrin intermediates. If the amine-substituted chain is indicated with a bar, would the reaction sequence be represented by $(\alpha_2\beta_2\gamma_2) \rightarrow (\alpha\beta\bar{\gamma}\cdot\alpha\beta\gamma) \rightarrow (\alpha_2\beta_2\bar{\gamma}_2)$ or by $(\alpha_2\beta_2\gamma_2) \rightarrow (\alpha_2\beta_2\bar{\gamma}_2)$, followed by $(\alpha_2\beta_2\bar{\gamma}_2) \rightarrow (\bar{\alpha}\beta\bar{\gamma}\cdot\alpha\beta\bar{\gamma}) \rightarrow (\bar{\alpha}_2\beta_2\bar{\gamma}_2)$ or by $(\alpha_2\beta_2\bar{\gamma}_2) \rightarrow (\bar{\alpha}_2\beta_2\bar{\gamma}_2)$?

As with the attack of thrombin on fibrinogen, also in the reaction of Factor XIII_a with fibrin, the aim is to define the stoichiometry of combination of the enzyme with the substrate in the Michaelis complex and to specify the chain at which this takes place. Could it be that four Factor XIII_a mole-

† If Factor XIII exists in plasma as an (ab) rather than an (a_2b_2) ensemble, only a single acyl-thrombin should be assumed.

Binding of F_1 to enzyme:

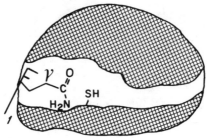

Binding of F_2 to acylenzyme:

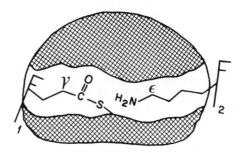

Regeneration of enzyme by aminolytic deacylation:

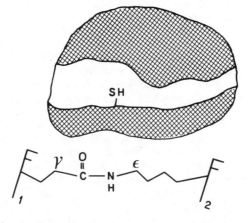

FIGURE 2. The production of γ-glutamyl-ϵ-lysine side chain peptides between two fibrin molecules (F_1 and F_2) as catalyzed by fibrinoligase (i.e., Factor XIII$_a$) with a cysteine active center, is thought to proceed by an acylation-deacylation pathway. If an amine such as dansylcadaverine was present,[10] this would compete against the ϵ-amino group of F_2 in the deacylation step. Fibrinogen reacts some tenfold slower than fibrin with the synthetic amine.[11, 12]

cules combine simultaneously with one fibrin and react with two γ and two α chain sites in a more or less concerted fashion, though at different turnover rates: $(\alpha_2\beta_2\gamma_2) \rightarrow (\overline{\alpha}_2\beta_2\overline{\gamma}_2)$?

Only painstaking further studies could resolve these issues. With Dr. Hanna Parnas of the Hebrew University in Jerusalem, we made a start for analyzing the reaction of Factor XIII_a with the γ chain sites using the following model (assuming that the two γ chains of fibrin are equivalent), where F represents monomeric fibrin, D dimeric fibrin, E the Factor XIII_a enzyme and R an amine such as dansylcadaverine:

$$1. \quad F + E \underset{k_{-1}}{\overset{k_1}{\rightleftharpoons}} F; E \overset{k_2}{\longrightarrow} F - E$$

$$2. \quad F - E + F \underset{k_{-3}}{\overset{k_3}{\rightleftharpoons}} F - E; F \overset{k_4}{\longrightarrow} D + E$$

$$3. \quad F - E + R \underset{k_{-5}}{\overset{k_5}{\rightleftharpoons}} F - E; R \overset{k_6}{\longrightarrow} F - R + E$$

This is essentially the same model we proposed in 1968 for the partitioning of the acyl-fibrin intermediate ($F - E$, in reaction 1) between the formation of fibrin dimer (as in reaction 2) and derivatization by the amine (as in reaction 3). Though still a great deal of data must be collected before the arrows can be decorated by actual kinetic constants, there can be little doubt that this scheme is in good agreement with all measurements thus far. A key point in this regard is the fact that there is an optimal fibrin concentration for amine incorporation, following which there is a decline in the velocity of the reaction (see FIGURE 6 in the article by Lorand, Chenoweth and Gray, 1972, Ann. N.Y. Acad. Sci. **202:** 155). Our computer analysis rests on the plausible assumption that k_2, k_4 and k_6, as chemical rate constants, are two to three orders of magnitude smaller than those controlled by diffusion in the scheme (k_1, k_{-1}, k_3, k_{-3}, k_5 and k_{-5}).

MOLECULAR DISEASES CONFIRM THE VALIDITY OF THE SCHEME FOR THE CLOTTING OF FIBRINOGEN

Whatever the outcome of the further analysis of the problems described above may be, there can be little doubt that the scheme presented in FIGURE 1 truly represents reality in terms of clotting in normal plasma. Best confirmation comes from nature's own experiments, in the variety of molecular diseases with specific relevance to each reaction step in the outline given. It is worth noting that this sequence of events was first reconstructed from *in vitro* studies, and progress in the clinical area could be made only after laboratory methods have been developed for examining individual reactions. Thus, abnormal fibrinogen molecules [13, 14] with defective susceptibility to thrombin or with impaired polymerization properties could be found only after techniques for measuring fibrinopeptide release [15, 16] and the reversible aggregation of fibrin monomers became available.[17-19] Similarly, when laboratory procedures could distinguish between covalently and noncovalently bonded fibrin clots, and when stages in the activation of the Factor XIII zymogen were worked out, differential diagnosis of a host of hemorrhagic diseases of fibrin stabilization became possible.[1] In addition to the autosomally inherited deficiency of Factor XIII, this "new" group of diseases is already known to comprise three other disorders with circulating inhibitors directed either against one of the steps in the production of the XIII_a enzyme, or against XIII_a activity in general, or against the cross-linking sites in fibrin itself.

The "feed-forward" regulation based on the interaction of the fibrinogen molecule with the thrombin-modified form of Factor XIII [20] is a very recent development, but it clearly brings up the possibility that a fibrinogen abnormality might exist which would be detrimental only to this regulatory function of the protein. We can predict that in such an individual the $a_2'b_2 \rightarrow a_2^* + b_2$ conversion of XIII' to XIII_a could not proceed normally at the physiological 1.5 mM concentration of Ca^{2+} in plasma.

WHY IS FIBRIN A PREFERRED SUBSTRATE FOR FIBRINOLIGASE (XIII_a)?

As shown in FIGURE 1, polymerization by covalent bonds is the biological *raison d'être* of the fibrinogen molecule. Mere clotting by aggregation, as in the n fibrin $\rightleftharpoons$ (fibrin)$_n$ step, does not suffice to stem hemorrhage. Patients with hereditary deficiency of Factor XIII or with acquired circulating inhibitors to fibrin stabilization usually have very severe bleeding tendencies.[1] Only a clot structure stabilized by γ-glutamyl-ϵ-lysine side chain peptide linkages is of real value for survival. Our recent measurements [20a] show that the addition of purified Factor XIII to a deficient plasma strengthens the clot by increasing its static viscoelastic modulus as much as fivefold.‡

It must be borne in mind, however, that it is fibrin—and not fibrinogen—which is the preferred substrate for fibrinoligase.[11, 12] Though this fact was known for some time, only the recent electronmicroscopic observations [21-23] allow us to suggest a possible explanation as to how the release of fibrinopeptides in the central E domain of the trinodular fibrinogen molecule might contribute to the unmasking of the amine-incorporating sites in the distally located D domains.

As illustrated in FIGURE 3, if some significant interaction between fibrinopeptides and the C-terminal halves of the α chains could be shown to exist, the thrombin-catalyzed removal of the N-terminal peptides might cause the swinging out of the wandering α chain domains from the central nodule of the molecule and the concomitant unmasking of the γ chain sites for easier access to Factor XIII_a. Such a transition in the molecule from a compact to a more extended configuration could be important from a regulatory point of view to ensure the preferential reactivity of the D domains in the extended form, without affecting the basic trinodular design of the protein.

As yet we have no concrete evidence to buttress the validity of this hypothesis. Nevertheless, we consider this idea sufficiently intriguing and of such far reaching significance that we are hard at work to devise experiments which would either prove or disprove it altogether.

‡ It is interesting that a fivefold reduction in this modulus is obtained when a competing substrate such as dansylcadaverine is added to normal plasma prior to clotting.

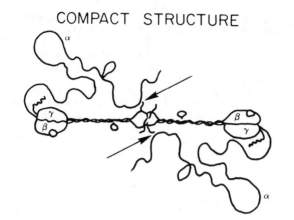

FIGURE 3. The compact structure of fibrinogen might be stabilized by some inter-action (see arrows) of the wandering α chain with the fibrinopeptide-containing central domain of the molecule. Cleavage of fibrinopeptides might thus lead to transformation to an extended form, with concomitant unmasking of amine-in-corporating sites (marked by X) in the end domains of γ chains.

REFERENCES

1. LORAND, L., M. S. LOSOWSKY & MILOSZEWSKI. 1980. Human Factor XIII: Fibrin stabilizing factor. *In* Progress in Hemostasis and Thrombosis. T. H. Spaet, Ed. **5:** 245–290. Grune and Stratton. New York.
2. KEZDY, F. J., L. LORAND & K. D. MILLER. 1965. Titration of active centers in thrombin solutions. Standardization of the enzyme. Biochemistry **4:** 2302–2308.
3. CHASE, T. C. JR. & E. SHAW. 1969. Comparison of the esterase activities of trypsin, plasmin and thrombin on guanidinobenzoate esters. Titration of the enzymes. Biochemistry **8:** 2212–2224.
4. FENTON, J. W. II, M. J. FASCO, A. F. STACKROW, D. L. ARONSON, A. M. YOUNG & J. S. FINLAYSON. 1977. Human thrombins. J. Biol. Chem. **252:** 3587–3598.
4a. JANUS, T. J., S. D. LEWIS, L. LORAND & J. A. SHAFER. 1983. The effect of

fibrinogen on the thrombin-catalyzed release of activation peptide from human Factor XIII. Fed. Proc. **42**: 1032 (Abst. 4335)

5. CURTIS, C. G., P. STENBERG, C.-H. J. CHOU, A. GRAY, K. L. BROWN & L. LORAND. 1973. Titration and subunit localization of active center cysteine in fibrinoligase (thrombin activated fibrin stabilizing factor). Biochem. Biophys. Res. Commun. **52**: 51–56.

6. CURTIS, C. G., K. L. BROWN, R. B. CREDO, R. A. DOMANIK, A. GRAY, P. STENBERG & L. LORAND. 1974. Calcium dependent unmasking of active center cysteine during activation of fibrin stabilizing factor. Biochemistry **13**: 3774–3780.

7. CURTIS, C. G., P. STENBERG, K. L. BROWN, A. BARON, K. CHEN, A. GRAY, I. SIMPSON & L. LORAND. 1974. Kinetics of transamidating enzymes. Production of thiol in the reactions of thiol esters with fibrinoligase. Biochemistry **13**: 3257–3262.

8. STENBERG, P., C. G. CURTIS, D. WING, Y. S. TONG, R. B. CREDO, A. GRAY & L. LORAND. 1975. Transamidase kinetics. Amide formation in the enzymic reactions of thiol esters with amines. Biochem. J. **147**: 155–163.

9. PARAMESWARAN, K. N. & L. LORAND. 1981. New thioester substrates for fibrinoligase (coagulation factor XIIIₐ) and for transglutaminase. Transfer of the fluorescently labelled acyl group to amines and alcohols. Biochemistry **20**: 3703–3711.

10. LORAND, L., N. G. RULE, H. H. ONG, R. FURLANETTO, A. JACOBSEN, J. DOWNEY, N. ONER & J. BRUNER-LORAND. 1968. Amine specificity in transpeptidation. Inhibition of fibrin cross-linking. Biochemistry **7**: 1214–1223.

11. LORAND, L. & D. CHENOWETH. 1969. Intramolecular localization of the acceptor cross-linking sites in fibrin. Proc. Natl. Acad. Sci. USA **63**: 1247–1252.

12. LORAND, L., D. CHENOWETH & A. GRAY. 1972. Titration of the acceptor cross-linking sites in fibrin. Ann. N.Y. Acad. Sci. **202**: 155–171.

13. HENSCHEN, A., F. LOTTSPEICH, M. KEHL & C. SOUTHAN. 1983. Ann. N.Y. Acad. Sci. **408**:. This volume.

14. MÉNACHÉ, D. 1983. Ann. N.Y. Acad. Sci. **408**:. This volume.

15. LORAND, L. 1951. Fibrino-peptide: New aspects of the fibrinogen-fibrin transformation. Nature **167**: 992.

16. LORAND, L. 1952. Fibrino-peptide. Biochem. J. **52**: 200–203.

17. LORAND, L. A study on the solubility of fibrin clots in urea. 1948. Hung. Acta Physiol. **1**: 192.

18. LORAND, L. & W. R. MIDDLEBROOK. 1952. The Action of Thrombin on Fibrinogen. Biochem. J. **52**: 196–199.

19. DONNELLY, T. H., M. LASKOWSKI, JR., N. NOTLEY & H. A. SCHERAGA. 1955. Equilibria in the fibrinogen-fibrin conversion. II. Reversibility of the polymerization steps. Arch. Biochem. Biophys. **56**: 369–

20. CURTIS, C. G., T. J. JANUS, R. B. CREDO & L. LORAND. 1983. Ann. N.Y. Acad. Sci. **408**:. This volume.

20a. SHEN, L. & L. LORAND. 1983. Contribution of fibrin stabilization to clot strength. Supplementation of Factor XIII-deficient plasma with the purified zymogen. J. Clin. Invest. In press.

21. MOSESSON, M. W., J. HAINFELD, R. H. HASCHEMEYER & J. WALL. 1981. Identification and mass analysis of human fibrinogen molecules and their domains by scanning transmission electron microscopy, J. Mol. Biol. **153**: 695–718.

22. WALL, J., J. HAINFELD, R. H. HASCHEMEYER & M. W. MOSESSON. 1983. Ann. N.Y. Acad. Sci. **408**:. This volume.

23. ERICKSON, H. P. & W. E. FOWLER. 1983. Ann. N.Y. Acad. Sci. **408**:. This volume.

FACTORS INFLUENCING FIBRIN GEL STRUCTURE STUDIED BY FLOW MEASUREMENT *

Masahisa Okada and Birger Blombäck

New York Blood Center
New York, New York 10021

INTRODUCTION

In order to form a gel, fibrinogen has to be activated. Physiologically, this event is governed by thrombin, which releases two peptides, fibrinopeptide A (FPA) and fibrinpeptide B (FPB), from fibrinogen in the course of its transformation to fibrin. It is believed that release of FPA leads to formation of linear polymers. The polymer molecules appear to constitute elongated structures, having a width double that of fibrinogen. They are formed by end-to-end and side-to-side association of fibrinogen units in a half-staggered fashion. These polymer molecules were first observed in thrombin-fibrinogen systems in which clotting was inhibited by addition of hexamethylene glycol or similar agents [1] (FIGURE 1). However, polymers do also form in clottable systems and they are present before the time of gelation.[3-6] The polymer molecules are very likely the result of interaction between polymerization sites in activated bifunctional fibrinogen molecules.[7, 8] The polymers formed when FPA only is released by the snake venom enzyme Batroxobin is different from those which are formed by release of both FPA and FPB by thrombin.[9] The polymers eventually cross-link † to form the infinite network structure which is the gel according to the definition by Flory.[10]

Ferry and Morrison [11] were the first to perform systematic studies on fibrin gel structures. They distinguished between two types of gels: i.e., the coarse and the fine types. In the fine type, more or less individual polymer strands were suggested to cross-link to form the gel. In the coarse gel, on the other hand, it was assumed that the polymers were laterally aggregated and cross-links occurred between these bundles of polymers. The hypothetical structure of the fine and coarse type of gel is shown in FIGURE 2.

Most information on fibrin gel structure has been obtained from studies of the optical and mechanical properties of the gels. We will present here studies we have performed using liquid and particle permeation of fibrin gels under a variety of experimental conditions. This approach was also used by Carr et al.[12] to study thickness of fibrin strands in fibrin gels. We have studied the flow properties of gels produced by Batroxobin and by thrombin. The former gels are called Fibrin I gels, and the latter Fibrin II gels.[7] The simple flow-device we have used is shown in FIGURE 3. The gels were prepared in special cups (FIGURE 3A) as previously described.[13]

* This research work was supported by a grant from the U.S. National Institutes of Health (HL27279–01).

† This cross-linking must be distinguished from Factor XIIIa-induced cross-linking.

233

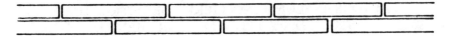

FIGURE 1. Polymer formed from fibrin monomer in a half-staggered fashion. (From Ferry et al.[2])

FIGURE 2. Schematic illustration of fibrin gel structure. *Left panel*: fine gel; *right panel*: coarse gel. (From Ferry and Morrison.[1])

A

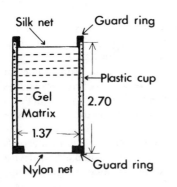

B

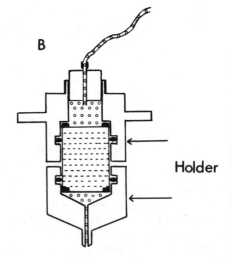

FIGURE 3. Apparatus for flow measurements. **A.** Transverse section of the cylindrical plastic gel cup. **B.** Transverse section of permeation apparatus containing the gel cup; OOOOOO, permeant; and ------, gel. Height and width in cm.

CLOTTING TIME AND FLOW PROPERTIES AT DIFFERENT pH, IONIC STRENGTH AND
FIBRINOGEN CONCENTRATION

Liquid permeation studies of fibrin gels which were formed in the presence of thrombin and in the presence of Batroxobin showed that those gels have distinctly different flow properties. The flow through the gels was viscous in all cases and consequently the Poiseuille's law applies. For both gels the permeability coefficient (Ks)‡ and average pore sizes ‡ varied depending on the conditions used for gel formation. Low pH and ionic strength favored high Ks and large pore sizes, whereas high pH and ionic strength produced gels with low Ks and small pore sizes. Parallel turbidity studies showed correlations between the optical properties of the gels and permeation data. Of particular importance was the finding that the clotting time (Ct) is directly related to Ks of the final gels. Thus events preceding gel formation determine the final gel structure. It was also found that Ks is inversely related to the fibrin concentration (C) in the gels. The ratio Ct/C (permeability index) is thus an important determinant for the gel structure. The flow properies of fibrin gels under different experimental conditions are summarizd in FIGURE 4.

Recently we studied the effect of temperature on gelation. As shown in FIGURE 5a, the clotting time decreases with increasing temperature. This may be explained by the increased rate of activation (release of fibrinopeptides) with increasing temperatures observed earlier.[4] The gels formed at the different temperatures were percolated at a constant temperature. The results are shown in FIGURE 5b. It is evident that the Ks values of the final gels are directly related to the clotting times of the gel forming system (cf. FIGURE 5a).

CLOTTING TIME AND ACTIVATION OF FIBRINOGEN

We also determined the release of FPA and FPB before and after gelation under different conditions. The results obtained at the gel point are shown in TABLE 1. These studies showed that the activation required for gelation decreased as the gelation time increased. After completion of gelation, quantitative release of peptides had occurred in all instances.

TABLES 2a and 2b show the activation of fibrinogen at different fibrinogen concentrations. With the exception of Batroxobin at the highest fibrinogen concentration, the initial rate of release of FPA increases with increasing fibrinogen concentrations (TABLE 2a). The initial rate of release of FPB

‡ *Darcy coefficient* (Ks) was calculated as follows:[14, 15]: $Ks = \dfrac{Q \cdot L \cdot \eta}{t \cdot A \cdot \Delta P}$ (1)

where Q is the volume of liquid (in cm³), having viscosity η (in poise), flowing through a column with height L (in cm) and area A (in cm²) in a given time t (in seconds) under a pressure differential, ΔP (in dyne/cm²). The dimension of the resulting Ks is cm².

Pore sizes were calculated assuming a cylindrical capillary system with the capillaries parallel to the direction of flow.[15–17] The equation was:

$$ r = \sqrt{\frac{8Ks}{\epsilon}} \qquad (2) $$

where r is the average pore radius (in cm), and ϵ is the fractional void volume of the gel.

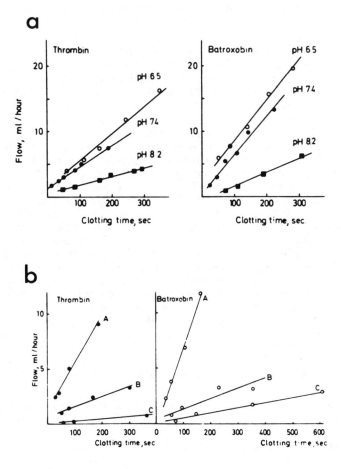

a

Thrombin

Flow, ml / hour

pH 6 5
pH 7 4
pH 8 2

Clotting time, sec

Batroxobin

pH 6 5
pH 7 4
pH 8 2

Clotting time, sec

b

Thrombin

Flow, ml / hour

A
B
C

Clotting time, sec

Batroxobin

A
B
C

Clotting time, sec

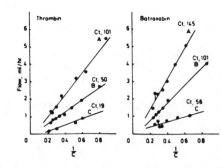

c

Thrombin

Flow, ml/hr

Ct, 101
A
Ct, 50
B
Ct, 19
C

$\frac{1}{t}$

Batroxobin

Ct, 145
A
Ct, 101
B
Ct, 56
C

$\frac{1}{t}$

FIGURE 4. Clotting time and flow properties at different pH, ionic strength and fibrinogen concentration. Gel formation: Tris-imidazole buffer containing 20mM CaCl₂, of ionic strength 0.21, temperature 22–25° C. Fibrinogen concentration; in panel a and b, 2g/1; in panel C, between 1.2 and 4.9 g/1. Panel **a:** different pH. Thrombin 0.04–1.32 NIH units/ml. Batroxobin 0.27–3.6 BU/ml. Panel **b:** different ionic strengths. A; 0.21, B; 0.26, C; 0.29. pH 7.4. Thrombin 0.09–0.78 NIH units/ml. Batroxobin 0.27–1.78 BU/ml. Panel **c:** different fibrinogen concentrations. pH 7.4. Thrombin: I,0,15; II, 0.76 and III, 2.5 NIH units/ml. Batroxobin: I, 0.44; II 0.78 and III 1.89 BU/ml. Average Ct at different enzyme concentrations are shown in figures. Permeation: same buffer as used for gel formation, temperature between 22–24° C. (From Blombäck & Okada.[13] By permission of *Thrombosis Research.*)

is more or less constant at increasing fibrinogen concentrations. The clotting times, at a given enzyme concentration, vary relatively little with increasing fibrinogen concentration (TABLE 2b). At a given enzyme concentration there is an increase of FPA at Ct with increasing fibrinogen concentrations. This is less pronounced at lower enzyme concentrations in case of thrombin. The amount of FPB at Ct is more or less constant.

INFLUENCE OF CALCIUM AND MAGNESIUM IONS ON CLOTTING TIME AND
FLOW PROPERTIES OF FIBRIN GELS

In all the studies described so far, the fibrin gels were formed in the presence of constant amounts of calcium. We have recently found that variations of the calcium concentrations greatly affects the gel structure.

Previous work has shown that calcium ions influence the fibrinogen transition. It is known that calcium shortens the clotting time of fibrinogen in the

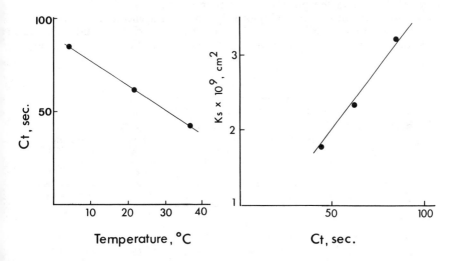

FIGURE 5. Ks and Ct at different temperatures. Gel formation: Tris-imidazole buffer containing 20mM CaCl₂, pH 7.4. Ionic strength 0.21. Temperature 4.5, 22 and 37° C. Fibrinogen concentration, 2.16 g/1. Thrombin, 0.15 NIH units/ml. Permeation: same buffer as used for gel formation. Temperature 22° C.

TABLE 1

RELEASE OF FPA AND FPB AT GEL POINT (Ct) *

pH	Ct (secs)	Thrombin FPA (nmole/ml)	FPB (nmole/ml)	$\frac{FPA}{FPB}$	Ct (secs)	Batroxobin FPA (nmole/ml)
	51	3.99	0.24	17	64	6.13
6.5	78	3.97	0.14	28	106	5.68
	160	3.28	0.12	27	220	2.99
	29	4.97	0.47	11	39	5.19
7.4	52	4.32	0.32	14	66	4.33
	89	3.28	0.20	16	119	3.82
	29	7.79	0.88	9	42	8.68
8.2	56	5.13	0.66	8	69	5.38
	115	4.48	0.56	8	127	4.69

* At each pH, three different enzyme concentrations were used: for thrombin, 0.15, 0.38 and 0.78 NIH units/ml and for Batroxobin, 0.44, 0.89 and 1.78 BU/ml. Samples were taken at ct. The peptides were determined as described previously. Ionic strength in all experiments was 0.21. The fibrinogen concentrations (μM) at pH 6.5, 7.4 and 8.2 were: 5.76 (11.52), 5.79 (11.58), and 6.03 (12.06), respectively. Figures within brackets denote chain concentrations. (From Blombäck & Okada. By permission of *Thrombosis Research*.)[13]

presence of thrombin.[18-20] This effect is not due to enhancement of the release of fibrinopeptides but more likely to an effect on the polymerization or gelation reactions.[7, 21-24] The pattern of transition from opaque to transparent clot is changed in the presence of calcium,[7] and the rigidity of gels is increased when they are formed in the presence of calcium.[25]

Removal of calcium from fibrinogen by chelating agents leads to prolongation of the thrombin clotting time,[26, 27] reversible structural changes in the molecule [28] and increased susceptibility to heat denaturation.[29] These findings

TABLE 2A

INITIAL RATE OF RELEASE OF FPA AND FPB
AT DIFFERENT FIBRINOGEN CONCENTRATIONS *

Fibrinogen (g/l)	Thrombin, NIH units/ml 0.035 FPA (p mole/ml/sec)	FPB (p mole/ml/sec)	0.15 FPA (p mole/ml/sec)	FPB (p mole/ml/sec)	Batroxobin, BU/ml 0.27 FPA (p mole/ml/sec)	1.78 FPA (p mole/ml/sec)
2.02	25	1.2	184	14	34	249
2.24	29	1.3	189	13	46	348
3.18	30	1.3	271	17	70	378
4.30	37	1.1	301	12	69	268

* Gelformation: Tris-imidazole buffer containing 20 mM $CaCl_2$, pH 7.4, ionic strength 0.21, temperature 26–27° C.

TABLE 2B

RELEASE OF FPA AND FPB AT GEL POINT (Ct) AT DIFFERENT FIBRINOGEN CONCENTRATIONS *

| Fibrinogen (g/l) | Thrombin, NIH units/ml | | | | | | | | Batroxobin, BU/ml | | | |
| | 0.035 | | | | 0.15 | | | | 0.27 | | 1.78 | |
	Ct (sec)	FPA (nmole/ml)	FPB (nmole/ml)	$\frac{FPA}{FPB}$	Ct (sec)	FPA (nmole/ml)	FPB (nmole/ml)	$\frac{FPA}{FPB}$	Ct (sec)	FPA (nmole/ml)	Ct (sec)	FPA (nmole/ml)
2.02	180	4.38	0.31	14	36	6.42	0.68	10	165	5.54	40	6.9
2.24	166	3.92	0.22	18	37	6.41	0.70	9	157	6.03	38	10.14
3.18	179	5.16	0.29	18	37	8.22	0.74	11	153	9.31	47	13.83
4.30	182	5.12	0.24	21	42	11.29	0.96	12	180	11.11	63	14.91

* Gel formation: see legend to TABLE 2A.

might suggest that calcium plays a role in stabilizing the fibrinogen molecule in a conformation that is favorable for its function in polymerization and gelation.

The binding of calcium to fibrinogen has been reported by several investigators.[23, 30-32] Marguerie *et al.*[30] found three high and several low affinity binding sites for calcium in fibrinogen. After fragmentation of fibrinogen and fibrin with plasmin, the high affinity binding sites appeared to remain in plasmin fragment D.[32]

FIGURE 6 shows the *Ks* values for gels formed with thrombin and Batroxobin at different calcium concentrations and at constant ionic strength. As shown in the figure, the *Ks* values increase with increasing calcium concentration. The activity of calcium ions (aCa) in the fibrinogen solution is also shown. For thrombin gels, the effect of increasing calcium concentrations on *Ks* appears to approach a maximum at about 20 mM. In the experiments using Batroxobin, the *Ks* values increased much slower with increasing calcium concentration.

In FIGURE 6 are shown the final turbidity of gels formed at different calcium concentrations. For both types of gels, the turbidity increases with increasing calcium concentrations.

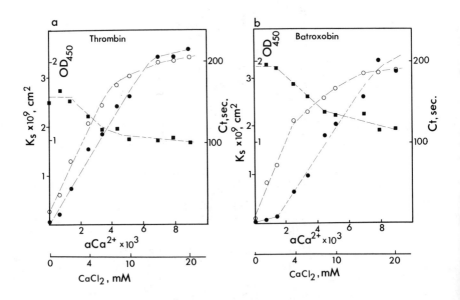

FIGURE 6. *Ks*, OD and *Ct* at different calcium concentrations. Gel formation: Tris-imidazole buffer, pH 7.4, ionic strength 0.14, temperature 22–23° C, fibrinogen concentration 1.95 g/l. Thrombin and Batroxobin concentrations were 0.06 NIH units/ml and 0.36 BU/ml, respectively. Permeation: same buffer as used for gel formation, temperature 21–22° C. Panels a and b: ●—●, *Ks*; ■—■ and ○—○ were *Ct* and OD, respectively. OD measured after two hr of gelation. Ion activity of calcium was determined by calcium selectrode Type F2112 Ca and a calomel electrode Type 401 (Radiometer, Copenhagen, Denmark). Ionic strength calculations were based on activity of the ions.

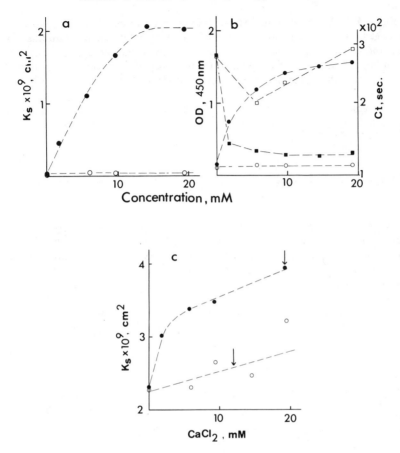

FIGURE 7. Ks, OD and Ct at different calcium and magnesium concentrations. Gelformation: Tris-imidazole buffer, pH 7.4, ionic strength, 0.18–0.24 in panel **a** and **b**; 0.12–0.14 (Ca) and 0.13–0.15 (Mg) in panel **c**. Temperature 22–24° C. Fibrinogen 2.03 g/l. Thrombin 0.06 NIH units/ml. Permeation: same buffer as used for gelformation, temperature 23–24° C. Panel **a**: Ks for Ca ●—● and for Mg ○–○. Panel **b**: Ct for Ca, ■–■ and for Mg, □–□; OD for Ca, ●–● and for Mg ○–○. OD measured after 2 hrs of gelation. Panel **c**: Ks for Ca alone ●–● and for Ca plus constant amounts of Mg (20 mM) ○–○. Ionic strength calculations in panel **c** were based on activity of the ions and arrows shown in figure indicate same ionic strength.

FIGURE 6 also shows the Ct of the fibrinogen solutions at different calcium concentrations. Concomitant with increasing Ks and turbidity, there is a decrease in Ct.

In other experiments, magnesium was substituted for calcium in the concentration range between 0–20 mM. As shown in FIGURE 7, no effect on Ks and turbidity was demonstrated for thrombin gels. However, magnesium did shorten Ct to some extent (FIGURE 7b). The latter effect reached a maximum at about 5 mM magnesium. Above this concentration, *Ct* was progressively prolonged. This is in contrast to the effect exerted by calcium (FIGURE 7b). Since the activity coefficient for magnesium is about the same as for calcium,[33] the prolongation cannot be explained on the basis of increasing ionic strength. Qualitatively, the same results as described above were obtained with Batroxobin.

When calcium is added to fibrinogen in the presence of constant amounts of magnesium, the effect exerted by calcium appears to be suppressed. This is shown in FIGURE 7c.

RELATIONSHIP BETWEEN CLOTTING TIME (*Ct*) AND *Ks* AT DIFFERENT CALCIUM CONCENTRATIONS

We have shown previously that there is a linear relationships between clotting time and flow rate in the presence of 20 mM calcium. FIGURE 8 shows that for thrombin this applies also for a situation where no calcium is present as well as at different calcium concentrations. The same results were obtained with Batroxobin.

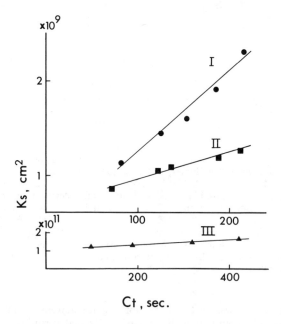

FIGURE 8. *Ks* versus *Ct* at different calcium concentrations. Gel formation: Tris-imidazole buffer, pH 7.40, ionic strength 0.18–0.24, temperature 22–23° C, fibrinogen concentration 2.09 g/l. Thrombin, 0.04–0.17 NIH units/ml. Permeation: same buffer as used for gel formation, temperature 22° C. Curves I, II and III were for calcium concentrations 19.3, 5.8 and 0 mM, respectively.

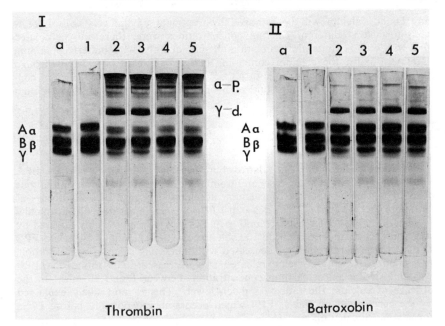

FIGURE 9. Factor XIIIa-induced cross-linking of fibrin gels at different calcium concentrations. Gel formation: Tris-imidazole buffer, pH 7.4, ionic strength 0.18–0.24, temperature 23° C, fibrinogen concentration 2.06 g/l. Thrombin, 0.06 NIH units/ml. After 2 hr of gelation, the samples were reduced and analyzed by SDS-gel electrophoresis on 7% polyacrylamide gels. Panel I and II: a, fibrinogen; 1–5 were for calcium concentrations 0, 1.9, 5.8, 9.7 and 19.3 mM, respectively.

FACTOR XIIIa-INDUCED CROSS-LINKING OF FIBRIN GELS IN PRESENCE AND ABSENCE OF CALCIUM

SDS-polyacrydamide gel electropherograms (SDS-PAGE) of reduced fibrin gels obtained with thrombin are shown in FIGURE 9. It is evident that no cross-linking occurred in the absence of calcium, whereas in the presence of calcium, cross-linking of both γ and α chains took place at all calcium concentrations used. SDS-PAGE was also performed on the fibrinogen solutions at the gel point (Ct). In none of the experiments, regardless of calcium concentration, was any cross-linking observed at Ct (results not shown in FIGURE 9). The SDS-PAGE of reduced fibrin gels obtained with Batroxobin are shown in FIGURE 9. In absence of calcium, no cross-linking occurred. Even at the lowest calcium concentration, complete or almost complete cross-linking of the γ chains had occurred. As compared with the thrombin gels, cross-linking of the α chain was far from complete. Analysis of fibrinogen samples at Ct showed, like in the case with thrombin, no cross-linking.

PERTURBATION OF GEL STRUCTURE WITH EDTA

It is possible that calcium is important for stabilization of fibrin gels after their formation. Since EDTA has a high affinity for calcium, it is likely that any bound calcium will be removed by treatment of the gels with EDTA. Gels formed at different calcium concentrations were, therefore, percolated first with buffers containing calcium and, subsequently, with buffers containing EDTA. Finally, the calcium containing buffer was again percolated through the gels. No substantial change in Ks occurred by increasing the calcium concentration in the permeation fluid or by percolating EDTA solutions through thrombin or Batroxobin gels.

ACTIVATION OF FIBRINOGEN AT DIFFERENT CALCIUM CONCENTRATIONS

With both thrombin and Batroxobin there is a linear release of FPA before Ct. FPB is, in the case of thrombin, released at a slower initial rate than FPA. In the absence of calcium, accelerated release of FPB occurs at or before Ct. No appreciable release of FPB was seen in the case of Batroxobin.

The initial rate of release of both peptides was not appreciably affected by increasing calcium concentration. TABLE 3 shows the amount of FPA and FPB at Ct. It is evident that as Ct decreased with increasing calcium concentration, the amount of peptides being released also decreased. The ratio FPA/FPB was almost constant at calcium concentrations between 6 and 20 mM. In the absence of calcium the ratio was much lower. This is most likely explained by the accelerated release of FPB which occurs at and possibly before Ct in the absence of calcium.

PROBING GEL STRUCTURE BY PERMEATION OF DIFFERENT SPHERICAL PARTICLES

In the above study we used the Poiseuille equation for calculation of pore sizes.[13, 16, 17] This equation applies to a capillary system with the capillaries parallel to the direction of liquid flow. It is, however, doubtful whether this simple model can be applied to the pores in fibrin gels. In order to obtain more information on the effective pore sizes in fibrin gels, we have performed permeation studies employing spherical latex particles of known particle size

TABLE 3

RELEASE OF FPA AND FPB AT GEL POINT (Ct)
AT DIFFERENT CALCIUM CONCENTRATIONS *

Ca (mM)	Thrombin				Batroxobin	
	Ct (sec)	FPA (nmole/ml)	FPB (nmole/ml)	$\dfrac{FPA}{FPB}$	Ct (sec)	FPA (nmole/ml)
0	248	8.5	1.8	5	350	8.9
6	156	5.6	0.4	13	190	7.8
10	152	4.2	0.4	12	179	6.9
19	178	4.4	0.4	12	182	5.2

* Gelformation: Tris-imidazole buffer, pH 7.4, ionic strength 0.18–0.24, temperature 23°C, fibrinogen 2.06 g/l. Thrombin and Batroxobin concentrations were 0.06 NIH units/ml and 0.36 BU/ml, respectively.

for probing the pores in the fibrin gels. Gels were prepared in the special cups mentioned before. Different amounts of thrombin were added to aliquots of fibrinogen solutions to obtain clotting times varying between about 25 seconds and 300 seconds. After complete gelation of the solution, permeation with buffer was performed and Ks values calculated (Equation 1). The correlation coefficients between clotting times and Ks were between 0.90 and 0.99. On the basis of the Ks values the theoretical average pore sizes of the gels were subsequently calculated (Equation 2). In our first experiments the latex particles were suspended in water and subsequently percolated through gels having different pore sizes.[34] However, these experiments were somewhat cumbersome because appreciable aggregation of the particles occurred on the walls of the upper part of the gel cup. Consequently, seldom more than 50% of particles were recovered in the eluates and sometimes anomalous filtration profiles [16] were observed i.e., decreasing concentration of particles in the eluate with time. In order to avoid this complication we have found it convenient to treat gels and particles in the following way: (1) The gels were treated with 1% glutardialdehyde (for about 1 hr). This treatment stabilizes the gels and diminishes the interaction between gel matrix and the particle preparation to be described below. The treatment only slightly affected the flow-rate of the gels. (2) The latex particle suspensions (10%, w/v) were first vigorously shaken (Vibromixer) for 10 minutes. The suspensions were drop-wise diluted with water to 0.2%. The particles were subsequently coated by addition of albumin to a final concentration of 0.01%. After coating, the suspension was shaken for 30 minutes. Finally one volume of the suspension was added drop-wise during stirring to nine volumes of 0.004–0.1% urea (or formamide) in 0.1 M glycine. The pH of the suspension was eventually adjusted to 8.

The suspended particles (concentration about 0.02%) were next percolated through the gels. FIGURE 10 shows a series of experiments with gels being permeable to varying degrees to a particular latex particle. The particles start to appear at the void volume of the gels and the turbidity reaches a maximum after passage of about 2–3 column volumes of permeation fluid. The concentration of particles in the effluent from the gel columns was determined by reading the turbidity (OD) of the effluent at 450 nm. The experiment shows that particles, which are permeable, pass through the gels without being retained by the gel matrix. Furthermore, as shown in the figure, the turbidity reached a constant value after percolation of about two column volumes. This indicates that neither does any appreciable blockage of capillaries inside the gels occur nor does aggregation on top of gel take place during filtration. The recovery of particles in the effluents from gels which should be completely permeable to the particles (cf. gel with 184 secs clotting time in FIGURE 10) was between 85–95%. The remaining particles (5–15%) were always recovered from the top of the gels by rinsing the surface. These nonpenetrable particles most likely represent irreversibly aggregated particles. This was confirmed by observation of the latex particle suspension in electronmicroscope. It was found that about 90% of the particles observed were monodisperse. The remainder of the particles were mainly present as dimers. A few trimers were also observed. A few particles were also seen which were much smaller than the majority of particles.

FIGURE 11 shows the penetration of different particles using gels having different theoretical pore sizes. As shown in FIGURE 11, the particles penetrated gels having theoretical pore sizes above a certain value, and did not

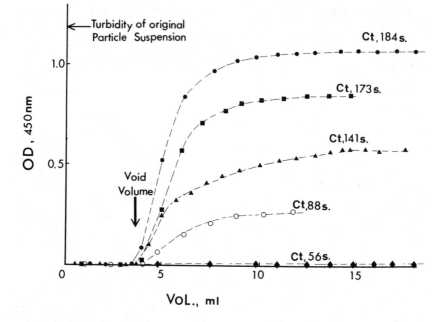

FIGURE 10. Profile of Latex particle permeation through fibrin gels. Gel formation: Tris-imidazole buffer containing 20 mM $CaCl_2$, pH 7.4. Ionic strength 0.24, temperature 24–25° C, fibrinogen concentration 2.3 g/l. Thrombin, 0.03–0.17 NIH units/ml. Permeation: Latex particles were suspended in 0.1 M glycine containing 0.001% bovine serum albumin and 0.004% urea, pH 8.0. Temperature, 23–25° C. For preparation of latex particle suspension, see text. Turbidity of original particle suspension (0.02%, w/v) was 1.18 at 450 nm. Ct of the gel forming system are shown in figure.

penetrate gels with pore sizes below a certain value. It is evident that the transition from maximum to minimum penetration occurred in a narrow pore size range. The theoretical pore size at 50% of maximum turbidity would represent the average size of pores and particles. Since the average diameter of the particle is known, the ratio between theoretical and effective pore size of the gels can be calculated. The effective pore size is about one order of magnitude less than the theoretically calculated pore size.

The difference in pore size between minimum and maximum permeation is a measure of the total variation in size of pores plus particles. The horizontal bars in FIGURE 11 represent the variation in particle diameters (average particles size ±3 SD). It is evident that the variation in particle size accounts for a considerable part of the total variation observed. By subtracting the variation in particle size from the approximate total variation we find that the variation in pore size is remarkably small (SD about ±5%). FIGURE 11 also shows the ratio theoretical/effective pore size at different ionic strengths and protein concentrations. The ratio seems not to be influenced by increasing the protein concentration in the system. On the other hand, an increase in ionic strength seems to decrease the ratio.

The large ratio between theoretical and effective pore size as determined with latex particles might possibly be due to artificially induced changes in pore size by e.g., the treatment of the gels with glutardialdehyde, urea or formamide. However, this is not so likely since the flow properties of the gels do not change appreciably after these treatments. In order to rule out this possibility, we performed permeation experiments on untreated gels using virus particles. We chose mouse-specific Sendai virus, since we had found that this virus did not appreciably interact with the gel matrix of untreated gels. Sendai virus is known to have a size distribution of about 90–200 nm. We

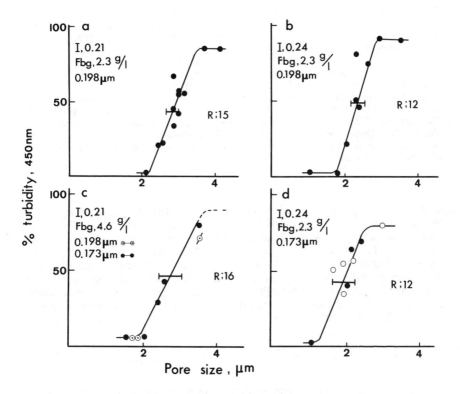

FIGURE 11. Penetration of latex particles in gels of different theoretical pore sizes. Gel formation: Tris-imidazole buffer containing 20 mM CaCl₂, pH 7.4. Ionic strength and fibrinogen concentrations are shown in figure. Thrombin concentrations were between 0.02 and 0.35 NIH units/ml. Temperature 23–25° C. Permeation: particle suspension in 0.1M glycine containing 0.01% BSA and urea or formamide, pH 8.0. Urea and formamide concentrations were as follows: 0.004% urea in panels **a** and **b**, 0.1% formamide (●–●) and 0.004% urea (⊙–⊙) in panel **c**, 0.004% urea (●–●) and 0.02% formamide (○–○) in panel **d**. Before permeation of the particles, the gels in the columns were equilibrated with 0.1M glycine, pH 8.0, containing BSA and urea or formamide. Latex particles: 0.198 ± 0.0036 μm and 0.173 ± 0.0068 μm were used. *R*: the ratio between theoretical and effective pore size of the gels. The turbidity of the effluent was expressed in % of that in the suspension applied to the gel columns.

TABLE 4

PERMEATION OF SENDAI VIRUS THROUGH FIBRIN GELS OF VARIOUS PORE SIZES *

	Properties of Gel		Amount of Sendai Virus				Total
	$Ks(cm^2)$	Pore nø, μm (effective)	In Effluent		On Top of the Gel		Recovery (%)
			HA Units	(%)	HA Units	(%)	
I	2.49×10^{-11}	0.028	0	0	128	100	100
II	1.66×10^{-9}	0.153	64.4	50.3	32	25	75.3
III	5.87×10^{-9}	0.289	121.8	95.2	nd	nd	95.2

* Gel formation: Tris-imidazole buffer rontaining 20 mM $CaCl_2$, pH 7.4. Experiment I: ionic strength ($\Gamma/2$) 0.29, fibrinogen concentration (C) 4.6 g/1, thrombin concentration 0.17 NIH units/ml, temperature 22°C. Experiments II and III: $\Gamma/2$, 0.21 C, 2.08 g/1, thrombin concentration 0.08 and 0.04 NIH units/ml, respectively, temperature 24°C. Permeation: for determination of Ks and average pore sizes the same buffer was used as for gel formation. After equilibration of the gels with phosphate saline buffer (PBS), pH 7 containing 1% bovine serum albumin, 0.2 ml of Sendai virus (640 HA units/ml) in PBS albumin buffer, pH 7.0, was percolated. The amount of Sendai virus was determined by hemagglutination of chicken red blood cells. Effective pore diameter was calculated from Ks values on the basis of the ratio theoretical/effective obtained from experiments in which latex particles were used. nd = not determined.

used a crude preparation of this virus and confirmed in the electronmicroscope that the sizes of the virus particles fell within the stated range. The virus particles were suspended in albumin containing buffer and percolated through gels having different theoretical pore sizes. The results of two experiments are shown in TABLE 4. Based on the ratio theoretical/effective diameter observed with latex particles we found, as is shown in the table, that the majority of virus particles are retained by gels having effective pore sizes less than about 30 nm. Partial permeation occurred for gels having pore sizes of about 150 nm. Essentially all particles penetrate gels with pore sizes of about 290 nm.

INTERACTION OF PROTEINS WITH THE FIBRIN GEL MATRIX

Fibrinogen is known to interact with fibrin monomers to form complexes.[35] Fibrinogen is also adsorbed to fibrinomonomers coupled to Sepharose.[36] We have studied the interaction between fibrin gels and fibrinogen. Gels containing about 8 mg of gel matrix were prepared using thrombin and Batroxobin in the presence of calcium at pH 7.4, ionic strength 0.21. When fibrinogen solutions containing 4 mg of protein were percolated through the gels no adsorption to the gel matrix was observed.

Fibronectin is known to interact with the fibrinogen in the cold[37] and to adsorb to fibrinmonomer-Sepharose conjugates.[38] We have found that when fibronectin is percolated through completely cross-linked gels, formed in the presence of thrombin, calcium and Factor XIIIa, no adsorption takes place whether calcium is present or not in the permeation fluid.

DISCUSSION

The permeability coefficient (*Ks*) reflects on the size and shape of the pores in a gel and provides information on the over-all gel structure. Of considerable importance is our observation that, under a variety of conditions, *Ct* of the gel-forming system is directly related to the flow rate and thus, also with *Ks*. This relationship holds for different pHs, different ionic strengths, and calcium and fibrin concentrations. Furthermore, activation and gelation at different temperatures give gels with flow properties in accordance with the clotting time of the system, i.e., decreasing temperatures gives longer clotting times and higher *Ks* values. This shows that the *Ct* determines the gel structure over a large temperature range.

Our findings give strong support for the conclusion that the over-all gel structure must be determined by events preceding gel formation. This conclusion applies to both Fibrin I and Fibrin II gels, although the flow properties are distinctly different for these gels, in the sense that the *Ct* versus *Ks* plots have different slopes for the two enzymes (Batroxobin and thrombin) reflecting the difference in their structure. We have proposed that the average lengths of polymers formed prior to gelation varies directly with *Ct*.[13] Shorter polymers give infinite networks (gels) which are tight and longer polymers provide for formation of infinite networks which are more porous. Since longer polymers have more cross-linking (gelation) sites per polymer unit, the activation required for gelation would be relatively lower for long polymers than for short ones. In fact, we found that the activation (release of FPA and FPB) required for gelation is less at long clotting times than at short ones.

Increasing fibrinogen concentration leads to tighter final gels and decrease in *Ks*, which is proportional to the increase in fibrinogen concentration. This suggests that the amount of matrix in the gels influences their porosity. As

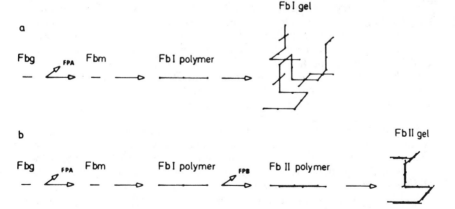

FIGURE 12. Schematic representation of polymerization and gelation. The picture shows alignment of activated fibrinogen molecules (Fbm) into polymers (FbI- or FbII polymers). These polymers, unlike Fbm, possess cross-linking sites, which are indicated as filled circles. Each site is thought to contain a set of complementary binding domains. We arbitrarily assume there exist three sites per polymer. Interaction between the domains yields an infinite new work (FbI or FbII-gels). Fbg: fibrinogen. (From Blombäck & Okada.[13] By permission of *Thrombosis Research*.)

shown in this study, the rate of activation (release of FPA) increased almost proportionally to the increase in fibrinogen concentration, without any decrease in clotting time (cf. TABLE 2). This finding seems to contradict our proposal relating clotting time to rate of activation.[13] However, it is possible that the expected shortening of clotting time due to the increased rate of activation is canceled out by shortening of polymer length when the fibrinogen concentration is increased. Such a mechanism would also explain the decrease in porosity with increasing fibrinogen concentration under our experimental conditions.

The gel structures for Fibrin I and Fibrin II we envisage is shown in FIGURE 12. In this model Fibrin I gels are in principle built up by linear polymers which are cross-linked at gelation sites. The lengths of these polymers are determined by Ct (inversely with enzyme concentration). The fibrin II structure is very similar, except that lateral alignment of polymers in this case provides for a different gel structure.

The present report has demonstrated that calcium ions have a pronounced influence on fibrin gel structure. The effect of calcium may be explained on the basis of binding of calcium to fibrinogen, fibrin monomer and/or fibrin polymers. Calcium seems to be of minor importance as a ligand for fibrin strands at crosspoints in the network, which is in agreement with the results of Ratnoff and Pott.[20] Several investigators have shown that fibrinogen from several species has high, as well as low affinity binding sites for calcium.[22, 29-31] The high affinity sites have Kd-values of about 10^{-6}M, whereas the low affinity sites have Kd-values of about 10^{-3}M. In case the structural changes demonstrated in this study are due to binding of calcium to fibrinogen, fibrinmonomer or fibrin polymers, one would expect that a low affinity site is involved.

Marguerie et al.[30] found that magnesium ions could compete with calcium for binding to the low affinity binding site. The effect by calcium on Ks is inhibited by magnesium. However, magnesium ions by themselves have no appreciable effect on the Ks values for either Fibrin I or Fibrin II gels, despite the fact that they shorten the Ct of the gel forming system. These findings suggest that magnesium interacts with fibrinogen (or monomers or polymers of fibrin), but its effect is clearly different from that exerted by calcium.

With both thrombin and Batroxobin, nearly complete cross-linking of the γ chains occurred in the final gels at all calcium concentrations. In the case of thrombin also the α chains were almost completely cross-linked. With neither enzyme was any cross-linking observed at Ct. We have demonstrated that there is at all calcium concentrations a linear relationship between Ks and Ct. As we have already discussed, this signifies the dependence of gel structure on events preceding Ct. It is therefore unlikely that differences in the degree of cross-linking for each type of gel can explain the variation in Ks at different calcium concentrations. On the other hand, the difference between Fibrin I and Fibrin II gels with regard to cross-linking may be an expression for the structural differences which exist in these gels.

The rate of release of FPA and FPB is not much affected by calcium. An increased rate of release is therefore not the explanation to the shortening of Ct, which can be demonstrated in the presence of calcium. The almost constant rate of release of fibrinopeptides coupled with shortening of clotting time, means that the activation required for gelation is decreased in he presence of calcium. The phenomenon may be explained by considering the mechanism for the formation of Fibrin I and Fibrin II gels we have just mentioned. The increasing Ks values and the lower activation required for gel formation in the presence of calcium, would thus mean that longer polymers are formed when calcium is present. The difference between Fibrin I and Fibrin II at increasing calcium concentrations may be explained by the proposition that the interaction between the complementary set of polymerization sites, B:b, which are operative on release of FPB,[7, 13] is enhanced in the presence of calcium, and that this interaction provides for formation of relatively longer polymers at the gel point.

The percolation of particles through gels having different clotting times supports our concept of the gel structure. Short clotting times clearly give finer lattice structures than long clotting times. However, the large ratio between theoretical and effective pore size, suggest that the pores in the gels are different from the pores in a conventional Poiseuille system. Furthermore, the difference in ratio between gels formed at different ionic strengths suggests that structural differences exist between gels formed under different conditions.

Our permeation experiments show that the pore structure in fibrin gels is remarkably uniform. It seems to us difficult to explain this uniformity unless we assume that the lattice structure is built up by uniform polymer units in an ordered fashion. A polymerization reaction similar to free radical polymerization [39] provides for information of a more or less homogeneous population of polymers at a given clotting time, having lengths inversely related to enzyme concentration. Such polymers would be expected to give a more regular pore structure of the final gels than would polymers formed by condensation polymerization.[39]

ACKNOWLEDGMENTS

The Sendai virus preparation was kindly provided by Dr. M. Wiebe, The New York Blood Center. We thank him for valuable advice regarding its handling and assay. For valuable assistance, we thank Mrs. Milla Block, Lisbeth Therkildsen, and Sonja Söderman.

REFERENCES

1. FERRY, J. D., S. SHULMAN, K. GUTFREUND & S. KATZ. 1952. The conversion of fibrinogen to fibrin. XI. Light scattering studies on clotting systems inhibited by hexamethylenglycol. J. Am. Chem. Soc. 74: 5709–5715.
2. FERRY, J. D., S. KATZ & I. TINOCO, JR. 1954. Some aspects of the polymerization of fibrinogen. J. Polymer Sci. 12: 509–516.
3. BLOMBÄCK, B. & T. C. LAURENT. 1958. N-terminal and light-scattering studies on fibrinogen and its transformation to fibrin. Arkiv Kemi 12: 137–146.
4. BLOMBÄCK, B. 1958. Studies on the action of thrombic enzymes on bovine fibrinogen as measured by N-terminal analysis. Arkiv Kemi 12: 321–335.
5. BACKUS, J. K., M. LASKOWSKI, JR., H. A. SCHERAGA & L. F. NIMS. 1952. Distribution of intermediate polymers in the fibrinogen-fibrin conversion. Arch. Biochem. Biophys. 41: 354–366.
6. CASASSA, E. F. 1956. The conversion of Fibrinogen to Fibrin. XIX. The structure of the Intermediate Polymer of Fibrinogen Formed in Alkaline Solutions. J. Am. Chem. Soc. 78: 3980–3985.
7. BLOMBÄCK, B., B. HESSEL, D. HOGG & L. THERKILDSEN. 1978. A two-step fibrinogen-fibrin transition in blood coagulation. Nature 257: 501–505.

8. KUDRYK, B. J., D. COLLEN, K. R. WOODS & B. BLOMBÄCK. 1974. Evidence for localization of polymerization sites in fibrinogen. J. Biol. Chem. **249:** 3322–3325.

9. LAURENT, T. C. & B. BLOMBÄCK. 1958. On the significance of the release of two different peptides from fibrinogen during clotting. Acta Chem. Scand. **12:** 1875–1877.

10. FLORY, P. J. 1941. Molecular size distribution in three dimensional polymers. I. Gelation. J. Amer. Chem. Soc. **63:** 3083–3090.

11. FERRY, J. D. & P. R. MORRISON. 1947. Preparation and properties of serum and plasma proteins. VIII. The conversion of human fibrinogen to fibrin under various conditions. J. Amer. Chem. Soc. **69:** 388–400.

12. CARR, M. JR., L. L. SHEN & J. HERMANS. 1977. Mass-length ratio of fibrin fibers from gel permeation and light scattering. Biopolymers **16:** 1–15.

13. BLOMBÄCK, B. & M. OKADA. 1982. Fibrin gel structure and clotting time. Thrombosis Research **25:** 51–70.

14. SIGNER, R. & H. EGLI. 1950. Sedimentation von Makromolekülen und Durchströmung von Gelen. Recueil des travaux chimiques de pay-Bas **69:** 45–58.

15. MADRAS, S., R. L. MCINTOSH & S. C. MASON. 1949. A preliminary study of the permeability of cellophane to liquids. Can. J. Res. Sect. B. **27B:** 764–779.

16. FERRY, J. D. 1936. Ultrafilter membranes and ultrafiltration. Chem. Rev. **18:** 373–455.

17. WHITE, M. L. 1960. The permeability of an acrylamide polymer gel. J. Phys. Chem. **64:** 1563–1565.

18. SEEGERS, W. H. & H. P. SMITH. 1942. Factors which influence the activity of purified thrombin. Am. J. Physiol. **137:** 348–354.

19. BOYER, M. H., J. R. SHAINOFF & O. D. RATNOFF. 1972. Acceleration of fibrin polymerization by calcium ions. Blood **39:** 382–387.

20. RATNOFF, O. D. & A. M. POTTS. 1954. The accelerating effect of calcium and other cations on the conversion of fibrinogen to fibrin. J. Clin. Invest **33:** 206–210.

21. KATZ, S., S. SHULMAN, I. TINOCO, JR., J. BILLICK, K. GUTFREUND & J. D. FERRY. 1953. The conversion of fibrinogen to fibrin. XIV. The effect of calcium on the formation and dissociation of intermediate polymers. Arch. Biochem. Biophys. **47:** 165–173.

22. LORAND, L. & K. KONISHI. 1964. Activation of the fibrin stabilizing factor of plasma by thrombin. Arch. Biochem. Biophys. **105:** 58–67.

23. ENDRES, G. F. & H. A. SCHERAGA. 1972. Equilibria in the fibrinogen-fibrin conversion IX. Effects of calcium ions on the reversible polymerization of fibrin monomer. Arch. Biochem. Biophys. **153:** 266–278.

24. MARGUERIE, G., Y. BENABID & M. SUSCILLON. 1979. The binding of calcium to fibrinogen: Influence on the clotting process. Biochim. Biophys. Acta **579:** 134–141.

25. SHEN, L. L., J. HERMANS, J. MCDONAGH, R. P. MCDONAGH & M. CARR. 1975. Effects of calcium ion and covalent cross-linking on formation and elasticity of fibrin gels. Thromb. Res. **6:** 255–265.

26. ROSENFELD, G. & B. JANSZKY. 1952. The accelerating effect of calcium on the fibrinogen-fibrin transformation. Science **116:** 36–37.

27. GODAL, H. C. 1960. The effect of EDTA on human fibrinogen and its significance for the coagulation of fibrinogen with thrombin. Scand. J. Clin. Lab. Invest. **12:** Suppl. 53, 3–20.

28. BLOMBÄCK, B., M. BLOMBÄCK, T. C. LAURENT & H. PERTOFT. 1966. Effect of EDTA on fibrinogen. Biochim. Biophys. Acta **127:** 560–562.

29. LY, B. & A. C. GODAL. 1972. Denaturation of fibrinogen, the protective effect of calcium. Haemostasis **1:** 204–209.

30. MARGUERIE, G., G. CHAGNIEL & M. SUSCILLON. 1977. The binding of calcium to bovine fibrinogen. Biochim. Biophys. Acta **490:** 94–103.
31. VAN RUIJVEN-VERMEER, I. A. M., W. NIEUWENHUIZEN & W. J. NOOIJEN. 1978. Calcium binding of rat fibrinogen and fibrin(ogen) degradation products. FEBS Letters. **93:** 177–180.
32. NIEUWENHUIZEN, W., A. VERMOND, W. J. NOOIJEN & F. HAVERKATE. 1979. Calcium binding properties of human fibrin(ogen) and degradation products. FEBS Letters. **98:** 257–259.
33. ROBINSON, R. A. & R. H. STOKES. 1955. Electrolyte solutions. Butterworths Scientific Publications. London, England.
34. BLOMBÄCK, B. & M. OKADA. 1982. On pores in fibrin gels. Thromb. Res. **26:** 141–142.
35. SHAINOFF, J. R. & I. H. PAGE. 1962. Significance of cryoprofibrin in fibrinogen-fibrin conversion. J. Exp. Med. **116:** 687–707.
36. HEENE, D. L. & F. R. MATTHIAS. 1973. Adsorption of fibrinogen derivatives on insolubilized fibrinogen and fibrin monomer. Thromb. Res. **2:** 137–154.
37. MOSESSON, M. W. 1978. Structure of human plasma cold-insoluble globulin and the mechanism of its precipitation in the cold with heparin or fibrin-fibrinogen complexes. Ann. N.Y. Acad. Sci. **312:** 11–29.
38. IWANAGA, S., K. SUZUKI & S. HASHIMOTO. 1978. Bovine plasma cold-insoluble globulin: Gross structure and function. Ann. N.Y. Acad. Sci. **312:** 56–73.
39. TANFORD, C. 1978. Physical chemistry of macromolecules. J. Wiley & Sons, Inc. New York.

DISCUSSION OF THE PAPER

J. D. FERRY (*University of Wisconsin, Madison*): In interpreting these very interesting experiments, it may be useful to focus attention on the thickness of the fibrin structures in the clot rather than on their length. In somewhat similar measurements of the flow of fluid through clots which we made, we used the Darcy formulation to calculate the thickness of fiber strands and showed how this increased as one progressed from a fine clot to a coarse clot. I think it may be of interest to use this alternative description of the structure. I do not know exactly what kind of formulas you use to calculate the pore size.

M. OKADA: We are using a conventional way of calculating the data. We have used the formula by Madras et al.[15] for the calculation of average pore size, assuming that we have a straight capillary system with the capillaries parallel to the direction of flow.

FERRY: The Darcy formulation will give you the fiber strand thickness.

OKADA: We do not know the fiber thickness or its effect on the liquid and particle permeation experiments. We assume, however, that the strands in the matrix are thicker at long clotting times than at short ones. The point in our experiments is the demonstration that the pore size in fibrin gels appears to be related to the clotting time and that the pore structure is very uniform.

H. A. SCHERAGA (*Cornell University, Ithaca, NY*): Are there any models yet to explain why pH and ionic strength will shift you over from a fine to a coarse clot?

OKADA: In this regard we are thinking along the same lines as Dr. Ferry.

FERRY: We used to think in terms of increasing electrostatic repulsion with increasing pH, and I do not think that we have ever progressed beyond that simple-minded idea.

FIBRINOPEPTIDE B IN FIBRIN ASSEMBLY AND METABOLISM: PHYSIOLOGIC SIGNIFICANCE IN DELAYED RELEASE OF THE PEPTIDE *

John R. Shainoff and Beatriz N. Dardik

Thrombosis Section
Department of Cardiovascular Research
The Cleveland Clinic Foundation
Cleveland, Ohio 44106

INTRODUCTION

As with fibrinopeptide A, release of fibrinopeptide B is capable of eliciting aggregation of fibrinogen in a staggered overlapping assembly as demonstrated by rapid γ chain cross-linking of the polymers by Factor XIIIa,[1] but the aggregation does not yield clot forming protofibrils at 37°C and clots formed at lower temperatures tend to be gelatinous rather than coarse.[2] Further, viscous solutions formed by the oligomers at 25°C tend to remain fluid when treated with thrombin to release fibrinopeptide A, but rapidly transform into a coarse clot when brought to 37°C. The question of what consequences might arise if B preceded A might be considered academic, because no mammal has as yet been found in which A does not precede B in course of blood coagulation.[3-6] We have recently proposed that the delayed release of B is important for rapid clearance of the small amounts of fibrin monomer that are continuously formed in the circulation, because monomers lacking A alone are cleared rapidly while those lacking both peptides A and B are cleared slowly.[7-9] The purpose of the present communication is to present studies on the extent to which each of the fibrinopeptides effects fibrinogen-fibrin interaction, and to consider ways in which these interactions effect oligomer dissociability, which we believe to be important for rapid uptake of circulating fibrin complexes from blood. It is suggested that a role of fibrinopeptide B release may be to prevent dissociation of the fibrin oligomers, which would open them to attack by mononuclear phagocytic cells.

EXPERIMENTAL

Fibrin lacking B alone (designated as β-fibrin) was prepared from human fibrinogen by treating with copperhead procoagulant enzyme [10] at 14°C, and was dissolved by warming to 37°C with PMSF added to inactivate the enzyme as described.[2] It was then further purified to remove an 8% contamination with species lacking A by passing concentrated solutions (>10 mg/ml)

* This work was supported in part by Grants HL-16361 and HL-19767 from the National Heart, Lung and Blood Institute.

through a column (20 ml/ml fibrin) of 2% agarose gel equilibrated with 0.3 M NaCl/TRIS-HCl (9/1) at pH 7.4 and 37°C, and retaining only the portion eluting at concentrations above six-tenths of the peak height. Preliminary separations with added trace amounts of radioiodinated monomer lacking both fibrinopeptides indicated that all but a small percentage eluted ahead of the β-fibrin peak, and enabled us to infer that monomers lacking A in addition to B comprised only 0.4% of the retained portion.

Monomers lacking A (designated α-fibrin) with only 8% loss of B were prepared with thrombin at pH 5.3, and the thrombin was inactivated with hirudin, as described.[11] Fibrin lacking both fibrinopeptides (designated $\alpha\beta$-fibrin) was prepared with thrombin at pH 7.4, and dissolved in 0.3 M TRIS HCl/acetate (9:1) with hirudin added, as described.[12]

The ultracentrifuge procedure for study of interaction of fibrinogen with β-fibrin was essentially the same as described before [2] for study of aggregation of β-fibrin itself. The chromatographic studies were also carried out as before,[11] and differed only with respect to the concentrations of unlabelled fibrinogen used for equilibration, and in the species of ^{125}I-monomer added along with the ^{131}I-fibrinogen subsequent to equilibration to track relative rates of transport of the monomer through the column. Fresh columns (ranging in size from 5–20 ml) were employed for each experiment, and were calibrated only on basis of known volume and known permeation ($V_e/V_t = 0.8$) of fibrinogen. Flow rates were maintained at 4 ml/cm^2/h by peristaltic pump.

Diffusion of β-fibrin aggregates was measured in a double sector, synthetic boundary cell by essentially the same ultracentrifuge procedure described by Ifft et al.[13] in which sucrose was added to the solution to stabilize boundary formation. Solutions of the β-fibrin at pH 7.4 and indicated concentrations in 0.15 M NaCl/TRIS-HCl (9:1) and 0.6% sucrose was loaded into the cell at 37°C, and overlaid with sucrose-free solution of more dilute (0.5 mg/ml), fibrin by centrifugal transfer from the reference chamber through the connecting channel. Velocity ultracentrifugation had indicated that the concentration of monomer equalled 0.2 mg/ml in all solutions over the concentration ranges studied, so that diffusion would in theory depend only on the difference in aggregate concentration in the overlaid solution. The measurements were carried out at 37°C as determined from the rotor thermistor. To minimize thermal convection in course of the run the room was closed off, and ambient temperature raised to 37°C. Also, the rotor chamber was preadjusted to maintain 37°C by setting refrigeration bleed to maintain that temperature at operating speed of 5600 rpm for a period of 24 hr prior to loading the fibrin solutions into the cells. Diffusion was measured from change in radial distance ($\triangle r$) between points corresponding to ¼ and ¾ of the absorbancy change (280 nm) across the protein boundary. Diffusion coefficients calculated from slopes of $(\triangle r)^2$ vs. t ($D = \text{slope}/3.64$) were adjusted for temperature (37°C) and viscosity to yield $D_{w, 20}$ which corresponded to 0.6211 times the observed value (D).

RESULTS

Ultracentrifuge Studies with β-Fibrin

To characterize the interaction between fibrinogen and monomeric β-fibrin we took advantage of the previous observation [2] that β-fibrin at 37°C dissociated fully into 8S monomers at dilutions below 0.2 mg/ml. When fibrinogen was added to the monomeric β-fibrin (0.15 mg/ml), complexing was discernible from formation of a 16S component (FIGURE 1). The sedimentation rate of the complex was the same as observed for the soluble aggregate formed by β-fibrin itself at concentrations above 0.2 mg/ml. The concentration of complex (0.10 mg/ml) formed with a mixture of 0.15 mg/ml fibrin and 0.2 mg/ml of fibrinogen was essentially the same as that (0.15 mg/ml) formed by β-fibrin itself at the same total protein concentration (0.35 mg/ml). Thus, it could be concluded that fibrinogen complexing and self aggregation by the β-fibrin involved equally avid interactions.

The sedimentation coefficient of the complex did not change on elevating fibrinogen concentrations 10 times to 2 mg/ml. Nor did the sedimentation coefficient of the aggregate formed by β-fibrin itself change over the same range of concentrations, as though aggregation and complex formation at

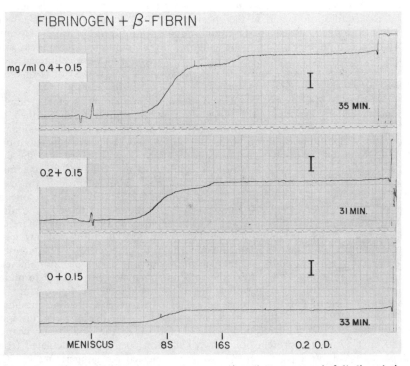

FIGURE 1. Ultracentrifuge patterns demonstrating (lower pattern) full dissociation of dilute (0.15 mg/ml) β-fibrin into monomers in buffered saline at 37°C, and incorporation of the monomers into 16S complexes (upper patterns) with added fibrinogen. The diagrams display scans of absorbency at 280 nm (ordinate) for the location and measurement of boundaries formed by protein sedimenting (left to right) in the centrifugal field (52,000 rpm). The position of boundaries corresponding to monomeric fibrinogen and fibrin with sedimentation coefficient ($s_{w,20}$) of 8S, and aggregates sedimenting at 16S are indicated in the diagram. The concentrations of 16S component formed corresponded to those anticipated if complexing of fibrinogen with β-fibrin and self-aggregation of β-fibrin [2] involved equally avid interactions.

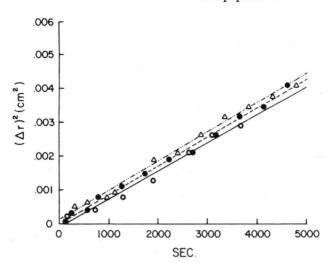

FIGURE 2. Measurements of boundary spread from diffusion of β-fibrin at 0.7 (——O——), 1.0 (– – ● – –), and 2.0 (– ·· △ ··) mg/ml into β-fibrin at 0.5 mg/ml. Ultracentrifugation as in FIGURE 4 indicated all solutions contained equal concentrations of monomer at 37° C, and differed only with respect to concentrations of 16S component. Constancy of the diffusion coefficients (given by slopes of the plots) with changing concentrations indicated the 16S component comprised a discrete species of aggregate, which from combined sedimentation and diffusion data corresponded to trimers.

37°C was limited to formation of a discrete structure. From measurement (FIGURE 2) of the diffusion coefficient at concentrations ranging from 0.7 to 2 mg/ml ($D_{w,20} = 1.46 \times 10^{-7}$), we concluded that the complex consisted of trimers, presumably with β-fibrin monomer linking two fibrinogen molecules (FIGURE 3), and a similar configuration for the trimers formed by β-fibrin itself. The diffusion constant determined [14] for fibrinogen (1.93×10^{-7} cm²/s) by the same procedure was essentially identical to the generally accepted value.[15]

Preceding studies on self association of β-fibrin into the 16S component could not readily be described in terms of a sequence involving stepwise addition of monomers to form dimers, then trimers, because monomer concentrations remained constant as β-fibrin concentrations were elevated to five times that (0.2 mg/ml) yielding 16S component. Because of absence of measurable change in monomer concentration, the association constant would be given by $K_\beta = 1/[\text{monomer}]$ at saturation or 1.6×10^6 M⁻¹.

There was no indication of formation of complexes with sedimentation rate between that of 8S monomers and 16S trimers. Dimeric fibrinogen linked by end to end (D-D) cross-linking had been shown [12] to sediment at 11S, and dimers formed by D-E interaction as anticipated for fibrinogen/monomer association would have three-fourths the length, and would sediment at rates only slightly faster. Since the diffusion and sedimentation measurements gave no indication of substantial formation of a dimeric component, the possibility could be considered that dimers were relatively unstable compared to trimers, and that complexing was reinforced in the trimers by cooperative interaction such as $\frac{D \cdot D}{E}$, where the dot represents an interaction between D-domains of two monomers linked by the B-dependent aggregation sites of the third (binding) monomer. If the interactions occurred as:

$$2f_1 \rightleftarrows f_2 \text{ (dimerization) } K_1 = [f_2]/[f_1]^2$$
$$f_1 + f_2 \rightleftarrows f_3 \text{ (trimerization) } K_2 = [f_3]/[f_1][f_2]$$

the reinforced assembly of f_3 would make $f_2 \ll f_3$ and f_1, and f_2 could be expressed as occurring in small proportion to f_3, or $[f_2] = c[f_3]$ where c would represent a stability constant for f_3). Then substituting for $[f_2]$, we'd obtain $K_2 = 1/c[f_1]$, which we suspect may be the reason for the constant monomer concentration over the range studied.

The interactions involved in monomer/trimer transition were not affected by Ca^{2+} in that identical results were obtained with solutions containing 0.1 mM EDTA or 2.5 mM Ca^{2+} (FIGURE 4). Althought Ca^{2+} had no effect on monomer/monomer association, it did accelerate clot formation by the β-fibrin at 14°C.

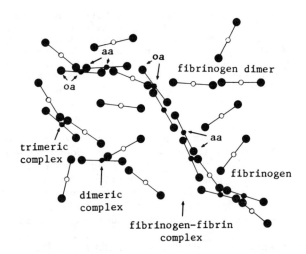

FIGURE 3. Perspective on the structure of fibrinogen-fibrin complexes. Fibrinogen is represented schematically as tridomainal structure with the central E-domain inactive (unshaded) in complexing, and fibrin monomer is depicted with functional E-domain (shaded) capable of interacting equivalently with the D-domains of either fibrinogen or fibrin. Monomer/monomer interactions are depicted as reinforced by duplicated D/E interactions between the monomers (labelled as "aa" couples), while such reinforcement is prohibited in monomer/fibrinogen interaction where binding is depicted with a single attachment point (labelled "oa"). Although the aa couple is depicted as fully expressed it is partially hindered in α-fibrin which sediments largely as 24S aggregates that are readily dissociable into trimers and monomers as evidenced by chromatography. Analogous coupling through the b-epitope is greatly hindered in β-fibrin. Fibrinogen dimers linked by the D-domains are not seen except when formed by Factor XIIIa crosslinking, but are believed to form transiently through a weak D·D interaction inferred to reinforce trimeric complexes and predispose the preferential formation of trimers over the relatively unstable dimeric complexes.

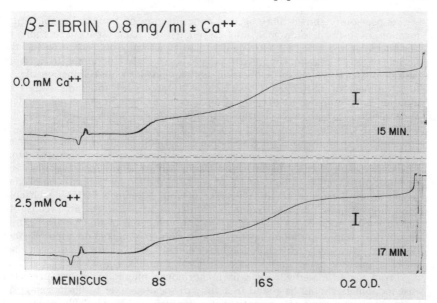

FIGURE 4. Ultracentrifuge patterns demonstrating that Ca^{2+} had no effect on self-aggregation of β-fibrin into trimers at 37°C. The observation warranted inference that acceleration of β-fibrin clot formation by Ca^{2+} at low temperatures was not due to an enhancement of monomer/monomer interaction, but to an effect on later stages of clot assembly.

As well known, α-fibrin was too insoluble to yield detectable levels of monomer in absence of fibrinogen, and relatively large (8- to 12-fold) molar excess of fibrinogen was required to maintain it in solution at 37°C.[16] The complexes formed in fibrinogen solution sedimented at 24S, as previously described,[17, 18] and presumably consisted of hybrid fibrinogen/fibrin protofibrils with structure corresponding to the intermediary polymers of fibrin described by Ferry.[19] Complexes of $\alpha\beta$-fibrin have the same sedimentation rate, and hence identical thickness as those formed by α-fibrin, but not necessarily the same length as indicated by Scheraga.[15] To approach study of the interaction of fibrinogen with these species of fibrin, we resorted to a chromatographic system where monomer dissociation was discernable but not directly measurable.

Chromatographic Assessment of Interactions through the A- and B-Dependent Sites

Dissociation of α-fibrin into monomer was discernable from its comigration with labelled fibrinogen in course of chromatography through agarose equilibrated with unlabelled fibrinogen at low concentrations,[20] and complexing of the monomers with fibrinogen was discernable from migration ahead of labelled fibrinogen in columns equilibrated with unlabelled fibrinogen at moderate or high concentrations.[11] We sought to compare strengths of association of fibrinogen with differing species of monomer by determining the concentration of fibrinogen required to elicit the same change in transport characteristics.

As previously shown, fibrinogen at concentration of 1 mg/ml enabled permeation of α-fibrin monomer into essentially the same space accessible to fibrinogen at the adopted chromatographic flow rate, but elevation of the fibrinogen to 2 mg/ml suppressed permeation of the monomer to a more restricted space which by calibration corresponded to that impermeable to linear dimers of fibrinogen [11, 12] which by inference of the same radius of gyration corresponded to trimeric complexes assembled in staggered overlap. With β-fibrin, a higher concentration of fibrinogen between 8 and 12 mg/ml was found (FIGURE 5) necessary to elicit the same shift in permeation. It was inferred from this that the association of fibrinogen with α-fibrin monomers to form trimeric complexes was four to six times stronger than the same type of association with β-fibrin. Since K_β for β-fibrin complexing was observed by ultracentrifugation to be near 1.6×10^6 M^{-1}, the K_α for α-fibrin complexing would be on the order of 10^7 M^{-1}. The 50-fold difference of fibrinogen concentrations required to elicit complex formation with β-fibrin as directly observed ($\simeq 0.2$ mg/ml) by ultracentrifugation, compared to that ($\simeq 10$ mg/ml) required to suppress monomer permeation in course of chromatography further illustrated why early chromatographic studies [20] with dilute fibrinogen had posed a question whether fibrinogen/α-fibrin complexes exist at 37°C.

From the estimated K_α for α-fibrin/fibrinogen interaction it was predicted that monomers should be found in a fully dissociated state in fibrinogen-fibrin solutions at 40 μg/ml. Examination of a mixture of 40 μg/ml fibrinogen and 4 μg/ml of α-fibrin that was labelled with heme-octapeptide [8] to allow tracking at 490 nm showed all of the fibrin sedimenting as monomers at 8S. Monomer itself at the same concentration formed insoluble strands, as anticipated.[21] The virtually full aggregation of dilute monomer in absence of fibrinogen was inter-

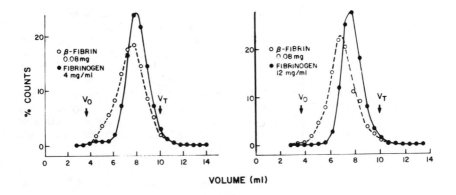

FIGURE 5. Elution profiles of one ml of labelled β-fibrin and fibrinogen applied to a 10 ml column of 2% agarose equilibrated with indicated concentrations of unlabelled fibrinogen at 37°C. Dissociability into monomers sufficed to allow the fibrin to permeate into the gel to a degree almost equivalent to fibrinogen with the fibrinogen at 4 mg/ml, but elevation of the fibrinogen to 12 mg/ml suppressed dissociation to a degree limiting permeation to space permeable to dimers and trimers. A fibrinogen concentration of 2 mg/ml sufficed to elicit a similar shift in permeation in previous studies [11] with α-fibrin, an indication that the interaction of fibrinogen with β-fibrin was six times weaker than its interaction with α-fibrin.

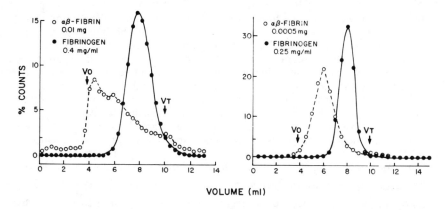

VOLUME (ml)

FIGURE 6. Very little permeation of $\alpha\beta$-fibrin into agarose occurred in chromatography with fibrinogen except when the fibrin was applied as a very dilute solution. Results qualitatively similar to that shown with 0.01 mg fibrin and 0.4 mg/ml fibrinogen were obtained when the fibrinogen concentration was raised to 20 mg/ml. Essentially full recovery (>90%) of the labelled fibrin and fibrinogen was obtained in all experiments, but required addition of albumin (10 mg/ml) in the equilibrating medium to minimize nonspecific adsorption in experiments with dilute solutions.

preted to result from reinforcement of fibrin oligomers from duplication of D-E interactions between monomers, depicted by the two attachment points labelled as "aa" in FIGURE 3. The D-E interaction between monomer and fibrinogen is not duplicated as in monomer/monomer interaction, because only the D-domain and not the E-domain of fibrinogen is involved in complex formation, depicted by the single attachment point labelled "oa" in FIGURE 3. Unhindered duplication of monomer interaction would function in a way analogous to chelation to suppress monomer dissociation from fibrin oligomers, while nonduplicated interaction with fibrinogen would serve to maintain monomer dissociability provided fibrinogen was present in large excess over fibrin. If there was no steric hindrance to duplicated interaction, dissociation of monomer in absence of fibrinogen would be as low as $K_\alpha K_\alpha = 10^{14}$ M^{-1}. Since only a 12-fold molar excess of fibrinogen suffices [16, 22] to maintain the fibrin in solution over wide ranges of concentration, the duplicated interaction is only partially expressed, and results in strengthening of monomer/monomer interaction by only a factor on the order of ten.

As observed by others,[23] permeation characteristics of $\alpha\beta$-fibrin differed greatly in that major portions applied to the columns eluted as nonpermeating aggregates in the void volume under conditions allowing permeation of α or β-fibrin as trimeric complexes. At 10 μg/ml substantial permeation of $\alpha\beta$-fibrin as trimers was not even observed when the fibrinogen in the equilibrating medium was raised to 20 mg/ml. To elicit permeation of this fibrin, we had to drop the concentration of the fibrin to a trace level of 0.5 μg/ml. At this level the $\alpha\beta$-fibrin permeated as trimers with columns equilibrated with fibrinogen at concentrations of 250 μg/ml or larger (FIGURE 6). Complex patterns with much elution of the fibrin in the void was obtained with fibrinogen below a 100-fold excess over the fibrin. When the fibrinogen concentration was dropped to 20 $\mu g/ml$ and fibrin to a trace level of 0.05 $\mu g/ml$, a complex elution pattern was obtained with about 30% of the fibrin eluting with fibrinogen as though a partial though slow dissociation into monomers was occurring. We estimated from the very high dilutions required to detect monomer dissociation that fibrinogen/monomer association with $\alpha\beta$-fibrin was at least a thousand times tighter than with α-fibrin. The A-dependent and B-dependent epitopes contributing to aggregation and fibrinogen complex formation were accordingly acting cooperatively as in a chelate rather than independently.

Cooperative rather than independent association through the A- and B-dependent sites accorded also with observation that complexes formed by $\alpha\beta$-fibrin sediment at the same rate and accordingly have the same thickness as those formed by α-fibrin. Thus, interaction of the D-domain of fibrinogen with the E-domain of $\alpha\beta$-fibrin would be represented by $D_{-b}^{-a} E_b^a$ (where a and b represent A- and B-dependent aggregation sites)—a chelated rather than extended interaction of the type $\begin{smallmatrix}D\text{-}a\\D\text{-}b\end{smallmatrix} E_b^a$. The coupled a and b interactions would be duplicated again in monomer/monomer assembly so that the attachment points labelled aa in FIGURE 3 might be reinforced by analogous bb complexes.

Since aggregation of β-fibrin at 37°C was shown to be limited to trimerization, the question arose whether the b-epitope contributes to duplicated interactions between monomers analogous to that labelled as "aa" in FIGURE 3. An affirmative answer was indicated by chromatography of $\alpha\beta$-fibrin in solution with β-fibrin (FIGURE 7). Whereas very dilute $\alpha\beta$-fibrin (0.5 $\mu g/ml$) was

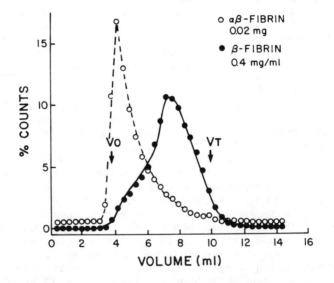

FIGURE 7. Chromatography of very dilute $\alpha\beta$-fibrin in columns equilibrated with β-fibrin (0.25 mg/ml) at 37°C failed to reveal dissociation to a degree evident with columns equilibrated with fibrinogen as in FIGURE 6. Identical results were obtained with fully dissociated β-fibrin at 0.2 mg/ml. The low dissociability of $\alpha\beta$-fibrin/β-fibrin complexes provided evidence of the bb couple depicted in FIGURE 8.

capable of dissociating adequately to permeate agarose as trimers in presence of fibrinogen at 0.2 mg/ml, it did not permeate in columns equilibrated with β-fibrin instead of fibrinogen at this concentration where the β-fibrin itself exists as monomer. The hindrance to duplication of interaction through the b-epitope which suppressed formation of large oligomers with B alone removed was at least partially overcome with both peptides A and B removed.

DISCUSSION

Fibrin Assembly and Dissociation

From preceding studies,[2, 24, 25] it is known that the fibrinopeptides A and B each mask penultimate segments (designated as a- and b-epitopes) which are individually capable of functioning as aggregation sites. The present study helps to define the individual contributions of each epitope, the degree of interaction required for protofibril generation, the extent to which the two epitopes function cooperatively in protofibril stabilization, and the manner in which fibrinogen helps to maintain fibrin in a soluble, dissociable state.

The analysis of monomer dissociation from α-fibrin and β-fibrin in fibrinogen solution indicate that interactions imparted by the a- and b-epitopes differ by a factor of six in strength, that aggregation into trimeric structures is preferred over dimeric units in course of protofibril generation, and that fibrinogen complexing and monomer aggregation involve equally avid interactions at the trimer stage of protofibril generation. The $K_\alpha = 10^7$ M^{-1} estimated for trimerization through the a-epitope is 15 times the association constant determined by Olexa and Budzynski[26] for interaction of α-fibrin with a peptide segment ($\gamma_{373-410}$) isolated from the D-domain as complementary site for the a-epitope. We suspect that the factor of 15-difference represents a weak D-D interaction helping to stabilize the trimeric structure, as proposed for the equilibrium involved in β-fibrin trimerization. The inference of a weak D-D interaction is supported by observation that concentrated fibrinogen undergoes cross-linking as rapidly as fibrin through formation of γ-γ dimers with Factor XIIIa.[12]

Olexa and Budzynski[27] suggested that a new aggregation site arises from alignment of the D-domains in the aggregation process, because cross-linked fibrin was found in chromatographic studies to bind isolated E-1 fragments while noncross-linked monomer did not bind the fragments. However, the method of study was insensitive to the binding ($K = 7 \times 10^5$ M^{-1}) with noncross-linked monomer observed later.[26] A large difference in binding would be expected to arise from cross-linking due to bifunctionality of cross-linked D-domains which provide for a coupled interaction with the E-domain. Such coupling is indicated by the low dissociability of the DXD-E complex isolated from cross-linked fibrin.[28] The D-D interaction proposed here for trimer stabilization is one that we believe exists naturally but is too weak to be of significance in the absence of aggregation promoted by fibrinopeptide release.

The suggested reinforcement favoring trimerization may be related to cooperative interaction suggested long ago by Donnelly et al.[29] Their light-scattering studies had indicated that monomer association was strengthened at degrees of polymerization beyond dimerization. From observation that β-fibrin aggregates only into trimers at 37°C, but forms clots at low temperatures of 20°–14°C, we can estimate the extent of the additional interaction required for generation of protofibrils. As judged from an enthalpy of 20 kcal/mole,[2] the interaction between β-fibrin becomes strengthened 5–10 times to 10^7 M^{-1} at the lower temperatures. Coincidentally, α-fibrin association is of this magnitude, and elicits clot formation in absence of fibrinogen, and large 24S aggregates in presence of fibrinogen at 37°C. However, some additional interaction may be inferred to contribute to protofibril generation by α-fibrin, because of indications from Landis and Waugh [21] that clot formation is favored at concentrations as low as 10^{-8} M. The added association involved in oligomer formation which would correspond to an increase in K_α by at least a factor of 10 over that measured here for trimerization probably arises from a partial coupling of D-E interactions from adjoining monomers, depicted as aa interactions in FIGURE 3 and 8. Analogous reinforcement through the b-epitope, although not evident in aggregation of β-fibrin itself, was evident from the chromatography of $\alpha\beta$-fibrin where little dissociation into trimers was evident in chromatography with monomeric β-fibrin as carrier under conditions where the fibrin equilibrated as trimers with fibrinogen as carrier.

The concerted interactions through the cooperative effect of the a- and b-epitopes (an "ab" couple) in linking individual D and E domains, together with the added reinforcement through the aa coupling resulting from the overlapping monomer assembly (FIGURE 8) can be viewed as analogous to a chelation. Stabilization of the overlapping assembly through an added bb couple can also occur. Although significant, the contribution of bb coupling to monomer aggregation is small, because the solubility of $\alpha\beta$-fibrin in fibrinogen solution is not a great deal less than the solubility of α-fibrin. Whereas full dissolution of α-fibrin requires an 8–12 molar excess of fibrinogen, $\alpha\beta$-fibrin requires a 30 molar excess.

As portrayed in FIGURE 3 and demonstrated in chromatography, oligomers formed by complexing of fibrinogen with α-fibrin are readily dissociable into trimers and, in turn, monomers due to (1) inability of fibrinogen to function in aa coupling and (2) weakness of the aa couple relative to the intrinsic, unassisted binding of fibrinogen with the a-epitope. Dissociation of $\alpha\beta$-fibrin into trimers is partially suppressed by added effect of the bb couple between monomers; thus, concentrations of $\alpha\beta$-monomer had to be dropped 5–10 times relative to the fibrinogen concentration to detect oligomer → trimer dissociation analogous to that observed in chromatography of α-fibrin in fibrinogen solution. From these observations it is apparent that the bb couple enhances monomer/monomer association by only a factor of 5 or so over the weak enhancement imparted by the aa couple. But dissociation of monomers from the trimeric complexes of $\alpha\beta$-fibrin is severely restrained by the ab couple by which fibrinogen becomes intrinsically linked to the monomer with affinity approaching the product ($K_\alpha K_\beta > 10^3 \, K_\alpha$) of association effected through each epitope.

Physiologic Significance

Because of the more rapid release of A in course of coagulation, there is no doubt that protofibrils are generated initially through interactions imparted by the a-epitope. The release of B is usually portrayed as functioning in promoting side-by-side cross-bridging of the protofibrils to form thick strands.[24] Although it is known that the release of B accelerates coalescence of protofibrils into fibrin strands, the effect may be an indirect result of the strengthening of protofibril structure. Evidence against direct function of the b-epitope in cross-bridging is provided by the identical sedimentation rate and hence thickness of soluble complexes formed by either αβ-fibrin or α-fibrin in fibrinogen solution, and by the evidence that the b-eptiope acts in concert with the a-epitope to reinforce the complexes against dissociation by a factor greater than 10^3.

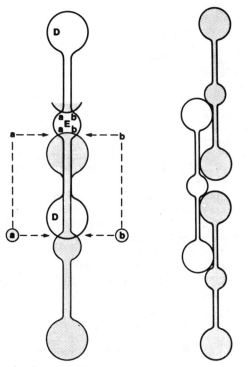

FIGURE 8. Schematic diagram depicting the staggered overlapping assembly yielding coupled interactions. The a- and b-epitopes in the E-domain are suggested to interact simultaneously and without appreciable hindrance to form a strong "ab" couple linking the αβ-monomers to the D-domain of either fibrinogen or another monomer. This "ab" couple is duplicated through "aa" couples (depicted by the arrows) in linkages between two monomers, but not in linkage of monomer to fibrinogen (the a- and b-epitopes being masked by fibrinopeptides in fibrinogen). Data provided suggest that the aa and bb couples, while contributing to reinforced profibril assembly, are hindered because dissociability in fibrinogen solution is far greater than would be anticipated for an unhindered quadridentate assembly.

Release of B before A could lead to defective clot formation, but only at low temperature as indicated from studies with copperhead procoagulant venom.[2] Considerations of large differences in clot stability and rates of clot formation after release of B led to a suggestion that the delayed release of B helps to prevent intravascular coagulation by enabling initial clots to undergo more rapid degradation by plasmin.[30] Recent studies on the mechanism of absorption of fibrin monomer by phagocytic cells [8] lead us to infer that the reason for delayed release of B may be more importantly related to a need to maintain soluble fibrin complexes in a highly dissociable state for rapid clearance from blood.

A saturable receptor for uptake of dissolved fibrin by phagocytic cells was implicated by Sherman and Lee [31] to function in clearance of fibrin complexes. We recently suggested that this receptor has high affinity for the a-epitope that is masked by fibrinopeptide A in fibrinogen, based on observation of rapid uptake of α-fibrin in a monomeric, nonaggregated form, and inhibition of this uptake by synthetic Gly-Pro-Arg.[8] To the extent that uptake depends on access to the a-epitope, rates of clearance of dissolved fibrin in blood would depend on dissociability of the fibrin, for the a-epitope would not be accessible while occupied in aggregation. Because of low dissociability of complexes formed by αβ-fibrin, absorption of these complexes would be expected to proceed slowly. As observed,[7, 9] intravenously injected monomers of αβ-fibrin are cleared from the circulation much more slowly than α-fibrin. Dependence on access to the a-epitope is indicated also by observation [7, 32] that β-fibrin has a half-life near that of fibrinogen, and cross-linking of complexes formed by α-fibrin impair their clearance. Thus, delayed activation of Factor XIII as well as the delayed release of B is essential to maintain dissociability of fibrin complexes for rapid clearance.

Release of B in the end stages of fibrin formation functions to stabilize the clot.[30] It may also function to confine transendothelial transport of fibrin oligomers, for as shown, the complexes formed by αβ-fibrin have low permeability, while complexes of α-fibrin can be transported out of space permeable to fibrinogen by virtue of their dissociability. The prospect may also be entertained that the release of B helps to prevent attack of the fibrin by mononuclear phagocytic cells by preventing exposure of the a-epitope through which the cells recognize fibrin.

SUMMARY

The delayed release of peptide B that accelerates towards the end of fibrin formation unmasks accessory (b-) epitopes for monomer interaction. Ultracentrifuge and chromatographic analysis of the composition and dissociation of soluble complexes formed by monomers in fibrinogen solution indicate that the b-epitope augments aggregation by acting cooperatively with the a-epitope to reinforce rather than cross-bridge oligomer assembly. Monomer/fibrinogen association by coordinated interactions through both epitopes is strengthened by an additional order of magnitude over associations (10^7 and 1.6×10^6 M^{-1}) through the a- and b-epitopes individually, without affecting oligomer thickness. It is suggested that the delayed release of B has purpose in allowing early complexes to dissociate for (1) rapid equilibration across interstitial fluids, and for (2) rapid uptake by phagocytic cells which depend on access to the a-epitope for monomer absorption. In late stages of coagulation, stabilization of oligomer assembly imparted by the b-epitope blocks both equilibration of fibrin concentrations and phagocytic clearance of the fibrin to localize deposition.

ACKNOWLEDGMENTS

Ms. Dolores Andrasic and Mr. David Dreshar assisted. Data processing was aided by the PROPHET computer system of the Biotechnology Resources Program, Division of Research Resources, NIH.

REFERENCES

1. DARDIK, B. N. & J. R. SHAINOFF. 1979. Thromb. Haemostas. **3:** 864–872.
2. SHAINOFF, J. R. & B. N. DARDIK. 1979. Science **204:** 200–202.
3. LAKI, K. & J. A. GLADNER. 1964. Physiol. Rev. **44:** 127–160.
4. BLOMBÄCK, B. & I. YAMASHINA. 1958. Arkiv. Kemi. **12:** 299–319.
5. SHAINOFF, J. R. & I. H. PAGE. 1960. Circ. Res. **8:** 1013–1022.
6. TEGER-NILSSON, A. C. 1967. Acta Chem. Scand. **21**(7)**:** 1879–1886.
7. DARDIK, B. N. & J. R. SHAINOFF. 1980. Am. Heart Assoc. **62:** 717.
8. GONDA, S. R. & J. R. SHAINOFF. 1982. Proc. Natl. Acad. Sci. USA **79:** 4565–4569.
9. MÜLLER-BERGHAUS, G., T. MAHN, G. KOVEKER & F. D. MAUL. 1976. Br. J. Haematol. **33:** 61–79.
10. HERZIG, R. H., O. D. RATNOFF & J. R. SHAINOFF. 1970. J. Lab. Clin. Med. **76:** 451–465.
11. SHAINOFF, J. R. & B. N. DARDIK. 1980. Thromb. Res. **17:** 491–500.
12. KANAIDE, H. & J. R. SHAINOFF. 1976. J. Lab. Clin. Med. **85:** 574–597.
13. IFFT, J. B., D. E. VOET & J. VINOGRAD. 1961. J. Phys. Chem. **65:** 1138–1145.
14. EDELSTEIN, C., L. A. LEWIS, J. R. SHAINOFF, H. K. NAITO & A. M. SCANU. 1976. Biochemistry **15:** 1934–1941.
15. SCHERAGA, H. A. 1961. *In* Protein Structure. **1:** 129–174. Academic Press. New York.
16. SHAINOFF, J. R. & I. H. PAGE. 1962. J. Exp. Med. **116:** 687–707.
17. SASAKI, T., I. H. PAGE & J. R. SHAINOFF. 1966. Science **152:** 1069–1071.
18. SHAINOFF, J. R., B. LAHIRI & F. M. BUMPUS. 1970. Thromb. Diathes. Haemorrh. (Suppl. 39): 203–217.
19. FERRY, J. D. 1954. Phys. Rev. **34:** 753–760.
20. MÜLLER-BERGHAUS, G., I. MAHN & W. KRELL. 1979. Thromb. Res. **14:** 561–572.
21. LANDIS, W. J. & D. F. WAUGH. 1975. Arch. Biochem. **168:** 498–511.
22. KIERULF, P. 1973. Thromb. Res. **3:** 613–630.
23. BROSSTAD, F. 1979. Thromb. Res. **15:** 563–568.
24. BLOMBÄCK, B., B. HESSEL, D. HOGG & L. THERKILDSEN. 1978. Nature (London) **275:** 501–505.
25. LAUDANO, A. P. & R. G. DOOLITTLE. 1978. Proc. Natl. Acad. Sci. USA **75:** 3085–3089.
26. OLEXA, S. A. & A. Z. BUDZYNSKI. 1981. J. Biol. Chem. **256:** 3544–3549.
27. OLEXA, S. A. & A. Z. BUDZYNSKI. 1980. Proc. Natl. Acad. Sci. USA **77:** 1374–1378.
28. FRANCIS, C. N., V. J. MARDER & S. E. MARTIN. 1980. Blood **56:** 456–446.
29. DONNELLY, T. H., M. LASKOWSKI, JR., N. NOTLEY & H. A. SCHERAGA. 1955. Arch. Biochem. Biophys. **56:** 369–387.
30. SHEN, L. L., J. HERMANS, J. MCDONAGH & R. P. MCDONAGH. 1977. Am. J. Physiol. **232:** H629–633.
31. SHERMAN, L. A. & J. LEE. 1977. J. Exp. Med. **145:** 76–85.
32. SHAINOFF, J. R., B. N. DARDIK & S. R. GONDA. 1981. Thromb. Haemostas. **46:** 953.

DISCUSSION OF THE PAPER

H. A. SCHERAGA (*Cornell University, Ithaca, NY*): You quoted 20 kilo-calories for the heat of association of the β fibrin. I presume that's exothermic?

J. SHAINOFF: Yes.

SCHERAGA: Do you have a corresponding value for the α fibrin and the α/β?

SHAINOFF: No, I don't.

J. McDONAGH (*University of North Carolina, Chapel Hill*): Have you examined how macrophages react to fibrin monomer lacking only peptide B?

SHAINOFF: We are in the process of doing that. We know, however, that β-fibrin is cleared from the circulation at a rate not much different from fibrinogen itself so we would anticipate that we will get only slow binding of this form.

A. BUDZYNSKI (*Temple University Health Sciences Center, Philadelphia, PA*): You have implied that there is interaction between D-domains of fibrin monomers which may reinforce the interactions. Is there any direct experimental evidence for this? We have had rather negative results in trying to aggregate D-dimers or fragments. In fibrin that is devoid of fibrinopeptide B only, one can partially uncover polymerization sites in the NH_2-terminal domain that would be sufficient to bind to fragment D domains of two different molecules forming a trimer, and that would be sufficient to explain formation of trimers preferentially over dimers.

SHAINOFF: One is not going to have preferential formation unless there is some reinforcement to promote trimerization, and this interaction need not be terribly strong: it could have been an association constant on the order of 10^2, which one is not going to see in affinity chromatography. The ability to detect association in gel filtration is 60 times less sensitive than the ability to detect it in the ultracentrifuge.

J. S. FINLAYSON (*Bureau of Biologics, NIH, Bethesda, MD*): You said that the rapid uptake of the α-fibrin is dependent on the availability of the Gly-Pro-Arg-Pro epitope. Is the take-home message that when the B peptide comes off, this is no longer available?

SHAINOFF: When the peptide comes off you have a reinforced aggregation and A is not accessible.

J. HERMANS (*University of North Carolina, Chapel Hill*): We seem to be having a nomenclature crisis; I count that there are eight different molecular species, starting with fibrin monomers. We need more precise names than fibrin-I and fibrin-II. One possibility would be to use Des-A, Des-AB, Des-2A fibrin and so on.

L. SHERMAN (*Washington University School of Medicine, St. Louis, MO*): Neither agreeing or disagreeing with your comments concerning the differences due to the availability of the Gly-Pro-Arg receptor, I would not lean too heavily on the time differences between our studies. You were using metabolically active cells wherein fibrin is actually taken up. In the studies that we reported, we were using metabolically inhibited cells.

FIBRINOPEPTIDE RELEASE FROM FIBRINOGEN *

Hymie L. Nossel, Anne Hurlet-Jensen,
Chung Y. Liu, James A. Koehn, and Robert E. Canfield

Department of Medicine
College of Physicians & Surgeons
Columbia University
New York, New York 10032

INTRODUCTION

Proteolysis of fibrinogen occurs in hemostasis, thrombosis, inflammation and many disease states. Every enzyme that attacks the molecule releases the fibrinopeptides in a distinct pattern. Hence measurement of the released peptides indicates not only formation of particular fibrinogen derivatives but reflects the action and hence formation of specific enzymes. Knowledge of the peptide bonds cleaved by a protease forms an essential basis for analyzing its action. Identification of these peptide bonds has generally been made by identifying newly exposed NH_2-terminal amino acids. The released peptides have been quantified by radioimmunoassay or by elution pattern on high performance liquid chromatography (HPLC). Before discussing thrombin and plasmin action on fibrinogen the action of some other enzymes will first be briefly considered.

ENZYMES OTHER THAN THROMBIN AND PLASMIN

Limited data have been reported on the patterns of fibrinopeptide release produced by trypsin, snake venom enzymes and by leukocyte-derived proteases. Initially trypsin releases both FPA and FPB at a similar rate but after release of about 0.5 mole of each, FPB is released more rapidly possibly consequent on partial polymerization of fibrin.[1,2] Reptilase has long been known to release FPB with considerable selectivity;[2,3] however, quite highly purified preparations of reptilase will also release about 0.02 to 0.06 mole FPB-containing peptide per mole of fibrinogen. It has not been established whether release of he FPB-containing peptide results from very slow action which will eventually be complete or represents release from a particular fibrinogen fraction. An extract of the procoagulant fraction of the Southern Copperhead snake, *Agkistrodon contortrix,* releases FPB more rapidly than FPA.[2,4] A crude extract of human leukocytes releases FPA-containing peptide more rapidly than FPB-containing peptide.[5] The active enzyme is probably elastase. In each instance the fibrinopeptide containing mtterial exhibits an approximately 100-fold increase in immunoreactivity with R2 antiserum when treated by thrombin.[6] This behavior implies that the elastase does not cleave the thrombin-

* This work was supported by research grants from the National Institutes of Health (HL15486 and HL21006). Dr. Hurlet-Jensen was supported by Training Grant in Thrombosis and Hemostasis HL07461.

269

sensitive Arg-Gly bonds on the Aα and Bβ chains but cleaves internal to these, probably between His (Aα 24) and Cys (Aα 28) on the α chain and between Lys (Bβ 21) and Cys (Bβ 65) on the β chain.

THROMBIN

Since the late 1950s it has been known that thrombin releases FPA more rapidly than FPB.[7] Blombäck and colleagues reported that FPB release is selectively delayed when fibrin polymerization is inhibited either by 2M urea or when formed from the congenitally abnormtl Fibrinogen Detroit in which serine is substituted for arginine at position A alpha 19.[8] On the basis of these findings they inferred that FPB release was greatly enhanced by polymerization and depended on prior release of FPA. Brosstad reported that very high concentrations of thrombin (100 U/ml) released FPB from fibrinogen dissolved in 2M urea implying that polymerization was not essential for FPB release.[9] Martinelli and Scheraga used HPLC to quantify fibrinopeptide release. They found FPB release at a low rate from the start of the reaction and thought that prior FPA release was not obligatory.[10] Hurlet-Jensen and colleagues studied the influence of fibrin polymerization on FPB release by thrombin.[11] They used the synthetic tetrapeptide Gly-Pro-Arg-Pro to inhibit fibrin polymerization. In a series of preliminary experiments conditions for preventing polymerization were determined. The tetrapeptide delays the onset of polymerization and slows its rate when it does occur but does not limit its extent. The length of time during which polymerization is prevented depends on the relative concentrations of tetrapeptide, thrombin and fibrinogen. Under appropriate conditions polymerization was delayed for more than 12 hours (2.3 nM tetrapeptide, 0.003 U/ml thrombin and 2.94 μM fibrinogen). Three techniques were used to indicate that polymerization had not occurred: measurement of optical density at 350 nm, light scattering monitoring and gel filtration behavior on Sephadex G-200. Light scattering was followed by dynamic measurements with an intensity fluctuation spectrometer utilizing an argon ion laser and a 60 channel computer controlled autocorrelator. This technique is quite sensitive enough to detect the one and a half-fold increase in apparent molecular size expected to result from fibrin dimer formation. Constant intensity and radius observed during prolonged periods of over 12 hours indicated that no significant polymerization (monomer concentration >90%) occurred in the presence of Gly-Pro-Arg-Pro. Gel filtration behavior was assessed by treating ^{125}I-fibrinogen (Ibrin-Amersham) with thrombin in the presence of excess Gly-Pro-Arg-Pro. More than 95% of the radioactivity eluted in a single peak at a point corresponding to 340,000 MW, which confirmed the inference that most of the fibrin molecules were in monomeric form.

In the absence of tetrapeptide the pattern of FPB release showed three phases: an initial slow phase succeeded by a second rapid phase and in turn followed by a third phase in which the rate declined progressively as the substrate concentration decreased. At concentrations which completely prevented polymerization for several hours the tetrapeptide had no effect on FPA release. The second accelerated phase of FPB release was completely abolished but the initial slow release phase was quite unaltered. If sufficient time was allowed, 2 moles FPB per mole of fibrinogen were released at the initial slow

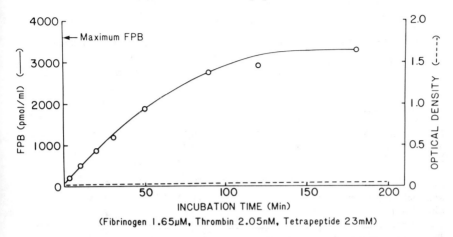

FIGURE 1. Fibrinopeptide B release from fibrin I monomer by thrombin (pH 7.4, 0.15 M tris NaCl, 37° C).

rate in the absence of polymerization (FIGURE 1). These data are interpreted to mean that polymerization is not essential for FPB release and to strongly suggest that FPA cleavage is not obligatory for release of FPB by thrombin since the release rate remains constant while the concentration of fibrin I steadily increases. The amount of FPB released by thrombin (0.02 U/ml) in the first minute from fibrinogen, fibrin I monomer and fibrin I polymer is shown in TABLE 1. These data indicate that FPB was released about twice as rapidly from fibrin I monomer as from fibrinogen and about 10 times as rapidly from fibrin I polymer. The release rate from the polymer is a minimal figure since we think that a thrombin concentration lower than 0.02 U/ml is required to permit measurement of the true initial rate. TABLE 2 lists some published data for K_m and K_{cat} for FPA and FPB release from fibrinogen by thrombin and also unpublished studies by Liu *et al.*[15] K_m and K_{cat} need to be determined for FPB release from fibrin I monomer and polymer. The data in TABLE 1 indicate that these will be different from the data for release from fibrinogen.

TABLE 1

INITIAL RATE OF CLEAVAGE OF FPB BY THROMBIN FROM FIBRINOGEN AND FIBRIN *

Substrate	Amount of FPB Released after 1 min. pmol
Fibrinogen	15.8 ± 5.8
Fibrinogen with tetrapeptide †	14.0 ± 3.6
Fibrin I monomer †	27.4 ± 3.4
Fibrin I polymer	164.4 ± 41.8

* The substrate concentration was 1.65 μM, thrombin concentration 0.02 U/ml and total volume 1 ml.
† Gly-Pro-Arg-Pro 2.3 μM. The data shown are the mean of five experiments.

TABLE 2

KINETIC PARAMETERS FOR FIBRINOPEPTIDE RELEASE FROM FIBRINOGEN BY THROMBIN
FIBRINOPEPTIDE A

	K_m (μM)	Kcat (M/M/sec)
Bando et al.[12]	5	
Hogg & Blombäck [13]	6.3	
Martinelli and Scheraga [10]*	9.2	73
Nossel et al.[14]	2.99	70
Fibrinopeptide B		
Martinelli and Scheraga [10]*	11.3	11.5
Liu et al.[15]	7.2	23

* Bovine reagents, pH 8.0, and human reagents in other studies.

Our interpretation of the data indicating different FPB release rates after addition of thrombin to fibrinogen is that they result from two separate changes in fibrinogen: release of FPA and polymerization of fibrin I, each of which increases the affinity of the Arg (Bβ 14)-Lys (Bβ 15) bond to thrombin with consequent decrease in K_m. Our data confirm the inference of Martinelli and Scheraga that FPB release occurs from fibrinogen. We suggest that fibrin

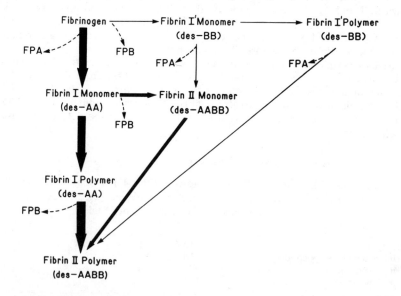

FIGURE 2. Fibrin formation—The thick lines represent the faster reactions. Evidence that fibrin I[1] (des BB) polymerizes is derived from the work of Shainoff [16] and observations on fibrinogen Metz, which polymerizes in the absence of FPA release.[17]

I and II formation occur simultaneously with a more rapid formation of fibrin I which results in the appearance of a sequential reaction (FIGURE 2). The K_ms for thrombin release of FPB from fibrin I monomer and polymer still need to be determined. On the basis of the data available, we suggest that other proteolytic reactions in the hemostatic system which appear to be sequential may also be simultaneous and that apparent sequential reactions result from varying release rates of different peptides. The potential significance of this concept is that the affinity of coagulation proteases for specific peptide bonds may be greatly altered when the protease is bound to a cell surface such as platelets or endothelial cells. Not only the rates but the apparent sequence of the reactions could be altered from those which occur in the fluid phase when cell surface bound proteases are active.

PLASMIN

Plasmin cleaves many peptide bonds in fibrinogen producing a considerable variety of large and small molecules. Here we will consider only the initial cleavages of the central and carboxy-terminal portions of the Aα chain and of the amino-terminal Bβ chain. The large size residual molecule is termed fragment X.[18–20, 28]

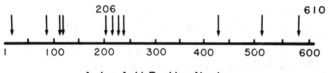

Amino Acid Residue Number

FIGURE 3. Aα chain showing plasmin cleavage sites (From Doolittle *et al.*[21] By permission of *Nature*.)

Formation of Fragment X

The detailed sequence of proteolyses which result in the formation of fragment X has not been established. On the basis of three types of information a model for the formation of fragment X is suggested. The essential basis for the interpretation of our data is the identification by Doolittle and colleagues of the peptide bonds cleaved by plasmin [21] (FIGURE 3). Other data are derived by characterization of fragment X produced by limited treatment of fibrinogen with plasmin.[22] Human fibrinogen (4 mg/ml) was treated with plasmin (0.05 U/ml) terminated in the presence of 1.8 mM $CaCl_2$ for 20 minutes at 37°C and the reaction with Trasylol. Following gel filtration over a Sephadex G-200 column (2.5 × 90 cm) five distinct protein peaks were eluted. Only peak I material (elution volume 76–88 ml) clotted on addition of thrombin. This material was termed fragment X and was analyzed for molecular size on SDS-PAGE (5%) and the content of various peptides measured immunochemically. On SDS-PAGE six bands migrating at different rates were identified. Two of these were quite narrow and molecular weight estimations based on R_f studies suggested 335,000 (X_1) and 267,000 (X_5) for these fractions, each of which constituted about 3% of the total protein. A dense band which migrated as an approximately 248,000 molecular weight protein (X_6) constituted about 33% of the total protein. Marder et al.[23] reported 260,000 for the molecular weight of fragment X and Pizzo et al. reported 248–252,000.[24] Immunochemical analysis of the fragment X preparation showed a content of 2 moles/mole FPA, 0.49 mole/mole FPB, absent Aα 540–554 and partial presence of Aα 253–268. Immunochemical study of α 540-554 and 253–268 was made with monoclonal antibodies to these two peptides generated by Joan Sobel and Paul Ehrlich of this department.[25] We infer that Aα 584–610 is released initially from each half of the dimer thus accounting for the 330–335,000 molecular weight species. On 7.5% SDS-PAGE a 40,000 molecular weight band appeared more rapidly than a 20,000 molecular weight band. On the basis of these data, we infer that the Aα segment 207–583 is released in at least two different reactions. The faster cleavage releasing Aα 207–583 and the slower cleavage releasing the Aα 425–583 followed by release of Aα

```
                   5                    10
    Pyr-Gly-Val-Asn-Asp-Asn-Glu-Glu-Gly-Phe-Phe-Ser-
         T P                           P
         ↓ 15                 20        ↓
    Ala-Arg-Gly-His-Arg-Pro-Leu-Asp-Lys-Lys-Arg-

         25                 30                    35
    Glu-Glu-Ala-Pro-Ser-Leu-Arg-Pro-Ala-Pro-Pro-Pro
                                P
                   40          ↓ 43
         Ile-Ser-Gly-Gly-Gly-Tyr-Arg-Ala
```

FIGURE 4. The primary structure of NH$_2$-terminal Bβ chain of human fibrinogen showing peptide bonds cleaved by thrombin (T) and plasmin (P) (From Iwanaga et al.[27] and from Koehn et al.[30])

207–424. In each instance we infer that release of a peptide from one half of the molecule is followed by more rapid release of the corresponding peptide from the other half of the molecule. This scheme accounts for an approximately 40,000 molecular weight fraction noted by many workers.[26] The rate of release of Bβ 1–42 appears to overlap the rate of release of the second major cleavage product of the Aα chain. Determination of the sequence of peptides released from the Aα and Bβ chain is under active investigation in our laboratory. We believe that at least six different fragments (X_{1-6}) and hence stages of the reaction can be identified.

NH$_2$-Terminal Bβ chain

The primary structure of the NH$_2$-terminal 43 amino acids of the Bβ chain is shown in FIGURE 4. The peptide bonds cleaved by thrombin and plasmin are also indicated. Takagi and Doolittle[28] reported that the Arg (Bβ 42)-Ala (Bβ 43) bond was the first bond cleaved on the NH$_2$-terminus of the Bβ chain, and we have confirmed this observation by directly isolating Bβ 1–42 from early plasmin digests of fibrinogen. Since studies on plasma samples from patients with thrombotic disease showed the presence of thrombin-increasable FPB immunoreactivity (TIFPB) [29] it was of interest to study the proteolysis by plasmin and thrombin of the isolated Bβ 1–42 segment. Treatment of this peptide with plasmin resulted in the progressive degradation of Bβ 1–42 and the appearance most rapidly of Bβ 22–42 and of Bβ 1–21. Bβ 1–21 in its turn began to decrease accompanied by the appearance of FPB.[30] These data indicate that plasmin cleaves the Lys (Bβ 21)-Lys (Bβ 22) bond and more slowly the Arg (Bβ 14)-Gly (Bβ 15) bond. Neither of these cleavages was altered in rate or amount by the presence of 0.34 units of hirudin in a mixture containing 0.16 units of plasmin and hence cannot be attributed to thrombin contamination of the plasmin.

The kinetics of release of Bβ 1–42 by plasmin from fibrinogen were then determined at pH 7.4, 0.15M at 37°C.[31] The maximum initial velocity of the reaction was 3.75×10^{-5} M/min/U of plasmin and the K_m of the reaction was 0.87 μM. Very similar kinetic data were obtained when the substrate was fibrin I monomer rather than fibrinogen. Fibrin I monomer was prepared by treating fibrinogen with reptilase in the presence of gly-pro-arg-pro. Observations were then made on Bβ 1–42 release from fibrin polymer. When plasmin was added to a fibrin I clot made by incubating reptilase with fibrinogen, the data did not fit the Michaelis-Menten equation primarily because of a lesser effect of plasmin at high concentrations of fibrin I. These data result at least in part from inhomogeneous distribution of the plasmin. When reptilase and plasmin were added to fibrinogen simultaneously Bβ 1–42 was released at a slower rate than when reptilase was omitted. Plasmin proteolysis of fibrin I polymer as the substrate clearly reflects a more complex situation involving two phases than is the case when the reactants are all in solution.

REFERENCES

1. MIHALYI, E. & J. E. GODFREY. 1963. Digestion of fibrinogen by trypsin. I. Kinetic studies of the reaction. Biochim. Biophys. Acta **67:** 73–89.
2. BILEZIKIAN, S. B., H. L. NOSSEL, V. P. BUTLER, JR. & R. E. CANFIELD. 1975. Radioimmunoassay of human fibrinopeptide B and kinetics of fibrinopeptide cleavage by different enzymes. J. Clin. Invest. **56:** 438–445.
3. BLOMBÄCK, B. & I. YAMASHINA. 1958. On the N-terminal amino acids in fibrinogen and fibrin. Ark. Kemi. **12:** 299–319.
4. HERZIG, R. H., O. D. RATNOFF & J. R. SHAINOFF. 1970. Studies on the procoagulant fraction of southern copperhead venom: The preferential release of fibrinopeptide B. J. Lab. Clin. Med. **76:** 451–465.
5. BILEZIKIAN, S. B. & H. L. NOSSEL. 1977. Unique pattern of fibrinogen cleavage by human leukocyte proteases. Blood **50:** 21–28.
6. NOSSEL, H. L., V. P. BUTLER, JR., G. D. WILNER, R. E. CANFIELD & E. J. HARFENIST. 1976. Specificity of antisera to human fibrinopeptide A used in clinical fibrinopeptide A assays. Thromb. Haemost. **35:** 101–109.
7. BLOMBÄCK, B. & A. VESTERMARK. 1958. Isolation of fibrinopeptides by chromatography. Ark. Kemi **12:** 173–182.
8. BLOMBÄCK, B., B. HESSEL, D. HOGG & L. THERKILDSEN. 1978. A two-step fibrinogen-fibrin transition in blood coagulation. Nature **275:** 501–505.
9. BROSSTAD, F., P. KIERULF, K. GRAVEN & H. C. GODAL. 1979. Evidence that thrombin releases fibrinopeptide B from non-polymerizing fibrin. Thromb. Haemost. **42:** 97 (Abstract 0222).

10. MARTINELLI, R. A. & H. A. SCHERAGA. 1980. Steady-stage kinetic study of the bovine thrombin-fibrinogen interaction. Biochemistry **19:** 2343–2350.
11. HURLET-JENSEN, A., H. Z. CUMMINS, H. L. NOSSEL & C. Y. LIU. 1982. Fibrin polymerization and release of fibrinopeptide B by thrombin. Thromb. Res. **27:** 419–427.
12. HOGG, D. M., B. BLOMBÄCK. 1978. The mechanism of the fibrinogen-thrombin reaction. Thromb. Res. **12:** 953–964.
13. BANDO, M., A. MATSUSHIMA, J. HIRANO & Y. INADA. 1972. Thrombin-catalyzed conversion of fibrinogen to fibrin. J. Biochem. **71:** 897–899.
14. NOSSEL, H. L., M. TI, K. L. KAPLAN, K. SPANONDIS, T. SOLAND & V. P. BUTLER, JR. 1976. The generation of fibrinopeptide A in clinical blood samples. J. Clin. Invest. **58:** 1136–1144.
15. LIU, C. Y. & H. L. NOSSEL. 1982. Unpublished observations. In preparation.
16. SHAINOFF, J. R., B. N. DARDIK. 1979. Fibrinopeptide B and aggregation of fibrinogen. Science **204:** 200–202.
17. HENSCHEN, A., C. SOUTHAN, J. SORIA, C. SORIA & M. SAMAMA. 1981. Structure abnormality of fibrinogen Metz and its relationship to the clotting defect. Thromb. Haemost. **46:** 103 (Abstract 0314).
18. MARDER, V. J., N. R. SHULMAN & W. R. CARROLL. 1969. High molecular weight derivatives of human fibrinogen produced by plasmin I. Physicochemical immunological characterization. J. Biol. Chem. **244:** 2111–2119.
19. BUDZYNSKI, A. W., V. J. MARDER & J. R. SHAINOFF. 1974. Structure of plasmic degradation products of fibrinogen. J. Biol. Chem. **249:** 2294–2302.
20. PIZZO, S. V., M. L. SCHWARTZ, R. L. HILL & P. A. MCKEE. 1972. The effect of plasmin on the subunit structure of human fibrinogen. J. Biol. Chem. **247:** 636–645.
21. DOOLITTLE, R. F., K. W. K. WATT, B. A. COTTRELL, D. D. STRONG & M. RILEY. 1979. The amino acid sequence of the alpha-chain of human fibrinogen. Nature **280:** 464–468.
22. LIU, C. Y., H. L. NOSSEL & K. L. KAPLAN. 1981. The binding of thrombin by clots formed from fragment X. Thromb. Haemost. **46** (548): 177.
23. MARDER, V. J., A. Z. BUDZYNSKI & H. L. JAMES. 1972. High molecular weight derivatives of human fibrinogen produced by plasmin. III. Their NH_2-terminal amino acid comparison with the "NH_2-terminal disulfide knot." J. Biol. Chem. **247:** 4775–4781.
24. PIZZO, S. V., M. L. SCHWARTZ, R. L. HILL & P. A. MCKEE. 1972. The effect of plasmin on the subunit structure of human fibrinogen. J. Biol. Chem. **247:** 636–645.
25. SOBEL, J. H., S. BIRKEN, P. EHRLICH, R. FRIEDMAN, Z. MOUSTAFA & R. E. CANFIELD. 1981. Characterization of a crosslink-containing fragment derived from the α polymer of human fibrin and its application in immunologic studies using monoclonal antibodies. Thromb. Haemost. **46:** 240 (Abstract 0758).
26. BLOMBÄCK, M., B. BLOMBÄCK, & H. HOLMQVIST. 1976. Immunological characterization of early fibrinogen degradation products. Thromb. Res. **8:** 567–577.
27. IWANAGA, S., P. WALLÉN, N. GRONDAHL, A. HENSCHEN & B. BLOMBÄCK. 1969. On the primary structure of human fibrinogen isolation and characterization of N-terminal fragments from plasmic digests. Eur. J. Biochem. **8:** 189–199.
28. TAKAGI, T. & R. F. DOOLITTLE. 1975. Amino acid sequence studies on plasmin-derived fragments of human fibrinogen: Amino terminal sequence of intermediate and terminal fragments. Biochemistry **14:** 940–946.
29. NOSSEL, H. L., J. WASSER, K. L. KAPLAN, K. S. LA GAMMA, I. YUDELMAN & R. E. CANFIELD. 1979. Sequence of fibrinogen proteolysis and platelet release after intrauterine infusion of hypertonic saline. J. Clin. Invest. **64:** 1371–1378.

30. KOEHN, J. A., A. HURLET-JENSEN, H. L. NOSSEL & R. E. CANFIELD. 1981.
Sequential plasmin proteolysis of the NH₂-terminus of the Bβ chain of
fibrinogen. Thromb. Haemost. **46:** 182 (Abstract 0563).
31. HURLET-JENSEN, A., J. A. KOEHN & H. L. NOSSEL. 1983. The release of
Bβ1–42 from fibrinogen and fibrin by plasmin. Thromb. Res. In press.

DISCUSSION OF THE PAPER

D. K. GALANAKIS (*State University of New York at Stonybrook*): As
you know, we reported on the sequence of plasmin digestion in early plasmic
digests. We studied a 90 percent clottable digest and we were able to isolate
a fraction that possessed intact β chains and no intact Aα chains by NH₂-terminal
analysis and by electrophoretic analysis of chains. Of the order of 40-plus
percent of all its α chains were 25,000 Daltons. Finding this product does
not quite agree with the sequence of having to cleave the Bβ 1 to 42 before
small α chains form.

H. NOSSEL: We found a significant amout of β release overlapping with
the Aα chain release. I suppose that a lot depends upon the sensitivity of the
assays for each alteration.

L. SHERMAN (*Washington University School of Medicine, St. Louis, MO*):
Dr. Nossel, a while ago we isolated and reported on an early derivative from a
plastic digest that had intact N terminal acids, both Aα and Bβ. To this
observation, I think it is simply a matter of whether one is analyzing a whole
digest or whether one is selecting out specific derivatives by one means or
another.

P. J. GAFFNEY: Blombäck and his colleagues have suggested that FPB
release and FPA release occur at the same rate in platelet-rich whole blood.
You suggested in your talk that this indeed actually might be true.

NOSSEL: We have some unpublished data suggesting that under the conditions of our experiments the K_m for the FPA release is about 1.2 micromole
and about 7 for FPB, and that the V_{max} are not very far apart for their
release. If that's correct, then the release of B relative to A would be more
rapid as the fibrinogen concentration increased. However, I do not believe they
would ever be equal to each other. In all our experiments, the A always
comes off faster than the B.

J. A. SHAFER (*University of Michigan Medical School, Ann Arbor*): Our
kinetic analysis, which I presented in a poster session, seems to speak for a
sequential release of fibrinopeptide B after fibrinopeptide A. By this I mean
that the velocity of release of B increases by more than an order of magnitude
and perhaps two orders of magnitude after fibrinopeptide A has been released.
I do not know if this increased rate of B release is due to polymerization. In any
case 90 percent of the molecules lose their B-peptides after they lose their
A-peptides. In fibrinogen Petovsky one half of the molecules have a histidine in
position 16, which causes a 100- to 200-fold decrease in the rate of FPA release
from the abnormal chain. In this situation, release of FPB is delayed to the
same extent as the release of the abnormal FPA. If release of FPB was truly a
process that occurred simultaneously with release of A, you would not expect a

delayed release of FPB to result from this amino acid replacement. Furthermore, we have found that the tetrapeptide Gly-Pro-Arg-Pro inhibits release of FPB.

Nossel: I can say that our interpretations disagree. With regard to your observations of fibrinogen Petovsky, they seem to be exactly the same as described for fibrinogen Detroit, and we would explain those data as being influenced by the polymerization reaction.

H. A. Scheraga: I agree with Dr. Nossel's interpretation. If you want to study a fundamental process, you should try to study it without complication with other processes. Thus, one should study the relative release of the A and B peptides without polymerization. In agreement with Dr. Nossel, our data show that where there is no polymerization release of the two peptides is a competitive reaction. When polymerization is taking place, the kinetics are much more complicated and we do see changes in the relative rates of FPB release.

J. Hermans (University of North Carolina, Chapel Hill): It is true that if one observes clotting, then one is absolutely sure there has been polymerization. On the other hand, there are conditions where one gets polymerization to staggered-overlap polymers, i.e., to protofibrils but there is no formation of a visible clot. Inhibitors like the tetrapeptide Gly-Pro-Arg-Pro and fragment D can delay gelation while permitting the formation of polymers. Thus, the question is whether fibrinopeptide B is released preferentially from soluble polymers and exclusion of visible clot formation in the experiment is therefore a necessary, but not sufficient, precaution.

CALCIUM ION FUNCTIONS IN
FIBRINOGEN CONVERSION TO FIBRIN *

James J. Hardy, Nadia A. Carrell,† and Jan McDonagh ‡

Department of Pathology
University of North Carolina School of Medicine
Chapel Hill, North Carolina 27514

The complex biochemical events which result in the conversion of fibrinogen to fibrin gel can be described by three general reactions: (1) proteolytic activation of fibrinogen with cleavage of the fibrinopeptides, (2) linear association of fibrin monomers to form long double stranded protofibrils, and (3) lateral association of protofibrils to form fibrin fibers of varying diameters.[1] When proteolysis is catalyzed by thrombin, the A peptides are cleaved rapidly and the B peptides more slowly.[2]

It has been well documented that calcium ions increase the efficiency of fibrinogen conversion with a resultant shortening of the clotting time while addition of chelating agents, such as ethylenediaminetetraacetate (EDTA), have the converse effect of prolonging the clotting time. However, the mechanisms for the Ca^{2+} and EDTA effect are not entirely clear. Both molecules bind to fibrinogen; per molecule of human fibrinogen there are three tight binding sites for calcium $(K_d = 1.9 \times 10^{-5}\ M)$[3] and 0.4 sites for EDTA $(K_d = 2.8 \times 10^{-5}\ M)$[4]. Hence, both molecules could affect fibrin formation through direct interaction with fibrinogen or by inducing a conformational change in fibrinogen or by some other indirect mechanism. Several investigators have found that Ca^{2+} accelerates the conversion of fibrinogen to fibrin by increasing the rate of fibrin monomer aggregation.[5-8] Boyer *et al.*[6] found that Ca^{2+} did not affect the initial rate of fibrinopeptide cleavage and indeed might be slightly inhibitory at later times. In contrast Marguerie *et al.*[9] on the basis of H^+ titrations at pH 8.6, found that Ca^{2+} enhances thrombin cleavage of fibrinogen. These investigators also postulated that there is Ca^{2+} dependent formation of a fibrinogen-fibrin monomer dimer as an intermediate in the polymerization process. The inhibitory activity of EDTA has been postulated to be due simply to removal of calcium ions [10] and also to a specific effect of EDTA on fibrinogen.[11]

Much of the earlier work on the effects of Ca^{2+} and EDTA was done before it was demonstrated that these molecules bind to fibrinogen. Although it is now clear that fibrinogen has specific Ca^{2+} binding sites and that bound Ca^{2+} protects fibrinogen from thermal denaturation [12] and proteolytic digestion,[13] it is less clear as to whether the Ca^{2+} effect on fibrin formation is related to the high affinity binding or to other interactions or both. Similarly, it is not clear whether the inhibitory activity of EDTA is due to removal of Ca^{2+} or to bound

* This work was supported in part by Grant HL29511 from the National Institutes of Health. J.M. is an Established Investigator of the American Heart Association.

† Present address: Department of Anatomy, Duke University, Durham, N.C.

‡ To whom correspondence should be addressed at: Department of Pathology, Harvard Medical School, Beth Israel Hospital, Boston, MA 02215.

0077–8923/83/0408–0279 $01.75/0 © 1983, NYAS

EDTA or to other mechanisms. We have carried out studies using EDTA and Chelex-treated fibrinogen in order to obtain further information on the Ca^{2+} and EDTA effects on fibrinogen.

MATERIALS AND METHODS

Human fibrinogen was prepared from cryoprecipitate-depleted Cohn Fraction I (Cutter Laboratory) by ammonium sulfate precipitation.[14] It was stored at $-80°$ in 0.3 M NaCl and diluted to appropriate concentrations in 0.15 M NaCl–0.05 M Tris, pH 7.5. The clottability was 90% or greater. Chelex-treated fibrinogen was prepared by incubation with excess Chelex (Biorad Laboratories) for 10 minutes or longer.[15] EDTA-treated fibrinogen was prepared by overnight dialysis at room temperature against 20 volumes of the appropriate EDTA solution. EDTA-dialyzed fibrinogen was further dialyzed against Tris-NaCl buffer. Dialysis bags were pretreated by boiling and washing in ethanol water. Purified human thrombin (1454 NIH units/mg) was provided by Dr. John Fenton, New York State Department of Health.

Fibrinopeptide release by thrombin was quantitated by reaction of peptide-bound arginine with phenanthrenequinone.[16] Two mg fibrinogen were activated by incubation with 0.2 units thrombin for various time periods (0–200 sec). The fibrinopeptides cleaved were isolated by either extraction after protein precipitation with trichloroacetic acid.

Thrombin time was defined as the point of visible fibrin formation after addition of 100 μl thrombin to a solution containing 600 μg fibrinogen in 300 μl. Fibrin assembly was monitored by recording turbidity at 350 nm of solutions containing 0.5 mg fibrinogen and 1 unit of thrombin; pH was 7.4 and ionic strength 0.15. When $CaCl_2$ was added, the final concentration was 2.5 mM.

Calcium ion concentration in fibrinogen solutions was measured by atomic absorption spectrometry and was performed by Dr. Miles Crenshaw, University of North Carolina School of Dentistry. $CaCl_2$ and Na_2EDTA stock solutions were prepared in resin-purified, distilled water.

RESULTS

In all experiments purified fibrinogen that had been extensively dialyzed against buffer but had not been treated with $CaCl_2$, Chelex, or EDTA was used as the reference (normal) fibrinogen. This fibrinogen retained its tightly bound Ca^{2+} but was free of any loosely associated or unbound calcium ions.[17] This material was assayed for Ca^{2+} by atomic absorption spectrometry. At a fibrinogen concentration of 4.05 mg/ml (12 μM) with three high affinity binding sites, the expected Ca^{2+} concentration would be 1.42 μg/ml. Experimentally 1.3 μg/ml were found. For Chelex-treated fibrinogen the Ca^{2+} concentration was 0.2 μg/ml, and for EDTA-treated, dialyzed fibrinogen it was 0.1 μg/ml. These results indicate that, within experimental error, normal fibrinogen contained 3 Ca^{2+} per molecule, all of which were removed by either Chelex or EDTA treatment.

The thrombin time was used to assess the effect of Ca^{2+} on the overall process of fibrin formation. Addition of Ca^{2+} to a fibrinogen solution con-

TABLE 1

EFFECT OF ADDED CA²⁺ ON THE RATE OF VISIBLE FIBRIN FORMATION

CaCl₂ mM	Thrombin Time (sec)
0	15.2
2.5	13.3
5.0	12.9
10.0	12.8
20.0	13.0

taining bound Ca²⁺ shortened the thrombin time in a manner which was not concentration dependent (TABLE 1). Similarly addition of Ca²⁺ to Chelex-treated fibrinogen shortened the thrombin time in a manner identical to that of normal fibrinogen (FIGURE 1 curves 2 and 4). However, removal of bound Ca²⁺ by Chelex had no effect on the thrombin time until a low thrombin concentration was used (FIGURE 1, curves 1 and 3). In the absence of added Ca²⁺,

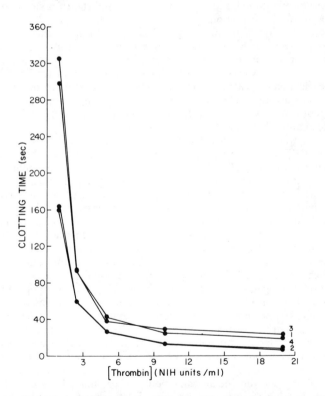

FIGURE 1. Effect of Ca²⁺ depletion by Chelex on the thrombin time. 1: Reference fibrinogen (containing structural Ca²⁺); 2: Fibrinogen plus 2.5 mM CaCl₂; 3: Chelex-treated fibrinogen; 4: Chelex fibrinogen plus 2.5 mM CaCl₂.

TABLE 2

EFFECT OF EDTA ON THE RATE OF VISIBLE FIBRIN FORMATION *

EDTA mM	Thrombin Time (sec)		
	(1) Before Dialysis	(2) After Dialysis	(3) +2.5 mM CaCl$_2$
0	12.7		9.5
1.0	169.9	23.6	9.4
2.5	175.2	68.8	9.6
5.0	179.6	72.9	9.7
10.0	184.9	109.2	9.1

* Samples were tested before (1) and after (2) dialysis to remove free EDTA and after addition of CaCl$_2$ to dialyzed material (3).

normal fibrinogen and Chelex-treated fibrinogen formed fibrin at the same rate until the thrombin:fibrinogen ratio was decreased to 0.1 unit thrombin: 1.5 mg fibrinogen. These results indicate that it is the added "free" Ca^{2+} in solution, not the three bound Ca^{2+}, which are important for decreasing the time required for fibrin formation.

The interaction of fibrinogen with EDTA has a pronounced, but reversible, inhibitory effect on the process of fibrin formation (TABLE 2). The effect of EDTA is concentration dependent, and both free and bound molecules appear to play a role. However, if the fibrinogen is dialyzed after EDTA treatment and then incubated with Ca^{2+}, the thrombin time returns to normal. The same observations are illustrated in FIGURE 2 with varying thrombin concentrations. Thus, removal of Ca^{2+} by EDTA produces a significantly different effect on fibrin formation than does Ca^{2+} depletion by Chelex.

Studies were also conducted to determine the effect of Ca^{2+}, Ca^{2+} depletion, and EDTA specifically on the first, proteolytic step in fibrin formation by measuring the initial rate of fibrinopeptide cleavage by thrombin (FIGURE 3). Peptide release was measured at constant thrombin concentration over a fourfold range in fibrinogen concentration. It can be seen that Ca^{2+} depletion with Chelex had no effect on the initial rate of release, nor within experimental error did addition of Ca^{2+}. From the data for the reference fibrinogen a K_m of 11.0×10^{-6} M was calculated (FIGURE 4). When the peptide release data for fibrinogen treated with Ca^{2+}, Chelex, or EDTA were incorporated into the Michaelis-Menton plot, the value for K_m remained the same. From these data it can be concluded that neither Ca^{2+} binding nor depletion nor EDTA binding to fibrinogen has any effect on the initial rate of fibrinopeptide cleavage.

The increase in turbidity which occurs during fibrin formation can be used as an approximate index of the overall assembly of molecules of fibrin into large fibers. Turbidity is a function of the size of molecules; the initial rate of increase is correlated with both linear and lateral association and is not separated into the component processes. FIGURE 5 and TABLE 3 show the effects of Ca^{2+} depletion and Ca^{2+} enrichment on the overall assembly process. It can be seen that both the intrinsically bound Ca^{2+} and added Ca^{2+} have a pronounced effect on the assembly process. Furthermore, replenishment of bound Ca^{2+} after its removal by either Chelex or EDTA does not reconstitute the assembly process.

Discussion

Fibrin formation is a kinetically controlled process in which the rates of the individual reactions determine the type of fibrin gel which is formed.[18, 19] The role of calcium ions in this process is complex and involves both the tightly bound ions as well as free or losely associated ions in solution. On the basis of our observations and also those of other investigators, we hypothesize that bound Ca²⁺ forms an integral part of fibrinogen structure while free Ca²⁺ has a regulatory role in fibrinogen conversion to fibrin. It is reasonable that the bound Ca²⁺ should not have a regulatory function since its dissociation constant is 100-fold lower than the free Ca²⁺ concentration in plasma. The structural Ca²⁺ maintains a polymerization contact site in the most appropriate conformation, but it does not form part of the contact site. Free Ca²⁺ helps to regulate the type of fibrin which is formed, probably by modulating the rate of lateral association.

Support for this hypothesis is provided by the experiments presented here. It is clear that neither Ca²⁺ enrichment nor depletion have any effect on the

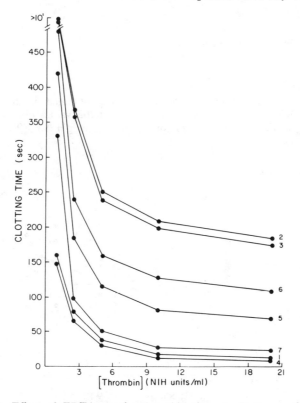

FIGURE 2. Effect of EDTA on the thrombin time. 1: Reference fibrinogen; 2: Fibrinogen treated with 10 mM EDTA; 3: Fibrinogen treated with 2.5 mM EDTA; 4: Sample 2, dialyzed, then treated with 2.5 mM CaCl₂; 5: Sample 3 dialyzed; 6: Sample 2 dialyzed; 7: Fibrinogen treated with 1.0 mM EDTA and dialyzed.

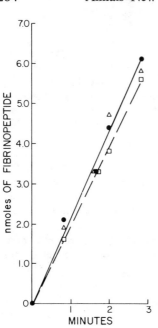

FIGURE 3. Initial fibrinopeptide cleavage by thrombin substrates—● reference fibrinogen; □ fibrinogen treated with CaCl₂; △ fibrinogen incubated with Chelex.

initial rate of fibrinopeptide cleavage. The K_m obtained in these studies (11.0×10^{-6} M) is in reasonable agreement with the values obtained earlier for A and B peptide cleavage of bovine fibrinogen at pH 8.6 (9.2×10^{-6} M

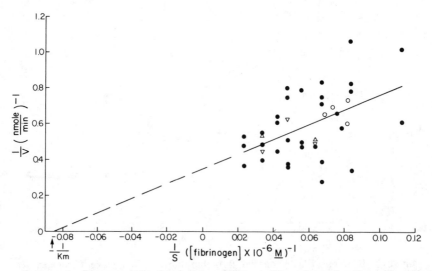

FIGURE 4. Lineweaver-Burk plot for fibrinopeptide release by thrombin substrates—● reference fibrinogen; △ cleavage with 5 mM CaCl₂ added; ▽ Ca²⁺ removed with Chelex; ○ treatment with 10 mM EDTA.

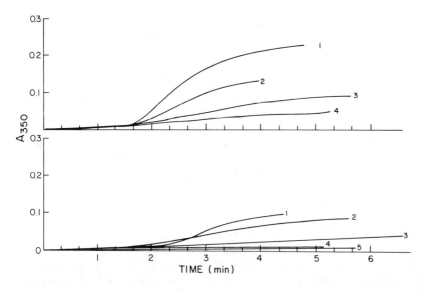

FIGURE 5. Effects of Ca²⁺ enrichment and depletion on fibrin aggregation. Upper graph–1: CaCl₂ added to reference fibrinogen; 2: Chelex-treated fibrinogen plus CaCl₂; 3: reference fibrinogen; 4: Chelex-treated fibrinogen. Lower graph–1: EDTA-treated fibrinogen (dialyzed) plus CaCl₂; 2: reference fibrinogen; 3: fibrinogen plus EDTA and CaCl₂; 4: EDTA-fibrinogen (dialyzed); 5: EDTA-fibrinogen.

for A and 11.0×10^{-6} M for B).[20] Binding of EDTA to fibrinogen also does not inhibit this reaction. Bound Ca²⁺ is not necessary for fibrin formation, since Chelex-treated fibrinogen can form fibrin fibers as rapidly as the reference fibrinogen when the thrombin concentration is not too low (FIGURE 1). However, the types of fibrin which form in the presence and absence of bound Ca²⁺ are different, and the conformational change induced by Ca²⁺ depletion is difficult to reverse fully (FIGURE 5, TABLE 3). Hence it can be concluded that

TABLE 3

EFFECTS OF Ca²⁺ ENRICHMENT AND DEPLETION AND EDTA
TREATMENT ON FIBRIN AGGREGATION *

Sample Treatment	Aggregation Rate
Control fibrinogen	0.22 ± 0.09
2.5 mM CaCl₂	0.84 ± 0.26
Chelex treatment	0.08 ± 0.02
Chelex, then CaCl₂	0.48 ± 0.01
EDTA incubation	0.02 ± 0.01
EDTA, then dialysis	0.03 ± 0.02
EDTA, dialysis, then CaCl₂	0.39 ± 0.02

* Defined as initial change in turbidity per unit time.

the structural role of Ca^{2+} in maintaining the functional integrity of a contact site is necessary for normal fibrin fibers to form but is not essential for formation of fibers with less well fitting or weaker contacts.

The functional role of free Ca^{2+} in regulating fibrin formation is apparent but is not clearly defined. We presume that this effect is largely on the process of lateral association. Ca^{2+} releases thrombin from fibrin monomer and polymer and thus can effectively increase the thrombin concentration.[21] Since increasing thrombin favors lateral association, this may be part of the mechanism for the observed Ca^{2+} effect. Ca^{2+} may also expose a contact site on the β-chain, following release of B peptide, which is also involved in lateral association.[22] Another possibility is that loosely associated Ca^{2+} may facilitate lateral association through neutralization of charge.

The marked inhibitory effect of EDTA on fibrin assembly is not due simply to the removal of Ca^{2+} since the effect of removal by Chelex is significantly less pronounced. It would appear, therefore, that one or more contact sites are significantly altered by interaction with EDTA. Whether or not this is due directly to binding of EDTA to fibrinogen is not known but seems likely.

The net effect of Ca^{2+} on fibrinogen conversion to fibrin is to increase the tensile strength of the fibrin fibers.[23] This would be the expected result if Ca^{2+} somehow enhances the fit of the contacts necessary for fiber formation. Our results indicate that both bound and free Ca^{2+} help to achieve the proper alignment of contact sites. In addition to the roles discussed here, Ca^{2+} is also necessary for the covalent cross-linking of fibrin by factor XIIIa, which further serves to increase the mechanical stability of the fiber network. It is thus clear that both the structural calcium ions bound to fibrinogen and free calcium ions in solution have several essential roles in the formation of a mechanically stable fibrin gel.

REFERENCES

1. HERMANS, J. & J. McDONAGH. 1982. Fibrin: Structure and Interactions. Semin. Thromb. Hemostas. **8:** 11–24.
2. BLOMBÄCK, B., B. HESSEL, D. HOGG & L. THERKILDSEN. 1978. A two-step fibrinogen-fibrin transition in blood coagulation. Nature **275:** 501–505.
3. NIEUWENHUIZEN, W., I. A. M. VAN RUIJVEN-VERMEER, W. J. NOOIJEN, A. VERMOND & F. HAVERKATE. 1981. Recalculation of calcium-binding properties of human and rat fibrin(ogen) and their degradation products. Thromb. Res. **22:** 653–657.
4. NIEUWENHUIZEN, W., A. VERMOND & J. HERMANS. 1981. Human fibrinogen binds EDTA and citrate. Thromb. Res. **22:** 659–663.
5. LORAND, L. & K. KONISHI. 1964. Activation of the fibrin stabilizing factor of plasma by thrombin. Arch. Biochem. Biophys. **105:** 58–
6. BOYER, M. H., J. R. SHAINOFF & O. D. RANOTFF. 1972. Acceleration of fibrin polymerization by calcium ions. Blood **39:** 382–387.
7. BRASS, E. P., W. B. FORMAN, R. V. EDWARDS & O. LINDAN. 1978. Fibrin formation: Effect of calcium ions. Blood **52:** 654–658.
8. ENDRES, G. F. & H. A. SCHERAGA. 1972. Equilibria in the fibrinogen-fibrin conversion. IX. Effects of calcium ions on the reversible polymerization of fibrin monomer. Arch. Biochem. Biophys. **153:** 266–278.
9. MARGUERIE, G., Y. BENABID & M. SUSCILLON. 1979. The binding of calcium to fibrinogen: Influence on the clotting process. Biochim. Biophys. Acta **579:** 134–141.

10. BITHELL, T. C.   1964.   A study of the inhibitory effect of ethylenediamine-tetraacetic acid on the thrombin-fibrinogen reaction. Biochem. J. **93:** 431–439.

11. GODAL, H. C.   The effect of EDTA on human fibrinogen and its significance for the coagulation of fibrinogen with thrombin. Scand. J. Clin. Lab. Invest. **12** (Suppl. 53): 3–20.

12. LY, B. & H. C. GODAL.   1973.   Denaturation of fibrinogen: the protective effect of calcium. Haemostasis **1:** 204–209.

13. HAVERKATE, F. & G. TIMAN.   1977.   Protective effect of calcium on the plasmin degradation of fibrinogen and fibrin fragments D. Thromb. Res. **10:** 803–812.

14. GRALNICK, H. R., H. M. GILVELBER, J. R. SHAINOFF & J. S. FINLAYSON.   1971. Fibrinogen Bethesda: A congenital dysfibrinogenemia with delayed fibrino-peptide release. J. Clin. Invest. **50:** 1819–1830.

15. MARGUERIE, G., G. CHAGNIEL & M. SUSCILLON.   1977.   The binding of calcium to bovine fibrinogen. Biochim. Biophys. Acta. **490:** 94–103.

16. YAMADA, S. & H. A. ITANO.   1966.   Phenanthrenequinone as an analytical reagent for arginine and other monosubstituted guanidines. Biochim. Biophys. Acta **130:** 538–540.

17. SHEN, L. L., J. HERMANS, J. McDONAGH, R. P. McDONAGH & M. CARR.   1975. Effects of calcium ion and covalent crosslinking on formation and elasticity of fibrin gels. Thromb. Res. **6:** 255–265.

18. HANTGAN, R. R. & J. HERMANS.   1979.   Assembly of fibrin: A light scattering study. J. Biol. Chem. **254:** 11272–11281.

19. HANTGAN, R. R., J. McDONAGH & J. HERMANS.   1982.   Fibrin assembly. This volume.

20. MARTINELLI, R. A. & H. A. SCHERAGA.   1980.   Steady-state kinetic study of the bovine thrombin-fibrinogen interaction. Biochemistry **19:** 2343–2350.

21. KAMINSKI, M. & J. McDONAGH.   1982.   Calcium regulates the interaction of thrombin with fibrin. Fed. Proc. **41:** 770.

22. LAUDANO, A. P. & R. F. DOOLITTLE.   1981.   Influence of calcium ion on the binding of fibrin amino terminal peptides to fibrinogen. Science **212:** 457–459.

23. SHEN, L. L., R. P. McDONAGH, J. McDONAGH & J. HERMANS.   1974.   Fibrin gel structure: Influence of calcium and covalent crosslinking on the elasticity. Biochim. Biophys. Res. Commun. **566:** 793–798.

CHEMICAL MODIFICATION OF FIBRINOGEN
AND THE EFFECT ON FIBRIN FORMATION

Y. Saito, A. Shimizu, A. Matsushima, and Y. Inada

Laboratory of Biological Chemistry
Tokyo Institute of Technology
Ookayama, Meguroku, Tokyo 152, Japan

INTRODUCTION

Isolation of various abnormal fibrinogens from patients has contributed much to understanding the mechanism of fibrin formation.[1-6] Chemical modifications of amino acid residues should also artificially generate a series of abnormal fibrinogens, but only if a limited number of specific amino acid residues can be modified. Studies along this line have been carried out since Laki and Mihalyi reported polymerization of iodinated fibrinogen in 1949.[7] Most studies tried so far, however, have failed to unambiguously identify specific amino acid residues directly involved in the fibrinogen polymerization. This lack of specific residue identification was probably due to the nonspecific modification of the amino acids. Although high specificity of modification in terms of the position as well as the kind of amino acid residues is desired, chemical modifications very often yield a mixture of heterogeneous products unlike genetic alterations. They can be a mixture of functional and dysfunctional molecules as well as a mixture of molecules whose various amino acid residues located at various positions are modified. If we could separate functional molecules from dysfunctional ones, comparison between them would help us to focus specifically on amino acid residues directly involved in the polymerization process. To see the change of small number of amino acid residues, it is desirable to work with the functional unit as little as possible.

The formation of fibrin clot from fibrinogen proceeds via two steps, (1) the release of fibrinopeptides A and B with thrombin and (2) the self-association of fibrin monomer. It is thus important to focus specifically on the latter step. Several studies revealed that the association occurs through the interaction of complementary binding sites on the N-terminal and the C-terminal domains in the fibrinogen molecule.[8-12] N-terminal disulfide knot activated with thrombin (N-DSK$_a$)[13] and Fragment D[14] have been isolated as functional units of the N-terminal and the C-terminal domains, respectively.

We have modified fibrinogen with photooxidation in the presence of methylene blue (and rose bengal), which does not introduce bulky substitution groups on any amino acid residues. Affinity chromatography with fibrin monomer-Sepharose strongly suggested that this modification is specifically directed to the N-terminal binding sites. In order to find the amino acid residues directly involved in the association step, we have then modified N-DSK$_a$ by the same technique. Functional and dysfunctional N-DSK$_a$ molecules were separated by affinity chromatography with fibrinogen-Sepharose. Amino acid analyses of functional and dysfunctional N-DSK$_a$ molecules, which were produced by the

0077–8923/83/0408–0288 $01.75/0 © 1983, NYAS

modification to varying degrees, have strongly suggested that histidine residue(s) located at specific position(s) is directly involved in the self-association of fibrin molecules.

MATERIALS AND METHODS

Fibrinogen (95% clottable), thrombin (12.5 NIH units/mg) and plasmin (0.5 casein units/mg) from humans were gifts of Green Cross Co., Japan. All other chemical reagents were of analytical grade. Fibrin monomer was obtained by dissolving noncross-linked fibrin polymer with 1 M NaBr in 50 mM acetate buffer (pH 5.3). This method was proven to give fibrin monomer solution by ultracentrifugal analysis by Donnelly *et al.*[15]

Preparation of Fibrinogen and Fibrin Monomer-Sepharose

Fibrinogen was immobilized on cyanogen bromide-activated Sepharose CL-6B.[8] A method was devised to prepare immobilized fibrin monomer. Unlike the original method introduced by Heene and Matthias,[16] in which immobilized fibrinogen was activated by thrombin, fibrin monomer (3 μM) was directly coupled to cyanogen bromide-activated Sepharose CL-6B in the presence of 1 M NaBr in 0.1 M borate buffer (pH 8.2). We found this method more reproducible and reliable than the original method.

Preparation of N-DSK$_a$ and Fragment D

N-terminal disulfide k not activated with thrombin (N-DSK$_a$) was obtained by directly treating noncross-linked fibrin monomer with cyanogen bromide, rather than by treating fibrinogen first and activating it later with thrombin.[13] The digested sample was purified by Sephadex G-100 (4.5 × 90 cm) [17] followed by Sephadex G-150 (1.4 × 90 cm) equilibrated with 5 mM Tris buffer (pH 8.0) containing 4 M urea. The final purification was performd with fibrinogen-Sepharose conjugate to eliminate materials which do not bind to the conjugate. This newly devised method produced homogeneous and functional N-DSK$_a$ molecules, and the yield was reproducibly high. Fragment D was obtained by plasmin digestion of fibrinogen in the presence of Ca ion [18] followed by DEAE ion exchange chromatography.[14] The obtained N-DSK$_a$ and Fragment D preparations were homogeneous and the molecular weight was determined to be about 60,000 and 95,000 by polyacrylamide gel electrophoresis with 0.1% sodium dodecyl sulfate, respectively.

Photooxidation of Fibrinogen, Fibrin Monomer, and N-DSK$_a$

A mixture of fibrinogen (or N-DSK$_a$) and methylene blue (or rose bengal) in 50 mM Tris buffer (or phosphate buffer) containing 100 mM NaCl (pH 7.0) was irradiated with a 300 W projector lamp. The final concentrations of fibrinogen, N-DSK$_a$ and methylene blue (or rose bengal) were 2 μM, 12 μM and about 15 μM (or 3 μM), respectively. During the irradiation, the sample was shielded from radiation heat by putting a flat bottle containing cold water between the light source and the sample. Fibrin monomers were also similarly photooxidized immediately after they were diluted from 1 M NaBr solution with the Tris buffer containing NaCl. When fibrinogen and N-DSK$_a$ were used for affinity tests after the photooxidation, they were dialyzed against 50 mM phosphate buffer containing 100 mM NaCl (pH 7.6) to eliminate the photosensitizer dye. Otherwise they were used immediately in the dark for the measurement of association. Photooxidation did not proceed under the conditions used for the measurement of association.

Measurement of Association of Fibrinogen and Fibrin Monomer

The association of fibrinogen (1 ml, 2 μM) was primed in 50 mM Tris buffer containing 100 mM NaCl (pH 7.6) by 50 μl of thrombin (12.5 units/ml) in 0.9% NaCl. The association of fibrin monomer (kept in 1 M NaBr in 50 mM acetate buffer, pH 5.3) was initiated by diluting it 15 times with 50 mM Tris buffer containing 100 mM NaCl (pH 7.2).[19, 20] The final pH value of the reaction mixture was 7.0. The association was monitored by measuring the increase of absorbance (turbidity) at 450 nm with a Shimadzu recording spectrophotometer model UV-200.

Affinity Test of Photooxidized Fibrinogen and N-DSK$_a$ with Fibrin Monomer-Sepharose and Fibrinogen-Sepharose

Fibrinogen and N-DSK$_a$ (before and after photooxidation) were applied to fibrin monomer-Sepharose (1.4 $\times$ 10 cm) and fibrinogen-Sepharose (1.5 $\times$ 16 cm) columns, respectively, equilibrated with 50 mM phosphate buffer containing 100 mM NaCl (pH 7.6). They were washed thoroughly with the buffer mentioned above and were eluted with 1 M NaBr in 10 mM phosphate buffer (pH 5.3). The bound materials and nonbound materials are functional and dysfunctional molecules, respectively.

Amino Acid Analysis of Modified Proteins

The native and modified proteins (separately obtained functional and dysfunctional molecules) were subjected to amino acid analysis with a JEOL amino acid analyzer JLC-6AH, after hydrolysis with 6 M HCl or 3 M mercaptoethanesulfonic acid for 22 hr at 110°C.[21]

RESULTS

The Effect of Photooxidation of Fibrinogen Molecules on Association

Fibrinogen was photooxidized with methylene blue for varied periods of time and the degree of association was studied with thrombin. As it is shown in FIGURE 1, the association decreased steadily as the period of irradiation time was increased, as Inada *et al.*[22] had reported previously. If we take the maximum absorbance increase as the measure of association, the irradiation for ten seconds was enough to inhibit the association by 50%, as depicted in the inset. This might suggest that a specific amino acid residue(s) located in binding sites is particularly sensitive to photooxidation. We are not sure at this stage, however, whether this dramatic decrease of the association is due to either the change of susceptibility of fibrinogen to the attack of thrombin

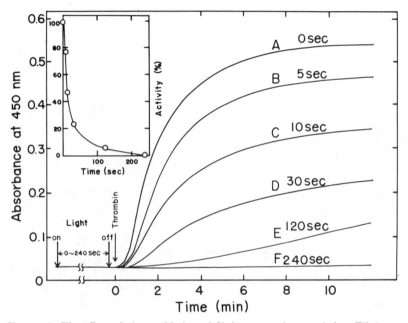

FIGURE 1. The effect of photooxidation of fibrinogen on its association. Fibrinogen in 50 mM Tris buffer with 100 mM NaCl (pH 7.0) was photooxidized in the presence of methylene blue, as described in MATERIALS AND METHODS, for 0(A), 5(B), 10(C), 30(D), 120(E) and 240 seconds(F). Thrombin was then added to see the association by the increase of absorbance at 450 nm. *Inset*: the effect of irradiation time on the association activity which was taken from the maximum absorbance increase of each curve.

or the real decrease of association of fibrin monomers. To test the possibility of the latter case, we have photooxidized fibrin monomer with methylene blue immediately after it was diluted with Tris buffer (pH 7.2) from 1 M NaBr solution (pH 5.3). As shown in FIGURE 2, the degree of association was indeed decreased by photooxidation. This indicates then the decrease of association of fibrinogen depicted in FIGURE 1 can be attributed to the step of association of fibrin monomers, although it does not mean that the step of peptide release is not affected at all by the modification. Qualitatively, similar effect was observed by photooxidation with rose bengal.

Binding Site Affected by Photooxidation

As there are complementary binding sites on the N-terminal and C-terminal domains in the fibrinogen molecule, we wanted to know which binding site (or both of them) of the fibrinogen molecule is affected by the modification. We have studied the affinity of the native and photooxidized fibrinogen with rose bengal to fibrin monomer using fibrin monomer-Sepharose. As it is shown in FIGURE 3, the modification does not affect the affinity at all. Since this

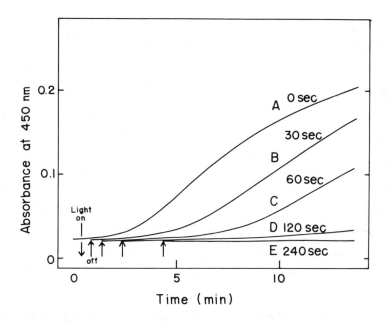

FIGURE 2. The effect of photooxidation of fibrin monomer on its association. Fibrin monomer was photooxidized in the presence of methylene blue for 0(A), 30(B), 60(C), 120(D) and 240 seconds(E) immediately after it was diluted 15 times with 50 mM Tris buffer containing 0.1 M NaCl (pH 7.2) from 1 M NaBr in 50 mM acetate buffer (pH 5.3). Association was then measured by the increase of absorbance at 450 nm.

affinity test depends on the association between the N-terminal binding site of the immobilized fibrin monomer and the C-terminal binding sites of the added fibrinogen molecule, the modification obviously does not affect the binding site present in the C-terminal domain of fibrinogen. This was further confirmed by using N-DSK$_a$-Sepharose (data not shown). It then implies that the real cause of the decrease of the association of fibrinogen by the photooxidation with rose bengal would be the destruction of the binding sites located in the N-terminal domain of the molecule.

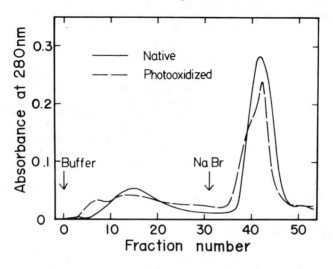

FIGURE 3. The effect of photooxidation of fibrinogen on its affinity with fibrin monomer. Fibrinogen in 50 mM phosphate buffer containing 0.1 M NaCl (pH 7.0) was photooxidized for 20 min in the presence of rose bengal as it is described in MATERIALS AND METHODS. It was then applied to fibrin monomer-Sepharose column (1.4 × 10 cm), washed with the above mentioned buffer and was eluted with 1 M NaBr in 10 mM phosphate buffer (pH 5.3). The wash and the eluate were collected at one ml per min into 2 ml each fraction.

Effect of Photooxidation of Fragment D and N-DSK$_a$ on Their Inhibitory Activity on Fibrin Formation

To confirm the above mentioned notion, we have decided to modify isolated Fragment D and N-DSK$_a$ by photooxidation. As we have shown previously,[20] isolated Fragment D inhibited the association of fibrin monomer when they were present simultaneously. This is because Fragment D binds to the N-terminal binding site of fibrin molecules and prevents the association between fibrin monomers. If the photooxidation does not affect the binding site of Fragment D (the C-terminal binding domain), the above mentioned inhi-

TABLE 1

INHIBITION OF FIBRIN FORMATION BY FRAGMENT D WITH AND
WITHOUT PHOTOOXIDATION

	Association Activity (%)
Fibrinogen	100
+ Fragment D	44
+ Fragment D *	46

* Irradiated for 10 min with rose bengal, molar ratio; Fragment D/fibrinogen = 10.

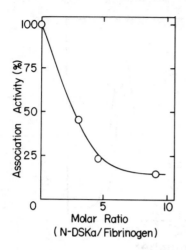

FIGURE 4. The effect of N-DSK$_a$ on the association of fibrinogen. The association of fibrinogen induced by thrombin was measured in the presence of varied amounts of N-DSK$_a$, as described in FIGURE 1. The association activity, which was given by the maximum absorbance increase at 450 nm, was expressed as values relative to those obtained in the absence of N-DSK$_a$.

bition should not be influenced by the modification. This was actually the case, as is illustrated in TABLE 1. The native and photooxidized Fragment D with rose bengal inhibited the association to the same extent. Based upon the analogous idea, N-DSK$_a$ inhibited the association of fibrinogen, as shown in FIGURE 4. The association was increasingly inhibited by increasing the amount of N-DSK$_a$ added. The introduction of mere threefold excess of N-DSK$_a$ relative to fibrinogen was enough to inhibit the association by 50%. We would predict that this inhibition would be relieved by the photooxidation of N-DSK$_a$ if the modification affects the binding sites of the molecule as we have discussed. This is indeed what we have observed, as shown in FIGURE 5. Under the condition described in the figure, the native N-DSK$_a$ added inhibits the association by about 90%, and the extent of inhibition was steadily decreased by increasing the period of irradiation time for photooxidation of

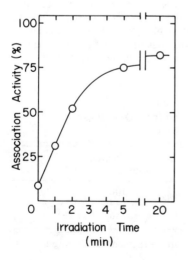

FIGURE 5. The effect of photooxidation of N-DSK$_a$ on the inhibition of the association of fibrinogen. The association of fibrinogen induced by thrombin was measured in the presence of N-DSK$_a$ which had been photooxidized with methylene blue for given periods of time. The amount of N-DSK$_a$ added was 9 molar excess relative to fibrinogen. The association activity, which was given by the maximum absorbance increase at 450 nm, was expressed as relative values to that obtained in the absence of N-DSK$_a$.

N-DSK$_a$ with methylene blue. The N-DSK$_a$ irradiated for five minutes inhibited the association by only 25%. In order to know the kind of amino acid residues oxidized under the experimental condition, the amino acid composition of N-DSK$_a$ photooxidized for varied periods of time was analyzed. The results were summarized in TABLE 2. All the amino acid tested except histidine stayed rather constant throughout the entire periods of photooxidation. This is further illustrated in FIGURE 6. The number of histidine residues decreased steadily with irradiation, while that of tyrosine residue stayed very constant. There are five kinds of amino acid residues altogether that can be photooxidized. They are histidine, tyrosine, tryptophan, methionine and cysteine.[23, 24] The

TABLE 2

THE EFFECT OF PHOTOOXIDATION ON THE AMINO ACID COMPOSITION OF N-DSK$_a$

Amino acid	Irradiation Time (Inhibition of association of fibrinogen)			
	DARK (91%)	1 min (69%)	2 min (48%)	5 min (25%)
Lys	29.8	30.6	29.7	30.3
His	7.84	5.63	4.78	2.34
Arg	26.0	26.8	26.5	26.9
Asp	50.4	50.8	50.3	50.3
Thr	19.1	19.4	19.1	19.0
Ser	32.8	33.2	32.8	32.8
Glu	55.7	56.4	55.2	55.6
Gly	23.4	23.0	23.0	22.9
Ala	29.5	29.5	29.5	29.3
Cys	8.82	8.12	8.43	8.41
Val	19.7	19.6	19.7	19.8
Met	–	–	–	–
Ile	15.3	15.2	15.4	15.4
Leu	32	32	32	32
Tyr	12.2	12.2	12.5	12.2
Phe	8.03	7.79	8.16	7.97
Pro	31.3	32.4	30.6	31.2

latter two amino acids are not present in N-DSK$_a$ molecule. The value of tryptophan was not reliable and was not included in TABLE 2 and FIGURE 6, due to HCl hydrolysis procedure employed. Since rose bengal tended to bind with fibrinogen too tenaciously to come off by dialysis, methylene blue was used instead throughout experiments shown below.

Amino Acid Residue in the N-Terminal Binding Site Directly Involved in the Association

In order to identify the kind of amino acid residues directly involved in the association and eventually to locate its amino acid residue in the molecule,

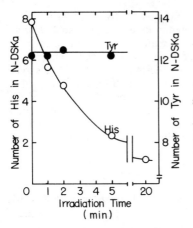

FIGURE 6. The effect of photooxidation of N-DSK$_a$ on its contents of histidine and tyrosine residues. N-DSK$_a$ was photooxidized for given periods of time with methylene blue, and the amino acid compositions were analyzed after hydrolysis with 6 M HCl for 22 hr at 110° C.

it may be advantageous to separate functional and dysfunctional molecules. Even if multiple kinds of residue are modified as a whole, for example, it might be possible to show that some of them are not necessarily involved in the association, provided that they are also modified in the functional molecules. We have separated functional N-DSK$_a$ from the dysfunctional one by fibrinogen-Sepharose affinity chromatography. The inhibition test illustrated above could not fulfill this purpose. As it is illustrated in FIGURE 7, the native N-DSK$_a$ bound almost quantitatively to the affinity column. This shows

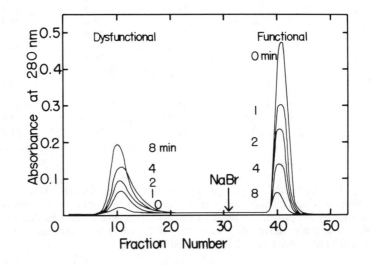

FIGURE 7. The effect of photooxidation of N-DSK$_a$ on its affinity with fibrinogen. N-DSK$_a$ in 50 mM Tris buffer containing 0.1 M NaCl (pH 7.0) was photooxidized for given periods of time in the presence of methylene blue. It was then applied to a fibrinogen-Sepharose column (1.5 × 16 cm), washed with the above-mentioned buffer, and eluted with 1 M NaBr in 10 mM phosphate buffer (pH 5.3). The wash and the eluate were collected at one ml per min into 3 ml fraction.

that the N-DSK$_a$ preparation used consisted entirely of functional molecules. The amount of dysfunctional N-DSK$_a$ was increased by increasing the period of the irradiation time. The irradiation for four minutes made 50% of the molecules dysfunctional. We have collected both functional and dysfunctional molecules after the affinity chromatography and they were subjected to amino acid compositional analysis with 3 M mercaptoethanesulfonic acid, which would yield reliable values for the content of tryptophan, too. FIGURE 8 summarizes the amino acid compositional analyses of the functional and dysfunctional molecules obtained from N-DSK$_a$ irradiated for varied periods of time. Tyrosine residues were oxidized neither in functional nor dysfunctional molecules. The importance of these residues in the association activity can not be ascertained by this technique. Tryptophan residues were apparently oxidized easily, and none of the dysfunctional samples retained these amino acid residues intact. It might be speculated that these amino acid residues are directly involved in the association step. The fact that no tryptophan residues were

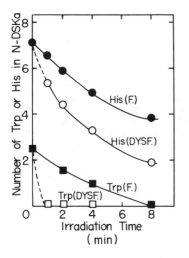

FIGURE 8. The effect of photooxidation of N-DSK$_a$ on the contents of histidine and tryptophan residues in functional and dysfunctional N-DSK$_a$ molecules. Photooxidized N-DSK$_a$ for given periods of time was fractionated into functional and dysfunctional molecules by the affinity chromatography shown in FIGURE 7. The amino acid compositions were analyzed after hydrolysis with 3 M mercaptoethanesulfonic acid for 22 hr at 110° C.

observed in functional N-DSK$_a$ molecules obtained from samples irradiated for eight minutes, however, strongly argues against this speculation. The amounts of histidine residues in functional molecules are always higher than in the corresponding dysfunctional molecules. At least some histidine residues are always found in the functional molecules. The results about histidine residues were, therefore, consistent throughout the experiment, permitting the conclusion that these amino acid residues are directly important for the association step.

An additional experiment was conducted to support the above mentioned conclusion. Fragment E (mostly E$_3$) was obtained by thorough digestion of fibrinogen by plasmin followed by DEAE ion exchange chromatography.[14] Fibrinogen used for the digestion had been photooxidized with methylene blue for varied periods of time. Although we have tried to isolate N-DSK fragments from those samples, the position of excision by cyanogen bromide was not fixed and fragments with varied length have been obtained after the modification. The compositions of amino acids in Fragment E obtained were analyzed.

FIGURE 9 summarizes the change of the contents of tryptophan and histidine residues in the fragment. The amount of tryptophan (and tyrosine) residues stayed constant but that of histidine residue decreased rather linearly with the modification. This result confirms the importance of histidine residue for the N-terminal binding sites to function.

DISCUSSION

Many chemical modification techniques employed so far have been unsuccessful in pointing out the involvement of any specific amino acid residues in the association process. As they often had to introduce so many bulky substitution groups in order to affect the association activity, gradual conformational alteration might indirectly affect the association. It is imperative to

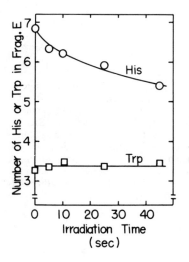

FIGURE 9. The effect of photooxidation of fibrinogen on the contents of histidine and tryptophan residues in Fragment E. Fibrinogen was photooxidized for given periods of time with methylene blue, and Fragment E was obtained from each sample, as described in the text. The amino acid compositions were analyzed after hydrolysis and 3 M mercaptoethanesulfonic acid for 22 hr at 110° C.

choose the reaction condition so as to ensure the specificity as high as possible. In order to augment this further, it is desirable to modify as little amino acid residues as possible and to separate dysfunctional molecules from the functional ones. Even then, multiple kinds of amino acid residues might be modified as in the case we have shown in this study. It would be possible, however, to come to the conclusion that some of the residues are not necessarily relevant to the function, if we could prepare a series of samples modified to varied degrees as we have done in this study.

Both tryptophan and histidine residues were oxidized to almost the same extent, and the dysfunctional molecules had always less of these two amino acid residues than the functional molecules did. It was, therefore, quite difficult to rule out any of them from the list of candidates. It was the almost complete absence of tryptophan residues in one of the functional samples that made us to conclude that tryptophan residues are not involved in the association process. As shown in FIGURE 8, tryptophan residues in the functional N-DSK$_a$ molecules obtained from samples irradiated for eight minutes were completely oxidized. An additional experiment shown in FIGURE 9 also indicated very strongly the importance of histidine residue(s) in the N-terminal binding site.

As the primary sequence of the fibrinogen molecule has been completely worked out by many systematic and enduring studies,[25-28] the position of histidine residues in N-DSK$_a$ can be easily pointed out. They are His[24] in α chain, His[16] and His[67] in β chain and His[48] in γ chain. Since the N-DSK$_a$ molecule has a dimeric structure, the total number of histidine residues present in the molecule is eight. The next obvious experiment we would like to do is to identify the critical histidine residue in N-DSK$_a$ molecule. If the molecule can be reasonably assumed to be symmetrical, the affinity chromatography employed in this study to test the association activity depends on the binding between one of the two identical binding sites of N-DSK$_a$ molecule and that of C-terminal domains of fibrinogen molecule. The destruction of only one of two binding sites does not make N-DSK$_a$ dysfunctional in this affinity test. It is, therefore, necessary to modify at least two of particular amino acid residues of N-DSK$_a$ to make it dysfunctional. If this is the case, the dysfunctional N-DSK$_a$ obtained from samples irradiated for one minute would be the ideal sample to use to identify the critical residues in the molecule. Since these dysfunctional molecules have exactly two histidine residues oxidized, only those critical two histidine residues should be modified uniformly among these molecules. If one uses other dysfunctional samples, histidine residues other than those two critical residues might also be heterogeneously modified, and their positions cannot be identified unambiguously. It is intriguing to point out that some of the functional N-DSK$_a$ molecules obtained from samples irradiated for longer period of time apparently have more than two histidine residues oxidized. This shows that those two critical histidine residues are not necessarily always oxidized preferentially in the N-DSK$_a$ molecule. It appears from the data shown in FIGURE 1, however, that those two critical histidine residues might be much more sensitive to the modification in the parent fibrinogen molecule. This might suggest some conformational difference between N-DSK$_a$ and fibrinogen molecules that remains to be elucidated.

REFERENCES

1. MAMMEN, E. F. 1974. Semin. Thromb. Haemostasis **1:** 184–201.
2. CRUM, E. D. 1977. *In* Hemostasis: Biochemistry, Physiology and Pathology. D. Ogston & B. Bennett, Eds.: 424–445. John Wiley and Sons. New York.
3. MORSE, E. E. 1978. Ann. Clin. Lab. Sci. **8:** 234–238.
4. BLOMBÄCK, B. & M. BLOMBÄCK. 1970. Nouv. Rev. Fr. Hematol. **10:** 671–678.
5. MOSESSON, M. W., D. L. AMRANI & D. MÉNACHÉ. 1976. J. Clin. Invest. **57:** 782–790.
6. HIGGINS, D. L. & J. A. SHAFER. 1981. J. Biol. Chem. **256:** 12013–12017.
7. LAKI, K. & E. MIHALYI. 1949. Nature **163:** 66.
8. KUDRYK, B., J. REUTERBY & B. BLOMBÄCK. 1973. Thromb. Res. **2:** 297–304.
9. KUDRYK, B. J., D. COLLEN, K. R. WOODS & B. BLOMBÄCK. 1974. J. Biol. Chem. **249:** 3322–3325.
10. YORK, L. L. & B. BLOMBÄCK. 1976. Thromb. Res. **8:** 607–618.
11. OLEXA, S. A. & A. Z. BUDZYNSKI. 1980. Proc. Natl. Acad. Sci. USA **77:** 1374–1378.

12. OLEXA, S. A. & A. Z. BUDZYNSKI. 1981. J. Biol. Chem. **256:** 3544–3549.
13. BLOMBÄCK, B., M. BLOMBÄCK, A. HENSCHEN, B. HESSEL, S. IWANAGA & K. R. WOODS. 1968. Nature **218:** 130–134.
14. DOOLITTLE, R. F., K. G. CASSMAN, B. A. COTTRELL, S. J. FRIEZNER & T. TAKAGI. 1977. Biochemistry **16:** 1710–1715.
15. DONNELLY, T. H., M. LASKOWSKI, N. NOTLEY & H. S. SCHERAGA. 1955. Arch. Biochem. Biophys. **56:** 369–387.
16. HENNE, D. L. & F. R. MATTHIAS. 1973. Thromb. Res. **2:** 137–154.
17. OLEXA, S. A. & A. Z. BUDZYNSKI. 1979. J. Biol. Chem. **254:** 4925–4932.
18. NIEUWENHUIZEN, W., A. VERMOND & F. HAVERKATE. 1981. Biochim. Biophys. Acta **667:** 321–327.
19. LAUDANO, A. P. & R. F. DOOLITTLE. 1978. Proc. Natl. Acad. Sci. USA **75:** 3085–3089.
20. MATSUSHIMA, A., H. TAKIUCHI, Y. SAITO & Y. INADA. 1980. Biochim. Biophys. Acta **625:** 230–236.
21. PENKE, E. B., R. FERENCZI & K. KOVACS. 1974. Anal. Biochem. **60:** 45–50.
22. INADA, Y., B. HESSEL & B. BLOMBÄCK. 1978. Biochim. Biophys. Acta **532:** 161–170.
23. MEANS, G. E. & R. E. FEENEY, Eds. 1971. In Chemical Modification of Proteins: 165–169. Holden-Day. San Francisco, CA.
24. SLUYTERMAN, L. A. 1962. Biochim. Biophys. Acta **60:** 557–561.
25. BLOMBÄCK, B., B. HESSEL, S. IWANAGA, J. REUTERBY & M. BLOMBÄCK. 1972. J. Biol. Chem. **247:** 1496–1512.
26. WATT, K. W. K., B. A. COTTRELL, D. D. STRONG & R. F. DOOLITTLE. 1979. Biochemistry **18:** 5410–5416.
27. WATT, K. W. K., T. TAKAGI & R. F. DOOLITTLE. 1979. Biochemistry **18:** 68–76.
28. HENSCHEN, A., F. LOTTSPEICH, E. TÖPFER-PETERSEN & R. WARBINEK. 1979. Thromb. Haemostas. (Stuttgart) **41:** 662–670.

DISCUSSION OF THE PAPER

R. F. EBERT (*Johns Hopkins Hospital, Baltimore, MD*): Have you done any experiments to show the effect of photooxidation on fibrinopeptide release?

Y. SAITO: Dr. Inada has done this experiment and showed that there was no effect on the peptide release.

H. A. SCHERAGA (*Cornell University, Ithaca, NY*): Do you have data for the number of histidines photooxidized in fibrinogen on the one hand and fibrin monomer on the other?

SAITO: In order to keep the fibrin monomeric one has to put them in 1 molar NaBr and we have not yet done this experiment.

J. L. YORK (*University of Arkansas College of Medicine, Little Rock*): Do you have evidence that it is a unique histidine? You may be statistically oxidizing a fraction of several or all histidines.

SAITO: I do not have this information yet, but this is the way I think you might get it. If you work with the original mixture of oxidized molecules, you will have oxidation in different places, but with affinity chromatography you might select molecules that have only two oxidized histidine residues, and these would be the very critical residues.

FIBRIN POLYMERIZATION SITES IN
FIBRINOGEN AND FIBRIN FRAGMENTS *

Andrei Z. Budzynski, Stephanie A. Olexa,† and Bharat V. Pandya

Thrombosis Research Center and Department of Biochemistry
Temple University School of Medicine
Philadelphia, Pennsylvania 19140

The inhibition of blood clotting by degradation products of fibrinogen reflects an interference with fibrin network formation. The original discovery that digestion of fibrinogen by plasmin resulted in the formation of fragments with anticoagulant activity [1,2] did not recognize the importance of inhibition of fibrin polymerization and was interpreted as an interference with thrombin action. During the past 25 years considerable progress has been achieved in the understanding of fibrin clot formation. The use of fibrinogen and fibrin fragments significantly contributed to the advancement of our knowledge, especially since many biologic functions of fibrinogen were recovered in various fragments.

Thrombin releases two molecules of fibrinopeptide A (FPA) and two molecules of fibrinopeptide B (FPB) from the Aα and Bβ polypeptide chains of the fibrinogen molecule, respectively. The cleavage of four arginylglycine peptide bonds converts fibrinogen into fibrin monomer, which polymerizes spontaneously forming a fibrin clot. This paper addresses the structure and localization of active contact sites with complementary affinities which are thought to specifically organize fibrin monomers and drive the polymerization reaction. Batroxobin, a protease from *Bothrops atrox* venom, cleaves FPA from fibrinogen. This event is sufficient to initiate clotting and formation of fibrin I, which lacks FPA but still contains FPB, and provides evidence that the loss of FPA alone is sufficient for fibrin gelation. Thrombin releases FPA at a much faster rate than FPB and the ultimate product, fibrin II lacks both fibrinopeptides.

POLYMERIZATION SITE ON FRAGMENT D

Fragment D, a plasmic degradation product of fibrinogen, contains the COOH-terminal of the Bβ and γ chains, and an Aα chain remnant. The conclusions inferred from chemical, physical and electron microscopic data clearly placed the origin of Fragment D in the outer terminal structures of the trinodular model of the fibrinogen molecule. Heterogeneity of Fragment D originates from the degradation of the γ chain remnant. It had been recognized that Fragment D inhibited polymerization of fibrin monomers [3,4] and caused the formation of defective clots.[5] Since Fragment D is thermolabile and precipitates at 60° while Fragment E is thermostable at this temperature, differential heat denaturation was used to separate the fragments from plasmic

---

* Supported by a Grant HL 14217 from the National Heart, Lung and Blood Institute, National Institutes of Health, Bethesda, Maryland.

† Present address: Rohm and Haas Co., Spring House, PA 19477.

0077–8923/83/0408–0301 $01.75/0 © 1983, NYAS

digests of fibrinogen and to demonstrate that Fragment E did not interfere with clot formation.[6] The development of affinity chromatography on insolubilized fibrinogen or fibrin monomer showed different binding properties of the two proteins, in particular a reversible binding of fibrinogen with insolubilized fibrin monomer.[7] This result was consistent with binding of fibrinogen to fibrin monomer in solution causing the formation of soluble fibrin complexes.[8] It appeared from these observations that a binding site, with specificity for fibrin monomer, was already exposed on the fibrinogen molecule without any action of thrombin.

The availability of structurally defined fragments obtained from fibrinogen after plasmin or cyanogen bromide degradation allowed the investigation of the localization of polymerization sites participating in the propagation of fibrin fibers. It was found that Fragment D bound with insolubilized fibrin monomer but not with insolubilized fibrinogen.[9] The binding capacity of Fragment D_1 was high; however, Fragment D_3 bound only minimally.[9, 10] Supportive evidence was provided by a study showing a strong inhibition of fibrin monomer polymerization by Fragment D_1 and poor inhibition by Fragment D_3.[11] Fragments D_1 and D_3 differ in the size of the γ chain remnants, of M_r 39,000 and 26,000, respectively; the α and β chain remnants are the same.[12] It appeared that after cleavage of the γ chain remnant in Fragment D_1 by plasmin, a polypeptide segment responsible for the binding function was either affected or removed. In order to resolve the question, Fragment D_1 from human fibrinogen was digested with plasmin in the presence of EDTA, since in this condition cleavages in the γ chain occurred. From the digestion mixture a peptide of approximate M_r 4,500 was isolated by affinity chromatography on insolubilized fibrin monomer; the amino acid sequence demonstrated that the peptide originated from the COOH-terminal of the γ chain encompassing the residues γ374Thr-411Val.[13] The peptide had no affinity for NH_2-terminal disulfide knot (NDSK), fibrinogen or Fragment D_1, but it bound to thrombin-treated NH_2-terminal disulfide knot (t-NDSK). It was concluded from these observations that the isolated peptide could be used as a functional probe for testing the polymerization sites in the NH_2-terminal domain after the cleavage of FPA and FPB. In order to determine which fibrinopeptide was critical for the exposure of a binding site, two types of fibrin monomers were studied. Batroxobin fibrin monomer was lacking FPA only, thrombin fibrin monomer was lacking both fibrinopeptides. The peptide inhibited polymerization of the two types of monomers (FIGURE 1). As the molar ratio of peptide to fibrin monomer increased, the maximum rate and extent of polymerization decreased proportionally, while the reaction lag time increased. The maximum rate of polymerization decreased to 50 percent at approximately 1:1 molar ratio of the peptide to fibrin monomer, regardless whether batroxobin or thrombin monomer was used. It was thus concluded that the cleavage of only FPA exposed a binding site in the NH_2-terminal domain, that the cleavage of FPA and FPB did not change the basic phenomenon, and that two moleclues of the peptide would combine with one molecule of either monomer.

A direct measurement of binding with fibrin monomers was complicated due to the formation of various polymers during equilibration with the peptide. Therefore, t-NDSK was used as a fragment containing thrombin-activable polymerization sites. In order to measure the number of binding sites and the strength of the interaction, the peptide was labeled with ^{125}I. A Scatchard analysis of binding data (FIGURE 2) gave the value of $K_d = 1.45 \times 10^{-6}M$

indicating a fairly tight binding. Approximately 2.1 binding sites for the peptide were found in t-NDSK, a result in excellent agreement with inhibition of fibrin monomer polymerization (FIGURE 1). Since Fragment D_1 had the same inhibitory effect as the peptide on polymerization reaction, it can be concluded that the functional polymerization site demonstrated in Fragment D_1 is contained in the amino acid sequence γ374Thr-411Val.

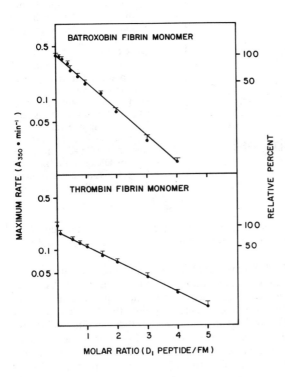

FIGURE 1. Inhibition of fibrin monomer polymerization by a 38-amino-acid-residue-long peptide from the COOH-terminal of the γ chain (γ374Thr-411 Val). Batroxobin fibrin monomer (top) lacked FPA only; thrombin fibrin monomer (bottom) lacked FPA and FPB; the monomers were prepared from human fibrinogen incubated with the respective enzyme. Various molar ratios of the peptide to fibrin monomer (abscissa) were tested. The formation of fibrin polymers was monitored in a spectrophotometer at 350 nm and for each run the maximum reaction rate was calculated (ordinate, left). The maximum rate of polymerization decreased by 50 percent at approximately a 1:1 molar ratio for both fibrin monomers. (From Olexa & Budzynski.[13] By permission of *Journal of Biological Chemistry*.)

POLYMERIZATION SITE ON THE NH$_2$-TERMINAL DOMAIN

A polymerization site, which will be called "a" site, is thus present in the last 38 amino acid residues of the COOH-terminal of the γ-chain. A complementary site, called "A" site, is not available in fibrinogen; however, after the cleavage of FPA only it is exposed in the NH$_2$-terminal of the fibrin molecule. This localization of polymerization sites was suggested after demonstration of Fragment D_1 binding to t-NDSK.[10, 14] Electron microscopic patterns of fibrin oligomers using either shadow casting[15] or negative staining[16] techniques showed a half-staggered association of the fibrin monomer molecules, suggesting that the lateral and central nodules may have contact sites. It should be pointed out that polymerizing molecules are not equivalent in all directions

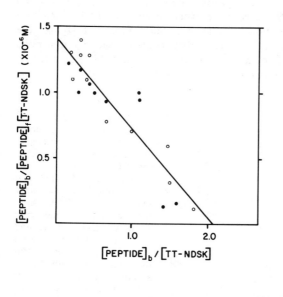

FIGURE 2. Binding of γ374Thr-411Val peptide with thrombin-treated NDSK. This fibrinogen derivative represents the NH₂-terminal domain and lacks both FPA and FPB. Binding was measured by either equilibrium dialysis (●) or the Sephadex gel equilibrium method (○). The ratio of peptide molecules bound per molecule of thrombin-treated NDSK (abscissa) is plotted versus the ratio of bound to free peptide (ordinate). The line was computed from a least square regression analysis of the data points and indicated approximately 2.1 binding sites on thrombin-treated NDSK molecule. (From Olexa & Budzynski.[18] By permission of *Journal of Biological Chemistry*.)

and the half-staggered association of monomers must occur face-to-face (FIGURE 3), otherwise the formation of fibrin sheets rather than fibers would be favored.

The cleavage of only FPA from the fibrinogen molecule is sufficient to initiate clotting.[17] This reaction is compatible with the exposure of "A" site. Its localization within amino acid sequence of the three polypeptide chains is still incompletely understood. There is evidence that the NH₂-terminal of the α chain of fibrin may be involved in this site. Synthetic peptide Gly-Pro-Arg-Val, corresponding to the amino acid sequence Aα17-20 of human fibrinogen, inhibited fibrin monomer polymerization and bound to fibrinogen with a K_d of 1×10^{-4} M [18]; a tetrapeptide Gly-Pro-Arg-Pro was even more effective and bound to two sites in fibrinogen or to one site in Fragment D.[19] The binding constant indicated rather weak affinity of these tetrapeptides for the "a" site. For instance, Gly-Pro-Arg-Pro did not inhibit polymerization at a molar ratio to fibrin monomer of 15:1 (FIGURE 4). Thus the amino acid sequence in the Aα chain following that of FPA seems to represent a small portion of the entire "A" site.

Since the "A" site was expected to be located in the NH₂-terminal domain of the molecule, the interaction of NDSK derivatives with fibrin monomer polymerization was tested. Unmodified NDSK and reduced NDSK did not affect the rate of polymerization at molar ratios of derivatives to fibrin monomer up to 18:1 (FIGURE 4). The cleavage of FPA and FPB from NDSK by thrombin resulted in the formation of t-NDSK which interfered with polymerization. At low molar ratios, of 1-3 moles of t-NDSK per mole of fibrin monomer, a small but significant increase of polymerization rate was consistently observed. However, at higher molar ratios the inhibition of polymerization was

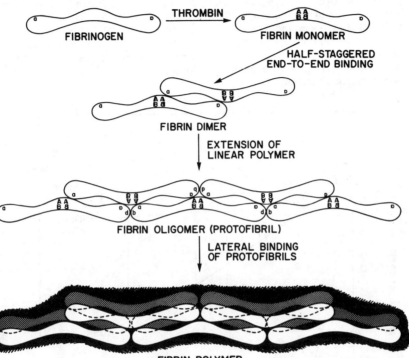

FIGURE 3. Sequence of events in human fibrin polymerization. Fibrinogen is a bivalent molecule depicted in a trinodular form with an available polymerization site "a" on each D domain of the molecule. Upon cleavage of FPA and FPB by thrombin, fibrin monomer is formed. By this reaction two sets of binding sites ("A" and "B") on the NH_2-terminal domain of the molecule become available. Polymerization sites on the two halves of the fibrinogen molecule are depicted to reflect the twofold rotational symmetry; one-half of the polymerization sites is on the opposite side of the molecule. The "A" sites are complementary to the "a" sites on the fragment D domain. The binding of these sites induces a half-staggered linear polymerization of the molecules, which are arranged in a face-to-face fashion. The contact between two D domains in linearly assembled fibrin monomers results in the formation of a new bivalent polymerization site "bb." The formation of crosslink bonds between two γ chains of the neighboring fibrin monomer molecules may stabilize the sites on aligned fibrin monomer molecules. The "b" site is complementary to the thrombin-activated site "B" on the NH_2-terminal domain of fibrin monomer. The two "B" sites on the second (upper) protofibril bind to the "bb" sites on the first (lower, shaded) one. Meanwhile, the alignment of the fibrin monomer molecules on the second (upper) protofibril results in the formation of "bb" binding sites that will enable the addition of a third fibrin protofibril. Therefore, the binding of "A" to "a" sites promotes linear polymerization of the fibrin monomer molecules as well as fibrin strand branching, whereas the binding of "BB" to "bb" sites allows lateral aggregation of protofibrils.

evident. This inconsistency can be explained by taking into consideration the presence of two "A" sites in t-NDSK (FIGURE 5) which can bind to complementary "a" sites on fibrin monomer molecules. At an equimolar ratio of t-NDSK and fibrin monomer the concentration of "A" sites was increased twofold, accelerating the reaction rate since t-NDSK can be incorporated into the growing heteropolymers. However, an excess of t-NDSK would bind to fibrin monomers in such a manner as to occupy both "a" sites preventing the polymerization reaction.

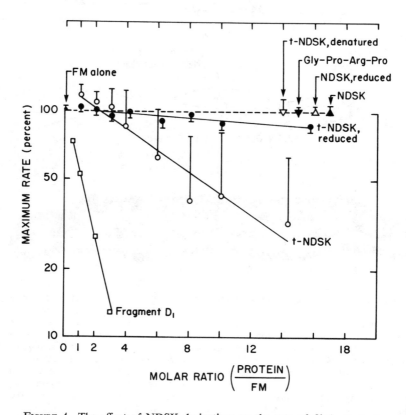

FIGURE 4. The effect of NDSK derivatives on the rate of fibrin monomer polymerization. A 0.1 ml aliquot of the tested preparation in 0.15 M Tris-HCl buffer, pH 7.55, was mixed with 0.8 ml of the same buffer in a 10mm-long quartz cell and the base line was recorded at 350 nm. At zero time the recorder was started, 0.1 ml of fibrin monomer (1.5 mg/ml in 0.02 M acetic acid) was added, mixed well and the increase of absorbance was recorded for 15 min. To measure the rate of polymerization of fibrin monomer alone (100% on hte ordinate), 0.1 ml of fibrin monomer was added to 0.9 ml of the buffer. The maximum rate of polymerization was calculated from the slope of the steepest part of the curve and expressed as percent of that for fibrin monomer alone. At the indicated molar ratio neither tetrapeptide Gly-Pro-Arg-Pro nor NDSK inhibited polymerization. After thrombin treatment, t-NDSK gained inhibitory activity, but reduced and carboxymethylated t-NDSK lost most of it; t-NDSK denatured in 6M guanidine hydrochloride and dialyzed in Tris-HCl buffer was totally inactive.

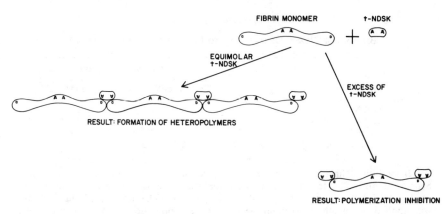

FIGURE 5. Model for t-NDSK interaction with fibrin monomer polymerization. Thrombin treatment of NDSK cleaves fibrinopeptides and exposes two "A" sites on the two halves of the dimeric molecule of t-NDSK. The derivative thus has a bivalent polymerization site. At equimolar ratio of t-NDSK and fibrin monomer a copolymerization occurs and the reaction rate increases due to higher substrate (= "A" sites) concentration. An excess of t-NDSK occupies "a" sites available on fibrin monomer molecules, which results in the inhibition of polymerization reaction.

The comparison of the kinetics of inhibition by Fragment D_1 and t-NDSK showed that the latter was significantly less potent (FIGURE 4). In addition, great variability between different batches of t-NDSK was noted and it was reflected by large standard deviations. One possible interpretation of the data would be that "A" sites on t-NDSK may have altered conformation and decreased affinity for the complementary polymerization sites due to the denaturing conditions (70 percent formic acid) used in its preparation. Theoretically one would expect that a fibrinogen derivative containing a monovalent fully active "A" site should decrease the polymerization rate to 50% at a molar ratio of 1:1. Further evidence supporting the involvement of polypeptide chain conformation in "A" site came from observations that denaturation of t-NDSK by reduction and carboxymethylation greatly decreased its inhibitory function; after treatment with 6M guanidine hydrochloride the inhibitory activity was totally abolished.

Affinity chromatography on insolubilized fibrinogen confirmed that the removal of fibrinopeptides from NDSK resulted in the appearance of a binding function (TABLE 1). However, this property was mostly lost after reduction and carboxymethylation of t-NDSK. Similar results were obtained in the presence of calcium ions, with the exception that all tested derivatives, including NDSK, bound in greater amounts. It seems that the effect of calcium ions may be independent of the interaction of polymerization sites and perhaps could introduce an additional link with the NH_2-terminal domain through the third calcium binding site in fibrinogen.

New evidence concerning the localization of "A" site was obtained by studying the structure of fibrinogen degradation products formed by Protease III isolated from *Crotalus atrox* venom.[20] This enzyme had an unusual proteolytic effect on fibrinogen, cleaving only the Bβ chain, whereas the Aα and

TABLE 1

BINDING OF NDSK DERIVATIVES TO INSOLUBILIZED FIBRINOGEN *

| Derivative | Protein Bound, nmoles | |
	Ca^{2+} absent	$+2mM\ Ca^{2+}$
NDSK	0	10.8
NDSK, reduced	0	16.8
t-NDSK	18.9	38.5
t-NDSK, reduced	3.8	19.2

* Insolubilized fibrinogen was prepared by coupling to cyanogen bromide-activated Sepharose 4B.[7] The tested proteins were dialyzed in a low-porosity tubing in 0.05 M Tris-HCl buffer containing 0.1 M NaCl, pH 7.4. The protein (5 mg; 0.2 ml) was applied to an affinity column (0.5 × 3 cm) containing 0.5 ml of packed resin equilibrated in the same buffer and was eluted with 10 ml of the buffer. The bound protein was eluted with 0.05 M acetic acid, 10 fractions 1 ml each collected and the total protein in the eluates determined spectrophotometrically at 280 nm.

γ chains were unaffected.[21] A fibrinogen derivative of M_r 325,000 isolated from the digest by salt precipitation (FIGURE 6) had impaired coagulability with batroxobin or thrombin in spite of the presence of FPA in the apparently

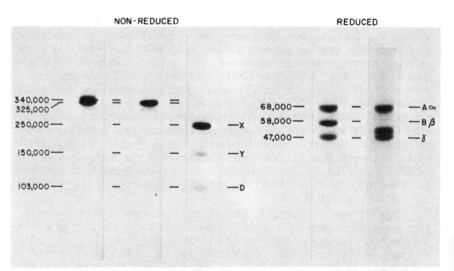

FIGURE 6. Electrophoretic patterns of M_r 325,000 derivative isolated from a *C. atrox* Protease III fibrinogen digest, by precipitation with ammonium sulfate (1.23 M final concentration). Polyacrylamide electrophoresis in the presence of 0.1% SDS was performed in 3.5% gels with non-reduced samples of fibrinogen, derivative and a standard mixture of Fragments X (M_r 250,000), Y (M_r 155,000) and D (M_r 103,000), or in 7% gels with 2-mercaptoethanol-reduced samples of fibrinogen and the derivative. In the latter preparation the Aα and the γ chains have the same electrophoretic mobility as those of fibrinogen, however, all Bβ chains have been degraded to a M_r 51,500 fragment.

undegraded Aα chain. The intactness of the NH$_2$-terminal of the Aα chain in M$_r$ 325,000 derivative was ascertained by the same rate of FPA cleavage by thrombin from fibrinogen and the derivative. This result casts a serious doubt on the notion that the NH$_2$-terminal of the α chain does contain the only polymerization region that would be complementary to "a" site.

It is known that the NH$_2$-terminal of the Bβ chain is very susceptible to the proteolytic degradation by thrombin, plasmin and trypsin, implying exposure on the surface of the fibrinogen molecule. On the other hand, the COOH-terminal of the Bβ chain is not accessible in native fibrinogen to proteolytic enzymes. Apparently, the cleavage by Protease III affected the NH$_2$-terminal of the Bβ chain and a M$_r$ 5,000 peptide was removed together with attached FPB as calculated from its amino acid composition. The peptide either contained a polymerization site or significantly contributed the function of a specific binding site. The peptide present in the supernate, after precipitation of the M$_r$ 325,000 derivative, was isolated using column gel filtration and had strong anticlotting activity (FIGURE 7). When tested in the fibrin monomer polymerization system it decreased the reaction rate by 50% at a molar ratio of approximately 1.5:1 of the peptide to fibrin monomer. The ratio is close to 1:1, as for the D$_1$ peptide, implying that the fibrin monomer molecule possesses two binding sites for the peptide. It has been shown that Gly-His-Arg-Pro, a peptide contiguous with FPB, binds with insolubilized fibrinogen with $K_d =$ 1.4 × 10^{-4}M but does not inhibit fibrin monomer polymerization.[18, 19] Thus, it is unlikely that the polymerization site in the M$_r$ 5,000 peptide is close to the NH$_2$-terminal.

The effect of the M$_r$ 325,000 derivative on the polymerization reaction resembled that seen with t-NDSK. At a low molar ratio of the derivative to fibrin monomer an acceleration of polymerization rate was observed, but at high ratios the derivative had an inhibitory effect. These phenomena can be explained assuming that the M$_r$ 325,000 derivative has two functionally operative "a" sites and significantly altered "A" sites with greatly decreased binding affinity (FIGURE 8).

The structure of "A" site appears to be complex. First, it seems to contain a high affinity binding region in the NH$_2$-terminal of the Bβ chain, most likely within the sequence Bβ15–58. Secondly, it involves the amino acid sequence in the Aα chain adjacent to FPA; however, its entire length is unknown at present. Thirdly, the segments of the Bβ and the Aα chains that participate in the structure of "A" site have specific conformation for the full expression of the binding function. The participation of the γ chain in this polymerization site cannot be excluded at the present time.

The location of a high affinity polymerization site in the NH$_2$-terminal of the Bβ chain is compatible with the hypothesis that the relative rates of proteolysis of this chain by thrombin and plasmin determine the occurrence of thrombosis.[22] Upon the cleavage by plasmin of Bβ1–42 from fibrinogen or from fibrin I monomer, which lacks FPA only, the "A" site seems to become impaired, formation of fibrin decreased, and clinical occurrence of thrombosis less frequent.

A NUCLEUS OF POLYMERIZATION SITES: (DD)E COMPLEX

An exquisite model for studies of polymerization sites is the (DD)E complex. It is formed as a predominant soluble plasmic degradation product of cross-linked fibrin according to the scheme: [23]

$$\text{cross-linked fibrin} \rightarrow (DD)E_1 \rightarrow (DD)E_2 \rightarrow DD + E_3$$

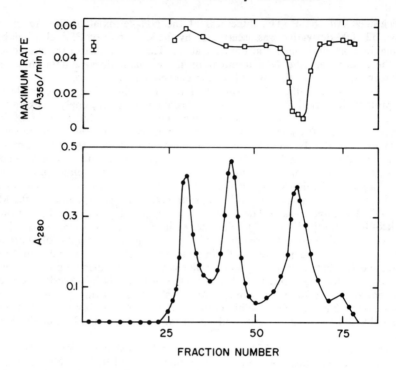

FIGURE 7. Isolation of a M_r 5,000 peptide that inhibits fibrin monomer polymerization. Human fibrinogen (1 g) was digested with Protease III (0.3 mg) from *C. atrox* venom at 37° for 90 min. High molecular weight derivatives were precipitated in 1.23 M ammonium sulfate; the supernate was dialyzed in a low-porosity tubing with a cutoff limit at M_r 3,500 and gel filtered on a BioGel P-10 column (1 × 100 cm) using 0.05 M ammonium bicarbonate, pH 7.0 (bottom). Eluted fractions were tested in fibrin monomer polymerization system, measuring the maximum rate as the increase of absorbance per minute ($A_{350} \cdot \min^{-1}$). Only the last peak, of approximate M_r 5,000, significantly inhibited the maximum polymerization rate (top).

Purified Fragments E_1 and E_2, but not Fragment E_3, reform the complex after incubation with purified Fragment DD.[24] Since t-NDSK also forms a stoichiometric complex with Fragment DD, it was accepted that the binding occurs between two "a" sites on Fragment DD aligned by crosslink bonds and two "A" sites on Fragment E_1 or E_2 or t-NDSK positioned specifically by the dimeric structure of these derivatives. Thus, the (DD)E complex represents a nucleus of polymerization sites holding together fibrin polymers. The determination of the amino acid sequence of Fragments E_2 and E_3 addressed the question of which parts of the polypeptide chains are lost along with the loss of binding to Fragment DD.[25] It was found that cleavage occurred at four sites: the most significant was the removal of Bβ15–53 and Aα17–19 in accordance with the sequences postulated to be involved in the "A" site; in addition Bβ121 and γ54–62 were split off. These data are in a good agreement with a possible localization of "A" site in the NH_2-terminals of the β and α chains of the fibrin monomer molecule.

It was proposed that the cleavage of FPA from fibrinogen by thrombin initiates an end-to-end polymerization; the removal of FPB would result in a lateral aggregation.[17] The latter notion has been supported by the demonstration that abnormal fibrinogen Detroit (Aα19Arg → Ser), with a defective "A" site due to the amino acid replacement, will eventually form a clot but only after the cleavage of FPB.[14] The role of FPB in clot formation is compatible with the absence of the (DD)E complex in plasmic digests of crosslinked fibrin derived from fibrinogen after batroxobin and factor XIIIa treatment.

The evidence supporting the second set of polymerization sites "b" and "B", was obtained by affinity chromatography on covalently bound oligomers of crosslinked fibrin II, lacking both FPA and FPB.[9] Fragments E_1 and E_2 did not bind to insolubilized fibrinogen or fibrin monomer but did bind to insolubilized crosslinked fibrin. This fact, taken together with binding of these fragments to Fragment DD but not to D_1, indicated a requirement for specifically aligned two sites on the two D domains in either Fragment DD or crosslinked fibrin. It was proposed that a new bivalent polymerization site "bb" was generated upon an end-to-end alignment of fibrin monomer molecules stabilized by crosslink bonds. The complementary site "B" could be exposed by the loss of FPB. The binding between "b" and "B" sites would allow lateral aggregation of protofibrils (FIGURE 3). Electron microscopic patterns did not show the addition of single fibrin monomer molecules to the growing protofibrils, and showed the identity of batroxobin- or thrombin-derived protofibrils.[26] The formation of thick fibrin fibers would occur by coalescence of protofibrils and side-to-side binding by the interaction of "b"

EQUIMOLAR

RESULT: FORMATION OF POLYMERS

EXCESS OF DERIVATIVE

RESULT: POLYMERIZATION INHIBITION

FIGURE 8. Model for M_r 325,000 derivative interaction with fibrin monomer polymerization. The derivative untreated with thrombin has 2 FPA per molecule; however, it lacks a M_r 5,000 fragment cleaved from the NH_2-terminal of the Bβ chain. The M_r 325,000 derivative has two operational "a" sites through which it copolymerizes with fibrin monomer when tested at an equimolar proportion. However, the derivative does not have fully functional "A" sites and used in excess inhibits polymerization by occupying "A" sites present on fibrin monomer molecules. Thrombin-treated M_r 325,000 derivative, which lost both FPA, interacts in the same way with fibrin monomer as the untreated counterpart.

and "B" sites. This interpretation, however, requires a caution since "b" and "B" sites' interaction may well serve to reinforce the binding strength established in protofibrils through "a" and "A" sites. Preferential cleavage of FPB did not result in normal polymerization unless FPA was also removed,[27] implying that FPB cleavage did not expose a primary polymerization site. The use of novel degradation products from fibrinogen and fibrin will contribute to better understanding of interactions between polymerization sites.

REFERENCES

1. NIEWIAROWSKI, S. & E. KOWALSKI. 1958. Un novel anticoagulant derive du fibrinogene. Rev. d'Hematol. **13:** 320–328.
2. TRIANTAPHYLLOPOULOS, D. C. 1958. Anticoagulant effect of incubated fibrinogen. Can. J. Biochem. Physiol. **36:** 249–259.
3. ALKJAERSIG, N., A. P. FLETCHER & S. SHERRY. 1962. Pathogenesis of the coagulation defect developing during pathological plasma proteolytic ("fibrinolytic") states. II. The significance, mechanism and consequences of defective fibrin polymerization. J. Clin. Invest. **41:** 917–934.
4. LATALLO, Z. S., A. P. FLETCHER, N. ALKJAERSIG & S. SHERRY. 1962. Inhibition of fibrin polymerization by fibrinogen proteolysis products. Am. J. Physiol. **202:** 681–686.
5. BANG, N. U., A. P. FLETCHER, N. ALKJAERSIG & S. SHERRY. 1962. Pathogenesis of the coagulation defect developing during pathological plasma proteolytic ("fibrinolytic") states. III. Demonstration of abnormal clot structure by electron microscopy. J. Clin. Invest. **41:** 935–948.
6. BUDZYNSKI, A. Z., M. STAHL, M. KOPEC, Z. S. LATALLO, Z. WEGRZYNOWICZ & E. KOWALSKI. 1967. High molecular weight products of the late stage of fibrinogen proteolysis by plasmin and their structural relation to the fibrinogen molecule. Binochim. Biophys. Acta **147:** 313–323.
7. HEENE, D. L. & F. R. MATTHIAS. 1973. Adsorption of fibrinogen derivatives on insolubilized fibrinogen and fibrinmonomer. Thromb. Res. **2:** 137–154.
8. SHAINOFF, J. R. & I. H. PAGE. 1960. Cofibrins and fibrin-intermediates as indicators of thrombin activity, *in vivo.* Circulation Res. **8:** 1013–1022.
9. OLEXA, S. A. & A. Z. BUDZYNSKI. 1980. Evidence for four different polymerization sites involved in human fibrin formation. Proc. Natl. Acad. Sci. USA **77:** 1374–1378.
10. KUDRYK, B. J., D. COLLEN, K. R. WOODS & B. BLOMBÄCK. 1974. Evidence for localization of polymerization sites in fibrinogen. J. Biol. Chem. **249:** 3322–3325.
11. DRAY-ATTALI, L. & M. J. LARRIEU. 1977. Fragments D—Correlation between structure and biological activity. Thromb. Res. **10:** 575–586.
12. PIZZO, S. V., L. M. TAYLOR, JR., M. L. SCHWARTZ, R. L. HILL & P. A. MCKEE. 1973. Subunit structure of Fragment D from fibrinogen and cross-linked fibrin. J. Biol. Chem. **248:** 4584–4590.
13. OLEXA, S. A. & A. Z. BUDZYNSKI. 1981. Localization of a fibrin polymerization site. J. Biol. Chem. **256:** 3544–3549.
14. BLOMBÄCK, B., B. HESSEL, D. HOGG & L. THERKILDSEN. 1978. A two-step fibrinogen-fibrin transition in blood coagulation. Nature **275:** 501–505.
15. KRAKOW, W., G. F. ENDRES, B. M. SIEGEL & H. A. SCHERAGA. 1972. An electron microscopic investigation of the polymerization of bovine fibrin monomer. J. Mol. Biol. **71:** 95–103.
16. FOWLER, W. E., R. R. HANTGAN, J. HERMANS & H. P. ERICKSON. 1981. Structure of the fibrin protofibril. Proc. Natl. Acad. Sci. USA **78:** 4872–4876.

17. LAURENT, T. C. & B. BLOMBÄCK. 1958. On the significance of the release of two different peptides from fibrinogen during clotting. Acta Chem. Scand. **12:** 1875–1877.
18. LAUDANO, A. P. & R. F. DOOLITTLE. 1980. Studies on synthetic peptides that bind to fibrinogen and prevent fibrin polymerization. Structural requirements, number of binding sites and species differences. Biochemistry **19:** 1013–1019.
19. LAUDANO, A. P. & R. F. DOOLITTLE. 1978. Synthetic peptide derivatives that bind to fibrinogen and prevent the polymerization of fibrin monomers. Proc. Natl. Acad. Sci. USA **75:** 3085–3089.
20. PANDYA, B. V., A. Z. BUDZYNSKI, R. N. RUBIN & S. A. OLEXA. 1981. Anticoagulant proteases from western diamondback rattlesnake venom. Fed. Proc. **40:** 1970.
21. PANDYA, B. V., R. N. RUBIN, S. A. OLEXA & A. Z. BUDZYNSKI. 1983. Unique degradation of human fibrinogen by proteases from western diamondback rattlesnake (*Crotalus atrox*) venom. Toxicon **21.** In press.
22. NOSSEL, H. L. 1981. Relative proteolysis of the fibrinogen Bβ chain by thrombin and plasmin as a determinant of thrombosis. Nature **291:** 165–167.
23. OLEXA, S. A. & A. Z. BUDZYNSKI. 1979. Primary soluble plasmic degradation product of human cross-linked fibrin. Isolation and stoichiometry of the (DD)E complex. Biochemistry **18:** 991–995.
24. OLEXA, S. A. & A. Z. BUDZYNSKI. 1979. Binding phenomena of isolated unique plasmic degradation products of human crosslinked fibrin. J. Biol. Chem. **254:** 4925–4932.
25. OLEXA, S. A., A. Z. BUDZYNSKI, R. F. DOOLITTLE, B. A. COTTRELL & T. C. GREENE. 1981. Structure of Fragment E species from human cross-linked fibrin. Biochemistry **21:** 6139–6145.
26. HANTGAN, R. R., W. E. FOWLER, H. P. ERICKSON & J. HERMANS. 1980. Fibrin assembly: A comparison of electron microscopic and light scattering results. Thromb. Haemostas. (Stuttgart) **44:** 119–124.
27. SHAINOFF, J. R. & B. N. DARDIK. 1979. Fibrinopeptide B and aggregation of fibrinogen. Science **204:** 200–202.

DISCUSSION OF THE PAPER

L. LORAND (*Northwestern University, Evanston, IL*): If you add calcium to reduce NDSK, do you get a dimer?

A. BUDZYNSKI: We did not measure that.

R. HANTGAN (*University of North Carolina, Chapel Hill*): As I understand it, your model of inhibition depends on the existence of what we may term single-stranded protofibrils that are stabilized strictly by end-to-end contacts between D domains. I should like to point out the alternative interpretation: inhibition occurs by the blocking of the further growth of staggered overlap polymers.

BUDZYNSKI: That formation of the protofibrils essentially is extended only by the cooperation between the "a" and "A" sites. It's sufficient to explain growth of protofibrils.

HANTGAN: But they should, in fact, see single stranded protofibrils according to that model.

BUDZYNSKI: No.

HANTGAN: The mode of inhibition that you showed shows protofibrils joined strictly end to end with the inhibitor sitting across where we would put a different contact or where you'd put those staggered overlap contacts, as I understand your model.

BUDZYNSKI: The model that I showed refers only to the formation of oligo-fibers or two-stranded protofibrils. I think that our interpretation of protofibril formation is consistent with the data of other investigators because there is a successive addition of individual fiber monomers in half-staggered overlap.

HANTGAN: I agree with that.

W. R. BELL (*Johns Hopkins Hospital, Baltimore, MD*): Do you think that persistence of the B fibrinopeptide would have some influence on some of the findings that you showed?

BUDZYNSKI: Yes, but we removed fibrinopeptide B together with a piece of the Bβ chain. Thus, fibrinopeptide B is absent in the M_r 325,000 derivative.

SYNTHETIC PEPTIDES MODELED ON
FIBRIN POLYMERIZATION SITES

Andrew P. Laudano,* Barbara A. Cottrell, and Russell F. Doolittle

*Department of Chemistry
University of California., San Diego
La Jolla, California 92093*

The transformation of fibrinogen into a fibrin clot is initiated by the proteolytic enzyme thrombin that cleaves fibrinopeptides A and B from the amino termini of the α and β chains of fibrinogen.[1] The resulting fibrin monomers spontaneously polymerize to form a noncovalently bound gel stabilized by intermolecular cross-linking in the presence of activated factor XIII and calcium.

Thirty years ago Ferry [2] proposed that polymerization proceeds through two associative processes involving, initially, an end-to-end interaction leading to the formation of intermediate polymers, and then, a lateral aggregation of these intermediate strands. He postulated that intermediate polymers two molecules thick arranged in a staggered overlapping fashion might form as a result of interactions involving the positive amino termini that are exposed by thrombin.

It was later found that all six amino termini of fibrinogen are located in a single disulfide-bonded cyanogen bromide fragment, the "amino-terminal disulfide knot" (NDSK),[3] which was found to be immunologically cross-reactive with the plasmin-derived fragment E.[4] Recent electron microscopic studies utilizing antibodies that are directed toward various domains of fibrinogen have supported previous notions that this amino terminal domain comprises the central domain [5-7] of the triglobular fibrinogen molecule first observed by Hall and Slayter,[8] and more recently by others.[9-11]

A decade ago it was proposed that release of fibrinopeptides from the central domain of a triglobular molecule allows reciprocal interactions of the central domains with the terminal domains of neighboring molecules.[12] This notion was subsequently verified by experiments in which fibrinogen and fragment D, the plasmin-derived fragment that approximates the terminal domain, bound to both fibrin monomer and thrombin activated NDSK immobilized on Sepharose.[12]

Amino acid sequence studies have revealed that the amino terminal regions of the α and β chains of fibrin that are exposed by fibrinopeptide release are highly conserved from species to species.[13-16] In particular, the α chain of fibrin begins with the same tripeptide sequence, glycyl-L-prolyl-L-arginine (Gly-Pro-Arg), in all species studied, including the primitive lamprey. It is likely that important functional constraints related to fibrin polymerization have led to the evolutionary persistence of this moiety. Furthermore, the variant human fibrinogen Detroit, which exhibits defective polymerization, has a replacement of arginine by serine at the third residue from the amino terminus

* Present address: Department of Biology, Massachusetts Institute of Technology, Cambridge, Massachusetts 02139.

0077–8923/83/0408–0315 $01.75/0 © 1983, NYAS

of the α chain of fibrin.[17] The thought that the amino termini of fibrin might themselves function as binding sites for polymerization seemed more than reasonable. In the fibrinogen molecule these regions are accessible to thrombin and contain polar residues. As such, they would likely be exposed to the solvent upon fibrinopeptide release, an advantageous position for them to function as sites of interaction with other molecules.

In order to test the proposition that the amino termini of fibrin exposed by thrombin are involved as binding sites for fibrin polymerization, we synthesized a series of peptides modeled after the amino termini of the α and β chains of fibrin and tested them for their ability to bind to fibrinogen and to inhibit fibrin formation.

PEPTIDE ANALOGUES OF THE AMINO TERMINUS OF THE FIBRIN α CHAIN

We found that peptides beginning with the sequence Gly-Pro-Arg, which is found at the amino terminus of the α chain of fibrin, bind to fibrinogen (2 mol/mol) and fragment D (1 mol/mol).[18, 19] Furthermore, these peptides inhibit fibrin formation. Both the binding affinity of different α chain analogues (FIGURE 1, TABLE 1) and their inhibitory activity were found to vary as a function of the carboxy-terminal amino acid. Glycyl-L-prolyl-L-arginyl-L-proline (Gly-Pro-Arg-Pro) and glycyl-L-prolyl-L-arginyl-sarcosine (Gly-Pro-Arg-Sar) have the highest affinity and are the most potent inhibitors. Gly-Pro-Arg is a weaker binder and a less effective inhibitor. Glycyl-L-prolyl-L-arginyl-L-valine (Gly-Pro-Arg-Val), the naturally occurring tetrapeptide sequence at the amino terminus of the α chain of human fibrin, was found to be the weakest binder and the least effective inhibitor of the Gly-Pro-Arg peptides tested. Studies of many peptide analogues demonstrated that only those peptides containing the ubiquitous sequence Gly-Pro-Arg at the amino terminus are able to exhibit both binding and inhibitory activity.

PEPTIDE ANALOGUES OF THE AMINO TERMINUS OF THE β CHAIN OF HUMAN FIBRIN

Glycyl-L-histidyl-L-arginyl-L-proline (Gly-His-Arg-Pro), the tetrapeptide found at the amino terminus of the β chain of human fibrin, binds to fibrinogen (2 mol/mol) (FIGURE 1, TABLE 1) and to fragment D (1 mol/mol), but does not inhibit polymerization. The tripeptide glycyl-L-histidyl-L-arginine (Gly-His-Arg), which binds to fibrinogen with a much weaker affinity than Gly-His-Arg-Pro (TABLE 1), does not inhibit polymerization either.

Competitive binding studies demonstrated that the binding sites for Gly-His-Arg-Pro, the tetrapeptide found at the amino terminus of the β chain of fibrin, are different from the binding sites for Gly-Pro-Arg-Val, the tetrapeptide found at the amino terminus of the α chain of fibrin.[19] Although both α and β chain peptides have separate and distinct binding sites, some cross-reactivity was observed. In particular, the peptide Gly-Pro-Arg-Pro was found to bind weakly to the Gly-His-Arg-Pro binding sites in addition to its major binding at the two binding sites for Gly-Pro-Arg type peptides.[19] This cross-reactivity is likely due to the structural similarities of the amino termini of the

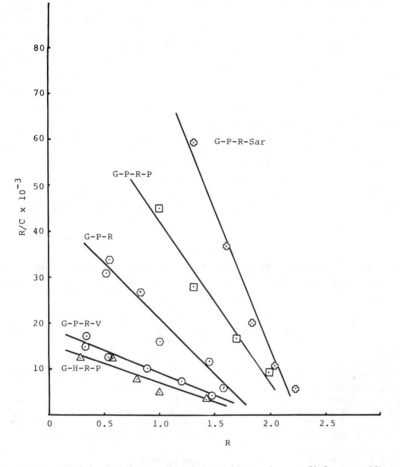

FIGURE 1. Binding of various radioactive peptides to human fibrinogen. All peptides were synthesized with [^{14}C]glycine. Gly-Pro-Arg-Sar (◒); Gly-Pro-Arg-Pro (⊡); Gly-Pro-Arg (⊙); Gly-Pro-Arg-Val (◎); Gly-His-Arg-Pro (△). R = number of moles of peptide bound per mole of fibrinogen; C = concentration of unbound peptide. All studies were performed at room temperature.

α and β chains of fibrin. The fact that the binding sites for peptides from the amino termini of both the α and β chains of fibrin are in fragment D is in complete harmony with a mode of polymerization in which structures exposed by thrombin at the amino terminal central domain interact with the terminal domains of overlapping neighboring molecules.[42]

STUDIES ON LAMPREY FIBRINOGEN

The clotting of lamprey fibrinogen by lamprey thrombin resembles mammalian clotting in that both fibrinopeptides A and B are released. The clotting of lamprey fibrinogen by bovine thrombin, however, proceeds by the exclusive removal of fibrinopeptide B.[20] The β chain of lamprey fibrin thus exposed begins with the sequence glycyl-L-valyl-L-arginyl-L-proline (Gly-Val-Arg-Pro),[14] differing from its mammalian counterpart by the substitution of valine for histidine at position 2. Accordingly, we compared the binding of this and other peptides to lamprey and human fibrinogen. We also tested the peptides for their ability to inhibit the clotting of lamprey and human fibrinogen.

Gly-Val-Arg-Pro, the peptide exposed at the amino terminus of the β chain of lamprey fibrin by bovine thrombin, bound to lamprey fibrinogen (FIGURE 2, TABLE 1) and exhibited a modest inhibitory effect on the clotting of lamprey fibrinogen by bovine thrombin.[19] It did not bind to or inhibit the clotting of human fibrinogen.[19] Gly-His-Arg-Pro, the peptide found at the amino terminus of the β chains of mammalian fibrins bound to both human and lamprey fibrinogen (TABLE 1). Surprisingly, the affinity of Gly-His-Arg-Pro for lamprey fibrinogen was much greater than its affinity for human fibrinogen and greater

TABLE 1

BINDING OF VARIOUS SYNTHETIC PEPTIDES TO HUMAN AND LAMPREY FIBRINOGENS *

Peptide	Human Fibrinogen †		Lamprey Fibrinogen ‡	
	$K(M^{-1})$	n	$K(M^{-1})$	n
Gly	does not bind			
Gly-L-Pro	does not bind			
Gly-L-Pro-L-Arg	2×10^4	1.8		
Gly-L-Pro-L-Arg-L-Pro	4×10^4	2.2	9×10^4	3.0
Gly-L-Pro-L-Arg-Ser	6×10^4	2.2		
Gly-L-Pro-L-Arg-L-Val	1×10^4	2.0		
Gly-L-His-L-Arg-L-Pro	7×10^3	1.8	1×10^5	1.8
Gly-L-Val-L-Arg-L-Pro	does not bind		2×10^4	2.2
Gly-L-Pro-L-Ser-L-Pro	does not bind		does not bind	
L-Ala-L-Pro-L-Arg-L-Pro	does not bind			
Gly-D-Pro-L-Arg	does not bind			
Gly-L-Pro-L-Lys-L-Pro	does not bind			
Gly-Gly-L-Pro-L-Arg-L-Pro	does not bind			
Acetyl-Gly-L-Pro-L-Arg-L-Pro	does not bind			
Acetyl-Gly-L-His-L-Arg-L-Pro	does not bind			
L-Pro-L-Arg-L-Pro	does not bind			
Bromoacetyl-L-Pro-L-Arg-L-Pro	does not bind			
Gly-L-His-L-Arg	$<2 \times 10^3$	§		
Gly-Gly	does not bind			
Gly-L-Pro-L-Arg-L-Pro-methyl ester	2×10^4	1.7		
Gly-L-Pro-L-Arg-methyl ester	2×10^4	1.5		
Gly-L-Pro-L-Arg-L-Val-methyl ester	7×10^3	1.6		

* Most, but not all, of these data were reported in reference 19.
† Temperature $= 22°$ C.
‡ Temperature $= 10°$ C.
§ Binding is too weak to determine n accurately.

than the affinity of Gly-Val-Arg-Pro for lamprey fibrinogen (TABLE 1). However, Gly-His-Arg-Pro does not inhibit the clotting of either lamprey or human fibrinogen by bovine thrombin.[19] Nonradioactive Gly-His-Arg-Pro in a twenty-fold excess markedly diminishes, but does not completely eliminate, the binding of Gly-Val-Arg-Pro to lamprey fibrinogen (FIGURE 2), indicating that these two peptides share common binding sites. The incompleteness of blocking by the more tightly binding Gly-His-Arg-Pro, however, suggests that Gly-Val-

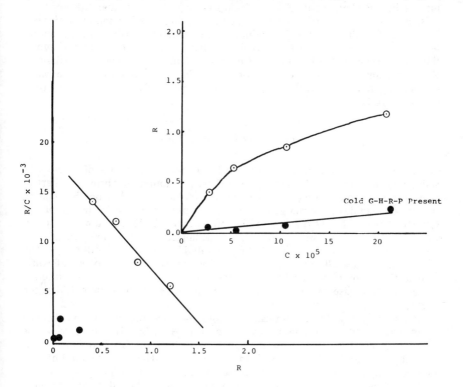

FIGURE 2. Effect of nonradioactive Gly-His-Arg-Pro on the binding of [¹⁴C]Gly-Val-Arg-Pro to lamprey fibrinogen. Inset: [¹⁴C]Gly-Val-Arg-Pro binding to lamprey fibrinogen in the absence of nonradioactive Gly-His-Arg-Pro (⊙). [¹⁴C]Gly-Val-Arg-Pro binding to lamprey fibrinogen in the presence of a 20-fold excess of nonradioactive Gly-His-Arg-Pro (●). The same data are shown in a Scatchard plot in the outer frame. The study was performed at 10° C.

Arg-Pro binds weakly to other sites in addition to its binding to the Gly-His-Arg-Pro binding sites.

Gly-Pro-Arg-Pro binds to lamprey fibrinogen as well as (actually, even better than) it does to human fibrinogen, although there appear to be more than two sites on lamprey fibrinogen (TABLE 1).[19] Peptides beginning with the sequence Gly-Pro-Arg were found to inhibit the clotting of lamprey fibrinogen, interestingly enough, to a greater extent than Gly-Val-Arg-Pro, the actual lamprey sequence exposed by bovine thrombin. The greater inhibition by Gly-

Pro-Arg peptides and the lack of any inhibition by Gly-His-Arg-Pro suggests that during the clotting of lamprey fibrinogen by bovine thrombin the orientation of polymerization is such that Gly-Val-Arg-Pro, the sequence exposed at the amino terminus of the β chain of fibrin, fits into the site that would normally bind to the amino terminus of the α chain of fibrin.

The less specific nature of the binding site for the amino terminus of the β chain in lamprey fibrinogen is evident from its ability to accommodate the mammalian β chain peptide Gly-His-Arg-Pro. It is likely that the apparent existence of more than two Gly-Pro-Arg-Pro binding sites in lamprey fibrinogen is because Gly-Pro-Arg-Pro is binding to sites that ordinarily bind the amino terminus of the β chain of lamprey fibrin. It may be that the common evolutionary ancestor of the lamprey and mammals had the sequence Gly-His-Arg-Pro at the amino terminus of the fibrin β chain, since Gly-His-Arg-Pro binds to lamprey fibrinogen with an association constant five times greater than Gly-Val-Arg-Pro.

EFFECT OF CALCIUM IONS

In view of the binding of calcium to fibrinogen,[21] the effects of calcium in stabilizing the structure of fibrinogen,[22-25] its promotion of fibrin polymerization [26, 27] and cross-linking,[28] we studied the effect of this ion on the binding of various peptides to fibrinogen. There is a tenfold increase in the affinity of Gly-His-Arg-Pro for fibrinogen in the presence of 2mM calcium ions (FIGURE 3).[29] The number of binding sites is unchanged, however; magnesium ions have virtually no effect on the binding of Gly-His-Arg-Pro (FIGURE 4). The α chain analogues Gly-Pro-Arg-Val and Gly-Pro-Arg-Pro do not significantly change their affinity for fibrinogen in the presence of calcium ions, although there is an increase in the number of binding sites for both peptides in the presence of calcium (FIGURE 3).[29]

These results indicate that calcium ions enhance the affinity of the site which binds to the amino-terminus of the fibrin β chain. The increase in the number of sites for the α chain analogues beginning with Gly-Pro-Arg . . . is evidently due to these peptides fitting into the Gly-His-Arg-Pro binding site as a consequence of the increased affinity of that site. This interpretation was confirmed by the finding that in the presence of a 20-fold excess of non-radioactive Gly-His-Arg-Pro, to prevent binding of α chain analogues to its site, there is no longer any significant increase in the number of binding sites for Gly-Pro-Arg-Pro in the presence of calcium (FIGURE 3).

The influence of calcium is therefore predominantly on the Gly-His-Arg-Pro (β chain type) binding site. In the absence of calcium this site is apparently much less accessible. Calcium ions have no influence on the binding site for α chain peptides. The increase in the number of binding sites for Gly-Pro-Arg-Pro in the presence of calcium is attributable to its binding to the β chain type site.

Recently Furlan et al.[30] have compared the effects of various fibrin amino-terminal peptides and calcium on the rates of fibrin polymerization induced by the enzymes thrombin, batroxobin and the *Agkistrodon contortrix* thrombin-like enzyme (ACTE). Thrombin removes both fibrinopeptides A and B, whereas batroxobin removes only fibrinopeptide A. In contrast, ACTE clots human fibrinogen at low temperatures by the preferential removal of fibrino-

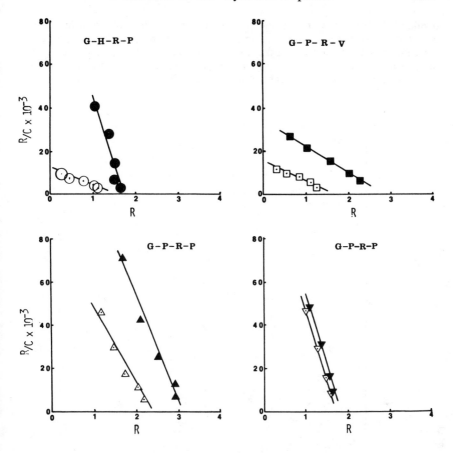

FIGURE 3. Effect of calcium on the binding of various radioactive peptides to human fibrinogen. Calcium-free fibrinogen solutions were prepared by dissolving human fibrinogen to a concentration of 13 mg/ml in 0.3 M NaCl, 0.005 M imidazole buffer, pH 7.2, containing 0.005 M EGTA. The solutions were dialyzed three times against 0.2 M NaCl, 0.01 M imidazole, pH 7.2; Hirudin (Sigma, Grade IV, 1400 U/mg) was added to a final concentration of 0.05 mg/ml. Aliquots of the fibrinogen solution (0.8 ml) were then dialyzed 10 hrs. versus 5 ml. of various peptide solutions prepared in 0.2 M NaCl 0.01 M imidazole, pH 7.2, both in the presence (closed symbols) and absence (open symbols) of 2 mM calcium chloride. Binding of radioactive peptides was determined by liquid scintillation counting of duplicate 0.2 ml aliquots from the inside and outside of each bag after equilibration for 10 hr. (Upper left) [^{14}C]Gly-His-Arg-Pro binding (●) 2 mM CaCl$_2$; (⊙) no CaCl$_2$. (Upper right) [^{14}C]Gly-Pro-Arg-Val binding (■) 2 mM CaCl$_2$; (⊡) no CaCl$_2$. (Lower left) [^{14}C]Gly-Pro-Arg-Pro binding (▲) 2 mM CaCl$_2$; (△) no CaCl$_2$. (Lower right) [^{14}C]Gly-Pro-Arg-Pro binding (▼) 2 mM CaCl$_2$, 20 × excess nonradioactive Gly-His-Arg-Pro; (▽) no CaCl$_2$, 20 × excess nonradioactive Gly-His-Arg-Pro. R = number of moles of peptide bound per mole of fibrinogen. C = concentration of unbound peptide. All studies were performed at room temperature (t = 22° C). (From Laudano & Doolittle.[20] By permission of *Science*.)

peptide B.[31] It has been found that calcium accelerates the clotting of fibrinogen by all three enzymes.[30] Since batroxobin does not remove fibrinopeptide B, the enhancement of fibrin reaggregation cannot be due solely to the increased binding affinity for the amino terminus of the β chain in the presence of calcium. Recent electron microscopic studies have demonstrated changes in the

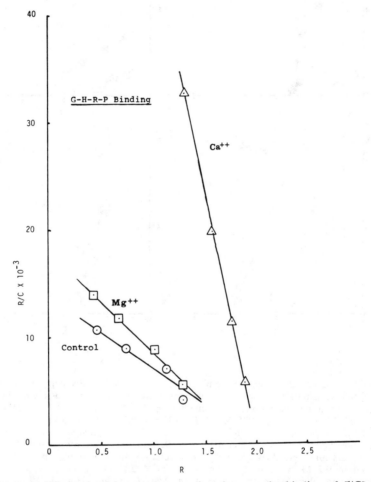

FIGURE 4. Effect of calcium and magnesium ions on the binding of [14C] Gly-His-Arg-Pro to human fibrinogen. (○) no calcium or magnesium; (△) 2 mM calcium ions; (▫) 2 mM magnesium ions.

structure of fibrinogen in the presence of calcium.[10] These structural alterations may be responsible for both the increased affinity for Gly-His-Arg-Pro and the acceleration of polymerization in the presence of calcium.

Furlan et al.[30] showed that at physiological concentrations of calcium ions, Gly-His-Arg inhibits the polymerization of fibrin in which only fibrinopeptide B is released by ACTE. This inhibition was expected since Gly-His-Arg is the tripeptide sequence that is exposed at the amino terminus of the β chain by ACTE. Neither Gly-His-Arg nor Gly-His-Arg-Pro inhibit the polymerization of fibrin induced by thrombin or batroxobin,[30] consistent with our earlier findings.[18, 19] Moreover, at low concentrations of calcium ions these peptides actually increased the rate of thrombin- and batroxobin-induced fibrin polymerization.[30] We had also observed an increase in initial turbidity when reaggregation was performed in the presence of Gly-His-Arg-Pro.[32] Furlan *et al.* have proposed that the binding of Gly-His-Arg induces a conformational change that facilitates polymerization.[30] An alternative explanation that stems from the observation that α chain peptides can bind to the Gly-His-Arg-Pro binding site is that the rate enhancement by Gly-His-Arg peptides at low calcium concentrations may be due to their preventing binding of the amino termini of the α chain of fibrin to the site to which the amino terminus of the β chain of fibrin ordinarily binds. By preventing binding of the amino terminus of the fibrin α chain to the "wrong site", Gly-His-Arg and Gly-His-Arg-Pro may promote the correct orientation for polymerization of fibrin monomers.

It was found, unexpectedly, by Furlan *et al.*[30] that Gly-Pro-Arg is a better inhibitor of ACTE-induced fibrin polymerization than is Gly-His-Arg or Gly-His-Arg-Pro, just as occurred in our experiments with the lamprey fibrinogen.[19] Since Gly-Pro-Arg is not exposed by ACTE, which preferentially removes fibrinopeptide B, it was suggested that the inhibitory effects of Gly-Pro-Arg may be due to the induction of a conformational change upon the binding of this peptide.[30] In view of the structural similarities of the amino termini of the α and β chains of fibrin, however, two other explanations merit consideration. First, some of the inhibitory activity of Gly-Pro-Arg is likely due to its fitting into the Gly-His-Arg binding site. Secondly, when the clotting of fibrinogen is unnaturally induced by enzymes which preferentially remove fibrinopeptide B, polymerization may proceed by a mode in which sites at the amino terminus of the β chain of fibrin interact with the sites that would normally bind to Gly-Pro-Arg at the amino terminus of the α chain of fibrin. This situation would account for the greater inhibitory activity of Gly-Pro-Arg peptides.

The role of fibrinopeptide B release in fibrin formation has been the subject of considerable investigation in the past. It was proposed long ago that the release of fibrinopeptide A may result in end-to-end polymerization leading to the formation of intermediate polymers, while subsequent removal of fibrinopeptide B could lead to lateral aggregation of these intermediate polymers.[33] Clotting by the exclusive removal of fibrinopeptide B was demonstrated initially in lamprey fibrinogen with the use of bovine thrombin,[20] and more recently in the case of human fibrinogen treated with ACTE at 14°.[31]

In order to assess the participation of the binding sites for the amino terminus of the β chain of fibrin when human fibrinogen is clotted by a mammalian thrombin, we tested the inhibitory activity of peptide analogues of the α and β chains of fibrin individually and in combination in thrombin-fibrinogen clotting assays both in the presence and absence of calcium (TABLE 2).

In accordance with previous observations, calcium shortens the clotting time, regardless of the presence of peptides. In the absence of calcium, however, Gly-Pro-Arg-Pro is a relatively more potent inhibitor than in its presence (TABLE 2). Although Gly-His-Arg-Pro by itself is not a significant inhibitor, whether or not calcium is present, when it is used in combination with Gly-Pro-Arg-Pro clotting is significantly delayed compared to when Gly-Pro-Arg-Pro alone is used (TABLE 2). This effect is more apparent in the presence of calcium. These results are consistent with the involvement of specific sites that bind to Gly-His-Arg-Pro during fibrin polymerization.

BINDING OF PEPTIDES TO FRAGMENTS D AND E

In order to localize the binding sites for the amino terminal peptides of the α and β chains of fibrin, Gly-Pro-Arg-Pro and Gly-His-Arg-Pro were tested for their ability to bind to the plasmin-generated fragments of fibrinogen. Neither peptide bound to fragment E, but both peptides bound to fragment D. We found, however, that the binding of Gly-Pro-Arg-Pro to fragment D only occurred when the fragment "D" was that species referred to variously as "big D" or "early D" or "D_1". This fragment is most readily generated by

TABLE 2

INHIBITION OF CLOTTING BY GLY-HIS-ARG-PRO AND GLY-PRO-ARG-PRO *

Peptides	Thrombin/Fibrinogen Clotting Time (Seconds)	
	2 mM CaCl₂ Present	CaCl₂ Absent
Control	13	20
Gly-His-Arg-Pro (10 mM)	15	18
Gly-Pro-Arg-Pro (1 mM)	49	180
Gly-Pro-Arg-Pro (1 mM) + Gly-His-Arg-Pro (10 mM)	113	200

* Final Conditions: Human fibrinogen $= 1 \times 10^{-5}$ M, Thrombin $= 5$ units/ml, 0.15M NaCl, 0.005M Imidazole, pH $= 7.2$. Temperature $= 22°$ C.

maintaining the presence of calcium throughout the preparation of fragment D. In contrast, the binding of Gly-His-Arg-Pro did not vary significantly when fragment D was prepared either in the presence or absence of calcium. Haverkate and Timan [25] have shown that calcium stabilizes the digestion of fibrinogen and fibrin by plasmin. They found that fragment D prepared in the presence of calcium has a molecular weight of 93,000 while fragment D prepared in the presence of EGTA has a molecular weight of only 80,000. The difference in size is due to the loss of a 13,000 molecular weight fragment from the carboxy terminus of the γ chain during plasmin digestion in the absence of calcium.

Fragment D prepared by the traditional Nussenzweig method,[35] in which calcium is neither added nor removed by chelating agents, exhibits some size heterogeneity and had one binding site for both Gly-Pro-Arg-Pro and Gly-His-Arg-Pro. In order to obtain large quantities of homogeneous fragments D_1 and D_3, we modified the original Nussenzweig method so as to include either calcium ions, on the one hand, or suitable chelating agents, on the other, throughout the preparation of D_1 and D_3, respectively.

The affinity of Gly-Pro-Arg-Pro for the larger fragment D_1 prepared in the presence of calcium is much greater than its affinity for the smaller D_3 prepared in the absence of calcium (FIGURE 5, TABLE 3). The number of binding sites for Gly-Pro-Arg-Pro also appeared to be greater for D_1 (n = 1.6)

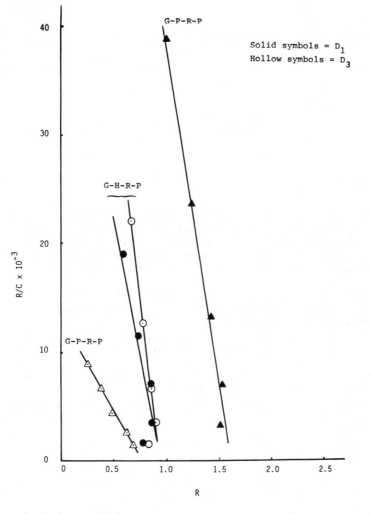

FIGURE 5. Binding of [^{14}C]Gly-Pro-Arg-Pro (▲ and △) and [^{14}C]Gly-His-Arg-Pro (● and ☉) to D_1 (solid symbols) and D_3 (hollow symbols). The study was performed at room temperature = 22° C in the presence of 2 mM calcium ions.

than for D_3 (n = 0.8). The weaker binding of Gly-Pro-Arg-Pro to D_3, however, was completely abolished when its binding was measured in the presence of excess nonradioactive Gly-His-Arg-Pro (FIGURE 6, TABLE 3) indicating that there are no Gly-Pro-Arg type binding sites on fragment D_3. Furthermore,

TABLE 3

BINDING PROPERTIES OF VARIOUS FORMS OF FRAGMENT D PREPARED
UNDER DIFFERENT CONDITIONS

Type of Fragment D	Preparative Conditions	[^{14}C]G-H-R-P Binding		[^{14}C]G-P-R-P Binding		[^{14}C]G-P-R-P Binding in the Presence of Nonradioactive G-H-R-P	
		$K(M^{-1})$	n	$K(M^{-1})$	n	$K(M^{-1})$	n
D	No Ca^{2+} or EGTA added	5.3×10^4	1.0	5.9×10^4	1.0		
D$_1$	Ca^{2+} added	5.2×10^4	0.9	6.2×10^4	1.6	1.1×10^5	1.0
D$_3$	EGTA added	8.3×10^4	0.9	1.7×10^4	0.8	No binding	

in the presence of nonradioactive Gly-His-Arg-Pro the number of Gly-Pro-Arg-Pro binding sites for fragment D$_1$ was reduced to a single site (FIGURE 6, TABLE 3). Therefore, both the weak binding of Gly-Pro-Arg-Pro to fragment D$_3$ and its additional binding over and above one site to fragment D$_1$ are attributable to its weak affinity for the Gly-His-Arg-Pro binding site.

In contrast, there was no decrease in either the affinity or binding of Gly-His-Arg-Pro to fragment D$_3$ (FIGURE 5, TABLE 3). If anything, the affinity of Gly-His-Arg-Pro for fragment D$_3$ is slightly greater than its affinity for D$_1$, although not dramatically so.

These studies provide clues as to the location of the binding sites for both peptides in the fragment D domain. The lack of a binding site for Gly-Pro-Arg-Pro in fragment D$_3$ indicates that the carboxy-terminus of the γ chain is critical to the binding of Gly-Pro-Arg type peptides.[34] In this regard, Olexa and Budzynski [36] have isolated a peptide from the carboxy-terminus of the γ chain of fragment D that is a potent inhibitor of fibrin polymerization and which binds to sites that are exposed by the release of fibrinopeptide A. Since the binding of Gly-His-Arg-Pro to D$_3$ is not appreciably different from its binding to D$_1$, the carboxy terminus of the γ chain is not likely to be involved in binding this peptide. In view of the structural similarities between Gly-His-Arg-Pro and Gly-Pro-Arg-Pro and the homology of the β and γ chains,[37, 38] it is likely that the Gly-His-Arg-Pro binding site is situated in the other subdomain of fragment D on a portion of the β chain which is homologous to the γ chain.

USES OF SYNTHETIC FIBRIN AMINO-TERMINAL PEPTIDES

Thus far we have focussed on the use of the peptides as probes for investigating the nature of the binding sites involved in fibrin polymerization. The approach that we have employed involves the binding of assorted peptide analogues to fibrinogen and various proteolytically derived fragments and the inhibitory properties of these peptides. Studies on the structure and physical properties of clots formed in the presence of Gly-His-Arg-Pro may provide further information as to the role of fibrinopeptide B release. These peptides

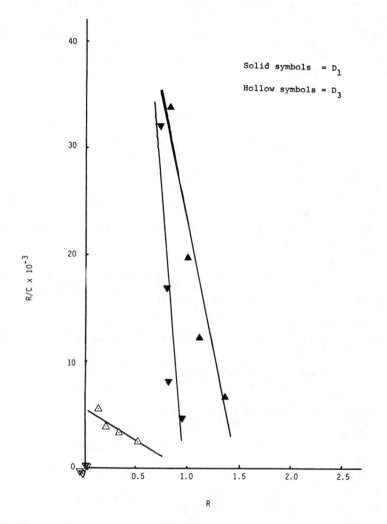

FIGURE 6. Effect of excess nonradioactive Gly-L-His-L-Arg-L-Pro on the binding of [^{14}C]Gly-L-Pro-L-Arg-L-Pro to D_1 (solid symbols and D_3 (hollow symbols) showing that primary binding site for G-P-R-P is completely lost during conversion of D_1 to D_3. [^{14}C]Gly-L-Pro-L-Arg-L-Pro binding to D_1 in the absence of nonradioactive Gly-L-His-L-Arg-L-Pro (▲). Binding of [^{14}C] Gly-L-Pro-L-Arg-L-Pro to D_1 in the presence of a 20-fold excess of nonradioactive Gly-L-His-L-Arg-L-Pro (▼). Binding of [^{14}C]Gly-L-Pro-L-Arg-L-Pro to D_3 in the absence of nonradioactive Gly-L-His-L-Arg-L-Pro (△). Binding of [^{14}C]Gly-L-Pro-L-Arg-L-Pro to D_3 in the presence of a nonradioactive Gly-L-His-L-Arg-L-Pro (▽). The study was performed at room temperature $= 22°$ C in the absence of calcium.

may also be useful for affinity labelling studies, and efforts in this direction have been undertaken.[39]

The α chain analogues, Gly-Pro-Arg-Pro in particular, have proved useful in the preparation of thrombin-degranulated platelets [40] and in studies of coagulation proteins such as factor VIII,[41] in which cases fibrin formation can interfere with the conduct of experiments. These synthetic peptides may also be useful in various clinical chemistry determinations on plasma where currently used anticoagulants are undesirable. Beyond that, their potential therapeutic use as anticoagulants should not be underestimated.

REFERENCES

1. BETTEHEIM, F. R. & K. BAILEY. 1952. Biochim. Biophys. Acta 9: 578–579.
2. FERRY, J. D. 1952. Proc. Natl. Acad. Sci. USA 38: 566–569.
3. BLOMBÄCK, B., M. BLOMBÄCK, A. HENSCHEN, B. HESSEL, S. IWANAGA & K. R. WOODS. 1968. Nature (London) 218: 130–134.
4. MARDER, V. J. 1971. Scand. J. Haematol. (Suppl.) 13: 21–36.
5. TELFORD, J. N., J. A. NAGY, P. A. HATCHER & H. A. SCHERAGA. 1980. Proc. Natl. Acad. Sci. USA 77: 2372–2376.
6. PRICE, T. M., D. D. STRONG, M. L. RUDEE & R. F. DOOLITTLE. 1981. Proc. Natl. Acad. Sci. USA 78: 200–204.
7. NORTON, P. A. & H. S. SLAYTER. 1981. Proc. Natl. Acad. Sci. USA 78: 1661–1665.
8. HALL, C. E. & H. D. SLAYTER. 1959. J. Biophys. Biochem. Cytol. 5: 11–16.
9. FOWLER, W. E. & H. P. ERICKSON. 1979. J. Mol. Biol. 134: 241–249.
10. WILLIAMS, R. C. 1981. J. Mol. Biol. 150: 399–408.
11. MOSESSON, M. W., J. HAINFELD, J. WALL & R. M. HASCHEMEYER. 1981. J. Mol. Biol. 153: 695–718.
12. KUDRYK, B. J., D. COLLEN, K. R. WOODS & B. BLOMBÄCK. 1974. J. Biol. Chem. 249: 3322.
13. IWANAGA, S., P. WALLEN, N. J. GRONDAHL, A. HENSCHEN & B. BLÖMBACK. 1967. Biochim. Biophys. Acta 147: 606–609.
14. COTTRELL, B. A. & R. F. DOOLITTLE. 1976. Biochim. Biophys. Acta 453: 426–438.
15. BIRKIN, S., G. D. WILNER & R. E. CANFIELD. 1975. Thromb. Res. 7: 599–610.
16. MURANO, G., D. WALZ, L. WILLIAMS, J. PINDYEK & M. W. MOSESSON. 1977. Thromb. Res. 11: 1–10.
17. BLOMBÄCK, M., B. BLOMBÄCK, E. F. MAMMEN & A. S. PRASAD. 1968. Nature 218: 134–137.
18. LAUDANO, A. P. & R. F. DOOLITTLE. 1978. Proc. Natl. Acad. Sci. USA 75: 3085–3089.
19. LAUDANO, A. P. & R. F. DOOLITTLE. 1980. Biochemistry 19: 1013–1019.
20. DOOLITTLE, R. F. 1965. Biochem. J. 94: 735–741.
21. MARGUERIE, G., G. CHAGNIEL & M. SUSCILLON. 1977. Biochim. Biophys. Acta 490: 94–103.
22. LY, B. & H. C. GODAL. 1973. Haemostasis 1: 204–409.
23. KOMENKO, A. K. & V. A. BELITSER. 1963. Ukr. Biokhim. Zh. 35: 829–833.
24. LUGUVSKOI, E. V., T. M. POZDNYAKOVA, T. V. VARETSKA, S. G. DERZSKAIA & V. M. TOLSTYCH. 1976. Ukr. Biokhim. Zh. 48: 743–747.
25. HAVERKATE, F. & G. TIMAN. 1977. Thromb. Res. 10: 803–812.
26. BOYER, M. H., J. R. SHAINOFF & O. D. RATNOFF. 1972. Blood 39: 382–387.
27. ENDRES, G. F. & H. A. SCHERAGA. 1972. Arch. Biochem. Biophys. 153: 266–278.
28. LORAND, L. & K. KONISHI. 1964. Arch. Biochem. Biophys. 105: 58–67.

29. LAUDANO, A. P. & R. F. DOOLITTLE. 1981. Science **212:** 457–459.
30. FURLAN, M., C. RUPP, E. A. BECK & L. SVENDSEN. 1982. Thromb. Haemostasis (Stuttgart) **47:** 118–121.
31. SHAINOFF, J. R. & B. N. DARDIK. 1979. Science **204:** 200–202.
32. LAUDANO, A. P. & R. F. DOOLITTLE. 1978. Proc. Natl. Acad. Sci. USA **75:** 3085–3089.
33. LAURENT, T. C. & B. BLOMBÄCK. 1958. Acta Chem. Scand. **12:** 1875–1877.
34. DOOLITTLE, R. F. & A. P. LAUDANO. 1980. Protides Biol. Fluids Proc. Colloq. **28:** 311–316.
35. NUSSENZWEIG, V., M. SELIGMAN, J. PELMONT & P. GRABER. 1961. Ann. Inst. Pasteur (Paris) **100:** 377–389.
36. OLEXA, S. A. & A. BUDZYNSKI. 1981. J. Biol. Chem. **256:** 3544–3549.
37. HENSCHEN, A. & F. LOTTSPEICH. 1977. Thromb. Res. **11:** 869–880.
38. WATT, K. W. K., T. TAKAGI & R. F. DOOLITTLE. 1978. Proc. Natl. Acad. Sci. USA **75:** 1731–1735.
39. HSIEH, K., M. S. MUDD & G. D. WILNER. 1981. J. Med. Chem. **24:** 322–327.
40. HARFENIST, E. J., M. A. GUCCIONE, M. A. PACKHAM & J. F. MUSTARD. 1982. Blood **59:** 952–955.
41. WEINSTEIN, M. J., Personal communication.
42. DOOLITTLE, R. F. 1973. Adv. Prot. Chem. **25:** 1–109.

DISCUSSION OF THE PAPER

S. NIEWIAROWSKI (*Temple University Health Science Center, Philadelphia, PA*): Is it possible using your peptides to stop polymerization at the level of oligomers?

A. LAUDANO: As assessed by ultracentrifugation, there is no dimerization if you use enough Gly-Pro-Arg-Pro. You need a high concentration in order to achieve that kind of effectiveness, however. Perhaps both peptides together may achieve a similar effect at lower concentrations.

A. BUDZYNSKI (*Temple University Health Science Center, Philadelphia, PA*): I have difficulty understanding the quantitative data on the binding of Gly-Pro-Arg-Pro to lamprey fibrinogen in the presence of calcium. From your binding data I surmise that there are three sites with essentially the same binding quality. One possibility is that, in excess, the peptide can bind to another binding site, for instance, Gly-His-Arg. But even if this might be the case, one would expect two combinations. How do you interpret three binding sites of the same class?

LAUDANO: I believe there are not really three binding sites: there are four for Gly-Pro-Arg-Pro. I think the problem comes when one does a Scatchard analysis. What happens is that one extracts a biphasic extrapolation. When both binding sites are acting together and are not sufficiently separated in terms of affinity—rather than achieve a biphasic extrapolation—what one gets is a sort of averaging that extrapolates to three binding sites.

INTERACTION OF THROMBIN AND FIBRINOGEN AND THE POLYMERIZATION OF FIBRIN MONOMER *

Harold A. Scheraga

Baker Laboratory of Chemistry
Cornell University
Ithaca, New York 14853

INTRODUCTION

Generation of a fibrin clot from purified thrombin and fibrinogen is initiated by a proteolytic reaction and proceeds by means of three *reversible* [1-5] steps: the thrombin-induced conversion of fibrinogen to fibrin monomer, the formation of intermediate polymers from fibrin monomer, and the incorporation of the intermediate polymers from fibrin gel network.[1-7] Thrombin is a proteolytic enzyme,[8-11] homologous to trypsin, chymotrypsin and elastase,[12-14] and this homology has been used to suggest a three-dimensional structure of thrombin.[15, 16] Several of its physical properties such as its molecular weight ($\sim$40,000), tendency toward aggregation, and titration behavior have been reported.[17-21] Physicochemical studies of fibrinogen [3, 6, 7, 22-24] indicate that its molecular weight is $\sim$340,000, and that it is an asymmetric elongated molecule with an axial ratio of $\sim$5:1. Electron microscopy [25-33] confirms this view, and provides the additional detail that the molecule consists of three interconnected linearly arranged spheres. The complete amino acid sequence of human fibrinogen [7, 34] and part of that of bovine fibrinogen [35-38] are known. The molecule is a disulfide-linked dimer of the form [24, 39-45] $(A\alpha, B\beta, \gamma)_2$, and Doolittle and his colleagues have used the sequence and electron-microscopy information to propose a detailed model of fibrinogen.[34]

Fibrinogen and fibrin can be degraded in recognizable stages by plasmin, and Francis and Marder have recently presented a molecular model for this process.[46] The results of these studies imply that the molecule consists of a series of linked domains, as had previously been concluded from differential scanning calorimetric observations on fibrinogen and on plasmin degradation products of fibrinogen.[47] This conclusion is consistent with the results of electron microscopy and with current thinking [34] about the possible three-dimensional structure of fibrinogen.

This paper is concerned with the kinetics of the proteolytic action of thrombin on fibrinogen to release fibrinopeptides A and B from the Aα and Bβ chains, respectively, and the mechanism of formation of the intermediate polymers from the resulting fibrin monomer. The latter is defined here as a fibrinogen molecule that lacks two fibrinopeptide A and two fibrinopeptide B segments. This definition differs somewhat from that of Nossel et al.[48] and Blombäck et al.,[49] who distinguish between clottable species devoid of fibrinopeptide A, and of both fibrinopeptides A and B, respectively.

* This work was supported by research grants from the National Heart, Lung and Blood Institute of the National Institutes of Health, U.S. Public Health Service (HL–01662) and from the National Science Foundation (PCM79–20279).

LIMITED PROTEOLYSIS

The reversibility observed [5] for the action of thrombin on fibrinogen appears to be a general property of limited proteolytic reactions. For example, reversibility is observed in the action of trypsin on both trypsin inhibitor [50] and hemoglobin.[51]

The mechanism of action of proteolytic enzymes has been studied by various techniques, e.g., by an active-site mapping approach.[52] This method has been used [53] to investigate the interactions between thrombin and the portions of the Aα and Bβ chains of fibrinogen that contain the Arg-Gly bonds that are hydrolyzed by the enzyme. The kinetics of hydrolysis of these bonds in intact fibrinogen [54] and in various fragments of fibrinogen [55-63] are expressed in terms of the kinetic constants k_{cat} and K_m. These quantities are shown in TABLE 1 for a selection of substrates investigated.[54-65]

It can be seen from these data that short peptides such as 15–17 are very poor substrates. The values of k_{cat} for the synthetic peptides do not approach that for the Aα chain of intact fibrinogen until a Phe is incorporated 9 residues on the N-terminal side of the Arg-Gly bond (cf. substrates 6, 11, 13, 14 and the Aα chain). The importance of this Phe residue had been suggested earlier by Blombäck,[44] and later by kinetic studies of the rate of the thrombin-catalyzed hydrolysis of chromogenic substrates [66] and by NMR studies of the structures of these substrates.[67] Comparison of substrates 6, 11, 13 and 14 of TABLE 1 indicates that residues on the C-terminal side of Pro do not influence k_{cat} to a great extent. However, it has been suggested [63, 68] that an Asp . . . Arg salt link (formed between residues outside of those of substrate 14) may contribute (see below). The influence of Phe on K_m is less clear since the values of K_m for those synthetic substrates with the higher values of k_{cat} are still larger than those for the Aα chain of intact fibrinogen.

Various hypotheses have been explored [53] to account for the apparently large size of the substrate (extending from at least Phe to Pro) that interacts with thrombin, and to explain the trend seen in the experimental values of K_m. A plausible one under current consideration is that there may be a bend in this portion of the Aα chain that brings the Phe residue close [67, 69] to the Arg-Gly bond. This suggestion is made because there is a type II β-turn in the Pro-Ala portion [69] (see FIGURE 1), and there may be a similar bend near the Gly-Gly-Gly segment. The possibility that a specific conformation is required for this portion of the Aα chain to interact with thrombin (and account for its high specificity) was examined [68] with an immunochemical technique. In this method, the conformation of peptides containing the hydrolyzable Arg-Gly bond was expressed in terms of an equilibrium constant, K_{conf}, which indicates the degree to which an excised peptide fragment has the same conformation that it had in the intact fibrinogen molecule. The antigenic determinant investigated by this procedure is localized among the residues of peptide 10 of TABLE 1.[68] The value of K_{conf} for this region in intact fibrinogen is very large, indicating that the antibodies used in this method recognize intact fibrinogen as a native molecule. As the molecule is degraded progressively to the disulfide knot (DSK),[64] to CNBr Aα, and to shorter peptides, the values of K_{conf} decreases, in a manner similar to the increase in K_m (see TABLE 1). This indicates that the character of the native antigenic determinant is disrupted to different degrees as the fragments decrease in size, and

TABLE 1

KINETIC CONSTANTS FOR HYDROLYSIS OF ARG-GLY BONDS BY THROMBIN (AT pH 8.0 AND 25°)* AND CONFORMATIONAL EQUILIBRIUM CONSTANTS (AT pH 8.3 AND 4° C)†

Substrate	$k_{cat} \times 10^{11}$ $M[(NIH\ U/L)s]^{-1}$ ‡	$K_m \times 10^6$ M ‡	K_{conf} ‡
1. Aα chain of intact fibrinogen	73	9	Very Large
2. Bβ chain of intact fibrinogen	12	11	
3. DSK §	¶	¶	
4. CNBr Aα §	48	47	4×10^{-3}
5. CNBr Bβ §	6	189	5×10^{-5}
6. Ac-Phe-Leu-Ala-Glu-Gly-Gly-Val-Arg-Gly-Pro-Arg-Val-Val-Glu-Arg-NHCH₃	11	680	
7. Ac-Leu-Ala-Glu-Gly-Gly-Val-Arg-Gly-Val-Arg-Gly-Pro-NHCH₃	‖	‖	
8. Ac-Ala-Glu-Gly-Gly-Val-Arg-Gly-Pro-Arg-Val-Val-Glu-Arg-NHCH₃	0.3	1560	
9. Ac-Gly-Gly-Val-Arg-Gly-Pro-Arg-Val-Val-Glu-Arg-NHCH₃	0.3	630	
10. H-Gly-Gly-Val-Arg-Gly-Pro-Arg-Val-Val-Glu-NHCH₃			6×10^{-7}
11. Ac-Phe-Leu-Ala-Glu-Gly-Val-Arg-Gly-Pro-Arg-Val-Val-Glu-NHCH₃	16	789	
12. H-Phe-Leu-Ala-Glu-Gly-Gly-Val-Arg-Gly-Pro-Arg-Val-Val-Glu-NHCH₃			11×10^{-7}
13. Ac-Phe-Leu-Ala-Glu-Gly-Gly-Val-Arg-Gly-Pro-Arg-Val-NHCH₃	20	633	
14. Ac-Phe-Leu-Ala-Glu-Gly-Gly-Val-Arg-Gly-Pro-NHCH₃	11	934	
15. H-Gly-Val-Arg-Gly-Pro-Arg-Leu-OH	0.5	3700	
16. H-Gly-Val-Arg-Gly-Gly-Arg-Leu-OH	0.2	9600	
17. H-Gly-Val-Arg-Gly-Pro-Gly-Leu-OH	0.1	15300	

* From references [54–63]. † From reference [68]. ‡ Though not shown here (but reported in the original papers), the experimental errors in these quantities are very large. § DSK is the disulfide knot produced by CNBr cleavage of fibrinogen.[64] Reduction and alkylation of the disulfide bonds of the DSK produces CNBr Aα, CNBr Bβ and CNBr γ. ¶ No values of k_{cat} or of K_m have been reported for the DSK although it is known that thrombin will release fibrinopeptide A from the DSK at nearly the same rate as it releases this peptide from the intact fibrinogen molecule.[65] ‖ While the values of k_{cat} and K_m could not be determined, this peptide is hydrolyzed by thrombin much more slowly than substrate 14.

suggests that a unique (native) conformation of this portion of the Aα chain is required for proper binding to thrombin. Long-range interactions (disrupted in these derivatives but present in the whole fibrinogen molecule) may be required to maintain this unique conformation. As indicated above, it is possible that a peptide, similar to peptide 13 but with an Asp residue inserted between Ac and Phe at the N-terminus, may adopt a hairpin-like conformation, stabilized by one of these long-range interactions, viz., a salt link between this additional Asp residue and the Arg residue near the C-terminus.

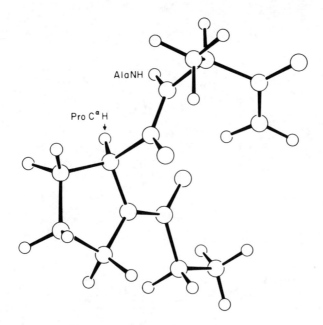

FIGURE 1. Illustration of the type II β-turn in the Pro-Ala segment.[69] In the series of peptides used for the NMR studies of conformation, Ala was substituted for Arg.[69]

FORMATION OF INTERMEDIATE POLYMERS

The limited proteolysis by thrombin releases fibrinopeptides A and B and presumably uncovers polymerization sites in the resulting fibrin monomer so that it can polymerize to form larger structures. It appears that the release of only fibrinopeptide A is sufficient to initiate polymerization, and that the release of fibrinopeptide B may uncover additional polymerization sites.[70-73] Immunoelectron microscopy [74] reveals that the site of thrombin action is in the central nodule (see FIGURE 2), and that residues 240–424 of the Aα chains [75] are in the two outer nodules (see FIGURE 3). Similar results have been reported by Price and coworkers [76] and by Norton and Slayter,[77] who used immunoelectron microscopy to localize the D and E domains to the outer and inner nodules, respectively.

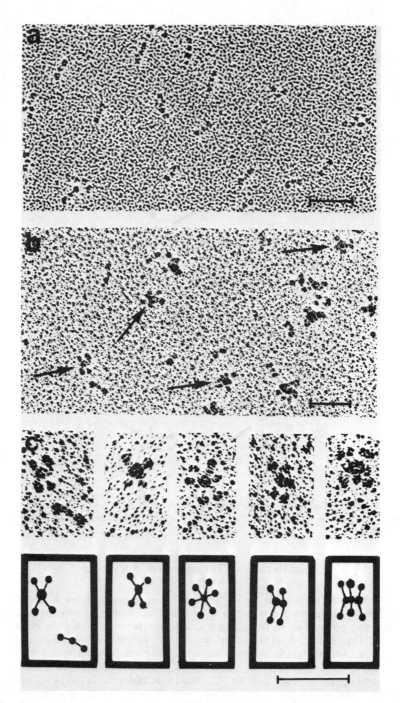

FIGURE 2. Electron micrographs [74] (bar = 1000Å). (a) Bovine fibrinogen. (b) Complexes of bovine fibrinogen with purified anti-DSK antibodies. Arrows indicate fibrinogen-antibody complexes. (c) Fibrinogen-antibody complexes, with an interpretive drawing given below each frame.

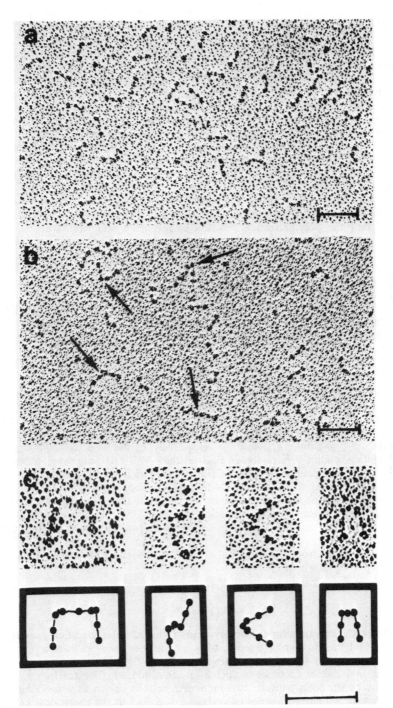

FIGURE 3. Electron micrographs [74] (bar = 1000Å). (a) Human fibrinogen. (b) Complexes of human fibrinogen with antibodies to a plasmin-degradation product from the C-terminal region of the Aα chain. Arrows indicate fibrinogen-antibody complexes. (c) Fibrinogen-antibody complexes showing individual fibrinogen molecules with attached antibody, and antibody-linked fibrinogen complexes, with an interpretive drawing given below each frame.

Physico-chemical studies [78] indicate that fibrin monomer is indistinguishable from fibrinogen in its overall size and shape, and electron micrographs [28, 31, 32] confirm this. In addition, physico-chemical studies indicate that fibrin monomer polymerizes to form rod-like structures with a staggered overlap,[2, 79-81] and electron micrographs [26-28, 32, 82-84] support this view. FIGURE 4 illustrates some of the polymers formed during the early stages of polymerization. A schematic representation of the staggered overlap (possibly involving hydrogen bond donor and acceptor groups; see below) is shown [85] in FIGURE 5. This figure also illustrates how a copolymer of fibrinogen (F) and fibrin monomer (f) can inhibit polymerization. This view is supported by Brass and coworkers [86] but not by Smith.[87]

These polymers can be dissociated by dilution and also by heating.[1, 2] This suggests that the bonding of the monomer units is weak, i.e. noncovalent, and that the polymerization reaction is reversible and exothermic. In fact, this is one of the few protein association reactions that is exothermic, most others being endothermic (i.e. entropy-driven). The exothermicity has been confirmed by calorimetry; [49, 88, 89] FIGURE 6 shows the calorimetric heat of association at two pHs, and a theoretical curve based on a model discussed below.

The exothermicity has been accounted for [88] by assuming that (in the absence of fibrin-stabilizing factor) polymerization involves hydrogen bond formation between donors (DH) on one monomer and acceptors on another, viz., according to the reaction

$$DH + A \rightleftharpoons DH \cdot \cdot \cdot A \tag{1}$$

Presumably either the donors or the acceptors (but not both) are free in the fibrinogen molecule (probably in the outer spheres) and either the acceptors or the donors are unmasked (in the central sphere) upon removal of the fibrinopeptides. Besides the location of the N- and C-termini in the central [74, 76, 77] and outer nodules,[76, 77] respectively, by immunoelectron microscopy, inhibition studies [90] have shown that residues 373–410 near the C-terminus of the γ chains contain a polymerization site that is complementary to the thrombin-activated site in the N-terminal region. Similarly, short peptides having the amino acid sequence of the N-terminal portion of the α-chain of fibrin also inhibit the polymerization reaction; [91, 92] this observation supports the conclusion that the N-terminus of the α-chain of fibrin is in a polymerization site.

The nonhydrogen-bonded donor and acceptor groups are themselves involved in equilibria of the type

$$DH \rightleftharpoons H^+ + D^- \tag{2}$$

$$AH^+ \rightleftharpoons H^+ + A \tag{3}$$

This model accounts for the observed pH range of polymerization (about pH 5 to 10). If the donors have a pK_a of ~9.5 and the acceptors have a pK_a of ~6 (consistent with these groups being tyrosine and histidine, respectively), then reaction 1 cannot occur below pH 5 or above pH 10. The fraction x_{ij} of any ij^{th} hydrogen bond between two monomer units is shown as a function of pH in FIGURE 6. The enthalpies of reactions 1–3 lead to the theoretical curve for ΔH, also shown in FIGURE 6. If the polymerizable species is not produced by thrombin but by a snake venom enzyme that liberates primarily fibrinopeptide

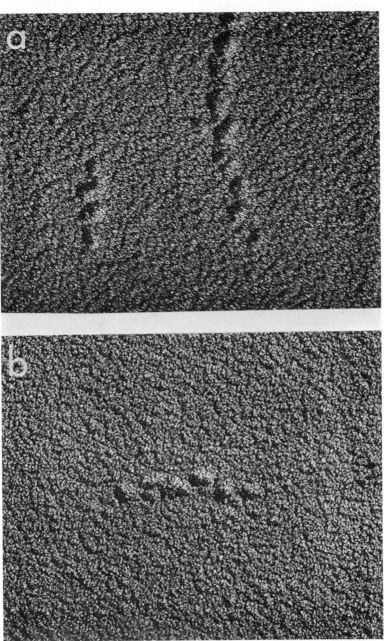

FIGURE 4. Electron micrographs.[28] **(a)** Fibrin dimer (left) and part of a higher polymer (right). **(b)** Fibrin tetramer.

B, then the enthalpy of polymerization is lower,[89] presumably because of the smaller number of available donor and acceptor sites, in accord with the above model.

Besides accounting for the pH range and the enthalpy of polymerization, this model also explains the release and uptake of protons during polymerization. A theoretical curve for q, the net number of protons released per ij^{th}

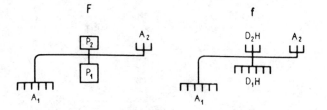

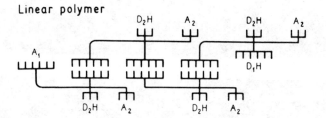

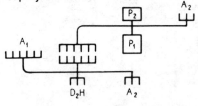

FIGURE 5. Idealized schematic representation[2, 85] of fibrinogen (F) and fibrin monomer (f) to show the functionality with respect to polymerization, and the staggered overlap structure of the polymers. The hydrogen bond donor and acceptor groups are represented by DH and A, respectively. A tetramer and an f-F copolymer are also shown.

hydrogen bond formed, is shown in FIGURE 6. The locations of the maximum and minimum depend on the pK_as of the acceptor and donor, respectively, and the crossing point (where $q = 0$) is a function of both pK_as. An experimental determination of the pK_as, according to this model is shown[93] in FIGURE 7. The location of the maximum suggests that the acceptors may be histidines, and the location of the crossing point suggests that the donors may

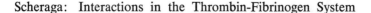

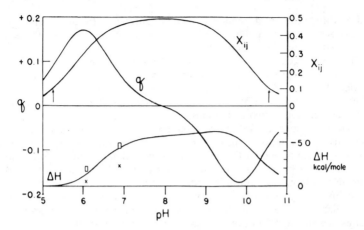

FIGURE 6. Curves [88] showing the pH dependence of x_{ij} (the fraction of the molecules containing a hydrogen bond between the ith donor and j^{th} acceptor), q (the net number of proteins released per ij^{th} pair during polymerization), and ΔT (the heat of polymerization). The experimental points at pH 6.08 and 6.88, indicated by the rectangles, are included for comparison with the theoretical curve. The crosses are theoretical points obtained with a different value of one of the parameters used to compute the theoretical curve.

be tyrosines. Iodination of fibrinogen changes the above properties [3] in the direction expected when the pK_a of tyrosine is lowered by iodination. Calcium ions (but not magnesium ions) appear to bind to fibrinogen and fibrin monomer,

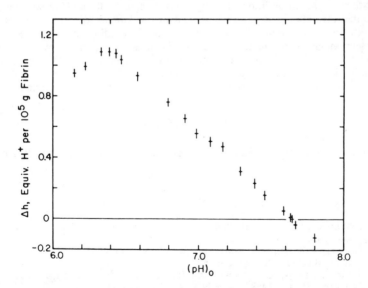

FIGURE 7. Proton release or absorption in the polymerization of fibrin monomer in 1.0M NaBr, 25.0°, as a function of the initial pH.[93]

and modify [94] the pH dependence of the quantities in FIGURE 6. It appears that the properties of the polymerization site at the N-terminus of the β chain of fibrin are influenced by calcium ions.[95]

The hydrogen-bonding model discussed above may involve two sets of donors and acceptors.[70-73] One of them may be generated by release of fibrino-peptide A and bind to the residues near the C-terminus of the γ chains. The other may be generated by release of fibrinopeptide B and bind to a site formed by the association of two D domains.

While the model implied by equations 1-3 seems plausible, and accounts for many observations, it must be regarded as a tentative one because of some inherent difficulties.[96, 97] The main difficulty is that a value for the enthalpy of formation of a hydrogen bond, somewhat higher than is normally observed in aqueous solution, is required in order to account for the enthalpy of polymeri-zation in the hydrogen-bonding model. Such a high value, however, can be rationalized by the plausible view that the intermolecular hydrogen-bonding groups are shielded from the aqueous solvent after the first one or two hydro-gen bonds are formed in the association of two fibrin monomers. Even if hydrogen bonding is the dominant interaction, other types of noncovalent interactions such as hydrophobic bonds or hydrogen bonds between non-ioniz-able groups might also occur.[97]

SOME PERSPECTIVES

While considerable information is available about the proteolytic and polym-erization steps in the thrombin-induced formation of fibrin monomer, and conversion of the latter to the fibrin clot, there is still much to be learned. First of all, it will be important to have x-ray crystal structures of thrombin and fibrinogen. In addition, investigations of chemically and/or enzymatically modified thrombins and fibrinogens, and of the kinetics of interaction of such species, will contribute further to our understanding of the mechanism. Nuclear magnetic resonance studies of fibrinogen-like peptides will give some indication of the role of conformation in the specificity of thrombin. The combination of chemical, physico-chemical, and immunochemical techniques should provide molecular details about the proteolytic and polymerization steps.

ACKNOWLEDGMENT

I am indebted to Dr. Janice A. Nagy for helpful comments on this manu-script.

REFERENCES

1. LASKOWSKI, M., JR., D. H. RAKOWITZ & H. A. SCHERAGA. 1952. J. Am. Chem. Soc. 74: 280.
2. DONNELLY, T. H., M. LASKOWSKI, JR., N. NOTLEY & H. A. SCHERAGA. 1955. Arch. Biochem. Biophys. 56: 369.
3. SCHERAGA, H. A. & M. LASKOWSKI, JR. 1957. Adv. Protein Chem. 12: 1.
4. SCHERAGA, H. A. 1958. Ann. N.Y. Acad. Sci. 75: 189.
5. LASKOWSKI, M., JR., S. EHRENPREIS, T. H. DONNELLY & H. A. SCHERAGA. 1960. J. Am. Chem. Soc. 82: 1340.

6. BLOMBÄCK, B. 1979. *In* Plasma Proteins. B. Blombäck & L. H. Hanson, Eds.: 221. John Wiley & Sons. New York.
7. DOOLITTLE, R. F., H. BOUMA, B. A. COTTRELL, D. STRONG & K. W. K. WATT. 1980. *In* The Chemistry and Physiology of the Human Plasma Proteins. D. H. Bing, Ed.: 77. Pergamon Press. New York.
8. BETTELHEIM, F. R. & K. BAILEY. 1952. Biochim. Biophys. Acta **9:** 578.
9. LORAND, L. & W. R. MIDDLEBROOK. 1952. Biochim. Biophys. Acta **9:** 581.
10. BETTELHEIM, F. R. 1956. Biochim. Biophys. Acta **19:** 121.
11. BLOMBÄCK, B., M. BLOMBÄCK, P. EDMAN & B. HESSEL. 1966. Biochim. Biophys. Acta **115:** 371.
12. HARTLEY, B. S. 1970. Phil. Trans. Roy. Soc. London **B257:** 77.
13. MAGNUSSON, S. 1971. *In* The Enzymes, 3rd Edit. P.D. Boyer, Ed. **3:** 277. Academic Press. New York.
14. MAGNUSSON, S., T. E. PETERSEN, L. SOTTRUP-JENSEN & H. CLAEYS. 1975. *In* Proteases and Biological Control. E. Reich, D. B. Riffin & E. Shaw, Eds.: 123. Cold Spring Harbor Laboratory. Cold Spring Harbor, N.Y.
15. ENDRES, G. F., M. K. SWENSON & H. A. SCHERAGA. 1975. Arch. Biochem. Biophys. **168:** 180.
16. BING, D. H., R. LAURA, D. J. ROBISON, B. FURIE, B. C. FURIE & R. J. FELDMANN. 1981. Ann. N.Y. Acad. Sci. **370:** 496.
17. SCHERAGA, H. A. & M. A. COHLY. 1961. Thromb. Diath. Haemorrh. **5:** 609.
18. COHLY, M. A. & H. A. SCHERAGA. 1961. Arch. Biochem. Biophys. **95:** 428.
19. SCHRIER, E. E., C. A. BROOMFIELD & H. A. SCHERAGA. 1962. Arch. Biochem. Biophys. Suppl. **1:** 309.
20. WINZOR, D. J. & H. A. SCHERAGA. 1964. J. Phys. Chem. **68:** 338.
21. WINZOR, D. J. & H. A. SCHERAGA. 1964. Arch. Biochem. Biophys **104:** 202.
22. SCHERAGA, H. A. 1961. Protein Structure. Chapter V. Academic Press. New York.
23. ENDRES, G. F. & H. A. SCHERAGA. 1971. Arch. Biochem. Biophys. **144:** 519.
24. DOOLITTLE, R. F. 1973. Adv. Protein Chem. **27:** 1.
25. HALL, C. E. 1949. J. Biol. Chem. **179:** 857.
26. HALL, C. E. & H. S. SLAYTER. 1959. J. Biophys. Biochem. Cytol. **5:** 11.
27. SIEGEL, B. M., J. P. MERNAN & H. A. SCHERAGA. 1953. Biochim. Biophys. Acta **11:** 329.
28. KRAKOW, W., G. F. ENDRES, B. M. SIEGEL & H. A. SCHERAGA. 1972. J. Mol. Biol. **71:** 95.
29. BACHMAN, L., W. W. SCHMITT-FUMIAN, R. HAMMEL & K. LEDERER. 1975. Die Makromolekulare Chemie **176:** 2603.
30. FOWLER, W. E. & H. P. ERICKSON. 1979. J. Mol. Biol. **134:** 241.
31. WEISEL, J. W., G. N. PHILLIPS, JR. & C. COHEN. 1981. Nature **289:** 263.
32. WILLIAMS, R. C. 1981. J. Mol. Biol. **150:** 399.
33. MOSESSON, M. W., J. HAINFELD, J. WALL & R. H. HASCHEMEYER. 1981. J. Mol. Biol. **153:** 695.
34. DOOLITTLE, R. F. 1981. Scientific American **245:** 126.
35. TIMPL, R., P. F. FIETZEK, E. WACHTER & V. VAN DELDEN. 1977. Biochim. Biophys. Acta. **490:** 420.
36. MARTINELLI, R. A., A. S. INGLIS, M. R. RUBIRA, T. C. HAGEMAN, J. G. R. HURRELL, S. J. LEACH & H. A. SCHERAGA. 1979. Arch. Biochem. Biophys. **192:** 27.
37. CHUNG, D. W., M. W. RIXON, R. T. A. MacGILLIVRAY & E. W. DAVIE. 1981. Proc. Natl. Acad. Sci. USA **78:** 1466.
38. CHUNG, D. W., M. W. RIXON, R. T. A. MacGILLIVRAY & E. W. DAVIE. 1981. Thromb. & Haem. **46:** Abstract No. 0313.
39. HENSCHEN, A. 1962. Acta Chem. Scand. **16:** 1037.
40. HENSCHEN, A. 1964. Ark. Kemi **22:** 1.
41. HENSCHEN, A. 1964. Ark. Kemi **22:** 375.

42. CLEGG, J. B. & K. BAILEY. 1962. Biochim. Biophys. Acta 23: 525.
43. McKEE, P. A., L. A. ROGERS, E. MARLER & R. L. HILL. 1966. Arch. Biochem. Biophys. 116: 271.
44. BLOMBÄCK, B. 1967. In Blood Clotting Enzymology. W. H. Seegers, Ed.: 143. Academic Press. New York.
45. BLOMBÄCK, B. & M. BLOMBÄCK. 1972. Ann. N.Y. Acad. Sci. 202: 77.
46. FRANCIS, C. W. & V. J. MARDER. 1982. Seminar in Thrombosis and Hemostasis 8: 25. Thieme-Stratton. New York, N.Y.
47. DONOVAN, J. W. & E. MIHALYI. 1974. Proc. Natl. Acad. Sci. USA 71: 4125.
48. NOSSEL, H. L., J. WASSER, K. L. KAPLAN, K. S. LaGAMMA, I. YUDELMAN & R. E. CANFIELD. 1979. J. Clin. Invest. 64: 1371.
49. BLOMBÄCK, B., B. HESSEL, M. OKADA & N. EGBERG. 1981. Ann. N.Y. Acad. Sci. 370: 536.
50. KOWALSKI, R. D. & M. LASKOWSKI, JR. 1976. Biochemistry 15: 1300.
51. NAGAI, K., Y. ENOKI, S. TOMITA & T. TESHIMA. 1982. J. Biol. Chem. 257: 1622.
52. SCHECHTER, I. & A. BERGER. 1967. Biochem. Biophys. Res. Commun. 27: 157.
53. SCHERAGA, H. A. 1977. In Chemistry and Biology of Thrombin. R. L. Lundblad, J. W. Fenton & K. G. Mann, Eds.: 145. Ann Arbor Science Publishers. Ann Arbor, MI.
54. MARTINELLI, R. A. & H. A. SCHERAGA. 1980. Biochemistry 19: 2343.
55. ANDREATTA, R. H., R. K. H. LIEM & H. A. SCHERAGA. 1971. Proc. Natl. Acad. Sci. USA 68: 253.
56. LIEM, R. K. H., R. H. ANDREATTA & H. A. SCHERAGA. 1971. Arch. Biochem. Biophys. 147: 201.
57. LIEM, R. K. H. & H. A. SCHERAGA. 1973. Arch. Biochem. Biophys. 158: 387.
58. LIEM, R. K. H. & H. A. SCHERAGA. 1974. Arch. Biochem. Biophys. 160: 333.
59. HAGEMAN, T. C. & H. A. SCHERAGA. 1974. Arch. Biochem. Biophys. 164: 707.
60. HAGEMAN, T. C. & H. A. SCHERAGA. 1977. Arch. Biochem. Biophys. 179: 506.
61. VAN NISPEN, J. W., T. C. HAGEMAN & H. A. SCHERAGA. 1977. Arch. Biochem. Biophys. 182: 227.
62. MEINWALD, Y. C., R. A. MARTINELLI, J. W. VAN NISPEN & H. A. SCHERAGA. 1980. Biochemistry 19: 3820.
63. MARSH, H. C., JR., Y. C. MEINWALD, S. LEE & H. A. SCHERAGA. 1982. Biochemistry 21: 6167.
64. BLOMBÄCK, B., B. HESSEL, S. IWANAGA, J. REUTERBY & M. BLOMBÄCK. 1972. J. Biol. Chem. 247: 1496.
65. HOGG, D. H. & B. BLOMBÄCK. 1974. Thromb. Res. 5: 685.
66. CLAESON, G., L. AURELL, G. KARLSSON & P. FRIBERGER. 1977. In New Methods for the Analysis of Coagulation Using Chromogenic Substrates. Proc. Symp. Deut. Ges. Klin. Chem., Titisee, Breisgau, July 1976. I. Witt, Ed.: 37–54. W. de Gruyter. Berlin, Federal Republic of Germany.
67. RAE, I. D. & H. A. SCHERAGA. 1979. Int. J. Peptide Protein Res. 13: 304.
68. NAGY, J. A., Y. C. MEINWALD & H. A. SCHERAGA. 1982. Biochemistry 21: 1794.
69. VON DREELE, P. H., I. D. RAE & H. A. SCHERAGA. 1978. Biochemistry 17: 956.
70. BLOMBÄCK, B., B. HESSEL, D. HOGG & L. THERKILDSEN. 1978. Nature 257: 501.
71. HANTGAN, R. R. & J. HERMANS. 1979. J. Biol. Chem. 254: 11272.
72. OLEXA, S. A. & A. Z. BUDZYNSKI. 1980. Proc. Natl. Acad. Sci. USA 77: 1374.
73. WILTZIUS, P., G. DIETLER, W. KÄNZIG, V. HOFMANN, A. HÄBERLI & P. W. STRAUB. 1982. Biophys. J. 38: 123.
74. TELFORD, J. N., J. A. NAGY, P. A. HATCHER & H. A. SCHERAGA. 1980. Proc. Natl. Acad. Sci. USA 77: 2372.

75. HARFENIST, E. J. & R. E. CANFIELD. 1975. Biochemistry **14:** 4110.
76. PRICE, T. M., D. D. STRONG, M. L. RUDEE & R. F. DOOLITTLE. 1981. Proc. Natl. Acad. Sci. USA **78:** 200.
77. NORTON, P. A. & H. S. SLAYTER. 1981. Proc. Natl. Acad. Sci. U.S. **78:** 1661.
78. ENDRES, G. F. & H. A. SCHERAGA. 1971. Arch. Biochem. Biophys. **144:** 519.
79. SCHERAGA, H. A. & J. K. BACKUS. 1952. J. Am. Chem. Soc. **74:** 1979.
80. BACKUS, J. K., M. LASKOWSKI, JR., H. A. SCHERAGA & L. F. NIMS. 1952. Arch. Biochem. Biophys. **41:** 354.
81. FERRY, J. D., S. KATZ & I. TINOCO, JR. 1954. J. Polymer Sci. **12:** 509.
82. HANTGAN, R., W. FOWLER, H. ERICKSON & J. HERMANS. 1980. Thrombosis and Haemostasis **44:** 119.
83. FOWLER, W. E., R. R. HANTGAN, J. HERMANS & H. P. ERICKSON. 1981. Proc. Natl. Acad. Sci. U.S. **78:** 4872.
84. FOWLER, W. E., H. P. ERICKSON, R. R. HANTGAN, J. McDONAGH & J. HERMANS. 1981. Science **211:** 287.
85. SCHERAGA, H. A. 1961. Protein Structure: 171. Academic Press. New York.
86. BRASS, E. P., W. B. FORMAN, R. V. EDWARDS & O. LINDEN. 1976. Thromb. Haemost. **36:** 37.
87. SMITH, G. F. 1980. Biochem. J. **185:** 1.
88. STURTEVANT, J. M., M. LASKOWSKI, JR., T. H. DONNELLY & H. A. SCHERAGA. 1955. J. Am. Chem. Soc. **77:** 6168.
89. SHAINOFF, J. R. & B. N. DARDIK. 1979. Science **204:** 200.
90. OLEXA, S. A. & A. Z. BUDZYNSKI. 1981. J. Biol. Chem. **256:** 3544.
91. LAUDANO, A. P. & R. F. DOOLITTLE. 1978. Proc. Natl. Acad. Sci. USA **75:** 3085.
92. LAUDANO, A. P. & R. F. DOOLITTLE. 1980. Biochemistry **19:** 1013.
93. ENDRES, G. F., S. EHRENPREIS & H. A. SCHERAGA. 1966. Biochemistry **5:** 1561.
94. ENDRES, G. F. & H. A. SCHERAGA. 1972. Arch. Biochem. Biophys. **153:** 266.
95. LAUDANO, A. P. & R. F. DOOLITTLE. 1981. Science **212:** 457.
96. ENDRES, G. F. & H. A. SCHERAGA. 1966. Biochemistry **5:** 1568.
97. ENDRES, G. F. & H. A. SCHERAGA. 1966. Biochemistry **7:** 4219.

DISCUSSION OF THE PAPER

L. LORAND (*Northwestern University, Evanston, IL*): Does calcium make a difference in your heat data and your protein?

H. A. SCHERAGA: We did not look at the ΔH, but we did look at the proton production and found that it is affected.

LORAND: From the number of protons that are produced during polymerization, can you get an estimate of how many pairs are involved?

SCHERAGA: The number we got is 19, but it is an uncertain number because it depends upon what we assume for the heat of formation of a hydrogen bond. This number can be anywhere from 8 to 10 up to 19, and that's another reason why I do not believe that there is only one histidine acceptor.

FIBRIN ASSEMBLY *

Roy Hantgan,† Jan McDonagh,‡ and Jan Hermans †

Departments of †Biochemistry and ‡Pathology
University of North Carolina
Chapel Hill, North Carolina 27514

INTRODUCTION

In order to arrive at a quantitative description of the molecular mechanism of fibrin assembly, we may begin with a minimum sequence of three consecutive reactions, a scheme that is based on the results of numerous investigators over the past three decades as well as our own contributions: activation, polymerization and lateral association.[1-5] *In vitro*, proteolytic cleavage of the Aα chain of fibrinogen at the Arg-Gly bond between residues 16–17 by thrombin or reptilase may be considered as the initial activation event.[6] The next assembly step, which involves only noncovalent protein-protein interactions, is a bimolecular polymerization of trinodular fibrin monomer molecules to form protofibrils,[5] two-stranded polymers in which the molecules of one strand are related to those of its complement by a stagger of 22.5 nm or one-half of a molecular length.[1, 7, 8] This linear growth dominates assembly until protofibrils greater than 600 nm are formed, at which point the second assembly step, lateral association of long protofibrils into thicker fibers, suddenly becomes favorable.[1, 5] Lateral fiber growth leads to fibers that are not only thick but also highly interconnected.[10-12] Fibrinopeptide B appears to be preferentially removed from the growing protofibrils by thrombin,[4, 5, 13] enhancing the rate and extent of lateral association, but this step is not required for gelation to occur.[14] Once all the fibers have been tied together, a network of fibrin fibers or gel, which is insoluble even in the absence of covalent crosslinking, is the result. Our results indicate that the structure of this fiber network is determined by the *rates* of these assembly steps, not by their equilibria, as polymerization proceeds essentially to completion and lateral association ceases only when the fibers are immobilized by the gel.[5]

These three *essential* assembly events are indicated by arrows in the following schematic description of fibrin assembly:

1. Fibrinogen $\xrightarrow[\substack{\text{Fibrinopeptide A Released:} \\ \text{Thrombin/Reptilase}}]{\text{Activation}}$ Fibrin Monomer

2. Fibrin Monomer $\xrightarrow{\text{Polymerization}}$ Protofibrils

2a. Factor XIIIa & Ca²⁺
    γ Chain Cross-linking:

2b. Fibrinopeptide B-released: Thrombin

3. Protofibrils $\xrightarrow{\text{Lateral Association}}$ Network of Fibrin Fibers

3a. α Chain cross-linking: Factor XIIIa & Ca²⁺

* This work was supported by National Institutes of Health Research Grants HL–20319 and HL–26309. J.M. is an Established Investigator of the American Heart Association.

0077–8923/83/0408–0344 $01.75/0 © 1983, NYAS

This scheme also includes three nonessential reactions, namely, release of fibrinopeptide B by thrombin and the two cross-linking reactions, which are catalyzed by Factor XIIIa in the presence of Ca^{2+}, and involve the formation of isopeptide bonds.[15] The first of these covalent stabilization steps results in the formation of a bond between γ chain residues Lys-405 and Gln-406 of adjacent fibrin monomers,[16, 17] both of which lie end-to-end in the same strand of the protofibril.[18] α chain polymers, which are probably formed between protofibrils in the assembled fibers, involve a less well-defined specificity but serve to reinforce the structural and mechanical integrity of the network of fibrin fibers.[19, 20]

We will now present the experimental evidence that supports this picture of fibrin assembly and show how an extension of this basic model provides an explanation of the anticoagulant properties of fibrinogen degradation products and small peptide inhibitors.

POLYMERIZATION

We have employed stopped-flow light scattering techniques to follow the time course of the first step of fibrin assembly, the polymerization of activated fibrin monomers to form protofibrils.[5] In its most basic form, light scattering is a technique in which an unpolarized beam of visible light is directed through a solution in a cuvette. Fluctuations in the refractive index of the solution, due to Brownian motion of the molecules, cause light to be scattered in all directions.[21] In turbidimetric techniques, the diminished intensity of the incident beam is measured in a spectrophotometer, while the more sensitive method, which we employ, involves detection of the scattered light by a photomultiplier whose position, in the plane of the incident beam, can be varied allowing measurements of intensity as a function of scattering angle. For molecules whose dimensions are small compared to the wavelength of the scattered light, measurements at a single fixed angle allow determination of the molecular weight of the particle. For molecules whose dimensions are comparable to the scattered light wavelength, one may obtain information about their size and shape from measurements of the intensity of light scattered as a function of the angle of observation.[21] The intensity of light scattered from solutions of fibrinogen and short fibrin oligomers is a function of concentration, molecular weight and dimensions of these macromolecules.[22, 23] As the length of the protofibrils increases, the scattering intensity reaches a plateau which is proportional to concentration and mass per unit length of the polymers.[5, 11, 24]

By initiating assembly with nonrate-limiting thrombin concentrations at a relatively high salt concentration we can clearly distinguish protofibril growth from the subsequent lateral association and gelation. Rapid mixing techniques have enabled us to observe the complete time course of protofibril polymerization over a sevenfold range of fibrinogen concentrations. The initial rate of the reaction was found to be proportional to the square of the initial fibrinogen concentration and the halftime to vary as the inverse first power of the fibrinogen concentration.[5] These results are consistent with a simple bimolecular polymerization[25] of bifunctional fibrin monomers. The stopped-flow trace of FIGURE 1 is typical of the results obtained; we note that the scattering intensity increases with time in a hyperbolic fashion with no evidence of the sigmoidal shape characteristic of a nucleation-controlled reaction.[26] Protofibril

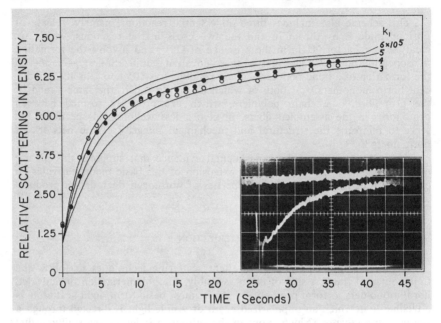

FIGURE 1. Experimental and calculated changes in light scattering intensity during protofibril formation in 0.5 M NaCl. The inset is a trace of a stopped-flow experiment carried out in 0.5 M NaCl, 0.05 M Tris-Cl, pH 7.4 at 0.25 mg/ml fibrinogen and 27 N.I.H. units/ml thrombin. Each division on the time axis (abscissa) equals 2 sec.; each division on the intensity axis (ordinate) corresponds to 2.1 times the scattering of the standard (benzene). The solid lines, which represent the intensity relative to monomeric fibrin as a function of time, were calculated from the theory of light scattering of rod-like particles for a bimolecular polymerization reaction with $k_1 = 6$, 5, 4 and 3×10^5 M^{-1} sec^{-1}. The points were obtained from experiments employing 35 and 27 NIH units/ml thrombin, open and closed circles, respectively. (From Hantgan and Hermans.[5] By permission of *Journal of Biological Chemistry*.)

growth thus stands out as an unusual example of a biological assembly system because in the presence of sufficient thrombin, rapid growth rather than finely tuned control is its distinguishing characteristic.

By modeling the fibrin monomer molecule as a long, thin rod with a polymerization site at each end, we have been able to describe the scattering intensity versus time curves (FIGURE 1) with a bimolecular kinetic model and a single rate constant for polymerization: [5]

$$df_1/dt = -k_1[f_1]^2 \tag{1}$$

The concentration of sites, f_1, remaining unpolymerized at a particular time is then:

$$f_1 = c_o/(1 + k_1 c_o t) \tag{2}$$

The maximum concentration of binding sites, c_o, is equal to twice the initial fibrinogen concentration. In 0.5 M NaCl, pH 7.4 we find this rate constant, k_1, to be equal to 5.0×10^5 $M^{-1}s^{-1}$. Under these conditions, protofibril growth continues until the scattering intensity approaches that of very long polymers whose width is twice that of the fibrinogen molecule. (The scattering intensity is independent of polymer length for polymers composed of more than 15 monomers).[5]

Results of stopped-flow experiments performed in 0.1 M NaCl, following thrombin and reptilase activation [5] are presented in FIGURE 2. The solid line is a scattering intensity versus time profile calculated from the bimolecular kinetic model and the scattering behavior of protofibrils, modeled as long, thin rods as previously described. The rate constant, k_1, was set equal to 1.2×10^6 $M^{-1}s^{-1}$ in this calculation.

The sharp rise in scattering intensity which occurs at 16 sec (thrombin) or 40 sec (reptilase) after initiation of the reaction corresponds to the formation of fibrin fibers composed of several protofibrils. These experiments demonstrate that the removal of fibrinopeptide A, alone, triggers the first step in fibrin assembly, the bimolecular polymerization of bifunctional fibrin monomer molecules to form protofibrils.

PROTOFIBRIL STRUCTURE

In collaboration with W. Fowler and H. Erickson, we have employed electron microscopy to study the fibrin assembly process and deduce details of

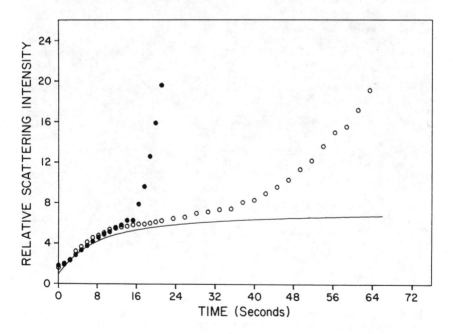

FIGURE 2. Experimental and calculated changes in light scattering intensity during fiber formation in 0.1 M NaCl. The experimental points were obtained in 0.1 M NaCl, 0.05 M Tris-Cl, pH 7.5 at 0.05 mg/ml fibrinogen activated with 5 NIH units/ml thrombin (solid circles) and reptilase (open circles). The solid line was calculated for protofibril growth as described in the text, using a value of $k_1 = 1.2 \times 10^6 M^{-1} sec^{-1}$.

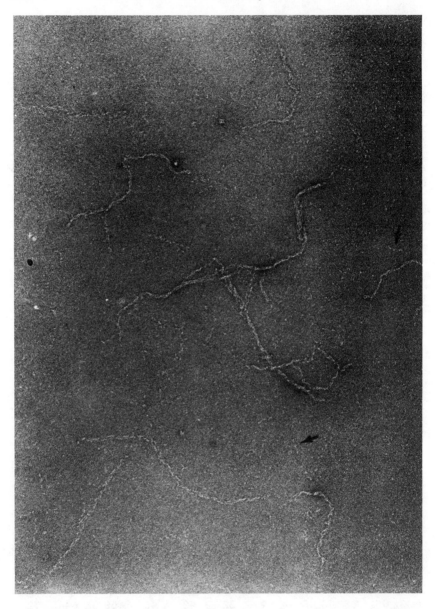

FIGURE 3. Species obtained by negative staining 1 min after thrombin addition to 0.01 mg/ml fibrinogen in 0.1 M NaCl, 0.005 M CaCl₂, 0.05 M Tri-Cl pH 7.4. Protofibrils as well as some thicker fibers are present. Individual trinodular molecules are indicated by black arrows. The presence or absence of CaCl₂ and Factor XIIIa did not alter the pattern of molecular species observed by electron microscopy during assembly. The electron micrographs of FIGURES 3, 4, 7, and 9 were prepared by Walter Fowler and are reproduced by permission of *Thrombosis and Haemostasis*, from Hantgan *et al.*[9]

protofibril structure.[8, 9, 18] The electron micrograph in FIGURE 3 shows a specimen negatively stained with uranyl acetate 60 sec after addition of thrombin to a solution of fibrinogen at 10 μg/ml in 0.1 M NaCl.[9] The light scattering kinetic profile of such a reaction mixture indicates that polymerization to protofibrils is nearly complete at this time but that lateral association of these long polymers has not yet taken place. The bimolecular kinetic model predicts the presence of a broad distribution of polymer lengths, with pentamers being the most probable species.

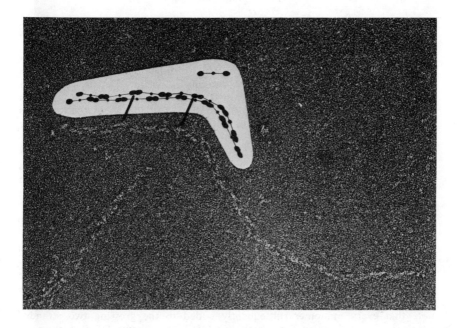

FIGURE 4. Selected region of FIGURE 3 at higher magnification containing a protofibril. An interpretative drawing, included for clarity, demonstrates the half-staggered overlap arrangement of the individual trinodular molecules. Further details of protofibril structure, deduced from electron micrographs, are discussed by Erickson and Fowler in this volume.[27] (From Hantgan *et al.*[9] By permission of *Thrombosis and Haemostasis.*)

Protofibrils are clearly the dominant species present in this micrograph. Although difficulties of nonuniform staining and sampling prevent quantitative length distribution measurements, we have ascertained that 90% of the polymers had lengths less than 800 nm.[9] Protofibrils, whose structure was indistinguishable from those in FIGURES 3 and 4, were also observed 60 sec after activation by reptilase; a maximum polymer length of 600 nm was observed in the electron micrographs. In 0.5 M NaCl, protofibrils ranging in length from 300–2000 nm were observed 15 min after thrombin activation.[9]

FIGURE 4 shows a selected region of the micrograph of the previous figure at a twofold higher magnification, as well as a schematic drawing to clarify our interpretation of its structure.[8, 9] Resolution of these polymers into two strands of trinodular molecules is apparent in the micrograph. Reference to the cross-linked fibrinogen dimers [18] and dimer:fibrin monomer complexes,[8] which are discussed in more detail in this volume by Erickson and Fowler,[27] has aided our interpretation of the half-staggered overlap arrangement of the trinodular molecules in one strand relative to those in the complementary strand of the protofibril.

PROTOFIBRIL MODEL

The schematic model of fibrinogen [28] in FIGURE 5a provides a good starting point for a description of the contact sites involved in protofibril formation. One distinguishing feature of this trinodular fibrinogen model (and, presumably any other acceptable model) is its twofold symmetry axis, which has been drawn through the center of the dimeric E-domain. That is, rotation of the molecule by 180° about the symmetry axis presents the viewer with an unchanged picture of the molecule. By contrast, rotation about the long axis by 180° would alter the steric features of the molecule presented, that is, the molecule's top side is distinct from its bottom.[28, 29]

Following thrombin activation, two fibrin monomer molecules interact first to form a half staggered overlap dimer (FIGURE 5b) with their contact sites facing each other and forming two identical sets of noncovalent bonds between the D-domain of one molecule and the E-domain of its complement.[8, 28] We have termed this noncovalent interaction a DE-staggered contact (DE-stag), as it is the essential structural element maintaining the half staggered overlap configuration of the protofibril.[8, 28]

As dimer formation involves only DE-stag contacts, this step could, in principle, occupy a unique position in the protofibril assembly scheme. Addition of a third fibrin monomer to a dimer forms a new contact (which we have termed DD-long),[8, 28] between D-domains of adjacent molecules in the same strand of the polymer in addition to another two DE-stag contacts.[5] We note that every monomer subsequently added to the growing two stranded polymer chain will again form both one DD-long and two DE-stag contacts. However, we find that the kinetics of protofibril growth are well described by a scheme with only *one* rate constant,[5] implying that the trimer and higher oligomers are formed at the same rate as the dimer, i.e., the DD-long contact does not contribute significantly to the rate of polymerization of these noncovalent complexes.

We have previously shown that the DD-long contact is at or near the site of γ-chain cross-linking,[18] a reaction catalyzed by Factor XIIIa, which results in formation of an isopeptide bond between Lys-405 and Gln-406, both of which are located on the γ-chains of the adjacent D-domains of molecules in the same strand of the protofibril. As discussed in detail in this issue by Erickson and Fowler,[27] a noncovalent complex involving a cross-linked fibrinogen dimer and an activated fibrin monomer [8] serves as a model of the repeating polymeric unit of the protofibril, as it involves both DD-long and DE-stag contacts. However, in that case we note that the D-domains of the monomer are frequently bent away from the E-domain of the dimer, probably due to steric or charge repulsions from the remaining fibrinopeptides present on the (unactivated) fibrinogen dimer.[8]

As noted, the addition of each subsequent monomer to the growing proto-fibril forms both DE-stag and DD-long contacts. The schematic of FIGURE 5d [28] shows the best model of the resultant protofibril according to our observations of its formation and structure by light scattering and electron microscopy both in 0.1 and 0.5 M NaCl, following either thrombin or reptilase activation.[5, 8, 9] FIGURE 5e depicts one arrangement by which protofibrils may be assembled

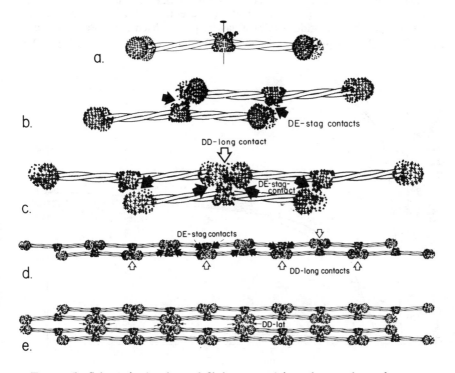

FIGURE 5. Schematic drawings of fibrinogen and its polymers drawn by a computer-controlled plotter, with excess structural details included in order to stress certain features of the molecular architecture. **(a)** fibrinogen: note the twofold symmetry axis that divides the central E-domain, the three stranded coiled-coil connectors and the dimeric, outer D-domains. **(b)** fibrin dimer: note two DE-stag contacts (dark arrows). **(c)** fibrin trimer: third monomer added forms both DE-stag (dark arrow) and DD-long (white arrow) contacts. **(d)** protofibril: formed by continued associations of monomers and oligomers via DE-stag and DD-long non-covalent bonds. **(e)** Possible arrangement of protofibrils in fibrin fibers, stabilized by DD-lat contacts. (From Hermans & McDonagh.[28] By permission of *Seminars in Thrombosis.*)

into fibrin fibers.[28, 29] Growth of protofibrils, as observed by electron micros-copy,[9] is approximately four times slower in 0.5 M NaCl, in accord with our light scattering kinetic model. Protofibrils ranging in length from 300–2000 nm have been observed in 0.5 M NaCl prior to gelation, while in 0.1 M NaCl maximum lengths of 600–800 nm were noted prior to the onset of lateral association.[9]

LATERAL ASSOCIATION

Scattering intensity vs. time profiles (FIGURE 6) show a second step in fibrin assembly as a rapid rise in intensity, due to a dramatic increase in fiber thickness.[5] Electron microscopy (FIGURE 7) confirms that lateral association of long protofibrils occurs in parallel with the rise in light scattering intensity.[9] Knowing that proteolytic cleavage of fibrinopeptide A triggers polymerization, we may now pose the question of what event initiates lateral association.

Although fibrinopeptide B release has frequently been associated with fiber growth,[4, 30] light scattering and electron microscopic data clearly show that

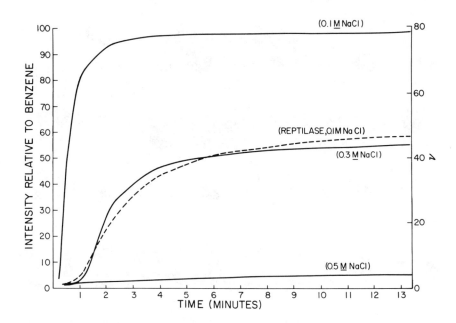

FIGURE 6. Changes in light scattering intensity due to fiber growth. The fibrinogen concentration was 0.05 mg/ml, solutions contained 0.05 M Tris-Cl, pH 7.4 and were activated by thrombin (solid curves) or reptilase (dashed line) at a final concentration of 5 NIH units/ml. The right hand ordinate, ν, is the number of fibrin molecules in the fiber cross-section calculated from the scattering intensity. (From Hantgan and Hermans.[5] By permission of *Journal of Biological Chemistry*.)

this step is not *required* for formation of thick fibers. Fibrinogen cleaved by reptilase does assemble to thick fibers [5, 14] as the data in FIGURE 6 demonstrate; electron microscopic observations [9] show that fibers form which are indistinguishable from those in FIGURE 7. The time at which fibrinopeptide B is cleaved by thrombin is therefore of special interest; we have shown that thrombin is capable of removing the B peptides from long protofibrils and that scattering intensity versus time measurements show that B peptide release, in the presence of sufficient thrombin, is complete prior to lateral association.[5]

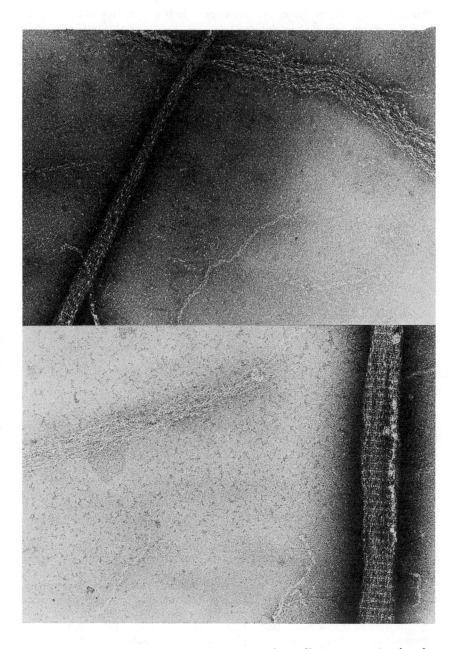

FIGURE 7. Intermediate polymers and cross-striated fibers present 2 min after thrombin addition to 0.01 mg/ml fibrinogen in 0.1 M NaCl, 0.005 M CaCl$_2$ 0.05 M Tris, pH 7.4. (From Hantgan *et al.*[9] By permission of *Thrombosis and Haemostasis*.)

Other investigators have shown that fibrinopeptide B release lags behind A peptide release,[4, 6, 13] and that inhibitors of polymerization delay fibrinopeptide B release.[31, 32] We have interpreted these results to mean that fibrinopeptide B is preferentially removed from the elongating protofibril, that this proteolytic removal of the negatively charged peptides accelerates lateral association and that this increased rate causes thicker fibers to be formed but that the structure of the resultant fibrin is not determined by the presence or absence of fibrinopeptide B.

Our results indicate that protofibril *length* is the key factor in determining the onset of lateral fiber growth; short protofibrils are incapable of lateral

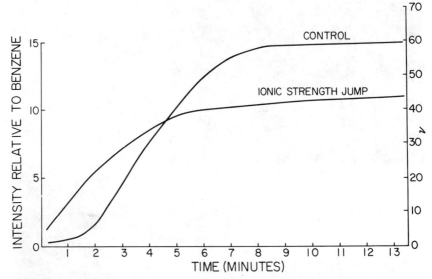

FIGURE 8. Results of an ionic strength jump experiment in which the NaCl concentration was lowered from 0.5 to 0.1 M 120 sec after activation with thrombin. For the experiment, the time scale refers to the time elapsed after the ionic strength jump; for the control the time is that elapsed after activation. Fibrinogen was at 0.05 mg/ml before dilution and 0.01 mg/ml after dilution and for control; thrombin was at 5 unit/ml before dilution and 1 unit/ml for control. (From Hantgan & Hermans.[5] By permission of *Journal of Biological Chemistry*.)

association because they cannot form a sufficient number of the (apparently weak) protofibril-protofibril contacts, which are necessary to stabilize the thick fiber structure.[5, 8, 9] Fortunately, this concept is directly testable by experiment. By initiating assembly in 0.5 *M* NaCl, we have obtained a solution of long, soluble protofibrils and prior to the gel point, diluted the solution to a final concentration of 0.1 *M* NaCl, 0.05 *M* Tris, pH 7.4. As the results in FIGURE 8 demonstrate, this ionic strength jump results in immediate, rapid lateral association with no intervening lag period.[5] As polymerization to long polymers was already complete prior to the salt jump, growth of thicker fibers could begin at once. Fibrinogen activated by reptilase would, according to this model, assemble more slowly because longer polymers are required to favor lateral association. One may therefore conclude from this that an increased number of inter-protofibril contacts is apparently needed to overcome the repulsive contributions of the remaining fibrinopeptide B. In a similar fashion, 0.5 M NaCl delays fiber growth because longer polymers are required to overcome the unfavorable effects of high ionic strength on assembly.[5, 9]

These observations also suggest that relatively weak, electrostatic forces are involved in the formation of contacts whose strength is diminished by the presence of a high concentration of ionic species. In our model of the protofibril, all possible DE-stag contacts are completed between the two strands of the polymer. Therefore, we have proposed the existence of a third noncovalent interaction involving the D-domains of the protofibrils comprising fibrin fibers and have termed these contacts DD-lat (FIGURE 5e), as they contribute to the stability of lateral interactions between protofibrils.[8] Such a contact is consistent with the experimental observation of the low density of fibrin, that is the fibers are approximately 80% water and 20% protein.[33] The conclusion that protofibrils pack *in register* to form fibers [29] follows from the electron microscopic and x-ray repeat patterns [7, 34] and is reinforced by the observations of "loose" fibers [9] (FIGURE 9) in which both individual protofibrils and the typical 22.5 nm repeating cross band pattern are clearly observed. Although details of fiber structure require further investigation, observations by neutron scattering indicate that it is reasonable to model the fiber as a regular, crystal-like arrangement of the protofibrils.[29, 35]

GELATION

We have observed that formation of an insoluble, three-dimensional gel network occurs prior to completion of lateral association of protofibrils, but that the rigidity of this network continues to increase well beyond the completion of changes in the intensity of scattered light.[5] In our model of fibrin formation, gelation is the necessary outcome of the essentially irreversible linear and lateral growth of protofibrils whose outer surface is capable of interacting with other protofibrils (or other fibers) to form thicker fibers.[5] These bonds between protofibrils and/or fibers necessary for mechanical strength need not be a new kind of contact, but may be formed by protofibrils or fibers that attach to more than one fibrin strand.[5, 8, 9] Such interfiber connections can be seen in FIGURE 9 and have been observed both before and after the gel point; [9] this is consistent with our postulate that fiber size is determined by the *rate* of assembly and ceases only when the inherently flexible fibers are prevented from achieving further contacts by the restrictions of the three dimensional gel.[5, 10, 12] These noncovalent interactions are reinforced by the Factor XIIIa-catalyzed formation of α chain polymers, which frequently involve multiple contacts between protofibrils [15] and cause a substantial increase in the rigidity of the gel.[20]

DELAYED FIBRIN ASSEMBLY

We are now in a position to ask the question: What scattering kinetics are observed if the rates of protofibril formation or lateral association are limited by the activation steps, i.e., at lower thrombin concentrations? As the fibrinogen concentrations employed in our studies are always less than the K_ms for fibrino-

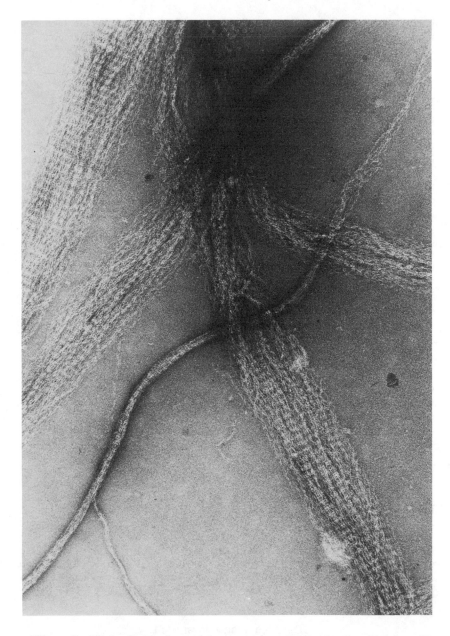

FIGURE 9. Fibrin fibers observed by negative staining 10 min after thrombin addition in 0.1 M NaCl, 0.005 M CaCl$_2$, 0.05 M Tris, pH 7.4. (From Hantgan et al.[9] By permission of *Thrombosis and Haemostasis.*)

peptide release (approximately 10^{-5} M),[13, 32, 37] considerable simplification of the Michaelis-Menten model results. For cleavage of fibrinopeptide A, the thrombin-limited model predicts that the concentration of uncleaved Aα

chains, [Aα], decreases at a rate which is governed by (k_{cat}/K_m), the enzyme and initial fibrinogen concentration:

$$[A\alpha] = c_o \; exp(-t[T]k_{cat}/K_m) \tag{3}$$

The binding sites which become available as a result of this cleavage react to form protofibrils at a rate determined by the polymerization rate constant k_1:

$$df_1/dt = (k_{cat}/K_m)[T]c_o \; exp(-t[T]k_{cat}/K_m) - k_1[f_1]^2 \tag{4}$$

Solution of equation 4 by a numerical method [36] yields a description of the concentration of binding sites remaining unpolymerized versus time. These

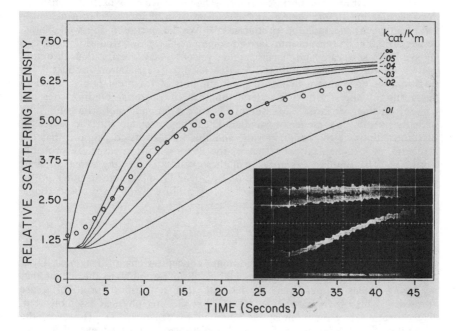

FIGURE 10. Experimental and calculated changes in light scattering intensity during thrombin-limited polymerization. The inset depicts an experiment carried out at 7.1 NIH units/ml thrombin; all other conditions are the same as in FIGURE 1. The solid lines were calculated from the delayed polymerization model with $k_1 = 5 \times 10^5 \; M^{-1} \; s^{-1}$, varying (k_{cat}/K_m). The uppermost line corresponds to nonrate-limiting thrombin concentrations; $(k_{cat}/K_m) = 0.05, 0.04, 0.03, 0.02$ and 0.01 for the lower curves.

combined results allow a calculation of the extent of reaction, defined by the fraction of Aα chains that have been cleaved and subsequently polymerized, as a function of time. With this information the scattering kinetic profile can be calculated.[5] (Note that if the thrombin concentration is sufficiently high, equation 4 reduces to equation 1.) The family of curves (solid lines) in FIGURE 10 have been calculated from the model for various values of k_{cat}/K_m and compared to experimental data. Using published values of K_m, we obtain an estimate of $k_{cat} \cong 2\text{--}4 \times 10^{-10}$ mol/unit/s for fibrinopeptide A release in reasonable agreement with values measured by other, more direct analytical techniques.[32, 37] In a similar fashion, light scattering measurements of the delay in fiber formation (lateral association) which occurs in 0.1 M NaCl, 0.05 M Tris, pH 7.4 due to low thrombin concentrations have enabled us to estimate k_{cat} for fibrinopeptide B release at $3\text{--}4 \times 10^{-10}$ mol/unit/s,[5] in accord with published values.[32, 37]

INHIBITION OF FIBRIN ASSEMBLY

Our basic three-step fibrin assembly scheme provides a useful framework for investigation of the molecular mechanism of inhibition of fibrin assembly by fibrin(ogen) degradation products.[38-42] We have chosen the well-characterized 100,000 dalton plasmin-derived fragment D for our studies.[40-42] Assays of synthetic substrate and fibrinopeptide release assays verify that thrombin activity is not significantly altered by this terminal degradation product.[42] In addition, we have shown that only small amounts of fragment D are bound to the resultant fibers, although a large number of potential binding sites are present on their surface.[42] Measurements of the amount of fibrinogen which actually clots in the presence of fragment D can be correlated with the decreased elastic modulus of the resultant gels, and this has led us to propose that the structure and mechanical properties of the fibers which do form are probably normal.[42] The simplest explanation of these results is that the *extent* of fiber formation is limited by the presence of fragment D; inhibition of either protofibril growth or lateral association (or both) could explain these observations.

KINETIC EXPERIMENTS

Under physiological buffer conditions, concentrations of fragment D in excess of 15 mol D/mol fibrinogen inhibit fibrin assembly to the extent that scattering intensity measurements indicate that only short protofibrils are formed.[42] This observation alone reveals that the first physical assembly step, linear protofibril growth, must be inhibited by fragment D. However, the possibility that lateral association is also disrupted cannot be excluded from such experiments. We have carried out two-step activation experiments to directly address this question.[43]

By activating fibrinogen at 0.01 mg/ml, the lag time, or period of protofibril growth is approximately 50–60 sec. In the absence of fragment D, fibers composed of 20–30 laterally associated protofibrils are formed, while the presence of 52 mol D/mol fibrinogen reduces the scattering to that of short fibrin oligomers. The addition of fragment D (to the same final concentration) 45 sec after thrombin activation did not impair the rate or extent of lateral association of protofibrils to form fibrin fibers.[43] Since long protofibrils were present in the solution at this time, these results indicate that fragment D does not significantly inhibit the rapid lateral association of these long polymers. This result has led us to propose that fragment D exerts its inhibitory effect primarily by blocking the bimolecular polymerization of activated fibrin monomers, thus inhibiting all subsequent steps.

We have carried out stopped-flow light scattering experiments in 0.5 *M* NaCl, pH 7.4 in the presence of fragment D in order to directly test this hypothesis.[43] The trace of FIGURE 11 was obtained at the same fibrinogen concentrations as those in FIGURES 1 and 10, but in the presence of 8.2 mol D/mol fibrinogen. We note that a quasi-hyperbolic curve typical of non-inhibited assembly results, but that the final intensity reached is approximately 25% less than in the absence of inhibitor. We can treat this case by an extension of our bimolecular polymerization model by postulating that an

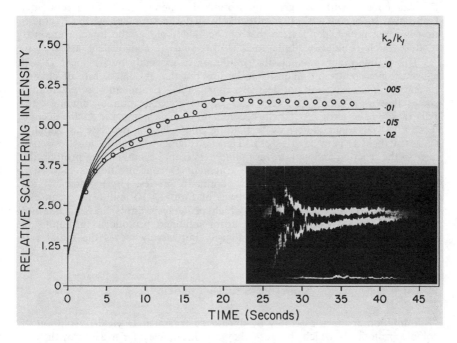

FIGURE 11. Experimental and calculated changes in light scattering intensity during polymerization in the presence of fragment D. The inset depicts an experiment carried out in the presence of 0.60 mg/ml fragment D, 0.25 mg/ml fibrinogen and 38 u/ml thrombin, buffer conditions as in FIGURE 1. The solid lines were calculated for the polymerization/inhibition model with $k_1 = 5.0 \times 10^5$ and k_2/k_1 = 0, 0.005, 0.010, 0.015 and 0.020 (top to bottom).

activated polymerization site may either react with another site, forming a polymer or it may bind fragment D forming one DE-stag contact and a dead-end complex incapable of further growth.[43] The extent of polymerization is then determined by the fraction of sites which have self-associated rather than bound fragment D. In the presence of a sufficient amount of thrombin, the following rate equations result:

$$df_1/dt = -k_1[f_1]^2 - k_2[f_1][D] \tag{5}$$

$$df^*/dt = k_2[f_1][D] \tag{6}$$

where k_1 is the polymerization rate constant, k_2 the inhibition rate constant,

$[f_1]$ the concentration of activated polymerization sites and $[D]$ the fragment D concentration, (treated as constant, to a first approximation); f^* represents the polymerization sites which have bound fragment D. Equations 5 and 6 were solved† [43] and a family of scattering intensity versus time curves were calculated by varying the value of the inhibition rate constant, k_2 (FIGURE 11). We find that the data can be described by a value of k_2 approximately 100-fold lower than the polymerization rate constant, k_1.[43] This decreased value of k_2 is presumably due in part to the fact that fragment D can form only one DE-stag contact with the growing protofibril, whereas each fibrin monomer forms two such contacts. It is also likely that the structure of the degradation product is not identical to that of the D-domain of the intact trinodular model, which could contribute to a slower rate of noncovalent association.

This simple model predicts that inhibition of assembly by fragment D will cause the final scattering intensity to be low but that the initial rate of change of scattering intensity will not decrease appreciably. Examination of equation 5 shows that as assembly proceeds and f_1, the concentration of binding sites falls rapidly, the term $k_1[f_1]^2$ will become very small, compared to the second term $k_2[f_1][D]$ which describes D binding and will not decrease as rapidly, since fragment D is in large excess. That is, fragment D will be most effective later in the reaction and will prevent oligomers from becoming long protofibrils. The model also predicts that as the fragment D concentration is increased, shorter oligomers will become the dominant species. By comparing the stopped-flow traces and calculated curves of FIGURES 10 and 11, we can distinguish delayed polymerization, in which proteolysis limits the rate but not the final extent of polymer growth from inhibited protofibril formation in which fragment D limits the final intensity but largely leaves the initial rate unaltered.

POLYMER SIZE DISTRIBUTION

Inhibition of fibrin assembly by fibrinogen fragment D yields a solution of soluble protofibrils which, because they are incapable of further growth, can be studied in detail by light scattering techniques. The angular dependence spectrum of light scattered from a solution of (large) polymers contains information about the concentration of components of different size in a manner completely analogous to the absorbance spectrum of a mixture of different chromophores.[21] If the distribution of polymer sizes and the particle scattering function of each oligomer are known, one may predict the scattering behavior as a function of angle. Conversely, experimental observations of the scattering intensity at, e.g., 11 angles can be used to derive estimates of the weight fractions of up to seven different components, differing only in length. Representative data, obtained in 0.1 M NaCl in the presence of two molar ratios of fragment D and fibrinogen [43] are shown in FIGURE 12; the solid lines were obtained with use of a nonlinear least squares fitting procedure.[44] The weight fractions of the seven oligomer species that resulted from this analysis are shown in the insert of FIGURE 12 in histogram form.[43, 44] The data show

$$\text{† } [f_1]/c_o = k_2[D] \, exp \, (-k_2[D]t)/[k_2[D]+k_1c_o(1-exp(-k_2[D]t)] \tag{7}$$

$$[f^*]/c_o = (k_2[D]/k_1) \, ln \, [(k_2[D]+k_1c_o(1-exp(-k_2[D]t))/k_2[D]] \tag{8}$$

that the most prevalent species at the lower fragment D concentration is a pentamer of fibrin monomer molecules whereas a fourfold increase in inhibitor concentration shifts the peak of the distribution to a trimer. These results agree with the bimolecular polymerization/inhibition model of fibrin assembly in that a broad distribution of inhibited oligomer sizes results and that increasing fragment D concentrations lead to shorter polymers.

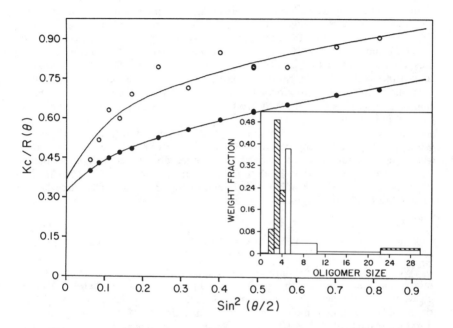

FIGURE 12. Angular dependence of the intensity of light scattered from fragment D inhibited protofibrils. The experimental data curve obtained in 0.1 *M* NaCl, 0.05 *M* Tris-Cl, pH 7.4 at a fibrinogen concentration of 0.16 mg/ml, 19 (open circles) and 68 (closed circles) moles fragment D/mole fibrinogen, activated at a final thrombin concentration of 3.3 NIH units/ml. The reciprocal scattering function (ordinate) is plotted versus the square of half the scattering angle. Solid lines were obtained from a nonlinear least squares fitting procedure. The inset shows the distribution of oligomer lengths obtained from this fitting procedure, plotted as a histogram of weight fraction versus number of monomers in the oligomer. The shaded bars correspond to the upper line of the main figure, 68 moles D/mole fibrinogen; the open bars to 19 moles D/mole fibrinogen.

REVERSIBILITY OF INHIBITION

In the bimolecular kinetic scheme proposed to explain the inhibitory properties of fragment D, we have treated polymerization and dead-end complex formation as irreversible steps.[5, 43] This postulate is supported by our observations that the inhibition of fibrin assembly by excess fragment D yields stable solutions of soluble fibrin which do not gel, even over a period of several hours.[43] If the dissociation of the noncovalently bound fragment D is sufficiently low, we may be able to isolate protofibrils that will remain soluble even in the absence of an excess of fragment D; Factor XIIIa will also be employed to covalently cross-link fragment D to the inhibited protofibrils, preventing dissociation.

Laudano and Doolittle have shown that the tetrapeptide Gly-Pro-Arg-Pro binds to fibrinogen with an affinity constant of 4×10^4 M^{-1}, inhibiting fibrin assembly.[45] We have carried out activation/inhibition experiments with a sample which they have kindly provided, and found results similar to those we have described for fragment D. That is, at a tetrapeptide concentration of 8×10^{-4} M, well above the dissociation constant,[45] gelation of thrombin activated fibrinogen $(5.5 \times 10^{-8}$ $M)$ was prevented and the scattering intensity was less than that of long protofibrils. Addition of the peptide to a solution of long protofibrils did not significantly inhibit lateral association. We may interpret this result, as with fragment D, to mean that the primary focus of inhibition of assembly by Gly-Pro-Arg-Pro is at the first assembly step, i.e., inhibition of protofibril growth.

One-step activation experiments carried out at a lower molar excess of tetrapeptide $(4 \times 10^3$ compared to $1.4 \times 10^4)$ resulted in delayed assembly, although the final intensity reached was comparable to control experiments. These results indicate that inhibition of fibrin assembly by Gly-Pro-Arg-Pro depends on a reversible binding step, in contrast to the (essentially) irreversible model discussed with fragment D.

CONCLUSION

We have proposed a comprehensive, quantitative model of the molecular interactions that govern the gelation of fibrin. In this model, proteolytic cleavage of fibrinopeptide A initiates the bimolecular polymerization of bifunctional fibrin monomer molecules to form protofibrils, which takes place with a rate constant of 1.2×10^6 $M^{-1}s^{-1}$ under physiological buffer conditions.[5] The individual trinodular molecules of these two-stranded polymers are related to each other by a twofold screw axis between the strands.[8] That is, if any molecule is rotated by 180° about an axis parallel to the long axis of the polymer and then translated along this axis for one-half a molecular length, its position will coincide with that of another molecule. Noncovalent interactions between the D-domain of one molecule and the E-domain of its complement in the adjacent strand are primarily responsible for the stability of the protofibril.[8, 28] An additional interaction occurs between D-domains of adjacent molecules of the same strand, and this DD-long contact is at or near the site of the Factor XIIIa-catalyzed γ chain cross-links.[18]

In our model further proteolytic cleavage is not required for fiber formation, as this step begins when protofibrils reach a sufficient length, which is of the order of 600–800 nm.[9] Lateral association of long protofibrils is postulated to involve a new noncovalent interaction, the DD-lat contact, which is considerably weaker than the DE-stag contact.[8, 28] The release of fibrinopeptide B occurs preferentially from fibrin oligomers or protofibrils; this removal of the negatively charged peptides contributes to the overall stability of the laterally associated protofibrils. As the effects of not removing the B peptide can be quantitatively mimicked by a modest increase in ionic strength,[5, 14] we propose that the loss of fibrinopeptide B does not expose a new binding site.

Electron microscopic observations confirm that fibers formed following thrombin or reptilase activation are structurally indistinguishable.[9]

Gelation occurs not by the formation of new contacts, but by the continued lateral association of long protofibrils and fibers, which inevitably make interfiber connections, resulting in a three dimensional network of fibrin, the gel. The size of individual fibers is kinetically controlled; fiber growth ceases only when interfiber connections have prevented the flexible fibers from diffusing together to form even more DD-lat contacts.[5, 12] Although noncovalent interactions alone will impart considerable mechanical strength to the network, α chain polymer formation further increases the gel's rigidity and stability.[20]

Our experimental data show that fragment D inhibits fibrin assembly by forming DE-stag contacts with the polymerization sites of growing protofibrils, resulting in blocked polymers incapable of further elongation.[42, 43] We find that fragment D binds to polymers in an essentially irreversible step at a rate 100-fold lower than the normal rate of polymerization.[43] These inhibited, short oligomers cannot form a sufficient number of DD-lat contacts, thus they are prevented from assembling into thicker fibrin fibers and do not gel. Measurements of the angular dependence of the intensity of light scattered from these inhibited solutions have yielded information on the distribution of polymer sizes and, in accord with our model, show that an increase of the inhibitor concentration shifts the distribution toward shorter oligomers.[43, 44] This model for inhibition can encompass the observation that fragment Y is a more potent inhibitor of gelation than fragment D, as the larger fragment Y can form two DE-stag contacts upon binding to the terminal fibrin monomer of a protofibril.[42] Our model is consistent with the findings of Olexa and Budzynski, who have experimentally defined residues 373–410 of the γ chain of fragment D as a polymerization site on the D-domain.[46]

Observations on inhibition of gelation by the peptide Gly-Pro-Arg-Pro, which Laudano and Doolittle[45] have shown to bind reversibly to the D-domain of fibrinogen, also fit with our postulated inhibition scheme. Our model predicts that Gly-Pro-Arg-Pro blocks the first assembly step, protofibril growth, and that the same DE-stag contact is involved here, although the tetrapeptide corresponds to a portion of the polymerization site on the E-domain which interacts with the D-domain in contrast to fragment D which binds to the E-domain. Our model would predict that antibodies directed against polymerization sites on either the D or E domain will also inhibit gelation by blocking protofibril growth.

References

1. FERRY, J. D. 1952. The mechanism of polymerization of fibrinogen. Proc. Natl. Acad. Sci. USA **38:** 566–569.
2. SCHERAGA, H. A. & M. LASKOWSKI, JR. 1957. The fibrinogen-fibrin conversion. Adv. Protein Chem. **12:** 1–131.
3. DOOLITTLE, R. F. 1973. Structural aspects of the fibrinogen to fibrin conversion. Adv. Protein Chem. **27:** 1–109.
4. BLOMBÄCK, B., B. HESSEL, D. HOGG & L. THERKILDSEN. 1978. A two-step fibrinogen-fibrin transition in blood coagulation. Nature **275:** 501–505.
5. HANTGAN, R. R. & J. HERMANS. 1979. Assembly of fibrin: A light scattering study. J. Biol. Chem. **254:** 11272–11281.

6. BLOMBÄCK, B. 1958. Studies on the action of thrombotic enzymes on bovine fibrinogen as measured by N-terminal analysis. Ark. Kemi. **12:** 321–335.
7. STRYER, L., C. COHEN & R. LANGRIDGE. 1963. Axial period of fibrinogen and fibrin. Nature **197:** 793–794.
8. FOWLER, W. E., R. R. HANTGAN, J. HERMANS & H. P. ERICKSON. 1981. Structure of the fibrin protofibril. Proc. Natl. Acad. Sci. USA **78:** 4872–4876.
9. HANTGAN, R., W. FOWLER, H. ERICKSON & J. HERMANS. 1980. Fibrin assembly: A comparison of electron microscopic and light scattering results. Thromb. Haemostas. **44:** 119–124.
10. FERRY, J. D. & P. R. MORRISON. 1947. Preparation and properties of serum and plasma proteins. VIII. The conversion of human fibrinogen to fibrin under various conditions. J. Am. Chem. Soc. **69:** 388–400.
11. CARR, M. E. JR., L. L. SHEN & J. HERMANS. 1977. Mass-length ratio of fibrin fibers from gel permeation and light scattering. Biopolymers **16:** 1–15.
12. ROSSER, W. R., W. W. ROBERTS & J. D. FERRY. 1977. Rheology of fibrin clots. IV. Darcy constants and fiber thickness. Biophys. Chem. **7:** 153–157.
13. MARTINELLI, R. A. & H. A. SCHERAGA. 1980. Steady-state kinetic study of the bovine thrombin-fibrinogen interaction. Biochemistry **19:** 2343–2350.
14. SHEN, L. L., J. HERMANS, J. McDONAGH & R. P. McDONAGH. 1977. Role of fibrinopeptide B release: Comparison of fibrins produced by thrombin and ancrod. Am. J. Physiol. **232:** 629–633.
15. McDONAGH, J. & R. P. McDONAGH. 1980. Factor XIII. *In* Clinical Laboratory Science. R. M. Schmidt, Ed. 3 Sect. I. (Hematology): 125–140. CRC Press. Boca Raton, FL.
16. LORAND, L., J. DOWNEY, T. GOTOH, A. JACOBSEN & S. TOKURA. 1968. The transpeptidase system which crosslinks fibrin by γ-glutamyl-ε-lysine bonds. Biochem. Biophys. Res. Commun. **31:** 222–230.
17. DOOLITTLE, R. F., R. CHEN & F. LAM. 1971. Hybrid fibrin: Proof of the intermolecular nature of γ-γ crosslinking units. Biochem. Biophys. Res. Commun. **44:** 94–100.
18. FOWLER, W. E., H. P. ERICKSON, R. R. HANTGAN, J. McDONAGH & J. HERMANS. 1981. Cross-linked fibrinogen dimers demonstrate a feature of the molecular packing in fibrin fibers. Science. **211:** 287–289.
19. DOOLITTLE, R. F., K. W. K. WATT, B. A. COTTRELL, D. D. STRONG & M. RILEY. 1979. The amino acid sequence of the α-chain of human fibrinogen. Nature **280:** 464–468.
20. SHEN, L. L., J. HERMANS, J. McDONAGH, R. P. McDONAGH & M. CARR. 1975. Effects of calcium ion and covalent crosslinking on formation and elasticity of fibrin gels. Thromb. Res. **6:** 255–265.
21. HUGLIN, M. B., Ed. 1972. Light Scattering from Polymer Solutions. Academic Press. New York.
22. HOCKING, C. S., M. LASKOWSKI, JR. & H. A. SCHERAGA. 1952. Size and shape of bovine fibrinogen. J. Am. Chem. Soc. **74:** 775–778.
23. FERRY, J. D., S. SHULMAN, K. GUTFREUND & S. KATZ. 1952. The conversion of fibrinogen to fibrin. XI. Light scattering studies on clotting systems inhibited by hexamethylene glycol. J. Am. Chem. Soc. **74:** 5709–5715.
24. CASASSA, E. F. 1955. Light scattering from very long rod-like particles and application to polymerized fibrinogen. J. Chem. Phys. **23:** 596–597.
25. FLORY, P. J. 1953. *In* Principles of Polymer Chemistry: 318–323. Cornell Univ. Press. Ithaca, NY.
26. OOSAWA, F. & M. KASAI. 1962. A theory of linear and helical aggregations of macromolecules. J. Mol. Biol. **4:** 10–21.
27. ERICKSON, H. P. & W. FOWLER. 1983. Ann. N.Y. Acad. Sci. **408:**. This volume.
28. HERMANS, J. & J. McDONAGH. 1982. Fibrin: Structure and interactions. Semin. Thromb. Hemostas. **8:** 11–24.

29. HERMANS, J. 1979. Models of fibrin. Proc. Natl. Acad. Sci. USA **76:** 1189–1193.
30. OLEXA, S. A. & A. Z. BUDZYNSKI. 1980. Evidence for four different polymerization sites involved in human fibrin formation. Proc. Natl. Acad. Sci. USA **77:** 1374–1378.
31. HURLET-JENSEN, A., H. Z. CUMMINS, H. L. NOSSEL & C. Y. LIU. 1981. The polymerization of fibrin and the clearance of fibrinopeptide B by thrombin. Thromb. Haemostas. **46:** 182.
32. SHAFER, J. A., D. L. HIGGINS & S. D. LEWIS. 1982. Steady state kinetics for the thrombin catalyzed sequential release of fibrinopeptides A & B from human fibrinogen. Fed. Proc. **41:** 654.
33. CARR, M. E. & J. HERMANS. 1978. Size and density of fibrin fibers from turbidity. Macromolecules **11:** 46–50.
34. WEISEL, J. W., G. N. PHILLIPS, JR. & C. COHEN. 1981. A model from electron microscopy for the molecular structure of fibrinogen and fibrin. Nature **289:** 263–267.
35. TORBET, J., J.-M. FREYSSINET & G. HUDRY-CLERGEON. 1981. Oriented fibrin gels formed by polymerization in strong magnetic fields. Nature **289:** 91–93.
36. WILKINS, C. L., C. E. KLOPFENSTEIN, T. L. ISENHOUR & P. C. JURS. 1975. *In* Introduction to Computer Programming for Chemists-BASIC Version: 288–292. Allyn & Bacon. Boston, MA.
37. BLOMBÄCK, B., B. HESSEL, D. HOGG & G. CLAESSON. 1977. Substrate specificity of thrombin on proteins and synthetic substrates. *In* Chemistry and Biology of Thrombin. R. L. Lundbald, J. W. Fenton and K. G. Mann, Eds.: 275–290. Ann Arbor Science. Ann Arbor, MI.
38. LATALLO, Z. S., A. P. FLETCHER, N. ALKJAERSIG & S. SHERRY. 1962. Inhibition of fibrin polymerization by fibrinogen proteolysis products. Am. J. Physiol. **202:** 681–686.
39. KOWALSKI, E. 1968. Fibrinogen derivatives and their biologic activities. Semin. Hematol. **5:** 45–59.
40. MARDER, V. J. & N. R. SHULMAN. 1969. High molecular weight derivatives of human fibrinogen produced by plasmin. II. Mechanism of their anticoagulant activity. J. Biol. Chem. **244:** 2120–2124.
41. HAVERKATE, F., G. TIMAN & W. NIEUWENHUIZEN. 1979. Anticlotting properties of fragments D from human fibrinogen and fibrin. Eur. J. Clin. Invest. **9:** 253–255.
42. WILLIAMS, J. E., R. R. HANTGAN, J. HERMANS & J. McDONAGH. 1981. Characterization of inhibition of fibrin assembly by fibrinogen fragment D. Biochem. J. **197:** 661–668.
43. WILLIAMS, J. E., R. R. HANTGAN, D. KNOLL, J. McDONAGH & J. HERMANS. The mechanism of inhibition of fragment assembly by fibrinogen fragment D. Manuscript in preparation.
44. KNOLL, D. A. 1983. Ph.D. thesis.
45. LAUDANO, A. P. & R. F. DOOLITTLE. 1980. Studies on synthetic peptides that bind to fibrinogen and prevent fibrin polymerization. Structural requirements, number of binding sites and species differences. Biochemistry **19:** 1013–1019.
46. OLEXA, S. & A. Z. BUDZYNSKI. 1981. Localization of a fibrin polymerization site. J. Biol. Chem. **256:** 3544–3549.

DISCUSSION OF THE PAPER

H. A. SCHERAGA (*Cornell University, Ithaca, NY*): Your distribution looks very similar to what both Ferry's group and our group observed many years ago in the ultracentrifuge: we had two peaks, which we interpreted as something around the monomer and then the higher polymer. Now the thing that puzzles me is you say that the polymerization is second order in fibrinogen, whereas polymerization was first order in our work. I wonder how you reconcile that?

R. HANTGAN: When I began this study some years ago, I at first used fibrin monomer, as in your experiments, rather than fibrinogen and thrombin. I also found first order behavior, which I attributed to a change in conformation of the fibrin monomer when pH and NaBr concentration are returned to normal. This reaction is slower than subsequent, presumably bimolecular, polymerization, and is absent in the experiments I reported today.

L. LORAND (*Northwestern University, Evanston, IL*): I did not quite understand your last statement about the lack of reversibility of D binding to your fibrin molecule?

HANTGAN: We conclude that the inhibition by fragment D is irreversible from this consideration: if fragment D were to again dissociate, then this would permit further polymerization; however, we find that in these inhibited solutions the light scattering intensity doesn't change over a period of up to 24 hours. Thus, dissociation of bound D followed by the joining of two polymerization sites is an extremely unlikely event.

A. BUDZYNSKI (*Temple University Health Sciences Center, Philadelphia, PA*): Your kinetic model does not appear to consider binding D to anything smaller than dimers of fibrin monomers. This will give a maximum stoichiometry of one molecule of D per fibrin monomer molecule. If I extrapolate from the data that I presented, I found that two fragments of D-1 will bind to one fibrin monomer. Why do you not consider the possibility that two Ds may bind to one fibrin monomer?

HANTGAN: I think that is a possibility, but we have not observed that species under our experimental conditions.

J. HERMANS: In relation to Dr. Budzynski's question: Our kinetic model consists of two parallel alternative reactions. Once a binding site becomes available after removal of the A peptide, then that site can either engage in a polymerization step or can *immediately* engage in an inhibition step. So in our model, a fibrin monomer can, in fact, be blocked by two molecules of fragment D.

V. MARDER (*University of Rochester School of Medicine and Dentistry, NY*): Could you explain the greater anti-coagulant effect of fragment Y. of which fivefold fewer molecules per molecule of monomer are needed to completely block clotting?

HANTGAN: I think a possible explanation is that fragment Y can form two rather than just one DE staggered contact with the free end of the growing polymer. Perhaps also the structure of fragment D is not as complete as that of the D domain of fragment Y and therefore D is a less effective inhibitor than Y.

THE STRUCTURE OF FIBRINOGEN AND FIBRIN: II. ARCHITECTURE OF THE FIBRIN CLOT *

John W. Weisel

Department of Anatomy
University of Pennsylvania School of Medicine
Philadelphia, Pennsylvania 19104

George N. Phillips, Jr. and Carolyn Cohen

Rosenstiel Basic Medical Sciences Research Center
Brandeis University
Waltham, Massachusetts 02254

INTRODUCTION

An understanding of the mechanism of clot formation requires detailed information about the three-dimensional structure of fibrin. Such information, together with biochemical data on the binding sites, will reveal the intermolecular interactions controlling clot formation. As outlined previously,[1] our strategy for the analysis of the structure of fibrinogen focuses on microcrystals and crystals of the modified molecule, and fibrin is considered to be but one other ordered assembly. Studies of these aggregates, which use coordinated electron microscopy and x-day diffraction, can reveal aspects of the molecular interactions in the clot. This novel approach to the molecular basis of clotting is possible because the formation of the fibrin clot is essentially a self-assembly process. All of the information necessary for fibrin formation is present in the structure of the fibrin(ogen) monomer so that under appropriate conditions, the molecules can assemble to form ordered arrays with specific interactions similar to those in the clot. A comparison of the modes of packing in various ordered fibrinogen aggregates reveal essential linkages of the molecule. Moreover, we have produced some arrays where the molecular packing is directly related to that of fibrin. In this paper, we describe studies on both fibrin itself and on more highly ordered arrays of modified fibrinogen that are related to fibrin and allow us to deduce aspects of its molecular arrangement.

MOLECULES IN FIBRIN ARE HALF-STAGGERED AND BONDED END TO END

Fibrin fibers formed *in vivo* were described from early light microscope observations; they are long (5μ or more) and thin (up to about 0.2μ) with occasional branch points. In the electron microscope, fibrin that has been shadowed by the evaporation of metal displays a cross-striation with a period of about 225 Å.[2] Since Hall and Slayter had observed by electron microscopy

* This work was supported by Grant AM17346–09 from the NIH to C.C. Some of the computing was supported by BRSG Grants RR–05415 and RR–07083 to J.W.W. J.W.W. is an Established Investigator of the American Heart Association. G.N.P. was a fellow of the Medical Foundation, Inc. of Boston.

0077-8923/83/0408-0367 $01.75/0 © 1983, NYAS

that individual shadowed molecules of fibrinogen were considerably larger than 225 Å (about 475 Å), they proposed that shrinkage of the molecules accounted for this period.[3] This proposal, however, ran counter to x-ray diffraction experiments. Wide angle x-ray diagrams of fibrinogen and fibrin both show the α pattern, indicating that these proteins, at least in part, have the same basic molecular plan.[4] While wdie-angle x-ray patterns reveal the polypeptide chain folding, small angle x-ray diagrams give information about molecular dimensions. The small angle patterns of fibrinogen (at high ionic strength) and fibrin are also very similar: both show an axial period of 225 Å and a similar intensity distribution. These results confirm the view that no large scale changes occur in fibrin formation.[5] On the basis of the small angle X-ray results, it was thus proposed that the 225 Å period of fibrin arises from half-staggering of molecules with a length of 450Å.[5] It should be emphasized that x-ray experiments allow measurements of the native, hydrated structure.

This basic conclusion has now been strengthened since we can describe the molecular packing in the fibrin clot in more detail. Several levels of organization may be distinguished in arrays of fibrinogen or fibrin. We have deduced a working model for the molecular shape of fibrinogen from coordinated x-ray diffraction and electron microscope studies of a variety of microcrystals and crystals of modified fibrinogens.[1] The shape appears to be similar in all aggregates. All of these forms are built from filaments of molecules bonded end to end. Although the crystalline aggregates that we have produced display many different packing arrangements, a basic parameter is the lateral stagger between filaments. Adjacent filaments in crystal forms of modified fibrinogen tend to be staggered by about one third of the molecular length of 450 Å, while those in microcrystals are often half-staggered. The molecular arrangement in these microcrystals is closely related to that of fibrin; these arrays will be described in detail below. Both chemical and electron microscope evidence presented at this meeting are also consistent with a half-stagger between adjacent filaments in fibrin.

FIGURE 1. (a)-(d) Various electron microscope images of bovine fibrin, as described below. In all cases the axial repeat is 225 Å.

(a, inset) Fibrin stained with phosphotungstic acid at low pH. The band pattern consists of a broad dark staining region alternating with a narrower dark band. This image probably arises from the effects of both the mass distribution and the charge along the fiber. (From Hall and Slayter.[8] By permission of Journal of Cell Biology.)

(a) Fibrin prepared by clotting Pseudomonas protease-modified fibrinogen with thrombin, negatively stained with uranyl acetate at low pH. Bright axial striations about 70 Å wide alternate with three narrower light bands. This image can be accounted for by the mass distribution along the fiber (see FIGURE 2). Note particularly that these fibers end in the middle of the broad, light striation—a result consistent with our structural analysis. The fibers also display some lateral order.

(b) Same as (a) above, but the lateral order here is better. Measurement of the lateral repeat from optical diffraction patterns of this image yields a period of about 180 Å.

(c) Positively stained by first negative staining with uranyl acetate, as in (a) and (b), and then washing off the excess stain with 8–10 drops of water.[14] The contrast of the band pattern is reversed compared to (a) and (b).

(d) High ionic strength precipitate (0.3M KF at neutral pH) of native fibrinogen, negatively stained with uranyl acetate. Both the repeat and the band pattern are indistinguishable from those of native fibrin (inset) and very similar to (a) and (b) above.

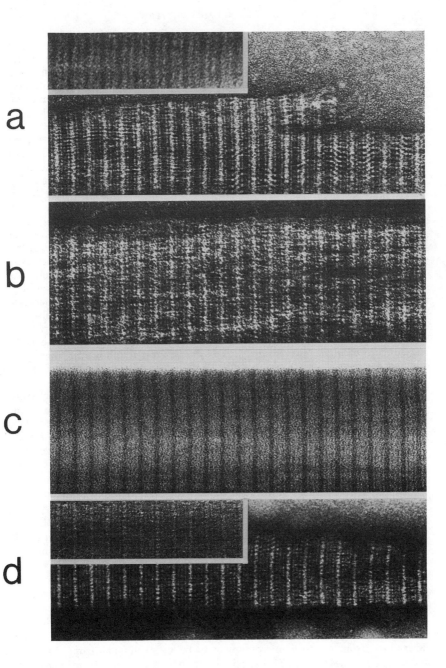

THE FIBRIN BAND PATTERN

Fibrin has been studied by electron microscopy since the early days of this technique, and several different images have been analyzed over the years. Since much of this research predated our present understanding of staining and contrast in the electron microscope, many reports contain spurious interpretations. With more recent knowledge of contrast in microscopy of proteins, as well as the structure of fibrinogen and aspects of the molecular packing in fibrin, we can now begin to account for these images. Shadow-cast fibrin observed in the electron microscope has prominent cross-striations with an axial repeat of 225 Å.[2] Since shadowing reveals features of the mass distribution of the specimen, the ridges indicate that there are regions of different density along the fiber. These images are of limited use, however, because of the poor resolution of this technique. Fibrin was first stained with heavy metals by the use of phosphotungstic acid at low pH.[6] These specimens display a band pattern with a repeat of 225 Å consisting of a broad dark region alternating with a narrower dark band (FIGURE 1a, inset). This image of fibrin was originally misinterpreted as a representation of the mass distribution along the polymer. We now know that these specimens have a good deal of positive staining [7] (see also below). Negatively contrasted fibrin, which does display a pattern representing the mass distribution along the fiber, was first observed by Cohen et al.[8, 9] In the technique of negative contrast, the relatively electron transparent protein is embedded in a continuous film of electron opaque stain. In such a micrograph, darker areas represent regions where stain accumulates because of low protein density, while light areas are stain-excluding domains of higher protein density. It should be noted, however, that these staining methods generally do not yield purely negative or positive contrast. Fibrins from all species that have been examined thus far have a similar appearance. Within each 225 Å repeat, there is one bright stain-excluding band about 70 Å wide and three narrow light bands, the middle one being somewhat less prominent or variable in tensity (FIGURES 1a, b and 2a). There are slight variations in these images of fibrin depending upon the intactness of the fibrinogen as well as the preparation for microscopy. For example, the central band is sometimes obscured in human or avian fibrins where proteolysis tends to occur. There have been numerous attempts to account for this image, although most have dealt mainly with the axial repeat rather than the details of the band pattern. One exception is the work by Kay and Cuddigan,[10] who proposed a model for fibrinogen with domains based on the size of the stain-excluding bands of fibrin. Although the molecular length and staggering of this model are incorrect, these authors recognized that the domains of fibrinogen must be considerably smaller than those of the Hall-Slayter trinodular model. A major point, however, is that the fibrin band pattern seen in negative contrast is not readily interpreted in terms of this simple model.

The heptad model that we have derived by coordinated electron microscopy and x-ray diffraction [1, 11] yields the observed band pattern simply and directly. Before discussing the modeling, it should be noted that the sizes of domains determined from electron microscopy may be overestimated from shadowed specimens because of metal deposition or underestimated from negatively stained specimens because of stain penetration. Thus, while relative dimensions of our model [1, 11] were determined from simulations of the electron microscope images, absolute sizes were calculated from the volume of the hydrated molecule.[12, 13] In FIGURE 2, a filtered electron micrograph of fibrin at the top (a and b) is compared with our computer simulation below (c). The inset shows that the arrangement of molecules that gives rise to this image consists of two half-staggered filaments of molecules bonded end to end. The juxtaposition of the end globular units of two molecules with the middle domain of an adjacent molecule gives rise to the bright axial striation. The narrower light bands on either side of this feature arise from the adjacent globular

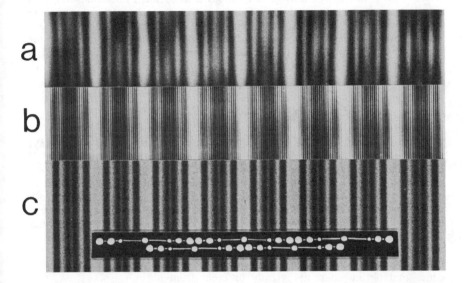

FIGURE 2. Simulation of electron microscope images of negatively stained fibrin. (a), (b) Computer-filtered electron micrographs of negatively contrasted fibrin. The axial repeat is 225 Å. In (a), the fibrin is derived from native fibrinogen clotted with thrombin, whereas in (b) *Pseudomonas* protease-modified fibrinogen has been clotted. Note that the weakest central band that arises from the smallest globular domain is less prominent in the fibrin from modified fibrinogen. (c) Video display of computer simulation of fibrin based on the molecular arrangement shown in the inset: that is, two half-staggered filaments of molecules bonded end-to-end. The derivation of this model for the fibrinogen molecule is described elsewhere.[1, 11]

domains in these molecules. The thinnest, central band in the period corresponds to the smallest globular domain in the molecule.

This simple model is also consistent with several other electron microscope images of fibrin. The model predicts that molecules will end in the middle of the broad light band of fibrin. Although the ends of fibrin fibers are usually tapered, occasionally we do see blunt ends that always terminate in the middle of a light band (FIGURE 1a).

A basic test of the validity of the interpretation of a negatively stained image is the reversal of contrast to a positively stained image: excess stain is washed away and the change in band pattern between light and dark regions can be observed.[14] This reversal does, in fact, occur and an example of positively stained fibrin is shown in FIGURE 1c. Such an image resembles a photographic negative of the usual negatively stained fibrin, but the contrast is lower since much of the stain has been removed. As described above, however, other methods of generating "positively stained" fibrin (e.g., with phosphotungstic acid) yield a somewhat different band pattern (FIGURE 1a, inset). A great many studies comparing these two types of images have been reported, but much of the early work was carried out before the methods, as well as the nature, of negative and positive contrast were well understood. Positive staining depends upon the charge distribution along the molecule, rather than the mass distribution as in negative staining. This is a rather complex matter but we have preliminary indications that these images are broadly consistent with our heptad model. In summary, the sizes and disposition of the domains in the molecular model now yield the correct dimensions for the band pattern. It will become clear below, however, that this picture is somewhat oversimplified.

FIBRIN-LIKE AGGREGATES OF FIBRINOGEN

Native fibrinogen can be precipitated at high ionic strength to form arrays that appear identical to fibrin in the electron microscope (FIGURE 1d).[8, 9, 15] These aggregates can be produced by precipitation of fibrinogen with divalent cations, fluoride,[15] protamine sulfate,[16] or citrate.[17] Whatever the exact conditions, in all cases the fibrin-like arrays are produced at high ionic strength where the charges on fibrinogen are shielded, allowing interactions between molecules that do not otherwise occur. In fact, the most common aggregates that we have observed in attempts to crystallize fibrinogen under a wide variety of conditions show a fibrin-like structure. Although these arrays are not sufficiently ordered for detailed structural analysis, they provide additional evidence that fibrinogen is not greatly different from fibrin.

THE ORTHOGONAL SHEET MICROCRYSTALS

One of the most remarkable and detailed arrays of modified fibrinogen is the orthogonal sheet form (FIGURE 3).[11, 12, 15] As noted early on,[15] this form reveals most directly the relationship between fibrinogen and fibrin. These arrays are produced by precipitation at intermediate ionic strength (about .15μ) of bovine fibrinogen which has been digested by treatment with bacterial protease. The motif here is a very complex dumbbell-shaped region about 400 Å long with a darker gap of about 50 Å. The motifs in adjacent rows separated by 450 Å are staggered by half of the lateral period of 90 Å so that the true axial repeat is 900 Å. Within the dark-staining gap there is a small globular domain as well as strands connecting the adjacent motifs. Superposition of images of the orthogonal sheet that are half-staggered generates a band pattern very similar to that of negatively stained fibrin. Like fibrin, this array consists of molecules bonded end to end and half-staggered. The exceptional detail of the thin orthogonal sheets, however, allows analysis of the molecular structure and lateral interactions between molecules.

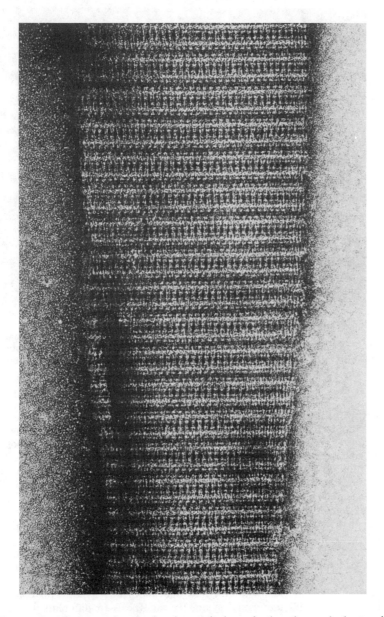

FIGURE 3. Electron micrograph of negatively stained orthogonal sheet microcrystal. The fibrinogen was slightly digested by the *Pseudomonas* protease and precipitated at intermediate ionic strength (about .15μ). The unit cell dimensions are 900 Å, plane group cmm. Note the regions of fibrin-like band pattern along the edges.

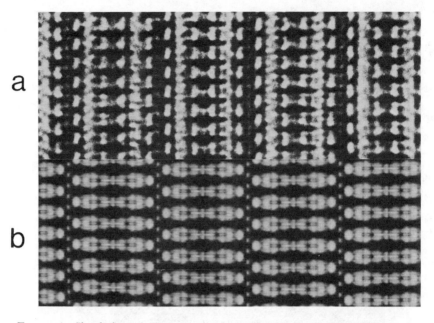

FIGURE 4. Simulation of electron microscope image of orthogonal sheet micro-crystals. (a) Filtered electron micrograph of the orthogonal sheet microcrystal (cf. the unfiltered image in FIGURE 3). (b) Video display of the computer simulation of the image in (a) using the bent heptad model, as described elsewhere.[1, 11] The molecules are both tilted and bent slightly.

Using the heptad model and methods described previously,[1, 11] we have deduced a probable packing arrangement for the orthogonal sheet (FIGURE 4). The high degree of symmetry of this structure limits considerably the possible modes of packing. As in fibrin, the axial striations which occur every 225 Å arise from the juxtaposition of the end globular domains of two molecules which form a pocket for the middle domain of a neighboring molecule. The pairs of smaller bright units on either side of the striations arise from the adjacent domains. The smallest domains are seen as bright nodes in the darker regions halfway between each bright striation. Both the angle of the thin strands seen in the gap region and the lateral disposition of the bright nodes require that the molecules be both tilted and bent slightly. Since this image is a projection of two layers of molecules, it is difficult to visualize the molecular packing.

Not allowing molecules to run into one another is a powerful constraint which, together with the crystal symmetry, allows us to differentiate the two layers of this structure (FIGURE 5a). It is immediately apparent by inspection of a single layer that the orthogonal sheet is made up of two-stranded proto-fibrils, i.e., filaments are paired. As suggested by Fowler et al.,[18] all molecules in one filament face the same direction and the two filaments face each other. The protofibril is a "closed structure" in the sense that all binding sites (two in the central domain exposed upon removal of the A fibrinopeptides and one complementary binding site in each end region) interact with molecules in the adjacent, half-staggered filament. (Note that this molecular packing contrasts with that of filaments with a one-third stagger in the macroscopic crystals where all filaments face in the same direction to form an open, or unlimited, structure.[1, 11]) This is the first time that these detailed features, proposed from the existence of the two-stranded protofibril and the complementary binding sites,[18] have been resolved. The twofold symmetry axes that generate the two-stranded protofibril are readily apparent also since they are symmetry elements of the microcrystal. The molecular twofold axis of fibrinogen described by Cohen *et al.*[1] is in the plane of the sheet and is not a crystallographic axis. Furthermore, it may be seen that adjacent protofibrils are not aligned but half-staggered. All of the protofibrils in any one layer are parallel, but are tilted slightly with respect to the long axis of the microcrystal; alternate

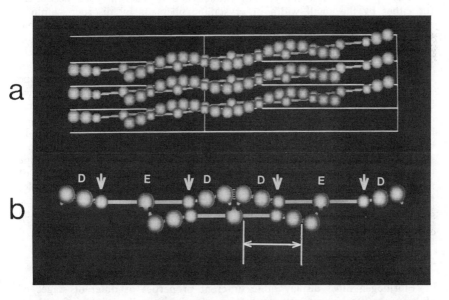

FIGURE 5. Model for fibrin assembly. (a) Computer model of a single layer of the orthogonal sheet microcrystal. Each layer is made up of parallel arrays of protofibrils, or paired filaments. Protofibrils in a second layer, not shown, are tilted in the opposite direction. Adjacent protofibrils are half-staggered. (b) Functional aspects of molecular model. Filaments made up of end-to-end bonded molecules are half-staggered and face each other. Binding sites exposed by the removal of the A fibrinopeptides in the central domain are represented as holes while the complementary binding sites in the end domains are represented as knobs; both sets lie along the same "face" of the molecule. Cleavage sites essential for fibrinolysis through fragmentation to the D and E products are shown by the vertical arrows. These regions were identified as the smallest globular domain and correspond to the plasmin-sensitive region in the middle of the coiled-coil determined from the amino acid sequence. (The predicted coiled-coil region [21] is represented by the horizontal arrow at the bottom right.) The D and E domains, localized on this basis, are also labeled. The Factor XIII ligation sites between the γ chains of two adjacent end domains are also represented schematically.

layers are tilted in opposite directions. The orthogonal sheet is a basic array that underlies many other microcrystal forms such as those in FIGURE 2 of Cohen et al.[1]). Many of these arrays can now be accounted for by modifications of the packing model presented here. This result emphasizes the close relationship of fibrin to the microcrystalline forms of modified fibrinogen.

IS FIBRIN A CRYSTAL?

The question of the degree of order in fibrin is important in order to relate its structure more directly to that of the orthogonal sheet microcrystals. Present evidence indicates that the molecules in fibrin do not form a truly crystalline lattice. The fibrin x-ray diagram, like that of other fibrous proteins, is characterized by a high degree of order in the axial direction, but relatively poor order laterally. The limited lateral order may be accounted for either by the small width of crystalline fibers or by disorder within a single fiber. Fibrin fibers that have been oriented in a strong magnetic field have also been studied by neutron diffraction.[19] Although this orientation method markedly improves the diffraction patterns, there are still relatively few reflections arising from lateral features. Electron microscopy may be of some use in interpreting the diffraction experiments. Electron microscope images of fibrin are generally ordered only in the axial direction, although we have occasionally observed fibers with some lateral order (FIGURE 1b). Optical diffraction of these micrographs shows weak reflections corresponding to spacings in the lateral direction of about 180 Å, similar to that measured in oriented specimens by neutron diffraction.[19] In contrast to the orthogonal sheet microcrystals, therefore, fibrin does not appear to display crystalline lateral associations although both forms have the same axial arrangement of two-stranded protofibrils. In summary, the precise arrangement of molecules to form protofibrils generates extensive axial order in fibrin; these protofibrils, however, appear to associate in a number of different ways, resulting in limited lateral order.

FUNCTIONAL REGIONS OF FIBRINOGEN

Fibrinogen must display quite different properties at different stages of clotting: its life history includes a stage of solubility, of rapidly induced insolubility and assembly, of stabilization and of dissolution. Certain domains of its structure may be associated with some of these functions (FIGURE 5b). Chemistry can first be related to morphology for the regions of the molecule involved in the assembly of fibrin. The small, highly charged fibrinopeptides cleaved by thrombin are located in the central domain. The outer globular domains contain the complementary binding sites for fibrin formation. The pair of A fibrinopeptides and their complementary binding sites appear to lie along one face of the molecule and are represented schematically in FIGURE 5b as knobs and holes. As described above, this location results in the formation of the two-standed protofibril. The B fibrinopeptides and associated binding sites cannot yet be located with certainty since their removal does not appear to be a prerequisite for aggregation of protofibrils. Covalent cross-linking of γ chains by fibrin-stabilizing factor involves end-to-end bonding between the outer globular domains of two molecules.[20]

The plasmin-sensitive portion of the molecule can be identified with the smallest globular domain which interrupts the regular run of the rod-like regions (FIGURE 5b). The α helical coiled-coil region was identified in the amino acid sequence and a distinctly nonhelical domain was predicted to be in the middle of the coiled-coil.[21, 22] The distance of this region from the center of the molecule corresponds to the location of the smallest globular region in our model. Cleavage here is crucial for generating the D and E fragments of fibrin. In our model of fibrin, all of these plasmin-sensitive domains are aligned so that dissolution of the clot by cleavage at this site is easily accomplished (FIGURE 5b). This conclusion, which is a good example of efficiency in a biological process, is worth restating. The enzymatic specificity of plasmin requires that each molecule of fibrinogen be cleaved in the same, minimal number of locations. A simple way to dissolve a fiber is to cut straight across it. To fulfill both of these requirements for a structure made up of half-staggered filaments of end-to-end bonded molecules, one would cleave the molecules at a distance of one quarter of the length from either end. We have found that the plasmin-sensitive region for the formation of the D and E fragments is in approximately this location. This finding also gives us some information about the lengths of fragments D and E because the distal ends of the molecules appear to be intact during the early stages of plasmin degradation (since D fragments still bind end to end). On the basis of these arguments, then, E would be about 225 Å long while D would be about half this length.

The smallest globular region appears slightly smaller in fibrin made by clotting our modified fibrinogen (FIGURE 2b) than in native fibrin (FIGURE 2a), suggesting that proteolytic treatment removes some mass from this region. We already know that the bacterial protease used to modify fibrinogen for crystallization removes a portion of the carboxy-terminal end of the Aα chain.[12] We also know that this same region is very susceptible to other proteases as well as a variety of other treatments. Correspondingly, the smallest globular domain is the most labile feature of the band pattern; it may decrease in prominence from protease digestion or various preparative treatments for microscopy. It is also possible that part of the smallest globular domain may comprise a piece of the C-terminus of the Aα chain. Such a proposal would be consistent with the suggestion of Mosesson *et al.*[23] that this chain is folded back toward the center of the molecule.

SUMMARY

Our present low resolution model for fibrinogen based on electron microscopy and x-ray diffraction data has been described by Cohen *et al.*[1] A unique aspect of the structural analysis of fibrous proteins is that the molecular packing in ordered arrays reflects biologically significant intermolecular interactions. We have shown that the orthogonal sheet microcrystals, which are closely related to fibrin, are made up of a highly regular arrangement of two-stranded protofibrils, and we have visualized aspects of both the substructure of the protofibrils as well as their packing to form the fibrin clot. By correlation of structural data with biochemical studies we have begun to identify certain functional regions of the fibrinogen model related to fibrin. Many aspects of fibrinogen's physiological activity remain to be related to its structure. As our present model is improved by higher resolution studies, we will see with increasing clarity molecular features critical for clot formation and fibrinolysis.

ACKNOWLEDGMENTS

We thank Drs. Cynthia Stauffacher and James Fillers for helpful discussions, Paul Norton for electron microscopy, and William Saunders and Vicki Ragan for photography.

REFERENCES

1. COHEN, C., J. W. WEISEL, G. N. PHILLIPS, JR., C. V. STAUFFACHER, J. P. FILLERS & E. DAUB. 1983. Ann. N. Y. Acad. Sci. 408:. This volume.
2. HAWN, C. V. Z. & K. R. PORTER. 1947. J. Exp. Med. 86: 285–292.
3. HALL, C. E. & H. S. SLAYTER. 1959. J. Biophys. Biochem. Cytol. 5: 11–16.
4. BAILEY, K., W. T. ASTBURY & K. M. RUDALL. 1943. Nature 151: 716–717.
5. STRYER, L., C. COHEN & R. LANGRIDGE. 1963. Nature 197: 793–794.
6. HALL, C. E. 1949. J. Biol. Chem. 179: 857–864.
7. KARGES, H. E. & K. KUHN. 1970. Eur. J. Biochem. 14: 94–97.
8. COHEN, C., J.-P. REVEL & J. KUCERA. 1963. Science 141: 436–438.
9. COHEN, C., H. SLAYTER, L. GOLDSTEIN, J. KUCERA & C. HALL. 1966. J. Mol. Biol. 22: 385–388.
10. KAY, D. & B. J. CUDDIGAN. 1967. Br. J. Haemat. 13: 341–347.
11. WEISEL, J. W., G. N. PHILLIPS, JR. & C. COHEN. 1981. Nature 289: 263–267.
12. TOONEY, N. M. & C. COHEN. 1977. J. Mol. Biol. 110: 363–385.
13. WEISEL, J. W., S. G. WARREN & C. COHEN. 1978. J. Mol. Biol. 126: 159–183.
14. COHEN, C., A. G. SZENT-GYORGYI & J. KENDRICK-JONES. 1971. J. Mol. Biol. 56: 223–237.
15. TOONEY, N. M. & C. COHEN. 1972. Nature 237: 23–25.
16. STEWART, G. J. & S. NIEWIAROWSKI. 1969. Biochim. Biophys. Acta 194: 462–469.
17. GOLLWITZER, R., W. BODE, H.-J. SCHRAMM, D. TYPKE & R. GUCKENBERGER. 1983. Ann. N. Y. Acad. Sci. 408:. This volume.
18. FOWLER, W. F., R. R. HANTGAN, J. HERMANS & H. P. ERICKSON. 1981. Proc. Natl. Acad. Sci. USA 78: 4872–4876.
19. TORBET, J., J.-M. FREYSSINET & G. HUDRY-CLERGEON. 1981. Nature 289: 91–93.
20. FOWLER, W. F., H. P. ERICKSON, R. HANTGAN, J. McDONAGH & J. HERMANS. 1981. Science 211: 287–289.
21. DOOLITTLE, R. F., D. M. GOLDBAUM & L. R. DOOLITTLE. 1978. J. Mol. Biol. 120: 311–325.
22. PARRY, D. 1978. J. Mol. Biol. 120: 545–551.
23. MOSESSON, M. W., J. HAINFELD, J. WALL & R. H. HASCHEMEYER. 1981. J. Mol. Biol. 153: 695–718.

DISCUSSION OF THE PAPER

R. C. WILLIAMS (*University of California at Berkeley*): I must say that I have never seen from my own material or from anyone else's material, a human thrombin-clotted fibrin that has a central band that is comparable to the central band that one finds in bovine fibrin.

J. WEISEL: One of the things that I did in looking at the band pattern of fibrin was to go back to the literature. In the mid-60s Kay and Cuddigan showed very distinctly the central band in human fibrin. It is there but it seems to be a more labile feature of the band pattern than it is in bovine fibrin.

H. ERICKSON (*Duke University Medical Center, Durham, NC*): How consistent and reproducible do you find the banding pattern in clotted fibrin and precipitated fibrinogen?

WEISEL: The band patterns are very similar and repeatable. I should emphasize though that we can see much more directly the domainal structure of the molecule from the crystal forms.

ORIENTATION OF FIBRIN IN STRONG MAGNETIC FIELDS

Gilbert Hudry-Clergeon and Jean-Marie Freyssinet

Laboratoire d' Hématologie
Centre d' Etudes Nucléaires
F- 38041 Grenoble, Cedex, France

James Torbet

Hochfeld-Magnetlabor des Max-Planck-Institutes
für Festkörperforschung
F- 38042 Grenoble, Cedex, France

Jean Marx

Laboratoire de Recherches Optiques
Faculté des Sciences
F-51062 Reims, Cedex, France

The magnetic orientation of some proteins can be attributed to the diamagnetic anisotropy of the molecules.[1] This anisotropy is due to oriented aromatic groups and to oriented peptide bonds. It seems that the average orientation of the aromatic groups is such that they have a small influence. In α helices the planar peptide bonds are oriented parallel to the helix axis. Therefore α helical structures orient with the α helices parallel to the magnetic field. In β sheets the planar peptide bonds are nearly parallel to the sheet and β structures also orient axially. The diamagnetic anisotropy $\Delta\chi$ for a structure containing N peptide bonds is 7.3×10^{-29} N $(J \times T^{-2})$ for an α helix compared to 1.8×10^{-29} N for a β pleated sheet.[1] Consequently, the diamagnetic anisotropy of α helical structures predominates. Orientation can occur only if the magnetic energy E is significant with respect to the thermal energy kT. For example, for an α helix containing N peptide bonds, $E = \frac{1}{2} B^2\Delta\chi = 3.6 \times 10^{-29} B^2 N$ (J) where B is the field strength.[1] Magnetic orientation of biological samples has been successful with membranes [2-4] and cylindrical viruses.[5, 6] In this paper, we report some recent results concerning the orientation of fibrin in strong magnetic fields. Fibrinogen is known to contain $\simeq 33\%$ α helix measured by ORD.[7] A significant part of this structure is arranged in two coiled-coils forming rodlike connectors between the D and E domains.[8] In fibrin, these coiled-coils are oriented along the axis of the fiber as predicted long ago by x-ray scattering.[9] This fact explains without doubt our success in experiments consisting in the orientation of the fibrin fibers during the polymerization process.[10-12] These experiments give interesting new data on the polymerization process and on the structure of fibrin.

MATERIALS AND METHODS

Fibrinogen was purified from bovine plasma [13] and thrombin was purchased from I.S.H., Paris. Oriented clots were obtained by slow polymerization in strong magnetic fields (10 to 20 teslas, 1 tesla $=$ 100,000 gauss) available at

0077–8923/83/0408–0380 $01.75/0 © 1983, NYAS

the Service National des Champs Intenses, CNRS, Grenoble.[10-12] The magnetically induced birefringence was measured using a combined photoelastic modulation and compensation technique.[14]

The neutron diffraction patterns were obtained on a small angle scattering camera at the Institut Laue-Langevin, Grenoble. The oriented clots were usually observed after dialysis against D_2O buffer. Raman spectroscopy was performed with an equipment available at the Faculté des Sciences, Reims. A substractive method was used to observe the Amide I vibration normally masked by the strong band of the H_2O bending mode.[15] Unless stated, all experiments were performed in 0.10 to 0.15 *M* NaCl solutions buffered with 0.05 *M* Tris or 0.01 *M* sodium phosphate (pH 7.5).

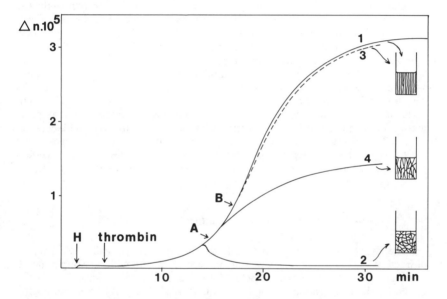

FIGURE 1. The change with time of the birefringence under conditions allowing full orientation of the fibrin fibers (curve 1). When the field is cut below point A (curve 2) the final clot is unoriented. Cutting the field above point B (approximately the gel point) results in a full orientation (curve 3). Different orientation degrees are obtained when the field is cut between A and B (curve 4).

RESULTS AND DISCUSSION

The Polymerization Process

The degree of orientation of the protein material that has been placed in the magnetic field can be estimated during the experiments by the optical birefringence Δn of the sample.[10-12] A typical experiment is shown in FIGURE 1. In the absence of thrombin, the orientation is very small because the energy of orientation is low compared to the thermal energy kT. After thrombin addition, a sigmoidal variation of Δn is observed (curve 1). The upsurge in the birefringence begins with polymerization. Now large aggregates are formed, and the magnetic orienting energy gives rise to significant orientation. The increase of Δn terminates in a plateau whose value depends on several parameters such as field strength, rate of polymerization, ionic strength, and protein concentration. If the experimental conditions are well chosen, a limiting value can be reached that corresponds to a full orientation of the fibers. Typically, fully oriented clots can be obtained under the following conditions using a magnetic field of 11T : fibrinogen concentration $\leqslant 5$ mg/ml, ionic strength $\leqslant 0.15$, clotting time $\geqslant 20$ mn (temperature $= 22°C$). The orientation is favored under conditions allowing the formation of thick fibers, i.e., low protein concentration, low ionic strength, traces of calcium, and long clotting times.[12]

The study of the birefringence curves shows that after approximately the gel point the polymerization reaction follows pseudo-first order kinetics.[12] Experiments involving cutting the magnetic field at different stages give new interesting data on the polymerization process. If the field is switched off below time A Δn decreases rapidly (FIGURE 1, curve 2). This indicates the disorientation of the first polymers by thermal motion. If the field is switched off above a critical time B, which corresponds approximately to the gel point, there is very little effect and a nearly full orientation is obtained (curve 3). This indicates that at time B a rigid matrix has already been formed and that polymerization continues using this matrix as a basis. Cutting the field between A and B gives intermediate orientation degrees (curve 4). This suggests that the final gel structure depends on early steps of the reaction : after a certain point (B in FIGURE 1) no new fibers are formed and polymerization continues by lateral growth of already existing fibers.[12]

The Crystalline Lattice of Fibrin

The structure of fully oriented gels can be studied by small angle neutron diffraction.[11, 12] FIGURE 2 shows the pattern obtained with a sample ($\simeq 10$ mg/ml) oriented in a magnetic field of 15 T and dialyzed against D_2O buffer to reduce the background level. The familiar 223 Å (± 5 Å) axial periodicity can be observed along the meridian. The sharpness of these reflections reveals a high degree of crystallinity along the fiber axis and their small arcing indicates that the fibers are highly oriented along the magnetic field direction. In addition to the meridional reflections the pattern shows equatorial and other Bragg reflections arising from a regular lateral arrangement of monomers and protofibrils. A detailed interpretation (see reference 11) shows that the neutron diffraction pattern can be indexed on a tetragonal unit cell ($a = b = 185$ Å and $c = 446$ Å). The true axial repeat of 446 Å agrees with the view that $\simeq 450$ Å long fibrin monomers form a half-staggered arrangement parallel to the fiber axis.[16] The small number of unit cells in the cross-section of a fiber can explain the breadth of the equatorial reflections and does not necessarily mean that lateral crystallinity is less than along the axis. Using a value of 6 Å^3 per dalton for the specific volume,[17] the unit cell would contain eight fibrin monomers or to be more precise, four monomers and eight half-monomers. Comparative experiments performed on oriented clots prepared in the presence or absence of calcium and on samples cross-linked by Factor XIII_a show no variation in the unit cell.[12]

The Secondary Structure of Fibrin

Fibrinogen contains a relatively high amount ($\simeq 33\%$) of α helix.[7] About two-thirds of this structure are arranged in two coiled-coils connecting the D and E domains.[8] These coiled-coils are oriented along the fiber axis according to X-ray scattering[9] and birefringence data.[12] The β sheet structure which is relatively low in fibrinogen ($\simeq 10\%$) increases significantly during the polymerization process,[15] as shown by Raman spectroscopy. This increase in β sheet

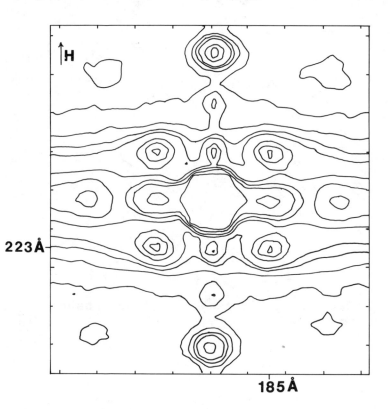

FIGURE 2. Low angle diffraction pattern from fibrin oriented in a magnetic field of 15 teslas (150 000 gauss). After orientation, the gel is dialyzed against a D_2O buffer to reduce the background level. Protein concentration $\simeq$ 10 mg/ml.

structure is favored under conditions allowing the formation of thick fibers.[18] For example, the β sheet content is $\simeq 20\%$ in fine clots formed at high ionic strength and $\simeq 30\%$ in coarse clots formed at low ionic strength. Hence the influence of the fiber diameter suggests that the β sheet structure is a result of lateral contacts between fibrin monomers. This statement is also supported by polarized Raman spectroscopy performed on clots oriented in high magnetic fields.[18] Two particular geometries of the directions of the incident and scattered light relative to the fiber axis are shown in FIGURE 3 with the cor-

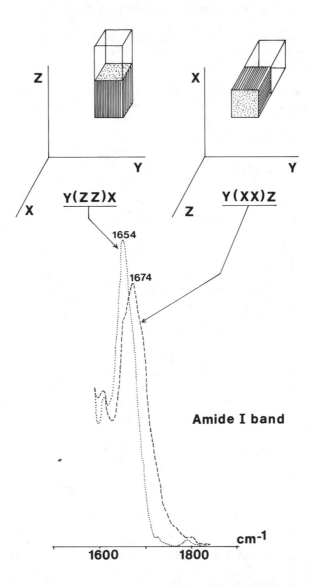

FIGURE 3. Polarized Raman spectroscopic study of oriented fibrin. Shown are two particular geometries, Y(ZZ)X and Y(XX)Z, of the directions of the incident and scattered light relative to the fiber axis and the two corresponding spectra of the Amide I band. The α helix vibration is increased and the β sheet vibration is decreased in the Y(ZZ)X spectrum compared to the Y(XX)Z one. This indicates an orientation of the $C = 0$ bonds of α helices parallel to the fiber axis and a perpendicular orientation of the $C = 0$ bonds of the β sheet structure.

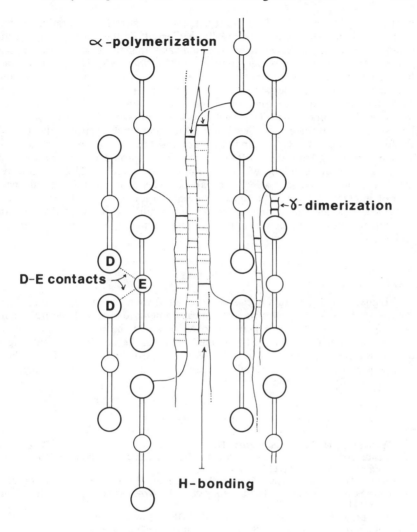

FIGURE 4. Schematic representation of hydrogen-bonding resulting in β sheet structure which could occur between the C-terminal parts of the α chains arranged nearly parallel to the fiber axis. This type of interaction adds to already known interactions between D and E domains, γ chain dimerization, and α chain polymerization.

responding Raman spectra in the Amide I region. The Amide I frequencies chiefly due to $C = 0$ stretching are determined by Φ and Ψ dihedral angles and are sensitive to H-bonding. The $C = 0$ stretching vibration in the α helix region (1654 cm^{-1}) is greatly increased in the $Y(ZZ)X$ spectrum compared with the $Y(XX)Z$ one. This indicates that the $C = 0$ bonds of the α helices are nearly parallel to the fiber axis, in agreement with crystallographic [9] and birefringence [12] data. On the contrary, the $C = 0$ stretching vibration in the β sheet region (1674 cm^{-1}) is greatly decreased in the $Y(ZZ)X$ spectrum compared with the $Y(XX)Z$ one. This indicates that the $C = 0$ bonds of the β sheet structure are nearly perpendicular to the fiber axis. Hence, this structure results from H-bonding between peptidic chains which are parallel to the fiber axis. Moreover, the positive effect of the fiber thickness on the β sheet content of fibrin indicates that these interactions develop laterally from one monomer to another one. The relatively high number of intermonomer H-bonds seems to exclude that they result from interactions between the D and E globular domains. Better candidates are the long C-terminal parts of the α chains emerging from the two D domains. These protuberances are involved in the α chain cross-linking which is also favored under conditions allowing the formation of coarse clots.[19] We can conceive that the role of this new type of interaction (schematic representation in FIGURE 4) results in strengthening the cohesion inside and between the protofibrils of the fibrin fiber in addition to interactions between the D and E domains, γ chain dimerization and α chain polymerization.

SUMMARY

Magnetic field orientation can be considered as a new valuable technique for the study of fibrin. Birefringence measurements during polymerization in the magnetic field enables a new approach to study the mode of association of fibrin monomers. Stable fully oriented clots can be obtained for investigation of the three-dimensional and the secondary structures of the fibrin fiber. Such experiments could be useful to assess the structure of other biopolymers diamagnetically anisotropic.

REFERENCES

1. WORESTER, D. L. 1978. Proc. Natl. Acad. Sci. USA **75:** 5475–5477.
2. GEACINTOV, N. E., R. VAN NOSTRAND, M. POPE & J. B. TINKEL. 1972. Biochim. Biophys. Acta **226:** 486–491.
3. SAIBIL, H. R., M. CHABRE & D. L. WORCESTER. 1976. Nature **262:** 266–270.
4. NEUGEBAUER, D.-CH., A. E. BLAUROCK & D. L. WORCESTER. 1977. FEBS Lett. **78:** 31–35.
5. TORBET, J. & G. MARET. 1979. J. Mol. Biol. **134:** 843–845.
6. TORBET, J. & G. MARET. 1981. Biopolymers **20:** 2657–2669.
7. MIHALYI, E. 1965. Biochim. Biophys. Acta **102:** 487–499.
8. DOOLITTLE, R. F., D. M. GOLDBAUM & L. R. DOOLITTLE. 1978. J. Mol. Biol. **120:** 311–325.
9. BAILEY, K., W. T. ASTBURY & K. M. RUDALL. 1943. Nature **151:** 716–717.
10. HUDRY-CLERGEON, G., J.-M. FREYSSINET & J. TORBET. 1980. *In* Proceedings of the 28th Colloquium on Protides of the Biological Fluids. H. Peeters, Ed. **28:** 317–320. Pergamon Press. Oxford and New York.
11. TORBET, J., J.-M. FREYSSINET & G. HUDRY-CLERGEON. 1981. Nature **289:** 91–93.
12. FREYSSINET, J.-M., J. TORBET, G. HUDRY-CLERGEON & G. MARET. 1983. Proc. Natl. Acad. Sci. USA. In press.
13. KEKWICK, R. A., M. E. MACKAY, M. M. NANCE & B. H. RECORD. 1955. Biochem. J. **60:** 671–683.
14. MARET, G. & K. DRANSFELD. 1977. Phys. B, C, **86:** 1077–1083.

15. MARX, J., G. HUDRY-CLERGEON, F. CAPET-ANTONINI & L. BERNARD. 1979. Biochim. Biophys. Acta **578:** 107–115.
16. FERRY, J. D. 1952. Proc. Natl. Acad. Sci. USA **38:** 566–569.
17. HERMANS, J. 1979. Proc. Natl. Acad. Sci. USA **76:** 1189–1193.
18. MARX, J., G. HUDRY-CLERGEON & M. BERJOT. To be published.
19. DOOLITTLE, R. F. 1973. Adv. Protein Chem. **27:** 1–109.

DISCUSSION OF THE PAPER

H. A. SCHERAGA (*Cornell University, Ithaca, NY*): If your interpretation of the amide-I frequency is correct you should be able to see that even without orienting the fibrin. If you are getting α-α C-terminal associations to form β sheets you should be able to see it in your Raman spectra even without orienting the fibrin. Do you?

G. HUDRY-CLERGEON: No, the association of α chains is only an hypothesis to explain the β sheet increase, but we do observe this increase even without orienting the fibrin.

J. D. FERRY (*University of Wisconsin, Madison*): But maybe you do not get the α chains lined up unless they are oriented.

HUDRY-CLERGEON: We suppose that α chains orient parallel to the fibrin axis in the two cases: with or without fiber orientation.

THE ABNORMAL CARBOHYDRATE COMPOSITION
OF THE DYSFIBRINOGENEMIA ASSOCIATED
WITH LIVER DISEASE *

Jose Martinez, P. M. Keane, and P. B. Gilman

Department of Medicine
Cardeza Foundation for Hematologic Research
Thomas Jefferson University
Philadelphia, Pennsylvania 19107

Joseph E. Palascak

Division of Hematology-Oncology
Department of Internal Medicine
University of Cincinnati College of Medicine
Cincinnati, Ohio 45267

INTRODUCTION

An acquired dysfibrinogenemia has been described in a variety of hepato-cellular disorders including alcoholic cirrhosis,[1,2] postnecrotic cirrhosis,[1,3] chronic active liver disease,[2] severe hepatitis,[4,5] acute hepatic necrosis following acetaminophen (paracetamol) overdose,[3,6] acute liver failure[2] and hepatoma.[7-9] In those patients in whom the purified abnormal fibrinogen has been studied, the functional defect has been demonstrated to be due to an impairment of fibrin monomer polymerization, i.e., an impairment in the second step of the thrombin-induced conversion of fibrinogen to fibrin; fibrinopeptide release, the first step in this thrombin-induced reaction, is normal.[1,9] The dysfibrinogenemia associated with liver disease has been further characterized by an increased sialic acid content[3,4,9] which directly correlated with the prolongation of its thrombin time (FIGURE 1).[3] Enzymatic desialylation of the abnormal fibrinogen resulted in an asialoderivative whose thrombin time was the same as that of the asialoderivative of normal fibrinogen.[3] These findings provide further evidence of the important influence of sialic acid on the functional properties of the fibrinogen molecule.[10] Periodic acid-Schiff reactivity was limited to the Bβ and γ chains of reduced normal and patient fibrinogens on sodium dodecyl sulfate-polyacrylamide gel electrophoresis (SDS-PAGE) suggesting that the carbohydrate moiety, and thus the sialic acid content, was located on these chains.[3,9]

To further evaluate the role of sialic acid in the dysfibrinogenemia, we studied the effects of removal of only the excess sialic acid residues of the abnormal fibrinogen on its thrombin time and fibrin monomer aggregation. The distribution of tritium-labeled sialic acid on the constituent chains of normal fibrinogen, patient fibrinogen and partially desialylated patient fibrinogen was studied on SDS-PAGE of the reduced proteins.

* This work was supported in part by National Institutes of Health Research Grant HL–20092.

MATERIALS AND METHODS

The abnormal fibrinogens studied were obtained from three patients with alcoholic cirrhosis and two with postnecrotic cirrhosis of undetermined etiology whose plasma thrombin times were 50% longer than normal controls.[12] Blood was collected and fibrinogen purified from patients and normal controls as previously described.[1, 3]

Total sialic acid content of the fibrinogens was measured by the thiobarbituric acid method [11] after acid hydrolysis or cleavage with *Vibrio cholerae*

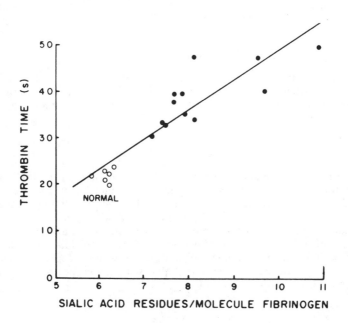

FIGURE 1. Comparison of sialic acid content vs. thrombin time of normal (○) and 12 patient (●) purified fibrinogens. Linear regression analysis of the data revealed a correlation coefficient of +0.91. The equation of the thrombin time vs. sialic acid content is $Y = 6.51 \times - 15.32$. (From Martinez *et al.*[3] By permission of *Journal of Clinical Investigation.*)

neuraminidase {Behringwerke AG, Marburg/Lahn, West Germany}.[10] Partially desialylated fibrinogens with a normal number of sialic acid residues remaining were obtained by incubating the fibrinogens with 50 units of neuraminidase per mg fibrinogen for 1½ to 2½ minutes, depending upon the amount of sialic acid to be removed. The neuraminidase was then rapidly removed from the reaction mixture by rabbit anti-neuraminidase antibody affinity chromatography.[12] The rabbit anti-neuraminidase antibody employed in these experiments was produced by injecting New Zealand white rabbits subcutaneously with *Vibrio cholerae* neuraminidase mixed with Freund's adjuvant.[12]

Labeling Studies

The sialic acid residues of normal fibrinogen and of patient fibrinogen, both with its full sialic acid complement and with its excess sialic acid residues removed, were labeled with tritium by the technique of Van Lenten and Ashwell.[12, 13] Two methods were employed to demonstrate restriction of the tritium label to the sialic acid analogues of fibrinogen. One method involved enzymatic cleavage of the sialic acid analogues by *Vibrio cholerae* neuraminidase [10, 12, 14] and the other involved cleavage of these analogues by acid hydrolysis.[12, 14] In both instances the protein was precipitated with a three volume excess of ethanol:chloroform (9:1), and the radioactivity was measured in the precipitates and supernatants.[12]

Functional Studies

Thrombin times of the purified normal, patient, and partially desialylated patient fibrinogens, before and after labeling with tritium, were performed as previously described.[1, 12] Fibrin monomer aggregation was studied by a modification of the method of Belitser *et al.*[12, 15, 16]

Electrophoretic Studies

Normal, patient, and partially desialylated patient fibrinogens, both unlabeled and tritium-labeled, were reduced with β-mercaptoethanol and run in SDS-PAGE. Densitometric scans of Coomassie blue stained gels were made with a Gilford gel scanner (Gilford Instrument Laboratories, Oberlin, OH). Gels of the reduced, labeled proteins were cut into 1 mm slices, and the protein was eluted with 6 M guanidinium chloride at 90°C overnight. The eluates were mixed with 15 ml of Triton X-100-toluene scintillation cocktail,[17] and radioactivity was measured in a Packard liquid scintillation-spectrometer (Packard Instrument Co., Downers Grove, IL).

RESULTS

The sialic acid content of all patient fibrinogens studied was greater than that of normal fibrinogen by at least 1.4 residues per molecule (FIGURE 2).

Partially Desialylated Fibrinogen

In incubation studies 0.1 ml of rabbit anti-neuraminidase antibody fraction neutralized the activity of 50 units of *Vibrio cholerae* neuraminidase.[12] Rapid passage of patient fibrinogen-neuraminidase incubation mixtures through anti-neuraminidase antibody affinity columns effectively removed all detectable neuraminidase activity from the mixtures and allowed for the production of partially desialylated fibrinogen derivatives with normal numbers of sialic acid residues (FIGURE 2).[12] The prolonged thrombin times of the patient fibrinogens with their full sialic acid complements were normalized after removal of the excess sialic acid residues (FIGURE 2). Patient fibrin monomer aggregation was, likewise, normalized by removal of excess sialic acid residues (FIGURE 3).

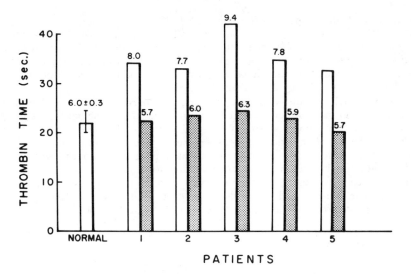

FIGURE 2. Thrombin times of purified normal and patient fibrinogens. Open bars represent untreated fibrinogen and crosshatched bars represent fibrinogen after removal of its excess sialic acid. Numbers above bars indicate sialic acid residues per molecule of fibrinogen. (From Martinez *et al.*[12])

(Sialyl-³H) Fibrinogen Studies

Ninety-five percent or more of the radioactivity of the tritium-labeled normal and patient fibrinogens was released by acid hydrolysis with only 4% to 5% of the radioactivity remaining with the protein.[12] Incubation with neuraminidase resulted in release of 85% of the protein's radioactivity.[12] These studies indicate that the radioactive label was restricted to the sialic acid

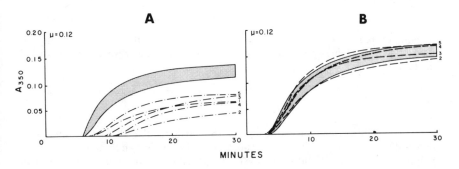

FIGURE 3. Aggregation of fibrin monomers of purified normal and patient fibrinogens. Shaded area represents the range of aggregation of five normal controls. The dashed lines represent the fibrin monomer aggregation of five patient fibrinogens with their full sialic acid complement (A) and after removal of their excess sialic acid (B). (μ = ionic strength). (From Martinez *et al.*[12])

analogues of the respective fibrinogens.[10, 12, 18] Incorporation of the radio-active label into normal, patient, and partially desialylated patient fibrinogens is shown in TABLE 1. Patient fibrinogens with their full complement of sialic acid incorporated more of the radioactive label than normal fibrinogen treated in the same manner. Partial desialylation resulted in a decreased incorporation of the radioactive label by the patient fibrinogens.[12]

SDS-PAGE of reduced normal, patient, and partially desialylated patient (sialyl-³H) fibrinogens demonstrated the presence of radioactivity on the Bβ and γ chains with no radioactivity detectable on the Aα chain (FIGURE 4). The distribution of radioactivity between the Bβ and γ chains was similar for the normal, patient, and partially desialylated patient fibrinogens with 60% of the radioactivity present in the Bβ chain and 40% in the γ chain. Coomassie blue stained gels run in parallel revealed normal mobility and amount of Aα, Bβ and γ chains with no evidence of proteolysis.[12] The [³H]borohydride labeling process did not appear to affect the biological properties of control,

TABLE 1

SIALIC ACID LABELING STUDIES

Sample	Sialic Acid Residues Molecule Fibrinogen *	(Sialyl-³H) Fibrinogen cpm/mg Fibrinogen
Control †	6.0 ± 0.3	8.7 × 10⁵ ± 1.5 × 10⁵
Patient 1	8.0	12.0 × 10⁵
Patient 1 ‡	5.4	8.3 × 10⁵
Patient 2	9.0	13.5 × 10⁵
Patient 2 ‡	5.0	9.5 × 10⁵
Patient 3	7.4	12.0 × 10⁵
Patient 3 ‡	5.0	6.5 × 10⁵

* Sialic acid residues were measured after acid hydrolysis by the thiobarbituric acid assay, as previously described.
† 6 Normal fibrinogens. The control is expressed as the mean ± 1 S.D.
‡ Indicates partially desialylated patient fibrinogens.

patient, or partially desialylated patient fibrinogens as the thrombin times of the labeled proteins were the same as those of the respective unlabeled proteins.[12]

DISCUSSION

The dysfibrinogenemia associated with hepatocellular disease [1-6] and with hepatoma [7-9] has a prolonged thrombin time due to impaired fibrin monomer polymerization. In addition, this dysfibrinogenemia has an increased sialic acid content [3, 9, 12] that correlates directly with the prolongation of its thrombin time.[3] Removal of the excess sialic acid results in normalization of the thrombin time and of the fibrin monomer polymerization.[9, 12] Periodic acid-Schiff reactivity [3, 9] and radioactivity after tritium borohydride labeling of sialic acid residues [12] of normal and patient fibrinogens were limited to the Bβ and γ

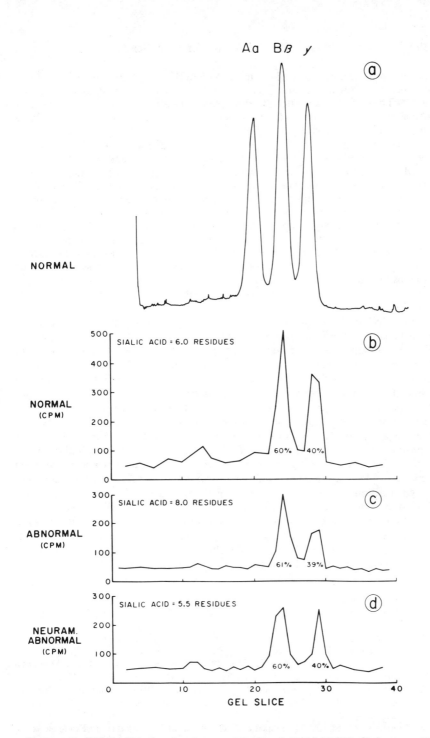

Aα Bβ γ

(a)

NORMAL

NORMAL
(CPM)

SIALIC ACID = 6.0 RESIDUES

(b)

60% 40%

ABNORMAL
(CPM)

SIALIC ACID = 8.0 RESIDUES

(c)

61% 39%

NEURAM.
ABNORMAL
(CPM)

SIALIC ACID = 5.5 RESIDUES

(d)

60% 40%

GEL SLICE

FIGURE 4. SDS-PAGE of reduced normal and patient (sialyl-³H) fibrinogens. (a) Densitometric scan of reduced normal (sialyl-³H) fibrinogen. The anode is to the right. Distribution of (³H)-labeled sialic acid analogues on the constituent chains of normal (b), patient (c) and partially desialylated patient fibrinogens (d). (From Martinez *et al.*[12])

chains of the reduced protein after SDS-PAGE. Similarly, the tritium label was restricted to the Bβ and γ chains of partially desialylated patient fibrinogen.[12] These studies indicate that the carbohydrate moieties of this functionally abnormal fibrinogen are attached to the Bβ and γ chains of the molecule as they are in normal fibrinogen.[19, 20, 22]

The observations on the abnormality of the carbohydrate moiety of the dysfibrinogenemia of liver disease have been further extended by the studies of Martinez et al.[21] reported elsewhere in this volume. These studies demonstrate an increased content of β-galactose, the penultimate sugar residue, exposed after enzymatic removal of sialic acid, in the dysfibrinogenemia. The mannose content of patient fibrinogen after stepwise enzymatic removal of sialic acid, galactose and N-acetylglucosamine was normal, however.[21, 22] These findings suggest that the branching of the oligosaccharide chain distal to mannose is increased in the dysfibrinogenemia of liver disease as compared to normal fibrinogen.[21] Thus, an abnormality of oligosaccharide processing may be responsible for the biochemical alteration of the fibrinogen molecule in liver disease, which is expressed functionally by abnormal fibrin monomer polymerization.

In addition to the finding of hypersialylated fibrinogen in some patients with hepatocellular disease and with hepatoma, a hypersialylated R-type B_{12}-binding protein has been demonstrated in the sera of several adolescents with hepatoma.[23, 24] A variant alkaline phosphatase has also been described in the sera of patients with hepatoma and in extracts of the tumor. This tumor-associated alkaline phosphatase has an increased negative charge which has been postulated to be related to an increased sialic acid content similar to that seen in placental alkaline phosphatase.[25] Human hepatoma tissue has also yielded an abnormal ferritin with an increased negative charge and a lower iron content than liver ferritin.[26] This hepatoma isoferritin is similar to the isoferritin found in fetal liver in early gestation.[26] Fetal fibrinogen has been reported to have an increased sialic acid content and abnormal fibrin monomer polymerization similar to the dysfibrinogenemia associated with hepatoma.[9, 27] The similarity between hepatoma isoferritin and fetal isoferritin and between the dysfibrinogenemia associated with hepatocellular disease and fetal fibrinogen[9, 12] suggests a reversion to a more primative evolutionary level of glycoprotein production and may provide clues to the mechanism of disordered glycoprotein synthesis in liver disease.[28]

REFERENCES

1. PALASCAK, J. E. & J. MARTINEZ. 1977. Dysfibrinogenemia associated with liver disease. J. Clin. Invest. **60:** 89–95.
2. GREEN, G., J. M. THOMSON, I. W. DYMOCK & L. POLLER. 1976. Abnormal fibrin polymerization in liver disease. Br. J. Haematol. **34:** 427–439.
3. MARTINEZ, J., J. E. PALASCAK & D. KWASNIAK. 1978. Abnormal sialic acid content of the dysfibrinogenemia associated with liver disease. J. Clin. Invest. **61:** 535–538.
4. SORIA, J., C. SORIA, M. SAMAMA, J. COUPIER, M. L. GIRARD, J. BOUSSER & G. BILSKI-PASQUIER. 1970. Dysfibrinogénémies acquises dans les atteintes hépatiques sévéres. Coagulation **3:** 37–44.

5. AIACH, M., J. ROGÉ, M.-F. BUSY, H. DURAND, N. GUÉROULT, CH. CHANRION, M. LECLERC & L. JUSTIN-BESANCON. 1973. Dysfibrinogénémies acquises et affections hépatiques. A propos de observations. Sem. Hôp. Paris. **49:** 183–197.
6. LANE, D. A., M. F. SCULLY, D. P. THOMAS, V. V. KAKKAR, I. L. WOLF & R. WILLIAMS. 1977. Acquired dysfibrinogenemia in acute and chronic liver disease. Br. J. Haemat. **35:** 301–308.
7. VON FELTEN, A., P. W. STRAUB & P. G. FRICK. 1969. Dysfibrinogenemia in a patient with primary hepatoma. First observation of an acquired abnormality of fibrin monomer aggregation. N. Engl. J. Med. **280:** 405–409.
8. VERHAEGHE, R., B. VAN DAMME, A. MOLLA & J. VERMYLEN. 1972. Dysfibrinogenemia associated with primary hepatoma. Scand. J. Haematol. **9:** 451–458.
9. GRALNICK, H. R., H. GIVELBER & E. ABRMAS. 1978. Dysfibrinogenemia associated with hepatoma. Increased carbohydrate content of the fibrinogen molecule. N. Engl. J. Med. **299:** 221–226.
10. MARTINEZ, J., J. PALASCAK & C. PETERS. 1977. Functional and metabolic properties of human asialofibrinogen. J. Lab. Clin. Med **89:** 367–377.
11. WARREN, L. 1959. The thiobarbituric acid assay of sialic acids. J. Biol. Chem. **234:** 1971–1975.
12. MARTINEZ, J., K. A. MACDONALD & J. E. PALASCAK. 1982. The role of sialic acid in the dysfibrinogenemia associated with liver disease. Distribution of sialic acid on the constituent chains. (Submitted for publication).
13. VAN LENTEN, L. & G. ASHWELL. 1971. Studies on the chemical and enzymatic modification of glycoproteins. A general method for the tritiation of sialic acid-containing glycoproteins. J. Biol. Chem. **246:** 1889–1894.
14. BUTKOWSKI, R. J., S. P. BAJAJ & K. G. MANN. 1974. The preparation and activation of (sialyl-^{3}H) prothrombin. J. Biol. Chem. **249:** 6562–6569.
15. BELITSER, V. A., T. V. VARETSKAJA & G. V. MALNEVA. 1968. Fibrinogen-fibrin interaction. Biochim. Biophys. Acta **154:** 367–375.
16. GRALNICK, J. R., H. M. GIVELBER, J. R. SHAINOFF & J. S. FINLAYSON. 1971. Fibrinogen Bethesda: A congenital dysfibrinogenemia with delayed fibrinopeptide release. J. Clin. Invest. **50:** 1819–1830.
17. MANN, K. G., R. YIP, C. M. HELDEBRANT & D. N. FASS. 1973. Multiple active forms of thrombin. III. Polypeptide chain location of active site serine and carbohydrate. J. Biol. Chem. **248:** 1868–1875.
18. SUTTAJIT, M. & R. J. WINZLER. 1971. Effect of modification of *N*-acetylneuraminic acid on the binding of glycoproteins to influenza virus and on susceptibility to cleavage by neuraminidase. J. Biol. Chem. **246:** 3398–3404.
19. GAFFNEY, P. J. 1972. Localization of carbohydrate in the subunits of human fibrinogen and its plasmin induced fragments. Biochim. Biophys. Acta. **263:** 453–458.
20. PIZZO, S. V., M. L. SCHWARTZ, R. L. HILL & P. A. McKEE. 1972. The effect of plasmin on the subunit structure of human fibrinogen. J. Biol. Chem. **247:** 636–645.
21. MARTINEZ, J., P. M. KEANE & P. B. GILMAN. Carbohydrate composition of normal fibrinogen compared to the abnormal fibrinogen of liver disease. Ann. N.Y. Acad. Sci. **408:**. This volume.
22. TOWNSEND, R. R., E. HILLIKER, Y-T. LI, R. A. LAINE, W. R. BELL & Y. C. LEE. 1982. Carbohydrate structure of human fibrinogen. J. Biol. Chem. **257:** 9704–9710.
23. BURGER, R. L., S. WAXMAN, H. S. GILBERT, C. S. MEHLMAN & R. H. ALLEN. 1975. Isolation and characterization of a novel vitamin B_{12}-binding protein associated with hepatocellular carcinoma. J. Clin. Invest. **56:** 1262–1270.
24. WAXMAN, S., C.-K. LIU, C. SCHREIBER & L. HELSON. 1977. The clinical and physiological implications of hepatoma B_{12}-binding proteins. Cancer Res. **37:** 1908–1914.

25. HIGASHINO, K., M. HASHINOTSUME, K. Y. KANG, Y. TAKAHASHI & Y. YAMAMURA. 1972. Studies on a variant alkaline phosphatase in sera of patients with hepatocellular carcinoma. Clin. Chim. Acta **40:** 67–81.
26. BULLOCK, S., A. BOMFORD & R. WILLIAMS. 1980. A biochemical comparison of normal human liver and hepatocellular carcinoma ferritins. Biochem. J. **185:** 639–645.
27. GALANAKIS, D. K., J. MARTINEZ, C. McDEVITT & F. MILLER. 1983. The role of sialic acid on human fetal fibrinogen function. Demonstration of low peptide A release rate corrected by partial removal of sialic acid. Ann. N.Y. Acad. Sci. **408:**. This volume.
28. MARTINEZ, J. & J. E. PALASCAK. 1982. Hemostatic alterations in liver disease. *In* Zakim, D. and T. D. Boyer, Eds. Hepatology. Saunders. Philadelphia, Pa. In press.

PLASMIN DEGRADATION OF
CROSS-LINKED FIBRIN *

Victor J. Marder and Charles W. Francis

Hematology Unit, Department of Medicine
University of Rochester School of Medicine and Dentistry
Rochester, New York 14642

This report describes a unified structural model of plasmic degradation of cross-linked fibrin, incorporating new concepts of macromolecular structure with known details of polymerization, binding sites, cross-linking and plasmic cleavage sites. Our model is based on the two-stranded protofibril as the structural unit of the clot [1-3] and explains the entire sequence of degradation from intact fibrin to the "terminal" lysate. The principles of fibrin structure [4-13] that must be accommodated in this model are as follows:

(1) Non-covalent D to E "stag" contacts between chains of the protofibril.
(2) Covalent γ chain cross-linking of D to D "long" contacts along each chain of the protofibril.
(3) Weaker, non-covalent "lateral" contacts between protofibrils.
(4) Cross-linked α polymers interlaced between all fibrin monomers and perhaps wrapped around the fibrin fiber.

The model of fibrin degradation also implies that the major plasmin-susceptible sites known to exist in fibrinogen are accessible within cross-linked fibrin as well.[14-26] Thus, in the cross-linked dimer, the sequence of relevant cleavage points is as follows:

(1) Cleavage at or near the emergence of the α chain polar appendage.
(2) Concordant cleavages of all three chains between the central and terminal domains, and
(3) Cleavage of γ chains near the cross-link site.

The γ-γ cross-links produce unique derivatives after plasmic action, as shown in the schematic diagram of a short protofibril "backbone" (FIG. 1). With cleavage of the coiled coils on either side of the two terminal domains, fragment DD is formed,[27-29] and under physiologic conditions, DD is bound noncovalently to a central domain (Fragment E) located in the adjacent fibrin polymer of the protofibril.[30, 31] This DD/E complex is held together by the same D to E "stag" contacts that promoted the initial half-overlap polymerization of fibrin monomers. In further considerations, we define fragments such as DD as derivatives that are not dissociable into smaller units by ionic detergents, and complexes such as DD/E as two or more fragments that are held in place by such non-covalent bonds.

We have noted [32] that particulate fibrin is degraded in an orderly sequence of proteolytic cleavages that can be correlated with physical changes in clot struc-

* This work was supported in part by Grants No. HL–18208, HL–21379 & HL–00733 from the National Heart, Lung and Blood Institute, National Institutes of Health, Bethesda, Maryland.

Fibrinogen

Thrombin

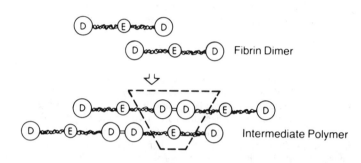

Fibrin Dimer

Intermediate Polymer

FIGURE 1. Schematic representation showing the polymerization of fibrin into a short protofibril, and the derivation of the DD/E complex by plasmic action. The definition of larger cross-linked fibrin derivatives is schematized in FIGURE 3 and TABLE 2. (Reproduced from Francis & Marder.[13])

ture (TABLE 1). The earliest plasmic cleavages leave all proteins still covalently bound together. Degradation to this extent is designated as phase 1 and is characterized by cleavage of α polymer chains, but relatively few cleavages of the coiled coils. Phase 2 describes the extent of degradation present in fibrin when degradation fragments are retained in the matrix by noncovalent bonding. By washing the phase 2 degrading clot with SDS, this coat of adherent protein can be removed. Gel electrophoresis after disulfide bond reduction shows that only a residual α polymer network remains, and electrophoresis without disulfide bond reduction demonstrates that the fragments are of very high molecular weight, the majority being greater than M_r 500,000. Further degradation of the coiled coils results in solubilization of complexes from the matrix (phase 3),

TABLE 1

DISTINGUISHING PROPERTIES OF THE FOUR PHASES OF PLASMIC
DEGRADATION OF CROSS-LINKED FIBRIN *

Phase 1 – Limited proteolysis but all D and E domains still attached covalently to other parts of the matrix.
Phase 2 – Some fragments are bound to the particulate matrix by noncovalent bonds and can be solubilized by ionic detergents such as SDS.
Phase 3 – Spontaneous release of soluble complexes into solution.
Phase 4 – Degradation of soluble complexes subsequent to release from the matrix.

* Table derived from Francis & Marder.[13]

resulting in a smaller degrading clot, and phase 4 represents degradation of the solubilized derivatives.

The polypeptide chain composition of the residual phase 1 fibrin matrix remains constant regardless of the incubation time with plasmin.[32] Such experiments show that the particulate fibrin contains high molecular weight polymers of α chains, intact γ-γ dimers, and small amounts of the largest /γγ remnants, with a preponderance of undegraded β chains relative to the /β remnant of 43,000. Although the fibrin clot becomes smaller during plasmin exposure, the polypeptide chain composition of the clot present at each time interval is the same. This polypeptide chain composition defines the extent to which fibrin degradation can occur without producing fragments that can be washed from the surface with SDS.

Just as for phase 1 particulate fibrin, those fragments which are washed free with SDS and the derivatives that are freely liberated into solution have an unchanging composition regardless of the prior exposure time of the clot to plasmin. The spontaneously liberated complexes are composed of fragments of molecular weight in the range of 195,000 to approximately 800,000 (14,33,34), held together by noncovalent bonds. Liberated derivatives are dominated by fragments at the higher molecular weight range.

The data are consistent with progressive proteolysis of accessible superficial portions of the clot, in which a circumscribed number of specific cleavages caused predictable alterations in the clot, beginning with those of phase 1 (TABLE 1, FIG. 2). Portions of the clot were converted to enormous noncovalently bound derivatives (phase 2), and these superficial regions were then liberated as large complexes (phase 3), leaving behind a smaller particulate fibrin clot that was in the process of undergoing the same sequence of dissolution. We postulate that the major difference between *in vivo* and *in vitro* fibrinolysis relates to the removal of large complexes by the flow of blood.[32] Such circulating complexes would undergo relatively minor proteolytic degradation, in contrast with their continued exposure to plasmin *in vitro*. The phase 4 cleavages lead to ever-smaller complexes that are eventually degraded to the terminal lysate that contains only the smallest complex, DD/E, and individual fragments DD and E.

This process of sequential degradation of soluble plasmic derivatives during phase 4 has been shown by electrophoretic analysis to start with a group of large complexes and to end with only the smallest complex (DD/E).[14] Four such complexes have been purified by elution through Sephacryl S-300,[14] subjected to analytical centrifugation to determine their molecular weights, and analyzed by two dimensional electrophoresis to determine the types of fragments present. The fragments have been similarly analyzed in a two dimensional electrophoresis system after disulfide bond reduction to determine their polypeptide chain composition. The identity of fragments was also predicted by analysis of single chain fibrin polymer models, and with this, the maximum size of component fragments within each complex could be determined.[14] These results are summarized in TABLE 2 (right), which schematizes the linked domains derived from a single fibrin polymer chain. Most fragments contain a DD or E at the end of the polymer, reflecting cleavage of a coiled coil. Fragments that terminate with a D domain, such as fragment XY, represent examples of a cleavage near the γ-γ cross-link site which splits the Factor XIII-ligated D to D domains.

TABLE 3 records the observed molecular weights of four complexes and compares these values with those predicted on the basis of two independent approaches. The first summates the molecular weights of the two largest fragments that were detected in each complex by two dimensional electrophoresis in SDS; the second assumes that the complexes represent longitudinal additions of

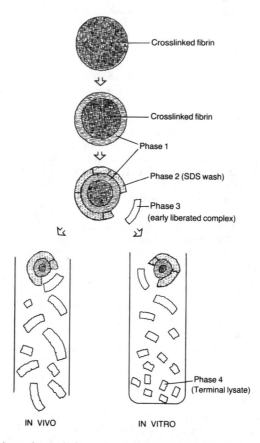

FIGURE 2. Schematic depiction of particulate fibrin dissolution by "outside-in" progressive alterations in clot structure. The substance of the clot is sequentially converted from a fully cross-linked form to a state of cleavage (see TABLE 1) that allows for spontaneous liberation of soluble complexes into solution. With that step, the clot becomes smaller and previously unexposed portions would now be accessible to binding and proteolytic cleavage by the plasminogen activator/plasmin enzyme system. It is postulated that further degradation of complexes in solution is mostly an *in vitro* phenomenon, this being limited *in vivo* by the entry of derivatives into the circulation, their rapid cellular clearance and inhibition of degradation by α_2-antiplasmin. (Reproduced from Francis *et al.*[14])

three domainal (DD/E) segments along the protofibril. The agreements are close, and warrant our schematic for the structure of complexes as shown in TABLE 2 (left).

Complexes are diagramed as antiparallel strings of domains which are

TABLE 2

PLASMIC DERIVATIVES OF CROSS-LINKED FIBRIN

Number	Composition	Complexes		Fragments		
		Molecular Weight	Structure	Name	Molecular Weight	Structure
				E*	44–55,000	
1	DD/E	228,000		D*	100,000	
				Y*	150,000	
2	DY/YD	465,000		DD	195,000	
				X*	247,000	
3	YY/DXD	703,000		DY	247,000	
				YY	285,000	
4	YXD/DXY	850,000		XD	334,000 or 365,000	
				XY	365,000 or 391,000	
				DXD	461,000	
				YXD	500,000	

* Fragments of very similar structure also derive from plasmic degradation of fibrinogen or noncross-linked fibrin.

TABLE 3

OBSERVED AND PREDICTED MOLECULAR WEIGHTS OF COMPLEXES RELEASED FROM
CROSS-LINKED FIBRIN BY PLASMIN

Complex	Observed (Equilibrium Sedimentation)	Predicted* Multiples of Complex 1	Sum of Component Fragments
1 (DD/E)	228,000	–	245,000 (+8%)
2 (DY/YD)	465,000	465,000 (−2%)	494,000 (+6%)
3 (YY/DXD)	703,000	684,000 (−3%)	746,000 (+6%)
4 (YXD/DXY)	850,000	912,000 (+7%)	1,000,000 (+15%)

* Values in parentheses represent % differnce between predicted and observed molecular weights on the basis of either simple multiples of complex 1 or the sum of observed largest component fragments. (From Francis & Marder.[13])

oriented so that each central domain is approximated to two terminal domains of the other fibrin polymer strand. Their structure is compatible with the two-stranded protofibril model of fibrin formation, in which three domainal clusters of DD associated with E are liberated singly as complex 1 or joined together by coiled coils as the larger complexes.

The liberation of complexes into the circulation has relevance at both molecular and physiologic levels. FIGURE 3 schematizes solubilization (phase 3)

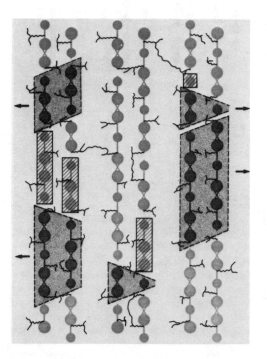

FIGURE 3. *In situ* formation of plasmic-derived fragments (lightly shaded) and complexes (heavily shaded) within a small segment of cross-linked fibrin, here depicted as three laterally associated protofibrils. The complexes represent portions of both strands of a given protofibril that have been freed of covalent (α polymer or coiled coil) attachments and are considered liable to spontaneous liberation from the fibrin. The fragments are single chain derivatives that are held in place by noncovalent attractive forces to other portions of the clot and which can be washed free with SDS. Further degradation of complementary coiled coils opposite these fragments would convert them to complexes. (This figure is modified from the scheme presented in reference 13.)

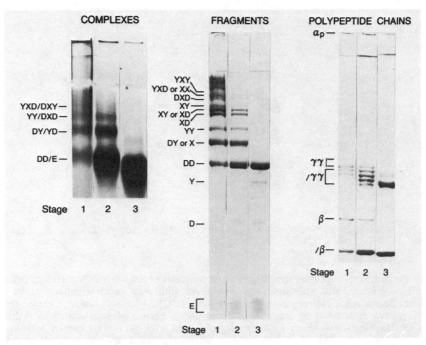

FIGURE 4. Polyacrylamide gel electrophoretic systems showing plasmic degradation of complexes, fragments and subunit polypeptide chains, after spontaneous liberation from cross-linked fibrin. Progressive degradation from larger to smaller derivatives is obvious ending in stage 3 with the smallest complex (DD/E) (left), composed of fragments E and DD (center), the latter of which contains primarily the smallest γ-γ dimer remnants (right). (Reproduced from Francis & Marder.[13])

of fibrin degradation. Solubilization follows cleavages at complementary sites of paired strands, so that free release of the two-stranded complexes occurs. These complexes contrast with the single-stranded fragments that are characteristic of phase 2 fibrin degradation. These latter fragments can be solubilized by SDS and, after appropriate complementary cleavage of the coiled coils, will themselves become part of a two-stranded complex. All four of the complexes shown in TABLE 2 are illustrated in their *in situ* position prior to spontaneous liberation. Phase 4 degradation of the soluble derivatives is shown in FIGURE 4, showing the noncovalent complexes, SDS-separated fragments, and disulfide-bond reduced polypeptide chains in their progressive conversion from large to small ("terminal") forms.

Since the largest complexes and fragments appear to constitute the bulk of newly liberated derivatives from fibrin, their detection in the blood of patients with thrombotic disorders becomes a prime objective. Thus, although the determination that γ-γ dimers or D dimer present in a sample is evidence of cross-linked fibrin dissolution, it seems that the ability to detect specific larger complexes or fragments will have relevance for understanding pathophysiology. In FIGURE 5, antifibrinogen immunoaffinity column eluates from one patient with DIC and another with deep vein thrombosis were electrophoresed in an

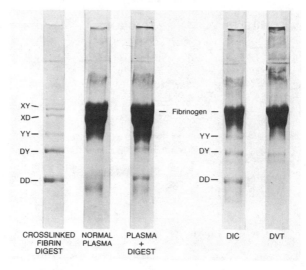

FIGURE 5. SDS-polyacrylamide gel electrophoresis of affinity purified eluates from plasma of a normal individual and of patients with deep vein thrombosis (DVT) and disseminated intravascular coagulation (DIC). Control samples show the fragments in a digest of cross-linked fibrin and a normal plasma sample to which such fragments have been added prior to elution over the antifibrinogen antibody affinity column. The fibrinogen band is excessively broad and dense in all of the plasma samples, obscuring the possible presence of any fragments of greater size than that of fragment YY. Nevertheless, fragments DD, DY and YY are clearly present in the pathologic samples, indicating the circulation of these cross-linked fibrin derivatives and suggesting that even larger fragments comparable to those present in stage 1 digests (FIG. 4) may also circulate.

SDS-polyacrylamide gradient gel. Bands are in the region not only of fragment DD, but also of DY and YY; larger fragments were obscured by the massive fibrinogen band, indicating the need for technical improvement to demonstrate these larger fragments.

Thus, the mass of derivatives that are liberated from fibrin by plasmin can be categorized logically, and relevance to diagnosis and especially to pathophysiology is evident. Although technical problems still exist in their detection in clinical samples, at least we now can appreciate the size and component structure of the derivatives that circulate.

SUMMARY

On the basis of structural studies of both degrading insoluble cross-linked fibrin and of soluble derivatives, we have developed a model to explain the principal structural and physical features of plasmic degradation of cross-linked fibrin *in vitro* from the completely intact matrix to terminally degraded soluble derivatives. The critical event of solubilization occurs only as the result of coincident cleavages at complementary sites in the basic two-stranded half-staggered overlap fibrin structure, resulting in the release of two-stranded complexes held together by noncovalent forces. The four smallest complexes that are released into solution have structures corresponding to DD/E, DY/YD, YY/DXD, and YXD/DXY. The protein initially solubilized has a constant composition with a predominance of large derivatives that are composed of at least one fragment from each of the two strands of the protofibril. Following their release into solution the larger complexes are converted *in vitro* to smaller ones by the continued action of plasmin, so that the complex found following prolonged digestion is DD/E. It is proposed that this newly defined group of complexes represents the major form of circulating plasmic derivatives of cross-linked fibrin.

REFERENCES

1. FERRY, J. D. 1952. Proc. Natl. Acad. Sci. USA **38:** 566–569.
2. KRAKOW, W., G. F. ENDRES, B. M. SIEGEL & H. A. SCHERAGA. 1972. J. Mol. Biol. **71:** 95–103.
3. HANTGAN, R., W. FOWLER, H. ERICKSON & J. HERMANS. 1980. Thromb. Haemost. **44:** 119–124.
4. LAUDANO, A. P. & R. F. DOOLITTLE. 1978. Proc. Natl. Acad. Sci. USA **78:** 3085–3089.
5. OLEXA, S. & A. Z. BUDZYNSKI. 1981. J. Biol. Chem. **256:** 3544–3549.
6. CHEN, R. & R. F. DOOLITTLE. 1971. Biochemistry **10:** 4486–4491.
7. MCKEE, P. A., P. MATTOCK & R. L. HILL. 1970. Proc. Natl. Acad. Sci. USA **66:** 738–744.
8. COTTRELL, B. A., D. D. STRONG, K. W. K. WATT, et al. 1979. Biochemistry **18:** 5405–5410.
9. HERMANS, J. 1979. Proc. Natl. Acad. Sci. USA **76:** 1189–1193.
10. HANTGAN, R. R. & J. HERMANS. 1979. J. Biol. Chem. **256:** 11272–11281.
11. FOWLER, W. E., E. P. ERICKSON, R. R. HANTGAN, et al. 1981. Science **211:** 287–289.
12. HERMANS, J. & J. MCDONAGH. 1982. Sem. Thromb. Hemost. **8:** 11–24.
13. FRANCIS, C. W. & V. J. MARDER. 1982. Sem. Thromb. Hemost. **8:** 25–35.
14. FRANCIS, C. W., V. J. MARDER & G. H. BARLOW. 1980. J. Clin. Invest. **66:** 1033–1043.
15. MILLS, D. & S. KARPATKIN. 1970. Biochem. Biophys. Res. Commun. **40:** 206–211.
16. MARDER, V. J., N. R. SHULMAN & W. R. CARROLL. 1967. Trans. Assoc. Am. Phys. **80:** 156–167.
17. MARDER, V. J., N. R. SHULMAN & W. R. CARROLL. 1969. J. Biol. Chem. **244:** 2111–2119.
18. PIZZO, S. V., M. L. SCHWARTZ, R. L. HILL & P. A. MCKEE. 1972. J. Biol. Chem. **247:** 636–645.
19. TAKAGI, T. & R. F. DOOLITTLE. 1975. Biochemistry **14:** 940–946.
20. GAFFNEY, P. J. 1977. Haemostasis: Biochemistry, Physiology, and Pathology. D. Ogston & B. Bennett, Eds. p. 105.
21. FURLAN, M. & E. A. BECK. 1972. Biochem. Biophys. Acta **263:** 631–644.
22. DOOLITTLE, R. F. 1975. The Plasma Proteins. Structure, Function and Genetic Control, Vol. 2. F. W. Putnam, Ed. p. 109. Academic Press. New York.
23. MARDER, V. J., C. W. FRANCIS & R. F. DOOLITTLE. 1982. Hemostasis and Thrombosis: Basic Principles and Clinical Practice. R. W. Colman, J. Hirsh, V. J. Marder & E. W. Salzman, Eds. p. 145. Lippincott Co. Philadelphia.
24. MIHALYI, E., R. M. WEINBERG, D. W. TOWNE, et al. 1976. Biochemistry **15:** 5372–5381.

25. NUSSENZWEIG, V., M. SELIGMANN, U. PELMONT & P. GRABAR. 1961. Ann. Inst. Pasteur 100: 377–389.
26. PIZZO, S. V., M. L. SCHWARTZ, R. L. HILL & P. A. McKEE. 1973. J. Biol. Chem. 248: 4574–4583.
27. KOPEC, M., E. TEISSEYRE, G. DUDEK-WOJCIECHOWSKA, M. KLOCZEWIAK, A. PANKIEWICZ & Z. S. LATALLO. 1973. Thromb. Res. 2: 283–291.
28. PIZZO, S. W., L. M. TAYLOR, JR., M. L. SCHWARTZ, R. L. HILL & P. A. McKEE J. Biol. Chem. 248: 4584–4590.
29. GAFFNEY, P. J. & M. BRASHER. 1973. Biochim. Biophys. Acta 295: 308–313.
30. HUDRY-CLERGEON, G., L. PATUREL & M. SUSCILLON. 1974. Pathologie et Biologie (Suppl) 22: 47–52.
31. GAFFNEY, P. J., D. A. LANE, V. V. KAKKAR & M. BRASHER. 1975. Thromb. Res. 7: 89–99.
32. FRANCIS, C. W., V. J. MARDER & G. H. BARLOW. 1980. Blood 56: 456–464.
33. ALKJAERSIG, N., A. DAVIES & A. FLETCHER. 1977. Thromb. Haemost. 38: 524–535.
34. REGANON, E., V. VILA & J. AZNAR. 1978. Thromb. Haemost. 40: 368–376.

———◆———

DISCUSSION OF THE PAPER

F. HAVERKATE (*Gaubius Institute, Leiden, the Netherlands*): I would like to point out that there is a difference between the *in vivo* and the *in vitro* situation; *in vitro* you have breakdown of fibrin and also of the degradation products formed in solution. However, *in vivo* as long as α-plasmin inhibitor is present, you have only action on the fibrin.

V. J. MARDER: I think you are entirely right. The largest derivatives, the ones that first come loose from fibrin, are really the ones that circulate, not only because they are washed away but because there is α-2 plasmin inhibitor present to impair further lysis once they are liberated from the clot.

THE OCCURRENCE AND CLINICAL RELEVANCE OF FIBRIN FRAGMENTS IN BLOOD

Patrick J. Gaffney

National Institute for Biological Standards and Control
London NW3 6RB, England

INTRODUCTION

The measurement of fibrin degradation products (FDP) in blood has been fashionable for some time and the availability of various methodologies has influenced the interpretation of the FDP assay. The term "FDP" has been subject to a deal of misinterpretation, ranging from a misconception about the source of the FDP fraction in blood to fundamental communication problems between those knowledgeable in fibrin-fibrinogen biochemistry and those who actually measure FDP in blood. The term "FDP" was often used to mean fibrinogen and fibrin degradation products; whereas this can indeed be true, it is now generally accepted that the FDP fraction mainly results from the interaction of plasmin with fibrin rather than fibrinogen.[1] It is well to keep in mind that the interaction of fibrin with other enzymes is also a possible source of the FDP fraction.[2] Thus in most cases of blood FDP the investigator is measuring the plasmin-mediated fragments derived from fibrin as it forms in the soluble state or is deposited, mostly in the microvasculature. A communication problem has been highlighted by the supposed measurement of fragment E rather than the whole FDP fraction, using tests based on antisera specific for the E domain of fibrinogen or of fibrin.[3] Staff in clinical laboratories may be under the impression that they can measure the E fragment, whereas in reality the early FDP fragments, such as X and Y, are the major source of the E antigen reacting in the immunologic assay. Thus "E antigen assays" are no more than modifications of existing FDP assays with an altered sensitivity and no real change in specificity. Since radioimmunoassays (RIA) are already adequately sensitive for the measurement of FDP in an undiluted plasma sample, any increase in sensitivity of these assays is disadvantageous[4]; it can be concluded that the "E antigen assay" in human serum is little more than a promotional modification that capitalizes on the erroneous impression that free fragment E is measured and that this may be clinically relevant.

Despite these problems about the source and meaning of the FDP fraction in blood, highlighted by the examples given above, a note of optimism was sounded in 1972, when it was predicted[5] that the cross-linked (XL) FDP fraction, notably D dimer,[6] might clarify our understanding of clinical fibrinolysis and serve as a useful diagnostic tool of the prethrombotic state. Ten years on the implementation of this hopeful prediction remains unfulfilled. However, during this time our understanding of the digestion of fibrin by plasmin has increased via the characterization of plasmin-mediated XL-FDP *in vitro*,[7, 8] and the presence of a whole family of XL derivatives has been established beyond doubt in various body fluids.[9] Limited characterization of XL-FDP found *in vivo* has been achieved, and an effort is made here to collate some of the more recent reports in this area. Efforts are also made to relate these FDP fractions to the type of fibrin from which it came and to access the

0077–8923/83/0408–0407 $01.75/0 © 1983, NYAS

relevance of the latter to the health of the vascular tree. This review covers the relevance of FDP as markers of fibrin deposition and possibly thrombosis and ignores what must be of great importance, namely, the effect of these fibrin fragments on platelet aggregability and other clot-initiating factors in blood.

We will divide this communication into three parts. One dealing with a rationalization of some of the biochemical and structural reactions associated with fibrin formation *in vitro*; two, drawing attention to probable structures of FDP found in clinical material, nearly always during disseminated intravascular coagulation (DIC); and three, a discussion on the relevance of this information to thrombosis and hemorrhage.

Fibrin Formation

It is becoming increasingly evident that an understanding of fibrin fragmentation depends on knowledge of how fibrin in blood is initiated and organized. Furthermore, since this communication will emphasize the relationships that exist between fibrin fragments and the originating fibrin, an appreciation of the structures of various forms of fibrin is necessary. Fibrin formation is initiated by the interaction of thrombin with fibrinogen, releasing two smaller peptides known as fibrinopeptides A and B (FPA and FPB) from the amino terminal ends of the fibrinogen molecule.[10] Indeed only the release of FPA seems to be required for the occurrence of primary polymerization leading to clot formation. The release of FPB has been demonstrated to be associated with the exposure of other polymerization sites [11] which may be related to aggregation of fibrin in a direction other than that associated with the primary (or FPA mediated) aspect of polymerization. Thrombin also activates plasma transglutaminase, known as Factor XIII, such that cross-linking is catalyzed between the γ chains of adjacent fibrin molecules and at a later stage between the α chains of several fibrin molecules.[12] The above reactions have been copiously reviewed in recent years,[13-16] and have also been covered in detail in other chapters of this book; thus it is the intention here to comment only on relationships between these reactions which contribute to our understanding of fibrin clot lysis. Areas of uncertainty are briefly highlighted.

It is generally accepted that FPA release precedes FPB release, though a recent paper has indicated [17] that both peptides are released at the same time but at different rates, these rates being altered significantly in platelet-rich plasma and whole blood.[18] Whether kinetics or sequential order of release govern FPA and FPB hydrolysis by thrombin (for review see ref. 19), it is accepted that elevated levels of FPA precede those of FPB during the fibrinogen-thrombin interaction. Since FPA release exposes a polymerization region in the E domain of fibrinogen-fibrin,[20] it seems reasonable to suppose that this is a bivalent site, fibrinogen being a dimeric molecule and containing two FPA peptides.[1] The question arises whether thrombin releases the two FPA peptides from the same fibrinogen at the same time or whether they are cleaved in a sequence that mediates the growth of protofibrils to form thicker fibers. The latter alternative is depicted schematically in FIGURE 1. The thinking that has generated the scheme in FIGURE 1 was fostered by experiments conducted to examine the constitutive effects of the various component reactions taking place during the well-known, spectrographically delineated aggregation curve of fibrinogen-thrombin mixtures at 350 mμ. Whereas the aggregation curve shown in FIGURE 2 is rather arbitrary, it does serve as a "template" from which to comment on the various reactions which take place during clotting. Kierulf [21] suggested that Factor XIII-mediated γ chain cross-links

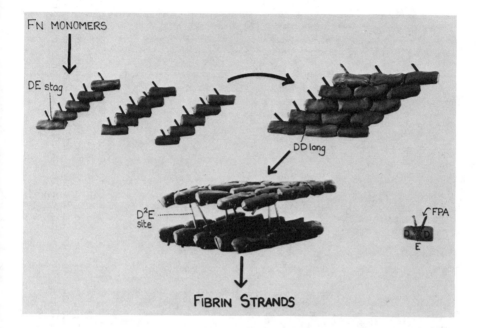

FIGURE 1. Three proposed steps during fibrin (FN) assembly. A fibrinogen model is shown at the extreme right of the diagram, indicating the locations of D, E structural domains and the two fibrinopeptides A (FPA). The first step shows the organization of linear polymers of FN via the "DE stag" polymerization sites; the individual FN monomers retain one FPA during this step. The second step involves the association of a number of linear polymers into fibrin strands via a secondary "DE stag" polymerization site exposed by the release of the second FPA. The D-D long association site forms between two adjacent D domains and Factor XIII cross-links (not shown) also occur between these domains. A further polymerization site (D²-E site) is exposed by the release of fibrinopeptides B(FPB) and this allows two stranded fibers to form that probably enhance the speed of "fiber bundling" during the formation of the ultimate thick fibrin strand. The three steps outlined here are compatible with the biochemical data relating to FPA and FPB release, Factor XIII cross-linking and the ultimate domainal composition of the cross-linked FDP fraction found *in vitro* and *in vivo*.

could form in fibrin in the soluble state whereas it had been previously assumed that "clot formation" or "aqueous insolubilisation" was a prerequisite for cross-linking to occur. The data in FIG. 2 indicate that (1) fibrin strands, physically removable as "miniclots," make the total contribution to the

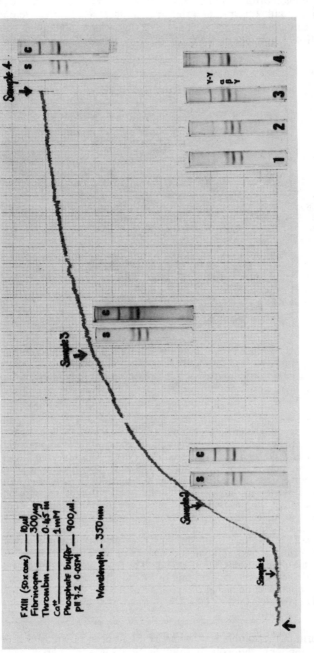

FIGURE 2. Fibrinogen to fibrin aggregation curve. During the lag phase (sample 1) and at various periods during aggregation samples (2 to 4) were examined for Factor XIII-induced cross-linking. This was achieved by inhibiting thrombin with a 10 × excess of hirudin, removing the optically dense fibers and examining these (C) and the supernatants (S) for γ chain cross-linking by SDS-PA gel electrophorsseis following reduction with β-mercaptoethanol. In each case the total sample (clot + supernatant) was also examined following reduction to its polypeptide chains, and the electrophoretic patterns are shown at the lower right of the figure.

absorbance at 350 mμ and that (2) only these fibrin strands contained Factor XIII-mediated γ chain cross-linking while the OD-negative "supernatant" fibrin was not cross-linked. This suggests that soluble fibrin aggregates cannot be γ chain cross-linked and the data of Kierulf [21] and other clinical findings of cross-linked soluble fibrin in plasma [22] can only be expressions of fibrin strand formation that may be easily confused with the truly soluble fibrin. The sudden increase in OD (Fig. 2) must be related with the association of already well-developed polymerized forms of fibrin (e.g. protofibrils), and Figure 1 suggests that this aggregation step could be associated with the dimeric expression of the E located polymerization site. This of course would also explain why the formation of γ chain cross-links between the adjacent fibrin subunits (marshalled together on the dimeric E domain of another fibrin subunit) is observed only when the fibrin strands, being optically active at 350 mμ, appear.

The release of FPB follows that of FPA and thus seems to take place only from fibrin strands or aqueously insoluble fibrin. Support for this is the evidence that FPB is not released during the lag phase of the "template" aggregation curve shown in Figure 2, while others [23] have indicated that FPB is released only after protofibril formation. Ever since the earliest suggestive evidence [24] that FPB release stimulated side-to-side aggregation of end-to-end fibrin polymers, clarification of the role of FPB release during the formation of fibrin has been sought. The terms end-to-end and side-to-side are difficult to visualize in terms of the half staggered fibrin arrangement first suggested by Ferry [25] and now clearly demonstrated in the electronmicroscope by Erickson and Fowler in this volume. That FPB release is associated with the exposure of a distinct set of polymerization sites [11] must be taken into account in any full interpretation of fibrin assembly. It should be remembered that the coagulant enzymes in some snake venoms, which only remove FPB from fibrinogen, can clot fibrinogen and that FPB release alone can cause clotting, albeit slowly, below 25°C.[26]

While the role of FPB release has been alluded to in a recent review by Hermans and McDonagh,[19] they allocate to it no clear status in the overall mechanism of fibrin assembly. These authors also do not explain why the earliest protofibrils, being two molecules in diameter and having the two D domains of adjacent fibrin molecules associated via their DD long contacts, have no Factor XIII-mediated γ-γ chain cross-links. Light scattering and electronmicroscopic data have formed the basis of the sequential steps proposed by these authors for fibrin assembly and little consideration has been given to the fact that the fibrinogen-thrombin molar ratio is of considerable importance in designating the sequence of events in fibrin assembly and the physico-chemical properties of the resultant gel matrix. Indeed, Hartgan and Hermans [23] have conducted all their experiments at very high thrombin where the enzyme concentration was not rate limiting. Light scattering data [23] suggest that FPB release contributes to fibrin fiber formation while EM data [27] do not confirm this, in that fibrin made with Batroxobin (a coagulant enzyme from the venom of *Botrops atrox* which only removes FPA) looks the same in the electronmicroscope as the fibrin resulting from thrombin action. Since the FPB-related polymerization sites play such an important role in defining the aggregated forms of plasmin-mediated fibrin fragments, it is necessary to allocate to such sites a role in the fibrin assembly process. In Figure 1 the half-stagger concept is integrated with the proposal that fibrin-AB$_2$ (i.e. fibrin retaining one FPA) aggregated to form protofibrils, which in turn form small "bundles" via the further loss of FPA; these latter "bundles" or strands are optically absorbing at 350 μm (FIG. 2), can cross-link with each other near the D-D long contacts, and are further associated into double stranded arrangements via the FPB-mediated polymerization sites. This sequence of events suggests that "insoluble" fibrin strands or mini-clots of at least two types, which are relevant to our subsequent discussion of the FDP fraction in plasma, can possibly form *in vivo*. The first would contain γ chain cross-links and retain its FPB moiety while the second would contain γ chain cross-links (and possibly α chain cross-links) and lack both FPA and FPB; some investigators would describe these as Fibrin I and Fibrin II, respectively. The scheme in FIGURE 1 is designed to relate more to the biochemical information concerning fibrin assembly and degradation rather than to the data obtained from the electronmicroscope. This preamble is meant to acquaint the reader with the origins of two types of cross-linked fibrin which may be formed *in vivo* and which may be the source of important FDP structures found in patient plasma. It is well to add that α chain-γ chain cross-linked fibrin is a likely occurrence *in vivo* (e.g. in thrombi, Ref. 28), but the α chain cross-links do not affect the structures of the high molecular weight fibrin fragments [29] and thus such fibrin is not discussed here. Similarly, the formation of single chain protofibril (lacking only FPA) are probably a frequent occurrence *in vivo*, their rapid destruction by plasmin yielding a non-cross-linked FDP fraction [30] which needs to be distinguished from the XL-FDP fractions.

Structures of FDP Found *in Vivo*

While a whole range of XL fibrin fragments has been classified in fibrin digests by reasonably rigid biochemical criteria (for review, see ref. 31), cross-linked fragments found *in vivo* have been classified with less surety. Comprehensive reviews have appeared recently [32-34] and little new knowledge is currently available. There are many reports in the literature describing elevated FDP in a variety of clinical conditions, [30, 35, *inter alia*] but the description of D dimer and Y-D (then described as "X") in patient plasma during clinically diagnosed amniotic fluid embolism [36] began the search for individual cross-linked fragments, which have already been described *in vitro*. Graeff and coworkers [9] first coined the term X-oligomers to describe the large fibrin fragments found in the plasma of obstetric patients with severe intravascular coagulation. Whether these fragments originate from the destruction of fibrin clots in the microvasculature or result from the digestion of systemic cross-linked soluble fibrin will be discussed later. X-oligomers and related cross-linked fragments are frequently found in the ascitic fluid of patients with ovarian cancer [34] and FIGURE 3 demonstrates the clear-cut classification of D-D, Y-D or D-Y and a heterogeneous X-oligomer fraction. Similar fragments have been found in the plasma of patients with severe intravascular coagulation associated usually with amniotic fluid embolism,[32, 36] while FIGURE 4 shows the separation and identification on SDS-PA gel of fibrin fragments found in the plasma of a patient with intravascular coagulation associated with pneumococcal sepsis.[37] It is well to draw attention to the fact that the technique normally used for identification, namely SDS-PA electrophoresis, disrupts noncovalent aggregates of these cross-linked complexes and, as Francis and

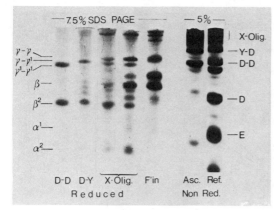

FIGURE 3. Electrophoretic patterns of the polypeptide chains in fragments D-D, D-Y and the X-oligomer fraction. Partly cross-linked fibrin chains (F'in) were used as a marker. The fragments were obtained by cutting a typical acrylamide separation of the cross-linked fibrin fragments found in ascitic fluid (Asc.) and non-reduced fibrin derivatives made *in vitro* were used as reference (ref.). (Taken from Graeff and Hafter [32] with permission.)

Marder [31] have indicated in their *in vitro* analysis of XL-FDP, all the fragments found clinically and described here probably exist in a double stranded form having a heterogeneous mixture of D and E domains making up the composition of each strand. Supporting evidence for this has been the finding that nearly all the D dimer found in some cases of intravascular coagulation was present as the D dimer-E complex.[37]

In an effort to more completely classify the X-oligomer fraction found in ascitic fluid, Hafter and colleagues [34] have assigned, by SDS-gel electrophoresis, molecular weights to the various components. From the molecular weight data the most likely domainal composition for each electrophoretic component has been proposed and the data are shown in FIGURE 5, using *in vitro* derived fragments for comparison.

While others [38-42] have demonstrated the presence of D-dimer in patient plasma during a variety of clinical conditions, the major biochemical examinations of the FDP fraction in patient plasma have been performed by Graeff

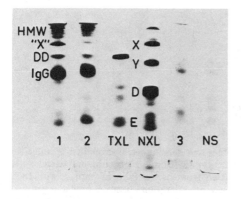

FIGURE 4. Sodium dodecyl sulfate polyacrylamide gel electrophoresis (5% gels) of patient sera immunoprecipitated with an antiserum to fibrinogen fragment E. Major degradation products observed were D dimer, "X" (now known to be cross-linked Y-D), the HMW fraction (X-oligomer) and fragment E. All these fragments were evident in two patients' sera (1 and 2) during DIC associated with pneumococcal sepsis while they were absent in convalescent serum (No. 3) and in normal serum (NS). Two reference gels show the lysates of totally cross-linked (TXL) and non-cross-linked (NXL) fibrin as marker mixtures. (Taken from Whitaker *et al.*[37] with permission.)

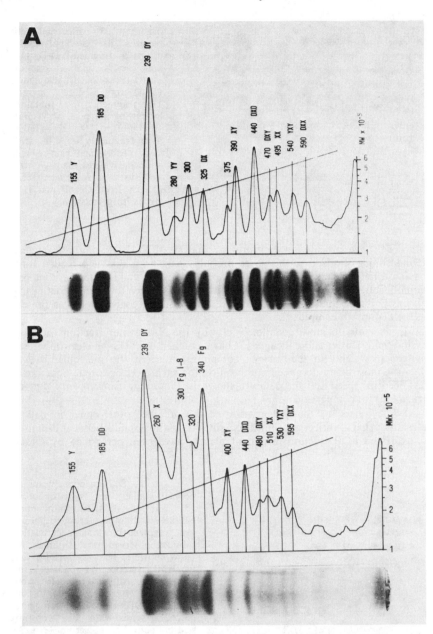

FIGURE 5. SDS-polyacrylamide gel profiles and densitometric scans of fibrin derivatives. (A) sample made by simultaneous action of thrombin, plasmin and Factor XIII on fibrinogen *in vitro;* (B) sample extracted from ascitic fluid of patient with liver cirrhosis. Molecular weights were assessed by migration distance in each gel and the most likely dominal composition is marked above each peak, together with the molecular weights $\times 10^{-3}$. (Taken from Hafter *et al.*[34] with permission.)

and Hafter (for review see ref. 32). These workers have suggested, despite their data shown in FIGURE 5, that the X-oligomers probably exist as homologues of the structure D-X-Y and not as symmetric structures like D-X-D and Y-X-Y.[33] This is supported by the finding that the purified X-oligomer fraction always contains two forms of the dimeric γ chains,[7] i.e. γ-γ and γ-γ' and this is not compatable with symmetrical structures such as Y-X-Y. This, indeed, may reflect the asymmetry that is associated with the interaction of plasmin with fibrinogen-like structures, which was first described by Marder and colleagues in their studies on fibrinogen-plasmin interactions.[43] In an effort to introduce a degree of simplicity into the classification of the XL-FDP frac-

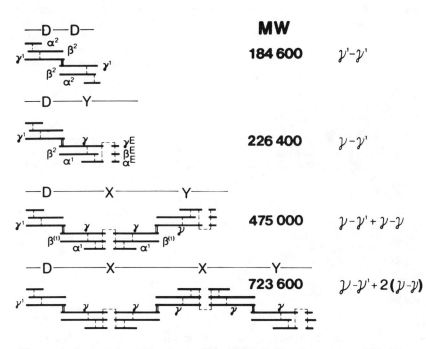

FIGURE 6. Structural models of cross-linked D-D, D-Y and two types of X-oligomer (D-X-Y, D-X-X-Y). The molecular weights and the form of cross-linked γ chains are shown at the right. (Taken from Hafter and Graeff[33] with permission.)

tion found *in vivo*, the general formula for the X-oligomer fraction might be $[D(X)_n Y]_2$, where the double stranded nature of fibrin contributes in a noncovalent manner to the molecular size under physiological conditions. This generalization for the *in vivo* XL-FDP is made despite reports of finding D-X-D and Y-Y in fibrin digests made *in vitro*.[44, 45] This will be discussed later when considering the clinical relevance of certain FDP structures. The known polypeptide structures of D dimer and Y-D have allowed Hafter and Graeff[33] to propose the series of structures in FIGURE 6 as the most likely for those fractions found *in vivo*. It can be seen from this Figure that the cross-linked γ'-γ' dimer is a marker of the D dimer fragment; however, it should be remembered that the D dimer frequently occurs *in vivo* during DIC as the (D-D)E complex [37] mirroring the two stranded nature of the originating fibrin. The cross-linked γ-γ′ dimer is a marker for the Y-D fragment while the ratio of γ-γ dimer to γ-γ′ dimer reflects both the ratio of X oligomers to the Y-D complex and the chain length of the X oligomers.

There are unexpected structural similarities between the cross-linked fibrin fragments (XL-FDP) derivd by plasmin *in vitro* from cross-linked fibrin with those fragments derived from the simultaneous interaction of thrombin, plasmin, and Factor XIII with fibrinogen. Similarities in polypeptide structures between the XL-FDP fragments found by these two distinct mechanistic procedures have been observed and are demonstrated in FIGURE 7. As suggested elsewhere,[46] it would seem that high molecular weight cross-linked fibrin fragments, described approximately as X-oligomers, Y-D and (D-D)E represent a family of fibrin derivatives that may be pivotal during the formation and destruction of fibrin *in vivo*. Both mechanisms generate cross-linked derivatives of higher (X-oligomers) and lower (D-Y, D-D) molecular weight than fibrinogen. While

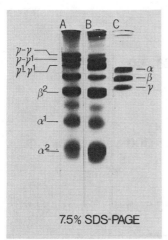

FIGURE 7. SDS-polyacrylamide gel electrophoretic patterns of the polypeptide subunits of fibrin lysates obtained as follows: (A) simultaneous action of plasmin, thrombin, Factor XIII on fibrinogen; (B) plasmin-induced digestion of preformed cross-linked fibrin. Pattern C is a reference marker of the fibrin chains α,β,γ, (Taken from Graeff and Hafter [82] with permission.)

our major consideration here is the characteristization of XL-FDP, there are many reports on the formation of soluble fibrin and its digestion as non-XL-FDP. In an excellent review, Bang and Chang [30] suggest that fibrin X, Y, D, and E can form large molecular weight complexes. Any assessment of the FDP fraction in blood must distinguish between what has been called the soluble fibrin monomer complexes (SFMC), non-XL-FDP and XL-FDP if these fractions are going to help in assessing the severity of the clinical conditions that stimulated their production.

Possibly owing to the tediousness of the methodology involved not many workers have reported on the structures of XL-FDP found *in vivo*. Nearly all samples obtained have come from patients with gross DIC-associated defibrination with clinical evidence of obstetric disorders,[32] disseminated cancer,[41] or thrombotic disease.[46] Vila and colleagues,[40] using columns of insolubilized antifibrinogen serum, demonstrated relatively large amounts of D dimer and fragment E in normal plasma, whereas others [41] have been unable to confirm this. Since normal plasma is unlikely to contain significant amounts of D dimer, the above conflict of data may be little more than a reflection of the handling of the plasma samples. During these studies it is necessary to take blood into both anticoagulant and antifibrinolytic solution and storage should be in liquid nitrogen if the samples are not processed immediately. Such precautions will be even more critical when sensitive tests are developed to examine the products of the combined action of thrombin and plasmin on circulating fibrinogen.

CLINICAL RELEVANCE OF STRUCTURAL FEATURES OF FDP

It is necessary to declare some assumptions about the generation of the FDP fraction in blood before suggestions can be made concerning the relevance of any particular form of the cross-linked FDP *in vivo*. The first assumption is that thrombin-mediated fibrin must be present to act as a source of the FDP fraction. Whereas this may not always be true, the assumption is reasonable in light of our knowledge about the molar concentrations of plasminogen and the fast-acting inhibitor of plasmin (α_2-antiplasmin) in blood (for review see ref. 47). Another assumption is that the plasmin-associated fibrinolytic system generates the FDP fraction in blood. While this may be quite acceptable when viewing the destruction of forming fibrin fibers, it is not fully logical when one thinks in terms of the digestion of organized and cross-linked thrombi. The latter situation (i.e. thrombus formation *in vivo*) can be interpreted as one in which the fibrinolytic system has failed to protect against the organization of fibrin fibers as a locus of thrombus formation—why then should this same fibrinolytic system be capable of digesting a fibrin matrix that is rapidly becoming more resistant to lysis via Factor XIII-mediated cross-linking, particularly of its α chains?[48, 49] Thus we need to postulate that enzymes other than plasmin, having their origin in the cellular fraction of thrombi, may be responsible for the slow digestion of clots.[50] This would not, of course, exclude a supportive role for the plasmin-associated fibrinolytic system in the lysis of established thrombi. For the sake of this review we are assuming that plasmin-associated fibrinolysis is the major source of the FDP fraction during DIC and that reactions similar to those which occur during overt DIC may also occur during events prior to thrombosis.

To comment on the clinical relevance of FDP occurrence in blood it is inevitable that a degree of controversy may surround one's remarks in that no acceptable consensus is available at the moment. Here we will try to collate some information about fibrin formation *in vitro* and the structures of fibrin fragments found *in vivo*. It is hoped that this approach may justify some speculations on aspects of the molecular pathology of DIC-associated hemorrhage and the role of the XL-FDP fraction in blood as a predictive marker of thrombosis. It is well here to attempt to clarify the term thrombosis.[51] Normally we think in terms of severe coronary artery disease complicated by acute clot formation in a partly occluded vessel. However, the occlusion of the microvasculature by thrombi is a condition of far broader aetiology and must be considered as the major precipitating event in acute forms of DIC-associated hemorrhage.

During severe cases of DIC it is clear that large amounts of thrombin are produced *in vivo*, sufficient to allow the conversion of fibrinogen to fibrin, activate Factor XIII, and cause the aggregation of platelets. It is evident

(FIG. 2) that fibrin strand formation *in vitro* is a prerequisite for Factor XIII cross-links to form in the fibrin. Since in many cases of DIC the presence of XL-FDP is evident,[32] one must assume that insoluble fibrin strands or clots had previously been formed *in vivo*. It is accepted that occlusion of certain sections of the microvasculature by fibrin and platelet thrombi can occur during DIC. The obvious result, if these deposits are not rapidly cleared, would be localized tissue death. These blockages of the microvasculature could cause rupture of small vessels with resultant bleeding. We must now ask whether the fibrin fragments found during DIC are any help in assessing the extent of both tissue death and hemorrhage which can occur during DIC.

Though cross-linked soluble fibrin has been reported *in vivo*,[22] it is here suggested that fibrin strands may evidence as seemingly soluble fibrin in the macrovasculature. These fibrin strands may, however, begin the process of occlusion in the microvasculature under unfavorable flow conditions. It is here speculated that the type of cross-linked fibrin or cross-linked FDP could indicate the severity of the occlusion of the microvasculature. Where the level of fibrinopeptide A (FPA) in blood was until recently regarded as a marker of thrombotic disease,[52] it has subsequently been suggested that fibrinopeptide B (FPB) release from fibrin may indicate a more pathological thrombotic state,[53] in that it would be a more reliable marker for the deposition of a more advanced form of fibrin *in vivo*. Thus cross-linked fragments and FPB seem to be emerging as markers of potential tissue damage during DIC. The release of FPB from fibrin has been demonstrated to expose a distinct set of polymerization sites in fibrin [11] that mediate some of the organization of fibrin fibers (FIG. 1); this type of fibrin has also been named Fibrin II to distiguish it from fibrin lacking only FPA, namely Fibrin I. Fibrin II, on digestion with plasmin, generates the D dimer-E complex while cross-linked Fibrin I forms D dimer and E,[11] these fragments not being associated by the FPB-mediated polymerization sites of the originating fibrin. We can conclude that cross-linked FDP fractions that are double stranded have their origin in Fibrin II, i.e. an advanced stage of fibrin that is both cross-linked and lacking FPA and FPB. Thus the XL-FDP fraction which, on treatment with detergents, reduces in molecular size owing to disruption of noncovalent D:E secondary polymerization sites may be a marker of advanced microvascular damage which can be accompanied by severe hemorrhage.

The generalized, and possibly over-simplified, formula for cross-linked FDP, e.g. $[D(X)_n Y]_2$, has already been discussed. While cross-linked γ'-γ' chains are a marker for D dimer (FIG. 6), the ratio of γ-γ/γ-γ' is a marker for the value of n in the above formula for XL-FDP. We have proposed that should it be possible to isolate sufficient of the FDP fraction from a clinical sample to achieve a polypeptide profile by SDS-PA gel electrophoresis, the ratios of γ-γ/γ-γ'/γ'-γ' may be quite revealing concerning the nature of fibrin formed and its susceptibility to clearance by the fibrinolytic system. A higher ratio of γ-γ/γ'-γ' may reflect poor fibrinolytic response and a slow clearance of fibrin from the occluded vessels while the converse would be suggested by a low ratio of γ-γ/γ'-γ'. Assays using specific antisera directed towards unique epitopes on the γ-γ, γ-γ' and γ'-γ' may also help in defining both the size of the X oligomer fraction and the extent of lysis. Nossel and his colleagues [53] have proposed that the ratio of the FPB level to the level of the plasmin mediated $B\beta$ chain remnant 1–42 may give an idea of the efficiency of the fibrinolytic system in destroying the fibrin which can accumulate during the prethrombotic state. Our above suggestions follows the same line of thinking but would yield more information on both fibrin formed and the vigour of the fibrinolytic response.

While our comments have dealt with situations where a disseminated condition exists in the patient and a large turnover of fibrinogen is involved, some consideration must be given to the localized condition of thrombosis where the level of FDP is marginally elevated as measured by conventional FDP assay. While it has been speculated that "thrombosis" of the microvasculature associated with DIC is an acute phenomenon, long-term atherosclerotic disease is the most likely precondition of an acute occlusive event in some of the larger vessels at the arterial side of the circulation.[51] This is not to say that subclinical microvascular damage may not also be a predisposing feature. The Rokitansky-Duguid[54, 55] hypothesis has suggested that fibrin deposition is associated with atheromatous lesions and, possibly, with coronary artery disease. Local hypercoagulability with impaired fibrinolysis at plaque-regions in the vessel wall can be expected. Should we accept that localized plasmin-thrombin-fibrinogen interactions follow similar mechanisms to those understood to take place systematically during DIC the generation of low levels of the X-oligomer fraction and possibly Y-D and D dimer may act as systemic markers of localized arterial disease and may indicate the coronary-prone individual. Validation of some of these thoughts awaits the availability of sensitive and specific assays for unique structural components of the XL-FDP fraction.

CONCLUDING REMARKS

A number of assumptions about fibrin formation and lysis *in vivo* have been made, some on the basis of observations *in vitro* and others based on current views expressed in the literature. Nearly all the data on clinical material presented in the literature suggested that cross-linked fibrin fragments (notably D dimer and Y-D) were present. Does this suggest that cross-linked fibrin fragments are the usual components of the FDP fraction in blood or, because of the insensitivity of assay techniques, have we limited ourselves to examining only those plasmas from patients whose FDP titer is extremely high (>100 μg/ml) and associated with advanced intravascular coagulation? The answer must lie in our future ability to measure sensitively the above described cross-linked fibrin fragments as distinct from non-cross-linked fragments. We have, as yet, no clear picture of the state of aggregation of non-cross-linked FDP, should they arise in blood (for review, see ref. 30). Our present understanding of *in vivo* formed cross-linked FDP is that they are held together by covalent γ-γ crosslinks and noncovalent polymerization bonds, the latter being mediated by FPB release. If γ-γ cross-linking and FPB release occur at about the same stage in the fibrin fiber organization (FIG. 2 and ref. 56, respectively), non-cross-linked FDP would be indicative of the formation and lysis of non-cross-linked soluble fibrin polymers, probably retaining their FPB moieties. The presence of these latter polymers and their resultant FDP fractions in blood would suggest a low level production of thrombin or a highly active fibrinolytic system or a combination of both. Considerable *in vitro* experimentation and further examination of clinical material is required before adequate substance can be given to a number of these speculations. Assay procedures of a sensitive and more specific nature need to be developed. The assays available currently have been adequately reviewed [32] but it might be well here to comment on some of these and try to point to directions of future assay development.

Polyacrylamide gel electrophoresis in detergent (SDS-PAGE) is the commonest method used to classify fibrin fragments found in vivo. The technique normally used is some modification of the method of Weber and Osborn [57] and it is usually necessary to partly purify and enrich the sample before application to electrophoresis. The latter is achieved by precipitation with β-alanine,[58, 59] glycine,[60] ammonium sulfate [61] or acid.[62] Agarose gel chromatography is widely used [32] as a purification step preceded by one of the above precipitation steps. Immunoadsorption using Sepharose-bound antibodies [36, 63] has proved a valuable tool to isolate fibrin fragments from plasma and other body fluids.

Though immunological methods for the detection of cross-linked derivatives are not yet available, efforts have been made to raise specific antisera to D dimer, the D dimer-E [64] complex and the X-oligomer fraction.[46] It was found [64] that the cross-link-related markers represent only a small proportion of the total antigenicity of fragment D-D whereas most of the antibodies reacted with fibrinogen fragment D. Antisera to highly purified X-oligomers have been raised in rabbits, which antisera to not react with fibrinogen but cross-react with non-cross-linked FDP (X, Y, D and E) in serum and plasma.[65] Such antisera need further development, possibly using the monocloncal approach to achieve more specificity for the conformationally related antigens on the high molecular weight XL-FDP. Antisera are not yet available specific for the Y-D or (Y-D)(D-Y) fractions. These latter antisera should be quite useful since it seems that cross-linked Y-D is one of the major components in fibrin digests in vitro and in plasma during DIC. (see FIG. 6).

The polypeptide chain compositions of the XL-FDP described here and elsewhere in this volume suggest that it would be useful to develop sensitive assays for unique epitopes on cross-linked γ-γ chains as distinct from those of the D dimer marker (γ'-γ'). Any assay which would indicate whether the XL-FDP fraction was single stranded or double stranded would give information on the formation of Fibrin II in the cirulation, the latter being an indication of an advanced stage of fibrin deposition in the microvasculature; levels of FPB or the ratio of FPA/FPB may yield similar information. Ratios of cross-linked γ chains to cross-linked γ chain remnants (γ-γ' or γ'-γ') should give information on the state of the hemostatic balance between thrombin and plasmin activity in vivo, which in turn may determine the outcome of any hypercoagulable episode.

While examinations of the XL-FDP during overt DIC have allowed some of the above suggestions about future assay development, there is an immediate need to develop sensitive assays for structural features of the FDP fraction when present in plasma at the nanogram/ml level during the subclinical hypercoagulable state. Such information may give an early warning of an acute event and allow the clinician to suggest both therapeutic and life-style interventions which may reduce both morbidity and mortality from a wide spectrum of cardiovascular disease.

REFERENCES

1. GAFFNEY, P. J. 1981. *In* Thrombosis and Haemostasis. A. L. Bloom & D. P. Thomas, Eds.: 198–224. Churchill-Livingstone. Edinburgh.
2. MOROZ, L. A. & N. J. GLIMORE. 1976. Blood **48:** 531–545.
3. COOKE, E. D., Y. B. GORDON, S. A. BOWCOCK, C. M. SOLA, M. F. PILCHER, T. CHARD, R. M. IBBOTSON & M. E. AINSWORTH. 1975. Lancet **2:** 51–54.
4. GAFFNEY, P. J., M. MAHMOUD, F. JOE, M. SPITZ & R. GAINES-DAS. 1981. *In* Progress in Chemical Fibrinolysis and Thrombolysis. J. F. Davidson, I. M. Nilsson, M. M. Samama & P. C. Desnoyers, Eds. Vol. 5: 379–383. Churchill-Livingston. Edinburgh.
5. GAFFNEY, P. J. 1972. Lancet **2:** 1422.
6. GAFFNEY, P. J. 1973. Thrombosis Res. **2:** 201–218.
7. GAFFNEY, P. J., F. JOE & M. MAHMOUD. 1980. Thrombosis Res. **20:** 647–662.
8. FRANCIS, C. W. & V. J. MARDER. 1982. Sem. Thromb. Haemost. **8:** 25–35.
9. GRAEFF, H., R. HAFTER & L. BACHMANN. 1979. Thrombosis Res. **16:** 313–328.
10. BAILEY, K. & F. R. BETTLEHEIM. 1955. Biochim. Biophys. Acta **18:** 495–503.
11. OLEXA, S. A. & A. Z. BUDZYNSKI. 1980. Biochemistry. **19:** 647–651.
12. McKEE, P. A., P. MATTOCK & R. L. HILL. 1970. Proc. Natl. Acad. Sci. USA **66:** 738–744.
13. DOOLITTLE, R. F. 1973. Adv. Protein Chem. **27:** 1–109.
14. FINLAYSON, J. S. 1974. Sem. Thromb. Haemost. **1:** 33–62.
15. BLOMBÄCK, B. & M. BLOMBÄCK. 1972. Ann. N.Y. Acad. Sci. **202:** 77–97.
16. BLOMBÄCK, B. 1967. *In* Blood Clotting Enzymology. W. H. Seegers, Ed.: 143–215 Academic Press. New York.
17. MARTINELLI, R. A. & H. A. SCHERAGA. 1980. Biochemistry **19:** 2343–2350.
18. BLOMBÄCK, B., B. HESSEL, M. OKADA & N. EGBERG. 1981. Ann. N.Y. Acad. Sci. **370:** 536–550.
19. HERMANS, J. & J. McDONAGH. 1982. Sem. Thromb. Haemost. **8:** 11–24.
20. KUDRYK, B. J., D. COLLEN, K. R. WOODS & B. BLOMBACK. 1974. J. Biol. Chem. **249:** 3322–3325.
21. KIERULF, P. 1974. Thrombosis Res. **4:** 183–187.
22. LY, B., P. KIERULF & E. JAKOBSEN. 1974. Thrombosis Res. **4:** 509–522.
23. HANTGAN, R. R. & J. HERMANS. 1979. J. Biol. Chem. **254:** 11272–11281.
24. LAURENT, T. C. & B. BLOMBÄCK. 1958. Acta Chem. Scand. **12:** 1875–1877.
25. FERRY, J. D. 1952. Proc. Natl. Acad. Sci. USA **38:** 566–569.
26. SHAINOFF, J. R. & B. N. DARDIK. 1979. Science **204:** 200–202.
27. HANTGAN, R. R., W. FOWLER, H. ERICKSON & J. HERMANS. 1980. Thromb. Haemost. **44:** 119–124.
28. GAFFNEY, P. J., M. BRASHER, K. LORD, C. J. L. STRACHAN, A. R. WILKINSON, V. V. KAKKAR & M. F. SCULLY. 1976. Cardiovasc. Res. **10:** 421–426.
29. GAFFNEY, P. J., D. A. LANE & M. BRASHER. 1975. Clin. Sci. Mol. Med. **49:** 149–156.
30. BANG, N. U. & M. L. CHANG. 1974. Sem. Thromb. Haemost. **1:** 91–128.
31. FRANCIS, C. W. & V. J. MARDER. 1982. Sem. Thromb. Haemost. **8:** 25–35.
32. GRAEFF, H. & R. HAFTER. 1982. Sem. Thromb. Haemost. **8:** 57–68.
33. HAFTER, R. & H. GRAEFF. 1982. *In* Fibrinogen. Recent Biochemical and Medical Aspects. A. Henschen, H. Graeff & F. Lottspeich, Eds.: 291–306. Walter de Gruyter. Berlin-Hawthorne, NY.
34. HAFTER, R., E. KLAUBERT & H. GRAEFF. 1983. *In* Fibrinogen. Structure, Functional Aspects, Metabolism. F. Haverkate, W. Nieuwenhuizen & A. Henschen, Eds. Walter de Gruyter. Berlin-Hawthorne, NY. In press.

35. WILNER, G. D. 1978. *In* Progress in Haemostasis and Thrombosis. T. H. Spaet, Ed. Vol. 4: 211–248. Grune and Stratton. New York.
36. GAFFNEY, P. J. 1975. Clin. Chim. Acta **65:** 109–115.
37. WHITAKER, A. N., E. A. ROWE, P. P. MASCI & P. J. GAFFNEY. 1980. Thrombosis Res. **19:** 381–391.
38. GRAEFF, H., R. HAFTER & R. VON HUGO. 1977. Thromb. Haemost. **38:** 724–726.
39. LANE, D. A., P. A. ROBBINS, M. W. RAMPLING & V. V. KAKKAR. 1977. Br. J. Haemostat. **36:** 137–148.
40. VILA, V., E. REGANON & J. AZNAR. 1978. Clin. Chim. Acta **87:** 245–252.
41. MERSKEY, C., A. J. JOHNSON, J. U. HARRIS, M. T. WANG & S. SWAIN. 1980. Br. J. Haemost. **44:** 655–670.
42. EDGAR, W., M. J. WARRELL, D. A. WARRELL & C. R. M. PRENTICE. 1980. Br. J. Haemost. **44:** 47–1481.
43. MARDER, V. J., N. R. SHULMAN & W. R. CARROLL. 1969. J. Biol. Chem. **244:** 2111–2119.
44. FRANCIS, C. W., V. J. MARDER & G. H. BARLOW. 1980. J. Clin. Invest. **66:** 1033–1043.
45. REGANON, E., J. VILA & J. AZNAR. 1978. Thromb. Haemost. **40:** 368–376.
46. GAFFNEY, P. J. 1981. *In* Fibrinogen. Recent Biochemical and Medical Aspects. A. Henschen, H. Graeff, F. Lottspeich, Eds.: 307–327. Walter de Gruyter. Berlin-Hawthorne, NY.
47. COLLEN, D. 1980. Thromb. Haemost. **43:** 77–89.
48. MCDONAGH, R. P. JR., J. MCDONAGH & F. DUCKERT. 1971. Br. J. Haemost. **21:** 323–332.
49. GAFFNEY, P. J. & A. N. WHITAKER. 1979. Thromb. Res. **14:** 85–94.
50. PLOW, E. F. & T. S. EDGINGTON. 1975. J. Clin. Invest. **56:** 30–38.
51. NEMERSON, Y. & H. L. NOSSEL. 1982. Ann. Rev. Med. **33:** 479–488.
52. NOSSEL, H. L. 1971. Proc. Natl. Acad. Sci USA **68:** 2350–2352.
53. NOSSEL, H. L. 1981. Nature **291:** 165–167.
54. ROKITANSKY, K., VON. 1844. *In* Handbuch der pathologischen Anatomie, 2. Braunmüller and Seidel. Vienna.
55. DUGUID, J. B. 1946. J. Pathol. Bact. **58:** 207–212.
56. BLOMBACK, B., B. HESSEL, D. HOGG & L. THERKILDSEN. 1978. Nature **275:** 501–505.
57. WEBER, K. & M. OSBORN. 1969. J. Biol. Chem. **244:** 4406–4412.
58. STRAUGHN, W. & R. H. WAGNER. 1966. Thromb. Diath. Haemost. **16:** 197–206.
59. GRAEFF, H. & R. VON HUGO. 1972. Thromb. Diath. Haemost. **27:** 610–618.
60. KAZAL, L. A., S. AMSEL, O. P. MILLER & L. M. TOCANTINS. 1963. Proc. Soc. Exp. Biol. Med. **113:** 989–994.
61. PARTFENTJEV, J. A., M. L. JOHNSON & E. E. CLIFTON. 1953. Arch. Biochem. Biophys. **46:** 470–480.
62. HAFTER, R., R. VON HUGO & H. GRAEFF. 1978. Hoppe Seyler's Z. Physiol. Chem. **359:** 759–763.
63. LANE, D. A., M. F. SCULLY & V. V. KAKKAR. 1976. Thromb. Res. **9:** 191–200.
64. BUDZYNSKI, A. Z., V. J. MARDER, M. E. PARKER, P. SHAMES, B. S. BRIZUELA & S. OLEXA. 1979. Blood **54:** 794–804.
65. GAFFNEY, P. J., F. JOE, M. MAHMOUD & M. SPITZ. 1981. *In* Progress in Fibrinolysis and Thrombolysis. J. F. Davidson, I. M. Nilsson, M. M. Samama & P. C. Desnoyers, Eds. Vol. 5: 399–405. Churchill-Livingstone. Edinburgh.

DISCUSSION OF THE PAPER

A. L. COPLEY (*Polytechnic Institute of New York, Brooklyn*): I do not believe that one can equate thrombus formation or thrombosis, which is, as you say, a disease, with intravascular coagulation of fibrin. We have known since almost the beginning of the century that you have both fibrin formation and fibrinolysis going on in the circulating blood all the time and that this is a part of a normal physiology.

P. J. GAFFNEY: I agree that the pathology of thrombosis and DIC differ. However, my intention in this paper was to equate some of the molecular reactions that take place prior to and during both clinical states. Though DIC can be accompanied by bleeding, an initiating biochemical event is fibrin formation. Fibrin deposition is also a structural feature of arterial thrombosis, and we have proposed above that, while the fibrinolytic response in DIC is obvious, a similar reaction may take place at a more discrete level during arterial thrombosis.

VASOACTIVE PEPTIDES DERIVED FROM
DEGRADATION OF FIBRINOGEN AND FIBRIN

Tom Saldeen

Institute of Forensic Medicine
University of Uppsala
75237 Uppsala, Sweden

Pulmonary microemboli seem to be of pathogenetic importance in a certain type of acute respiratory distress, the microembolism syndrome.[1] In the delayed form of this syndrome the fibrin component of the microemboli seems to be significant in the pathogenesis. Different observations have indicated that fibrin plays a role other than that of causing mechanical blockage of the vessels. Thus, there is evidence that fibrin degradation products released from persistent pulmonary microemboli into the lung tissue may be involved. Protein-rich interstitial and alveolar edema is a characteristic finding in the delayed microembolism syndrome, and both lysates from plasmin and elastase degradation of fibrin are strong inducers of increased microvascular permeability with edema as a result.[2, 3]

Our studies have indicated that the permeability-increasing effect is due to some fibrin-derived peptides. We have isolated two such peptides, obtained by plasmin degradation of human fibrin or fibrinogen, namely a pentapeptide and an undecapeptide, that increase microvascular permeability,[2] and recently also two elastase degradation products from fibrin or fibrinogen with a stronger effect.[3]

The pentapeptide and both the peptides from elastase degradation belong to the Bβ chain of the fibrinogen molecule. The undecapeptide is situated in the Aα chain.

The pentapeptide corresponds to residues 43–47 of the Bβ chain and the undecapeptide to residues 220–230 of the Aα chain. The following procedure was used for isolation of these peptides. Human fibrin or fibrinogen was incubated for 24 hours at physiological pH and at 37°C in a lysis system containing plasmin. The low molecular weight degradation products formed were separated from the bulk of the material by membrane filtration. After concentration and subsequent chromatography on a Bio-Gel P-6 column, which gave 12 distinct fractions, fraction 6 showed the highest permeability-increasing activity (FIG. 1).

This major activity fraction was further separated into eight subfractions by column zone electrophoresis at pH 5 (FIG. 2); two of these subfractions had a permeability-increasing effect. These fractions, called 6A and 6D, respectively, were further purified to a homogeneous state by electrophoresis at pH 1.9 (FIG. 3). The recovery of the active peptides was about 60%.

Peptide 6A is part of an undecapeptide that is split from the Bβ-chain on rather limited fibrinolysis. Since we have observed that comparable amounts of peptide 6A are produced after 1 and 24 hours of fibrinolysis, it is reasonable to assume that the undecapeptide is further split at a fairly high rate to release 6A. Peptide 6D from the Aα chain is also released from fibrinogen in the initial stage of plasmin degradation. It originates from a highly exposed region

0077–8923/83/0408–0424 $01.75/0 © 1983, NYAS

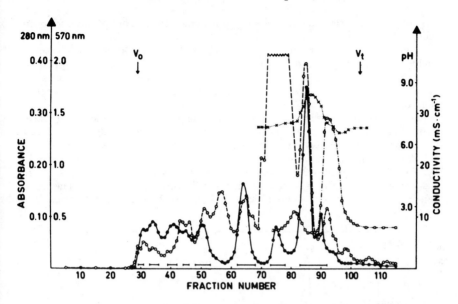

FIGURE 1. Chromatography of the low molecular weight degradation products from plasmin degradation of fibrinogen on Bio-Gel P–6. ●———● = A₂₈₀; ○– – – –○ = A₅₇₀; –·—·—·– = specific conductance at 20° C; x ···· x = pH |—| = pooled fractions.

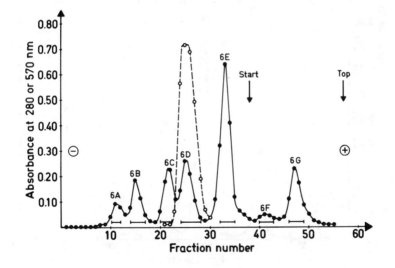

FIGURE 2. Column zone electrophoresis of fraction 6 from the run shown in FIGURE 1. Column dimensions = 1 × 90 cm (V₀ = 61 ml). Electrode buffer: 0.05 M pyridine/0.05 N acetic acid, pH 5.0. Duration of run: 16 hours at 1000 V and 7 mA. The run was made at +8° C. The separated zones were eluted at a flow rate of 20 ml/h and 1 ml fractions were collected. ●– – – –● = A₂₈₀; ○———○ = A₅₇₀ |—| = pooled fractions.

linking an easily released hydrophilic peptide chain to the core of the fibrinogen molecule.[4]

Peptide 6E (Thr-Ser-Glu-Val-Lys, residues 54–58 of the γ-chain) showed slight, but significant vasoconstrictor activity. This peptide is part of fragment E released from fibrinogen. This fragment is less sensitive to plasmin attack.[5] The resulting slow release of peptide 6E suggests that its concentration and thereby its vasoconstrictor activity *in vivo* must be rather low. The peptide is, however, released much more quickly after degradation of fibrin than after degradation of fibrinogen,[6] a finding that might be of some interest in the differentiation between fibrinogenolysis and fibrinolysis.

The two fragments from elastase degradation of fibrinogen correspond to Bβ 13–43 and Bβ 30–43, respectively. The following purification procedure was used for the elastase fragments: Human fibrin(ogen) was made plasminogen-free by affinity chromatography on lysine-Sepharose prior to use. The

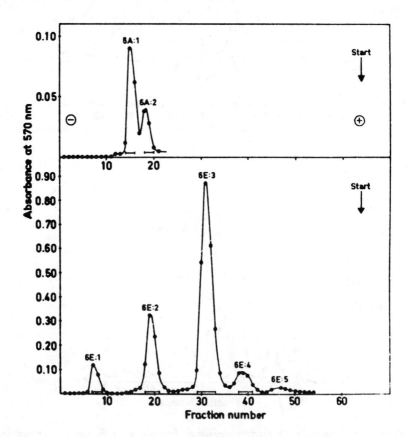

FIGURE 3. Further purification of subfractions 6A and 6E by column zone electrophoresis at pH 1.9. Column dimensions: 1×95 cm ($V_o = 60$ ml). Electrode buffer: 0.54 M formic acid/1.30 M acetic acid, pH 1.9. Duration of run: 16 h at 1000 V and 15 mA. The run was made at $+8°$ C. The separated zones were eluted at 20 ml/h and 1 ml fractions were collected. |—| = pooled fractions.

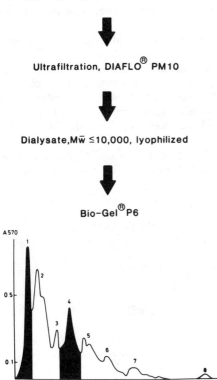

Human leukocyte elastase mediated lysis of human fibrinogen

FIGURE 4. Chromatography on Bio-Gel P–6 of the low molecular weight degradation products from elastase degradation of fibrinogen.

human leukocyte elastases were isolated from extracts of lysosome-like granules of human leukemic cells by a combination of gel filtration, affinity chromatography and preparative agarose gel electrophoresis. The fibrinogen and the leukocyte elastase (in a molar ratio of 100:1) were incubated together for 48 hours at $+37°C$ at pH 8.5.

The lysis was stopped by rapid freezing. After ultrafiltration, the dialysate, with a MW below 10 000, was lyophilized and subjected to chromatography on Bio-Gel P-6. Eight peaks were obtained, peaks 1 and 4 containing the permeability-increasing activity (FIG. 4). Those fractions were further purified by column zone electrophoresis at pH 5. Fraction 1 gave six peaks and fraction 4 five peaks, and in both cases the first peak was active (FIG. 5). The active peak of fraction 1 was further purified to homogeneity by HPLC, which gave six peaks of which the last one was active (FIG. 5). The active peak of fraction 4 was further purified by column zone electrophoresis at pH 1.9, which gave two peaks of which the second and smallest one was active (FIG. 5).

Amino acid analysis of the active peak from HPLC indicated position 13–43 in the Bβ-chain and the peak from the electrophoresis at pH 1.9 indicated position 30–43 of the Bβ-chain (FIG. 6). Both these peptides were more active than peptide 6A. The smaller one was at least 15 times and the larger one about twice as active as 6A.

Peptides 6A and 6D and the 14-residue peptide from elastase degradation have been synthesized and exhibit full potency.

Thirty-three analogues of peptide 6A have been synthesized for structure-function studies (TABLE 1).

We have found that there are certain definite structural prerequisites for biological activity of the pentapeptide 6A.[7, 8] The existence of such firm requirements indicates that the peptide interacts with other molecules in the tissues in a manner for which the three-dimensional structure, the position of certain important functional groups and the charge of the molecule all seem to be important.

A tetrapeptide with basic amino acids at both ends and a proline residue adjacent to the NH$_2$-terminal amino acid is essential for a strong effect on permeability.

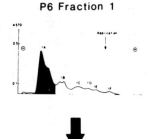

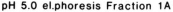
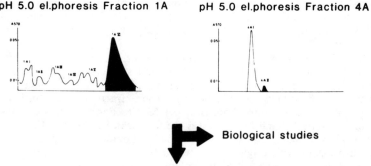

FIGURE 5. Cellulosa column zone electrophoresis at pH 5.0 of fractions 1 and 4 from the run shown in FIGURE 4. The active peak from fraction 1 was further purified by HPLC.

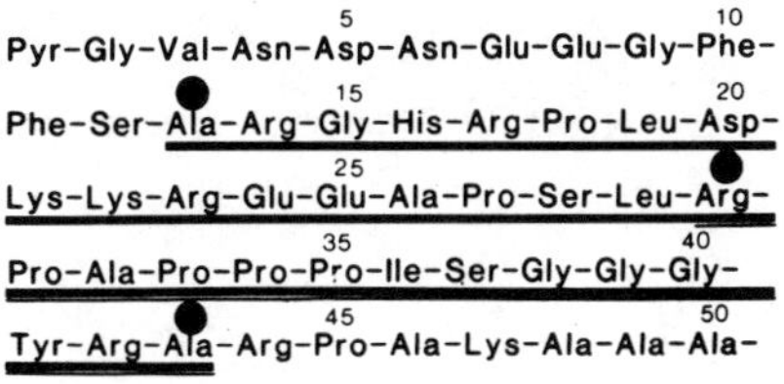

Pyr-Gly-Val-Asn-Asp-Asn-Glu-Glu-Gly-Phe-
(5) (10)
Phe-Ser-Ala-Arg-Gly-His-Arg-Pro-Leu-Asp-
(15) (20)
Lys-Lys-Arg-Glu-Glu-Ala-Pro-Ser-Leu-Arg-
(25)
Pro-Ala-Pro-Pro-Pro-Ile-Ser-Gly-Gly-Gly-
(35) (40)
Tyr-Arg-Ala-Arg-Pro-Ala-Lys-Ala-Ala-Ala-
(45) (50)

FIGURE 6. NH$_2$-terminal part of the Bβ-chain of human fibrinogen, showing the two vasoactive peptides derived from elastase degradation of fibrinogen (Bβ 13–43 and Bβ 30–43) and the vasoactive peptide 6A derived from plasmin degradation of fibrinogen (Bβ 43–47).

The restrictions seem to reside more or less completely within the NH$_2$-terminal part of the molecule, whereas the COOH-terminal region can be extensively modified without loss of activity.[9, 10] Neither the carboxyl nor the ε-amino group is necessary for full activity. The former in fact seems to be a slight disadvantage, since in all cases tested so far amidation leads to enhanced potency. The great difference between the norleucine analogue and the corresponding amide especially suggests that the net charge of this part of the molecule rather than the presence of a basic group is important for full activity. It seems that this region must not be negatively charged. A free carboxyl group, if present, has to be neutralized by a neighboring basic moiety or, possibly, be positioned at such a distance as to render its influence negligible.

The necessity for an L-Arg-L-Pro sequence has been confirmed.[9] Reversal of the configuration of either of these residues reduces the effect drastically, as does insertion of a third amino acid between them. The rigidity associated with the presence of proline is important. Replacement by other amino acids, even dehydroproline, again reduces the activity. Not unexpectedly, L-arginine can be replaced by L-lysine, but substitution of L-norleucine results in markedly diminished potency. One might argue that this decrease again reflects the importance of the net charge of the molecule, but the restrictions on the chirality of the arginine and proline residues rather contradict this view. It appears that it is not only the presence but also the orientation of the basic side chain that is important for full activity.[10]

The results of TABLE 1 and, especially, the fact that Arg-Pro-Lys (peptide 9) retains most of the activity of peptide 6A, directly led us to the four naturally occurring peptides of TABLE 2, whose activities are well interpretable qualitatively in terms of this model. The fact that substance P, neurotensin and bradykinin, but not tuftsin, all exhibit activities 25–100 times higher than that of peptide 6A may perhaps mean that the former peptides, in addition to their message for vasoactivity, carry information with a potentiating effect. Such an interpretation has been proposed by Schwyzer[11] in an attempt to rationalize the function of different parts of the ACTH molecule.

The structural similarities just discussed open the way for at least two sets of further speculations:

(1) The similarities seem to indicate that receptors for the peptides in question have one phylogenetic origin.

(2) It is possible that peptide 6A actually interacts with one or more types of receptors that are created by nature for other more specific biologically active peptides.

TABLE 1

EFFECT OF VARIOUS SYNTHETIC ANALOGUES OF THE PENTAPEPTIDE
ALA-ARG-PRO-ALA-LYS ON RAT SKIN PERMEABILITY *

Peptide No.							Activity (%)
6A		Ala	Arg	Pro	Ala	Lys	100
1	Gly	Ala	Arg	Pro	Ala	Lys	50
2		D-Ala	Arg	Pro	Ala	Lys	15
3		C₂H₅CO	Arg	Pro	Ala	Lys	15
4		Phe	Arg	Pro	Ala	Lys	40
5			Arg	Pro	Ala	Lys	100
6		Arg	Ala	Pro	Ala	Lys	15
7			D-Arg	D-Pro	Ala	Lys	10
8			Arg	Pro	Gly	Lys	10
9			Arg	Pro	D-Pro	Lys	70
10			Arg	Pro	Pro	Lys	30
11			Arg	Pro	Lys	Lys	35
12			Arg	Pro	Pro	Lys	75
13			Arg	Ala	Ala	Lys	100
14			Arg	D-Pro	Ala	Lys	20
15			Arg	Ala	D-Pro	Lys	15
16	Arg		Pro	Ala	Ala	Lys	80
17	Ala		Lys	Pro	Ala	Lys	70
18	Ala		Nle	Pro	Ala	Lys	20
19	Ala		Lys	Pro	Ala	Arg	200
20	Ala		Arg	Gly	Ala	Lys	10–15
21	Ala		Arg	Dhp	Ala	Lys	50
22	Ala		Arg	Ala	Pro	Lys	35
23	Ala		Arg	Pro	D-Ala	Lys	70
24	Ala		Arg	Pro	Lys	Lys	100
25	Ala		Arg	Pro	Gly	Lys	55
26	Ala		Arg	Pro	Pro	Lys	75
27	Ala		Arg	Pro	Ala	Nle	25

28	Ala	Arg	Pro	Ala	Arg		100
29	Ala	Arg	Pro	Ala	Lys-NH₂		200
30	Ala	Arg	Pro	Ala	D-Lys		100
31	Ala	Arg	Pro	Ala	Nle-NH₂		200
32	Ala	Arg	Pro	Ala	D-Lys-NH₂		200
33	Ala	Arg	Pro	Ala	Lys	Gly	50

* The results are expressed as per cent of the activity of the native peptide, whose effect was quantitatively determined simultaneously. In each case, the relative activities of the modified peptides were calculated from a dose-response curve, which is a routine procedure in our assay system. The activity obtained in each case is the average of at least four independent determinations performed on adjacent areas of the skin of four different rats. In the assay procedure adopted, the detection limit corresponds to a response evoked by injection of about 2–4 nmoles of native unmodified peptide. Since the usual dose injected is 40 nmoles, the detection limit thus corresponds to about 10 per cent.

TABLE 2

RELATIVE ACTIVITY OF VARIOUS BIOLOGICALLY ACTIVE PEPTIDES ON MICROVASCULAR PERMEABILITY AS COMPARED TO THE EFFECT OF PEPTIDE 6A

		Relative Activity Compared to Peptide 6A
Peptide 6A	Ala-Arg-Pro-Ala-Lys	1
Peptide 6D	Ser-Gln-Leu-Gln-Lys-Val-Pro-Pro-Glu-Trp-Lys	1
Tuftsin	Thr-Lys-Pro-Arg	1
Substance P	Arg-Pro-Lys-Pro-Gln-Gln-Phe-Phe-Gly-Leu-Met-NH₂	50
Neurotensin	<Glu-Leu-Tyr-Glu-Asn-Lys-Pro-Arg-Arg-Pro-Tyr-Ile-Leu	100
Bradykinin	Arg-Pro-Pro-Gly-Phe-Ser-Pro-Phe-Arg	25–50
Neurotensin fragment	Lys-Pro-Arg-Arg-Pro-Tyr	10
Substance P fragment	Arg-Pro-Lys-Pro-Gln	1

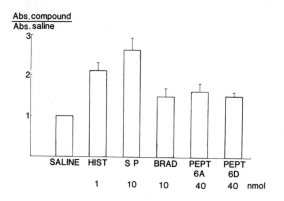

FIGURE 7. Evans blue dye leakage: cat trachea. Effects on protein leakage of various substances injected into the cat trachea.

The peptides induce increased vascular permeability in rat skin (our screening system) and in human skin.[12]

These peptides have a short half-life in the circulation, which is probably due to rapid enzymatic degradation, *e.g.* by angiotensin-converting enzyme and carboxypeptidase, and intravenous infusion of these peptides alone did not seem to induce a significant increase in the vascular permeability in the lungs. However, subpleural injection of peptide 6D led to pulmonary edema that was more severe than that following injection of histamine or bradykinin.[13] Injection of peptides 6A and 6D into the cat trachea resulted in an increased leakage of protein (FIG. 7), and inhalation of peptide 6A caused an increased permeability to protein in the lungs (FIG. 8).[14]

Fibrin peptides might be able to induce contraction of endothelial cells, since the permeability-increasing effect of these peptides was counteracted by the β-adrenoreceptor stimulants terbutaline and isoprenaline.[15]

Such substances have been postulated to counteract endothelial contraction by a direct effect on the endothelium.[16, 17]

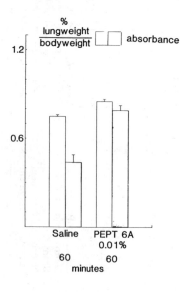

FIGURE 8. Aerosol exposure: guinea pig. Effect of peptide 6A (Ala-Arg-Pro-Ala-Lys) on protein leakage after inhalation of the peptide.

Our view on the effect of the peptides on vascular permeability during states of fibrinolysis inhibition, based on electron microscopic studies, is presented in FIGURE 9.

When the fibrinolytic system is functioning normally (FIG. 9A), fibrin and the peptides rapidly leave the lungs and do not cause any damage to the vascular endothelium. If the fibrinolytic system is inhibited (FIG. 9B-F), fibrin remains in the lungs for a longer time and escapes between the endothelial cells to the extravascular space, where the clearance of the peptides is much

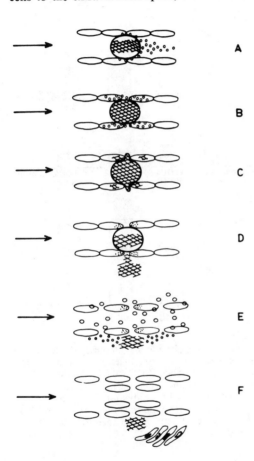

FIGURE 9. Effect of fibrinolysis inhibition on accumulation of peptides derived from the breakdown of fibrin in the lungs. See text.

slower than intravascularly. The reduced fibrinolytic activity in the circulating blood thus favors the persistence of fibrin deposits in the lungs and local liberation of peptides under the influence of tissue plasminogen activators and other proteases, such as elastase, in the lungs.

Part of the permeability-increasing effect of the peptides seems to be due to interference with other vasoactive substances. Peptide 6A is an inhibitor of angiotensin converting enzyme (I_{50} $1.9 \times 10^{-5}M$) and prevents the breakdown of bradykinin.[18] This peptide potentiates the permeability-increasing effect of bradykinin [19] (FIG. 10). Peptide 6D is a much weaker inhibitor of angiotensin-converting enzyme (I_{50} 1.0×10^{-4}). This peptide, however, induces histamine release from mast cells in a much lower concentration than peptide 6A.

Peptide 6A induces vasodilation of bovine arteries and also an increase in cyclic AMP (FIG. 11) but not in cyclic GMP in these vessels.[20] Both changes can be abolished by pretreatment with indomethacin, indicating that the cyclo-oxygenase system may be involved. However, indomethacin did not reduce

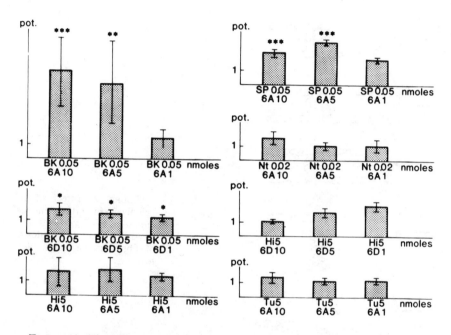

FIGURE 10. The effects of peptides 6A and 6D (10, 5 and 1 nmoles) on the activity of various permeability-increasing substances, namely 6A together with bradykinin (Bk) 0.05 nmoles (n = 72 for each column), substance P (SP) 0.05 nmoles (n = 24), histamine (Hi) 5 nmoles (n = 36), and neurotensin (Nt) 0.02 nmoles (n = 24) and tuftsin (Tu) 5 nmoles (n = 24), and 6D together with bradykinin 0.05 nmoles (n = 48) and histamine 5 nmoles (n = 12). The potentiations shown on the ordinate represent the ratio of obtained activity to expected activity and therefore describe the amplification of the response. $* = p < 0.05$, $** = p = < 0.01$, $*** = p < 0.001$.

the permeability-increasing effect of 6A, but enhanced it. This might possibly be due to a shift in the arachidonic acid metabolism from the cyclo-oxygenase to the lipoxygenase pathway. Lipoxygenase products are known to have a strong permeability-increasing effect. The arachidonic acid cascade might thus be involved as one mechanism underlying the effect of the peptides.

Peptides derived from degradation of fibrinogen or fibrin also have other interesting biological activities.

We found that both peptides 6A and 6D are immunosuppressive *in vitro*. They inhibited thymidine uptake by mitogen-stimulated lymphocytes in

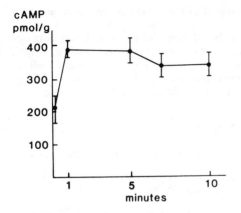

FIGURE 11. Effect of peptide 6A (25 nmol/ml) on cyclic AMP in bovine mesenterial arteries *in vitro*.

concentrations that had no cytotoxic effect.[21] Low molecular weight fibrinogen degradation products also depressed cell-mediated immunoreactivity *in vivo* and promoted tumour growth and metastasization in mice with an immunogenic tumour.[22] Peptides 6A, 6D, and 6E have been tested for their effect on phagocytosis.[23] TABLE 3 presents the results of a study on the influence of peptides on phagocytosis of [³H]thymidine-labeled *S. aureus* by peritoneal macrophages of BALB/c mice. Tuftsin, a tetrapeptide (Thr-Sys-Pro-Arg) cleaved by trypsin as well as by granulocyte protease(s) from a particular fraction of serum immunoglobulins, was included in the investigation for comparison. Tuftsin is considered to be a strong natural stimulant of phagocytotic activity of macrophages and blood granulocytes.[24, 25] It can be seen that the phagocytosis-stimulating activity of peptide 6A is of the same degree as that of tuftsin. Peptide 6E in concentrations above 10 nmol/ml produced slight inhibition—and at a dose of 40 nmol/ml pronounced inhibition—of phagocytosis. Peptide 6D had no effect.

Low molecular weight fibrinogen degradation products are also cytotoxic to endothelial cells [26] and rabbit kidney cells [27] in culture.

TABLE 3

EFFECTS OF PEPTIDES ON PHAGOCYTOSIS OF [³H]THYMIDINE-LABELED *S. aureus* BY PERITONEAL MACROPHAGES *

Peptide nmol/ml	Percent Change Related to Control		
	Tuftsin	6A	6E
1.0	+30.4	+28.9	0
2.5	+47.0	+37.0	0
5.0	+56.8	+61.4	0
10.0	+55.0	+86.3	−17
20.0	+51.9	+50.1	−16
40.0		+61.1	−50

* Results are expressd as mean values obtained in 5 samples.

Fragments from the NH_2-terminal part of the $B\beta$-chain are chemotactic for leukocytes, and fibrinopeptide B is a vasoconstrictor.

In conclusion, one can state that the human fibrinogen molecule contains several vasoactive peptides that are released by various proteolytic enzymes. These peptides also seem to have other biological effects and might well be involved in certain pathological conditions associated with extravascular fibrin deposition and high local concentrations of these peptides.

REFERENCES

1. SALDEEN, T. 1976. The microembolism syndrome. Microvasc. Res. **11:** 227–259.
2. BELEW, M., B. GERDIN, J. PORATH & T. SALDEEN. 1978. Isolation of vasoactive peptides from human fibrin and fibrinogen degraded by plasmin. Thrombos. Res. **13:** 983–994.
3. WALLIN, R., M. BELEW, K. OHLSSON & T. SALDEEN. 1981. Purification of vasoactive peptides from human fibrinogen degraded by human leucocyte elastase. Thrombos. Haemostas. **46:** 52.
4. TAKAGI, T. & R. F. DOOLITTLE. 1975. Amino acid sequence studies on the alpha-chain of human fibrinogen. Location of four plasmin attack points and a covalent cross-linking site. Biochemistry **14:** 5149–5156.
5. TAKAGI, T. & R. F. DOOLITTLE. 1975. Amino acid sequence studies on plasmin-derived fragments of human fibrinogen: Amino terminal sequences of intermediate and terminal fragments. Biochemistry **14:** 940–946.
6. WALLIN, R., M. BELEW, B. GERDIN & T. SALDEEN. 1981. Difference in plasmin degradation of fibrin and fibrinogen derived fragment E. *In* Progress in fibrinolysis. Vol. 5: 208–211. Churchill Livingston. Edinburg.
7. BELEW, M., B. GERDIN, G. LINDEBERG, J. PORATH, T. SALDEEN & R. WALLIN. 1979. Structure-activity relationships of vasoactive peptides derived from fibrin or fibrinogen degraded by plasmin. Biochim. Biophys. Acta **621:** 169–178.
8. BELEW, M., B. GERDIN, L. E. LARSSON, G. LINDEBERG, U. RAGNARSSON, T. SALDEEN & R. WALLIN. 1980. Structure-activity studies on synthetic analogs to vasoactive peptides derived from human fibrinogen. Biochim. Biophys. Acta **632:** 87–94.
9. GERDIN, B., G. LINDEBERG, U. RAGNARSSON, T. SALDEEN & R. WALLIN. 1981. Structural similarities between synthetic analogs to a vasoactive peptide derived from human fibrinogen and other biologically active peptides. *In* Progress in Fibrinolysis. Vol. 5: 383–389. Churchill Livingstone. Edinburgh.
10. GERDIN, B., G. LINDEBERG, U. RAGNARSSON, T. SALDEEN & R. WALLIN. 1983. Structural requirements for microvascular permeability-increasing ability of peptides: Studies on analogues of a fibrinogen pentapeptide fragment. Biochim. Biophys. Acta. In press.
11. SCHWYZER, R. 1977. ACTH: A short introductory review. Ann. N.Y. Acad. Sci. **297:** 3–25.
12. GERDIN, B., L. JUHLIN & T. SALDEEN. 1981. Cutaneous reaction to two fibrin-derived peptides. Acta Derm-Venereol. **61:** 558–560.
13. GERDIN, B., M. BELEW, O. LINDQUIST & T. SALDEEN. 1979. Effect of a fibrin derived peptide on pulmonary microvascular permeability. *In* The Microembolism Syndrome. T. Saldeen, Ed.: 233–239. Almqvist & Wiksell International. Stockholm.
14. ERJEFÄLT, I., C. PERSSON & T. SALDEEN. 1981. Induction of pulmonary edema following administration of substances with short half-life in plasma. (Swe.) Acta Soc. Med. Suec. Hyg., Stockholm. **90:** 497.
15. GERDIN, B. & T. SALDEEN. 1978. Effect of fibrin degradation products on microvascular permeability. Thrombos. Res. **13:** 995–1006.

16. GREEN, K. L. 1972. The anti-inflammatory effect of catecholamines in the peritoneal cavity and hind paw of the mouse. Brit. J. Pharmacol. **45:** 322.

17. SVENSJÖ, E., C. G. A. PERSSON & F. RUTILI. 1977. Inhibition of bradykinin induced macromolecular leakage from post-capillary venules by a β_2-adrenoreceptor stimulant, terbutaline. Acta Physiol. Scand. **101:** 504.

18. SALDEEN, T., J. W. RYAN & P. BERRYER. 1981. A peptide derived from fibrin(ogen) inhibits angiotensin converting enzyme and potentiates the effects of bradykinin. Thrombos. Res. **23:** 465–470.

19. ERIKSSON, M., B. GERDIN & T. SALDEEN. 1981. Enhancement of the microcirculatory leakage due to decreased extravascular inactivation of bradykinin by a peptide derived from fibrinogen. Thromb. Haemostas. **46:** 695.

20. ANDERSSON, R. G. G., K. SALDEEN & T. SALDEEN. 1983. A fibrin(ogen) derived pentapeptide induces vasodilation, prostacyclin release and an increase in cyclic AMP. Thrombos. Res. In press.

21. GERDIN, B., T. SALDEEN, W. ROSZKOWSKI, S. SZMIGIELSKI, J. STACHURSKA & M. KOPEĆ. 1980. Immunosuppressive effect of vasoactive peptides derived from human fibrinogen. Thrombos. Res. **18:** 461–468.

22. ROSZKOWSKI, W., J. STACHURSKA, B. GERDIN, T. SALDEEN & M. KOPEĆ. 1981. Peptides cleaved from fibrinogen by plasmin enhance the progression of L–1 sarcoma in BALB/c mice. Eur. J. Cancer **17:** 889–892.

23. KOPEĆ, M., E. ROSZKOWSKI, B. GERDIN & T. SALDEEN. 1982. Effects of peptides derived from fibrinogen proteolysis on immunereactions and tumour growth. In Fibrinogen-Recent Biochemical and Medical Aspects. A. Henschen, H. Graeff & F. Lottspeich, Eds.: 355–360. Walter de Gruyter & Co. Berlin, New York.

24. NAJJAR, U. A. & K. NISHIOKA. 1970. "Tuftsin": A natural phagocytosis stimulating peptide. Nature **228:** 672–673.

25. NISHIOKA, K., A. CONSTANTOPOWLOS, P. S. SATOK & V. A. NASSAR. 1972. The characteristics, isolation and synthesis of the phagocytosis stimulating peptide tuftsin. Biochem. Biophys. Res. Commun. **47:** 172–179.

26. BUSCH, C. & B. GERDIN. 1981. Effect of low molecular weight fibrin degradation products on endothelial cells in culture. Thrombos. Res. **22:** 33–39.

27. STACHURSKA, J., M. JANIK, M. KOBUS, M. LUCZAK, S. SZMIGIELSKI, M. ROSZKOWSKI, B. GERDIN, T. SALDEEN & M. KOPEĆ. 1983. Effects of peptides cleaved from human fibrinogen by plasmin on rabbit kidney cells in culture. Thrombos. Res. **29:** 419–424.

INTRODUCTORY REMARKS

Gerald M. Fuller

Department of Human Biological Chemistry and Genetics
University of Texas Medical Branch
Galveston, Texas 77550

The structure and function of fibrinogen has occupied the attention of scores of investigators for the better part of a century. Indeed in the past two days we have heard our colleagues of "fibrinogenology" relate their findings on the physical and chemical subtilities of this complicated and fascinating molecule. The primary amino acid sequence of fibrinogen is finally complete after 5–7 years of intense study and work in two different laboratories. What has emerged from this immense undertaking are fascinating details of amino acid sequence homologies indicating the evolutionary relatedness of the three chains as well as the elucidation of complex disulfide bridging and the discovery of specific regions rich in secondary structure. As expected, our knowledge of the primary structure has added significantly to further understanding the conformation of the protein. The three-dimensional shape of the protein has also come under careful scrutiny, and although the details are still only vaguely known it now appears quite evident that the protein is fundamentally trinodular. The diffraction patterns of x-ray crystals of partially degraded fibrinogen give images of the overall architecture that reinforce the depiction of fibrinogen as three spheres connected by two strands of approximately equal length. Much more needs to be learned before the full three-dimensional picture of fibrinogen is known, but what is obvious now is that basic dimension and spatial arrangements of subunits are firmly established.

This session addresses a fundamentally related but altogether different set of questions concerning the fibrinogen molecule. That is, what are processes involved in regulating the biosynthesis of fibrinogen and precisely how are the six polypeptide chains assembled into an intact and functional molecule? The details of these mechanisms have not as yet been shown, but new and important information is being discovered. Although we have known that the liver was the principal site of fibrinogen production for some time, we are still somewhat uncertain about what induces its synthesis. It does seem to be under a particular type of constitutive control system. For example, the protein is synthesized at a fairly steady rate under normal conditions; however, under suitable provocation (an acute inflammatory challenge) the synthesis can be increased four to fivefold. Some of the factor(s) involved in this process will be discussed in this session.

As would be expected, the study of the regulation of fibrinogen synthesis has been approached from several directions. The most thoroughly documented approach so far involves the demonstration that several hormones can influence its production. Dr. Gerd Grieninger will present material on the quantitative role exerted by steroid and thyroid hormones on fibrinogen synthesis. Dr. David Ritchie and I have been investigating what endogenous factors, in addition to the well-established regulator molecules (hormones), influence fibrinogen production. He will present some studies showing that a monocyte derived regulator protein can increase fibrinogen synthesis severalfold in hepatocyte cultures. Fibrinogen has been thought to be under some type of autoregulatory control ever since Barnhart and her colleagues demonstrated an increase in fibrinogen production following infusion or fibrin or fibrinogen peptides (derived from plasmin cleavage). Dr. William Bell has pursued these fascinating observations and will report how one of the core peptides (Fragment D) appears to stimulate fibrinogen production *in vivo*.

Certainly some of the most exciting discoveries in the past two years have been the cloning and partial chemical characterization of the cDNAs for bovine and rat fibrinogens. In this session we will hear of some of the molecular details of the genes coding for the fibrinogen chains. Thus, in a matter of a few days we have moved from the amino acid sequence of fibrinogen subunits to the nucleotide sequence of the gene that code for them. I should point out that recently it has been shown that there are two different γ subunits for fibrinogen. Dr. Gerald Crabtree will describe studies which show that the two γ chains mRNA arise from a single gene utilizing a unique splice pattern during RNA processing. Dr. Dominque Chung will discuss the nucleotide sequence of the fibrinogen subunit including the arrangement of the coding and noncoding regions. Finally, I will present some of the molecular events taking place during translation of fibrinogen chains. It is to be hoped this session will provide new and provocative information on the regulation of fibrinogen synthesis and assembly.

TRANSLATIONAL AND COTRANSLATIONAL EVENTS IN FIBRINOGEN SYNTHESIS *

Gerald M. Fuller, John M. Nickerson,† and Mark A. Adams

*Department of Human Biological Chemistry and Genetics
The University of Texas Health Science Center
Galveston, Texas 77550*

*† Laboratory of Molecular Genetics
National Institutes of Health
Bethesda, Maryland 20205*

TRANSLATIONAL AND COTRANSLATIONAL EVENTS DURING FIBRINOGEN SYNTHESIS

Ample evidence exists implicating the hepatocyte as the principal cell responsible for making fibrinogen.[1-4] It is also now known that certain hepatoma cells can synthesize and secrete fibrinogen at levels fairly close to that of their nontransformed counterpart.[5-7] Many steps involved in the mRNA directed synthesis of liver-derived proteins are known in general; however, much less is known about details of cotranslational and posttranslational modifications of particular proteins. Because of the structural complexity of fibrinogen one might expect it to require a complex series of modifications prior to formation of a fully functional protein.

In order to get the proper perspective of the complexities of the problem faced by the hepatocyte as it synthesizes and assembles fibrinogen, we need to remind ourselves of some of the better known steps required to make the fibrinogen molecule. First, there are at least three different genes that must be coordinately transcribed and the resultant mRNAs processed and delivered to the cytoplasm. Once in the cytoplasm, each mRNA must be translated and the nascent peptide translocated, proteolytically processed, glycosylated, folded and covalently cross-linked with five other subunits to finally make a functional protein. Following these complex events the molecule must then be packaged and secreted. This is truly an impressive feat.

Attempting to elucidate the details of these events has occupied a good deal of our interests for the past few years. We have been fortunate to be working in an era in which cell culturing systems have vastly improved and thus we have capitalized on the techniques involved in hepatocyte and hepatoma culture methodologies. We have concentrated our efforts on using this culture system to explore the various steps leading to the production of an intact functional molecule.

FIBRINOGEN SUBUNITS HAVE SIGNAL PEPTIDES

We believed that an important first question to ask was whether the Aα, Bβ and γ subunits of fibrinogen have leader or signal peptides on them. Our initial

* These studies were supported in part by DHHS NIH, 16445.

approach was to evaluate fibrinogen chains under various conditions of synthesis and posttranslational modification. We examined fibrinogen chains isolated from (1) the plasma of rats,[8] (2) the medium of hepatoma cells treated with tunicamycin, (3) a mRNA directed *in vitro* translating system, and (4) poly-ribosomes taken from rat liver. Each of these types played an important role in helping to deduce that each of the fibrinogen subunits was indeed derived from "prepeptide." Our basic approach was to use sodium dodecylsulfate polyacrylamide gels to measure the mass of each of the polypeptides taken from the different systems. One of the complicating factors in this type of analysis, i.e., measuring molecular mass by migration of reduced and alkylated polypeptides in detergent gels, is the anomalous R_f exhibited by almost all glycoproteins. Since both the Bβ and γ chains of fibrinogen have carbohydrate clusters an additional set of experiments had to be carried out before we were convinced that we had an accurate estimate of the mass of each subunit. Thus to have a degree of certainty in the determination of protein mass by this method we needed to devise a way to isolate nonglycosylated forms of Bβ and γ chain. Since both subunits link to their respective carbohydrate through an N-glycosidic bond we were able to use the antibiotic tunicamycin to block the glycosylation event.[23]

An analysis of fibrinogen translation products is shown in FIGURE 1. Two types of *in vitro* translation products (B & C) are compared with the subunits of fibrinogen that has been isolated and purified from plasma (A). The translational products were isolated by immunoprecipitation as we have previously described.[6] Basically, what is shown in this experiment are the relative migration distances of three types of fibrinogen subunits. A close evaluation of patterns show that all the bands in Lane B migrate somewhat more slowly when compared to their counterparts in Lanes A or C. This is what one would expect since the chains were derived from an mRNA directed *in vitro* translating system using mRNA or liver polysomes without the components of posttranslational modification. It is well known that the Aα chain is not glycosylated. Therefore a direct size correlation can be made between the Aα chain in Lanes B and C. Careful measurment of several experiments consistently showed that the Aα chain synthesized in the mRNA directed synthesis (shown in Lane C) was larger than that of either the Aα isolated from plasma fibrinogen or from polyribosomes that had been added to an *in vitro* translating system. This is clear evidence for the presence of a signal or leader peptide for this subunit.

The analysis of gel migration patterns of Bβ and γ chains require additional explanation since both of these chains have carbohydrate clusters. To get around the problem two additional experiments were performed. In the first we simply added an antibiotic pactomycin to our *in vitro* polysome translating system. This antibiotic blocks the initiation reaction and thus only polypeptide elongation can occur. Analysis of the gel pattern in Lane C suggests that the preponderance of γ chain may have already been glycosylated since it migrates quite similarly to the already glycosylated γ chain in Lane A. The Bβ pattern, however, is somewhat less definitive. A reminder should be made here, that is, when the total polyribosomes are isolated from the liver using our previously described procedures,[9] the nascent chain on the polysomes has already been passed through the endoplasmic reticular and thus would have had an amino terminally situated signal peptide already removed. The slightly faster migration pattern suggests that a leader or signal peptide has been removed. Because the carbohydrate on the Bβ chain is linked to an asparagine that is closer to the

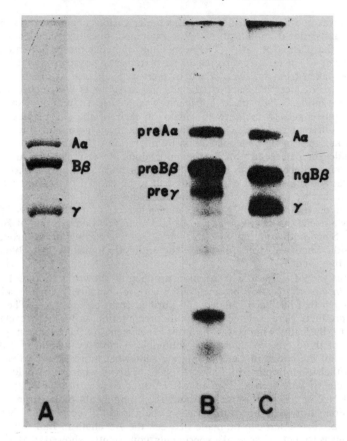

FIGURE 1. Sodium dodecylsulfate polyacrylamide gel patterns of fibrinogen translation products. **A,** purified fibrinogen from rat plasma, stained with Coomassie R 250; **B,** mRNA translated *in vitro*, fibrinogen immunoprecipitate; **C,** rat liver polysomes translated in the presence of 3 μM pactomycin, fibrinogen immunoprecipitate. The prefix "pre" refers to a precursor form of a chain. The prefix "ng" refers to a chain missing an asparagine linked carbohydrate.[6]

COOH-terminal end of the molecule it is not already glycosylated when the polysomes are isolated. The second experiment involved treating hepatoma cells with tunicamycin which blocks the N-linked glycosylation process. Isolation of fibrinogen from the hepatoma cell medium and then an examination of the subunits on SDS gels provided us with some additional information on the migration patterns of the glycosylated (Bβ and γ) subunits. We were able to perform direct measurements of molecular mass relating the molecular weight versus distance migrated on nonglycosylated Bβ and γ chains and demonstrate that these two chains are also synthesized with signal peptides attached. Interestingly, if we treated the nonglycosylated protein with thrombin full cleavage of the fibrinopeptides occurred. Moreover, by comparing the molecular mass of thrombin treated translation products (from an mRNA directed *in vitro* translation) with those from tunicamycin treated subunits we showed that the signal peptides are located on the NH_2-terminal ends of the chains. Although this is not unexpected, it was important to show this since it has been reported that the signal peptide can be located in the middle of the polypeptide chain.[10]

Additional analyses along these lines have convinced us that each subunit of fibrinogen has a single peptide, which is in keeping with the general scheme of how secretory proteins are selected for binding to the ER membrane.[12] TABLE 1 gives an estimate of molecular weight of the translation products for rat fibrinogen. The presence of signal peptides on fibrinogen subunits from the dog has also been shown.[13]

STOICHIOMETRY OF INTRACELLULAR FIBRINOGEN SUBUNITS

Identifying that signal peptides are present on all three chains of fibrinogen reveals several important features about the assembly of the molecule. Although our previous studies involving analysis of fibrinogen specific polysomes provided indirect evidence that separate mRNAs exist for each subunit these studies provide stronger, more direct evidence. It should be pointed out that each of the mRNAs have now been isolated and several of their chemical features determined.[14, 15] In a broader sense, however, the finding that each subunit of fibrinogen behaves as an independently operated protein implies that for assembly to take place each of the mRNAs must be translated and translocated in some sort of coordinated fashion. All three genes must be regulated fairly closely. Thus, transcription, processing, translation and translocation would be on-going simultaneously for all three chains.

Just how stringent a control exists for the transcription of the three mRNAs is not as yet known. One way to assess how tightly the translation of the subunits are linked would be to look for the presence of unassembled chains. In order to determine whether there are pools of unassembled fibrinogen subunits or whether one or more of the subunits is translated more rapidly than the other, we performed a set of pulse-chase experiments using primary hepatocytes in culture. Our basic strategy was to pulse the cells with a radioactive label (^{35}S-methionine) using first a 5-minute and then a 10-minute label followed by a chase of 25 and 20 minutes, respectively. The chase period was chosen after we had determined that the transit time for labeled fibrinogen to exit from the cell was approximately 40 minutes. Following the chase with unlabeled methionine, we lysed the cells in detergent solution to insure solubilization of the intracellular membrane systems. We then added monospecific polyclonal anti-

ESTIMATED MOLECULAR WEIGHTS FROM TRANSLATION PRODUCTS

From Total mRNA		From Polysomes+ Pactamycin		Differences
				(Putative Signal Peptides)
Chain	Molecular Weight	Chain	Molecular Weight	
PreAα	66,000	Aα	65,400	600
PreB	57,600	Bβ	56,500	1,100
Preγ	53,800	γ	50,800	3,000

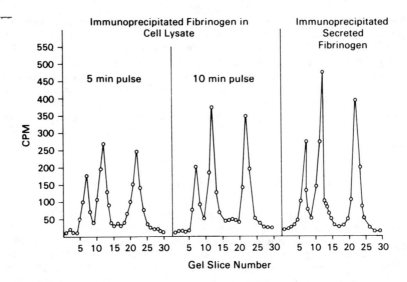

FIGURE 2. Pulse-chase experiments showing stoichiometry of fibrinogen subunits. Hepatocytes were stimulated with a monocyte-derived stimulating factor (see Ritchie and Fuller, these proceedings) then pulsed with either ^{35}S-methionine or ^{3}H-leucine. Secreted radiolabeled fibrinogen, and total intracellular fibrinogen were immunoprecipitated using monospecific antibodies. The labeled fibrinogen was released and electrophoresed in a sodium dodecylsulfate polyacrylamide gel system. The gel was sliced and the radioactivity of each slice determined.

body to the lysate in order to bind all intact fibrinogen as well as any unassembled subunits. The antibody-antigen complex was removed from the lysate by adding formalized *Staphylococcus aureus* cells.[16] This complex was washed exhaustively and then boiled in the presence of a reducing agent and detergent. Samples were separated on acrylamide gels that were then cut into 1 mm slices, and the radioactivity was determined for each slice. The results of these experiments are shown in FIGURE 2.

Since the stiochiometry of the chains in the secreted fibrinogen is known to be 2 : 2 : 2, we could compare the ratio of the radioactivities of the secreted protein with those from the intracellular pool to give us some information whether there were pools of unassembled chains within the cell. We reasoned that if any one of the chains was made in a greater amount than the others we could see this by comparing the radioactivity in subunits isolated from intracellular fibrinogen with those of secreted fibrinogen. Our results indicate that there are no obvious pools of chains. That is to say, all three chains appear to be synthesized at about the same rate, moreover, no large pools of previously existing but unassembled chains are evident in the cell. Thus, our original contention that some types of coordinated control for synthesis and assembly seems to hold true. Obviously we are a long way from knowing the assembly processes but what appears likely from these experiments is that (1) all three mRNAs exist in relatively equal amounts, (2) they are translated at about the same rate, and (3) no very obvious pools of unassembled fibrinogen subunits exist within the cell. We believe at this time that assembly occurs either during translation or very shortly thereafter. It is important to point out that it has been reported in an *in vivo* pulse-chase type of experiment using rabbits, that the Bβ subunits appeared to be synthesized more rapidly than the other two chains.[7] Furthermore these investigators suggest that the Bβ may be involved in assembly of the other two chains. At the present time no clear assembly mechanism has been elucidated; however, the systems for probing this area of the fibrinogen puzzle do exist and hopefully new information will be forthcoming soon.

TEMPORAL SEQUENCE OF GLYCOSYLATION OF Bβ AND γ SUBUNITS

We were interested in determining precisely when the two fibrinogen chains are glycosylated, since several investigations of secretory proteins showed that there is a tight coupling among polypeptide translation, translocation and core glycosylation.[18-20] From our studies identifying the presence of signal peptides we believed that core glycosylation of the γ chain may be an early cotranslational event and core glycosylation of Bβ may be an extremely late translational event. The strategy we employed was to translate in a cell free system (1) isolated fibrinogen mRNAs, (2) total liver polysomes and (3) isolated rough microsomes, i.e., polysomes still attached to endoplasmic reticular membranes. The elongated chain from the polysomes and microsomes and the mRNA translation products were chromatographed on a lectin column (Conconavalin A bound to Sepharose, (Con A-Sepharose).

It has been shown that immunoglobulin heavy and light chains could receive their core carbohydrates almost as soon as the current asparagine residue

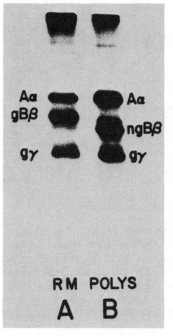

FIGURE 3. Comparison of rough microsome and total polysome translation products. Rat liver rough microsomes (RM) or total polysomes (POLYS) were added to the translation system; subsequently, fibrinogen chains were immunoprecipitated and electrophoresed. There is a substantial difference in the mobilities of the polysome elongated Bβ chains and the rough microsome Bβ chains. The rough microsome product Bβ chain migrated slower than the total polysome Bβ chain by an amount corresponding to about 3000 daltons. This migration was attributed to a glycosylated rough microsome Bβ chain and a nonglycosylated total polysome Bβ chain. **A**, Rough microsome products; **B**, total polysome products. The "g" refers to the putative glycosylated state, and "n" refers to the nonglycosylated state.[24]

appeared on the cisternal side of the endoplasmic reticulum.[21, 22] We reasoned this is also likely for the γ subunit. The carbohydrate of the γ subunit is located on Asn-52 (in human), which would of course, appear quite early and be available for carbohydrate attachment while the chain was still being elongated. The Bβ chain, however, has its single carbohydrate attachment position at Asn-364 quite some distance into the molecule. Thus this subunit may not be modified (glycosylated) until after release into the cisternal space. That this is in fact the case is shown in FIGURE 3.

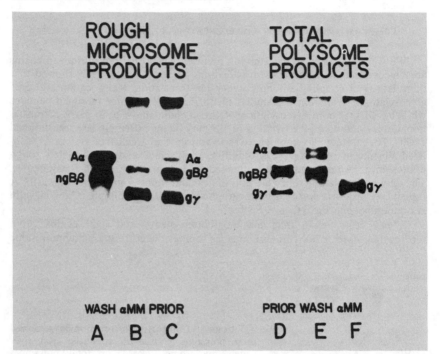

FIGURE 4. Con A-Sepharose chromatography of rough microsome and total polysome translation products. Rough microsomes or total polysomes from rat liver were added to the translation system, and fibrinogen products were immunoprecipitated. ⁔ fibrinogen chains were chromatographed on Con A-Sepharose. Unbound chains ⁔H) and specifically eluted chains (α-MM; 0.4 M α-methyl mannoside) were ⁔resed with material that was not chromatographed (PRIOR). Lanes A, B, ⁔rived from total microsomes, and lanes D, E, and F were derived from The "g" refers to glycosylated, and the "ng" refers to nongly-

⁔f isolated polysome were compared with trans-
⁔crosomes. We showed that chains isolated
⁔lly processed with signal peptides removed
⁔e A of FIGURE 3). The polysomal profile shows
⁔migrates similar to the products from the rough
⁔wever, shows a large amount of nonglycosylated
⁔s precisely what we would predict since the Bβ acceptor

asparagine is closer to the COOH-terminal end of the peptide. Apparently the elongated Bβ chain had not reached the ER at the time the polysomes were isolated.

To verify that the carbohydrates were attached to the subunits we passed the translation products through a Con A-Sepharose column. Those chains that were nonglycosylated did not bind to the column, whereas those which had their carbohydrates attached, bound. We released the bound fraction and analyzed both the bound and unbound fractions as shown in FIGURE 4. The first panel (Lanes A, B and C) shows translated fibrinogen polypeptides that did not bind to the column. The Aα chain of course served as an excellent control since it would not bind to the lectin column. The presence of Bβ in the unbound (wash) fraction also indicates that a fair proportion of this chain has not been glycosylated. When α-methylmannoside was added to the column the γ chain was eluted from the column together with a small amount of Bβ. Both of these subunits apparently have their core carbohydrate clusters attached. Analysis of total polysomal products (Panel B) on the other hand, also reveal that both the Aα and Bβ do not bind (Lane E), whereas all the γ does (Lane F). Thus we believe that the γ chain is modified while it is being elongated, whereas the Bβ does not receive its carbohydrate either until very late in the elongation process or quite soon after released into the cisternal space.

SUMMARY

What we have shown in these studies are some of the first steps in the assembly of fibrinogen. We believe that using the hepatocyte culture system together with molecules capable of stimulating an increase in synthesis holds much promise in helping to learn how the cell puts together this elegant and complicated molecule. Moreover, these studies may also provide clues to mechanisms operant for other complex proteins produced by the liver.

ACKNOWLEDGMENT

We thank Ms. Rose Byrdlon for her excellent secretarial help.

REFERENCES

1. DRURY, D. R. & P. O. MCMASTER. 1929. J. Exptl. Med. **50:** 569–578.
2. BARNHART, M. I., D. C. CRESS, S. M. NOONAN & R. T. WALSH. 1970. Thromb. Diath. Haemorrh. Suppl. **39:** 143–159.
3. GRIENINGER, G. & S. GRANICK. 1978. J. Exptl. Med. **147:** 1806–1823.
4. RUPP, R. G. & G. M. FULLER. 1977. Exptl. Cell Res. **118:** 23–30.
5. ADKISSON, V. T. & G. M. FULLER. 1982. J. Cell Biol. **95**, 202a.
6. NICKERSON, J. M. & G. M. FULLER. 1981. Proc. Natl. Acad. Sci. USA **78:** 303–307.
7. KNOWLES, B. B., C. C. HOWE & D. P. ADEN. 1980. Science **209:** 497–499.
8. BOUMA, H. III & G. M. FULLER. 1975. J. Biol. Chem. **250:** 4678–4683.
9. BOUMA, H., III, S.-W. KWAN & G. M. FULLER. 1975. Biochemistry **14:** 4787–4792.
10. LINGAPPA, V. R., J. R. LINGAPPA & G. BLOBEL. 1979. Nature **281:** 117–121.
11. BLOBEL, G. & B. DOBBERSTEIN. 1975. J. Cell Biol. **67:** 835–851.

12. BLOBEL, G., P. WALTER, C. N. CHANG, B. M. GOLDMAN, A. H. ERICKSON & V. R. LINGAPPA. 1979. Symposia of the Soc. of Exp. Biol. **33:** 9–36.
13. YU, S., C. M. REDMAN, J. GOLDSTEIN & B. BLOMBÄCK. 1980. Biochem. Biophys. Res. Commun. **96:** 1032–1038.
14. CRABTREE, G. R. & J. A. KANT. 1981. J. Biol. Chem. **255:** 9718–9723.
15. CHUNG, D. W., M. W. RIXON, R. T. A. MACGILLIVRAY & E. W. DAVIE. 1981. Proc. Natl. Acad. Sci. USA **78:** 1466–1470.
16. KESSLER, S. W. 1976. J. Immunol. **117:** 1482–1490.
17. ALVING, B. M., S. I. CHUNG, G. MURANO, D. B. TANG & J. S. FINLAYSON. 1982. Arch. Biochem. Biophys. **217:** 1–9.
18. LINGAPPA, V. R., J. R. LINGAPPA, R. PRASAD, K. E. EBNER & G. BLOBEL. 1978. Proc. Natl. Acad. Sci. USA **75:** 2338–2342.
19. HORTIN, G. & I. BOIME. 1978. J. Biol. Chem. **255:** 8007–8010.
20. ROSEN, J. M. & D. SHIELDS. 1980. *In* Testicular Development, Structure and Function. A. Steinberger & E. Steinberger, Eds. p. 343. Raven Press. New York.
21. BERGMAN, L. W. & W. M. KUEHL. 1977. Biochemistry **16:** 4490–4497.
22. BERGMAN, L. W. & W. M. KUEHL. 1979. J. Biol. Chem. **254:** 5690–5694.
23. TKACZ, J. S. & J. O. LAMPEN. 1975. Biochem. Biophys. Res. Commun. **65:** 248–257.
24. NICKERSON, J. M. & G. M. FULLER. 1981. Biochemistry **20:** 2818–2821.

DISCUSSION OF THE PAPER

L. LORAND (*Northwestern University, Evanston, IL*): In most of your gels there is a high molecular weight material. Would you care to comment on this?

G. M. FULLER: We feel that there is a co-precipitating material that comes down with the *Staphylococcus aureus*. In every gel that was immunoprecipitated with either anti-fibrinogen or anti-albumin we saw a high molecular weight band right near the top of the gel.

LORAND: I was hoping you would say it was fibrinogen.

FULLER: Well, it isn't, because these are all reduced gels.

N. U. BANG (*Lilly Laboratory for Clinical Research, Wishard Memorial Hospital, Indianapolis, IN*): Did you say that you saw non-glycosylated protein secreted into the medium in the tunicamycin experiments? Does that mean that secretion is possible without glycosylation?

FULLER: Yes. In a number of systems glycosylation is not a necessary event for secretion of hepatocyte-derived proteins.

UNIDENTIFIED SPEAKER: Do you have any information regarding the number of polypeptide chains that might be on a single messenger RNA molecule? In other words could a single messenger RNA molecule code for the α, β and γ chains?

FULLER: On the basis of antibody binding to *nascent* chains and then centrifuging them in a sucrose gradient one can get a general approximation of the size of the polysome unit; none of them were ever large enough to code for more than one of the messages.

D. L. AMRANI (*Mt. Sinai Medical Center, Milwaukee, WI*): Have you done any studies to look at the presence of the γ/γ' chain heterogeneity in your *in vitro* translation studies?

FULLER: No.

CLONING OF FIBRINOGEN GENES
AND THEIR cDNA *

Dominic W. Chung, Mark W. Rixon, Benito G. Que, and Earl W. Davie

Department of Biochemistry
University of Washington
Seattle, Washington 98195

Comparisons of the amino acid sequences of the three chains of fibrinogen reveal that the three chains are highly homologous and suggest that they are derived from a common ancestral gene.[1] Subsequent studies on the biosynthesis of fibrinogen in cell-free systems provide evidence that each polypeptide is encoded by distinct messenger RNA species, confirming the notion that the synthesis of fibrinogen is the result of concerted expressions of three genes.[2-4] In order to understand the structure and regulation of these genes, we have employed recombinant DNA techniques for the cloning of the cDNA for human and bovine fibrinogen α and β chains.[2, 5] Some of the properties of these cDNA clones and the isolation and characterization of the gene for the human β chain are described in this report.

ISOLATION OF THE GENE FOR THE β CHAIN

The identification and characterization of a cDNA for the β chain of bovine fibrinogen enabled us to predict the complete amino acid sequence of the bovine fibrinogen β chain.[5] Amino acid sequence comparisons with the human β chain indicated that the overall homology was about 75%. This suggested that the nucleotide sequence homology between the human and bovine β chains was also very high. Under conditions of reduced stringency of hybridization, the bovine cDNA was found to hybridize to discrete fragments of human genomic DNA generated by various restriction endonucleases. This specific hybridization indicates that the bovine cDNA could be used as a hybridization probe for the isolation of the gene for the human fibrinogen β chain.

A human genomic DNA library containing random fragments of the human genome constructed by Maniatis and coworkers,[6] was screened using a radiolabeled cDNA from the bovine β chain of fibrinogen. Screening of the recombinant phage plaques was carried out by a modification of the method of Benton and Davis,[7] as described by Woo.[8] Seven out of two million recombinant phages hybridized specifically with the probe. These phages were purified by plaque dilution and the DNA extracted for analyses. Restriction endonuclease mapping and Southern hybridization [9] indicated that they contained varying amounts of human DNA, including a common segment that contained nucleotide sequences homologous to the bovine cDNA probe. This segment of DNA, which was presumed to contain the gene for the β chain of human fibrinogen, was about 10 kilobases in length. It was contained within four

* This work was supported in part by Research Grant HL 16919 from the National Institutes of Health.

Eco RI restriction endonuclease sites. Further hybridization analyses on this segment using more restricted regions of the bovine β chain cDNA defined the direction of the coding sequences.

ELECTRON MICROSCOPIC HETERODUPLEX MAPPING

The overall structure of the gene for the β chain of human fibrinogen was visualized by electron microscopic heteroduplex mapping studies, performed in collaboration with Dr. Myles Mace at Baylor College of Medicine in Houston. Cloned genomic DNA containing the gene for the β chain was denatured and hybridized to total human liver mRNA under conditions which favored DNA-RNA hybridization to that of DNA-DNA reassociation. The hybrids were then examined by electron microscopy. Intervening sequences, which are not represented in mature mRNA, remained as single-stranded loops connecting segments of duplex structural sequences. Results from these experiments indicated seven apparent intervening sequences. Using the appropriate single-stranded and double-stranded DNA standards, we estimated the sizes of the eight structural segments, or exons, to be 0.15, 0.2, 0.2, 0.2, 0.15, 0.3, 0.3, and 0.4 kilobases in length, and the sizes of the seven intervening sequences, or introns, were about 1.5, 0.2, 0.7, 0.3, 0.6, 0.1, and 0.4 kilobases in length. A schematic representation of this structure is shown in FIGURE 1. We were unable to discern any nonhybridized poly-A protrusions at either ends of the hybrid molecules. Hence, the orientation of the gene could not be assigned from these studies.

CHARACTERIZATION OF THE GENE FOR THE β CHAIN OF HUMAN FIBRINOGEN

To further characterize the gene for the β chain, the cloned genomic DNA was excised from the recombinant phage by restriction endonuclease *Eco* RI and the contiguous fragments subcloned into the single *Eco* RI site of the

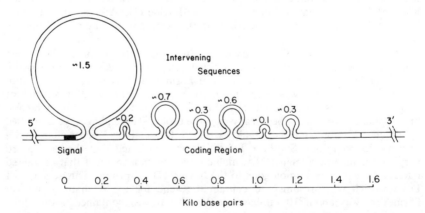

FIGURE 1. Schematic representation of the genomic DNA for the β chain of human fibrinogen as observed by electron microscopic heteroduplex mapping.

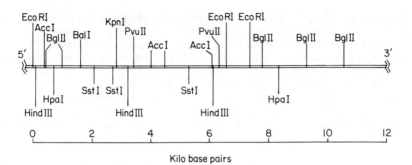

FIGURE 2. Restriction endonuclease map of the gene for the β chain of human fibrinogen. The locations of restriction enzyme cleavage sites with six-base recognition sequences are shown.

plasmid vector pBR322. Each of the subcloned *Eco* RI fragments was characterized by digestion with restriction endonucleases. The positions of restriction enzyme cleavage sites were determined by estimating the lengths of DNA segments generated by digestion to completion by a single enzyme or a combination of two or more enzymes. In some cases, partial digests of end-labeled DNA fragments were also performed. A partial restriction map showing cleavage sites of restriction enzymes with six-base recognition sequences is shown in FIGURE 2.

Using the detailed restriction map, the nucleotide sequence of the gene for the β chain was determined by the partial chemical degradative method of Maxam and Gilbert.[10] To date, over 65% of the nucleotide sequence of the gene has been determined. Specifically, these results confirm that the cloned DNA in fact codes for the β chain and that the cloned fragments contain the entire gene. The positions of the seven intervening sequences have been located and the direction of the gene was established. The size and distribution of these intervening sequences are in good agreement with those observed by electron microscopic heteroduplex mapping. The nucleotide sequence for the amino terminal and its surrounding region is shown in FIGURE 3. The first intervening sequence occurs between amino acid residues 8 and 9 within fibrinopeptide B. The size of this intervening sequence, determined from the restriction map, was about 1.5 kilobases in length, and it is the largest of the seven intervening sequences. The first amino acid of the β chain, as predicted from the genomic sequence, is glutamine which presumably undergoes cyclization to form pyroglutamic acid, which is found in the β chain of mature plasma fibrinogen.

Three potential translation initiation codons (ATG), located 16, 27 and 30 amino acids immediately upstream from the NH_2-terminal glutamine residue, have been identified. The predicted amino acid sequence for this region shows striking similarities to that of a typical signal peptide or leader sequence (FIG. 3). It contains a cluster of hydrophobic residues (six leucines) flanked by hydrophilic residues on either side, and an uncharged amino acid, in this case serine, at the putative cleavage site of signal peptidase. Comparison of the size of the mature β chain with its precursor synthesized in a cell-free translation system indicates that the length of the signal peptide is about 20 to 30 amino acids in length. This is in agreement with the predicted maximal length of 30 amino acids determined from the genomic DNA sequence.

The second intervening sequence occurs between amino acid residues 72 and 73, immediately preceding the disulfide knot region. The third intervening sequence occurs between amino acid residues 133 and 134, and is located in the middle of the coiled-coil region. This region is particularly sensitive to cleavage by plasmin. The fourth intervening sequence occurs between amino acid residues 209 and 210, about 12 amino acid residues past the coiled-coil domain. The remaining intervening sequences occur between

```
                                        -30
                          ***        Met Lys Arg Met Val Ser Trp Ser
5'- - - - - C T C A A G T T A A G T C T A C A T G A A A A G G A T G G T T T C T T G G A G C
          1030                                   1050


                -20
  Phe His Lys Leu Lys The Met Lys His Leu Leu Leu Leu Leu Leu Leu
                                                          -10
T T C C A C A A A C T T A A A A C C A T G A A A C A T C T A T T A T T G C T A C T A T T G
  1070                              1090                            1110


                                   1                              8
  Cys Val Phe Leu Val Lys Ser Gln Gly Val Asn Asp Asn Glu Glu
T G T G T T T T T C T A G T T A A G T C C C A A G G T G T C A A C G A C A A T G A G G A G
                1130                              1150


                                                           9
←─────────────────Intervening Sequence A ─────────────────→ Gly Phe Phe Ser
G T G A A T T T T T T A A A · · ·1.5 kilobases · · · · T G T A G G G T T T C T T C A G T
                1170                                        2700


                            20
  Ala Arg Gly His Arg Pro Leu Asp Lys Lys Arg Glu Glu Ala Pro
G C C C G T G G T C A T C G A C C C C T T G A C A A G A A G A G A G A A G A G G C T C C C
                2720                              2740


        30
  Ser Leu Arg
A G C C T G A G G - - - - - 3'
        2760
```

FIGURE 3. Nucleotide sequence of a portion of the gene for the β chain of human fibrinogen coding for the NH₂ terminal region of the polypeptide. The numbering of the nucleotide sequence refers to the distance from the 5′ *Eco* RI site. *** indicates the position of an in-phase stop codon proximal to the potential initiation sites.

amino acid residues 247 and 248, 289 and 290, and 384 and 385. A number of the nucleotide sequences at the intron-exon junctions on the β chain gene have been determined. A shown in TABLE 1, they agree well with the consensus junction splice sequence derived from the analyses of similar junction sequences in other eucaryotic genes.[12] Intervening sequences can be classified according to their positions in the triplet codon they interrupt: type 0 intervening sequences interrupt the coding region between triplet codons; type I interrupt between the first and second nucleotide, and type II between the second and third nucleotide of the triplet codons. Intervening sequences of all three types are present in the gene for the β chain of human fibrinogen.

TABLE 1

NUCLEOTIDE SEQUENCE AT INTRON:EXON JUNCTIONS IN THE GENE FOR THE β CHAIN OF HUMAN FIBRINOGEN

Intervening Sequence	Location (AA residues)	Exon	Intron	Exon	Type
A	8–9	GAG GTGAAT	AGGTTGTAG	GG	0
B	72–73	TTG GTGGGT			0
C	133–134		TTTTTCCAG	AT	I
D	209–210		CATTTGCAG	AA	I
E	248–249	GAG GTAAGC			I
F	289–290		TTCTTTTAG	GT	I
G	384–385	CTG GTATGT	TTCTTTCAG	GT	II
Consensus Sequence *		C/A AG GT A/G AGT	TTTTT T / CCCCC C – AG	G –	–

* The consensus sequence for splice junctions as reported by Mount.[12]

It was proposed by Gilbert [13] that a possible function of intervening sequences may be to segregate a gene into structural regions corresponding to the structural domains of the protein. Analyses of the β globin and γ immunoglobulin gene support this notion. Although we presently do not know all the exact structural or functional domains on the β chain of fibrinogen, the positions of the first four intervening sequences are consistent with this hypothesis.

Preliminary cell-free transcription studies in HeLa cell extracts supplemented with polymerase II show that there is a strong promoter for initiation of transcription located about two to three hundred bases upstream from the putative signal peptide. The nucleotide sequence in this region is rich in A and T residues and several potential promoter sequences or TATA sequences have been located. Experiments are in progress to define the true promoter and transcription initiation sites.

```
                      590
      Asp Glu Ala Gly Ser Glu Ala Asp His Glu                 Gly Thr His
Human  5' G A T G A G G C C G G A A G T G A A G C C G A T C A T G A A - - - - - - G G A A C A C A T
Bovine    G A T G A G G C C G A A A G C C T A G A G G A T C T T G G C T T T A A A G G A G C A C A C
      Asp Glu Ala Glu Ser Leu Glu Asp Leu Gly Phe Lys Gly Ala His


              600
      Ser Thr Lys Arg Gly His Ala Lys Ser Arg Pro Val  Arg  Gly  Ile
Human  A G C A C C A A G A G A G G T C A T G C T A A A T C T C G C C C T G T C A G A G G T A T C
Bovine G G C A C C C A G A A A G G C C A T A C T A A A G C T C G T C C T G C C A G A G G T A T C
      Gly Thr Gln Lys Gly His Thr Lys Ala Arg Pro Ala  Arg  Gly  Ile


                          620                          625
      His  Thr  Ser  Pro  Leu  Gly  Lys  Pro Ser Leu Ser Pro Stop
Human  C A C A C T T C T C C T T T G G G G A A G C C T T C C C T G T C C C C C T A G A C T A A G
Bovine C A C A C T T C T C C T T T G G G G G A G C C T A C C C T G A C C C C C T A G A C T A A G
      His  Thr  Ser  Pro  Leu  Gly  Glu  Pro Thr Leu Thr Pro Stop


Human  T T A A A T A T T T C T G C A C A G T G T T - C C C A T G G C C C 3'
Bovine T T C A T T A C T C C T G C A A C G T G T C T C C C A T A G C A C
```

FIGURE 4. Comparison of the nucleotide sequences around the termination codons in the cDNAs for the human and bovine α chain. Blanks are introduced to display maximum homology. The nucleotide sequence of the bovine α chain cDNA is taken from reference 2.

STUDIES ON THE CDNA FOR THE α CHAIN OF HUMAN FIBRINOGEN

A cDNA clone coding for the carboxyl 202 amino acids of the bovine α chain has been characterized.[2, 14] Unlike the β chain, the bovine α chain shares significantly less amino acid sequence homology with the corresponding region in the human α chain. A COOH-terminal extension of 15 amino acids has been identified. A hypervariable region that consists of 53 amino acids and shows no homology with the human chain is present in the bovine chain. This hypervariable region is flanked on both sides by regions of moderately good homology (70% and 56%, respectively). A portion of this cDNA, corresponding to a region of high homology was isolated and used as a hybridization probe for the human α chain. Preliminary hybridization experiments of this fragment with restriction fragments of total human genomic DNA show discrete bands of hybridization, indicating the adequacy in specificity of this probe for the human α chain. However, because of the restricted size, the possibility of isolating the entire gene for the human α chain from the human genomic library is decreased. Alternatively, we have used this probe to isolate and identify a cDNA clone for the human α chain, which will then be used to isolate the human α chain gene. Accordingly, we have screened a collection of five thousand human cDNA clones prepared from total human liver mRNA. Fifty-six independent cDNA clones were positively identified. The inserts from these clones were mapped and the longest one we have sequenced to date is about 1.3 kilobases in length. It was observed that the codon for the COOH-terminal amino acid (Val 610) was not immediately followed by a stop codon. Instead, the mRNA apparently codes for an additional 15 amino acids, ending in proline as its COOH-terminal residue. As shown in FIGURE 4, this region is highly homologous to the putative bovine COOH-terminal extension predicted from its cDNA nucleotide sequence. These results suggest that the primary translation product of the α chain is probably longer than the mature form found in plasma. Fibrinogen subpopulations of varying degree of solubility have been isolated and characterized by Mosesson and coworkers.[15] Some of the heterogeneities have been accrued to proteolytic cleavage of the COOH-terminus of the α chain.[16-18]

SUMMARY

Cross-species hybridizations have enabled us to isolate and clone the gene for the β chain of human fibrinogen. Highlights of the gene for the β chain revealed by nucleotide sequence analyses, particularly in areas that have a direct bearing on defining the overall organization of the gene, have been presented. Nucleotide sequence determination has confirmed the presence of seven intervening sequences. The positions where several of these intervening sequences interrupt the coding region appear to be related to the functional domains of the polypeptide. A putative signal peptide has been identified. Studies on the cDNA for the human α chain indicate that the α chain polypeptide may be synthesized in a precursor form with a COOH-terminal extension of 15 amino acids as compared to the α chain present in the mature molecule found in plasma.

We are in the process of isolating the genes for the α and γ chains by a similar approach. We are hopeful that these studies will provide information as to how they are regulated and how they have undergone changes in the course of evolution.

REFERENCES

1. DOOLITTLE, R. F., K. W. K. WATT, B. A. COTTRELL, D. D. STRONG & M. RILEY. 1979. Nature (London) **280:** 464–468.
2. CHUNG, D. W., M. W. RIXON & E. W. DAVIE. 1982. The Biosynthesis of Fibrinogen and the Cloning of its cDNA. *In* Proteins in Biology and Medicine. R. A. Bradshaw, R. L. Hill, J. Tang, L. Chih-chuan, T. Tien-chin & T. Chen-lu, Eds. Academic Press. New York. pp. 309–328.
3. NICKERSON, J. M. & G. M. FULLER. 1981. Proc. Natl. Acad. Sci. USA **78:** 303–307.

4. CRABTREE, G. R. & J. A. KANT. 1981. J. Biol. Chem. **256:** 9718–9723.
5. CHUNG, D. W., M. W. RIXON, R. T. A. MACGILLIVRAY & E. W. DAVIE. 1981. Proc. Natl. Acad. Sci. USA **78:** 1466–1470.
6. MANIATIS, T., R. C. HARDISON, E. LAZY, J. LANER, C. O'CONNELL, D. QUON, G. K. SIM & A. EFSTRATIADIS. 1978. Cell **15:** 687–702.
7. BENTON, W. D. & R. W. DAVIS. 1977. Science **196:** 181–182.
8. WOO, S. L. C. 1979. Methods Enzymol. **68:** 389–395.
9. SOUTHERN, E. M. 1975. J. Mol. Biol. **98:** 503–517.
10. MAXAM, A. M. & W. GILBERT. 1980. Methods Enzymol. **65:** 499–560.
11. TAKAGI, T. & R. F. DOOLITTLE. 1975. Biochemistry **14:** 940–946.
12. MOUNT, S. M. 1982. Nuc. Acid Res. **10:** 459–472.
13. GILBERT, W. 1982. Nature (London) **271:** 501.
14. CHUNG, D. W., M. W. RIXON, R. T. A. MACGILLIVRAY & E. W. DAVIE. 1981. Thrombos. Haemostas. **46:** 103.
15. MOSESSON, M. W., N. ALKJAERSIG, B. SWEET & S. SHERRY. 1967. Biochemistry **6:** 3279–3287.
16. MOSESSON, M. W., J. S. FINLAYSON, R. A. UMFLEET & D. GALANAKIS. 1972. J. Biol. Chem. **247:** 5210–5219.
17. FINLAYSON, J. S., M. W. MOSESSON, T. J. BRONZERT & J. J. PISANO. 1972. J. Biol. Chem. **247:** 5220–5222.
18. MOSESSON, M. W., J. HAINFELD, R. H. HASCHEMEYER & J. WALL. 1981. Blood **58** (Suppl. 1): 222a.

DISCUSSION OF THE PAPER

G. GRIENINGER (*New York Blood Center, New York*): How can you do a transcription assay with HeLa cells? Do they make fibrinogen?

D. W. CHUNG: We used HeLa cell extracts to provide necessary enzymes and co-factors. We then added the cloned DNA containing the putative promoter regions to produce *in vitro* transcription. If you cut the DNA into a specific size you can see that it terminates from that site.

L. J. WANGH (*Brandeis University, Waltham, MA*): I wonder if Gerry Fuller might comment on whether or not in translation of rat fibrinogens the pre-α's might still have an extension on the COOH-terminus and would therefore be expected to be larger than the secreted α?

G. FULLER (*University of Texas Medical Branch, Galveston*): We did not notice it, but I think that we will go back and look at that question again.

WANGH: I might add that in our examination of *xenopus* fibrinogens the pre-α that we observe is the same size as the secreted α.

REGULATION AND CHARACTERIZATION OF THE mRNAs FOR THE Aα, Bβ AND γ CHAINS OF FIBRINOGEN

Gerald R. Crabtree, Jeffrey A. Kant, Albert J. Fornace, Jr.,
Carol A. Rauch, and Dana M. Fowlkes

Laboratory of Pathology
National Institutes of Health
Bethesda, Maryland 20205

INTRODUCTION

A variety of local and systemic injuries are accompanied by an increase in the plasma levels of a group of proteins collectively known as the acute phase reactants. The levels of about 20 proteins are affected, including α_2 macroglobulin, haptoglobin, and fibrinogen. These proteins are produced in the liver, and *in vitro* labeling experiments have indicated that the rate of synthesis of these proteins is increased after experimental induction of the acute phase reaction. Our interest in this response arises from the fact that it resembles developmental processes where complex groups of genes are coordinately activated or inactivated.

The acute phase reaction appears to be mediated by a small polypeptide released from leukocytes [1] (see also Ritchie *et al.,* this volume). Ritchie and coworkers have found that rat hepatocytes cultured in the absence of serum increase the production of fibrinogen as well as several other acute phase proteins following the addition of a 15,000 dalton polypeptide, which they call hepatocyte-stimulating factor (HSF). This factor is derived from the supernatants of leukocytes that have been activated by exposure to the plasmin degradation products of fibrinogen. This suggests that HSF may be part of an indirect feedback mechanism controlling fibrinogen synthesis.

The intracellular events controlling the production of the acute phase proteins are still obscure; in fact, this is true of virtually all eukaryotic proteins. Recent developments in molecular biology have opened this complex level of biologic regulation to experimental approaches. To explore the control of fibrinogen expression we have constructed cDNA clones for each of the three fibrinogen chains in the rat. Thus far we have used these probes to examine the structure of the genes and to measure mRNA levels. Our results indicate that the increase in fibrinogen synthesis which occurs after defibrination as well as during the acute phase response is related to an increase in mRNA for fibrinogen, and it is likely that this regulation occurs at least in part at the transcriptional level.

RESULTS AND DISCUSSION

In Vitro *Translation of Fibrinogen*

Fibrinogen can be synthesized *in vitro* using extracts prepared from rabbit reticulocytes or wheat germ.[2-6] FIGURE 1 (*lane a*) shows the proteins produced by *in vitro* translation of hepatic RNA from rats injected with the venom of

0077–8923/83/0408–0457 $01.75/0 © 1983, NYAS

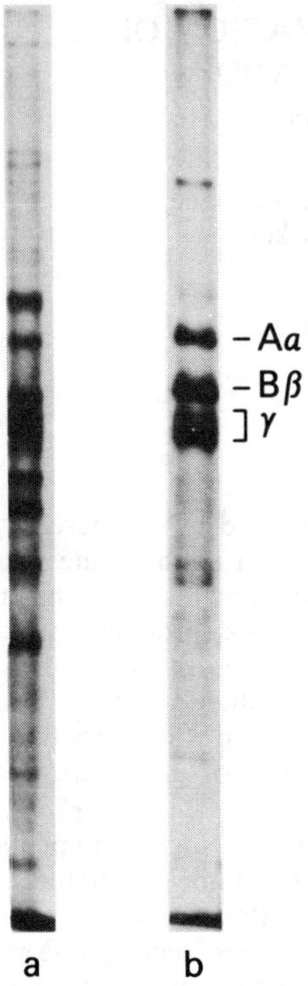

−Aα
−Bβ
] γ

FIGURE 1. Translation products of hepatic mRNA obtained from a rat defibrinated with 200 μg of Malayan pit viper venom. RNA was prepared, translated and the translation products immunoprecipitated and run on 10% polyacrylamide gels as previously described.[3, 4] The identities of the α, β, and γ chains were established using chain-specific antisera and later confirmed by translation of purified mRNA's. *Lane a:* total translation products of rat liver mRNA. *Lane b:* immunoprecipitates of the translation products using antibodies prepared to the denatured fibrinogen chains.

a   b

the Malayan pit viper. Under optimum conditions, about 100 separate bands can be discerned. When the total translation products are immunoprecipitated using specific antifibrinogen antiserum and staphylococcal protein A three dominant bands are visualized (FIG. 1, *lane b*). These can be identified as pre-Aα, pre-Bβ and pre-γ fibrinogen using chain-specific antiserum. The molecular weights of the chains are 66,000, 57,000, and 55,000 compared to 64,000, 57,000, and 55,000 for ^{125}I-labeled fibrinogen chains prepared from plasma fibrinogen. The difference in molecular weights between these two forms reflect the presence of NH_2-terminal signal polypeptides[2] and carbohydrates.[7]

Defibrination and the Acute Phase Reaction Are Accompanied by an Increase in the Level of Translatable Fibrinogen mRNA

Envenomation by a variety of pit vipers, including the western diamondback rattlesnake[8] and the Malayan pit viper,[9] is accompanied by defibrination. In many cases the low levels of fibrinogen occurring after envenomation are rapidly replenished and may return to normal or greater than normal by 24 hours.[8] This rapid replenishment is remarkable since the venom may persist and produce continued defibrination. These findings suggested to us that there might be a brisk increase in fibrinogen synthesis, and possibly mRNA levels for fibrinogen following defibrination with Malayan pit viper venom.[10, 11] FIGURE 2A illustrates the changes in the *in vitro* translation products of rat liver mRNA that accompany defibrination by Malayan pit viper venom. Eight out of 90 visible bands are increased in intensity 8 hours after injection of venom. When these translation products were immunoprecipitated with anti-fibrinogen antiserum four bands could be attributed to fibrinogen. One of the bands, indicated by the asterisk, is an early termination product of the α chain mRNA.[3, 4] In contrast, at least two of the bands visible in FIGURE 2A were

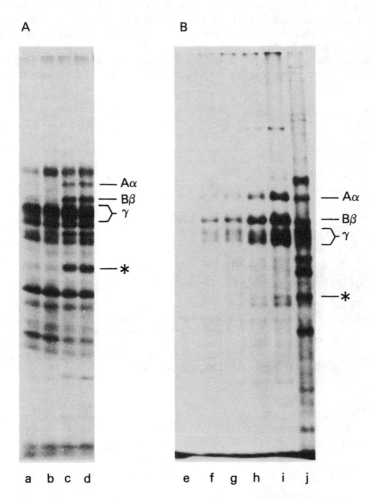

FIGURE 2. Changes in the level of translatable fibrinogen mRNA in the rat liver after intramuscular injection of 200 μg Malayan pit viper venom. *Lanes a and b:* total translation products prior to the injection of venom. *Lanes c and d:* total translation products 8 hr after injection of venom. *Lanes e-i:* immunoprecipitated translation products 0, 1, 2, 4 and 8 hrs after injection of venom. *Lane j:* Total translation products 8 hr after injection of venom. The asterisk indicates the position of a premature termination product of the α fibrinogen mRNA (from reference 4).

found to be reduced in intensity after injection of venom. One of these is likely to be albumin, since it is the proper molecular weight and is an abundant translation product. It appears then that the response to defibrination is complex and involves several proteins that may not be related to the coagulation system.

FIGURE 2B shows a time-course of the changes in translatable fibrinogen mRNA occurring after defibrination with Malayan pit viper venom. The immunoprecipitates of translation products of mRNA prepared 0, 1, 2, 4 and 8 hours after defibrination are shown in lanes e–i. An increase in fibrinogen mRNA can be detected as early as 1 to 2 hours and generally reaches maximal levels by 8 to 16 hours. By 48 hours the levels return to near normal values (not shown in this figure).

Similar changes in translatable fibrinogen mRNA are seen following induction of the acute phase reaction with turpentine. FIGURE 3 compares the trans-

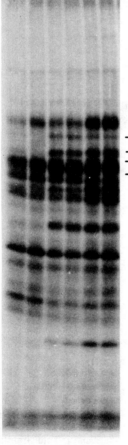

$-A\alpha$
$-B\beta$
$]\gamma$

FIGURE 3. Comparison of the translation ·products of mRNA obtained from uninjected animals (*lanes a and b*), animals injected with 200 μg Malayan pit viper venom (*lanes c and d*), or with 100 μl turpentine (*lanes e and f*).

a   b   c   d   e   f

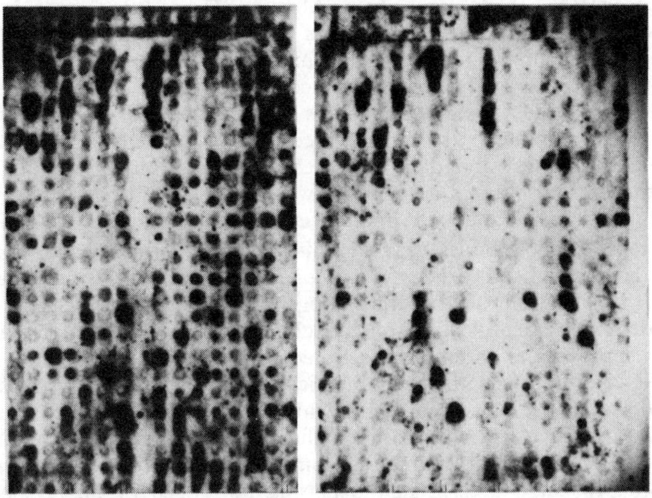

Induced ³²P cDNA   Non-Induced ³²P cDNA

FIGURE 4. Changes in the populations of mRNAs in the rat liver following defibrination. A library of cloned rat liver cDNAs was grown on duplicate nitrocellulose filters and hybridized to ³²P cDNA prepared from RNA of venom-injected animals (on the left) or from noninjected animals (on the right). A point-to-point correspondance exists between the colonies on the left and those on the right (from reference 3).

lation products of mRNA prepared from control noninjected animals (*lanes a and b*), venom-injected animals (*lanes c and d*), and turpentine-injected animals (*lanes e and f*). It is clear that the proteins induced in venom-injected and turpentine-injected animals are very similar. This is consistent with the notion put forth by Ritchie *et al.* that a single peptide is responsible for both responses.

Molecular Cloning of cDNA for the α, β, and γ Chains of Fibrinogen

In order to learn more about the mechanisms underlying these changes in translatable fibrinogen mRNA and to study the structure of the fibrinogen genes, we have constructed cDNA clones for the three chains of rat fibrinogen. This work has been described in detail [3] and only a few points not made in the original publication will be discussed. A report by Chung *et al.* has appeared describing the characterization of a cDNA clone for bovine fibrinogen.[12] The strategy we used for cloning made use of the induction of fibrinogen mRNA following defibrination. At maximal induction fibrinogen mRNA accounts for 10% of the total liver cell translation products compared to 0.5 to 1% in noninjected control animals. This response allowed us to pick prospective fibrinogen cDNA clones out of a library of cloned cDNAs prepared from induced rat liver mRNA as illustrated in FIGURE 4. Clones were grown on duplicate nitrocellulose filters and hybridized to either [³²P]cDNA prepared from induced mRNA (FIG. 4, *left*) or to [³²P]cDNA prepared from noninduced mRNA (FIG. 4, *right*). About 12% of the clones hybridized more to the induced than the noninduced cDNA. About 50% of these clones were found to be fibrinogen cDNAs. Thus, as expected from the translation results, several other mRNAs were induced which were not fibrinogen. Also about 3% of the cDNA clones from this bank of rat liver cDNAs hybridized less intensely to induced than noninduced rat liver cDNA, indicating that these mRNAs are less abundant after defibrination. Here again these results parallel those from protein translation which suggested that certain translation products were less abundant in the liver after defibrination.

Fibrinogen cDNA clones were identified using hybrid selection and translation.[3] The purified cDNA insert was used to specifically select complementary mRNA. Identification of the translation products of this selected mRNA allowed definitive identification of the cDNA clones. FIGURE 5 shows the translation products of the mRNA selected by three cloned cDNAs. A dominant band comigrating with the authentic pre-α, β and γ chains was present along with several less intense bands which most likely represent premature termination products. In the case of the γ chain a less intense band was present about 2000 daltons above the dominant band. This band corresponds to the γ′ or $γ_B$ chain present in the plasma fibrinogen of many species,[12-17] thus adding additional evidence that this chain is not a post-translational modification. As theorized by Wolfenstein-Todel and Mosesson,[17] we have shown that alternative mRNA splice patterns are responsible for the production of the major γ chain and the minor γ′ or $γ_B$ chain (Crabtree, G. R. and Kant, J. A., manuscript submitted).

Characterization of the Rat and Human Fibrinogen mRNAs

The size of the mRNAs coding for the fibrinogen chains was determined using Northern blotting. Cross-hybridization between the rat cDNAs and the human mRNAs permitted the sizing of both the human and the rat fibrinogen mRNAs using the rat cDNA probes. These results are shown in FIGURE 6. Comparing the mobility of the fibrinogen chains to ribosomal size standards allowed us to determine that the α fibrinogen mRNA is about 2400 nucleotides, the β fibrinogen mRNA about 2000 nucleotides and the γ fibrinogen mRNA about 1800 nucleotides: note that under the conditions used there is no cross hybridization between the cDNAs for the three chains. The lack of significant cross-hybridization under stringent conditions has been further documented using purified mRNAs.[3] However, if the conditions of hybridization are reduced in stringency by increasing the salt concentration and/or lowering the temperature the homology between the three fibrinogen mRNAs can be investigated. This approach revealed minimal homology between the γ and β fibrinogen mRNAs and little or no homology between the α and the other two mRNAs. These findings are consistent with the amino acid sequences of the human proteins.[18, 19]

Homologies between the human and the rat mRNAs can be inferred from the results shown in FIGURE 6. The human and rat α fibrinogen mRNAs cross-hybridize weakly. On the other hand, both human β and γ fibrinogen mRNAs cross-hybridize well with the rat β and γ cDNAs. These results indicate that the β and γ chains are more evolutionarily conserved than the α chain.

Changes in Hybridizable Fibrinogen mRNA Accompanying Defibrination

FIGURE 7, A and B, shows a time course of fibrinogen mRNA levels after defibrination of rats with 200 μg of Malayan pit viper venom. Here mRNA levels have been estimated by hybridization to the cloned cDNAs for the α,

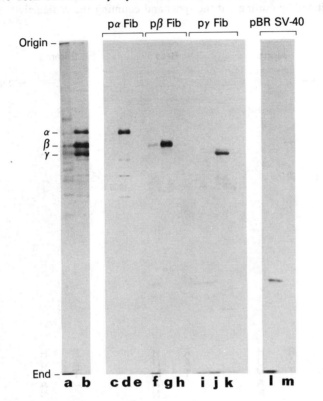

FIGURE 5. Identification of cDNA clones for the α, β and γ chains of rat fibrinogen. Immobilized cDNA was used to select mRNA under highly stringent conditions.[3] The selected mRNA was eluted, translated *in vitro* and run on polyacrylamide gels as described.[3] *Lane a:* Total translation products of rat liver polyadenylated RNA. *Lane b:* Immunoprecipitate of the total translation products shown in lane a. *Lane c:* Total translation products of the mRNA selected by the cDNA clone pα fib. *Lane d:* Immunoprecipitated translation products as in lane d but with 10 μg of the purified α chain added to the antibody before precipitation. *Lane f:* Total translation products of the mRNA selected by the clone pβ fib. *Lane g:* Immunoprecipitate of the translation products in lane f. *Lane h:* Immunoprecipitate of the translation products in lane f except with 10 μg of the purified β chain added to the antibody. *Lane i:* Total translation products of the mRNA selected by the clone pγ fib. *Lane j:* Immunoprecipitated translation products of the mRNA selected by p γfib. *Lane k:* Immunoprecipitated translation products of lane i except with 10 μg of purified γ chain added to the antibody. *Lane 1:* Translation products of mRNA selected by the short Pst I fragment of SV–40 inserted into Pst I site of pBR322 by G-C tailing. *Lane m:* Immunoprecipitate using fibrinogen specific antibodies of the translation products shown in lane 1.

β, and γ chains of rat fibrinogen. Under the conditions used there was no significant cross-hybridization between any of the probles. FIGURE 5A shows dot blots in which hepatic mRNA prepared at various times after defibrination was fixed to nitrocellulose and hybridized to [32P]cDNA for each chain. The amount of cDNA hybridized reflects the amount of specific mRNA and can be quantitated by cutting out the spots and counting the radioactivity as shown

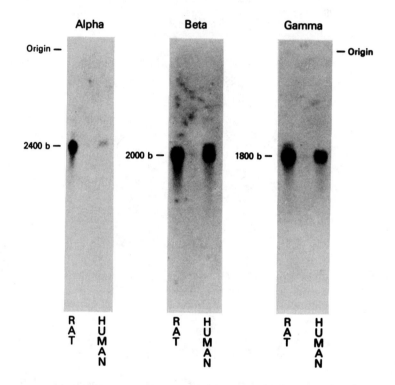

FIGURE 6. Estimation of the size of the α, β and γ mRNA of human and rat fibrinogen. mRNA was electrophoresed on a 1% denaturing gel and transferred to nitrocellulose.[22] To detect the human mRNA with the rat fibrinogen probes the stringency was reduced to 5× SSC, 50% formamide, 50 mM sodium phosphate pH 6.5 at 37° C. The nitrocellulose filters were washed with 2× SSC 4 times at 25° C and exposed for 48 hours with intensifying screens. The lanes labeled rat represent 2 μg total RNA, while the lanes labeled human represent 2 μg poly-adenylated RNA. The rat total RNA was about 1/50 polyadenylated RNA.

in FIGURE 5B. The most notable feature of the time-course is the similarity of the induction kinetics for the three mRNAs. The lag time, rate, and extent of induction are similar for each of the three fibrinogen mRNAs. Although such highly coordinated synthesis might be expected, since it would be in the interest of economy to balance production of each chain, similar induction kinetics have not been found for either the globin[21] or the casein[22] mRNAs. Simi-

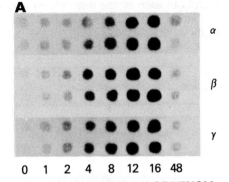

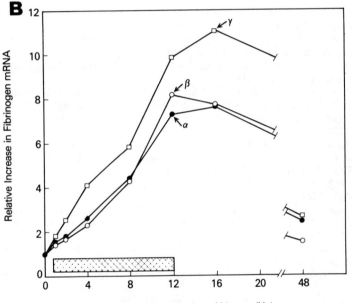

FIGURE 7. Changes in hybridizable fibrinogen mRNA after defibrination. Rats were injected with 200 μg of Malayan pit viper venom at time 0. At the indicated times hepatic mRNA was prepared and the amount of fibrinogen mRNA estimated by dot-blot hybridization.[22] Under the conditions used for the hybridization (50% formamide, 5 × SSC, 50 mM sodium phosphate pH 6.5, 42° C) and wash[22] there was no significant cross-hybridization between the three mRNAs. **A**. Autoradiograms of dot-blots prepared by spotting 4 μg of hepatic mRNA onto nitrocellulose and hybridizing with ^{32}P-labeled cDNA inserts for the α, β, and γ chains of rat fibrinogen. **B**. Relative changes in fibrinogen mRNA assayed by scintillation counting of the spots shown in A above. The results are expressed relative to the cpm hybridizing at time 0.

larities in induction kinetics of each of the three chains may point to common structural features in the mRNAs or genes which could represent sites of interaction with regulatory molecules. It is likely that this response is at least in part transcriptionally regulated, since the relative abundance of high molecular weight fibrinogen mRNA precursors parallels the increase in mature cytoplasmic mRNA seen after defibrination. However, definitive pulse-labeling studies of nuclear RNA will be essential to show an increase in transcriptional rates of these genes.

REFERENCES

1. RITCHIE, D. G., B. A. LEVY, M. A. ADAMS & G. M. FULLER. 1982. Proc. Natl. Acad. Sci. USA **79:** 1530–1534.
2. NICKERSON, J. M. & G. M. FULLER. 1981. Proc. Natl. Acad. Sci. USA **78:** 303–307.
3. CRABTREE, G. R. & J. A. KANT. 1981. J. Biol. Chem. **256:** 9718–9723.
4. CRABTREE, G. R. & J. A. KANT. J. Biol. Chem. In press.
5. YU, S., C. M. REDMAN, J. GOLDSTEIN AND B. BLOMBACK. 1980. Biochem. Biophy. Res. Commun. **96:** 1032–1038.
6. CHUNG, D. W., R. T. A. MACGILLIVARY & E. W. DAVIE. 1980. Ann. N. Y. Acad. Sci. **343:** 210–215.
7. NICKERSON, J. M. & G. M. FULLER. 1981. Biochemistry **20:** 2818–2821.
8. SIMON, T. L. & T. G. GRACE. 1981. New Engl. J. Med. **305:** 443–447.
9. REGOECZI, E., J. GERGELY & A. S. McFARLANE. 1966. J. Clin. Invest. **45:** 1202–1212.
10. ESNOUF, M. P. & G. W. TUNNAH. 1967. Brit. J. Haemat. **13:** 581–590.
11. CHUNG, D. W., M. W. RIXON, R. T. A. MACGILLIVARY & E. W. DAVIE. 1981. Proc. Natl. Acad. Sci. USA **78:** 1466–1470.
12. MOSESSON, M. W., J. S. FINLAYSON & R. A. UMFLEET. 1972. J. Biol. Chem. **247:** 5223–5227.
13. MOSHER, D. F. & E. R. BLOUT. 1973. J. Biol. Chem. **248:** 6896–6903.
14. FRANCIS, C. W., V. J. MARDER & S. E. MARTIN. 1980. J. Biol. Chem. **255:** 5599–5604.
15. WOLFENSTEIN-TODEL, C. & M. W. MOSESSON. 1980. Proc. Natl. Acad. Sci. USA **77:** 5069–5073.
16. WOLFENSTEIN-TODEL, C. & M. W. MOSESSON. 1981. Biochemistry **20:** 6146–6149.
17. LEGRELE, C. D., C. WOLFENSTEIN-TODEL, Y. HURBOURG & M. W. MOSESSON. 1982. Biochem. Biophy. Res. Commun. In press.
18. DOOLITTLE, R. F., K. W. K. WATT, B. A. COTTRELL, D. D. STRONG & M. RILEY. 1979. Nature **280:** 464–468.
19. WATT, K. W. K., T. TAKAGI & R. F. DOOLITTLE. 1978. Proc. Natl. Acad. Sci. USA **75:** 1731–1735.
20. ORKIN, S. H., D. SWAN & P. LEDER. 1975. J. Biol. Chem. **250:** 8753–8760.
21. HOBBS, A. A., G. PRABHAKAR, D. J. KESSELER & J. M. ROSEN. 1981. J. Supramol. Struct. Cell. Biochem., Suppl. **5:** 426.
22. THOMAS, P. S. 1980. Proc. Natl. Acad. Sci. USA **77:** 5201–5205.

DISCUSSION OF THE PAPER

R. F. DOOLITTLE (*University of California, San Diego, La Jolla*): I wonder if I could ask about the γ chain extension, or γ'. In particular it was not clear to me when you showed the slide whether you had only sequenced the rat γ' or whether you had also done the human. Let me explain the basis of the question. On Wednesday, we heard Dr. Francis talk about two different kinds of variant in the γ chain and the implication, I believe, was that these might have all been separate gene products or spliced products, whereas, Wolfenstein-Todel and Mosesson clearly have evidence for only two. If you had the human, as opposed to the rat, you would be able to see whether there was another possibility for splicing, as opposed to some artifact involving carbohydrate, etc.

The question is: Have you done the human sequence? Also, since you reported the α chain extension here and you have a sequence for that, and looking at the way your introns are lined up vis-à-vis the γ', it is clear that in your α chain extension the 15 residues could be homologous to either the rat or the human γ' extension.

G. R. CRABTREE: We have not really tried that. We do not have the sequence of the human γ 7th intron so that we can not say whether there were other alternative splice sites within that intron that might allow it to be spliced in other ways.

My speculation about that is that there may very well be other ways. In the rat, the last intron, is eliminated in three steps. You saw that there was a very high molecular message of about 2400 base pairs. There is an intermediate of about 2000, and then below that there is one of 1800 that represents the dominant message. Introns can be eliminated in a directional manner. This has been found for collagen genes for example, in one of the introns. It may well be that another fibrinogen chain appears as a result of a different type of splice junction.

DOOLITTLE: Is there any possibility that in the α chain there could be an equivalent splice at the COOH-terminal valine that goes directly to a termination, instead of running to an expressable product? Is that possible?

CRABTREE: I see. Thus, you would actually have two termination sites for the α chain as well as for the γ chain. I think Dominic Chung really has the best answer to that, because he has looked for heterogeneity of the messengers. We have not.

D. CHUNG (*University of Washington School of Medicine, Seattle*): From a sequence of a cDNA coding for body extension it does not look like there is an AGGT, which would be a splice point immediately within the vicinity of the valine 610. If it is a splice point, it would have a very different recognition sequence from the consensus sequence.

CRABTREE: It certainly seems that the more abundant form of the message would have to be this one with the 15 amino acid extension. Dr. Chung has found it in his clones, we found it in our human α chain, we found it in the rat, and it is found in the cow also. I think it is very likely that this is a dominant form of the message. But it should be pointed out that if the heterogeneity that one finds in the plasma at the COOH end of the α chain is due to proteolysis, then it is due to proteolysis other than the typical trypsin-like enzyme, because that is not an arginine or lysine COOH terminal. It is a Val-Arg-Gly. It is as though maybe an Arg-Gly bond was cleared and then a carboxy peptidase-like activity removed the arginine.

D. L. AMRANI (*Mt. Sinai Medical Center, Milwaukee, WI*): If I understood correctly what you said, you found approximately 10 hybridizable γ chain messenger RNAs for every one γ'?

CRABTREE: That is right.

AMRANI: You know that I have recently reported that the circulating rat fibrinogen molecule has approximately three γ chains to each γ'. I was wondering if your isolated translation and *in vitro* translation of your messenger RNAs are isolated by the hybridization? Do you have any idea of the portion that is being translated?

CRABTREE: I meant to mention that when I showed the translation of the purified message. There it is about ten to one also, so that what you are saying would suggest that there is some transcriptional or translational regulation, and I think that this is very possible. However, translational regulation would have to occur *in vivo* and might not be found in the *in vitro* translation.

W. R. BELL (*Johns Hopkins Hospital, Baltimore, MD*): I wonder what your thoughts are on how the venom of the Malayan pit viper induces the message?

CRABTREE: As far as the mechanism goes, I believe that there will be a nice answer to that in D. Ritchie's and G. Fuller's work with the hepatocyte-stimulating factor. The proteins that are induced after the injection of the venom are very much like the acute phase proteins.

HORMONAL REGULATION OF FIBRINOGEN
SYNTHESIS IN CULTURED HEPATOCYTES *

Gerd Grieninger,† Patricia W. Plant, T. Jake Liang,
Robert G. Kalb, David Amrani,‡ Michael W. Mosesson,‡
Kathe M. Hertzberg, and Johanna Pindyck

*Lindsley F. Kimball Research Institute
The New York Blood Center
New York, New York 10021*

INTRODUCTION

Plasma levels of fibrinogen are very responsive to physiological changes *in vivo*. They are elevated, for example, in pregnancy [1, 2] and during the body's response to injury and stress (the acute phase reaction).[3] The central role of the liver as the organ responsible for fibrinogen production has been clearly established, and the specific site of synthesis has been localized to the hepatic parenchymal cell.[4-7]

Investigation of the nature of the key regulatory agents governing fibrinogen production has been a major concern of this and other laboratories. The hepatic output of fibrinogen is apparently under humoral control, since there is no evidence for neural control. Considerable support for this thesis has come from changes in fibrinogen synthesis observed with whole animals, following surgical removal of endocrine glands and/or administration of a variety of hormones and other factors. Among the agents that reportedly stimulate fibrinogen production in these experiments are glucocorticoids,[8, 9] certain sex steroids,[10] adrenocorticotropic hormone,[11] thyroid hormones,[12] growth hormone,[13] prostaglandins,[14] endotoxin,[15] leukocytic endogenous mediator,[16, 17] and fibrinogen degradation products.[18, 19] However, the fact that fibrinogen is an acute phase reactant, coupled with the great number of variables inherent in experiments with whole animals, has tended to complicate the interpretation of many of these studies.

To determine the signals that *directly* affect production of fibrinogen by the liver, the hepatic tissue must be separated from the uncontrolled influence of products from other organs in the body. This can be achieved with several model systems, including the isolated perfused liver, liver slices, and isolated hepatocytes in culture.

Our laboratory has developed the primary monolayer culture of embryonic chick hepatocytes for dissection of the complex events that influence the liver's

* This research was supported by Grants HL–28444 and HL–09011 from the National Institutes of Health. P.W.P. was supported by an Institutional Grant NRSA 5 T32 HL–07331 from the National Institutes of Health. Part of this work was done during the tenure of an Established Investigatorship to G.G. from the American Heart Association.

† To whom correspondence should be addressed.

‡ Present address: Mt. Sinai Medical Center, Milwaukee, Wisconsin 53233.

0077–8923/83/0408–0469 $01.75/0 © 1983, NYAS

synthesis of fibrinogen. Our studies of chicken fibrinogen have shown it to be structurally homologous and functionally similar to mammalian fibrinogen.[20, 21] The advantages of this avian system over mammalian hepatocyte culture systems for studying hormonal control of plasma protein synthesis have been recently reviewed.[22] Briefly stated, the cultured chick embryo hepatocytes are capable of providing a uniquely well-controlled baseline against which to evaluate the effects of added hormones since they do not require serum, hormone or other macromolecular supplement to the culture medium for efficient plating or viability. Under such minimal culture conditions, the hepatocytes display stable, submaximal rates of fibrinogen synthesis for several days and are fully responsive to physiological levels of a number of added hormones. Using this system, we have demonstrated both the effects of individual hormones on fibrinogen synthesis as well as complex interactions between agents, which are reflected in additivity, suppression, or masking of the individual effects. This paper summarizes these observations, and indicates the versatility of the chick embryo hepatocyte cultures as a model for exploring the control of fibrinogen biosynthesis.

MATERIALS AND METHODS

Primary hepatocyte monolayer cultures were prepared from 16-day-old chicken embryos as described.[23-25] In short, hepatocytes were plated and maintained in modified Ham F-12 *without* hormones and serum supplement so that the cells are cultured from the onset in a completely chemically defined medium, free of added macromolecules. Culture medium was replaced with an equal volume of fresh medium every 24 hr unless otherwise indicated. When medium was intended for fibrinogen assay, heparin sodium salt (Fisher Scientific Co., Fairlawn, NJ; 156 units/mg) was added at a concentration of 15 μg/ml. Inclusion of this compound has no effect on the control of plasma protein synthesis in this system (unpublished observation).

The methodology and utilization of primary cultures of embryonic chicken hepatocytes for studies on the regulation of plasma protein synthesis have been reviewed recently by our laboratory.[22] In this model system, the rate at which plasma proteins are synthesized is generally reflected in their rate of secretion, which is determined by measuring extracellular accumulation with electroimmunoassays.[23] Fibrinogen and other secreted plasma proteins were measured in 3 μl samples of unconcentrated culture medium using electroimmunoassays with monospecific antisera.[26] Crossed immunoelectrophoresis was performed as described [27] on samples of culture medium that were concentrated by ultrafiltration with supported YM-10 membranes (Amicon).

Pertinent details of the experiments reported in this paper are specified in the legends to the figures and tables. A full description of the labeling of hepatocytes with [^{35}S] methionine, preparation of cell extracts using detergents and protease inhibitors, as well as immunoprecipitation of fibrinogen and other plasma proteins using monospecific antibodies and Protein A-Sepharose CL-4B (Pharmacia) as adsorbent are described by Plant *et al.*[28] The labeled plasma proteins were separated by SDS-polyacrylamide gel electrophoresis according to Laemmli [29] with a 7.5 to 11.4% linear gradient; [^{35}S]methionine incorporation into individual polypeptides was determined by liquid scintillation measurements of excised gel bands. Total cellular RNA was prepared from hepatocyte monolayers using a high salt extraction procedure modified from that of Deeley *et al.*,[30] and poly(A)-containing RNA was then isolated by affinity chromatography. Wheat germ lysates were prepared according to Roberts and Paterson,[31] and the conditions of the cell-free translation reaction were similar to those reported by Tse and Taylor.[32] Immunoprecipitation of fibrinogen and other plasma proteins from the cell-free translation products followed the procedure of Goldman and Blobel.[33]

3,3',5-Triiodo-L-thyronine (sodium salt) was obtained from Sigma as was insulin (bovine pancreas, crystalline; 24.3 units/mg). Dexamethasone was purchased from Steraloids, Wilton, NJ. Fetal bovine serum was obtained from Grand Island Biological Co., Grand Island, NY. Crude rabbit leukocytic endogenous mediator (LEM) was provided by Dr. Philip Sobocinski, Fort Detrick, MD; 1 ml contained the amount of LEM obtained from 10^8 peritoneal exudate cells prepared as described.[34] Supernatant fluid from lipopolysaccharide-stimulated mouse peritoneal macrophages was supplied by Dr. William Vine,

TABLE 1

PLASMA PROTEIN SYNTHESIS UNDER BASAL CULTURE CONDITIONS

	Fibrinogen	α_1-Globulin "M"	Transferrin	Albumin
	(μg per ml per day)			
Day 1	1.6 ± 0.2	0.30 ± 0	16 ± 1	22.7 ± 0.9
Day 2	1.5 ± 0.1	0.89 ± 0.04	26 ± 2	1.5 ± 0.1
Day 3	1.4 ± 0.1	1.07 ± 0.06	24 ± 2	0.19 ± 0.02
Day 4	1.5 ± 0	1.21 ± 0.09	19 ± 2	0.20 ± 0.02

Cells were plated and maintained in 5 ml of medium without supplement in 55-mm diameter dishes. Secreted plasma proteins were determined by electroimmunoassay of medium samples taken every 24 hr, immediately before fresh medium was given. Samples for determination of fibrinogen were taken from two separate sets of dishes that had received fresh medium containing heparin, beginning either at 3 hr after plating (for day 1 synthesis) or at 24 hr (for days 2, 3 and 4). Values represent the mean ± SD of triplicate dishes (data from Liang and Grieninger [25]).

Rockefeller University. Interleukin-1 was partially purified by gel filtration from endotoxin (100 ng/ml)-stimulated human monocytes and contained 300 LAF units/μl [35]; it was a gift from Dr. David Wood, Merck Sharp and Dohme Research Laboratories, Rahway, NJ.

RESULTS AND DISCUSSION

Fibrinogen Synthesis in the Absence of Hormone Supplement

Primary cultures of chick embryo hepatocytes synthesize fibrinogen and a wide range of plasma proteins when maintained, from the onset of culture, in a chemically defined medium devoid of culture supplements such as serum, hormones or other macromolecules (FIG. 1A, see also FIG. 6A and C). The rate at which fibrinogen and most plasma proteins are synthesized under these basal culture conditions is steady, or even increasing (TABLE 1), so that their

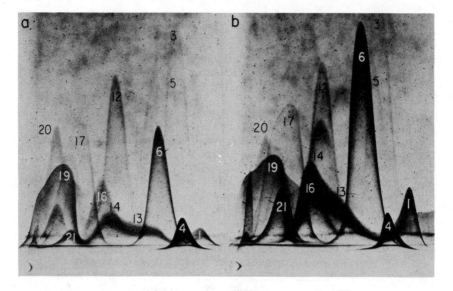

FIGURE 1. Secreted plasma proteins synthesized in the presence and absence of 20 nM T₃. Cells were given fresh medium with or without T₃ 7 hr after plating in hormone-free medium. Fifteen hr later, the medium was again replaced with fresh medium, with or without T₃ as before, but now containing heparin. Medium was collected 24 hr later, concentrated 300-fold and analyzed by crossed immunoelectrophoresis. Samples (3 μl) were applied in the wells in the lower left corner of each panel. Electrophoresis in the first dimension was performed from left to right and in the second dimension from bottom to top. The second-dimension gel (antibody-containing) contained immunoglobulin fractions prepared from the antisera by ammonium sulfate precipitation: 19.5 mg equivalent to 22.8% anti-adult chicken serum (i.e. 32 μl/cm²) and 0.96 mg equivalent to 1.1% anti-chicken fibrinogen (i.e 1.5 μl/cm²). Immunoplates were stained with Coomassie blue. In this assay, the amount of each plasma protein secreted is reflected by the intensity and area of its respective peak.[27] In order to match the corresponding peaks in each panel, the samples of culture medium were compared by tandem crossed immunoelectrophoresis.[28] Several peaks have been identified with the use of specific antisera or purified antigens and are numbered according to references 23 and 24: peak 4, albumin; peak 6, α_1-globulin "M"; peaks 13 and 14, lipoproteins; peak 17, α-antitrypsin; peak 19, transferrin; peak 21, fibrinogen. Unlike the anti-chicken serum used in the above references, this antiserum did not recognize prealbumin "C." (a) control; (b) 20 nM T₃. (From Hertzberg et al.,[36] used with permission.)

syntheses may be considered to be a "constitutive" hepatocellular function. The one notable exception to this rule is albumin; an investigation of the selective decrease of its synthesis in culture is described elsewhere.[28]

The fibrinogen molecule synthesized and secreted in culture is fully assembled and has a molecular weight of about 320,000, similar to the molecular weight of circulating chicken plasma fibrinogen.[37] In FIGURE 2, the polypeptide chains of fibrinogen isolated from chicken plasma are compared with those from fibrinogen synthesized in culture. Separation of both fibrinogens by SDS-polyacrylamide gel electrophoresis yields similar patterns of Aα, Bβ, and γ

chains, although the γ chain of the plasma form migrates as a doublet. Whether both γ chains are synthesized by the cultured hepatocytes is currently under investigation.

The Stimulatory Effects of Specific Hormones

The addition of certain hormones, individually and in physiological concentrations, to the cultures of chick embryo hepatocytes elicits an increase in fibrinogen production. The effects are characteristic for each agent and vary with regard to the dose of the hormone and minimum exposure required. Differences also exist in the time elapsed between initial exposure and onset of the response as well as in the extent of accompanying effects on the synthesis of other plasma proteins. Our laboratory has characterized the effects of glucocorticoids, thyroid hormones, and insulin on plasma protein synthesis in culture.[24, 25, 36] In the following subsections, the stimulation of fibrinogen synthesis by these hormones is described in the context of their effects on three other major plasma proteins: albumin, transferrin, and α_1-globulin "M", which, like fibrinogen, is a major acute phase protein in the chicken.[38]

Thyroid Hormones

When triiodothyronine (T_3) is added to the culture medium, the synthesis of fibrinogen as well as that of α_1-globulin "M" is selectively increased about threefold (TABLE 2 and FIG. 1). In contrast, only slight, if any, increases are

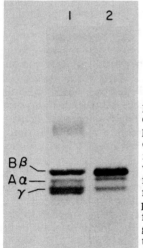

FIGURE 2. Comparison of the polypeptide chains of plasma fibrinogen with those of fibrinogen synthesized in culture. The preparation of fibrinogen from chicken plasma was as described.[20] To obtain fibrinogen from cultured hepatocytes, the cells were maintained in hormone-free medium for 24 hr and then labeled for 90 min with [^{35}S]methionine. Immunoprecipitation of fibrinogen from an extract of the cells and the separation of the fibrinogen chains by SDS-polyacrylamide gel electrophoresis are described elsewhere.[37] *Lane 1*—chicken plasma fibrinogen, stained with Coomassie blue; *lane 2*—fibrinogen from cultured hepatocytes, autoradiographed. Since the Bβ chain contains twice as much methionine as the other chains,[21] it incorporates more label.

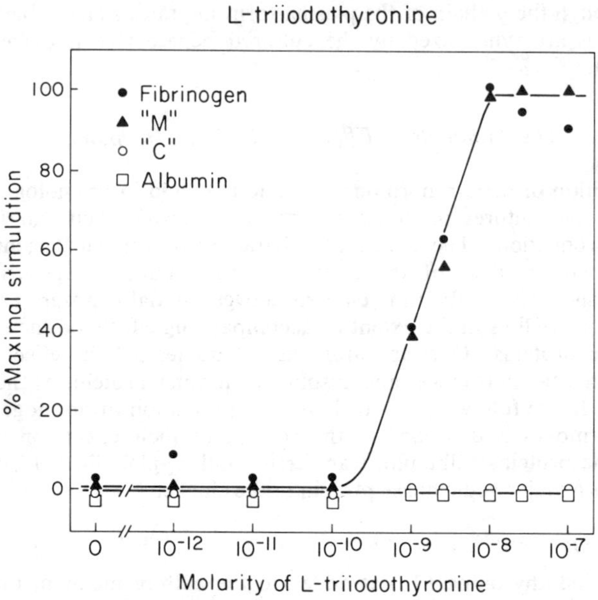

FIGURE 3. Effect of increasing concentrations of T_3 on plasma protein synthesis. Twenty-four hr after plating, the cells were given fresh medium containing heparin and various concentrations of T_3 as indicated on the abscissa. Medium was collected 24 hr later and the secreted plasma proteins, fibrinogen, α_1-globulin "M," prealbumin "C," and albumin, were determined by electroimmunoassay.

found in the synthesis of albumin, transferrin and a variety of other plasma proteins. The response of fibrinogen synthesis to increasing concentrations of T_3 is shown FIGURE 3. Half-maximal stimulation is elicited by 1 nM T_3, a value well within the physiological range. Diminished potency is associated with less iodination of the hormone and is consistent with the relative biological activity of such derivatives in a variety of other systems. A marked reduction is noted with the diiodinated derivative and no stimulation occurs at all with either mono- or non-iodinated thyronine.[36]

The stimulation of fibrinogen synthesis by T_3 can be detected about 6 hours after initial addition of the hormone (FIG. 4). However, exposure of the cells for only 30 minutes is nearly as effective as continuous exposure in eliciting the ultimate response (TABLE 3). Apparently, critical changes occur in the cells almost immediately upon exposure to the hormone, initiating a series of events that no longer require the presence of the hormone. These early events have yet to be elucidated.

Glucocorticoids

The effects of glucocorticoids on fibrinogen synthesis are similar to those of thyroid hormones in terms of selectivity and degree of stimulation. Dexamethasone (1 nM), a synthetic glucocorticoid, stimulates synthesis of fibrinogen and α_1-globulin "M" about threefold while scarcely affecting synthesis of

TABLE 2

EFFECTS OF HORMONES AND SERUM ON PLASMA PROTEIN SYNTHESIS

Additions	Fibrinogen	α_1-Globulin "M" (degree of stimulation *)	Transferrin	Albumin
1. None	1.0	1.0	1.0	1.0
2. T$_3$	2.5	3.0	1.4	1.1
3. Dexamethasone	2.8	2.8	1.4	1.3
4. Insulin	0.9	4.0	1.4	2.8
5. Insulin + dex	2.5	4.1	1.3	2.7
6. Insulin + T$_3$	1.4	6.2	1.4	2.8
7. Dex + T$_3$	4.8	5.3	1.5	1.1
8. Dex + T$_3$ + insulin	2.7	7.2	1.4	2.6
9. Serum	4.4	5.4	1.6	2.5
10. Serum + dex	5.5	6.7	1.6	2.5

* The basal level of synthesis in the absence of hormones has been normalized to a value of 1.0 for each of the individual plasma proteins for better comparison. Degree of stimulation is expressed relative to this basal level. Absolute values of synthesis in this experiment are similar to those presented in TABLE 1.

Cells were plated in 55-mm diameter dishes in 5 ml of medium. Twenty-two hours later, the cells were given fresh heparin-containing medium, with and without hormones and/or serum. Addition of agents, whether singly or in combination, were made at the following concentrations: 10 nM T$_3$, 1 nM dexamethasone, 35 nM insulin, and 2% (vol/vol) fetal bovine serum. After a 6 hr incubation, fresh medium was given, with additions as in the last medium change. The cells were incubated for an additional 18 hr prior to sampling of spent culture medium for analysis by electroimmunoassay.

transferrin and albumin (TABLE 2 and FIG. 5). Stimulation is detectable 5 hours after initial exposure.[39] Expression of the effect apparently requires an hydroxyl group at the 11-position of the steroid nucleus; comparable fibrinogen stimula-

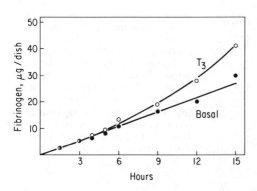

FIGURE 4. Time course of the T$_3$-induced stimulation of fibrinogen synthesis. Cells were plated in dishes 100 mm in diameter and maintained in 12 ml of medium. After 43 hr, the medium was replaced with one-fourth volume of fresh medium plus heparin (basal) or fresh medium with heparin and T$_3$ at 50 nM. Samples of culture medium (25 μl) were taken at the designated times after the culture medium change, and secreted levels of fibrinogen were determined by electroimmunoassay. Data from a matched pair of control and hormone-treated dishes are plotted.

TABLE 3

FIBRINOGEN SYNTHESIS FOLLOWING EXPOSURE TO T_3*

Conditions	Fibrinogen
	μg/dish
Control	2.8 ± 0.4
Group A, cells exposed for only 30 min	5.8 ± 1.0
Group B, cells exposed continuously	6.1 ± 0.8
Full stimulation remaining after hormone removal	91%

* Cells were plated in dishes 35 mm in diameter and maintained in 2 ml of hormone-free medium. The culture medium was replaced with fresh hormone-free medium (controls) or medium containing 20 nM T_3 25.5 hr after plating. After 30 min of further incubation, the medium was removed and the cells were gently rinsed with three changes of hormone-free medium in rapid succession. Three control dishes as well as three dishes that had been exposed to hormone were then incubated with hormone-free medium in the presence of heparin (Group A); the remaining three pre-exposed dishes were given fresh hormone-containing medium now also with heparin (Group B). This format was continued 4 hr later when the medium was replaced with one-half volume of fresh medium. After another 20 hr of culture, the medium was collected and fibrinogen was determined by electroimmunoassay. The results are expressed as the mean ± SD of triplicate dishes. The percentage of full stimulation remaining after hormone removal is calculated from the difference between Group A and the controls divided by the difference between Group B and the controls. Previously published data from this experiment (Hertzberg et al.[36]) contained a systematic calculation error, which has been rectified in this table.

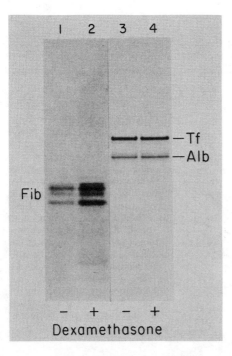

FIGURE 5. Effect of dexamethasone on synthesis of fibrinogen, transferrin and albumin. Cells were given fresh heparin-containing medium, with or without 1 nM dexamethasone, 20 hr after plating. Ten hours later, the cells were labeled for 20 min with [35S]methionine and cell extracts prepared. The three plasma proteins were immunoprecipitated from these extracts, separated by SDS-polyacrylamide gel electrophoresis, and autoradiographed. Quantitation of radioactivity in the respective polypeptides indicated that dexamethasone stimulated the synthesis of fibrinogen 2.5-fold whereas synthesis of both transferrin and albumin was increased only 1.3-fold. Full experimental details are described in Amrani et al.[37]

tion is not observed with cortisone (the 11-keto derivative) or cortexolone (the 11-deoxy derivative) at levels up to 100 nM. Aldosterone, which is structurally related to the glucocorticoids and which contains a hemiacetal on the 11-position, is significantly more effective in this concentration range than the 11-keto or 11-deoxy derivatives.[24]

Unlike the effect of thyroid hormones, however, enhanced fibrinogen synthesis by dexamethasone depends on the continued presence of the hormone.[24] Several other lines of evidence also indicate that thyroid hormones and glucocorticoids exert their effects via different mechanisms. While the glucocorticoid-induced stimulation of fibrinogen synthesis is largely abolished by the transcriptional inhibitor, actinomycin D,[24] the T_3-induced stimulation is only partially reduced (unpublished observations). Furthermore, when the cells are exposed to optimal concentrations of both T_3 and dexamethasone simultaneously, the degree of stimulation elicited is nearly twice that produced by either hormone alone (see below).

Two of the plasma proteins most responsive to dexamethasone in culture, fibrinogen and α_1-globulin "M", are also acute phase plasma proteins. A recently conducted study of the time course of the turpentine-induced acute phase reaction in the chicken has revealed a close correlation between high glucocorticoid levels and increased *in vivo* synthesis of these proteins.[38] The highest levels of corticosterone—the natural glucocorticoid found in chickens— occur approximately 24 hours after turpentine treatment, just when the plasma levels of fibrinogen and α_1-globulin "M" are increasing most rapidly. This study provides further support for the thesis that glucocorticoids contribute to the regulation of the production of acute phase proteins.

Insulin

The polypeptide hormone insulin has not been suspected of affecting fibrinogen synthesis on the basis of whole animal experiments or studies with relatively short-lived *ex vivo* systems such as the isolated perfused liver and isolated hepatocyte suspensions.[40-42] Indeed, even in the chick embryo hepatocyte cultures, a 24-hour exposure to insulin (35 nM) has no effect on fibrinogen synthesis, although it does increase synthesis of α_1-globulin "M" and albumin severalfold (TABLE 2). We have found, however, that prolonged exposure to the hormone results in a significant increase in fibrinogen production, on the order of three- to fourfold (FIG. 6). The concentration of insulin required for eliciting this delayed response of fibrinogen synthesis is 10 times higher than the dose (3.5 nM) to which albumin synthesis readily responds (FIG. 7 and ref. 25). The physiological implications of these observations have yet to be explored.

The Effects of Other Hormones

The direct effect on fibrinogen synthesis of other individual hormones has also been examined, but their importance remains unclear.

Epinephrine

Epinephrine has been studied because of the possibility that it may play a role in the hyperfibrinogenemic response that occurs in some physiological conditions of stress. Epinephrine has been reported to stimulate fibrinogen

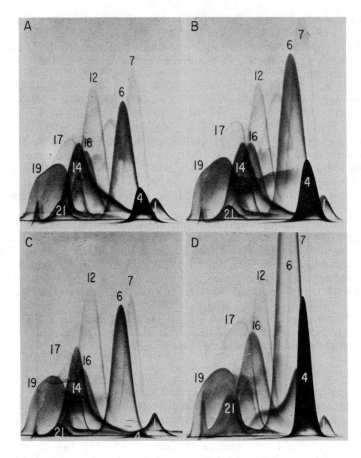

FIGURE 6. Secreted plasma proteins synthesized in the presence and absence of 35 nM insulin. Twenty-four hr after plating in hormone-free medium, the cells were given fresh heparin-containing medium, with or without insulin. The medium was collected 24 hr later—i.e., at the end of the second day of culture—and the cells were reexposed to fresh medium as before, with or without insulin. After an additional 24-hr incubation—i.e., at the end of the third day of culture—the medium was again collected. Medium samples, concentrated 400-fold, were applied (8 μl) in the appropriate well in the lower left corner of each panel. Crossed immunoelectrophoresis, staining, and identification of peaks were performed as in FIGURE 1 using 27.8 mg, equivalent to 32.6% anti-adult-chicken serum (i.e., 46 μl per cm²) and 1.92 mg, equivalent to 2.2% anti-chicken fibrinogen (i.e. 3 μl per cm²). Plasma protein synthesis on the first day of exposure is shown in controls (A) and in the presence of insulin (B); synthesis on the second day of exposure is shown in controls (C) and in the presence of insulin (D). (From Liang and Grieninger,[25] used with permission.)

synthesis in hepatocyte suspensions isolated from adrenalectomized rats.[41] However, the hormone concentrations found effective were unphysiologically high, depressing general secretory protein synthesis considerably. Preliminary experiments with chick embryo hepatocyte monolayers in our laboratory have revealed no effect on production of fibrinogen or α_1-globulin "M" due to this hormone in the concentration range of 0.01 to 100 nM.

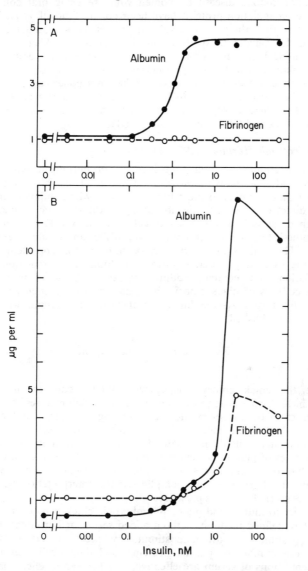

FIGURE 7. Effect of increasing concentrations of insulin on the synthesis of fibrinogen and albumin. Twenty-four hr after plating, the cells were exposed to various concentrations of insulin, as indicated on the abscissa, in fresh medium containing heparin. After another 24 hr, samples were taken and incubation in fresh medium, with hormone as before, was continued for a second 24-hr interval at the end of which samples were again taken. Secreted plasma proteins were determined by electroimmunoassay and are expressed as the mean of duplicate dishes in μg per ml medium. (A) first 24-hr exposure to hormone; (B) second 24-hr exposure to hormone. (From Liang and Grieninger,[25] used with permission.)

Estrogens

In thromboembolic disease in women who are using oral contraceptives, high levels of coagulation proteins have been noted. Pindyck *et al.*[43] advanced the hypothesis that these high levels might reflect a stimulatory effect of estrogen on the hepatic biosynthesis of these proteins. Their studies, using serum-containing hepatocyte monolayers derived from chick embryo, reveal a modest stimulation of fibrinogen production (less than twofold) in response to several synthetic estrogens. However, other hormones and/or factors in addition to the estrogens appear to be involved, since almost no stimulation is produced when these estrogens are supplied alone to the cells in serum-free medium (Pindyck and Grieninger, unpublished).

Adrenocorticotropic Hormone (ACTH)

In adrenalectomized rabbits, an increase in fibrinogen synthesis occurs several hours after administration of ACTH.[11, 44] Thus this hormone, which provides a major stimulus to the adrenal gland in whole animals, appears also to have a direct effect on hepatic fibrinogen synthesis in the adrenalectomized animals. In no case, however, has a direct effect of ACTH on hepatic plasma protein synthesis been demonstrated *ex vivo*. Miller and Griffin [45] evaluated the effect of ACTH on the synthesis of fibrinogen, α_1-acid glycoprotein, and haptoglobin in the isolated perfused rat liver and found no stimulation. Nor was any stimulation observed upon addition of ACTH alone (1 to 100 nM) to monolayer cultures of chick embryo hepatocytes in experiments in which the synthesis of fibrinogen, α_1-globulin "M", transferrin and albumin were examined (Grieninger, unpublished).

The Effects of Other Agents

Serum

Unlike the chick embryo hepatocyte system, cultures of mammalian liver cells have a general requirement for serum which exposes these cells to a complex mixture of hormones, hormone-binding proteins, and factors as yet unidentified. Unless the medium is supplemented with serum (usually 5 to 15%), mammalian cells do not plate efficiently, sustain long-term viability, or retain high rates of plasma protein synthesis.[41, 42, 46-48] The complex and poorly defined nature of serum, as well as the potential for interaction between serum components and hormones, has complicated the interpretation of many hormone studies carried out in its presence.

The ability to initiate and maintain embryonic chick hepatocyte monolayers in completely defined medium, without added serum or hormones, has enabled the demonstration of a dramatic stimulatory effect of serum itself on the synthesis of a number of plasma proteins, including fibrinogen (TABLE 2). Low concentrations of serum are effective, e.g. 1% serum elicits half-maximal stimulation of fibrinogen synthesis (FIG. 8). Such a finding suggests that when serum is present in culture, even at low concentrations, it may elevate baseline synthesis of fibrinogen as well as other plasma proteins sufficiently to obscure the effects of added hormones. This is indeed the case for the stimulation of fibrinogen synthesis by dexamethasone (TABLE 2, lines 3, 9, and 10). In the absence of serum, there is a 180% increase in the synthesis of fibrinogen whereas, in the presence of serum, the increase is only 25%.

Fibrinogen Degradation Products

Fibrinogen degradation products have been reported to increase fibrinogen production in the intact animal.[18, 19] However, experiments directly comparing serum and plasma in chick embryo hepatocyte cultures suggest that a direct effect of the degradation products is unlikely; no significant differences in the stimulation of fibrinogen were observed,[49] even though serum should contain a higher concentration of these products than plasma. Ritchie *et al.*[50] confirmed and extended this result in rat hepatocyte monolayers by adding a set of defined fragments produced by plasmin treatment of homologous rat fibrinogen. No effect on fibrinogen synthesis was observed. However, when these fragments are incubated with human peripheral leukocytes, the leukocytes increase their production of factors that dramatically stimulate the synthesis of fibrinogen by rat hepatocyte monolayers. Thus, fibrinogen degradation products themselves appear capable of regulating fibrinogen synthesis only via an indirect leukocyte-mediated pathway.

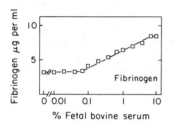

FIGURE 8. Effect of various concentrations of serum on the synthesis of fibrinogen. After 24 hr of culture, the cells were given fresh medium plus heparin, some dishes without serum and others containing fetal bovine serum at the concentrations indicated. Twenty-four hr later, culture medium was collected, and secreted fibrinogen was determined by electroimmunoassay. The values represent the mean of duplicate dishes. (From Plant *et al.*,[49] used with permission.)

Leukocytic Endogenous Mediator (LEM)

Because of the reported stimulatory effects of LEM on fibrinogen synthesis, both *in vivo* [16, 17] and *in vitro*,[48] we have tested several different heterologous preparations in embryonic chick hepatocyte cultures: crude LEM from rabbits (0.1 to 100 μl/ml culture medium), the supernatant fluid from stimulated mouse macrophage cultures (10 to 300 μl/ml), and partially purified interleukin-1 derived from human monocytes (0.7 to 5 μl/ml). No stimulatory effect of these preparations was observed on the synthesis of the two acute phase proteins, fibrinogen and α_1-globulin "M", either in the presence or absence of 1 nM dexamethasone. The effectiveness of a homologous LEM preparation remains to be investigated.

The Effects Produced by Combinations of Hormones

TABLE 2 summarizes the effects of T_3, dexamethasone, and insulin, alone and in various combinations. In the mixtures, each hormone is present at a concentration which is optimally effective for that hormone alone (see FIG. 3 and 7; see also ref. 24). The data indicate that certain stimulatory effects produced by the hormones individually are additive in the mixtures, whereas others are reduced.

When dexamethasone is combined with T_3, the stimulation of both fibrinogen and α_1-globulin "M" synthesis is greater than with either hormone alone (compare lines 2, 3, and 7). The additive effect of these hormones on α_1-globulin "M" synthesis—but not that of fibrinogen—is further enhanced by inclusion of insulin (lines 2, 3, 7, and 8).

Insulin, which has no effect on fibrinogen synthesis in the first 24 hours of exposure, does not alter the dexamethasone-induced stimulation of this protein (lines 3, 4, and 5). The stimulatory effects of insulin and dexamethasone on α_1-globulin "M" synthesis, on the other hand, are not additive, i.e. the nearly threefold increase in α_1-globulin "M" synthesis due to dexamethasone alone is obscured when this hormone is evaluated in the presence of insulin (lines 4 and 5). Thus, the specificity of the glucocorticoid effect for fibrinogen synthesis is most apparent in insulin-containing medium, a fact that has been noted previously [24]: dexamethasone, in the presence of insulin, produces only a slight stimulation in the synthesis of 3 minor proteins—out of the more than 20 plasma proteins monitored by crossed immunoelectrophoresis—in addition to its threefold increase of fibrinogen synthesis.

Curiously, when insulin is combined with T_3, the 2.5-fold stimulation of fibrinogen synthesis due to T_3 is reduced, despite the fact that synthesis of α_1-globulin "M" is enhanced to a greater degree than with either hormone alone (lines 2, 4, and 6).

It becomes apparent from these few examples that many of the cooperative effects (negative and positive) of hormones on the synthesis of fibrinogen and other plasma proteins are not only complex but selective. The mechanism(s) which confer the selectivity of hormone action on fibrinogen synthesis are the subject of intense investigation in our laboratory.

Further Studies

To deepen our understanding of the way in which hormones alter the production of fibrinogen, our laboratory has explored the following issues:

Cellular Localization of Fibrinogen Synthesis in Culture

The cellular distribution of fibrinogen synthesis in culture was evaluated by indirect immunofluorescence staining. This approach is possible in culture since, as in vivo, there is no intracellular storage of plasma proteins. Synthesis and secretion are continuous, and intracellular plasma protein levels correlate directly with the rate of secretion.[23] When the cultures are tested with anti-serum directed against fibrinogen, all the parenchymal cells fluoresce (FIG. 9). The fluorescence is distributed diffusely in the cytoplasm and is of similar density throughout the monolayer. (Since the cells originally attach to the surface of the culture dish in aggregates and only gradually spread out to form confluent monolayers, their spatial arrangement has some influence on fluorescence distribution. Cells at the center of aggregates are less flattened than those at the perimeter, hence they appear to fluoresce more brightly and their nuclei appear slightly stained.)

When fibrinogen synthesis in culture is stimulated by addition of serum, increased fluorescence is evidenced by all cells stained for fibrinogen.[51] Thus, increased fibrinogen synthesis appears to be due to a universal cellular event rather than to recruitment of previously nonsynthesizing cells. The hepatic parenchymal cells in culture apparently constitute a homogeneous population with respect not only to the synthesis of fibrinogen but also to the synthesis of several other plasma proteins.[51] It has been recently demonstrated using rat liver, that a similar homogeneity exists *in vivo*.[52, 53]

Fibrinogen Messenger RNA Levels during Stimulation by Hormones

To explore the mechanisms by which hormones control plasma protein synthesis, our laboratory has developed methods to determine the levels of

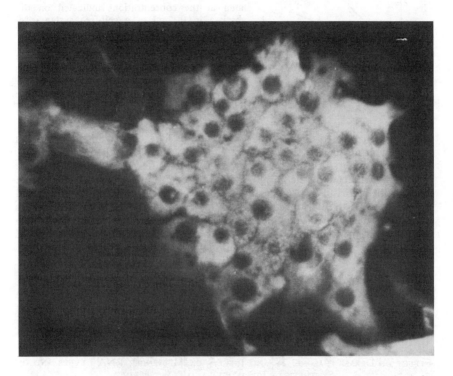

FIGURE 9. Immunofluorescence micrograph of cultured hepatocytes stained for fibrinogen. Cells were plated and maintained in medium supplemented with 10% fetal bovine serum. After 24 hr of culture, cells were fixed and fibrinogen was localized by indirect immunofluorescence staining as described by Kalb and Grieninger.[51] Magnification ×600.

mRNAs coding for several different plasma proteins in cultured hepatocytes.[28] In one case, a nearly threefold increase in the level of functional fibrinogen mRNA was found in response to hormones. Poly(A)-containing RNA was prepared from hepatocyte monolayers cultured in the absence or presence of a hormone mixture containing dexamethasone, T_3 and insulin. The RNA was then used to direct the cell-free synthesis of fibrinogen in a wheat germ translation system (FIG. 10). Translation of fibrinogen was linear with up to at least 8 μg of RNA from either the controls or the hormonally stimulated cells, with the latter reflecting a 170% increase in the amount of functional messenger

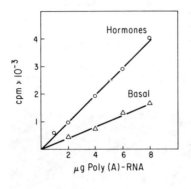

FIGURE 10. Fibrinogen synthesis in a cell-free translation system directed by RNA derived from cultured hepatocytes. Cells were plated and maintained in hormone-free medium. At 24 hr of culture, the medium was changed, with and without addition of a combination of hormones. After an additional 24 hr of incubation, poly(A)-containing RNA was prepared from both groups. The RNA was translated at the concentrations indicated on the abscissa in 0.1 ml of a cell-free system prepared from wheat germ and fibrinogen was immunoprecipitated and separated by SDS-polyacrylamide gel electrophoresis. [³⁵S]methionine incorporation into the fibrinogen polypeptides was quantitated and plotted on the ordinate. Basal—no addition; hormones—1 nM dexamethasone, 10 nM T₃, and 35 nM insulin.

RNA over the controls. Thus, the stimulation of fibrinogen synthesis correlates directly with an increased level of its functional messenger RNA (TABLE 4). In contrast, synthesis of transferrin and its functional messenger RNA level were only slightly affected by the hormones. It will require further study to determine whether the increase in fibrinogen messenger RNA level is the result of increased transcription of the fibrinogen genes in this case as well as when each of the hormones is examined individually. A transcriptional mechanism has been proposed for the stimulation of fibrinogen by glucocorticoids on the basis of kinetic studies and experiments with inhibitors of RNA synthesis.[24]

Intracellular Transit Time

Following its synthesis on the ribosomes, the fibrinogen molecule requires a finite time for intracellular transit along the secretory pathway. A series of

TABLE 4

EFFECT OF DEXAMETHASONE, T₃, AND INSULIN ON MESSENGER RNA LEVELS AND ON SYNTHESIS OF FIBRINOGEN AND TRANSFERRIN *

| | % Increase over Control | |
	Synthesis of Protein	Messenger RNA Level
Fibrinogen	160	170
Transferrin	30	20

* Cells were the same as described in FIG. 10. Synthesis was evaluated by determining with electroimmunoassay the amount of fibrinogen and transferrin secreted between the 24th and 48th hour of culture in the presence and absence of the hormones. The level of functional fibrinogen messenger RNA was determined by cell-free translation of poly(A)-containing RNA as described in FIG. 10. The increase due to the hormones was derived by comparing the slopes of the curves plotted in that figure and taking into account a 20% increase in total cellular RNA in the presence of the hormones. The level of functional transferrin messenger RNA was determined in a like fashion.

recent experiments indicates that hormones and other agents known to stimulate fibrinogen synthesis can, in addition, shorten this transit time. Under basal culture conditions, the average time required for synthesis, assembly, and secretion of the fibrinogen molecule amounts to about 2 hours (Grieninger and Plant, manuscript in preparation). This is considerably longer than the transit time for the population of plasma proteins as a whole; within 30 minutes of their synthesis, 80% of the plasma proteins are secreted.[23] While this 2-hour period is required for transit of fibrinogen in the absence of hormones and macromolecular supplement, the time can be reduced by a factor of 2 upon exposure of the cells to the simultaneous presence of dexamethasone, T_3, and insulin (as in TABLE 2) or by as much as a factor of 3 upon exposure to 2% serum. Thus, these agents appear to enhance not only the rate of fibrinogen synthesis but also the intracellular "maturation" of the fibrinogen molecule. The mechanisms underlying these phenomena are currently under investigation.§

SUMMARY AND CONCLUSIONS

Most of what was originally known of the effects of hormones on fibrinogen synthesis was based, as noted above, on experiments involving surgical removal of endocrine glands. Some caution should be exercised when using such *in vivo* experiments to derive the hormonal requirements of fibrinogen synthesis, however, since multiple hormonal alterations often occur in these animals. The development of a variety of *ex vivo* systems has allowed investigators to more carefully control the hepatocellular environment. The work of several laboratories, including our own, has now made it clear that hormones and other agents directly stimulate hepatocellular synthesis of fibrinogen. From the studies summarized here, using chick embryo hepatocytes as a model, several generalizations emerge:

• Fibrinogen synthesis may be considered to be a "constitutive" liver function, since hepatocytes cultured without serum, hormones or other macromolecular supplements synthesize this protein at a basal rate for several days.

• Addition of certain hormones (e.g. T_3, dexamethasone, insulin), individually and in physiological concentrations, elicits an increase in fibrinogen production, varying with each agent in onset, dose, minimum exposure required and accompanying effects on the synthesis of other plasma proteins. Glucocorticoids and thyroid hormones are similar in the selectivity of their stimulation (neither affects albumin or transferrin synthesis) but differ in that thyroid hormones need to be present for just a short "triggering" period. The stimulation of fibrinogen synthesis by insulin occurs only following prolonged exposure to concentrations 10-times higher than the very low doses to which albumin synthesis responds rapidly.

§ **Note added in proof:**
Another interpretation, not mentioned in this review, deserves serious consideration in light of more recent pulse-labeling experiments. It was determined that 30 minutes following synthesis fibrinogen molecules do begin to appear in the medium but that, in the absence of hormones, the exported fibrinogen represents only 20% of the molecules synthesized. The remainder is apparently degraded within the cell by a process selective for fibrinogen. Certain hormones and serum factors increase the proportion of secreted fibinogen molecules by interfering with this degradative process.

- Other hormones such as epinephrine, estrogens and ACTH, and agents, such as LEM, which have been suggested as capable of increasing fibrinogen production on the basis of whole animal studies, were found to be ineffective in our system when tested individually. These agents do not appear to have a direct effect on the hepatocyte, although it is possible that a second hormone or factor is required for stimulation ("permissive effect").

- Serum, frequently used as medium supplement in hepatocyte cultures, has been shown to have a dramatic stimulatory effect on fibrinogen synthesis. When serum is present in culture, it elevates baseline synthesis of fibrinogen sufficiently to obscure the effects of added hormones. For example, cultures of hepatocytes in serum-free medium evidence a generally greater sensitivity, as well as a greater response to glucocorticoids than that seen in serum-containing medium.[24, 41, 54, 55]

- Cooperative effects (positive and negative) of hormones on fibrinogen synthesis are selective, as revealed by using combinations of dexamethasone, T_3 and insulin.

- Stimulation of fibrinogen synthesis is a universal cellular event in culture, all cultured hepatocytes reacting uniformly to the addition of agents.

- The mechanisms underlying hormone-induced changes in fibrinogen synthesis can be studied at the molecular level in the controlled environment of the chick embryo hepatocyte culture. Messenger RNA levels for fibrinogen as well as other plasma proteins have been determined in the presence and absence of hormones. We have shown that, during stimulation with a combination of hormones (dexamethasone, T_3, insulin), increased fibrinogen synthesis correlates directly with an increased level of fibrinogen mRNA. The selectivity of hormone action at the level of plasma protein synthesis is reflected at the mRNA level, since biosynthesis of transferrin as well as the amount of its mRNA are only slightly affected by these hormones.

- The time required for synthesis, assembly and secretion of the fibrinogen molecule is considerably longer than that of other major plasma proteins. Preliminary experiments indicate that hormones and other agents known to stimulate fibrinogen production can in addition shorten the time required for intracellular transit along the secretory pathway (see note §).

Much remains to be learned about the mechanisms by which the body orchestrates control of fibrinogen synthesis by the liver cell. The liver *in situ* is subject to the simultaneous effect of a multitude of hormones and other agents, and the ratios of available hormones and nutrients fluctuate in both physiological and pathological states.

In the chick embryo hepatocyte system, the hepatocyte's environment has been simplified to the point where the response to single hormones can be evaluated and the direct role of these hormones assessed. These relatively "endocrine deficient" cultures have also served well as a base for the analysis of cooperative effects, that is, the way in which one hormone influences the effect of another. They have the advantage that, in culture experiments—unlike whole animal studies—large arrays of variables can be systematically manipulated. Eventually, by extending this approach to simulate fluctuations in hormone and nutrient ratios in culture, it may be possible to determine the critical balance of factors upon which the hormonal milieu *in vivo* is built, and so to understand the regulation of fibrinogen synthesis.

ACKNOWLEDGMENTS

We acknowledge the skilled technical assistance of Mary Ann Chiasson and Peter Golikov in the conduct of recent experimental work.

REFERENCES

1. GILLMAN, T., S. S. NAIDOO & M. HATHORN. 1959. Lancet **2:** 70–71.
2. REGOECZI, E. & K. R. HOBBS. 1969. Scand. J. Haematol. **6:** 175–178.
3. KOJ, A. 1974. *In* Structure and Function of Plasma Proteins. A. C. Allison, Ed. Vol. **1:** 73–131. Plenum Press. New York.
4. DRURY, D. R. & P. O. McMASTER. 19292. J. Exp. Med. **50:** 569–578.
5. MILLER, L. L., C. G. BLY & W. F. BALE. 1954. J. Exp. Med. **99:** 133–153.
6. STRAUB, P. W. 1963. J. Clin. Invest. **42:** 130–136.
7. BARNHART, M. I. & W. B. FORMAN. 1963. Vox Sang. **8:** 461–473.
8. JEEJEEBHOY, K. N., A. BRUCE-ROBERTSON, J. HO & U. SODTKE. 1972. Biochem. J. **130:** 533–538.
9. SELIGSOHN, U., S. I. RAPAPORT & A. ZIVELIN. 1973. Thromb. Diath. Haemorrh. **29:** 76–86.
10. GILLMAN, T., S. S. NAIDOO & M. HATHORN. 1958. Clin. Sci. **17:** 393–408.
11. SELIGSOHN, U., S. I. RAPAPORT & P. R. KUEFLER. 1973. Am. J. Physiol. **224:** 1172–1179.
12. CHEN, Y.-H. & E. B. REEVE. Thromb. Haemostas. **37:** 243–252.
13. JEEJEEBHOY, K. N., A. BRUCE-ROBERTSON, U. SODTKE & M. FOLEY. 1970. Biochem. J. **119:** 243–249.
14. CARLSON, T. H., D. C. FRADL, B. D. LEONARD, S. H. WENTLAND & E. B. REEVE. 1977. Am. J. Physiol. **233:** H1–H9.
15. STAFFORD, B. T., S. I. RAPAPORT & S. M.-C. SHEN. 1975. Thromb. Diath. Haemorrh. **34:** 159–168.
16. KAMPSCHMIDT, R. F. & H. F. UPCHURCH. 1974. Proc. Soc. Biol. Med. **146:** 904–907.
17. BORNSTEIN, D. L. & E. C. WALSH. 1978. J. Lab. Clin. Med. **91:** 236–245.
18. BARNHART, M. I., D. C. CRESS, S. M. NOONAN & R. T. WALSH. 1970. Thromb. Diath. Haemorrh. **39** (Suppl.): 143–159.
19. KESSLER, C. M. & W. R. BELL. 1980. Blood **55:** 40–47.
20. PINDYCK, J., M. W. MOSESSON, D. BANNERJEE & D. GALANAKIS. 1977. Biochim. Biophys. Acta **492:** 377–386.
21. MURANO, G., D. WALZ, L. WILLIAMS, J. PINDYCK & M. W. MOSESSON. 1977. Thromb. Res. **11:** 1–10.
22. GRIENINGER, G. 1982. *In* Plasma Protein Secretion by the Liver. H. Glaumann, T. Peters & C. Redman, Eds. Academic Press. New York. In press.
23. GRIENINGER, G. & S. GRANICK. 1978. J. Exp. Med. **147:** 1806–1823.
24. GRIENINGER, G., K. M. HERTZBERG & J. PINDYCK. 1978. Proc. Natl. Acad. Sci. USA **75:** 5506–5510.
25. LIANG, T. J. & G. GRIENINGER. 1981. Proc. Natl. Acad. Sci. USA **78:** 6972–6976.
26. GRIENINGER, G., J. PINDYCK, K. M. HERTZBERG & M. W. MOSESSON. 1979. Ann. Clin. Lab. Sci. **9:** 511–517.
27. GRIENINGER, G. & S. GRANICK. 1975. Proc. Natl. Acad. Sci. USA **72:** 5007–5011.
28. PLANT P. W., R. G. DEELEY & G. GRIENINGER. Manuscript submitted.
29. LAEMMLI, U. K. 1970. Nature **227:** 680–685.
30. DEELEY, R. G., J. I. GORDON, A. T. H. BURNS, K. P. MULLINIX, M. BINA-STEIN & R. F. GOLDBERGER. 1977. J. Biol. Chem. **252:** 8310–8319.

31. ROBERTS, B. E. & B. M. PATERSON. 1973. Proc. Natl. Acad. Sci. USA **70:** 2330–2334.
32. TSE, T. P. H. & J. M. TAYLOR. 1977. J. Biol. Chem. **252:** 1272–1278.
33. GOLDMAN, B. M. & G. BLOBEL. 1978. Proc. Natl. Acad. Sci. USA **75:** 5066–5070.
34. MAPES, C. A. & P. Z. SOBOCINSKI. 1977. Am. J. Physiol. **232:** C15–C22.
35. WOOD, D. D. 1979. J. Immunol. **123:** 2400–2407.
36. HERTZBERG, K. M., J. PINDYCK, M. W. MOSESSON & G. GRIENINGER. 1981. J. Biol. Chem. **256:** 563–566.
37. AMRANI, D., P. W. PLANT, J. PINDYCK, M. W. MOSESSON & G. GRIENINGER. 1983. Biochim. Biophys. Acta **743:** 394–400.
38. PINDYCK, J., G. BEUVING, K. M. HERTZBERG, T. J. LIANG, D. AMRANI & G. GRIENINGER. 1983. Ann. N.Y. Acad. Sci. **408:**. This volume.
39. GRIENINGER, G., K. M. HERTZBERG, T.-Y. LIANG & J. PINDYCK. 1979. *In* The Liver. Quantitative Aspects of Structure and Function. R. Preisig & J. Bircher, Eds. Vol. **3:** 118–125. Editio Cantor. Aulendorf. West Germany.
40. JOHN, D. W. & L. L. MILLER. 1969. J. Biol. Chem. **244:** 6134–6142.
41. CRANE, L. J. & D. L. MILLER. 1977. J. Cell Biol. **72:** 11–25.
42. JEEJEEBHOY, K. N., J. HO, G. R. GREENBERG, M. J. PHILLIPS, A. BRUCE-ROBERTSON & U. SODTKE. 1975. Biochem. J. **146:** 141–155.
43. PINDYCK, J., M. W. MOSESSON, M. W. ROOMI & R. D. LEVERE. 1975. Biochem. Med. **12:** 22–31.
44. ATENCIO, A. C., P.-Y. CHAO, A. Y. CHEN & E. B. REEVE. 1969. Am. J. Physiol. **216:** 773–780.
45. MILLER, L. L. & E. E. GRIFFIN. 1975. *In* Biochemical Actions of Hormones. G. Litwack, Ed. Vol. **3:** 159–186. Academic Press. New York. London.
46. SIRICA, A. E., W. RICHARDS, Y. TSUKADA, C. A. SATTLER & H. C. PITOT. 1979. Proc. Natl. Acad. Sci. USA **76:** 283–287.
47. FOUAD, F. M., R. SCHERER, M. ABD-EL-FATTAH & G. RUHENSTROTH-BAUER. 1980. Eur. J. Cell Biol. **21:** 175–179.
48. RUPP, R. G. & G. M. FULLER. 1979. Exp. Cell Res. **118:** 23–30.
49. PLANT, P. W., T. J. LIANG, J. PINDYCK & G. GRIENINGER. 1981. Biochim. Biophys. Acta **655:** 407–412.
50. RITCHIE, D. G., B. A. LEVY, M. A. ADAMS & G. M. FULLER. 1982. Proc. Natl. Acad. Sci. USA **79:** 1530–1534.
51. KALB, R. G. & G. GRIENINGER. 1979. Biochim. Biophys. Acta **563:** 518–526.
52. LEBOUTON, A. V. & J. P. MASSE. 1980. Anat. Rec. **197:** 183–194.
53. LEBOUTON, A. V. & J. P. MASSE. 1980. Anat. Rec. **197:** 195–203.
54. FOUAD, F. M., M. ABD-EL-FATTAH, R. SCHERER & G. RUHENSTROTH-BAUER. 1981. Z. Naturforsch. **36:** 350–352.
55. JEEJEEBHOY, K. N., J. HO, R. MEHRA, J. JEEJEEBHOY & A. BRUCE-ROBERTSON. 1977. Biochem. J. **168:** 347–352.

DISCUSSION OF THE PAPER

L. WANGH (*Brandeis University, Waltham, MA*): What percent contamination by non-parenchymal cells do you have in your parenchymal cell fraction?

G. GRIENINGER: I would say not more than 3 to 4 percent.

WANGH: Do you purify them through a density gradient?

GRIENINGER: We do not purify them. But other cells do not contribute to the synthesis of plasma proteins although they may contribute some other factors.

WANGH: Have you looked for increased DNA synthesis in the long-term insulin studies?

GRIENINGER: Yes, we did look at DNA synthesis, and we found no increase in either DNA or protein content. The cell mass is unchanged.

B. M. ALVING (*Walter Reed Army Institute, Washington, D.C.*): I am interested in the prolonged time it takes between label addition and the emergence of label in the medium. We found in our studies that if we give a whole animal [^{75}Se]methionine and then look at the plasma fibrinogen at time as early as 15, 20, 30, 45 minutes after injection we can extrapolate that the label would appear about 12 to 15 minutes after injection; this is true in both stimulated and nonstimulated states. Do you have any thoughts about this transit time?

GRIENINGER: I indeed have thoughts about it. When we measure the transit time with serum, which I think is the situation you would see in the intact animal, the average transit time is about 45 minutes. Peters has examined this question with albumin. The average transit time is about 30 minutes; he can see a few molecules after only 12 minutes, but these are the fast albumins. I think our studies are very similar. In a stimulated state, i.e., in the presence of serum, you have a transit time similar to the one observed *in vivo*. Under our endocrine-deficient conditions, we can see, maybe for the first time, that there is a longer transit time, which has not been seen before, because people have always worked with intrinsically stimulated liver cells.

G. D. QURESHI (*Richmond, VA*): Did you see similar stimulation with corticosterone or hydrocortisone in your cultures?

GRIENINGER: Yes, we reported previously that corticosterone has essentially the same effect as dexamethasone, as does hydrocortisone.

W. R. BELL (*Johns Hopkins Hospital, Baltimore, MD.*): What is your plating efficiency with the chicken hepatocytes? And what is the frequency of cell death per day?

GRIENINGER: Our plating efficiency is around 90%, about as high as is achieved in any culture system. Cell death, even under basal conditions, is less than 5% over a period of 4 days.

HEPATOCYTE-STIMULATING FACTOR: A MONOCYTE-DERIVED ACUTE-PHASE REGULATORY PROTEIN

David G. Ritchie and Gerald M. Fuller

Department of Human Biological Chemistry and Genetics
The University of Texas Medical School at Galveston
Galveston, Texas 77550

INTRODUCTION

Peripheral subcutaneous infections undergo a well-defined sequence of clinically noticeable changes known collectively as acute inflammation. For example, following infection, changes in vascular permeability give rise within hours to the rapid biphasic influx of polymorphonuclear leukocytes followed by monocytes into the injured site. Eight to 10 hours after the initial infection, serum levels of specific acute-phase responsive plasma proteins (i.e., α_1 antitrypsin, fibrinogen, α_2 macroglobulin, α_1 acid glycoprotein, C3) increase. While numerous studies reported during the past century have explored the roles of leukocytes in bacterial killing, little is known about hepatocyte stimulation during the acute phase response. The factor(s) responsible for this stimulation, their mechanism(s) of action, site(s) of origin, and the regulation of their own production following infection have, until recently, remained a most difficult and controversial subject for lack of a suitable experimental model system.

Studies initiated by Homburger,[1] using leukocytes obtained from a turpentine-induced abscess, demonstrated the ability of these leukocytes to secrete soluble factors, which, when injected back into dogs, caused an increase in serum fibrinogen levels. This same protocol, with minor changes, was closely followed by subsequent investigators during the 1960s and '70s. Taken together, these studies all demonstrated that symptoms mimicking those found during the acute inflammatory response (i.e., increased hepatic amino acid flux, increased hepatic protein synthesis, fever, and decreased serum iron and zinc levels) could be passively conferred upon experimental animals by factor(s) secreted from whole leukocyte preparations.[2-6] Thus, while these studies provided no definitive data to any of the aforementioned questions regarding hepatocyte stimulation, they succeeded in fostering the notion that leukocytes found at the site of an infection are capable of releasing modulating factors in addition to their bactericidal activity.

We have been particularly interested in identifying and characterizing those leukocyte-derived factors responsible for regulating the synthesis of hepatic acute-phase responsive plasma proteins. Initial studies in our laboratory demonstrated the feasibility of using cultured fetal hepatocytes as a bioassay for hepatocyte-stimulating factor (HSF). In these studies, the rate of fibrinogen secretion (as measured by an enzyme-linked immunosorbent assay, ELISA) was found to be dependent on HSF concentration.[7] More recently, we have developed a quantitative HSF bioassay (using cultured adult hepatocytes) that can be used to determine the specific activity (i.e., units/mg protein) of any

490

0077-8923/83/0408-0490 $01.75/0 © 1983, NYAS

HSF sample.[8] The present work describes our studies on the production, partial characterization and regulation of leukocyte HSF.

MATERIALS AND METHODS

Hepatocyte-Stimulating Factor Bioassay

Optimal conditions and rationale for the bioassay of HSF using cultured adult rat hepatocytes has been described in detail elsewhere.[8] Briefly, samples containing undetermined amounts of HSF are added at several dilutions to individual 35 mm culture dishes containing confluent hepatocyte monolayers. Twenty-four hours later, the media are removed and assayed for fibrinogen by an enzyme-linked immunosorbent assay. Absorbances obtained from these assays are converted directly into μg/ml of fibrinogen by a program written for a TI-59 programmable desk top calculator.[9] The percent maximal response is then calculated for each unknown sample and, from a standard dose-response curve, the concentration of HSF (units/ml) is determined.

Monocyte Culture

Details pertaining to the preparation and culture of human monocytes have been described elsewhere.[10] In brief, "buffy coats" obtained from healthy human donors were separated over Ficoll-Paque (Pharmacia). The monocyte-lymphocyte fraction found at the gradient interface was further fractionated by plating into 60 mm dishes with McCoy's 5A medium containing 10% fetal bovine serum. After one hour the nonadherent lymphocytes were aspirated and fresh serum-free McCoy's 5a medium added. These cultures were 95 to 98% lymphocyte free as judged by nonspecific esterase staining.[11]

Measurement of ^{3}H-Fibrinogen Synthesis

Hepatocytes cultured for 18 hours in 60 mm tissue culture dishes, were washed 3 times with phosphate-buffered saline (PBS). Fresh serum-free Williams Essential medium containing leucine (57 μM) was then added. At the appropriate 2-hour intervals ^{3}H-leucine (50 μCi/dish, 50Ci/mmol) was added for 30 minutes. The incorporation of ^{3}H-leucine into newly synthesized intracellular fibrinogen was linear for up to 40 minutes. After each 30-minute pulse, the dishes were washed three times again with PBS. A solubilization buffer containing 5 mM leucine, 1 mM EDTA, 25 mM NaCl, 1% Triton X-100, 0.5% sodium deoxycholate, 0.1% sodium lauryl sulfate in 10 mM sodium phosphate pH 7.4, was then added (1.5 ml) to each dish. The cells were scraped from the surface of the dishes, then triturated with a Pasteur pipet to help solubilize the cells which were then transferred to centrifuge tubes. At the conclusion of the experiment, the samples were all centrifuged at 18,000 rpm for 60 minutes. One ml of supernatant from each sample was removed and added to microfuge (2 ml) tubes containing 0.05 ml of Sepharose-coupled goat anti-rat fibrinogen (GARF-sepharose, 12). The GARF-Sepharose containing samples were then gently shaken overnight at 4° C to assure complete binding of the ³H-fibrinogen. The GARF-Sepharose bound ³H-fibrinogen was pelleted by a brief 30-second centrifugation (10,000 rpm). After aspirating the supernatants, the Sepharose pellets were washed 5 times with 1.5 mls of a borate-saline pH 8.6 buffer to remove nonspecifically bound radioactivity. The washed GARF-Sepharose containing ³H-fibrinogen was then resusupended in borate-saline and duplicate 0.5 ml aliquots from each sample were counted in aquasol using a liquid scintillation spectrometer. Data were calculated as cpm ³H-fibrinogen/mg protein/30 min.

Purification of Plasminolytic Fragments D and E from Fibrinogen and Fibrin

Fibrinogen (Fragments D and E) and fibrin (Fragments D' and E') were prepared as described previously.[12]

RESULTS

Hepatocyte-Stimulating Factor Mediated Stimulation of Hepatocyte Plasma Protein Synthesis and Secretion

The addition of HSF to cultured adult rat hepatocytes results in increased synthesis and secretion of a number of well-known, acute-phase-responsive plasma proteins.[8, 13] FIGURE 1a demonstrates the effect of HSF on fibrinogen secretion. It is apparent that, following a 4-hour lag period, the rate of fibrinogen secretion from HSF-treated cells increases dramatically during the next 16 hours as compared to untreated cells. We have recently determined the time course for HSF-mediated changes in the rates of fibrinogen synthesis and secretion. Preliminary studies indicated that the incorporation of ³H-leucine into newly synthesized fibrinogen was linear for up to 40 minutes. For this study a 30-minute pulse of ³H-leucine was administered at each time point in order to measure the rate of fibrinogen synthesis (see MATERIALS AND METHODS). Not unexpectedly, the rate of fibrinogen synthesis is 240% higher than control after 4 hours (FIG. 1b) and the lag period is reduced to less than 2 hours, suggesting that HSF must affect fibrinogen synthesis within 1–2 hours after coming into contact with hepatocytes.

Regulation of Monocyte Hepatocyte-Stimulating Factor Production

Previous studies from our laboratory have demonstrated that HSF is derived from peripheral blood monocytes.[10] Highly enriched monocyte cultures, maintained in serum-free McCoy's 5A medium, can synthesize and secrete about 220 units of HSF/ml during the first 24 hours in culture. HSF secretion is also completely inhibited by cycloheximide (100 μM).

To better understand the role played by HSF during inflammation, we first chose to study the effects of specific plasminolytic fragments of fibrin and fibrinogen on HSF production. Plasmin is known to activate both the classical and alternate (via C3) complement pathways and the kinin system as well as fibrinolysis during acute inflammation. When fibrin (or fibrinogen) is incubated in the presence of plasmin, a series of specific enzymatic cleavages occurs which ultimately results in the degradation of the parent molecule to one molecule termed E fragment (M_r 45,000) that comprises the central NH_2-terminal domain and two molecules of D fragment (M_r 80 to 100,000) that comprise the COOH-terminal portions of the parent molecule (see FIG. 2). When purified D or E fragments from either fibrinogen or fibrin were added directly to cultured hepatocytes, no increase in the rate of fibrinogen secretion was observed. Intact fibrinogen or fibrin also had no effect. However, the

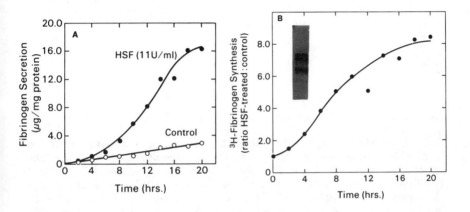

FIGURE 1. A. The effect of HSF on hepatocyte fibrinogen secretion. Hepatocytes were cultured for 18 hr in 35 mm tissue culture dishes. The media were removed and fresh medium, either with (●) or without (○) HSF, was added. Media from duplicate dishes were then removed at each time point and fibrinogen measured by ELISA. The remaining cells were washed twice with PBS and dissolved in 0.2 N NaOH for protein analysis.

B. The effect of HSF on hepatocyte fibrinogen synthesis. Hepatocytes were cultured for 18 hr in 60 mm tissue culture dishes. The media were removed and fresh medium either with or without HSF was added (see MATERIALS AND METHODS). At each time point 50 μCi of ^{3}H-leucine was added. After 30 min, the cells were washed then solubilized in a detergent buffer to release the labeled intracellular proteins. ^{3}H-fibrinogen was removed using GARF-Sepharose, and its radioactivity was counted in a liquid scintillation spectrometer. The results shown are mean values obtained from two experiments. *Inset:* Hepatocytes, labeled for 10 min with ^{35}S-methionine then chased for 30 min with cold methionine, were washed, lysed and treated as described above. The resulting radioactivity was electrophoresed on a SDS polyacrylamide gel. Major bands comigrated with authentic fibrinogen Aα, Bβ and γ chains.

addition of either of these fragments to leukocyte suspensions resulted in a dose-dependent (between 1.0 and 10 μM) increase in the rate of HSF production. Again, intact fibrinogen had no effect on monocyte HSF production suggesting that plasmin digestion may have opened up previously "hidden" regulatory sites within the parent molecule. On the basis of these data, we have drawn an indirect feedback pathway for the regulation of fibrinogen synthesis during acute inflammation (see FIG. 3). In this scheme, the production of plas-

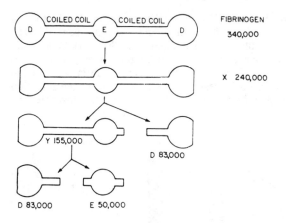

FIGURE 2. Schematic representation of the plasminolytic cleavage of fibrinogen. Exhaustive treatment with plasmin results in the formation of 2 D fragments and 1 E fragment per mole of fibrinogen.

minogen activator by endothelial cells and macrophages at the site of injury results in the production of plasmin which degrades fibrin and thereby forms the D and E fragments (fdp). With localized injury, the fdp's would be produced in the vicinity of numerous macrophages already present at the wound site. The presence of fragments D and E at the site of injury as well as their presence (with time) in the peripheral circulation will lead to stimulation of macrophage (and most likely Kupffer cell) HSF production. As blood levels of HSF increase (as little as one unit/ml can cause a four-to-fivefold increase in hepatocyte fibrinogen secretion) hepatic secretion of select plasma proteins, including fibrinogen, increases. We believe that this indirect pathway, involving FDP/fdp intereacting with and stimulating monocyte (and possibly Kupffer

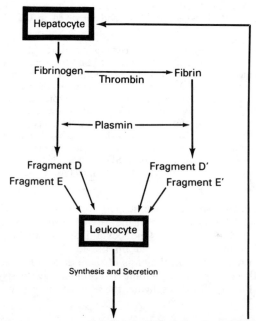

FIGURE 3. A suggested indirect feedback pathway for fibrinogen and other plasma protein synthesis. The plasmin-derived fragments of fibrinogen or fibrin stimulate the production of monocyte HSF, which is transported via the blood to the liver where it signals hepatocytes to increase their synthesis and secretion of selected plasma proteins. Kupffer cells also make HSF; however their specific response to fibrinogen degradation products has not as yet been tested.

cell) HSF synthesis and secretion, is of major importance in signaling liver hepatocytes to increase fibrinogen synthesis following tissue injury.[12]

Mononuclear phagocytes contribute much to infection and inflammation by their antimicrobial activities. Resistance to certain microorganisms, including mycobacteria, lysteria and some viruses and protozoa, is dependent upon the "activation" of macrophages by lymphocytes.[14] Certain properties of activated macrophages (i.e., enhanced spreading on glass or plastic, increased activity of the hexose monophosphate shunt, and increased release of superoxide anion (O_2) and hydrogen peroxide (H_2O_2)) can be found in macrophages that have also been pre-treated with thioglycollate, endotoxin (bacterial lipopolysac-

TABLE 1

EFFECTS OF MONOCYTE (ACTIVATION) AGENTS ON
HEPATOCYTE-STIMULATING FACTOR PRODUCTION *

Cell Type	Substance Added	Concentration	HSF Secretion (units/dish)	% Control
Experiment 1 Human monocyte	No addition	—	79.9 ± 15.5	100
	Lipopolysaccharide (LPS)	1.0 µg/ml	134.0 ± 11	167
	Phorbol myristate acetate (PMA)	1.0 µg/ml	77.5 ± 33	97
Experiment 2 Human monocyte	No addition	—	16.1 ± 2.8	100
	LPS plus	1.0 µg/ml	26.1 ± 3.2	162
	PMA	1.0 µg/ml		
Experiment 3 Human monocyte	No addition	—	84.4 ± 6.3	100
	Fragment E	5.0 µM	153.0 ± 12.9	181

* LPS, PMA, Fragment E had no effect on the fibrinogen product when incubated directly with hepatocyte monolayers. *Experiment 1*—Monocytes incubated in 60 mm dishes. *Experiments 2 and 3*—Monocytes incubated in 35 mm dishes. Cell incubations were carried out at 37° C in 5% CO_2/95% air for 6 hr.

charides, LPS) or Bacille Calmette-Guerin (BCG).[15] Moreover, these primed macrophages can be further activated or triggered by such surface-acting agents as phorbal myristate acetate (PMA), a fatty acid ester of a polyfunctional diterpene alcohol.[16] We have begun to study the relationship between monocyte activation and HSF production by using LPS and PMA as activating and triggering agents respectively. Neither agent had any direct effect of hepatocyte fibrinogen production. However, as shown in TABLE 1, LPS (1.0 µg/ml), but not PMA (1.0 µg/ml), stimulated monocyte HSF production. The addition of both agents simultaneously resulted in the same stimulation (i.e., 162%) as LPS alone. Fragment E, obtained by treating fibrinogen with plasmin, was also effective in stimulating monocyte HSF production as expected.

Hepatocyte-Stimulating Factor: Structure-Activity Relationship

We have attempted to gain some insight into the structure of HSF by observing the activity profile of crude HSF from Sephadex G-100 and G-75. In this way, we hoped to answer a number of fundamental questions regarding HSF-hepatocyte activation including: (1) How many different active hepatocyte stimulating factors exist? (2) What is the approximate molecular weight of HSF? (3) Is freshly secreted monocyte HSF further degraded by polymorphonuclear leukocyte proteases? (4) Does HSF contain disulfide bonds or free sulfhydryls that are important to its structure and/or activity?

One-thousand units of monocyte HSF containing several protein molecular markers was pumped into a reverse-flow Sephadex G-100 column. Two-hundred microliters of each 10 ml fraction was removed and assayed for HSF activity. Only one biologically active peak (see FIG. 4), which co-eluted with chymotrypsinogen A (MW 24,500), was found. When a similar quantity of HSF was preincubated for 3 hours at 37° C with leukocytes (10^8/ml), then chromatographed, an identical peak was obtained, suggesting that HSF is not further modified by neutrophil secreted enzymes. When crude HSF was treated with 2 mM N-ethyl maleimide (NEM) to block free-SH groups (a procedure that did not affect its biological activity) then chromatographed on Sephadex G-75, again only one biologically active peak (M_r 30,000) was found (see FIG. 4). Finally, treatment of HSF with dithiothreitol (a strong disulfide reducing agent) followed by NEM again failed to alter HSF biological activity. Chromatography of this sample on G-75 also resulted in one peak (M_r 30,000, see FIG. 4).

Relationship between Hepatocyte-Stimulating Factor and Interleukin-1

Currently in addition to HSF, there are a number of monocyte-derived "hormone-like" factors being investigated. Among the more prominent of these are Interleukin (IL-1[17]), serum amyloid A–stimulating factor (SAASF[18]),

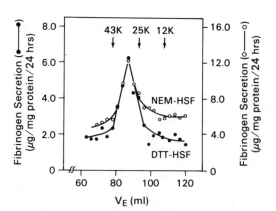

FIGURE 4. Effect of reducing and blocking agents on HSF G–75 elution profiles. Two thousand units of HSF (2 ml) was treated with dithiothreitol (DTT 50 mM) dialyzed against 0.5 mM DTT then blocked with NEM (2 mM). Another sample was treated with NEM only. Each sample was chromatographed separately on a calibrated Sephadex G–75 column (0.9 × 90 cm) equilibrated with PBS. Three ml fractions were collected and 0.1 ml from each fraction was assayed for hepatocyte-simulating activity. Molecular weight markers used were Ovalbumin (M_r 43,000); Chymotrypsinogen A (M_r 24,500); and Cytochrome c (M_r 11,700).

TABLE 2

COMPARISON BETWEEN HEPATOCYTE-STIMULATING FACTOR AND INTERLEUKIN–1*

Parameter	Response	
	HSF	IL–1
Monocyte derived	Yes	Yes [17]
Kupffer derived	Yes	Yes [17]
P–388D$_1$ derived	No	Yes [20]
PMA stimulates	No	No [20]
Cycloheximide inhibits	Yes	No [20]; Yes [27]
LPS stimulates	Yes	Yes [27]
Forms large MW complexes	No	Yes
Molecular weight	25–30,000	12–16,000 [17]
Isoelectric point	5.1–5.3	5.0–5.4 [20]
		7.0 [27]
Stable at 56° C	Yes	Yes [17]
Stimulates hepatic acute-phase proteins	Yes (fibrinogen)	Yes (SAA;[21]) No (fibrinogen †)

* Various parameters relating to IL–1 were obtained from the referenced literature and compared to our own data on HSF.

† IEF-purified IL–1 was supplied to us by Dr. L. Lachman.

and endogenous pyrogen (EP[19]). Because of the extreme difficulties in obtaining large quantities of any of these factors, only IL-1 has thus far been purified sufficiently for amino acid compositional analysis.[20] For the most part, questions pertaining to structural similarities between these three factors have relied heavily on the inherent biological activities (both *in vivo* and *in vitro*) of crude or only partially pure preparations. Thus the demonstration that (1) IL-1 and EP can induce SAA,[21] (2) that all three factors are monocyte derived, (3) that LPS stimulates their production, and (4) that they all have a molecular weight (as determined by Sephadex chromatography) of 12–16,000, has lead to the suggestion that IL-1 may be capable of modulating the activities of hepatocytes and hypothalamic cells as well as thymocytes.[22] Our results, however, suggest otherwise (see FIG. 4 and TABLE 2). Table 2 compares the properties of HSF and IL-1. While many similarities exist, several significant differences are also apparent. Most important is the difference in molecular weight. We have chromatographed HSF (obtained from human monocytes and rat Kupffer cells) on Sephadex G-100 and G-75 and never have we obtained fibrinogen stimulating activity in the 12–16,000 molecular weight range. Secondly, we have been unable to detect HSF activity in conditioned medium from P-388D$_1$ cells. Finally, we were unable to demonstrate any dose-dependent fibrinogen stimulation when IEF purified IL-1 was added to our cultured hepatocyte system (see TABLE 2).

DISCUSSION

We have attempted to answer a number of fundamental questions relating to leukocyte-mediated hepatocyte stimulation during periods of acute infection and inflammation. Our results have been faciliated by the development of tissue culture systems with which cells responsive to HSF (i.e., hepatocytes) and cells producing HSF (i.e., monocytes and Kupffer cells) can be manipulated under well-controlled conditions. For example, the preparations and culture of enriched monocyte cultures has not only eliminated the problem of cell aggregation and lysis that occurred when incubating rabbit peritoneal leukocytes in PBS (a procedure still in use by some investigators preparing LEM) but, by increasing the number of cells/ml that are actually involved in HSF production as well as the time of incubation (from 3 to 48 hours), large quantities of high activity (450 units/ml) HSF can now be produced on a routine basis. Similarly, our hepatocyte bioassay has enabled us to study, selectively, only those monocyte-derived factors capable of stimulating hepatocyte plasma protein biosynthesis.

The mechanisms by which hepatocyte acute phase proteins are regulated by HSF are unknown. We have begun to approach this problem by following the effect of HSF on fibrinogen synthesis and secretion (FIG. 1, a and b). Our studies demonstrate that the HSF-mediated stimulation of fibrinogen biosynthesis is fairly rapid (less than 2 hours). The requirement for glucocorticoids [23] as well as the well known rapid effects of glucocorticoids on $\alpha 2_u$ globulin synthesis,[24] further suggest that the hepatocyte response to HSF is modulated by intracellular glucocorticoid levels. Nevertheless, to better understand the primary signal responsible for HSF-mediated hepatocyte stimulation we must first determine whether hepatocytes contain a specific HSF receptor and, if so, whether the expression of this receptor can be modulated by glucocorticoids. We also do not know whether HSF can be rapidly internalized by hepatocytes or whether HSF can cause membrane modifications (i.e., methylation or phosphorylation) or rapid divalent cation (i.e., Ca^{2+}) movements by binding to is putative membrane receptor.

Plasmin plays a central role in acute inflammation by regulating blood clot (fibrin) dissolution, complement activation, and the kinin system. The production and secretion of plasminogen activator by activated monocytes [25] results in the accumulation of plasminolytic fragments of fibrin, fibrinogen, complement (i.e., C3a) and bradykinin in the vicinity of these monocytes. Our results demonstrating the ability of plasminolytic fragments of both fibrinogen and fibrin, but not the parent molecules themselves, to regulate monocyte HSF production in a dose-dependent manner has provided one possible means by which HSF production can be increased or decreased as needed (see FIG. 3, and ref. 12). While this regulatory mechanism is not unique,[26] it suggests that plasminolytic fragments of other proteins may also be capable of stimulating monocyte HSF production.

Direct chemical comparisons between HSF and other monocyte-derived factors such as IL-1, SAASF, and EP have met with limited success owing to the relatively small quantities of generally impure material available for sharing between investigators. Our studies on the chemistry of HSF have revealed that (1) HSF activity elutes as a single peak (M_r 25–30,000) from Sephadex G-100 or G-75. No high molecular weight (85,000) or low molecular weight (12–16,000) forms of HSF exist as is true for IL-1,[27] (2) newly secreted HSF cannot be further degraded by incubation with neutrophils, and (3) treatment with disulfide reducing and -SH blocking agents does not interfere with biological activity. These observations suggest that HSF is a single entity whose biological activity does not require free -SH groups and whose size (M_r 25–30,000) is significantly different from EP or IL-1. We are currently purifying HSF in order to obtain sufficient quantities for amino acid composition analysis of the molecule, which can then be compared with that already published for IL-1.[20]

In this report we have pointed out and answered a number of fundamental questions relating to the regulation and action of HSF. Nevertheless we are only beginning to understand the complex relationships existing between monocytes, inflammatory mediators, and liver function.

REFERENCES

1. HOMBURGER, F. 1945. A plasma fibrinogen-increasing factor obtained from sterile abesses in dogs. J. Clin. Invest. **24:** 43–45.
2. KAISER, H. K. & W. B. WOOD. 1962. Studies on the pathogenesis of fever. IX. The production of endogenous pyrogen by polymorphonuclear leukocytes. J. Exp. Med. **115:** 27–36.
3. KAMPSCHMIDT, R. F. & H. UPCHURCH. 1969. Lowering of plasma iron concentration in the rat with leukocytic extracts. Am. J. Physiol. **216:** 1287–1291.
4. KAMPSCHMIDT, R. F. & H. F. UPCHURCH. 1970. The effect of endogenous pyrogen on the plasma zinc concentration of the rat. Proc. Soc. Exp. Biol. Med. **134:** 1150–1152.
5. PEKAREK, R. S., M. C. POWANDA & R. W. WANNEMACHER. 1972. The effect of leukocytic endogenous mediator (LEM) on serum copper and ceruloplasmin concentration in the rat. Proc. Soc. Exp. Biol. Med. **141:** 1029–1031.
6. PEKAREK, R. S. & W. R. BEISEL. 1971. Characterization of the endogenous mediator(s) of serum zinc and iron depression during infection and other stresses. Proc. Soc. Exp. Biol. Med. **138:** 728–731.
7. RUPP, R. G. & G. M. FULLER. 1979. The effects of leukocytic and serum factors on fibrinogen biosynthesis in cultured hepatocytes. Exp. Cell. Res. **118:** 23–30.
8. RITCHIE, D. G. & G. M. FULLER. 1981. An *in vitro* bioassay for leukocytic endogenous mediator(s) using cultured rat hepatocytes. Inflammation **5:** 287–299.
9. RITCHIE, D. G., J. M. NICKERSON & G. M. FULLER. 1981. Two simple programs for the analysis of data from enzyme-linked immunosorbent assays (ELISA) on a programmable desk top calculator. Anal. Biochem. **110:** 281–290.
10. FULLER, G. M. & D. G. RITCHIE. 1982. A regulatory pathway for fibrinogen biosynthesis involving an indirect feedback loop. Ann. N.Y. Acad. Sci. **389:** 308–322.
11. LI, E. Y., K. W. LAM & L. T. YAM. 1973. Esterases in human leukocytes. J. Histochem. and Cytochem. **21:** 1–12.
12. RITCHIE, D. G., B. A. LEVY, M. A. ADMAS & G. M. FULLER. 1982. Regulation of fibrinogen synthesis by plasmin-derived fragments of fibrinogen and fibrin: An indirect feedback pathway. Proc. Natl. Acad. Sci. USA **79:** 1530–1534.
13. FOUAD, F. M., R. SCHERER, M. ABD-EL-FATTAH & G. RUHENSTROTH-BAUER. 1980. Biosynthesis of plasma proteins in serum-free medium by primary monolayer culture of rat hepatocytes. Eur. J. Cell Biol. **21:** 175–179.
14. MACKANESS, G. B. 1980. The mechanism of macrophage activation. In Infectious Agents and Host Reactions. S. Mudd, Ed.: 61. W. B. Saunders Company. Philadelphia.
15. NATHAN, C. F. & R. K. ROOT. 1977. Hydrogen peroxide release from mouse peritoneal macrophages. J. Exp. Med. **146:** 1648–1662.

16. JOHNSTON, R. B., JR., C. A. GODZIK & Z. A. COHN. 1978. Increased super-oxide anion production by immunologically activated and chemically elicited macrophages. J. Exp. Med. **148:** 115–127.

17. OPPENHEIM, J. J., S. B. MIZEL & M. S. MELTZER. 1979. In Biology of the Lymphokines. S. Cohen, E. Pick & J. J. Oppenheim, Eds.: 281–323. Academic Press. New York.

18. SIPE, J. D., S. N. VOGEL, J. L. RYAN, K. P. W. J. McADAM & D. L. ROSEN-STREICH. 1979. Detection of a mediator derived from endotoxin stimulated macrophages that induces the acute-phase serum amyloid A response in mice. J. Exp. Med. **150:** 597–606.

19. DINARELLO, C. A., N. P. GOLDIN & S. M. WOLFE. 1974. Demonstration and characterization of two distinct human leukocytic pyrogens. J. Exp. Med. **139:** 1369–1381.

20. MIZEL, S. B. & D. MIZEL. 1981. Purification to apparent homogeneity of murine interleukin-1. J. Immunol. **126:** 834–837.

21. SZTEIN, M. B., S. N. VOGEL, J. D. SIPE, P. A. MURPHY, S. B. MIZEL, J. J. OPPENHEIM & D. L. ROSENSTREICH. 1981. The role of macrophages in the acute-phase response: SAA inducer is closely related to lymphocyte activating factor and endogenous pyrogen. Cellular Immunol. **63:** 164–176.

22. OPPENHEIM, J. J. & I. GERY. 1982. Interleukin-1 is more than an interleukin. Immunol. Today **3:** 113–119.

23. THOMPSON, W. L., F. B. ABELES, F. A. BEALL, R. E. DINTERMAN & R. W. WANNEMACHER. 1976. Influence of the adrenal glucocorticoids on the stimulation of synthesis of hepatic ribonucleic acid and plasma acute-phase globulins by leukocytic endogenous mediator. Biochem. J. **156:** 25–32.

24. CHEN, C. C.-L. & P. FEIGELSON. 1978. Glucocortocid induction of α_{2u} globulin protein synthesis and its mRNA in rat hepatocytes in vitro. J. Biol. Chem. **253:** 7880–7885.

25. GORDON, S., J. C. UNKELESS & Z. A. COHN. 1974. Induction of macrophage plasminogen activator by endotoxin stimulation and phagocytosis. J. Exp. Med. **140:** 995–1010.

26. GRAVES, C. B., T. W. MUNNS, T. L. CARLISLE, G. A. GRANT & A. W. STRAUSS. 1981. Induction of prothrombin synthesis by prothrombin fragments. Proc. Natl. Acad. Sci. USA **78:** 4772–4776.

27. LACHMAN, L. B., S. O. PAGE & R. S. METZGAR. 1980. Purification of human lymphocyte activating factor—Interleukin-1. J. Supramol. Struct. **13:** 457–466.

DISCUSSION OF THE PAPER

L. LORAND (*Northwestern University, Evanston, IL*): There is evidence that when fibrinogen goes up other factors go up as well. Since you showed the differential stimulation of fibrinogen, I was wondering if you looked at plasminogen or other protein.

D. RITCHIE: We have examined plasminogen, and it seems that almost nothing stimulates plasminogen production.

LORAND: Any of the clotting factors?

RITCHIE: None of those that we have studied.

T. EDGINGTON (*Research Institute of Scripps Clinic, La Jolla, CA*): The monocyte, particularly under your conditions, can be stimulated to produce a number of arachidonic acid derivatives, particularly PGE_2, which are well known to influence other cell functions. Have you tried any experiments using indomethacin, or have you tried to use PGE_2 or some of the other prostaglandin to see whether this might be the pathway?

Secondly, do you have any idea whether some of the proteases that are released from the monocyte or macrophage upon interaction might be your factor?

RITCHIE: With respect to the first question, we have tried several experiments using indomethacin and aspirin, looking at the effect both on hepatocytes and on the hepatocyte response to the factor. In fact, we do get some decrease in the responsiveness in the presence of both indomethacin and aspirin. But we have not looked at what either of these drugs does as far as the monocyte production of this factor.

EDGINGTON: How about the proteases?

RITCHIE: As far as the proteases are concerned, we know that macrophages are activated during acute phase response and under activating conditions they do secrete a number of different plasminogen activators; there are at least three that I am aware of. At any rate this could be another feedback mechanism for for stimulating hepatocyte-stimulating factor production.

EDGINGTON: The plasminogen activator seems to be predominantly of a urokinase class, although there are a number of different molecular weights sizes. Do you know whether any further active degradation products are formed? I mean you could simply incubate them with the monocytes to see whether in fact the D or E fragment would generate a peptide or something that would in fact represent your factor.

RITCHIE: As I mentioned, that is one area that we would like to look into.

L. SHERMAN (*Washington University School of Medicine, St. Louis, MO.*): Could you tell me exactly what the concentration is of degradation products that result in the stimulatory release? Were those fibrinogen or fibrin degradation products?

RITCHIE: We have used both of them and both seem to be equally active.

SHERMAN: That is interesting, because both Dr. Bell and we have found that while fibrinogen products in fairly high amounts are stimulatory in the whole animal, fibrin products are not.

RITCHIE: We have prepared our fibrin products by first incubating the fibrinogen with thrombin and then going about degrading the fibrin. That may have something to do with it.

D. GALANAKIS (*State University of New York, Stony Brook*): Did you evaluate crosslinked fibrin digests?

RITCHIE: No.

A. BUDZYNSKI (*Temple University Health Sciences Center, Philadelphia, PA.*): Since high concentrations of degradation products are required for the reaction maybe it is not physiologic.

RITCHIE: Let me say that fibrinogen is present in grams per milliliter and the quantities I am working are present in terms of several hundred micrograms per milliliter. I do not know what the concentration of fragments are *in vivo.*

UNIDENTIFIED SPEAKER: In your system the monocyte-induced hepatocyte-stimulating factor stimulates fibrinogen synthesis, but what happens to the other proteins?

RITCHIE: We demonstrated that haptoglobin was certainly stimulated, albumin was decreased and some other unidentifiable proteins were also increased, so the factor does stimulate more than just fibrinogen.

P.J. GAFFNEY (*National Institute for Biological Standards and Control, London*): Since most people would agree that the presence of fragments D and E *in vivo* is quite rare, I am surprised you have not looked at things like fragments X or Y

RITCHIE: We were very aware of all the studies on the D and E fragments, and we felt that we had a system that could now approach the D and E problem directly, so we decided to start with that particular aspect.

THE BIOSYNTHESIS OF FIBRINOGEN:
IN VIVO STUDIES *

William R. Bell

Department of Medicine, Division of Hematology
The Johns Hopkins University School of Medicine
Baltimore, Maryland 21205

REGULATION OF FIBRINOGEN BIOSYNTHESIS

Fibrinogen, the thrombin coagulable glycoprotein circulating in the blood and its polymer fibrin, the essence of all thrombi, was first described and designated as such in 1861 by Schmidt.[1,2] At that time the origin of this glycoprotein was unknown. The studies by Drury and McMasters in 1929[3] were the first to suggest that this unique substance was produced by the liver. Employing ^{35}S-labeled methionine in intact normal *versus* hepatectomized dogs it was concluded that the liver was the only site where fibrinogen was produced.[4] This conclusion was further supported by studies using the isolated perfused rat liver.[5] Although these studies established the liver to be the organ system where fibrinogen was produced, the cellular type where this protein was synthesized remained unresolved. Approximately 60% of the liver cell population are parenchymal cells (hepatocytes), about 30% are Kupffer cells and the remaining 5–10% consist of fibroblasts, bile duct, lymphatic duct and vascular cells. Studies by Gitlin, Landing and Whipple,[6] which attemtped to identify the responsible cell, resulted only in the demonstration of extracellular fibrin. Histochemical studies of human hepatic biopsy specimens revealed that 1% of parenchymal cells and 70% of Kupffer cells contained fibrinogen.[7] These studies did not enable one to decide whether the described cellular localizations represented synthesis, storage, or accumulation of fibrinogen-fibrin breakdown products. Employing specific fluorescent antifibrinogen, Barnhart and Forman[8,9] established that the hepatic parenchymal cells (hepatocytes) were the site of fibrinogen biosynthesis.

Since then several studies have been performed to identify the mechanism(s) involved in regulation of the synthesis of this glycoprotein. Because of considerable confusion and conflicting reports, we performed studies in a single animal model to examine possible mechanisms regulatory for the synthesis of fibrinogen. In these studies, homology within the rabbit model of all biologic substances was rigidly maintained to eliminate the possibility of any nonspecific interspecies reaction.

METHODS

Animals

Healthy male New Zealand white rabbits, weighing between 2.0 and 2.9 kg were caged individually for at least 3 days before use. Baseline fibrinogen

* Supported in part by Research Grants HL–0714303, HL–06188, HL–24898 from the National Heart, Lung and Blood Institute of the National Institutes of Health and The Whitehall Foundation. William R. Bell is a Hubert E. and Ann E. Rogers Scholar in Academic Medicine.

0077–8923/83/0408–0503 $01.75/0 © 1983, NYAS

concentrations and hematocrit values as well as white blood cell and platelet counts were within normal ranges in all control and experimental animals.

Purification of Thrombin for Infusion

Purification of rabbit thrombin for infusion was performed according to modifications of the techniques of Lundblad [10] and Mann et al.[11] Prothrombin was obtained by repeated barium chloride precipitation of rabbit plasma. The resulting precipitate was resuspended in a trisodium citrate-saline solution (0.155M sodium chloride-0.2M trisodium citrate, diluted 1:9 (v/v) with distilled water) at 4° C. The third resuspended precipitate was dialyzed serially at 4° C against 0.2M EDTA pH 7.4, 0.2M trisodium citrate—0.155M sodium chloride, and distilled water (1:1:8, v/v/v); and 0.148M sodium chloride and 17M imidazole-distilled water (1:9, v/v) over at least 15 hr with frequent changes of dialysate. The solution then was activated at 37° C by the addition of thromboplastin (Difco two-stage reagent; Difco Laboratories, Detroit, MI.) and AC-globulin (Difco), or 0.85M trisodium citrate, in the presence of 0.25M calcium chloride. After maximum clotting activity was reached, the thrombin solution was dialyzed against 0.2M EDTA (pH 7.0)–0.2M trisodium citrate–0.155M sodium chloride distilled H_2O (1:1:1:8, v/v/v/v) for 4 hr at 4° C and then against 0.025M sodium phosphate, pH 6.5 for at least 15 hr at 4° C. Further purification was achieved by stepwise ion-exchange column chromatography employing C-50 sulfopropyl Sephadex (Pharmacia Fine Chemicals, Uppsala, Sweden) and 80 to 100 ml volumes of 0.025M, 0.1M, 0.25M, and 0.5M sodium phosphate, pH 6.5. The single peak eluted in the 0.25M fraction was placed on a Sephadex G-100 column and eluted with 0.1M sodium phosphate, pH 6.5. The resulting thrombin contained at least 2100 NIH U of activity per milligram of protein.

The clotting activity of purified thrombin was determined by measuring thrombin times in fresh-frozen plasma from rabbits. The results were converted into arbitrary units by means of a standard curve derived from thrombin clotting times of unpurified commercial bovine thrombin of known activity.

Diisopropylfluorophosphate Inactivation of Thrombin

Purified bovine thrombin (1,200 US U in 4 ml of 0.5M sodium phosphate buffer, pH 6.5) was incubated with 0.076 ml of 0.54M diisopropylfluorophosphate (DFP) in 2-propanol.[12] Coagulant activity of thrombin was undetectable at 2.5 hr, at which time DFP-thrombin was frozen at −70° C until further use within 6 weeks.

Purification of Fibrinogen

Rabbit fibrinogen was purified from citrated plasma according to the technique of McFarlane,[13] using repeated precipitation by 25% ammonium sulfate saturation. Cryofibrinogen was removed by the method of Senior et al.[14] following the second precipitation. Coagulability of the resulting fibrinogen was 89% to 98%. Purity was assessed also by gel electrophoresis [15] and immunodiffusion.[16] Preparations were lyophilized and maintained at −20° C until used.

^{125}I-Labeled Fibrinogen

Fibrinogen was purified by the method of McFarlane [13] with one modification from the technique of Senior et al.[14] After the initial precipitation of fibrinogen with 25% saturated ammonoium sulfate, the precipitate was dissolved in sodium chloride-sodium citrate solution. Ammonium sulfate was added to 12% saturation at 4° C. The cyroprecipitate was removed by centrifugation and the fibrinogen in the supernatant was precipitated with ammonium sulfate made up to 23% saturation. The fibrinogen was precipitated once more before dialysis. After dialysis, the purified fibrinogen (92%–95% coagulable) was labeled with ^{125}I (Amersham/Searle Corp., Arlington Heights, IL) by the technique of McFarlane.[13] The labeled fibrinogen was dialyzed at 4° C against 1 liter of sodium chloride-sodium citrate that was changed twice within 24 hours. After removal of the free ^{125}I by dialysis, the labeled fibrinogen (sp act 14 μCi/mg) was stored at $-20°$ C. Each rabbit received 0.67 mg of fibrinogen for catabolism studies.

Preparation of FDP in Vitro

Digestion of purified rabbit fibrinogen was accomplished at 37° C with 0.003 casein units of plasmin (Thrombolysin; Lot #0960W Merck, Sharp & Dohme Research Laboratories, West Point, PA) per milligram of lyophilized fibrinogen dissolved in 30 ml aliquots of sterile water to a concentration of 1 mg/ml. The pH of the digestion mixture was approximately 7.0 and was maintained during fibrinogenolysis by the periodic addition of 0.5 N sodium hydroxide. The reaction was terminated by the addition of ε-ACA, 0.2M final concentration,[17, 18] at 30 min, 40 min, or 15 hr to provide optimal production of predominantly stage 1 fragment X, stage 2 intermediate fragments X, Y, D, and E, and stage 3 final fragments D and E, respectively. To exclude the presence of residual fibrinolytic activity following addition of epsilon aminocaproic acid (ε-ACA), aliquots from all digestion mixtures were examined on heated fibrin plates (Enzo-Diffusion fibrin plate test; Hyland Products, Costa Mesa, CA).

The effectiveness of ε-ACA (0.2M) as an in vitro inhibitor of fibrinogen proteolysis by activated plasmin was compared to the termination of fibrinogenolysis in additional digestion mixtures with either DFP (Sigma Chemical Co., St. Louis, MO) or phenylmethanesulfonyl fluoride (Sigma), both potent serine protease inhibitors;[19-21] these agents were incubated in a five- to tenfold molar excess over the amount of plasmin added.

The FDP (70 to 150 μg of protein from each preparation) were identified in their unreduced form by SDS-polyacrylamide gel electrophoresis in 5% gels.[15] Protein standards for molecular weight estimation included aldolase (M_r 158,000); human immunoglobulin G (M_r 150,000); bovine serum albumin (M_r 140,000 to 70,000); carboxypeptidase (M_r 33,000); chymotrypsin (M_r 25,000); and myoglobin (M_r 17,000).

Solutions of FDP containing either ε-ACA or DFP (hydrolyzed over 24 hr at 4° C) were infused into rabbits, as described previously. At the termination of the infusions, small portions of each digestion mixture were subjected to polyacrylamide electrophoretic analysis, to ensure that no further proteolysis of FDP had occurred during the infusion period. Control rabbits were given equal volumes of 0.155M sodium chloride-0.013M sodium citrate solution, pH 7.5, containing 0.09 casein units of plasmin and 0.2M ε-ACA or 0.60 mM DFP, hydrolyzed over 24 hr. [75]SeM was injected 4 hr after infusion, and its incorporation into circulating fibrinogen was determined.

Immunoprecipitation of FDP Formed in Vitro

In an attempt to substantiate the specificity of FDP to stimulate fibrinogen synthesis in rabbits, purified homologous rabbit fibrinogen degradation products (FDP), generated *in vitro* were removed from the infusion solution by immunoprecipitation. This was accomplished by constructing a full precipitin curve in which either goat anti-rabbit fibrinogen antibody (Cappel Laboratories, Downington, PA) or specific canine anti-human D and E antibodies (generously provided by Dr. James P. Chen, University of Tennessee School of Medicine, Knoxville) were added to various dilutions of stage 2 or stage 3 digests of fibrinogen.[22] The equivalence point (the point at which maximum precipitation occurs) provided the approximate ideal antibody-antigen ratio necessary to avoid either antigen or antibody excess in the supernatants. To confirm this, the sample from the equivalence point was subjected to immunodiffusion. No antibody excess was detected; however, evidence of slight antigen excess persisted, possibly reflecting soluble antigen-antibody complexes. The Ouchterlony double-immunodiffusion technique [16] demonstrated immunological cross-reactivity between the specific canine antihuman D and E antibodies and the rabbit fibrinogen D and E fragments. This confirms the observations of others,[22-24] which indicate immunological cross-reactivity among heterologous mammalian fibrinogens and their terminal digestion fragments.

For infusion of the FDP-absorbed material into rabbits, the antibody was incubated with FDP derived from 30 mg of homologous purified fibrinogen, in a ratio corresponding to the equivalence point, and the previous experimental model was followed.

Separation of Components in Stage 3 Digests—Fragments D and E

The results of a stage 3 plasmic digest of rabbit fibrinogen were placed on a DE-52 DEAE cellulose column (2.5×38 cm) that was previously equilibrated with 0.01M $NaHCO_3/Na_2CO_3$ buffer, pH 8.9. Following application of the sample (200 or 300 mg protein) it washed into the column with starting buffer 15 ml 0.01M $NaHCO_3/Na_2CO_3$ pH 8.9. Employing descending technique, we started a sequential two part gradient elution. The first portion of the gradient was achieved with 400 ml of the column buffer (0.01M $NaHCO_3/Na_2CO_3$ pH 8.9) in chamber 1 of the gradient mixer and 400 ml 0.01M $NaHCO_3/Na_2CO_3$ 0.09M NaCl pH 8.9 in chamber 2. The second portion of the gradient was accomplished with 150 ml 0.01M $NaHCO_3/Na_2CO_3$ 0.17M NaCl pH 8.9. The third portion of the gradient was achieved with 150 ml 0.01M $NaHCO_3/Na_2CO_3$ 0.17M NaCl pH 8.9 in chamber 1 and 150 ml 0.01M $NaHCO_3$ 0.34M NaCl pH 8.9 in chamber 2. The fourth portion of the gradient was achieved with 150 ml 0.01M $NaHCO_3/Na_2CO_3$ 0.5M NaCl in chamber 2. The limit buffer solution was 200 ml 0.01M $NaHCO_3/Na_2CO_3$ 0.5M NaCl pH 8.9. The flow rate at 25° C was 55–60 ml/hour with a fraction size of 6 ml.

Fractions corresponding to peak D2 and peak E were pooled, dialyzed extensively in 20 volumes of distilled water at 4° C with changes of the distilled water every 8 hours for 2 days. The dialyzed peaks were then lyophilized.

Fragment D2 (hereafter referred to as Fragment D) and E were also identified by employing SDS-PAGE and the Ouchterlony double-immunodiffusion technique using specific canine antihuman D and E antibodies.

Following identification of Fragment D and Fragment E, quantities of the lyophilized digests were dissolved in 0.155 M sodium chloride 0.013 M sodium citrate pH 7.5 up to a volume of 30 ml. These solutions were then assayed for protein content [25] and fibrinogen-fibrin degradation products (FDP-fdp) by the tanned red cell hemagglutination inhibition immunoassay [26] modified for use in rabbits.[27]

Determination of FDP-fdp Titers

Titers of FDP-fdp were determined by the tanned red cell hemagglutination-inhibition immunoassay [26] modified for use in rabbits.[27] Normal values in our laboratory are $<1:8$ and represent the highest serum dilution that prevented agglutination of fibrinogen-coated red cells by anti-rabbit fibrinogen antibodies. The sensitivity of this technique detects less than 1.0 μg of fibrinogen per milliliter of plasma.

Preparation of fdp in Vitro

Noncrosslinked fibrin degradation products (fdp) were prepared by clotting 30 ml aliquots of rabbit fibrinogen (1 mg/ml) with either 3 U of ancrod (Venacil; Abbott Laboratories, North Chicago, IL) or 2 U of purified homologous thrombin [28] in the presence of 10^{-3}M EDTA. After a 4-hr incubation at 22° C, the clots were wound around Teflon or glass rods and digested, as previously described,[28] for 15 hr at 37° C in 0.155M sodium chloride-0.013M sodium citrate containing 0.09 casein units of plasmin (Thrombolytsin; Merck, Sharp & Dohme Research Laboratories, West Point, PA) per milligram of fibrinogen. The reaction was terminated by the addition of ϵ-ACA to a final concentration of 0.2M.[17, 18] No residual fibrinolytic activity was detected on heated fibrin plates following addition of ϵ-ACA. Residual coagulant activity of ancrod and thrombin was excluded by incubating 0.1 ml of each digestion mixture with 0.3 ml of fresh rabbit plasma at 37° C and observing for clot formation over 24 hr.

Production of crosslinked fdp from thrombin-formed clots was accomplished similarly, except that 0.8M calcium chloride replaced EDTA in the initial incubation period.

Ancrod

Ancrod was obtained as Venacil Lot #23-126DH from Abbot Laboratories in concentrations of 100 U/ml. All rabbits infused with ancrod received 2 U/kg body mass in 30 ml of isotonic saline intravenously over a period of 1 hr in the same manner as thrombin.

Quantification of Haptoglobin Levels

Nonhemolyzed serum was obtained from each rabbit immediately before infusion of the dialyzed stage 3 fragments and at 0, 3 and 5 hr following the injection of [75]SeM, for determination of haptoglobin levels by the immuno-diffusion technique of Mancini *et al.*[29] as modified by Fahey and McKelvey.[30] Several samples were evaluated simultaneously with the techniques outlined by Mouray *et al.*[31] for determination of haptoglobin concentration and incorporation of labeled amino acid into haptoglobin.

After 18 hr of diffusion areas of agar corresponding to the circumference of the immunoprecipitin rings that had developed around each sample well were measured for quantification,[25, 29] carefully cut out, placed in a gamma-well scintillation counter (Picker Nuclear Autowell II; Intertech, Inc., North Haven, CT) set for an energy range of 350 to 450 Kev, and counted over 15 min to estimate haptoglobin radioactivity.

Isolation in Vitro *of FPA and FPB from Fibrinogen*

Preparation of fibrinopeptides was accomplished according to modification of the technique described by Blombäck and Vestermark.[32] In 30 ml 0.005M phosphate buffer-0.15M sodium chloride, pH 7.0, 1.0 gm of lyophilized purified rabbit fibrinogen was dissolved and clotted with purified homologous thrombin [28] adequate to produce a clotting time of 18 to 20 min. After 6 hr at 22° C, the clot was wound on a Teflon rod, pressed in gauze, and washed with small volumes of distilled water thrice, with all washings added to the original clot liquor. After being acidified to pH 3 with formic acid, the clot liquor was applied to a Dowex 50 X-2 (200 to 400 mesh), BioRad Laboratories, Richmond, CA) column (2.5 by 15 cm) equilibrated with ammonium formate and eluted with 0.2M ammonium acetate, pH 5.5. The peptide-containing fractions were detected according to protein content or pH (between pH 3 and pH 6), pooled, and lyophilized; they were then redissolved and lyophilized three more times. The fibrinopeptides were identified subsequently by high-voltage paper electrophoresis at pH 2.0 and ultraviolet light (366 mU) analysis after staining with phenanthrenequinone and ninhydrin reagents.[33] The total yield of peptide material was 10 to 12 mg/gm of fibrinogen following chromatographic isolation. This is comparable to the results of Blombäck *et al.*[34]

Exactly 3 mg of lyophilized FPA and FPB were dissolved in 30 ml of sterile saline and infused over 1 hr into experimental animals. Control rabbits received equal volumes of sterile saline alone. [75]SeM was injected 4 hr following the completion of FPA and FBP infusions and its incorporation into fibrinogen was determined hourly for 5 hr.

Urokinase extracted from human urine (Winkinase; Winthrop Laboratories, New York NY) was administered intravenously to rabbits as a loading dose of 4400 CTA U/kg over 15 min, followed by continuous infusion of 4400 CTA U/kg/hr for 4 hr through a 23-gauge needle (Butterfly-23 infusion set; Abbott Laboratories, North Chicago, IL) inserted into the right marginal ear vein. Ancrod (Venacil; Abbott Laboratories, North Chicago, IL) supplied in concentrations of 100 U/ml, was mixed in 30 ml of sterile isotonic saline and infused in the same manner. Rabbits received either 2 U/kg body mass for 1 hr to achieve complete defibrinogenation or 0.75 U/kg body mass for 1 hr to obtain partial defibrinogenation. Control animals were given equal volumes of saline or saline followed by 150 mg homologous fibrinogen.

Limulus Lysate Assay

The Limulus lysate assay for endotoxin contamination according to the method of Levin and Bang [35] was performed on each preparation infused in this study. The results consistently indicated 10 ng or less of contamination per milliliter of sample. This is below the lowest concentrations of endotoxin observed to affect the rate of fibrinogen synthesis.[36] Aerobic and anaerobic cultures of all infusates were negative for bacterial growth.

Rabbit Model

The basic model for all studies to evaluate the stimulatory capacity of all compounds examined and the rate of newly synthesized fibrinogen was as follows: All experimental and control animals were studied for normal baseline values including hematocrit value, WBC count, platelet count, normal body temperature, plasma fibrinogen concentration. After only healthy animals were selected, the animals were infused with the compound under study in identical volumes over 1 hour by means of a continuous infusion pump (B. Brown Perfusion; Quigley-Rochester Inc., Rochester, NY). In all instances after completion of the infusion the animals were returned to their cages and rested for 5–6 hours. At the end of this time [75]SeM was injected and its incorporation into fibrinogen and other proteins was measured hourly for 6 hours.

In each experiment a minimum of 10 treated animals and 2–3 concomitant control rabbits were employed.

Determination of Newly Synthesized Fibrinogen

In vivo incorporation of [75]SeM fibrinogen and other proteins was measured serially after the injection of 20 μCi of [75]SeM into the marginal ear vein. Blood specimens were obtained form the opposite ear for measurement of fibrinogen concentration and radioactivity and were collected in siliconized tubes-containing 0.13M sodium citrate and 0.02M ϵ-ACA (1:9 v/v). Hematocrit values were determined simultaneously with heparinized capillary tubes. Plasma was prepared by centrifugation of anticoagulated whole blood at 5500 $\times$ g for 10 min at 4° C.

Fibrinogen concentrations and radioactivity were determined from plasma samples according to the method of Lerner et al.[37] [75]SeM-labeled fibrinogen was separated as thrombin-coagulable protein and was counted for 15 min in a gamma-well scintillation counter (Picker Nuclear Autowell II; Intertech, Inc., North Haven CT.) after the washed clot had been dissolved in alkaline urea solution. Radioactivity in fibrinogen was expressed as percent of the administered dose of [75]SeM incorporated into newly synthesized circulating coagulable protein, according to the following equation:

$$\% \, ^{75}\text{SeM incorporated} = (\text{cpm/mg fibrinogen-background}) \times \frac{\text{mg total circulation fibrinogen} \times 100}{\text{total cpm } ^{75}\text{SeM injected}}$$

Plasma volume for rabbits was assumed to be 40 ml/kg body mass.[38] This method of calculation was adapted to determine the radioactivity of total noncoagulable proteins from defibrinogenated plasma. Hematocrit values were not altered significantly during the 6 hr sampling period.

The absorbance of the dissolved fibrin (fibrinogen) was measured at 282 nm against an alkaline urea blank, and the fibrinogen concentration was calculated from the absorption coefficient of 1.617 for rabbit fibrin dissolved in alkaline urea.[39]

The incorporation of [75]SeM into plasma proteins was distinguished from nonspecific binding by treating portions of the plasma samples obtained after administration of [75]SeM with 1.0N sodium bisulfite to destroy nonspecific seleno-sulfide bonds. The proteins were precipitated with 20% TCA and washed repeatedly with cold 10% TCA; the radioactivities of the supernatant, washings and protein precipitate were then determined. A similar procedure was performed for serum samples.

Fibrinogen contamination by other radioactive proteins, generated during the experiments was assessed by incubating unlabeled rabbit plasma with an equal volume of serum from a rabbit administered [75]SeM 24 hr previously. The fibrinogen was coagulated and dissolved in alkaline urea, and its radioactivity was measured. Similar studies were conducted in unlabeled plasma to which [75]SeM had been added to give 10,000 cpm of [75]SeM per milliliter of plasma. In addition, plasma samples from rabbits injected with 75 SeM 6 hr previously were immunoprecipitated with optimal quantities of goat anti-rabbit fibrinogen antibody, as determined by a precipitin curve. After centrifugation and washing, the radioactivities of the supernatant and precipitin were determined.

Fibrinogen was precipitated from the plasma of rabbits administered [75]SeM 6 hr previously, by 25% saturated ammonium sulfate. Following SDS-poly-acrylamide electrophoresis [15] of the protein in reduced and unreduced forms, each gel band was cut and measured for radioactivity.

Statistical Evaluation

Mean, standard deviation, standard error, and comparative group analysis were computed by conventional methods.[40]

RESULTS

Incorporation of [75]SeM into fibrinogen after injection of 20 μCi of [75]SeM into rabbits, samples of blood were obtained hourly for 6 hr and the radioactivity in the circulating fibrinogen was determined. In 40 control rabbits that received either [75]SeM alone or 0.85% sodium chloride-sodium citrate infusion before [75]SeM; the incorporation of [75]SeM into fibrinogen reached a maximum of 0.61 $\pm$ 0.06 (SE)% of the injected dose at 3 hours and was unchanged at 24 hours.

To determine the degree of contamination of fibrinogen by other radioactive proteins, 0.3 ml of unlabeled rabbit plasma was incubated for 1 hour at 37° C with 0.3 ml of radioactive serum from a rabbit that 24 hours earlier had received 20 μCi of [75]SeM. The unlabeled fibrinogen was coagulated, washed, dissolved in alkaline urea and measured. The unlabeled fibrin clot was associated with less than 4% of the radioactivity of the same amount of labeled fibrinogen from rabbits that had received identical quantities of [75]SeM.

To determine if the assay system was sensitive in detection of inhibition of fibrinogen synthesis, we measured the amount of [75]SeM incorporated into fibrinogen during a 3-hr period in a rabbit in which the hepatic circulation had been completely bypassed. In this animal the concentration of labeled fibrinogen was barely detectable and always less than 1/10 the amount in animals with intact hepatic circulation.

Effect of in Vivo Elevated Plasma Fibrinogen Concentration on Fibrinogen Synthesis

In several groups of animals infused with purified homologous fibrinogen to raise the plasma concentration from 100–300% above normal there was no alteration in fibrinogen synthesis when compared with control animals with normal plasma fibrinogen concentration. No differences were detected in different groups of animals who had elevated levels of fibrinogen maintained for 1 day or for 6 consecutive days.

Effect of Severe Hypofibrinogenemia on Fibrinogen Synthesis

In groups of rabbits who received ancrod (purified fraction of crude venom from *Agkistroden rhodostoma*) 2 U/kg body mass during 1 hour fibrinogen values declined from 2.33 ± 0.14 gm/liter preinfusion to essentially unmeasurable levels after infusion. Fibrinogen synthesis measured at hourly, then at 6 hour intervals up to 26 hours did not differ from the normal control animals.

Also in these animals the catabolism of ^{125}I-labeled fibrinogen injected 5, 12, or 18 hr after ancrod infusion was measured for 120 hr and compared to controls that received ^{125}I-fibrinogen 18 hr after 0.85% NaCl. Analysis of covariance revealed that the regression slopes of fibrinogen radioactivity in the ancrod-treated groups did not differ (half-life of ^{125}I-fibrinogen was 42 hours) from the control animals.

Effect of Homologous Thrombin on Fibrinogen Synthesis

In groups of rabbits receiving either 100 or 200 NIH U of purified rabbit thrombin over 1 hour, a three- to fivefold increase in fibrinogen synthesis above the control animals was observed. It was demonstrated that the rate of synthesis was directly dose dependent on the amount of thrombin infused. It was observed in all animals that received thrombin infusions that large quantities of fibrinogen-fibrin degradation products (FDP-fdp) were present in the circulating blood. The quantities of these FDP-fdp correlated with the magnitude of increase in fibrinogen synthesis observed.

If the animals who received the purified homologous thrombin were pretreated with ε-ACA to give a blood level of 0.2M concentration, the accelerated synthesis of fibrinogen induced by thrombin was prevented and the appearance of FDP-fdp was inhibited. The same concentration of ε-ACA given to control animals did not disturb the normal rate of fibrinogen synthesis.

Effect on Fibrinogen Synthesis of Homologous FDP Generated in Vitro

The effects on fibrinogen synthesis of homologous FDP were examined by infusion of variously timed digests of purified fibrinogen. Nine controls received solutions of sodium citrate, plasmin and either ε-ACA or DFP hydrolyzed over 24 hr. Because similar changes in qualitative and quantitative fibrinogen synthesis occurred, whether the FDP digestion material contained ε-ACA or DFP as the inhibitor of fibrinogenolysis, the results obtained from both groups were considered collectively. The percent incorporation of ^{75}SeM into fibrinogen in nine rabbits that received stage 2 intermediate fibrinogen fragments X, Y, D, and E (FDP titer $\geq$ 1:512) was 1.74% $\pm$ 0.15 (SE) 5 hr after isotope injection, compared to 2.02% $\pm$ 0.17 for five animals that received final fibrinogen fragments D and E (FDP titer $\geq$ 1.256). These levels were significantly higher (p < 0.025) than the percent ^{75}SeM incorporation, 0.95% $\pm$ 0.10 (SE) produced by early fragment X in six animals. ^{75}SeM incorporation into fibrinogen after administration of fragment X was not statistically different from control values (0.79% $\pm$ 0.12). Fibrinogen concentrations rose 2.68 $\pm$ 0.23 and 2.95 $\pm$ 0.08 gm/L 9hr after infusions of intermediate or final FDP (5 hr after ^{75}SeM injection), respectively, as compared to their baseline levels of 2.09 $\pm$ 0.14 g/L.

The titers of FDP-fdp ranged between 1:128 and 1:512 immediately following administration of intermediate and final fibrinogen digests, as compared to the normal titers after infusion of fragment X.

When FDP were given to anmals pretreated with ε-ACA, there was stimulation of fibrinogen synthesis to the same degree as seen in the absence of ε-ACA. It is apparent that ε-ACA did not interfere with the stimulation of fibrinogen synthesis induced by FDP.

Effect of Removal of FDP by Immunoprecipitation

After immunoprecipitation of intermediate and late fibrinogen digests by goat anti-rabbit fibrinogen antibody and specific anti-human D and E antibodies, respectively, the supernatants, containing titers of degradation products between 1:16 and 1:32, were infused. Residual anti-fibrinogen antibody in the supernatants was not detected by immunodiffusion. Immunoprecipitation of the intermediate and late FDP significantly reduced the accelerated fibrinogen synthesis induced by their native unadsorbed counterparts. The percent incorporation 5 hr following injection of ^{75}SeM reached 0.98% $\pm$ 0.15 (S.E.) in the four animals receiving the supernatants of immunoprecipitated fibrinogen fragments X, Y, D, and E and 1.15% $\pm$ 0.04 in the five animals infused with supernatants of immunoprecipitated fragments D and E. Stimulation of fibrinogen synthesis was inhibited when compared with the stimulation produced by the respective native fragments (p < 0.005). Fibrinogen concentrations remained unaffected.

When compared to the controls, the fibrinogen synthesis associated with supernatants adsorbed from D and E fragments remained enhanced slightly. This may reflect the activity of residual rabbit fragments D and E not completely removed by the specific anti-human D and E antibodies.

Potential effects of contaminating anti-rabbit fibrinogen antibody in the infused supernatants following immunoprecipitation were evaluated. Five rabbits received 30 ml infusions of a sodium chloride-sodium citrate ε-ACA-plasmin solution that contained goat anti-rabbit fibrinogen antibody added in amounts adequate to immunoprecipitate and remove the fragments of fibrinogen digestion (as determined by the equivalence points on precipitin curves). Five hours following [75]SeM injection, the percent incorporation of isotope into fibrinogen was 1.70% ± 0.14 (S.E.) a result significantly different (p < 0.005) from control values of 0.79% ± 0.12. The titers of FDP-fdp in the experimental animals ranged from 1:16 to 1:32.

Fibrinogen Synthesis and the Influence of LMW Peptides of Fibrinogenolysis

In vitro generation of terminal stage 3 digests of fibrinogen (FDP-DE) yielded 0.544 mg/ml protein (69% recovery) before dialysis and 0.387 mg/ml protein after extensive dialysis. Corresponding titers of FDP were 1:1024 to 1:2048 before dialysis and 1:512 to 1:1024 after dialysis.

The ability of dialyzed and undialyzed FDP to stimulate fibrinogen synthesis was compared. The data demonstrate that fibrinogen production was influenced only by the presence of FDP-DE and not by the LMW peptides derived during fibrinogenolysis. The rates of fibrinogen synthesis stimulated in the FDP-DE groups were similar and were statistically significant (p < 0.05) when compared to the controls. The disparity of stimulation observed in the isotope incorporation curves of the FDP-DE groups 7 hr following completion of infusion (3 hr after injection of [75]SeM) may represent loss of inhibitors during dialysis; however, it is most likely secondary to sampling variability. Twenty-four hours following the infusion fibrinogen concentrations had increased to 150% of baseline values in the FDP-DE groups, whereas levels in the control animals remained unchanged.

The titers of FDP immediately after administration of both types of stage 3 digests ranged between 1:128 to 1:512, with normal titers in the controls.

The specificity of stimulation of fibrinogen synthesis by FDP-DE was evaluated by analyzing [75]SeM incorporation into nonclottable serum proteins and serum haptoglobin. Although the levels of [75]SeM-labeled fibrinogen (clottable protein) were altered significantly by infusions of dialyzed and undialyzed FDP-DE, the incorporation of isotope into nonclottable serum proteins and haptoglobin was unaffected. Haptoglobin concentrations were 365 ± 35 mg/dl in rabbits receiving FDP-DE preparations vs 308 ± 22 in the controls. The results obtained by the immunodiffusion method and the technique of Mouray et al.[31] were not significantly different. In addition, the radioactivity of plasma samples 6 hr following [75]SeM administration was 93% precipitable by trichloroacetic acid and 96% precipitable by anti-rabbit fibrinogen antibody.

Effect of Urokinase on Fibrinogen Synthesis

Urokinase (4400 CTA U/kg loading dose followed by continuous infusion of urokinase 4400 CTA U/kg/hr for 4 hr) was administered intravenously to a group of 6 rabbits. There was a slight, short-lived decline in the plasma fibrinogen levels following completion of the infusion; however, within 5 hr, the plasma fibrinogen concentration approached baseline values. After an additional 5 hr, plasma fibrinogen had decreased significantly to less than

1.40 g/liter. This degree of hypofibrinogenemia disappeared by 24 hr, as indicated by return to baseline levels in the 24-hr sample. In the 18 control animals given saline, the plasma fibrinogen concentration remained stable and within the normal range.

Abnormally elevated titers of FDP-fdp were detected in the experimental group within the first hour after initiating urokinase infusions. These titers subsequently peaked in the range of 1:128–1:256 at the completion of infusion and gradually decreased thereafter to normal levels by 5 hr, just prior to injection of ^{75}SeM. In the control group, the FDP-fdp remained normal throughout the study period.

Analysis of ^{75}SeM incorporation into clottable protein (fibrinogen) was performed in each animal. Fibrinogen synthesis was enhanced significantly 6 hr after terminating urokinase infusion (1 hr postinjection of isotope). Incorporation of ^{75}SeM into fibrinogen continued to rise over 24 hr. In contrast, levels of ^{75}SeM-labeled fibrinogen in control rabbits were maximal 4–5 hr after ^{75}SeM injection (9–10 hr following saline infusions) and remained constant over the next 24 hr.

The specificity of stimulation of fibrinogen synthesis following infusion of urokinase was evaluated by ^{75}SeM incorporation into total nonclottable serum proteins and serum haptoglobin. Although levels of ^{75}SeM-labeled fibrinogen in rabbits that received urokinase were about three times those of control animals, the incorporation of isotope into nonclottable serum proteins and haptoglobin was unaffected. The percentages of incorporation of ^{75}SeM into nonclottable serum proteins and haptoglobin 5 hr after injection of isotope were 4.98 ± 0.32 (SE) and 0.36 ± 0.07 (SE), respectively, in the urokinase group and 5.10 ± 0.26 (SE) and 0.32 ± 0.04 (SE) in the controls. Abnormal bleeding was not observed in animals receiving urokinase.

Effect of Homologous Noncrosslinked fdp Generated from Fibrin Formed by Ancrod or Thrombin

The response to infusions of noncrosslinked fdp was studied in two groups of normal rabbits that received digests of fibrin formed by ancrod or thrombin in the absence of calcium chloride. Infusates contained fdp titers ranging between 1:256 and 1:512. Basal fibrinogen synthesis was unaffected in both groups as compared to controls, and the fibrinogen levels 24 hr after infusion remained unchanged from baseline values $(2.15 \pm 0.25$ g/L). These results were similar to control fibrinogen concentrations (2.59 ± 0.40).

The serum titers of FDP-fdp immediately after infusions of the noncrosslinked fdp ranged from 1:64 to 1:128 in the ancrod fdp group and from 1:128 to 1:512 in the thrombin fdp group. Four hours later, just prior to isotope injection, FDP-fdp titers remained elevated, ranging from 1:128 to 1:512 in the former group and 1:32 to 1:64 in the latter group. The rise in titers of FDP-fdp in the ancrod fdp animals 4 hr following infusions is difficult to explain. Residual ancrod activity in the infusate might have produced such results; however, no concomitant decreases in fibrinogen concentrations occurred, and no clot formation was detected during incubation of the infusate with normal rabbit plasma. Titers of FDP-fdp in the controls were normal throughout the study.

Effect of Crosslinked fdp from Thrombin Fibrin

Effects of crosslinked fdp (FDP-fdp titer 1:512) are presented as follows: The maximum percent incorporation of ^{75}SeM into *de novo* fibrinogen (0.88% ± 0.04 (SE)) was not significantly different from control values (0.84% ± 0.06). Fibrinogen concentrations were comparable in both groups.

Titers of FDP-fdp were elevated in the range of 1:128 to 1:512 immediately after infusion of crosslinked fdp and gradually fell to 1:32 to 1:64 4 hr later (just prior to injection of ^{75}SeM). The titers of FDP-fdp in the control group remained normal.

Effects of FPA and FPB Generated in Vitro

The effects on fibrinogen synthesis of homologous FPA and FPB were examined. The maximum percent incorporation of ^{75}SeM into fibrinogen 5 hr after isotope injection reached 0.83% ± 0.09 (SE) in the rabbits administered FPA and FPB; the difference from control values (0.76% ± 0.07) was not statistically significant.

The titers of FDP-fdp remained within normal limits for both groups throughout the study.

Effect of Individual Fragment D and Fragment E on Fibrinogen Synthesis

Infusion of varying quantities (4 mg–12 mg) of Fragment D and Fragment E separately into different groups of rabbits revealed that each fragment can stimulate fibrinogen synthesis. For each fragment it was observed that the degree of stimulation was dose dependent. When equal amounts of each fragment were infused into different groups of animals, Fragment D produced a fourfold greater increase in fibrinogen synthesis than did Fragment E above control animals. There was no increase in the amount of radioactivity in haptoglobin and no increase in haptoglobin concentration following infusion of either Fragment D or Fragment E.

Quantification of the sialic acid content in each of these fragments was nearly identical.

DISCUSSION

The results of earlier studies [27, 41-43] as well as those conducted in the present series of experiments firmly support the contention that severe reductions or striking elevations of plasma fibrinogen concentration do not influence the basal rate of fibrinogen synthesis. This is in sharp contrast to almost all known plasma proteins, hormones, or elements.

Likewise infusions of large quantities of fibrinogen do not stimulate the production of this glycoprotein.[27] Increases, delayed by 16 hours, seen in earlier experiments [44] following infusions of large quantities of fibrinogen may have been due to an increase in metabolic fibrinogen breakdown products (the reason for the delay) and not fibrinogen.

Initially it appeared that thrombin which normally exists in the blood could be the naturally occurring regulator of fibrinogen biosynthesis.[45] We decided to evaluate this agent because of the difficulties involved in the interpretation of previous studies.[8, 27, 46-52] Although stimulation may have been present following thrombin, a number of possibilities besides the thrombin *per se* were present that could explain the stimulation. Frequently the thrombin was heterologous, and a nonspecific species interaction could not be excluded. The thrombin was not pure but contaminated with a number of substances including nonprotease enzymes. The doses of thrombin were not always clearly stated.

As we explored this agent, employing homologous purified thrombin (α thrombin), mechanistically it became apparent that thrombin was an intermediate in the stimulatory process. Each time thrombin was infused, FDP-fdp appeared in the blood; the quantities of which correlated with the magnitude in increase of fibrinogen synthesis above control. When thrombin in a wide range of doses was infused in the presence of ϵ-ACA (0.2M which by itself did not alter normal fibrinogen synthesis) no FDP-fdp could be detected and no increase in fibrinogen synthesis was observed. The increase in fibrinogen synthesis may have resulted from the FDP-fdp and not directly from thrombin.

Employing homologous purified fibrinogen and fibrinogen degradation products (FDP) in several different studies [28, 53-55] we were able to demonstrate that FDP-D was the most potent (of the classic FDP's X, Y, D, and E) stimulator of fibrinogen synthesis in the intact animal.[54, 55] No stimulation was observed via FDP-X or Y and minimal stimulation was not following infusion of FDP-E.[55]

Instructive were the observations that homologous purified fibrin degradation products (fdp) cross-linked or noncrosslinked, did not stimulate the production of fibrinogen synthesis. Identical observations have been made by other groups.[56, 57] The reason why FDP are capable of stimulating fibrinogen synthesis and fdp are not remains puzzling and unexplained.

Concerning the stimulatory capacity of FDP the results of the above studies agree with those of other investigators [41, 44, 56, 58] and have been demonstrated in rabbits, rats and dogs. Recently Franks et al.[59] have confirmed our earlier report [55] regarding the stimulatory properties of FDP-D. There is only one report [60] where an increase in fibrinogen synthesis was not observed following infusion of FDP. Many authors have suggested that another group [56] also failed to observe an increase in fibrinogen synthesis following infusion of FDP. These authors specifically refer to animal group IV in the quoted report.[56] It is critical to recognize that group IV rabbits received degradation products prepared from fibrin, namely, fdp. These animals who failed to demonstrate an increase in fibrinogen synthesis did not receive FDP but only fdp, an observation confirmed by several investigators.[54, 55, 57]

Whether fibrinogen degradation products that are present normally in small quantities in mammalian blood are the physiologic regulators of fibrinogen synthesis remains to be established. Like thrombin it is possible that these fragments FDP-D and FDP-E are also intermediates in the stimulatory process.

Recently we have established a method for the isolation of individual rabbit hepatocytes and placement into primary tissue culture. We have demonstrated that these hepatocytes exist in a serum free tissue culture environment for 10–14 days. Under the proper conditions these hepatocytes are capable of synthesizing fibrinogen, transferrin, haptoglobin, and albumin at physiologic rates reported for the intact animal. We are now prepared to study the effect of FDP-D and

FDP-E at the cellular level to establish if these fragments directly stimulate fibrinogen synthesis.

ACKNOWLEDGMENTS

These studies were possible with the help of Drs. H. G. Klein, B. M. Alving, C. M. Kessler, and B. L. Evatt.

REFERENCES

1. SCHMIDT, A. 1861. Uber den Faserstoff und die Ursachen seiner Gerinnung. Physiol. p. 545.
2. SCHMIDT, A. 1862. Weiteres uber den Faserstoff und die Ursachen seiner Gerinnung Archiv. f. Anat. u. Physiol. p. 428.
3. DRURY, D. R. & P. D. MCMASTER. 1929. The liver as the source of fibrinogen. J. Exp. Med. **50:** 569–678.
4. TARVER, H. & W. O. REINHARDT. Methionine labeled with radioactive sulfer as an indicator of protein formation in the hepatectomized dog. J. Biol. Chem. **167:** 395–400.
5. MILLER, L. & W. F. BALE. 1954. Synthesis of all plasma protein fractions except gamma globulin by liver. J. Exp. Med. **99:** 125–132.
6. GITLIN, D., B. H. LANDING & A. WHIPPLE. 1953. Localization of homologous plasma proteins in tissues of young human beings as demonstrated with fluorescent antibodies. J. Exp. Med. **97:** 163–177.
7. HAMASHIMA, Y., J. G. HARTER & A. H. COONS. 1962. Cellular site of albumin and fibrinogen production in human liver. Fed. Proc. **1:** 304.
8. BARNHART, M. I. & W. B. FORMAN. 1963. The cellular localization of fibrinogen as revealed by the fluorescent antibody technique. Vox Sang. **8:** 461–473.
9. FORMAN, W. B. & M. I. BARNHART. 1964. Cellular site for fibrinogen synthesis. J. Am. Med. Assoc. **187:** 168–172.
10. LUNDBLAD, R. L. 1971. A rapid method for the purification of bovine thrombin and the inhibition of the purified enzyme with phenylmethyl-sulfonyl fluoride. Biochemistry **10:** 2501.
11. MANN, K. G., C. M. HELDEBRANT & D. N. FASS. 1971. Multiple active forms of thrombin J. Biol. Chem. **246:** 5994.
12. MILLER, K. D. & H. VAN VUNAKIS. The effect of DFP on the proteinase and esterase activities of thrombin and on prothrombin and its activators. J. Biol. Chem. **223:** 227–236.
13. MCFARLANE, A. S. 1963. In vivo behavior of ^{131}I-fibrinogen. J. Clin. Invest. **42:** 346.
14. SENIOR, R. M., L. A. SHERMAN & E. T. YIN. 1974. Effects of hyperoxia on fibrinogen metabolism and clotting factors in rabbits. Am. Rev. Respir. Dis. **109:** 156–161.
15. WEBER, K. & M. OSBORN. 1969. The reliability of molecular weight determinations by dodecylsulfate-polyacrylamide gel electrophoresis. J. Biol. Chem. **244:** 4406–4412.
16. OUCHTERLONEY, O. 1958. Diffusion-in-gel method for immunological analysis. Prog. Allergy **5:** 1–78.
17. MARDER, V. J. & N. R. SHULMAN. 1969. High molecular weight derivatives of human fibrinogen produced by plasmin. J. Biol. Chem. **244:** 2111–2119.
18. KAPLAN, A. P. & K. F. AUSTEN. 1972. The fibrinolytic pathway of human plasma. J. Exp. Med. **136:** 1378–1393.
19. FAHRNEY, D. E. & A. M. GOLD. 1963. Sulfonyl fluorides as inhibitors of esterase. I. Rates of reaction with acetylcholinesterase, α-chymotrypsin, and trypsin. J. Am. Chem. Soc. **85:** 997–1000.

20. SUMMARIA, L., B. HSIEH, W. R. GROSKOPF, et al. 1967. The isolation and characterization of the S-carboxymethyl β(light) chain derivative of human plasmin. J. Biol. Chem. **242:** 5046.

21. SUMMARIA, L., L. ARZADON, P. BERNABE, et al. 1970. The activation of plasminogen to plasmin by urokinase in the presence of the plasmin inhibitor trasylol. J. Biol. Chem. **250:** 3988.

22. CHEN, J. P. 1977. Unique immunological cross-reactivity between fragments D and/or E of three heterologous mammalian fibrinogens. Thromb. Res. **11:** 31–42.

23. BAUER, K. 1969. Immunological investigations on the evaluation of fibrinogen and plasminogen. Humangenetik **7:** 260–262.

24. NUSSENZWEIG, V. & E. A. SOUZA. 1962. Cross reaction studies in a protein-antiprotein system. The immunochemical relationship between two isolated components of digested human fibrinogen with bovine fibrinogen. Int. Arch. Allergy Appl. Immunol. **21:** 294–304.

25. LOWRY, O. H., N. J. ROSEBROUGH, A. L. FARR, et al. 1961. Protein measurement with the Folin phenol reagent. J. Biol. Chem. **193:** 265–275.

26. MERSKEY, C., P. LALEZAKI & A. J. JOHNSON. 1969. A rapid, simple, sensitive method for measuring fibrinolytic split products in human serum. Proc. Soc. Exp. Biol. Med. **131:** 871–875.

27. ALVING, B. M., W. R. BELL & B. L. EVATT. 1977. Fibrinogen synthesis in rabbits: Effects of altered levels of circulating fibrinogen. Am. J. Physiol. **232:** H478–484.

28. KESSLER, C. M. & W. R. BELL. 1979. The effects of homologous thrombin and fibrinogen degradation products on fibrinogen synthesis in rabbits. J. Lab. Clin. Med. **93:** 768–782.

29. MANCINI, G., A. O. CARBONARA & J. F. HEREMANS. 1965. Immuno-chemical quantitation of antigens by single radial immunodiffusion. Immunochemistry **2:** 235–254.

30. FAHEY, J. L. & E. M. McKELVEY. 1965. Quantitative determination of serum immunoglobulins in antibody-agar plates. J. Immunol. **94:** 84–90.

31. MOURAY, H. J. MORETTI & M. F. JAYLE. 1964. Incorporation in vivo des acides amines marques dans les proteines plasmatiques du lapin. CR Acad. Sci. (D) **258:** 4871–4874.

32. BLOMBÄCK, B. & A. VESTERMARK. 1958. Isolation of fibrinopeptides by chromatography. Arkh. Kemi **12:** 173–182.

33. YAMADA, S. & H. ITANO. 1966. Phenanthrenequinone as an analytical reagent for arginine and other monosubstituted guanidines. Biochim. Biophys. Acta **130:** 538–540.

34. BLOMBÄCK, B., P. EDMAN, et al. 1966. Human fibrinopeptides: Isolation, characterization and structure. Biochim. Biophys. Acta **115:** 371–396.

35. LEVIN, J. & F. B. BANG. 1968. Clottable protein in Limulus: Its localization and kinetics of its coagulation by endotoxin. Thromb. Diath. Haemorrh. **19:** 196–197.

36. ALVING, B. M., B. L. EVATT, J. LEVIN, W. R. BELL, R. B. RAMSEY & F. C. LEVIN. 1979. Platelet and fibrinogen production: relative sensitivities to endotoxin. J. Lab. Clin. Med. **93:** 437–448.

37. LERNER, R. G., S. I. RAPAPORT, J. R. SIEMSEN, et al. 1968. Disappearance of fibrinogen-^{131}I after endotoxin: effects of a first and second injection. Am. J. Physiol. **214:** 532–537.

38. BOCCI, V. & A. VITI. 1966. Plasma and blood volumes estimated by the serum ^{131}I-proteins method in normal rabbits of varying body weight. Q. J. Exp. Physiol. **51:** 27–32.

39. ANTENCIO, A. C., D. C. BURDICK & E. B. REEVE. 1965. An accurate isotope dilution method for measuring plasma fibrinogen. J. Lab. Clin. Med. **66:** 137–145.

40. SNEDECOR, G. W. & W. G. COCHRAN. 1965. Statistical methods. Iowa State University Press. Ames.
41. BARNHART, M. I., D. C. CRESS, S. M. NOONAN & R. T. WALSH. 1970. Influence of fibrnolytic products on hepatic release and synthesis of fibrinogen. Thromb. Diath. Haemorrh. Suppl. **39:** 143–159.
42. BARNHART, M. I. & S. M. NOONAN. 1973. Cellular control mechanism for blood clotting proteins. Throm. Diath. Haemorrh. Suppl **54:** 59–82.
43. ATENCIO, A. C., K. JOINER & E. B. REEVE. 1969. Experimental and control system studies of plasma fibrinogen regulation in rabbits. Am. J. Physiol. **216:** 764–772.
44. BOCCI, V. & A. PACINI. 1973. Factors regulating plasma protein synthesis II. Influence of fibrinogenolytic products on plasma fibrinogen concentration. Thromb. Diath. Haemorrh. **29:** 63–75.
45. SHUMAN, M. A. & P. W. MAJERUS. 1976. The measurement of thrombin in clotting blood by radioimmunoassay. J. Clin. Invest. **58:** 1249–1258.
46. PICKART, L. R. & L. O. PILGERAM. 1967. The role of thrombin in fibrinogen biosynthesis. Thromb. Diath. Haemorrh. **17:** 358–364.
47. MONKHOUSE, F. C. & S. MILOJEVIC. 1960. Changes in fibrinogen levels after infusion of thrombin and thromboplasin. Am. J. Physiol. **199:** 1165–1168.
48. RATNOFF, O. D. & C. L. CONLEY. 1950. The defibrinating effect on dog blood of intravenous injection of thromboplastic material. Bull. Johns Hopkins Hosp. **88:** 414–424.
49. PILGERAM, L. O. & L. R. PICKART. 1968. Control of fibrinogen biosynthesis: The role of free fatty acid. J. Atherosclerosis Res. **8:** 155–166.
50. PICKART, L. R. & M. THALER. 1976. Free fatty acids and albumin as mediators of thrombin-stimulated fibrinogen synthesis. Am. J. Physiol. **230:** 996–1002.
51. ALVING, B. M., B. L. EVATT & W. R. BELL. 1977. Stimulation of fibrinogen synthesis by thrombin in rabbits with ancrod-induced afibrinogenemia. Am. J. Physiol. **233:** H565–H567.
52. JOHN, D. W. & L. L. MILLER. 1969. Regulation of net biosynthesis of serum albumin and acute phase plasma proteins. J. Biol. Chem. **244:** 6134–6142.
53. KESSLER, C. M. & W. R. BELL. 1979. Regulation of fibrinogen biosynthesis: Effect of fibrin degradation products, low molecular weight peptides of fibrinogenolysis and fibrinopeptides A and B. J. Lab. Clin. Med. **93:** 758–767.
54. KESSLER, C. M. & W. R. BELL. 1980. Stimulation of fibrinogen synthesis: A possible functional role of fibrinogen degradation products. Blood **55:** 40–47.
55. KESSLER, C. M., W. R. BELL & R. R. TOWNSEND. 1978. Stimulation of fibrinogen synthesis by fibrinogen degradation products. Blood **52**(Suppl 1): 186.
56. KROPATKIN, M. L. & G. IZAK. 1968. Studies on the hypercoagulable state. Thromb. Diath. Haemorrh. **19:** 547–555.
57. ITTYERAH, T. R., N. WEIDNER, D. WOCHNER & L. A. SHERMAN. 1979. Effect of fibrin degradation products and thrombin on fibrinogen synthesis. Brit. J. Haematol. **43:** 661–668.
58. YOUNG, M. C. & S. N. KOLMEN. 1970. Recovery of fibrinogen in fibrinolytic dogs. Thromb. Diath. Haemorrh. **23:** 50–57.
59. FRANKS, J. J., R. E. KIRSCH, L. O. FRITH, L. R. PURVES, W. T. FRANKS, J. A. FRANKS, P. MASON & S. J. SAUNDERS. 1981. Effect of fibrinogenolytic products D and E on fibrinogen and albumin synthesis in the rat. J. Clin. Invest. **67:** 575–580.
60. OTIS P. T. & S. I. RAPAPORT. 1973. Failure of fibrinogen degradation products to increase plasma fibrinogen in rabbits. Proc. Soc. Exp. Biol. Med. **144:** 124–129.

DISCUSSION OF THE PAPER

P. J. GAFFNEY (*National Institute for Biological Standards and Control, London*): Can one conclude from your lecture that the infusion of thrombin generates fibrinogenolysis?

W. R. BELL: Yes, I think that that is a reasonable conclusion, since we were able to show that degradation products do appear in large quantities and these can be completely inhibited.

A. L. COPLEY (*Polytechnic Institute of New York, Brooklyn*): What do you think is responsible for this difference between fibrinogen and fibrin in the stimulation? Do fibrinopeptides perhaps play a role?

BELL: No, we looked at both fibrinopeptides A and B in the rabbit model and they were not stimulatory.

INTERACTION OF FIBRINOGEN WITH STAPHYLOCOCCAL CLUMPING FACTOR AND WITH PLATELETS *

J. Hawiger,† M. Kloczewiak, and S. Timmons

Departments of Pathology and Medicine
Vanderbilt University
Nashville, Tennessee 37232

D. Strong and R. F. Doolittle

Department of Chemistry
University of California at San Diego
La Jolla, California 92093

INTRODUCTION

The main role of plasma fibrinogen is to serve as a substrate for the proteolytic action of thrombin, which initiates a transformation of soluble fibrinogen into insoluble fibrin. There is, however, another important function of fibrinogen—that of a cellular agglutinin. Fibrinogen interacts with several types of cells, such as staphylococci, streptococci, and platelets.[1-6] These interactions involve binding of fibrinogen to specific cell receptors with concomitant agglutination (clumping) of bacteria or aggregation of platelets. Research on the interaction of fibrinogen with cell surfaces has been focused on two central aspects: (1) localization of the binding region on fibrinogen that reacts with cellular receptors and (2) determination of the nature of cellular receptors for fibrinogen. Our studies during the past few years were directed toward the characterization of the binding regions(s) on fibrinogen for staphylococcal clumping receptor and for human platelet receptors. Attendant with the recent elucidation of the overall covalent structure of human fibrinogen [7-9] we were able to complete a series of investigations, reported recently elsewhere,[10-13] that pinpointed not only the region of human fibrinogen interacting with cell wall receptor of staphylococci responsible for their clumping, but also recognizing human platelet receptors for fibrinogen involved in their aggregation induced by ADP, epinephrine, and thrombin.

Thus, the problem of interaction of fibrinogen with prokaryotic cells such as staphylococci and eukaryotic cell fragments such as human platelets is approaching its solution from the standpoint of fibrinogen structure and function. The most recent evidence accumulated by us [10-13] will be reviewed below.

INTERACTION OF FIBRINOGEN WITH STAPHYLOCOCCI

In 1908 the German bacteriologist Much observed that some staphylococci when added to plasma form thick clumps visible to the naked eye.[1] He suggested

* This work was supported by grants from the National Institutes of Health HL 25,935, HL 25,107, HL 27,560, HE 12,759, HE 18,576, and GM 17,702.

† Present address: Division of Experimental Medicine, New England Deaconess Hospital, 185 Pilgrim Rd., Boston, MA 02215.

521

that this phenomenon can be related to the presence of fibrinogen in plasma. It took, however, almost 50 years until it was clearly established by Duthie [2] that fibrinogen was solely responsible for the staphylococcal clumping phenomenon. Subsequently, Allington [14] and Lipinski et al.,[15] established that the staphylococcal clumping reaction is also caused by fibrinogen derivatives such as fibrin monomers, large molecular weight products of fibrinogen degradation (FDP) present at very early stages of plasmin digestion, and soluble complexes formed between fibrin monomer and FDP. By standardizing the preparation of staphylococci it was possible to develop a rapid, simple, and sensitive method to measure fibrinogen and fibrin degradation products with a limit sensitivity between 0.5 to 1.0 μg of fibrinogen or its equivalent per ml of tested material.[16]

The high sensitivity of the staphylococcal clumping reaction toward fibrinogen, exemplified by the fact that it occurs when 20 molecules of fibrinogen per one staphylococcus are present,[17] indicates that the affinity of fibrinogen toward staphylococcal clumping factor must be high. Indeed, using ^{125}I-human fibrinogen and a steady-state binding system,[10] we determined that staphylococci bind (M ± SD) 2,130 ± 48 molecules of fibrinogen per cell with dissociation constant (K_d) of $9.9 \times 10^{-9} \pm 0.02 \times 10^{-9}$.

We have observed that degradation of fibrinogen with plasmin rapidly destroys the clumping reactivity of human fibrinogen.[16] We measured the binding of fibrinogen at different stages of limited digestion with plasmin, and compared that binding with the ability of digested fibrinogen to clump staphylococci. Binding decreased rapidly during the first 3–10 minutes and essentially paralleled a decrease in the clumping titer (from 512 to 64). Between 10 and 240 minutes of plasmin action the binding remained within the same reduced range, whereas the clumping titer was negative. Thus, the decrease in the clumping activity of fibrinogen closely paralleled a decline in binding.

Originally, we attributed the rapid decline in the clumping ability of fibrinogen treated with plasmin to the proteolytic removal of α chain portions known to be most susceptible to plasmin action.[18] Furthermore, by reducing and carboxymethylating ^{125}I-fibrinogen and adding staphylococci in the presence of 3M urea to keep chains in solution we observed association of α chains and β chains with staphylococci.[17] However, when isolated and well-characterized α chain fragments were used for binding studies in aqueous solution without urea no measurable association of the α chain fragments with staphylococci could be detected.[10] This fact prompted us to change our general approach to the study of the interacton of fibrinogen with staphylococci. Instead of using poorly soluble fragments of fibrinogen or of its chains for direct binding experiments, we chose to use antibodies specific for distinct fibrinogen regions to explore the possibility that one of those regions must bear a site recognizing the staphylococcal clumping receptor. Hence, antibody specific for such a region will block its interaction with staphylococci and thereby inhibit the staphylococcal clumping reaction. Monospecific Fab antibody fragments against complete fibrinogen, the central domain of fibrinogen (fragment E), the terminal domain (fragment D), and against the α chain fragment encompassing residues 241–476 were tested. Among these antibodies only those directed against fragment D, and against fibrinogen were effective (FIG. 1). Antibodies against fibrinogen inhibited 83% of ^{125}I-fibrinogen binding and abolished clumping reaction; antibodies against fragment D inhibited 79% of ^{125}I-fibrinogen binding and blocked the clumping reaction.[10]

We gained from these experiments a very strong indication that the structural feature of fibrinogen responsible for interaction with staphylococcal clumping is associated with fragment D, which is present in each of the two terminal domains of the fibrinogen molecule.[18] Since each of these terminal domains should possess at least one region interacting with staphylococcal clumping receptor the following predictions could be made: (1) the intact fibrinogen molecule with two binding regions on two opposite terminal domains will clump (agglutinate) staphylococci, (2) isolated fragment D representing terminal domain with one binding region will not clump staphylococci but will bind to its receptor, and (3) excess of fragment D bound to staphylococci prior to addition of fibrinogen will inhibit its binding and clumping. Our experiments have proven these predictions correct with some qualifications.

Among the family of fragments D generated during digestion of fibrinogen with plasmin larger fragments such as fragment D_1 (M_r 90,000) not only inhibited binding of ^{125}I-fibrinogen to staphylococci but also their clumping. In contrast fragment D_3 (M_r 80,000) was without effect (TABLE 1). Furthermore, Fragment D_1 reversed immunoinhibition of the staphylococcal clumping reaction by anti-fragment D antibody whereas fragment D_3 was inactive. In direct binding experiments ^{125}I-fragment D_1 showed a sixfold higher binding to staphylococci than ^{125}I-fragment D_3.[10]

When the subunit composition of both fragments was compared, the apparent differences between fragments D_1 and D_3 was primarily related to the loss of a sizable portion of γ chain (a decrease in apparent M_r of 13,000). Since fragments D_1 and D_3 had identical amino-terminal profiles, corresponding to Asp-105 of the α chain, Asp-134 of the β chain, and Ser-86 of the γ chain,[10] the loss of a portion of the γ chain of approximately M_r 13,000 occurred at its COOH-terminal region during the transition from fragment D_1 to D_3. Thus, the change in a sizable portion of the COOH-terminal segment of the γ chain

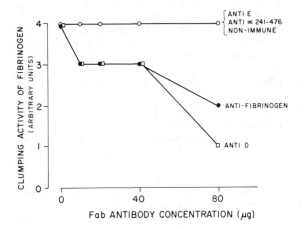

FIGURE 1. The effect of antibody Fab fragment on the clumping activity of fibrinogen toward staphylococci. Different concentrations of rabbit Fab antibody fragments were incubated with human fibrinogen (2 μg). Its reactivity toward staphylococci was tested in the clumping assay as described.[10] Only Fab antibody fragments against fibrinogen and fragment D inhibited the clumping reaction.

TABLE 1

INHIBITION OF BINDING AND CLUMPING ACTIVITY OF HUMAN FIBRINOGEN TOWARD
STAPHYLOCOCCI BY FRAGMENTS D *, †

System	^{125}I-Fibrinogen Bound to Staphylococci (ng/10^8 cells)	Clumping Reaction
Buffer + staphylococci + ^{125}I-labeled fibrinogen	23	4
Fragment D_1 + staphylococci + ^{125}I-labeled fibrinogen	5	2
Fragment D_3 + staphylococci + ^{125}I-labeled fibrinogen	25	4

* Adapted from reference.[10]

† Purified fragment D_1 (360 μg) and fragment D_3 (300 μg) were added to staphylococci prior to the addition of ^{125}I-labeled fibrinogen (4 μg). The binding assay and the staphylococcal clumping reaction were performed as described.[10]

encompassing 109 residues correlated with the ability of fragment D to interact with the staphylococcal clumping receptor.

The above conclusion was confirmed in a series of experiments with isolated γ chain. The addition of γ chain to anti-fragment D Fab antibody neutralized those antibody molecules that were blocking the fibrinogen region interacting with the staphylococcal clumping factor, thus resulting in a positive clumping reaction. Furthermore, γ chains without added antibody fragments interacted directly with staphylococci. Instead of solubilizing chains of fibrinogen in urea or other unfolding agents, isolated chains were dispersed in 0.05M ammonium bicarbonate, pH 7.8, and sonicated to assure a uniform suspension. Such suspensions of dispersed chains in the form of multimers in ammonium bicarbonate buffer, pH 7.8, appeared to be suitable for measuring their clumping activity with remarkable consistency and effectiveness. The γ chain was especially active toward the staphylococcal clumping factor, and it caused clumping of staphylococci at concentrations of 140 nM, which was of the same order as that of human fibrinogen (60 nM) (FIG. 2). Alpha chains showed 150-fold lower reactivity toward staphylococci than the γ chain.

The possibility of cross-contamination of the α chain exists. However, this seemed unlikely when one considers that β chain, which elutes between α and γ chains during isolation, remained inactive. It should be noted that a relatively high reactivity of a mixture of carboxymethylated chains toward staphylococci was observed by Stemberger and Horman,[19] although when isolated α, β, and γ chains were tested, they did not exhibit reactivity toward staphylococci. Since the chains in that study were twice dissolved in 8M urea and then dialyzed, this procedure could affect chain reactivity toward the staphylococcal receptor. The γ chain multimers effecting a positive clumping reaction with staphylococci were susceptible to plasmin digestion and to staphylococcal protease which reduced the clumping reactivity of γ chain multimers to very low values.[10]

The next series of experiments undertaken to pinpoint the interactive site on the γ chain more precisely were reported by Strong et al.[11] Since the γ chain has eight methionine residues, their cleavage with cyanogen bromide (CNBr) resulted in nine fragments.[20] After their fractionation on Sephadex G-75, only one pool designated "III + IV" was active in blocking the staphylococcal clumping reaction. This pool was resolved by paper electrophoresis at pH 2.0 into three bands which were eluted: only one showed a blocking effect on the staphylococcal clumping reaction induced with fibrinogen. The active material contained the well-characterized 27-residue CNBr COOH-terminal fragment.[11] This purified fragment lost its blocking activity after digestion with trypsin and chymotrypsin. In contrast, digestion with staphylococcal protease, which cleaves

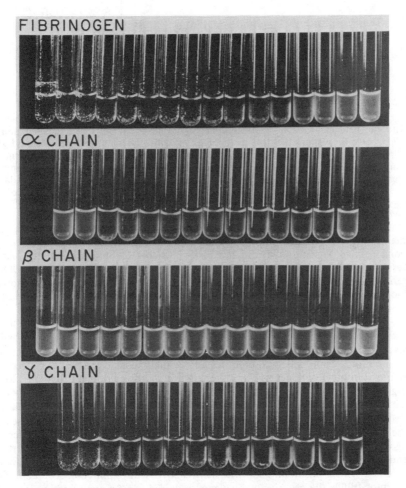

FIGURE 2. Direct interaction of fibrinogen and its isolated chains with staphylococci measured by the clumping reaction. Equivalent concentrations of fibrinogen and its chains [10] were prepared in twofold dilutions. Note the positive clumping reaction in tubes with fibrinogen and with γ chain multimers. Tubes containing α and β chain multimers show a lack of clumping reactivity towards staphylococci.

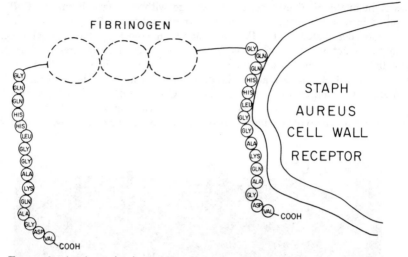

FIGURE 3. A schematic view of the three domainal molecule of human fibrinogen with the extended COOH-terminal segment of the γ chain with a detailed amino acid sequence responsible for its interaction with the clumping receptor on the staphylococcal cell wall. Only one γ chain recognition site for the staphylococcal clumping receptor is shown here interacting with the staphylococcus. An identical segment of the γ chain on the opposite half of the fibrinogen molecule is available to bind another staphylococcus, thus fulfilling the divalency requirement for the cell agglutinating function of fibrinogen.

mainly the bond on the carboxy side of glutamic acid and to a lesser degree the bond next to aspartic acid, did not cause any loss of blocking activity. As a result of this digestion, a 15-residue peptide corresponding to the COOH-terminal segment of the γ chain (residues 397–411) was isolated and it showed blocking activity toward staphylococcal clumping receptor when it was tested with fibrinogen.[11]

When it became apparent that the pentadecapeptide representing the COOH-terminal segment of the γ chain inhibits the interaction of fibrinogen with staphylococcal clumping receptor, a series of synthetic peptides corresponding to the same region was made and tested for their blocking activity.[11] Preparation of synthetic peptides of variable length demonstrated again that the blocking activity was confined to the 15-amino acid residues encompassing γ397–411.

A three-dimensional representation of the fibrinogen molecule [18] indicates that two symmetrical COOH-terminal segments of the γ chain are exposed at the extremities of the tridomainal structure, as postulated more than a decade ago.[21] The COOH-terminal segment of the γ chain is an integral part of fragment D, and two fragments D are recognized within one fibrinogen molecule.[23-25] Such positioning of the site reactive with staphylococcal clumping receptor fulfills the criteria for bivalency required for the agglutinating function of fibrinogen in staphylococcal clumping phenomenon (FIG. 3). On the other hand, monovalent fragment D_1 or isolated pentadecapeptide (γ 397–411) interact with the staphylococcal clumping receptor but without causing agglutination. Instead, they block the clumping function of fibrinogen. However, when monovalent synthetic pentadecapeptide (corresponding to the sequence of γ397–411) was conjugated with bovine serum albumin, which is not reactive by itself with staphylococcal clumping receptor, a polyvalent molecule was formed and was fully capable of causing a specific clumping reaction.[11] This experiment clearly demonstrated that the synthetic pentadecapeptide can not only block the agglutinating function of fibrinogen but also can provide a functional arm to an inert carrier molecule, albumin, to endow it with specific cell agglutinating properties toward staphylococci.

Thus, the localization of the binding region for the staphylococcal clumping factor to the 15-amino acid sequence at the COOH-terminus of the γ chain gives solid structural evidence for the cell agglutinating function of fibrinogen in regard to staphylococci.

INTERACTION OF FIBRINOGEN WITH HUMAN PLATELETS

The adhesion of platelets to the inner surface of the injured blood vessel and to each other plays a fundamental role in hemostasis and thrombosis.[26] These interactions require not only von Willebrand factor but also fibrinogen, because in the congenital deficiency of either of these glycoproteins, a hemorrhagic diathesis is observed in which the capillary bleeding time is prolonged.[27, 28]

The essential role of fibrinogen in the adhesion and aggregation of platelets has gained considerable evidence from the recent demonstration that fibrinogen binds to specific receptor sites on platelets. This binding occurs only when platelets are stimulated with certain agonists such as ADP, epinephrine, and thrombin.[6, 29-31] These agonists initiate the process of exposure of specific fibrinogen receptors otherwise not accessible. In contrast to staphylococci whose clumping factor receptor is always available, platelet receptors must be exposed and require calcium for their functions. Receptor exposure is regulated by changes in cyclic AMP reported elsewhere in this volume.[32] Until recently the localization of the region on the fibrinogen molecule responsible for interaction with the exposed platelet receptor remained unknown; however, evidence reported from our laboratories established which structural subunits of fibrinogen participate in its interaction with receptors on human platelets and what is the precise location of the main site recognizing these receptors.[12-13]

Because aggregation of platelets induced by ADP requires fibrinogen [5, 33-35] and is concomitant with its binding to the platelet receptor,[29, 36, 39] our basic system employed ADP-induced platelet aggregation. Human platelets were separated from plasma fibrinogen and, under such conditions, addition of fibrinogen in a concentration of 0.4 mg/ml (3.7 μM) resulted in platelet aggregation. When reduced and carboxymethylated fibrinogen composed of a mixture of separated chains dispersed in ammonium bicarbonate, pH 7.8, was tested, aggregation was somewhat less than with intact fibrinogen. However, isolated γ chains (7.5 μM) produced aggregation equivalent to that of intact fibrinogen. Alpha chains in an equimolar concentration produced a weaker effect, and β chains were inactive (FIG. 4).

The effect of isolated γ chains was concentration-dependent in terms of aggregation parameters (Tmax). The definite aggregating effect was still observed with 1.5 μM γ chain per 10^8 platelets (FIG. 5). In comparison, α chain gave measurable aggregation at 7.5 μM per 10^8 platelets.

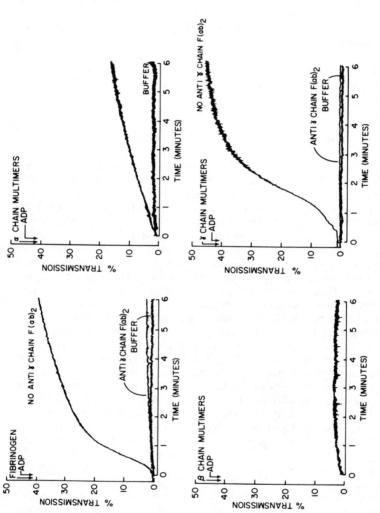

FIGURE 4. Aggregation of human platelets induced by ADP in the presence of fibrinogen or its isolated chains. Note the full aggregation in the presence of fibrinogen or γ chain multimers. The α chain multimers at an equivalent concentration were much less active and β chain multimers were inactive. Antibody F(ab)₂ fragments against γ chain inhibited ADP-induced platelet aggregation in the presence of fibrinogen and of γ chain (for details see ref. 12).

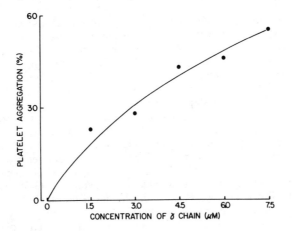

FIGURE 5. Concentration-dependent aggregation effect of the γ chain of fibrinogen on human platelets treated with ADP. The γ chain multimers were added to human platelets which were separated from plasma proteins (for details see ref. 12).

The β chain multimers were inactive toward platelets (FIG. 4). This means that, in spite of a significant homology between β and γ chains,[8, 22] the latter possess a highly chain-specific region interacting with platelets. The most significant difference between β and γ chains is that the β chain lacks the similar COOH-terminal segment containing cross-linking sites that are present in the γ chain.[37] Lack of reactivity by the β chain also indicates a lack of contamination of this chain by γ chain.

Because the platelet-aggregating activity of carboxymethylated fibrinogen was somewhat less than that of isolated γ chain, we examined the possibility of interference by β chain. Random reassociation of γ chain in the presence of β chains diminished the reactivity of the γ chain multimers with the platelet receptor.[12] This may explain the relatively weaker reactivity of the mixture of the α, β, and γ chains in the carboxymethylated fibrinogen preparation.

Since fibrinogen has a dimeric structure and is composed of two γ chains, in addition to pairs of α chains and β chains,[18] the aggregation of platelets by the γ chains would require dimers or larger multimers. Fortuitously, owing to the poor solubility of isolated γ chains in an aqueous solution, we worked with dispersed multimers of this chain and fulfilled requirements for multivalent γ chain arrangement. Alpha and β chains demonstrated a similar pattern of self-association in aqueous solution.[12] However, when the γ chains were separated twice on CM cellulose in such a way as to retain a monomeric structure they did not aggregate platelets treated with ADP.

It was apparent from these experiments that the direct interaction of the γ chain multimers with ADP-treated platelets resulted in their aggregation. However, we sought another line of evidence to prove that the γ chain of human fibrinogen is responsible for its interaction with the platelet receptor. Again, as before in the case of staphylococci, an immunologic approach has been used. Monospecific rabbit antibody F(ab)₂ fragments against γ chains were prepared in our laboratory. Such antibody fragments against γ chain inhibited binding of ^{125}I-fibrinogen to its platelet receptor by 60% and blocked platelet aggregation induced by ADP in the presence of purified plasma fibrinogen (FIG. 4). The antibody fragments also inhibited aggregation of ADP-treated platelets by the γ chain multimers. Accumulated lines of evidence firmly indicated that among three distinct chains (α, β, γ) of human fibrinogen the γ chain bears the main site reactive with human platelet receptor.[12] The analogy to the structural situation with the site for staphylococcal clumping factor was striking, and we decided to follow a similar approach in order to pinpoint the precise location of the site on the γ chain interacting with the platelet receptor.

Because platelets required a much higher concentration of fibrinogen and its chains than staphylococci, we modified[13] the procedure employed previously in work with the staphylococcal binding site.[11] The γ chain peptides obtained after CNBr cleavage were fractionated on a Sephadex G-50 column in 1% formic acid. The same peptide pool that inhibited the staphylococcal clumping reaction blocked the binding of ^{125}I-fibrinogen to human platelets treated with ADP in the experimental system used previously.[31] The pool containing inhibitory peptide(s) was then fractionated on a preparative HPLC column. Again, among several peaks the one that blocked the staphylococcal clumping reaction also inhibited binding of ^{125}I-fibrinogen to the platelet receptor. The inhibitory peptide produced one peak in analytical reverse phase HPLC. Amino acid analysis revealed a composition corresponding to the 27 residue CNBr COOH-terminal peptide. The NH_2-terminal sequence was Lys-Ile-Ile, corresponding to the γ385–411 peptide.[13]

The isolated 27-residue peptide inhibited binding of ^{125}I-fibrinogen to ADP-treated platelets in a concentration-dependent manner reaching full inhibition at 30 μM. The concentration of the peptide required for 50% inhibition of binding of ^{125}I-fibrinogen was 7 μM (FIG. 6). This represented a 44-fold molar excess in regard to ^{125}I-fibrinogen. Thus, the blocking peptide isolated from γ chain in our laboratory[13] was one order of magnitude more active than fragment D in a similar binding system and only one order of magnitude weaker than intact unlabeled fibrinogen.[38] In parallel, aggregation of ADP-treated platelets was completely inhibited by the peptide at 0.4 mM corresponding to a 2500-fold molar excess and representing more than enough to inhibit platelet aggregation by 50%.

The 27-residue peptide completely lost its blocking properties after digestion with trypsin (FIG. 7). Digestion with trypsin generated four peptides, among which the largest one (γ392–406) has 15 residues and is apparently devoid of blocking activity. However, when digestion with staphylococcal protease was performed only two peptides, a dodecapeptide (γ385–396) and a pentadecapeptide (γ397–411) were generated. The mixture of both peptides fully retained its blocking activity toward the platelet receptor as well as toward the staphylococcal clumping factor. Because trypsin-generated pentadecapeptide (γ392–406) was devoid of blocking activity and staphylococcal protease-generated peptides retained blocking activity toward both receptors, it appears likely that the pentadecapeptide (γ397–411) is required for interaction with the platelet receptor (FIG. 8). This segment of the γ chain is composed of highly hydrophilic amino acids without any predominant type of secondary structure.[13]

In addition to the site interacting with the platelet receptor and the staphylococcal clumping receptor, this segment of the γ chain possesses the cross-linking donor and acceptor sites.[21] It is susceptible to proteolysis by

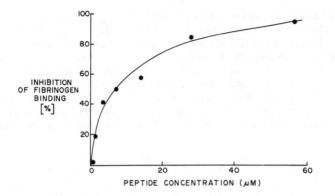

FIGURE 6. The inhibitory effect of the isolated 27 residue COOH-terminal segment of the γ chain (γ385–411) on binding of ^{125}I-fibrinogen to its receptor on human platelets treated with ADP (for details see ref. 13).

plasmin and this may explain the progressive loss of fibrinogen's ability to interact with platelet receptors after plasmin attack.

Whereas the γ chain bears the main site structurally and functionally identified as responsible for interaction with receptors on human platelets,[12, 13] there is still a problem of the α chain site. We have shown in direct experiments with isolated chains, that the α chain of human fibrinogen also interacts with ADP-treated human platelets.[12] This interaction is significantly weaker than those of the γ chain or intact fibrinogen. The existence of a site(s) interacting with human platelets on the α chain has been postulated by Niewiarowski and

```
       CNBr          TR   TR          SP                    TR        CPA
      ↓ K-I-I-P-F↓N-R↓L-T-I-G-E↓G-Q-Q-H-H-L-G-G-A-K↓Q-A-G-D↓V-COOH
         25          20          15          10          5
```

| K-I-I-P-F-N-R-L-T-I-G-E-G-Q-Q-H-H-L-G-G-A-K-Q-A-G-D-V-COOH | CNBr POOL4

G-Q-Q-H-H-L-G-G-A-K-Q-A-G-D-V-COOH   SP2

SP1   K-I-I-P-F-N-R-L-T-I-G-E

TR1   K-I-I-P-F

TR2   N-R

TR3   L-T-I-G-E-G-Q-Q-H-H-L-G-G-A-K

Q-A-G-D-V-COOH   TR4

FIGURE 7. Amino acid sequence of the 27 residue CNBr COOH-terminal peptide of the γ chain and of its enzyme derived fragments. Only the 27 residue peptide designated CNBr Pool IV (enclosed by a solid line) and the 15-residue peptide derived from digestion with staphylococcal protease and designated SPII (enclosed by a dashed line) were involved in binding of ^{125}I-fibrinogen to its platelet receptor;[13] the remaining peptides were inactive.

his colleagues on the basis of a loss of platelet aggregating activity of fibrinogen during the early stages of plasmin digestion.[35, 36] Whether early proteolysis with plasmin is limited to the COOH-terminal zone of the α chain or affects also the COOH-terminal segment of the γ chain bearing the main site interacting with platelet receptors, remains to be established. Nevertheless, our direct demonstration of the α chain reactivity toward platelets [12] in addition to the γ chain–specific site [13] introduces a greater degree of complexity to the fibrinogen interaction with platelet receptors. The reported existence of at least two classes of fibrinogen receptors of high and low affinity [36, 39] and the demonstration of

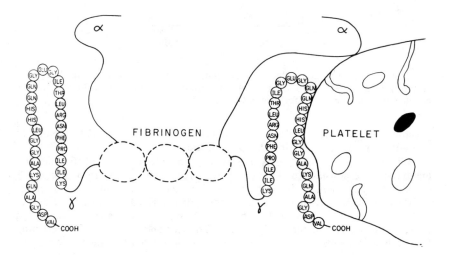

FIGURE 8. Schematic presentation of the γ chain fragment of human fibrinogen interacting with the platelet receptor. Fibrinogen is outlined here as a three domainal structure with an extended COOH-terminal segment of γ chain. The detailed amino acid sequence of the γ chain [13] is responsible for recognition of the human platelet receptor of γ chain specificity. The extended COOH-terminal segment of the α chain is shown also as interacting with a platelet receptor on the basis of its ability (albeit much weaker than γ chain) to participate in platelet aggregation.[12] COOH-terminal segments of the γ chain and of the α chain are also available on the opposite half of the fibrinogen molecule to interact with a corresponding set of receptors on another platelet.

fibrinogen binding to thrombospondin [40] bring about the distinct possibility of recognition of receptors on platelets as fibrinogen receptors of γ chain specificity and of α chain specificity. Such a functional distinction will provide the mechanism for multiple site interactions, which in a cooperative manner will enhance the ability of intact fibrinogen to interact with platelets.

SUMMARY

Fibrinogen, a clottable plasma glycoprotein, participates in cell adhesion phenomena involving prokaryotic cells, e.g. staphylococci, and eukaryotic cell fragments, e.g. platelets. Among the three chains (α, β, γ) of human fibrinogen, the γ chain bears the main site recognizing the staphylococcal clumping receptor and human platelet receptor induced by ADP. The platelet receptors are also recognized, albeit less avidly, by a site associated with the α chain. The γ chain site recognizing staphylococcal clumping factor exists on the COOH-terminal segment of this chain encompassing the 15 residues (γ397–411) including the COOH-terminal valine. The location of the γ chain site interacting with the human platelet receptor had been pinpointed to the 27 residue CNBr COOH-terminal segment (γ385–411). The results of enzymatic degradation of the 27-residue peptide indicate that the continuity of the last 15 amino acid residues at the COOH-terminal end of the γ chain of human fibrinogen seems to be essential for its interaction with human platelets. The sequence of the γ chain interacting with the platelet receptor (γ385–411) indicates that this segment is a unique region of fibrinogen endowed with three important functions: cross-linking of fibrin, clumping of staphylococci, and aggregation of platelets.

[**Note added in proof:** Recently we obtained evidence that dodecapeptide γ393–411 fully retains platelet receptor recognition site (Kloczewiak *et al.* 1983. Clin. Res **31:** 534A.)]

REFERENCES

1. Much, H.   1908.   Biochem. Z. **14:** 143–155.
2. Duthie, E. S.   1955.   J. Gen. Microbiol. **13:** 383–393.
3. Hawiger, J.   1976.   *In* Proceedings of the Workshop on Animal Models of Thrombosis and Hemorrhagic Diseases. : 127–131. National Academy of Science. Washington, D.C.
4. Hawiger, M. M., S. Timmons & J. Hawiger.   1979.   Clin. Res. **27:** 345A.
5. Born, G. V. R. & M. H. Cross.   1964.   J. Physiol. **170:** 397–414.
6. Mustard, J. F., M. A. Packham, R. L. Kinlough-Rathbone, D. W. Perry & E. Rogoeczi.   1978.   Blood. **52:** 453–567.
7. Henschen, A. & F. Lottspeich.   1977.   Hoppe-Seyler Z. Physiol. Chem. **358:** 935–938.
8. Watt, K. W. K., T. Takagi & R. F. Doolittle.   1978.   Proc. Natl. Acad. Sci. USA **75:** 1731–1735.
9. Doolittle, R. F., K. W. K. Watt, B. A. Cottrell, D. D. Strong & M. Riley.   1979.   Nature **280:** 464–468.
10. Hawiger, J., S. Timmons, D. D. Strong, B. A. Cottrell, M. Riley & R. F. Doolittle.   1982.   Biochemistry **21:** 1407–1413.
11. Strong, D. D., A. P. Laudano, J. Hawiger & R. F. Doolittle.   1982.   Biochemistry **21:** 1414–1420.
12. Hawiger, J., S. Timmons, M. Kloczewiak, D. D. Strong & R. F. Doolittle.   1982.   Proc. Natl. Acad. Sci. USA **79:** 2068–2071.
13. Kloczewiak, M., S. Timmons, J. Hawiger.   1982.   Biochem. Biophys. Res. Commun. **107:** 181–187.
14. Allington, M. J.   1967.   Br. J. Haemotal. **13:** 550–567.
15. Lipinski, B., J. Hawiger & J. Jeljaszewicz.   1967.   J. Exp. Med. **126:** 979–998.
16. Hawiger, J., S. Niewiarowski, V. Gurewich & D. P. Thomas.   1970.   J. Lab. Clin. Med. **75:** 93–108.
17. Hawiger, J., D. K. Hammond, S. Timmons & A. Z. Budzynski.   1978.   Blood. **51:** 799–812.
18. Doolittle, R. F., H. Bouma, B. A. Cottrell, D. Strong & K. W. K. Watt.   1979.   *In* The Chemistry and Physiology of the Human Plasma Proteins. D. H. Bing, Ed.: 77–96. Pergamon Press. Elmsford, NY.

19. STEMBERGER, A. & H. HORMAN. 1974. Thromb. Res. **4:** 753–756.
20. SHARP, J. J., K. G. CASSMAN & R. F. DOOLITTLE. 1972. FEBS. Letters **25:** 334–336.
21. CHEN, R. & R. F. DOOLITTLE. 1971. Biochemistry **10:** 4486–4491.
22. HENSCHEN, A. & F. LOTTSPEICH. 1977. Throm. Res. **11:** 869–880.
23. TAKAGI, T. & R. F. DOOLITTLE. 1975. Biochemistry **14:** 940–946.
24. MARDER, V. J. 1970. Thromb. Diath. Haemorrh. Suppl. **39:** 187–195.
25. PIZZO, S. W., M. L. SCHWARTZ, R. L. HILL & P. A. McKEE. 1973. J. Biol. Chem. **248:** 4574–4583.
26. SIXMA, J. J. & J. WEBSTER. 1977. Semin. Hematol. **14:** 265–299.
27. HOYER, L. W. 1976. *In* Progress in Hemostasis and Thrombosis. T. H. Spaet, Ed. Vol. **3:** 231–288. Grune & Stratton. New York.
28. WEISS, H. J. & J. ROGERS. 1971. N. Engl. J. Med. **285:** 369–374.
29. MARGUERIE, G. A., E. F. PLOW & T. S. EDGINGTON. 1979. J. Biol. Chem. **254:** 5357–5363.
30. BENNET, J. S. & G. VILAIRE. 1979. J. Clin. Invest. **64:** 1393–1401.
31. HAWIGER, J., S. PARKINSON, S. TIMMONS. 1980. Nature (Lond.) **283:** 195–197.
32. GRABER, S. & J. HAWIGER. 1983. Ann. NY Acad. Sci. **408:** This volume.
33. McLEAN, J. R., R. E. MAXWELL & D. HERTLER. 1964. Nature (Lond.). **202:** 605–606.
34. BRINKHOUS, K. M., M. S. READ & R. G. MASON. 1965. Lab. Invest. **14:** 335–342.
35. NIEWIAROWSKI, S., A. Z. BUDZYNSKI & B. LIPINSKI. 1977. Blood **49:** 635–644.
36. NIEWIAROWSKI, S., A. Z. BUDZYNSKI, T. A. MORINELLI, T. M. BRUDZYNSKI & G. J. STEWART. J. Biol. Chem. **256:** 917–925.
37. TAKAGI, T. & R. F. DOOLITTLE. 1975. Biochem. Biophys. Acta **386:** 617–622.
38. PLOW, E. F. & G. A. MARGUERIE. 1982. Fed. Proc. **41:** 1119.
39. PEERSCHKE, E. I., M. B. ZUCKER, R. A. GRANT, J. J. EGAN & M. M. JOHNSON. 1980. Blood **55:** 841–847.
40. NACHMAN, R. L. & L. L. K. LEUNG. 1981. Blood **58:** 199a.

DISCUSSION OF THE PAPER

J. McDONAGH (*University of North Carolina School of Medicine, Chapel Hill*): Do D dimers either bind or inhibit fibrinogen binding to staphylococci or to platelets?

J. HAWIGER: Yes they do, in the case of staphylococci.

McDONAGH: Do they have the same affinity? Is there any difference between the dimer and the monomeric form?

HAWIGER: We did not quantitate exactly in terms of the molar concentrations, but it appears as if the D dimer is as active as monomeric fragment D.

McDONAGH: Is the fibrinogen site for platelets and staphylococci the same?

HAWIGER: Well, from the work that I presented it appears that this is an essentially identical segment.

P.J. GAFFNEY (*National Institute for Biological Standards and Control, London*): Our work published two years ago suggested that fibrinogen supported ADP-induced platelet aggregation, fragment X did also but to a lesser degree, and fragments Y D, and E obviously did not. The suggestion we made at that time was that the dimeric structure may be involved in some way. My question is, does that fit into your scheme of things? Do you need a double-D to actually support platelet-induced aggregation? Would you agree with this notion?

HAWIGER: Yes and no. It is not only a matter of possessing the crucial segment of the γ chain, because we did not obtain aggregation supported by double-D dimers there; we are talking about two D's linking to platelets.

Consider, however, the conjugate that was prepared in Doolittle's laboratory where an active synthetic peptide was coupled to albumin. This synthetic clumping molecule has full clumping activity with staphylococci. Obviously, you need to have bivalency or polyvalency. In some fragments X, it is possible that the crucial COOH-terminal of the γ chain is present on both sides of the molecule. In some fragments X and in fragment Y, it will be absent.

GAFFNEY: There is an important point here. You did say that D dimer does not support platelet aggregation?

HAWIGER: Yes.

GAFFNEY: Okay, but remember D-dimer has the component parts coming from two distinctly separate molecules.

HAWIGER: Yes, but the geometry is completely different.

FIBRINOGEN INTERACTION WITH PLATELET RECEPTORS [*][†]

Stefan Niewiarowski,[‡] Elizabeth Kornecki, Andrei Z. Budzynski,[§]
Thomas A. Morinelli,[‡] and George P. Tuszynski

Thrombosis Research Center
Departments of Physiology [‡] and Biochemistry [§]
Temple University Health Sciences Center
Philadelphia, Pennsylvania 19140

INTRODUCTION

It has been accepted for many years that the major physiological role of fibrinogen is related to its conversion to fibrin. More recent studies indicate the importance of fibrinogen in platelet aggregation. In 1964, Cross [1] and McLean *et al.*[2] observed that fibrinogen plays the role of an essential cofactor in aggregation of platelets induced by ADP. This observation has been subsequently confirmed by a number of other investigators.[3-8] Tollefsen and Majerus [9] demonstrated that the monovalent Fab fragment of antifibrinogen antibody blocks thrombin-induced platelet aggregation and suggested that platelet fibrinogen released by thrombin subsequently participates in this process. In 1978, Mustard and colleagues [10] demonstrated that binding of ^{125}I-fibrinogen to washed platelets occurred during ADP-induced platelet aggregation. More recently, a number of investigators demonstrated that fibrinogen receptors are not available on intact platelets but can be exposed following treatment with ADP [11-14] and other aggregating agents such as thrombin,[15, 16] epinephrine,[12, 17] and prostaglandin endoperoxides.[18, 19] Moreover, Greenberg *et al.*[20] demonstrated that platelets treated by chymotrypsin or plasmin are aggregated directly by fibrinogen and showed that calcium is necessary in the incubation mixture. We have shown [21] that ADP and proteolytic enzymes (pronase and chymotrypsin), acting by different mechanisms, expose very similar fibrinogen receptors on the platelet surface and that platelet aggregation depends on the occupancy of these platelet fibrinogen receptors. In contrast to aggregation of platelets induced by ADP, fibrinogen-induced aggregation of platelets treated with pronase or chymotrypsin was irreversible. Platelets treated with proteolytic enzymes aggregated with fibrinogen directly even in the absence of ADP and in the presence of metabolic inhibitors.[21] Calcium is essential for fibrinogen binding to ADP-stimulated [12, 13, 22] and to chymotrypsin-treated platelets [21] and for fibrinogen-induced platelet aggregation. This article will focus on two points: (1) a discussion of the nature of the

* Supported by Grants HL–15226, HL–14217, HL–28149 from the National Institutes of Health by Training Grant HL–05976, and by National Research Service Award HL–06356. The present address of E.K. is: Department of Psychiatry, University of Vermont, Burlington, VT. Reprint requests should be sent to S. Niewiarowski, Thrombosis Research Center, Temple University Health Science Center, Philadelphia, PA 19140.

† Abbreviations: Fragment X (st. 1)—Fragment X (stage 1); GPIIb—glycoprotein IIb; GPIIIa—glycoprotein IIIa; SDS—sodium dodecyl sulfate.

platelet surface receptor for fibrinogen and (2) the localization of the binding region of the fibrinogen molecule which reacts with the platelet fibrinogen receptors.

EXPOSURE OF FIBRINOGEN RECEPTORS

Fibrinogen receptors are latent on intact unstimulated platelets thus preventing their spontaneous aggregation by plasma fibrinogen. The molecular mechanisms by which ADP or chymotrypsin expose fibrinogen receptors may possibly involve a 100,000 M_r polypeptide on the platelet surface. Figures *et al.*[23] have postulated that this polypeptide can be altered either by ADP-induced conformational changes or by proteolysis resulting in the exposure of specific fibrinogen receptors. As yet there is no agreement regarding the mechanism by which other aggregating stimuli expose fibrinogen receptors on the platelet surface. Bennett *et al.*[18] showed that exogenously added PGH_2 stimulated the exposure of fibrinogen receptors by a mechanism that was independent of secreted platelet ADP. Accordingly, Harfenist *et al.*[24] showed that arachidonate induced the binding of fibrinogen to thrombin-degranulated rabbit platelets independently of released ADP. However, other investigators have presented evidence that shows that the exposure of fibrinogen receptors by epinephrine [17] and thrombin [16] was mediated by ADP. Our data provide evidence that the exposure of fibrinogen receptors by stable prostaglandin endoperoxide analogues may be mediated by low concentrations of ADP, those which can not be detected by enzymatic methods.[19] The role of ADP can be demonstrated by a number of different methods: (1) the use of enzymes that remove this nucleotide from the system under study (i.e., apyrase or creatine phosphokinase in the presence of creatine phosphate); (2) the use of ATP, a competitive antagonist of ADP; and (3) 5'parafluorosulfonylbenzoyl adenosine (FSBA), an affinity label used for the study of ADP binding sites. In our own experimental system, all these agents mentioned here blocked both platelet aggregation and the binding of ^{125}I-fibrinogen to platelets stimulated by stable prostaglandin endoperoxide analogues.[19]

CHARACTERISTICS OF FIBRINOGEN RECEPTORS EXPOSED ON THE PLATELETS

Fibrinogen-binding sites on the platelet surface have been characterized by many investigators; however, the issue of their heterogeneity remains controversial. Marguerie *et al.*[11] estimated that the number of fibrinogen receptors which specifically bound ^{125}I-fibrinogen on an ADP-stimulated platelet was about 4600 ± 200 with a dissociation constant of 1.3×10^{-7} M. However, working in a system with optimalized calcium and magnesium concentration, these authors reported 49,800 fibrinogen receptors per platelet.[22] More recently, Marguerie *et al.*[25] studied the binding of radiolabeled fibrinogen to platelets in a milieu of plasma from a patient with congenital afibrinogenemia. By Scatchard plot analysis they found 32,000 fibrinogen molecules bound per platelet to a single class of receptors with a dissociation constant of 3.6×10^{-7} M. Bennett and Vilaire [12] found one class of binding sites (45,000 sites/ platelet) on ADP stimulated platelet with a K_d of 0.8×10^{-7} M. Hawiger *et al.*[15] estimated that 50,000 molecules of fibrinogen were bound per platelet to ADP-stimulated platelets with an apparent dissociation constant of 1.5×10^{-7} M. On the other hand, Peerschke et al.[13] obtained a curvilinear Scatchard plot when ^{125}I-fibrinogen binding to platelets were studied. They concluded that this may either be the result of negative cooperativity between a single class of receptors or that their results may be due to the presence of two or more distinct classes of binding sites. Assuming the presence of two classes of fibrinogen receptors, the Scatchard plot was resolved into two components illustrating a high affinity binding site with an apparent dissociation constant of 1.3×10^{-7} M and 7,000 receptors per platelet and a second, lower affinity site (K_d 4.3×10^{-7} M) with approximately 9,000 receptors per platelet. We detected 1,860 high affinity binding sites ($K_d = 1.9 \times 10^{-8}$ M) and 88,000 low affinity binding sites per platelet ($K_d = 3.8 \times 10^{-6}$ M) on chymotrypsin-treated platelets.[21] On ADP-stimulated platelets, the number of high affinity binding sites was 1,300 with a K_d of 3.2×10^{-8} M and there were 80,000 low affinity binding sites with a $K_d = 5.6 \times 10^{-6}$ M.[14]

High affinity fibrinogen-binding sites on the platelet surface may remain undetectable in certain preparations of platelets. For instance, we detected one class of fibrinogen-binding sites on human washed platelets stimulated by prostaglandin endoperoxide analogues that cause secretion of α-granule proteins.[19] Platelets stimulated by these compounds secrete fibrinogen which may bind to high affinity receptors competing with the labeled fibrinogen. Our observations indicate that the process of washing platelets by gel filtration may induce secretion of platelet fibrinogen which interferes subsequently with the detection of high affinity binding sites. Platelets isolated by differential centrifugation at 37° C with an excess of apyrase [8] appear to be most suitable for fibrinogen-binding studies. In nine experiments, we estimated fibrinogen concentration by means of the staphylococcal clumping assay,[26] and we found that the level of fibrinogen in the supernatants of the suspension of human washed platelets [8] and in the supernatants of the suspension of gel-filtered platelets [11] amounted to 0.84 ± 0.5 µg/ml and 5.06 ± 6.35 µg/ml, respectively. We consider that fibrinogen present in gel-filtered platelets as well as fibrinogen (10–11 µg/ml) present in afibrinogenemic plasma [25] may interfere with the detection of high affinity fibrinogen-binding sites on the platelet surface.

There is general agreement that calcium or magnesium are essential for binding of fibrinogen to its receptors both on ADP-stimulated and on chymotrypsin-treated platelets.[13, 21, 22] Binding of ^{125}I-fibrinogen is maximal at 1 to 3 mM for both calcium and magnesium.[22]

The interaction of fibrinogen with its receptors is time dependent. Time required to reach binding equilibrium varies depending on the preparation of platelets and fibrinogen used by various investigators.[12, 21–23] Equilibration of fibrinogen is much more rapid in suspensions of chymotrypsin-treated than in suspensions of ADP-stimulated platelets. ^{125}I-fibrinogen incubated with platelets for a short period of time (up to 5 min) can be easily displaced by an excess of unlabeled fibrinogen. However, prolonged incubation of ^{125}I-fibrinogen with ADP-stimulated platelets results in a formation of irreversible (i.e., nondisplaceable by unlabeled fibrinogen) but noncovalent bonds between fibrinogen and its platelet receptor.[22]

In experiments performed over wide ranges of temperature (4° C–37° C), similar values were obtained for the binding constants and number of fibrinogen receptors using both chymotrypsin-treated platelets [14] and ADP-stimulated platelets.[27] This suggests that the fluidity of platelet membranes is not a major factor in the fibrinogen–platelet receptor interaction. Accordingly, degradation of platelet membrane glycoproteins by chymotrypsin or pronase does not result in any alteration of platelet membrane fluidity as determined by a polarization of a fluorescent probe (G. Y. Stewart, S. Niewiarowski, Y. Whitin, E. Simons, T. A. Morinelli, and G. P. Tuszynski, unpublished observations).

By studying both the binding of ^{125}I-fibrinogen to ADP-stimulated platelets and the velocity of platelet aggregation in the presence of ADP and increasing concentrations of fibrinogen, Marguerie *et al.*[11] found that the dissociation constant obtained from binding studies was close to the value obtained from analysis of platelet aggregation curves. In our experimental system, K_m values for fibrinogen calculated on the basis of the rate of aggregation were 1.8×10^{-7} M for ADP-stimulated platelets and $1.0–5.0 \times 10^{-7}$ M for chymotrypsin-treated platelets.[14, 21] At higher concentrations of fibrinogen, platelet aggregation was inhibited. The inhibition constants (K_i) for fibrinogen in case ADP stimulated platelets and for chymotrypsin-treated platelets were 2.6×10^{-5} M and 8.5×10^{-5} M, respectively.[14]

DEFICIENCY OF FIBRINOGEN RECEPTORS IN GLANZMANN'S THROMBASTHENIA

Glanzmann's thrombasthenia is an inherited blood disorder characterized by a long bleeding time and delayed clot retraction. Thrombasthenic platelets do not aggregate in the presence of ADP, collagen, thrombin, and epinephrine.[28, 29] Thrombasthenic platelets do not bind ^{125}I-fibrinogen after ADP stimulation and hence, do not expose fibrinogen receptors. Thrombasthenic platelets are deficient in platelet membrane glycoproteins IIb (GPIIb) and IIIa (GPIIIa).[30, 31] This indicates that GPIIb and GPIIIa may be associated with fibrinogen receptors.

During our studies on four patients with Glanzmann's thrombasthenia, we found that chymotrypsin treatment of their platelets leads to the exposure of high affinity fibrinogen receptors (900–1,200 receptors per platelet, $K_d = 10^{-7}$ M) but it does not expose low affinity receptors.[14] In our hands, chymotrypsin-treated thrombasthenic platelets showed a slight but significant aggregation induced by low concentrations of fibrinogen.[14] Higher concentrations of fibrinogen ($>10^{-6}$ M) blocked this process.[14] On the other hand, Mustard *et al.*[32] did not observe aggregation of such platelets treated with chymotrypsin upon addition of fibrinogen.

Recently we determined that the level of ^{125}I surface–labeled GPIIIa on the platelets of three thrombasthenic patients studied in our laboratory amounted to 5–7% of the normal values. We also had an opoprtunity to study a thrombasthenic patient whose platelets contained less than 1% of the normal GPIIIa level. The platelets of this patient did not expose fibrinogen receptors following treatment with chymotrypsin. These data suggest that high affinity fibrinogen receptors detected on thrombasthenic platelets may be derived from residual GPIIIa (S. Niewiarowski, E. Kornecki, G. P. Tuszynski, C. Soria, J. Soria, F. Dunn, and unpublished observations).

ASSOCIATION OF FIBRINOGEN RECEPTORS WITH
PLATELET MEMBRANE GLYCOPROTEINS

Recent data provide further evidence that fibrinogen receptors may be associated with glycoproteins IIb and IIIa on the platelet membranes. It appears that these glycoproteins form a complex linked by calcium ions.[33] Nachman and Leung [34] demonstrated that partially purified GPIIb and GPIIIa formed a complex with the purified fibrinogen passively absorbed to the wells of the plastic microtitration plates. The formation of this complex was calcium dependent, fibrinogen specific and saturable. More recently, Bennett et al.[35] demonstrated incorporation of fibrinogen, modified with a photoreactive and cross-linking reagent, into a 105,000 molecular weight membrane polypeptide that also contained the P1[A1] antigen. They suggested that this polypeptide was identical with glycoproteins IIIa and that it constituted at least one component of the platelet fibrinogen receptor. However, antisera raised in different laboratories against glycoprotein IIb and IIIa did not show any consistent effect on platelet aggregation and on fibrinogen binding to platelets. Kaplan and Nachman [36] showed that antiplatelet membrane Fab fragments inhibited platelet aggregation and granular secretion, although an antibody directed against a major membrane glycoprotein (possibly GPIII) had no effect. Using polyclonal antibodies raised in rabbits against GPIIb and GPIIIa, two groups of investigators showed either no effect [37] or only a slight inhibitory effect on platelet aggregation.[38] Monoclonal hybridoma antibodies directed against GPIIb and GPIIIa either did not block platelet aggregation [39] or inhibited this process.[40-42] An alloantibody (designated IgG-L) which was obtained from the serum of a thrombasthenic patient who had received numerous platelet transfusions, precipitated GPIIb and GPIIIa from solubilized platelet membranes and it blocked platelet aggregation and ^{125}I-fibrinogen binding to ADP-stimulated platelets.[43]

In order to characterize further fibrinogen receptors on the platelet surface, we raised in rabbits antisera against membranes which were prepared either from intact or from chymotrypsin-treated or from pronase-treated human platelets. TABLE 1 shows that these antibodies inhibited fibrinogen-induced aggregation of ADP stimulated and chymotrypsin-treated platelets. They also blocked binding of ^{125}I-fibrinogen to these platelets. Inhibition was complete upon addition of 300 μg of Fab fragments of these antibodies.

FIGURE 1 shows the effect of chymotrypsin and pronase previously incubated with platelets on the immunoprecipitation of ^{125}I surface–labeled platelet membrane components and on ^{125}I-labeled fibrinogen binding and platelet aggregation. Antiplatelet membrane antibody acting on ^{125}I surface–labeled intact platelets precipitated essentially two components: GPIIb and GPIIIa. These platelets did not bind ^{125}I-fibrinogen and did not aggregate with fibrinogen. Incubation of platelets with chymotrypsin (500 μg/ml) or pronase (500 μg/ml) resulted in a significant increase of their reactivity with fibrinogen estimated on the basis of ^{125}I-fibrinogen binding to platelets and platelet aggregation. This correlated with the appearance of a surface component on platelet membranes that was labeled with ^{125}I and was immunoprecipitated with antiplatelet membrane antibody. This component migrated with an apparent molecular weight of 66,000 in a reduced system and 60,000 in a nonreduced system. Chymotrypsin and pronase also caused partial proteolytic degradation of GPIIb and GPIIIa.

Recent data obtained in our laboratory suggest that the 66,000 M_r component may originate as a result of proteolysis of GPIIIa. It may represent a fragment of GPIIIa that reacts directly with fibrinogen molecule.[43a] Intact thrombasthenic platelets that have low amounts of GPIIb and GPIIIa also

TABLE 1

INHIBITION OF PLATELET AGGREGATION AND ^{125}I-FIBRINOGEN BINDING BY ANTIPLATELET
MEMBRANE ANTIBODY FAB FRAGMENTS *

Concentration of Fab µg/ml	Percent Inhibition			
	ADP-stimulated Platelets		Chymotrypsin-treated Platelets	
	Platelet Aggregation	^{125}I-Fibrinogen Binding	Platelet Aggregation	Binding ^{125}I-Fibrinogen
400	92.7	100	100	100
300	84.2	—	—	—
250	—	91.5	—	—
200	45.2	—	83.3	—
150	—	54.9	—	86.2
100	12.8	25.0	69.5	—
50	—	—	48.7	51.4
Preimmune Fab				
500	0.0	0.0	0.0	0.0

* Platelet membranes were prepared from washed intact human platelets by the glycerol lysis technique of Barber and Jamieson.[44] Anti-intact-platelet membrane antibody were raised in rabbits by the method of Vaitukaitis *et al.*[45] Fab fragments were prepared by papain digestion of the immunoglobulin fraction of rabbit serum. Human washed platelets were prepared by the method of Mustard *et al.*[8] Chymotrypsin-treated platelets were prepared as described previously. Fibrinogen was labeled with ^{125}I by the method of Macfarlane [46] (specific radioactivity of ^{125}I-fibrinogen was 80,000 cpm/µg; the clottability was 95%). Platelet aggregation and binding of ^{125}I-fibrinogen to ADP-stimulated and to chymotrypsin-treated platelets were assayed as described previously.[14, 21]. The final concentrations of unlabeled fibrinogen (used in aggregation studies) and of ^{125}I-fibrinogen (used in binding studies) were 500 µg/ml and 100 µg/ml, respectively. The final concentration of ADP was 100 µM. The platelets were preincubated with Fab fragments before addition of the fibrinogen and ADP for 5 min. Percentages of the inhibition of ^{125}I binding to platelets refer to the specific binding.

Antibody raised in rabbits injected with platelet membranes prepared either from intact platelets, or from pronase-treated or chymotrypsin-treated platelets produced similar inhibitory effects on platelet aggregation and ^{125}I-fibrinogen binding to platelets.

exposed comparably lower amounts of the 66 K component on their surface following treatment with pronase or chymotrypsin. The relationship between platelet membrane components and fibrinogen receptors on normal and on thrombasthenic platelets requires further study.

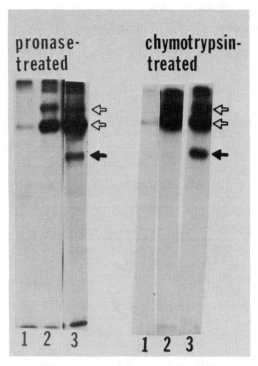

FIGURE 1. An autoradiogram showing the comparison of the immunoprecipitation by anti-platelet membrane antibody of ^{125}I-labeled surface components on intact, chymotrypsin- and pronase-treated platelets. Platelets were treated with chymotrypsin or pronase as previously described.[21] Intact platelets did not bind ^{125}I-fibrinogen nor were they aggregated by fibrinogen. The extent of fibrinogen-induced aggregation of pronase (500 μg)-treated and chymotrypsin (500 μg)-treated platelets was 27 and 22 light transmission units, respectively, using 100 μg fibrinogen and 4×10^8 platelets per ml. Platelet suspensions were labeled with ^{125}I in the presence of iodogen.[47] A 15–20% incorporation of ^{125}I into tyrosine-containing platelet proteins occurs during this procedure. Subsequently, ^{125}I surface–labeled proteins were solubilized with 0.5% Triton X-100 buffer containing 10 mM EDTA, 1 mM PMSF and 1 mM DFP and immunoprecipitated by the addition of anti-serum and suspensions of Staphylococcus aureus.[48] SDS polyacrylamide gel electrophoresis was performed in a non-reduced system according to Laemmli[49] using 10% acrylamide slab gels and the gels were exposed for 2 days at −70° C to Kodak X-ray film.

Immunoprecipitation was carried out as follows: Under the title *Pronase-treated*, (1) pre-immune serum + ^{125}I-labeled intact platelets, (2) anti-human pronase-treated platelet membrane antiserum + ^{125}I-labeled intact platelets, and (3) anti-human pronase-treated platelet membrane antiserum + ^{125}I-labeled pronase-treated platelets. Under the title *Chymotrypsin-treated* (1) pre-immune serum + ^{125}I-labeled intact platelets, (2) anti-human pronase-treated platelet membrane anti-serum + ^{125}I-labeled intact platelets, and (3) anti-human pronase-treated platelet membrane antiserum + ^{125}I-labeled chymotrypsin-treated platelets. None of the characteristic components (GPIIb, GPIIIa or 66K component) was precipitated following incubation of pre-immune serum with ^{125}I-labeled chymotrypsin-treated or pronase-treated platelets. Open arrows indicate GPIIb and GPIIIa; solid arrows indicate 66K component.

RELATIONSHIP BETWEEN FIBRINOGEN STRUCTURE AND FUNCTION IN PLATELET AGGREGATION

Our early data indicated that during plasmic digestion, fibrinogen loses its ability to act as a cofactor for ADP-induced platelet aggregation in suspensions of washed platelets.[50] Fragment X (stage 1) and Fragment X (stage 2) were less potent than fibrinogen in promoting ADP-induced platelet aggregation, whereas Fragments Y and D did not support ADP-induced platelet aggregation at all.[51] We concluded that the intact fibrinogen molecule was essential for ADP-induced platelet aggregation and that initial plasmic degradation of the COOH-terminal part of the Aα-chain alters this function of fibrinogen. More recently, we reported similar observations by studying the effect of fibrinogen and its fragments on the aggregation of chymotrypsin-treated platelets.[21] Holt *et al.*[52] observed decreased platelet-aggregating activity of Fragment X as compared to fibrinogen. However, the major loss of platelet-aggregating activity occurs as Fragment X is converted to Fragment Y and D.

Tomikawa *et al.*[53] studied the effect of various fibrinogen derivatives on fibrinogen-induced platelet aggregation in the presence of ADP. They found that fibrinogen derived Fragment E and the NH$_2$-terminal disulfide knot (NDSK) are the strongest inhibitors of platelet aggregation and suggested that the active sites of fibrinogen which may interact with platelets in the presence of ADP are located in the NH$_2$-terminal region of the molecule. These observations have not been confirmed in this study and by other investigators.[54]

TABLE 2 compares the effect of fibrinogen and its early degradation products on platelet aggregation. It can be seen that limited proteolysis of the fibrinogen

TABLE 2

ABILITY OF FIBRINOGEN AND ITS EARLY DEGRADATION PRODUCTS TO STIMULATE PLATELET AGGREGATION *

Protein Added	Aggregation, LTU/min	
	ADP-stimulated	Chymotrypsin-treated
Fibrinogen	40	22
Fragment X (stage 1)	17.5	15
Fragment X (stage 2)	7.5	3.5
Fragment Y	0	0
Fragment D	0	0

* Platelet aggregation was performed in Payton aggregometer (Scarborough, Ont.). In case of ADP stimulated platelets aliquots of 0.8 ml platelets (5×10^8/ml) were incubated with 100μl protein for one minute at 37° C. Then aggregation was triggered by addition of 100μl ADP (5×10^{-4} M). In the case of chymotrypsin-treated platelets [21] aggregation was induced directly by the addition of fibrinogen or its fragments. Light transmission units (LTU) refer to the arbitrary light transmission units, i.e., the light transmission change recorded during one minute of aggregation. Human fibrinogen (Kabi, Stockholm, Sweden) was further purified as described previously.[21] Fragments X (stage 1), X (stage 2), Y and D were purified from plasmic digest of fibrinogen.[55, 56] Platelet aggregation was tested as described.[21, 51] The concentration of protein tested varied from $1.2–1.4 \times 10^{-6}$ M.

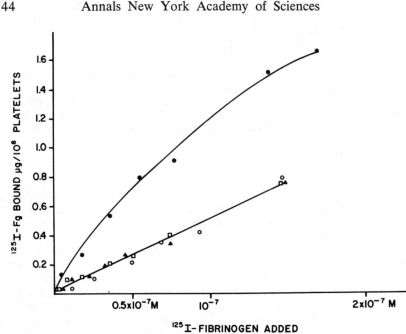

FIGURE 2. Inhibition of binding of ^{125}I-fibrinogen to chymotrypsin-treated platelets by unlabeled fibrinogen, Fragment X (stage 1) and Fragment X (stage 2). The same preparations of fibrinogen and both Fragments X (stage 1) and X (stage 2) were used as described in TABLE 1. Experiments were performed at 0–4° C. Aliquots of 300μl platelet suspension (10^9/ml) and 100μl Tyrode buffer (pH 7.5) or 100μl unlabeled fibrinogen or Fragment X (stage 1) or Fragment X (stage 2) were incubated for 5 min. Then aliquots of 100μl ^{125}I-fibrinogen were added. After further incubation for 5 min, 300μl samples of the incubation mixtures were applied to silicone oil and centrifuged for 1 min using Eppendorff centrifuge. The values of ^{125}I-fibrinogen bound per platelet were plotted against the concentration of total ^{125}I-fibrinogen in a double reciprocal plot. ●—● Tyrode buffer + ^{125}I-fibrinogen; ○—○ unlabeled fibrinogen (10^{-6} M) + ^{125}I-fibrinogen; □—□ Fragment X (stage 1) (1.5×10^{-6} M) + ^{125}I-fibrinogen; △—△ Fragment X (stage 2) (1.6×10^{-6} M) + ^{125}I-fibrinogen. All values refer to the final concentrations. Inhibition constants (K_i) calculated by means of double reciprocal plot for fibrinogen, Fragment X (stage 1) and Fragment X (stage 2) were 1.9×10^{-7} M, 2.8×10^{-7} M and 2.4×10^{-7} M, respectively.

molecule results in a significant impairment of platelet aggregation activity assessed using both ADP stimulated platelets and chymotrypsin-treated platelets. Fragments X (stage 1) and X (stage 2) promoted platelet aggregation but to a much lower degree than intact fibrinogen. Fragments Y and D did not support platelet aggregation. This experiment indicates that the bivalent structure of fibrinogen is required for platelet aggregation. Subsequently, we compared competition of intact fibrinogen, Fragment X (stage 1) and Fragment X (stage 2) with ^{125}I-fibrinogen for the binding sites on chymotrypsin-treated platelets. FIGURE 2 shows that fibrinogen and two Fragments X had similar inhibitory effects on the binding of ^{125}I fibrinogen to chymotrypsin-treated platelets. The last two experiments demonstrate that limited degradation of fibrinogen affects platelet aggregation more than the binding affinity to the receptor. It appears that the bivalent structure of fibrinogen, critical for platelet aggregation, is not necessary for binding of this protein to the receptor. Therefore, it can be expected that fibrinogen fragments that retain the ability to bind to platelets may block fibrinogen-induced platelet aggregation.

We decided to study the effect of various fibrinogen fragments on the fibrinogen-induced aggregation of chymotrypsin-treated platelets and on the binding of ^{125}I-fibrinogen to chymotrypsin-treated platelets. The results of the experiments were somewhat variable and depended on the particular preparations of platelets and on the preparation of the fibrinogen fragment. However, in all experiments Fragments P_{45} (COOH-terminal portion of the Aα chain), Fragments E and NDSK at the concentration of 10^{-5} M had no effect on the aggregation of chymotrypsin-treated platelets induced by fibrinogen ($10^{-8} - 6 \times 10^{-7}$ M) (FIG. 3). In the majority of experiments, fragment D_1 at the concentration of 5×10^{-6} M $- 5 \times 10^{-5}$ M inhibited fibrinogen-induced platelet aggregation (FIG. 4). This inhibition was not competitive.

Fragment D_1 (M_r 103,000) is a late plasmic degradation product of fibrinogen. However, upon prolonged incubation with plasmin, Fragment D_1 can be further digested to fragment D_3 (M_r 86,000). This digestion is associated entirely with the loss of the COOH-terminal portion of the γ chain. Recent

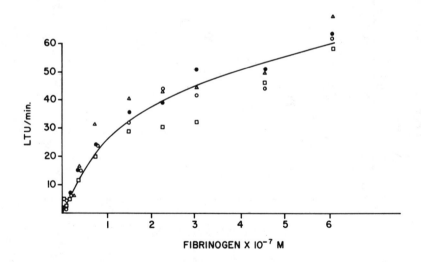

FIGURE 3. Effect of various fibrinogen fragments on fibrinogen-induced aggregation of chymotrypsin-treated platelets. Aliquots of 50μl fibrinogen fragments or Tyrode buffer were incubated with 400μl chymotrypsin-treated platelets for 1 min. Then 50μl of fibrinogen was added and aggregation was recorded in a Payton aggregometer. ●—● Control sample with buffer; ○—○ Sample with P_{45} (10^{-5} M) □—□ Sample with Fragment E, (10^{-5} M); △—△ Sample with NDSK, (5×10^{-6} M). Fragment P_{45} (COOH-terminal portion of Aα chain) was prepared according to the method of Budzynski *et al.*,[57] Fragment E—according to Olexa and Budzynski,[58] NDSK—according to Garlund *et al.*[59] All fragments were dissolved in 0.05 M Tris —0.15 M NaCl buffer, pH 7.6.

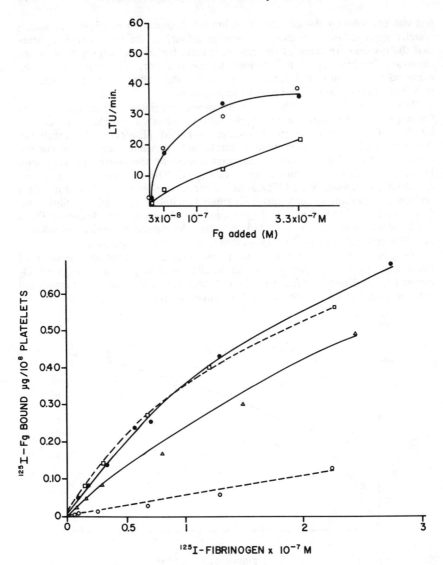

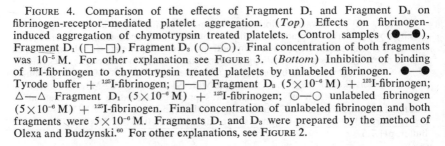

FIGURE 4. Comparison of the effects of Fragment D_1 and Fragment D_3 on fibrinogen-receptor–mediated platelet aggregation. (*Top*) Effects on fibrinogen-induced aggregation of chymotrypsin treated platelets. Control samples (●—●), Fragment D_1 (□—□), Fragment D_3 (○—○). Final concentration of both fragments was 10^{-5} M. For other explanation see FIGURE 3. (*Bottom*) Inhibition of binding of ^{125}I-fibrinogen to chymotrypsin treated platelets by unlabeled fibrinogen. ●—● Tyrode buffer + ^{125}I-fibrinogen; □—□ Fragment D_3 (5×10^{-6} M) + ^{125}I-fibrinogen; △—△ Fragment D_1 (5×10^{-6} M) + ^{125}I-fibrinogen; ○—○ unlabeled fibrinogen (5×10^{-6} M) + ^{125}I-fibrinogen. Final concentration of unlabeled fibrinogen and both fragments were 5×10^{-6} M. Fragments D_1 and D_3 were prepared by the method of Olexa and Budzynski.[60] For other explanations, see FIGURE 2.

studies indicate that sites such as the fibrin-polymerizing site,[60] the sites for fibrinogen crosslinking,[61, 62] sites interacting with staphylococcal clumping factor [63, 64] and high affinity calcium-binding sites [65] are all located in the COOH-terminal section of the γ chain. The most COOH-terminal part of the γ chain (γ 374–411) appears to be exposed on the surface of fibrinogen because functionally active sites are located in this region. Antibodies raised against site (epitope) γ 385–410 interact with the intact fibrinogen molecule suggesting that this epitope is exposed.[66, 67]

For the above reasons we decided to compare effect of Fragments D_1 and D_3 on the fibrinogen-induced aggregation and on binding of ^{125}I-fibrinogen to chymotrypsin-treated platelets (FIG. 4 a,b). The data from this particular experiment show that Fragment D_3 (10^{-5} M), in contrast to D_1 (10^{-5} M), did not inhibit fibrinogen-induced aggregation of chymotrypsin-treated platelets. Fragment D_1 (at concentration 5×10^{-6} M) inhibited partially ^{125}I-fibrinogen (concentration $0.2 - 2 \times 10^{-7}$ M) binding to the surface of chymotrypsin-treated platelets. Its effect was much less pronounced than the effect of fibrinogen. In this experiment Fragment D_3 did not cause any inhibition of ^{125}I-fibrinogen binding in this system, in other experiments, however, some inhibitory activity was detected in other Fragment D_3 preparations. These findings suggest that at least some platelet-binding sites may be located in the COOH-terminal portion of the γ chain, which is removed during proteolytic conversion of Fragment D_1 to Fragment D_3.

Independently of our study, Marguerie *et al.*[54] reached similar conclusions by studying the effect of various fibrinogen fragments on ADP-induced platelet aggregation and on ^{125}I-fibrinogen binding to ADP-stimulated platelets. According-ing to these authors, fragment D_1 (at concentrations $5 \times 10^{-5} - 10^{-4}$ M) inhibited platelet aggregation and fibrinogen binding whereas fragment E and D_3 were inactive. Kloczewiak *et al.*[68] isolated the COOH-terminal peptide of the γ chain (γ 385–411) and found that it was an inhibitor of platelet aggregation stimulated by ADP. The peptide at a concentration of 4×10^{-4} M completely inhibited the aggregation of platelets induced by ADP (5×10^{-6} M) and by fibrinogen (1.6×10^{-7} M). The complete inhibition of binding of ^{125}I-fibrinogen (1.6×10^{-7} M) to ADP-stimulated platelets occurred at the 3×10^{-5} M con-centration of the peptide. All these data suggest that active fibrinogen fragments cause inhibition of platelet interaction with fibrinogen at a molar concentration that is higher by 2–3 orders of magnitude than the concentrations of fibrinogen required to produce effects on platelets. These data suggest significance of the tertiary structure of fibrinogen in its interaction with platelets or point to the existence of numerous platelet-binding sites at the surface of the fibrinogen molecule.

A hypothetical mechanism by which fibrinogen causes platelet aggregation is presented in FIGURE 5. We postulate that fibrinogen causes platelet aggregation by bridging adjacent platelets together. The experimental evidence supporting this hypothesis is as follows: (1) the bivalent symmetrical structure of the fibrinogen molecule is a requirement for fibrinogen-induced aggregation to occur, both in ADP-stimulated platelets or in chymotrypsin-treated platelets; [21, 51] (2) limited degradation of the fibrinogen molecule significantly impairs fibrino-gen's ability to support platelet aggregation, but it does not interfere with its ability to bind to the platelets (TABLE 2, FIG. 2); (3) observations that an excess of fibrinogen blocks platelet aggregation [14] such that at increasing concentrations of fibrinogen, the binding sites on adjacent platelets would be saturated and therefore no fibrinogen bridges would form and platelets would not be held together.

We postulate that fibrinogen has numerous platelet-binding sites, some of which are located on the COOH-terminal portions of the γ chain (γ 374–411) and the Aα chain (Aα 250–610). The evidence for the participation of γ 374–411 is supported by the comparative studies on both the blocking effects of Fragment D_1 and D_3 in platelet aggregation and from other functional and immunochemical studies indicating that this portion of the molecule is exposed on the native molecule as well as from studies with isolated peptide fragments which inhibit platelet aggregation.[68] We suggest that the COOH-terminal portion of the Aα chain also participates in platelet aggregation because conversion of fibrinogen to Fragment X resulting in a loss of platelet aggregating activity is primarily associated with the degradation of the Aα chain. The γ chains remain intact in Fragment X (stage 1) and (stage 2). This portion of the Aα chain is also exposed on the surface of the fibrinogen molecule as documented by a number of immunochemical studies.[67] We were not able to demonstrate the blocking effect of Fragment P_{45} (COOH-terminal portion of the Aα chain) on platelet aggregation. This can be due to an instability of this fragment and to its susceptibility to the traces of proteolytic enzymes that may be present on the surface of chymotrypsin-treated platelets.

It is interesting that calcium-binding sites are located in the COOH-terminal portion of the γ chain.[65] Calcium is also critical for the stability of GPIIb and GPIIIa complex present on the surface of platelet membranes [33] and for the binding of fibrinogen to ADP-stimulated and chymotrypsin-treated platelets.[21, 22] It is conceivable that calcium is essential to the formation of links between fibrinogen receptors and COOH-terminal portion of the γ chains.

Recently, Hawiger et al.[69] demonstrated that whereas purified and partially polymerized γ and α chains of fibrinogen caused aggregation of platelets stimulated by ADP, β chain did not cause platelet aggregation. These findings are compatible with our observations discussed above and with the model of platelet-fibrinogen interaction presented in FIGURE 5.

ROLE OF PLATELET-FIBRINOGEN INTERACTION IN HEMOSTASIS

Analysis of patients with congenital afibrinogenemia may contribute to the understanding of the significance of platelet-fibrin interaction for hemostasis. The level of plasma fibrinogen measured in those patients by immunological techniques varied from 4–100 μg/ml; it is $10^{-8} - 3 \times 10^{-7}$ M.[25, 70–73] This level is too low to provide for clot formation, but it appears to be adequate for partial support of platelet aggregation. The continued ability of platelets to aggregate with fibrinogen may be the reason for the relatively moderate bleeding tendency observed in some of these patients.[70, 71] To our knowledge, there is no evidence that patients with a complete absence of fibrinogen do survive. Although platelets of patients with Glanzmann's thrombasthenia show deficiency of fibrinogen receptors there is an evidence that they may interact normally with polymerizing fibrin at its late stage.[74] On the basis of these observations, we suggest that the function of fibrinogen in platelet aggregation may be most critical for the formation of a hemostatic plug and even more important than its function in the formation of blood clots.

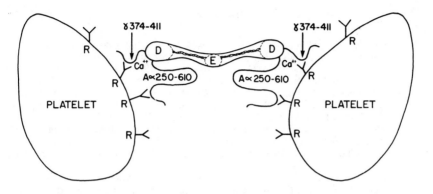

FIGURE 5. Hypothetical model of platelet-fibrinogen interactions. R indicates either high or low affinity fibrinogen receptors. The size of platelets and that of fibrinogen molecule do not reflect their real dimensions.

SUMMARY AND CONCLUSIONS

In summary:

(1) Incubation of platelets with ADP or proteolytic enzymes (chymotrypsin or pronase) results in an exposure of two classes of specific binding sites on platelet surface: low and high affinity fibrinogen receptors.

(2) Fibrinogen interaction with these receptors results in platelet aggregation.

(3) High affinity fibrinogen receptors are not exposed on thrombasthenic platelets stimulated by ADP but are rendered available on chymotrypsin-treated thrombasthenic platelets; low affinity receptors cannot be exposed by ADP or chymotrypsin on these platelets. Availability of high affinity fibrinogen receptors on thrombasthenic platelets may depend on the residual glycoprotein IIIa.

(4) Fibrinogen receptors appear to be associated with glycoproteins IIb, IIIa and a 66,000 M_r platelet membrane component that is exposed during proteolysis of platelet membranes.

(5) Some of the platelet-binding sites on the fibrinogen molecule appear to be associated with the COOH-terminal portion of the γ chain (γ 374–411). Additional binding sites may also be located in the COOH-terminal portion of the Aα chain. The conformation of the fibrinogen molecule may be important in its interaction with platelets.

(6) Platelet aggregation may result from bridging platelets by fibrinogen molecule in the presence of bivalent cations.

(7) In conclusion, platelet interaction with fibrinogen is a complex process involving different binding sites of the fibrinogen molecule.

Our own data and review of literature suggest that platelet-interaction with fibrinogen is of major significance in hemostasis.

REFERENCES

1. CROSS, M. J. 1964. Effect of fibrinogen on the aggregation of platelet by adenosine diphosphate. Thromb. Diath. Haemorrh. **12:** 524–527.
2. McLEAN, J. R., R. E. MAXWELL & D. HERTLER. 1964. Fibrinogen and adenosine diphosphate-induced aggregation of platelets. Nature **202:** 605–606.
3. BRINKHOUS, K. M., M. S. READ & R. G. MASON. 1965. Plasma thrombocyte agglutinating activity and fibrinogen. Synergism with adenosine diphosphate. Lab. Invest. **14:** 335–342.
4. SOLUM, N. O. & H. STORMORKEN. 1965. Influence of fibrinogen on the aggregation of washed human platelets induced by adenosine diphosphate, thrombin, collagen and adrenaline. Scand. J. Clin. Lab. Invest. **17**(Suppl. 84): 170–182.
5. KOPEC, M., A. Z. BUDZYNSKI, J. STACHURSKA, Z. WEGRZYNOWICZ & E. KOWALSKI. 1966. Studies on the mechanism of interference by fibrinogen degradation products (FDP) with the platelet function. Role of fibrinogen in platelet atmosphere. Thromb. Diath. Haemorrh. **15:** 476–490.
6. TANGEN, O., H. J. BERMAN & P. MURPHY. 1971. Gel-filtration. A new technique for separation of blood platelets from plasma. Thromb. Diath. Haemorrh. **25:** 268–287.
7. NIEWIAROWSKI, S., E. REGOECZI & J. F. MUSTARD. 1972. Platelet interaction with fibrinogen and fibrin: Comparison of the interaction of platelets with that of fibroblasts, leukocytes and erythrocytes. Ann. N.Y. Acad. Sci. **201:** 72–83.
8. MUSTARD, J. F., D. W. PERRY, N. G. ARDLIE & M. A. PACKHAM. 1972. Preparations of suspensions of washed platelets from humans. Br. J. Haematol. **22:** 193–204.
9. TOLLEFSEN, D. M. & P. W. MAJERUS. 1975. Inhibition of human platelet aggregation by monovalent antifibrinogen antibody fragments. J. Clin. Invest. **55:** 1259–1268.
10. MUSTARD, J. F., M. A. PACKMAN, R. L. KINLOUGH-RATHBONE, D. W. PERRY & E. REGOECZI. 1978. Fibrinogen and ADP-induced platelet aggregation. Blood **52:** 453–466.
11. MARGUERIE, G. A., E. F. PLOW & T. S. EDGINGTON. 1979. Human platelets possess an inducible and saturable receptor specific for fibrinogen. J. Biol. Chem. **254:** 5357–5363.
12. BENNETT, J. A. & G. VILAIRE. 1979. Exposure of platelet fibrinogen receptor by ADP and epinephrine. J. Clin. Invest. **64:** 1393–1401.
13. PEERSCHKE, E. I., M. B. ZUCKER, R. A. GRANT, J. J. EGAN & M. M. JOHNSON. 1980. Correlation between fibrinogen binding to human platelets and platelet aggregability. Blood **55:** 841–847.
14. KORNECKI, E., S. NIEWIAROWSKI, T. A. MORINELLI & M. KLOCZEWIAK. 1981. Effects of chymotrypsin and adenosine diphosphate on the exposure of fibrinogen receptors on normal human and Glanzmann's thrombasthenic platelets. J. Biol. Chem. **256:** 5696–5701.
15. HAWIGER, J., S. PARKINSON & S. TIMMONS. 1980. Prostacyclin inhibits mobilization of fibrinogen-binding sites on human ADP and thrombin-treated platelets. Nature (London) **283:** 195–197.
16. PLOW, E. F. & G. A. MARGUERIE. 1980. Participation of ADP in the binding of fibrinogen to thrombin-stimulated platelets. Blood **56:** 553–555.
17. PLOW, E. F. & G. A. MARGUERIE. 1980. Induction of the fibrinogen receptor on human platelets by epinephrine and the combination of epinephrine and ADP. J. Biol. Chem. **255:** 10971–10977.
18. BENNETT, J. S., G. VILAIRE & J. W. BURCH. 1981. A role for prostaglandins and thromboxanes in the exposure of platelet-fibrinogen receptors. J. Clin. Invest. **68:** 981–987.

19. MORINELLI, T. A., S. NIEWIAROWSKI, E. KORNECKI, W. R. FIGURES, Y. WACHT-FOGEL & R. W. COLMAN. 1983. Platelet aggregation and exposure of fibrinogen receptors by prostaglandin endoperoxide analogues. Blood **61:** 41–49.

20. GREENBERG, J. P., M. A. PACKHAM, M. A. GUCCIONE, E. J. HARFENIST, J. L. ORR, R. L. KINLOUGH-RATHBONE, D. W. PERRY & J. F. MUSTARD. 1979. The effect of pretreatment of human or rabbit platelets with chymotrypsin on their response to human fibrinogen and aggregating agents. Blood **54:** 753–765.

21. NIEWIAROWSKI, S., A. Z. BUDZYNSKI, T. A. MORINELLI, T. M. BRUDZYNSKI & G. J. STEWART. 1981. Exposure of fibrinogen receptor on human platelets by proteolytic enzymes. J. Biol. Chem. **256:** 917–925.

22. MARGUERIE, G. A., T. S. EDGINGTON & E. F. PLOW. 1980. Interaction of fibrinogen with its platelet receptor as part of a multistep reaction in ADP-induced platelet aggregation. J. Biol. Chem. **255:** 154–161.

23. FIGURES, W. R., S. NIEWIAROWSKI, T. A. MORINELLI, R. F. COLMAN & R. W. COLMAN. 1981. Affinity labeling of human platelet membrane protein with 5'p-fluorosulfonylbenzoyl adenosine. J. Biol. Chem. **256:** 7789–7795.

24. HARFENIST, E., M. A. GUCCIONE, M. A. PACKHAM, R. L. KINLOUGH-RATHBONE & J. F. MUSTARD. 1982. Arachidonate induced fibrinogen binding to thrombin degranulated platelets is independent of released ADP. Blood **59:** 956–962.

25. MARGUERIE, G. A., N. THOMAS-MAISON, M. J. LARRIEU & E. F. PLOW. 1982. The interaction of fibrinogen with human platelets in a plasma milieu. Blood **59:** 91–95.

26. HAWIGER, J., S. NIEWIAROWSKI, V. GUREWICH & D. P. THOMAS. 1970. Measurement of fibrinogen and fibrinogen degradation products in serum by staphylococcal clumping test. J. Lab. Clin. Med. **75:** 93–108.

27. MARGUERIE, G. A. & E. F. PLOW. 1981. Interaction of fibrinogen with its platelet receptors: Kinetics and effect of pH and temperature. Biochemistry **20:** 1074–1080.

28. CAEN, J. P., P. A. CASTALDI, J. C. LECLERC, S. INCEMAN, M. J. LARRIEU, M. PROBST & J. BERNARD. 1966. Congenital bleeding disorders with long bleeding time and normal platelet count: Glanzmann's thrombasthenia. Report of fifteen patients. Am. J. Med. **41:** 4–26.

29. HARDISTY, R. M., K. M. DORMANDY & R. A. HUTTON. 1964. Thrombasthenia. Br. J. Haematol. **10:** 371–387.

30. PHILLIPS, D. R. & P. AGIN. 1977. Platelet membrane defects in Glanzmann's thrombasthenia. J. Clin. Invest. **60:** 535–545.

31. HAGEN, I., A. NURDEN, O. J. BJERRUM, N. O. SOLUM & J. P. CAEN. 1980. Immunochemical evidence for protein abnormalities in platelets with Glanzmann's thrombasthenia and Bernard-Soulier syndrome. J. Clin. Invest. **65:** 722–731.

32. MUSTARD, J. F., R. L. KINLOUGH-RATHBONE, M. A. PACKHAM, D. W. PERRY, E. J. HARFENIST & K. R. M. PAI. 1979. Comparison of fibrinogen association with normal and thrombasthenic platelets on exposure to ADP or chymotrypsin. Blood **54:** 987–993.

33. POLLEY, M. J., L. L. K. LEUNG, F. Y. CLARK & R. L. NACHMAN. 1981. Thrombin-induced platelet membrane glycoprotein IIb and IIIa complex formation. An electron microscope study. J. Exp. Med. **154:** 1058–1068.

34. NACHMAN, R. L. & L. L. K. LEUNG. 1982. Complex formation of platelet membrane glycoproteins IIb and IIIa with fibrinogen. J. Clin. Invest. **69:** 263–269.

35. BENNET, J. S., G. VILAIRE & D. B. CINES. 1982. Identification of fibrinogen receptor on human platelets by photoaffinity labeling. J. Biol. Chem. **257:** 8049–8054.

36. KAPLAN, K. L. & R. L. NACHMAN. 1974. The effect of platelet membrane antibodies on aggregation and release. Br. J. Haematol. **28:** 551–560.

37. LEUNG, L. L. K., T. KINOSHITA & R. L. NACHMAN. 1981. Isolation, purification and partial characterization of platelet membrane glycoproteins IIb and IIIa. J. Biol. Chem. **256:** 1994–1997.
38. JENKINS, C. S. P., E. F. ALI-BRIGG & K. J. CLEMENTSON. 1981. Antibodies against platelet membrane glycoproteins II. Influence on ADP and collagen-induced platelet aggregation, crossed immunoelectrophoresis studies and relevance to Glanzmann's thrombasthenia. Br. J. Haematol. **49:** 439–447.
39. McEVER, R. P., N. L. BAENZIGER & P. W. MAJERUS. 1980. Isolation and quantitation of the platelet membrane glycoprotein deficient in thrombasthenia using a monoclonal hybridoma antibody. J. Clin. Invest. **66:** 1311–1318.
40. DIMINNO, G., P. THIAGARAJAN, P. PERUSSIA, B. MARTINEZ, S. S. SHAPIRO & S. MURPHY. 1983. Exposure of fibrinogen binding sites by collagen, arachidonic acid and ADP: Inhibition by a monoclonal antibody to glycoproteins IIb-IIIa complex. Blood **61:** 140–148.
41. COLLER, B. S., E. I. PEERSCHKE, L. E. SCUDDER & C. A. SULLIVAN. 1983. A murine monoclonal antibody that completely blocks the binding of fibrinogen to platelets produces a thrombasthenic-like state in normal platelets and binds glycoprotein IIb or IIIa. J. Clin. Invest. In press.
42. KORNECKI, E., H. LEE, G. P. TUSZYNSKI & S. NIEWIAROWSKI. 1982. Preliminary characterization of hybridoma antibodies that inhibit platelet-fibrinogen receptor interactions. Fed. Proc. **41**(3):1504 (abstract).
43. LEE, H., A. T. NURDEN, A. THOMAIDIS & J. P. CAEN. 1981. Relationship between fibrinogen binding and the platelet glycoprotein deficiencies in Glanzmann's thrombasthenia type I and type II. Brit. J. Haematol. **48:** 47–57.
43a. KORNECKI, E., G. P. TUSZYNSKI & S. NIEWIAROWSKI. 1983. Inhibition of fibrinogen receptor mediated platelet aggregation by heterologous anti-human platelet membrane antibody: Significance of a 66,000 MW protein derived from glycoprotein IIIa. J. Biol. Chem. In press.
44. BARBER, A. J. & G. A. JAMIESON. 1970. Isolation and characterization of plasma membranes from human blood platelets. J. Biol. Chem. **245:** 6357–6365.
45. VAITUKAITIS, J., J. B. ROBBINS, E. NIESCHLAG & G. T. ROSS. 1971. A method for producing specific antisera with small doses of immunogens. J. Clin. Endocrinol. **33:** 988–991.
46. MACFARLANE, A. S. 1956. Labelling of plasma proteins with radioactive iodine. Biochemical J. **62:** 135–143.
47. TUSZYNSKI, G. P., L. C. KNIGHT, E. KORNECKI & C. SRIVASTAVA. 1983. Labeling of platelet surface proteins with ^{125}I-Iodine by the iodogen method. Anal. Biochem. In press.
48. KESSLER, S. W. 1976. Cell membrane antigen isolation with the staphylococcal protein A—antibody absorbent. J. Immunol. **117:** 1482–1489.
49. LAEMMLI, U. K. 1970. Cleavage of structural proteins during the assembly of the head of bacteriophage T_4. Nature **227:** 680–685.
50. NIEWIAROWSKI, S., V. GUREWICH, A. F. SENYI & J. F. MUSTARD. 1971. The effect of fibrinolysis on platelet function. Thromb. Diath. Haemorrh. (Suppl.). **47:** 99–111.
51. NIEWIAROWSKI, S., A. Z. BUDZYNSKI & B. LIPINSKI. 1977. Significance of the intact polypeptide chains of human fibrinogen in ADP induced platelet aggregation. Blood **49:** 635–644.
52. HOLT, J. C., M. MAHMOUD & P. GAFFNEY. 1979. The ability of fibrinogen fragments to support ADP induced platelet aggregation. Thromb. Res. **16:** 427–435.

53. TOMIKAWA, M., M. IWAMOTO, S. SODERMAN & B. BLOMBÄCK. 1980. Effect of fibrinogen on ADP-induced platelet aggregation. Thromb. Res. **19:** 841–855.
54. MARGUERIE, G. A., N. ARDAILLOU, G. CHERE & E. F. PLOW. 1982. The binding of fibrinogen to its platelet receptor. Involvement of the D domain. J. Biol. Chem. **257:** 11872–11875.
55. MARDER, V. J., N. R. SHULMAN & S. CARROLL. 1969. High molecular weight derivatives of human fibrinogen produced by plasmin I. Physicochemical and immunochemical characterization. J. Biol. Chem. **244:** 2111–2119.
56. MARDER, V. J., H. L. JAMES & S. SHERRY. 1969. The purification of fibrinogen degradation products by Pevikon block electrophoresis. Thromb. Diath. Haemorrh. **22:** 234–239.
57. BUDZYNSKI, A. Z., R. T. JOSEPH, S. A. OLEXA & S. NIEWIAROWSKI. 1979. Binding function of fibrinogen molecule. Fed. Proc. **38:** 996 (abstract).
58. OLEXA, S. A. & A. Z. BUDZYNSKI. 1979. Binding phenomena of isolated unique plasmic degradation products of human cross-linked fibrin. J. Biol. Chem. **254:** 4925–4932.
59. GARLUND, B., B. HESSEL, G. MARGUERIE, G. MARIANO & B. BLOMBÄCK. 1977. Primary structure of human fibrinogen. Characterization of disulfide containing cyanogen-bromide fragments. Eur. J. Biochem. **77:** 595–610.
60. OLEXA, S. A. & A. Z. BUDZYNSKI. 1981. Localization of a fibrin polymerizing site. J. Biol. Chem. **256:** 3544–3549.
61. CHEN, R. & R. F. DOOLITTLE. 1971. γ-Crosslinking sites in human and bovine fibrin. Biochemistry **10:** 4486–4491.
62. HERMANS, J. & J. MCDONAGH. 1982. Fibrin: Structure and interactions. Seminars Thromb. Haemost. **8:** 11–24.
63. HAWIGER, J., S. TIMMONS, D. D. STRONG, B. A. COTTRELL, M. RILEY & R. F. DOOLITTLE. 1982. Identification of a region of human fibrinogen interacting with staphylococcal clumping factor. Biochemistry **21:** 1407–1413.
64. STRONG, D. D., A. P. LAUDANO, J. HAWIGER & R. F. DOOLITTLE. 1982. Isolation, characterization and synthesis of peptides from human fibrinogen that block the staphylococcal clumping reaction and construction of a synthetic clumping particle. Biochemistry **21:** 1414–1420.
65. NIEWENHUIZEN, W., A. VERMOND, W. J. NOOIJEN & F. HAVERKATE. 1979. Calcium binding properties of human fibrin(ogen) and degradation products. FEBS Lett. **98:** 257–259.
66. PURVES, L. R., G. G. LINDSEY & J. J. FRANKS. 1980. Sites of D-domain interaction in fibrin-derived D-dimer. Biochemistry **19:** 4051–4058.
67. PLOW, E. F. & T. S. EDGINGTON. 1982. Surface markers of fibrinogen and its physiologic derivatives revealed by antibody probes. Seminars Thromb. Haemost. **8:** 36–56.
68. KLOCZEWIAK, M., S. TIMMONS & J. HAWIGER. Localization of a site interacting with human platelet receptor on carboxy-terminal segment of human fibrinogen γ chain. Biochem. Biophys. Res. Commun. **107:** 181–187.
69. HAWIGER, J., S. TIMMONS, M. KLOCZEWIAK, D. D. STRONG & R. F. DOOLITTLE. 1982. γ and α Chains of human fibrinogen possess sites reactive with human platelet receptors. Proc. Natl. Acad. Sci. USA **79:** 2068–2071.
70. ALEXANDER, B., R. GOLDSTEIN, L. RICH, A. G. LEBOLLOCH, R. L. DIAMOND & W. BORGEN. 1954. Congenital afibrinogenemia. A study of some basic aspects of coagulation. Blood **9:** 843–865.
71. NIEWIAROWSKI, S., J. KOZLOWSKA, A. GULMANTOWICZ & E. PELCZARSKA-KASPERSKA. 1962. Afibrinogenemie congenitale. Etude biologique de deux cas. Hemostase (Paris) **2:** 191–202.

72. GUGLER, E. & E. F. LUSCHER. 1965. Platelet function in congenital afibrino-
    genemia. Thromb. Diath. Haem. **14:** 361–373.
73. WEISS, H. J. & J. ROGERS. 1971. Fibrinogen and platelets in the primary
    arrest of bleeding. New Engl. J. Med. **283:** 369–374.
74. NIEWIAROWSKI, S., S. LEVY-TOLEDANO & J. P. CAEN. 1981. Platelet interac-
    tion with polymerizing fibrin in Glanzmann's thrombasthenia. Thromb. Res.
    23: 457–463.

DISCUSSION OF THE PAPER

N.A. CARRELL (*Duke Medical Center, Durham, NC*): Have you examined the inhibition of fibrinogen binding to platelets with fibronectin?

S. NIEWIAROWSKI (*Temple University, Philadelphia, PA*): We have documented the fact that the binding of fibrinogen to the platelet receptor is fibronectin independent.

E.F. PLOW: I have two questions for Dr. Niewiarowski with regard to the two classes of receptors. First of all, I noticed that at the Federation Meeting, Peerschke, who also interpreted her Scatchard plots in terms of two receptors, recently indicated that there was probably one class of receptor with negative cooperativity. Did you consider that possibility? Secondly, in your model of fibrinogen binding to platelets, where you show only one site of interaction, the γ chain or the Aα chain, you do not consider both classes of receptors. Do you consider that both classes of receptors bind to that same set of sites?

NIEWIAROWSKI: I consider that negative cooperativity is a possibility. Secondly, we found that thrombasthenic platelets have high affinity receptors and do not have low affinity receptors. It cannot be excluded that different platelet receptors interact with the sites on γ chains or Aα chain, but we do not have any experimental data to support this hypothesis.

R.F. DOOLITTLE (*University of California at San Diego, La Jolla*): I have a comment. Although there appears to be a remarkable commonality to the last three papers, there were some subtle differences. In particular where Dr. Hawiger's work shows that he can bring about the inhibition of this interaction with a 15-residue peptide that corresponds to the COOH-terminal, it should be made very clear that Gly-Pro-Arg-Pro should not have an influence on such a system. So there's a difference here, and it's possible that the Gly-Pro-Arg-Pro activity that you are reporting could be interacting with some other element, either the platelet, the platelet factor, or whatever. It should not automatically be concluded that the results with a shortened D, and inhibition by Gly-Pro-Arg-Pro are manifestations of the same thing. I do not think that Gly-Pro-Arg-Pro has a complementary site in the last 15 residues of the γ chain. All current data indicate that that complementary site is much further in, and of course your data include a section that extends much further in also.

NIEWIAROWSKI: In our hands, a Gly-Pro-Arg-Pro peptide is a strong inhibitor of fibrin monomer polymerization and a weak inhibitor of fibrinogen receptor-mediated platelet aggregation.

J. HAWIGER (*Vanderbilt University School of Medicine, Nashville, TN*):
I would like to re-emphasize the basic differences between the receptor for
staphylococci and the platelet receptor in terms of their availability. Staphylococcal clumping involves a receptor that is available all the time; it does not
require mobilization, calcium dependency, etc. The binding is straightforward.
In platelets the process is metabolically active, and is regulated by cyclic AMP.
However, the fibrinogen γ chain 15 residue COOH-terminal segment binds both
receptors with about the same affinity.

THE FIBRINOGEN-DEPENDENT PATHWAY OF PLATELET AGGREGATION *

Gerard A. Marguerie

*Institut de Pathologie Cellulaire
Hôpital de Bicêtre, Inserm U 143
94270 Le Kremlin-Bicêtre, France*

Edward F. Plow

*Department of Molecular Immunology
Research Institute of Scripps Clinic
La Jolla, California 92037*

The capacity of platelets to interact with one another is fundamental to their function in the hemostatic process. In the resting state, platelets circulate freely as nonadhesive entities; but in response to a variety of stimuli, they develop the capacity to interact and form aggregates. Thus, the platelet exemplifies a cell in which the acquisition of adhesive properties permits cell-cell contact and the expression of biological activity. The initial event in the induction of platelet aggregation is the encounter of the cell with an appropriate stimulus, and several independent receptor systems for specific stimuli have been identified (i.e. thrombin, ADP, and epinephrine). The subsequent events, however, which ultimately lead to aggregation of the stimulated cells have not been defined at a molecular level. With ADP as the platelet stimulus, some insight into the nature of these events has been provided. Fibrinogen is a cofactor for ADP-induced platelet aggregation,[2-4] and participation of the molecule in cell-cell interaction is mediated by its binding to specific receptor sites.[5-6] A number of correlates between aggregation and fibrinogen binding have been identified, suggesting that the molecule may mediate aggregation of ADP-stimulated platelets.[5-8] The role of fibrinogen in platelet aggregation is not restricted to the ADP stimulus. Fibrinogen may also regulate platelet aggregation induced by a variety of stimuli, and this is achieved by the binding of the molecule to its platelet receptor. Thus, the existence of a common mechanism in which fibrinogen participates in platelet function—a fibrinogen-dependent pathway of platelet aggregation—can be hypothesized.

PARTICIPATION OF FIBRINOGEN IN ADP-INDUCED PLATELET AGGREGATION IS RECEPTOR MEDIATED

The identification of fibrinogen as an essential cofactor for ADP-induced platelet aggregation was derived from studies of washed platelets [9] and platelets in afibrinogenemic [10-11] and defibrinogenated plasma.[12] In these suspending media, platelets fail to aggregate in response to ADP, but normal aggregation

* This work was supported by Grants NIH-HL–16411, NATO 1852, and INSERM 80 5029.

ensues upon addition of fibrinogen. Mustard *et al.*[13] initially presented evidence of fibrinogen associated with platelets during ADP-induced aggregation. Subsequently, we [5, 7] and Bennett and Vilaire [6] have provided a molecular basis for these observations by demonstrating that the binding of fibrinogen to the platelet was specific and was mediated by discrete receptor sites. These conclusions were derived from studies in which radiolabeled fibrinogen was incubated with washed human platelets, and the platelet-bound from free ligand was separated by centrifugation through sucrose or nonaqueous media. The platelet-bound ligand was identified as native fibrinogen from its constituent chains and its fibrinopeptide A content.[5] The fibrinogen receptor system was not expressed by the resting platelet but was induced by ADP, and the extent of fibrinogen binding was dependent upon ADP dose with 10 μM ADP maximally supporting the interaction. The association of the molecule with ADP-stimulated platelets exhibited an absolute requirement for extracellular divalent ions and was supported by calcium or magnesium.[6, 7] In our laboratories, Scatchard analysis of specific binding isotherms suggested a single class of binding sites. Approximately 38,000 molecules were maximally bound per platelet in the presence of 1 mM calcium and 20,000 molecules in the presence of magnesium. The apparent affinity constant derived at equilibrium was $K_a = 2 \times 10^6$ M^{-1} and was independently confirmed from the ratio of the rate constants of association and dissociation derived from kinetic analysis which yielded a K_a of 3×10^6 M^{-1}. Several other laboratories have also provided evidence for a single class of binding sites with association constants ranging from 12.5 to 0.7×10^6 M^{-1}.[6,15, 16] Peerschke and Zucker [8] and Niewiarowski *et al.*[17] recently postulated the existence of high and low affinity receptor systems. We have, however, again found evidence for only a single class of binding site in a study assessing the binding of fibrinogen to platelets in afibrinogenemic plasma, which precluded the isolation and washing of the cells.[18]

The binding of fibrinogen to the platelet was initially reversible, but ultimately the bound ligand became nondissociable from the platelet surface by excess nonlabeled fibrinogen or ADP scavengers. On the basis of these observations, we postulated that the stabilization of platelet-bound fibrinogen was the final step in the interaction sequence and preceded by the reversible binding of fibrinogen to its receptor.[7] The reversible binding was the rate-determining and divalent ion-dependent step.[14] Although it may be argued that it is inappropriate to apply Scatchard-type analyses under conditions where the bound fibrinogen ligand is nondissociable, it should be noted that kinetic analyses measured at early time points when binding was fully reversible yielded a K_a very similar to that obtained from Scatchard plots. This indicates that the rate constant of the irreversible reaction does not significantly alter the rate constant of dissociation and suggests that K_a derived from Scatchard plots validly describes the affinity of the interaction.

The correlates relating the reversible and irreversible steps of fibrinogen binding with platelet function are multiple: (1) an apparent association constant of 3×10^6 M^{-1} was approximated from the initial rates of platelet aggregation, a value very similar to the K_a of the reversible binding of fibrinogen to the platelet; [5] (2) fibrinogen binding and platelet aggregation exhibited identical requirements for divalent ions; [6, 7] (3) when the binding of fibrinogen to the platelet was reversible, ADP scavengers [7] or ADP analog [19] dissociated the platelet-bound ligand and induced disaggregation; and (4) when platelet-bound fibrinogen was stabilized, ADP scavengers failed to dissociate fibrinogen and aggregation was irreversible.[7] Stirring was required to induce platelet aggregation, but fibrinogen binding did not require stirring, indicating that binding was not a consequence of platelet aggregation. Thus, all available information points to a direct participation of platelet receptor fibrinogen interaction in aggregation induced by ADP.

INDUCTION OF THE PLATELET RECEPTOR FOR FIBRINOGEN BY STIMULI OTHER THAN ADP

To assess the capacity of fibrinogen to participate in platelet aggregation induced by stimuli other than ADP, a selected panel of platelet agonists have been utilized. Collagen and thrombin are potent triggers of platelet secretion and aggregation and may represent two of the primary physiological stimuli of platelet function. Platelet-activating factor (PAF-acether) has been recently identified as acetyl-glyceryl-ether-phosphoryl-choline [20-21] and is capable of inducing platelet secretion and aggregation at doses as low as 10^{-10} M. Ionophore A23187 induces platelet aggregation and secretion by influencing calcium flux,[22] although this view has recently been challenged.[23] Finally, epinephrine is an example of a hormonal agonist that may be distinguished from the other stimuli by being capable of inducing aggregation in the absence of apparent shape change and dense granule secretion.[24] In response to low doses of these stimuli as well as to ADP, washed platelets in Tyrode's buffer did not aggregate or aggregated weakly in the absence of fibrinogen, even with calcium present. In all cases, however, significant aggregation was observed upon addition of fibrinogen (FIG. 1). These observations indicate that fibrinogen may modulate the aggregation of platelets induced by low concentrations of a variety of stimuli.

In considering a molecular basis for these results, the capacity of the stimuli to expose the platelet receptor for fibrinogen was evaluated. All stimuli tested supported the binding of ^{125}I-fibrinogen to the platelet (FIG. 2). Under nonstirring conditions, the binding was time dependent, and with ADP, thrombin, calcium ionophore and PAF, maximal binding was reached within 20 minutes after the addition of the stimulus. Collagen and epinephrine supported binding with prolonged time courses, and approximately 90 minutes were required to reach equilibrium. With collagen, a lag phase was observed, whereas with epinephrine binding occurred with prolonged linear time course consistent with previous reports.[25] Differences between the kinetics of fibrinogen binding and the rate of platelet aggregation with the various stimuli are apparent from comparison of the data shown in FIGURES 1 and 2. We have recently observed that platelet aggregation induced by stirring markedly increases the apparent rate constant of association of fibrinogen with the stimulated platelets (manuscript in preparation). We have chosen to perform the analyses shown in FIGURE 2 under nonstirring conditions since this accentuates the differences in fibrinogen binding with the various stimuli. This permits us to readily see the prolonged time course of fibrinogen binding with epinephrine and the lag phase with collagen. Both of these phenomena are reflected in the aggregation patterns.

Several lines of evidence indicate that fibrinogen binding is mediated by the same receptor system. First, as shown in TABLE 1, the binding induced by all stimuli was divalent ion dependent, and both calcium and magnesium supported the interaction. Second, the binding of fibrinogen was saturable with respect to fibrinogen concentration with all stimuli, and Scatchard analyses of

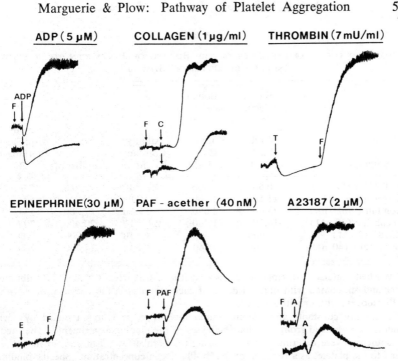

FIGURE 1. Aggregation of washed human platelets by various stimuli. Platelets (2×10^8 cells/ml) were suspended in Tyrode's buffer at pH 7.4 containing 1 mM calcium and 2% albumin. The final concentration of fibrinogen (F) was 0.2 mg/ml.

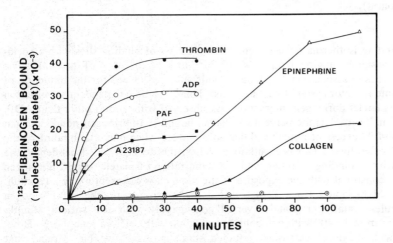

FIGURE 2. Time course of the association of ^{125}I-fibrinogen with platelets stimulated by various inducers. The washed platelets (2×10^8 cells/ml) were suspended in Tyrode's — 2% albumin buffer containing 1 mM calcium and 0.17 μM ^{125}I-fibrinogen. The cells were stimulated under nonstirring conditions at 22°C. The concentrations of stimuli utilized were: AP, 10 μM; thrombin, 2 mU/ml; epinephrine, 30 μM; collagen, 3.8 μg/ml; A 23187, 2 μM; PAF-acether 40 nM. The binding of ^{125}I-fibrinogen to nonstimulated platelets ($\odot$) is also illustrated for comparison.

TABLE 1

SUMMARY OF THE CHARACTERISTICS OF THE BINDING OF FIBRINOGEN TO PLATELETS
INDUCED BY DIFFERENT STIMULI *

| Stimulus | ^{125}I-Fibrinogen Bound (molecules/platelet) | | Affinity Constant K_a (M^{-1}) | Number of | Non-dissociable Fibrinogen % |
	With Ca^{++} (1 mM)	Without Ca^{++} (< 1 μM)			
A D P (10 μM)	30,500	300	2.2×10^6	38,000	63
Collagen (4 μ/ml)	11,500	1,500	ND	ND	62
Thrombin (2mU/ml)	25,600	2,050	2.0×10^6	37,000	73
Epinephrine (30 μM)	18,500	1,300	2.1×10^6	20,400	79.6
A 23187 (2 μM)	26,000	490	4×10^6	80,000	ND
PAF-acether (40 nM)	14,600	140	2.8×10^6	24,500	ND

* Washed platelets (10^8/ml) were suspended in a Ca^{++} free Tryode—2% albumin buffer and stimulated with optimal doses of each stimulus in the presence of 0.5 μM ^{125}I-Fibrinogen, with or without Ca^{++}.

The affinity constants and the maximum number of receptors induced by each stimulus were derived from Scatchard analyses of the binding experiments. The binding isotherms were determined by measuring the number of ^{125}I-fibrinogen molecules bound to the platelet as a function of the ^{125}I-fibrinogen concentration. Specific binding was obtained by subtracting the observed binding from the binding measured in the presence of a 100-fold molar excess of cold fibrinogen.[5] The nondissociable fibrinogen represents the platelet bound ^{125}I-fibrinogen, which is not displaced by cold fibrinogen, added when maximal binding of the radiolabeled ligand had been attained. (ND = not determined).

the binding isotherms revealed only a single class of binding sites with association constants ranging from 2×10^6 M^{-1} to 4×10^6 M^{-1} (TABLE 1). The similarities in association constants provide strong evidence for the induction of the same receptor system by all agonists. Third, the time-dependent stabilization of the platelet-fibrinogen interaction was observed with stimuli other than ADP. Listed in TABLE 1 is the percent of platelet bound ^{125}I-fibrinogen nondissociated by 100-fold excess of nonlabeled fibrinogen added when maximal binding of the radiolabeled ligand had been attained. Thus, the divalent ion dependency, the similarity in binding affinities, and the stabilization of platelet-bound fibrinogen are all consistent with the exposure of the same class of receptors by all stimuli. It should be noted from TABLE 1 that the maximum number of fibrinogen molecules bound per platelet varied considerably with the various stimuli, ranging from 10,400 with epinephrine to 80,000 with calcium ionophore. However with the most extensively studied stimulus, ADP, we have found that maximal fibrinogen binding also ranged from 15,400 to 82,500 molecules per cell. The variability was observed with platelets from different individuals as well as the same donor drawn on multiple occasions. Despite this variation on the number of sites, the affinity of the interaction was consistently the same.

Therefore, while we cannot exclude that stimuli induce different numbers of the same receptor sites, it seems more likely that these differences reflect donor variability.

With these data pointing to a single receptor system for fibrinogen, we considered the possibility that a common mechanism might be evoked for exposure of the receptor. Such a common mechanism could entail mobilization of platelet-associated ADP by stimulation of the cell. Thus, the inhibition of fibrinogen binding by apyrase, which degrades ADP to AMP, and creatinine phosphate/creatine phosphokinase, which converts ADP to ATP, was assessed. As shown in TABLE 2, these agents inhibited the binding of fibrinogen with all stimuli. Control experiments verified that products generated by these ADP scavenger enzymes did not inhibit fibrinogen binding, that creatinine phosphate or creatine phosphokinase separately did not inhibit binding, and that higher concentrations of calcium (5 mM) did not overcome the inhibition. These results are consistent with a common role of ADP in supporting the binding of fibrinogen to the stimulated cell. The presence of free contaminating ADP within the washed platelet suspension was excluded by washing the platelets in the presence of ADP scavengers (see ref. 26).

The synergistic effect of low levels of various stimuli on platelet aggregation is a well-recognized phenomenon. The capacity of a combination of a low concentration of ADP and the various stimuli to potentiate fibrinogen binding is demonstrated in TABLE 3. At doses of collagen and epinephrine that individually failed to support fibrinogen binding, addition of either one in combination with a low dose of ADP produced significant association. Similarly, doses of ADP as low as 10 nM significantly enhanced fibrinogen binding supported by 0.5 mU/ml of thrombin. With other stimuli such as PAF and calcium ionophore an additive rather than synergic effect was observed.

TABLE 2

INHIBITION OF THE BINDING OF FIBRINOGEN
TO STIMULATED PLATELETS BY ADP SCAVENGERS *

		ADP Removal by	
Stimulus	Control	CP/CPK	Apyrase
ADP (10 μM)	27,100	1,600	3,500
Collagen (4 μg/ml)	17,500	1,280	1,800
Thrombin (2 mU/ml)	38,690	4,780	5,350
Epinephrine (30 μM)	18,500	2,800	ND
A 23187 (2 μM)	19,000	672	ND
PAF-acether (40 nm)	23,300	1,608	ND

* Washed platelets (10^8/ml) were suspended in Tyrode 2% albumin solutions and were stimulated with optimal doses of the different inducers in the presence of 1 mM calcium and 0.5 μM ^{125}I-fibrinogen. Inhibition of the binding was achieved in the presence of CP/CPK (11 mg/ml/0.5 mg/ml) or apyrase (80 μg/ml). (ND = not determined).

TABLE 3

^{125}I-FIBRINOGEN BINDING TO PLATELETS, INDUCED BY LOW DOSES
OF STIMULI INDIVIDUALLY OR IN COMBINATION WITH ADP *

Stimuli	^{125}I-Fibrinogen Bound (molecules/platelet)	Potentiation of Effect
Collagen (1 μg/ml)	6,100	
ADP (0.5 μM)	4,500	synergic
Collagen+ ADP	16,000	
Thrombin (0.5 mU/ml)	450	
ADP (0.01 μM)	2,650	synergic
Thrombin + ADP	24,700	
Epinephrine (5 μM)	1,800	
ADP (0.5 μM)	2,500	synergic
Epinephrine + ADP	14,700	
A 23187 (1.1 μM)	3,600	
ADP (0.5 μM)	4,300	additive
A 23187 + ADP	7,800	
PAF-acether (1.8 nM)	1,200	
ADP (0.5 μM)	1,830	additive
PAF + ADP	3,200	

* In all experiments the platelets (2×10^8/ml) were stimulated with the selected inducers in the presence of 0.17 μM ^{125}I-fibrinogen and 1mM calcium. The number of ^{125}I-fibrinogen bound per cell was measured after a 30-minute incubation.

MODEL SEQUENCE FOR THE BINDING
OF FIBRINOGEN TO THE PLATELET

Taken together, these observations indicate that receptor exposure and fibrinogen binding occur with a variety of stimuli. Thus, a common pathway of fibrinogen-dependent aggregation may be considered with all stimuli and is schematically illustrated in FIGURE 3. Four basic steps are proposed to describe the reaction sequence. Reaction 1 illustrates the transition of the platelet from a resting state to an activated form. A key event associated with this transition may be the mobilization of platelet-associated ADP. Clearly, platelet secretion provides a mechanism for ADP mobilization; and all stimuli inducing secretion including prostaglandin derivatives,[27] should trigger this pathway. Epinephrine may represent a unique stimulus that is capable of mobilizing ADP in the absence of serotonin secretion.[25] In addition, it is postulated that individual differences between stimuli are established at the level of this initial reaction. Thus, the prolonged time course of fibrinogen binding with collagen and epinephrine-stimulated platelets probably reflects the rate of formation of the activated cell. Reaction 2 describes the induction of the fibrinogen receptor on the activated platelet. At low doses of stimuli, this reaction appears to require the continuous presence of ADP as suggested by the capacity of ADP scavengers [7] or ADP analogs,[19] to inhibit binding. The capacity of the initial stimulus or products derived from the stimulated platelet such as endoperoxides or thromboxanes to act synergistically with ADP must also be considered. Reaction 3 represents the reversible binding of fibrinogen to the cell in a divalent ion-dependent process. Reaction 4 describes the stabilization of the fibrinogen-platelet complex. Aggregation of the platelets and reaction 3 are concomitant events, whereas reaction 4 stabilizes the platelet aggregates.

The proposed model suggests a common pathway by which fibrinogen participates in platelet aggregation with all stimuli. The number of independent pathways which lead to platelet aggregation has been the topic of considerable interest and discussion. Our data implicating ADP as the essential inducer of the fibrinogen receptor with the various platelet stimuli were derived at low doses of the stimuli. This does not exclude the possibility that ADP-independent

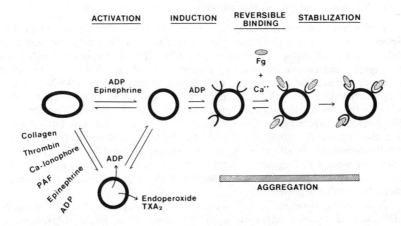

FIGURE 3. Schematic representation of the fibrinogen-dependent pathway of platelet aggregation. In this model sequence, it is hypothesized that aggregation of platelets by a variety of stimuli is dependent upon and concomitant with the binding of fibrinogen to its platelet receptor. This binding occurs in the presence of calcium (or magnesium) ions. Induction of this platelet receptor requires activation of the cell, and this activation can proceed through the secretion of platelet components such as ADP or prostaglandin derivatives. These components are themselves potent aggregating agents which can act synergistically or individually on other circulating platelets. ADP and epinephrine are unique inducers, since it has been possible to show the binding of fibrinogen with these stimuli in the absence of detectable secretion of serotonin. With all stimuli it was possible to demonstrate the association of fibrinogen with the platelets and it is postulated that this interaction regulates platelet aggregation.

mechanisms for the induction of the fibrinogen receptor may exist, particularly at higher doses of the platelet agonists. Clearly, washed platelets may aggregate in the absence of added fibrinogen, and this is usually achieved at high doses of the platelet agonists. The contribution of fibrinogen secreted from within the platelets under these conditions must, however, be considered. Tollefsen and Majerus [28] found that fragments of anti-fibrinogen inhibited thrombin-induced platelet aggregation in the absence of exogenously added fibrinogen, and we have made similar observations with collagen as the platelet stimulus. In addition, *in vivo* it is likely that the platelet will be exposed to low concentrations of multiple stimuli, and capacity of stimuli to act additively or synergistically in supporting fibrinogen binding enhances the physiological role of the fibrinogen-dependent pathway. In support of this proposal is the well-documented observation that platelets from patients with Glanzmann's thrombasthenia, which do not aggregate in response to a variety of stimuli,[29] fail to bind fibrinogen.[6]

THE SITES WITHIN FIBRINOGEN THAT MEDIATE BINDING TO THE PLATELET

The sites within fibrinogen that are recognized by its platelet receptor remain to be definitively localized. The following observations, however, are pertinent to this key issue and permit the proposal of a tentative hypothesis:

1. We have isolated a D fragment of 100,000 MW from prolonged plasmic digests in the presence of calcium which inhibits ^{125}I-fibrinogen binding to the platelet and ADP-induced platelet aggregation. The E fragment isolated from this digest is ineffective in inhibiting fibrinogen binding. When the D fragment is further digested in the absence of calcium to generate an 80,000 molecular weight derivative, its inhibitory capacity is abolished. These data provide evidence that the D domain contains an essential contact site for fibrinogen with the platelet. Corroborating this conclusion, we have found that Fab fragments of anti-D inhibit fibrinogen binding to the platelet, whereas Fab fragments of anti-E are much less effective. Since conversion of the 100,000 to the 80,000 molecular weight D fragment results primarily from proteolysis of the COOH-terminal region of the γ chain, the integrity of this region must be important for interaction with the platelet.[30]

2. Recently, Hawiger et al.[31] have found that aggregated γ chains support ADP-induced platelet aggregation in the absence of added fibrinogen and that Fab fragments of anti-γ chain inhibit fibrinogen binding and platelet aggregation. These results implicate the γ chain in the interaction of fibrinogen with its platelet receptor.

3. Recently, we have shown that the tetrapeptide, Gly-Pro-Arg-Pro, inhibits fibrinogen binding and platelet aggregation.[32] This peptide is an analog of the NH$_2$-terminal aspects of the α and β chains of fibrin, and Laudano and Doolittle [33] have shown that this peptide inhibits fibrin polymerization by interacting with the D domain of fibrinogen. Olexa and Budzynski [34] have shown that a fibrin polymerization site resides in the COOH-terminal region of the γ chain of fibrinogen. It has not been directly verified, but it can be reasonably hypothesized that Gly-Pro-Arg-Pro inhibits fibrin formation by binding to the polymerization site in the COOH-terminal region of the γ chain. Our data with this peptide suggest that it inhibits fibrinogen binding to the platelet by interacting with the fibrinogen ligand rather than the platelet receptor.

Thus, the present observations are consistent with involvement of the γ chain component of the D domain in the binding of fibrinogen to its platelet receptor. Moreover, the extreme COOH-terminal aspects of the γ chain appear to be critical for an interaction of high affinity. Whether the platelet binding site for fibrinogen resides specifically within this COOH-terminal aspect or whether its removal or interaction with Gly-Pro-Arg-Pro perturbs the binding site within adjacent regions remains to be elucidated. The γ chain of the D domain may not be the exclusive contact site within fibrinogen for the platelet. Limited plasmic proteolysis of fibrinogen results in a marked decrease in its capacity to bind to platelets [35] and support aggregation.[36] Aα chain aggregates of fibrinogen support platelet aggregation but more weakly than γ chain multimers.[31] A 240-fold molar excess of the 100,000 molecular weight D fragment as compared to non-labeled fibrinogen is required for similar inhibition of ^{125}I-fibrinogen binding.[31] Thus, additional contact sites, perhaps located in the Aα chain, may participate. The involvement of such multiple sites could provide a basis for the time dependent stabilization of platelet-bound fibrinogen. Once fibrinogen is bound to the platelet surface through its primary contact site, secondary interactions may occur such that free fibrinogen, even in large excess, cannot readily dissociate the bound ligand.

ROLE OF PLATELET-BOUND FIBRINOGEN IN AGGREGATION

While the data and the proposed model provide a molecular basis for the binding of fibrinogen to the platelet surface, the subsequent events that permit fibrinogen to directly induce or influence platelet aggregation remain uncertain. Some potential mechanisms may be considered at a speculative level, and fall into two general categories. First, the binding of fibrinogen molecules to the platelet surface may directly lead to platelet aggregation. It can be hypothesized that fibrinogen alters membrane characteristics such as charge properties which permit the platelets to aggregate. Another possibility for such a direct role of fibrinogen is a bridging function between platelets. Fibrinogen is a dimeric molecule, and it can be envisioned that one fibrinogen molecule may directly interact between two platelets. Alternatively, two adjacent platelets may both have fibrinogen bound to their surfaces, and aggregation may be mediated by associations between fibrinogen molecules in a fibrin polymerization-like interaction. Finally, in a physiological milieu, it must also be considered that other plasma or platelet-derived molecules interacting with fibrinogen may participate in bridging the platelets. Second, fibrinogen may modulate some basic pathway(s) of the platelet with the end result being aggregation. Modulation of the cystokeletal organization would fall into this category and would attractively explain the stabilization of platelet-bound fibrinogen. At present, there is no basis to distinguish between these two categories. Ultimately, a clear understanding of the mechanism by which fibrinogen bound to the platelet surface regulates cell-cell contact is fundamental to the understanding of the fibrinogen-dependent pathway of platelet aggregation.

SUMMARY

Fibrinogen participates in platelet aggregation induced by ADP and this participation is receptor mediated. The role of fibrinogen in platelet function is not restricted to the ADP stimulus as the molecule may regulate aggregation induced by the direct interaction of the protein with a single, specific receptor system. Thus, a common fibrinogen-dependent mechanism leading to platelet aggregation is hypothesized.

ACKNOWLEDGMENTS

We thank Dr. J. Benveniste for kindly providing platelet-activating factor (PAF-acether) and gratefully acknowledge the secretarial assistance of Sandy Thompson and Catherine Jacobson.

REFERENCES

1. MILLS, D. C. B. & D. E. MACFARLANE. 1976. *In* Platelets in Biology and Pathology. J. L. Gordon, Ed. Vol. 1: 159–202. North Holland Publishers. Amsterdam.
2. CROSS, M. J. 1964. Thromb. Diath. Haemorrh. **12:** 524–527.
3. MUSTARD, J. F., D. W. PERRY, R. L. KINLOUGH-RATHBONE & M. A. PACKHAM. Am. J. Physiol. **228:** 1757–1765.
4. NIEWIAROWSKI, S., A. Z. BUDZINSKI & B. LIPINSKI. 1977. Blood **49:** 635–644.
5. MARGUERIE, G., E. F. PLOW & T. S. EDGINGTON. 1979. J. Biol. Chem. **254:** 5357–5363.
6. BENNETT, J. S. & G. J. VILAIRE. 1979. J. Clin. Invest. **64:** 1393–1401.
7. MARGUERIE, G., T. S. EDGINGTON & E. F. PLOW. 1980. J. Biol. Chem. **255:** 154–161.
8. PEERSCHKE, E. I., M. B. ZUCKER, R. A. GRANT, J. J. EGAN & M. M. JOHNSON. 1980. **55:** 841–847.
9. MUSTARD, J. F., D. W. PERRY, N. G. ARDILE & M. A. PACKHAM. 1972. Br. J. Haematol. **22:** 193–204.
10. INCEMANN, S., J. CAEN & J. BERNARD. 1966. J. Lab. Clin. Med. **62:** 21–32.
11. WEISS, H. J. & J. ROGERS. 1971. New Engl. J. Med. **285:** 369–374.
12. SOLUM, N. O. & H. STORMORKEN. 1965. Scand. J. Clin. Lab. Invest. **17** (Suppl. 84): 170–182.
13. MUSTARD, J. F., M. A. PACKHAM, R. L. KINLOUGH-RATHBONE, D. W. PERRY & E. REGOEZCI. 1978. Blood **55:** 453–466.
14. MARGUERIE, G. & E. F. PLOW. 1981. Biochemistry. **20:** 1074–1080.
15. HAWIGER, J., S. PARKINSON & S. TIMMONS. 1980. Nature. **283:** 195–197.
16. HARFENIST, E. J., M. A. PACKHAM & J. H. MUSTARD. 1980. Blood **56:** 189–198.
17. NIEWIAROWSKI, S., A. Z. BUDZINSKI, T. A. MORINELLI, T. M. BUDZINSKI & G. J. STEWART. 1981. J. Biol. Chem. **256:** 917–925.
18. MARGUERIE, G., N. THOMAS-MAISON, M. J. LARRIEU & E. F. PLOW. 1982. Blood **59:** 91–95.
19. FIGURES, W. R., S. NIEWIAROWSKI, T. A. MORINELLI, R. F. COLMAN & R. W. COLMAN. 1981. J. Biol. Chem. **256:** 7789–7795.
20. BENVENISTE, J., M. TENCE, P. VARENNE, J. BIDAULT, G. BOULLET & J. POLONSKY. 1979. C.R. Acad. Sci, Paris **289:** 1017–1021.
21. DEMOPULOS, C. A., R. N. PINCKARD & D. J. HANAHAN. 1979. J. Biol. Chem. **254:** 9355–9358.
22. FEINMANN, R. D. & T. C. DETWILLER. 1974. Nature **249:** 172–174.
23. HOLMSEN, H. & C. A. DANGELMAIER. 1981. J. Biol. Chem. **256:** 10449–10452.
24. MILLS, D. C. B. 1973. Nature New Biol. **243:** 220–222.
25. PLOW, E. F. & G. MARGUERIE. 1980. J. Biol. Chem. **255:** 10971–10977.
26. PLOW, E. F. & G. MARGUERIE. 1980. Blood **56:** 553–55.
27. BENNETT, J. S., G. VILAIRE & J. W. BURCH. 1981. J. Clin. Invest. **68:** 981–987.
28. TOLLEFSEN, D. M., J. R. FEAGLER & P. W. MAJERUS. 1974. J. Biol. Chem. **249:** 2646–2651.
29. CAEN, J. 1972. J. Clin. Haematol. **1:** 383–391.
30. TAKAGI, T. & R. F. DOOLITTLE. 1975. Biochemistry **14:** 940–946.
31. HAWIGER, J., S. TIMMONS, M. KLOCZEWIAK, D. D. STRONG & R. F. DOOLITTLE. 1982. Proc. Natl. Acad. Sci. USA **79:** 2068–2071.
32. PLOW, E. F. & G. MARGUERIE. 1982. Proc. Natl. Acad. Sci. USA. In press.
33. LAUDANO, A. P. & R. F. DOOLITTLE. 1980. Biochemistry **19:** 1013–1019.
34. OLEXA, S. A. & A. Z. BUDZINSKI. 1981. J. Biol. Chem. **256:** 3544–3549.
35. NIEWIAROWSKI, S., A. Z. BUDZINSKI, T. A. MORINELLI, T. M. BUDZINSKI & G. J. STEWART. 1981. J. Biol. Chem. **256:** 917–925.
36. NIEWIAROWSKI, S., A. Z. BUDZINSKI & B. LIPINSKI. 1977. Blood **49:** 635–644.

REGULATION OF FACTOR XIIIa GENERATION BY FIBRINOGEN *

C. Gerald Curtis,† Todd J. Janus
R. Bruce Credo ‡ and Laszlo Lorand

Department of Biochemistry, Molecular and Cell Biology
Northwestern University
Evanston, Illinois 60201

By forming intermolecular γ-glutaminyl-ϵ-lysyl cross-links between fibrin molecules, activated coagulation factor XIII (FXIIIa) performs an essential hemostatic function.[1-3] However, the sequence of events leading to its activation at the surface of a wound and the contributions of the cellular components enmeshed in the unstabilized clot, have not as yet, been fully established. Unravelling some of the molecular interactions in the activation of purified FXIII *in vitro* has been possible using a combination of assay systems including the unmasking of the active center thiol to [^{14}C]iodoacetamide titration,[4, 5] the dissociation of the heterologous subunits monitored by polyacrylamide gel-electrophoresis under non-denaturing conditions,[5, 7] and generation of esterase and transamidase activities.[8-10]

Two experimental pathways are known by which FXIII zymogen (a_2b_2) may be activated.

Thrombin-independent Pathway [11, 11a, 11b]

In this pathway, activation does not require proteolytic cleavage of the zymogen but occurs by the addition of sufficiently high concentrations of calcium ions (Ca^{2+}). High (>100mM) Ca^{2+} concentrations cause dissociation of the zymogen, unmasking of titratable cysteine and the generation of enzyme activity. This thrombin-independent pathway is accelerated by chaotropic anions, the relative efficacy being p-toluenesulfonate $>$ thiocyanate $>$ iodide $>$ bromide. The Ca^{2+}-requirement for activation can be lowered significantly to about 50mM Ca^{2+}, in the presence of chaotropic agents. Moreover, the kinetics of acyl group transfer reactions catalyzed by the thiocyanate and Ca^{2+}-activated species (a_2^0 in Equation 1) is indistinguishable from those catalyzed by the more conventional and physiological thrombin-activated species, when measured on small synthetic substrates.

$$a_2b_2 \xrightarrow[\text{Ca}^{2+}]{\text{thiocyanate}} a_2^0 + b_2 \qquad \boxed{1}$$

Thrombin-dependent Pathway

In this pathway of activation, the limited proteolysis of the a_2 subunits in the zymogen,[12] is followed by the Ca^{2+}-dependent dissociation of the

† Present address: Department of Biochemistry, University College, Cardiff CF1 1XL, Wales.
‡ Present address: Abbott Laboratories, North Chicago, IL 60064.
* This work was aided by a USPHS Research Career Award (HL–03512) and by a grant from the National Institutes of Health (HL–02212).

0077–8923/83/0408–0567 $01.75/0 © 1983, NYAS

heterologous molecule and the concomitant unmasking of the catalytic centers
in the a subunits (Equation 2).

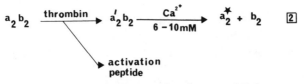

The cleaved $a_2'b_2$ species (FXIII') in the absence of Ca^{2+} cannot be titrated
with [^{14}C]iodoacetamide and is without significant catalytic activity when
measured in a fully synthetic substrate assay.[9] Thus it would appear that the
thrombin-catalyzed step, like the chaotropic solutes, reduces the interaction
between the subunits so that significantly less Ca^{2+} (i.e. 6–10mM Ca^{2+} at pH
7.5, $\mu = 0.15$, 37° C) can dissociate a_2' from b_2 and unmask the active center
thiol generating the catalytic a_2^* species.

ROLE OF FIBRINOGEN IN THE ACTIVATION OF FXIII' ($a_2'b_2$)

At normal plasma concentrations of free Ca^{2+} (1.5mM) it is unlikely that
thrombin and Ca^{2+} alone are responsible for FXIII activation, unless the
Ca^{2+} concentration is elevated at the site of coagulation by release from
platelets. By contrast, thrombin-activated platelet FXIII, which lacks the b subunit,[13] is fully activated at 1.5mM Ca^{2+}. This would suggest that a relatively high
Ca^{2+} concentration is required to dissociate the purified a_2b_2 complex and that
in vivo other components of the blood may promote the dissociation and consequently reduce the Ca^{2+} requirement for unmasking the active center thiol.
FIGURE 1 demonstrates that plasma concentrations of fibrinogen may fulfill this

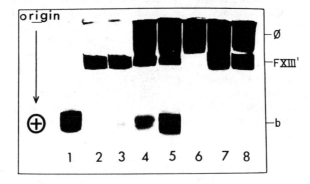

FIGURE 1. Fibrinogen-assisted dissociation of thrombin-modified factor XIII as
seen by polyacrylamide gel electrophoresis under nondenaturating conditions at pH
7.8. Factor XIII' (9.6×10^{-7} M) was incubated for 30 min at 37° C in a 0.1-ml
solution of 0.05 M Tris-HCl, pH 7.5, containing 2.16×10^{-5} M iodoacetamide, 4
units/ml hirudin, NaCl to maintain $\mu = 0.15$, and, as required, one or both 1.5 mM
$CaCl_2$ and 8.9×10^{-6} M fibrinogen (Kabi lot 32 20). The factor XIII zymogen itself
(9.6×10^{-7} M) and purified b subunit (9.6×10^{-7} M) served as markers. Aliquots
of 50 μl were taken for electrophoresis. *Tracks:* (1) b subunit; (2) factor XIII;
(3) factor XIII' and $CaCl_2$; (4) factor XIII' and fibrinogen; (5) factor XIII' with
$CaCl_2$ and fibrinogen; (6) fibrinogen and $CaCl_2$; (7) factor XIII and fibrinogen; and
(8) factor XIII with $CaCl_2$ and fibrinogen. Positions of fibrinogen (O), free b subunit, and the catalytic subunit containing factor XIII' species are indicated.

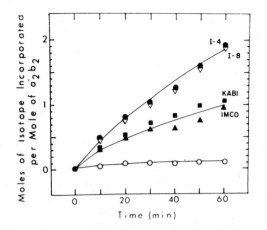

FIGURE 2. Alkylation of thrombin-modified factor XIII with iodoacetamide in the presence of fibrinogen, but without CaCl₂. The reaction was performed at 37° C in 0.1 ml solutions of 0.05 M Tris-HCl, pH 7.5, containing 1.26×10^{-6} M factor XIII', 3 units/ml hirudin, NaCl to maintain $\mu = 0.15$ (○), and also one of the following additions as required: 8.9×10^{-6} M IMCO fibrinogen (▲, lot F-144), 8.9×10^{-6} M Kabi (■, lot 32120), 1.1×10^{-5} M 1-4 (●), or 1.2×10^{-5} M]-8 (▽). The ordinate denotes protein-bound isotope.

role. The release of free b subunits from $a_2'b_2$ is promoted by the presence of both fibrinogen and Ca²⁺ (FIG. 1, *track 5*) when compared with incubations in which either fibrinogen (*track 3*) or Ca²⁺ (*track 4*) are omitted. Interestingly, fibrinogen causes significant dissociation of b subunits from the $a_2'b_2$ complex even when precautions are taken to remove Ca²⁺ by exhaustive dialysis. Moreover, this novel function of fibrinogen is not limited to the dissociation of $a_2'b_2$ because unmasking (though rather sluggishly) of titratable groups can be achieved simply by allowing $a_2'b_2$ to interact with fibrinogen but without Ca²⁺ (FIG. 2). Unmasking is more rapid when Ca²⁺ is added; with certain fibrinogen preparations nearly total unmasking of iodoacetamide reactive sites (i.e. 2 equivalents per $a_2'b_2$) could be achieved in 10 minutes with 1.5mM Ca²⁺ and less than 10^{-5} M fibrinogen which corresponds to their concentrations in plasma (FIG. 3).

The hitherto unsuspected role of fibrinogen in weakening the interaction between the a_2' and b_2 subunits may therefore be summarized as shown in Equation 3, with Ca²⁺ apparently shifting the equilibrium to the right.

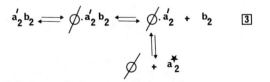

Although fibrinogen itself is a substrate for the a_2^* species there is no evidence to suggest that the formation of a covalent acylenzyme intermediate (Equation 4) contributes to the observed phenomenon. On the contrary, when the FXIIIa-sensitive glutamine residues in fibrinogen were blocked by reaction with hydroxylamine, the modified fibrinogen was as effective as intact fibrinogen in promoting the alkylation of $a_2'b_2$ with iodoacetamide.[7]

$$a_2^\star \rightleftharpoons \emptyset \cdot a_2^\star \longrightarrow \emptyset\text{–}a_2^\star \qquad \boxed{4}$$

LOCUS OF THE REGULATORY DOMAIN ON FIBRINOGEN

The specificity of four different fibrinogen preparations in modulating the reaction of $a_2'b_2$ with iodoacetamide is shown in FIGURE 4. These fibrinogens differ only in the $A\alpha$ chains, which in the Kabi preparation were smaller than in either I-4 or IMCO, whereas in I-8 the $A\alpha$ chains were cleaved into large fragments.[14] Perhaps the variability of response associated with the fibrinogen preparations could be explained by the differences in the integrity of the $A\alpha$ chains. The critical domain for contact with $a_2'b_2$ may therefore be located on these chains which can undergo rather extensive cleavages but still retain the ability to modify the $a_2'b_2$ complex. Similarly, plasmin digests of

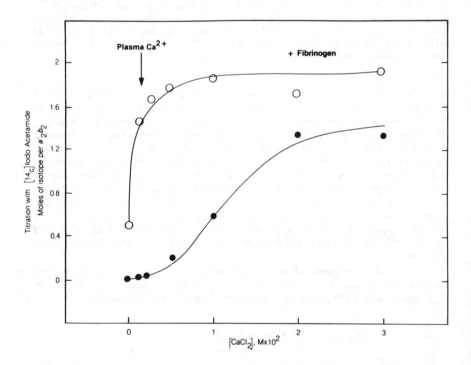

FIGURE 3. Calcium requirement for the alkylation of thrombin-modified factor XIII ($a_2'b_2$) with iodoacetamide in the absence (●) and presence (○) of 9 μM fibrinogen. Alkylation was performed for 10 min at 37° C in 0.1ml solutions of 0.05M Tris-HCl, pH 7.5, containing 1.2×10^{-6} M FXIII′, 3.4 units/ml hirudin, 4.4×10^{-5} M iodo[1-^{14}C-]acetamide, 0–30 mM CaCl$_2$, and NaCl to maintain $\mu = 0.15$. The ordinate denotes protein bound isotope.

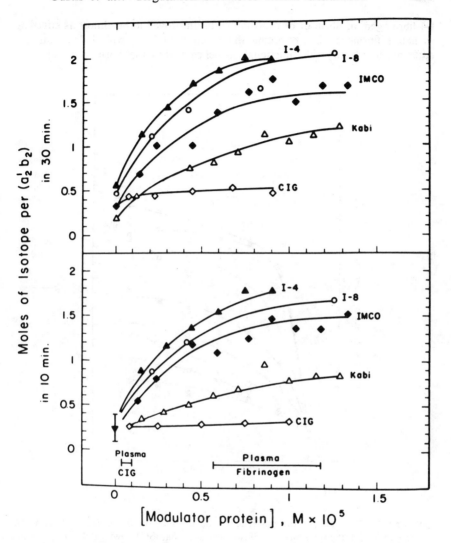

FIGURE 4. Dependence of the alkylation of thrombin-modified factor XIII with iodoacetamide on the concentration of protein modulators. Reaction was carried out for 10 (*bottom panel*) and 30 min (*top panel*) at 37° C in 0.1 ml solutions of 0.5 M Tris-HCl, pH 7.5, containing 9.6×10^{-7} M factor XIII′, 2.16×10^{-5} M iodo[1-¹⁴C] acetamide, 2.28 units/ml hirudin, 1.5 mM CaCl₂, NaCl to maintain $\mu = 0.15$, and $0–1.5 \times 10^{-5}$ M of protein modulators (abscissa) as follows: (▲) fibrinogen 1–4, (○) I–8, (♦) IMCO (Lot No. F–146), (△) Kabi (1024, lot 32120) and (◇) plasma fibronectin (i.e., cold-insoluble globulin or CIG). The ordinate denotes protein-bound isotope. Horizontal bars show the physiological range of concentrations for fibrinogen and CIG in human plasma.

I-4 fibrinogen, of fibrin or of cross-linked fibrin proved to be almost as effective as native fibrinogen in promoting the reaction of $a_2'b_2$ with iodoacetamide. Attempts to identify the functional sequence within the fibrinogen molecule

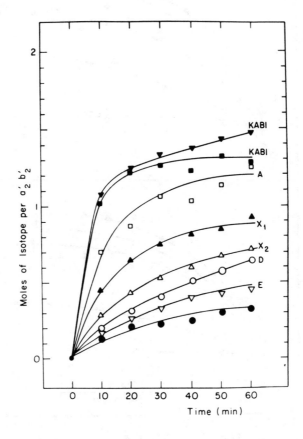

FIGURE 5. Time course for the reaction of thrombin-modified factor XIII with iodoacetamide in the presence of fibrinogen and that of known, isolated plasmin degradation products. Alkylations were performed at 37° C in 0.1 ml solutions of 0.05 M Tris-HCl, pH 7.5, containing 1.26×10^{-6} M factor XIII', 4.4×10^{-5} M iodo[1-^{14}C]acetamide, 1.5 mM $CaCl_2$, NaCl to maintain $= 0.15$ (●), and either 9.2×10^{-6} M Kabi fibrinogen (▼, ■), 1.1×10^{-5} M fragment A (□), 9.2×10^{-6} M fragment X of a stage 1 plasmin digest (▲), 1.8×10^{-5} M fragment D (○), or 1×10^{-5} M fragment E (▽). Protein-bound isotope is shown on the ordinate.

were made by testing purified plasmin fragments (A,D,E,X and the Hi2DSK fragment derived from the Aα chain by cleavage with cyanogen bromide). The positive results with fragment A {or Pl 21,[15]} and with Hi2DSK (Figs. 5

and 6) indicate that the area of contact is located in the midsection of the Aα chain, between residues 242–424 for fragment A and residues 243–476 for H12DSK.

Details of the molecular interactions between the intact fibrinogen molecule and the heterologous thrombin-activated FXIII have yet to be established, but there is little doubt that fibrinogen has at least three functional domains which are important in synchronizing the last stages of coagulation. One domain is located at the NH$_2$-terminal regions of the Aα and Bβ chains. The thrombin-catalyzed removal of the fibrinopeptides A and B [16-18] is necessary for unmask-

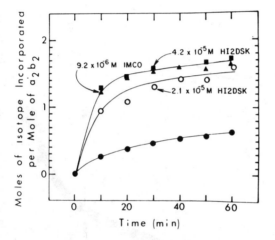

FIGURE 6. Time course for the alkylation of thrombin-modified factor XIII with iodoacetamide in the presence of the CNBr fragment from the Aα chain of fibrinogen, Hi2DSK. Reactions were performed at 37° C in 0.1 ml solutions of 0.05 M Tris-HCl, pH 7.5, containing 1.26×10^{-6} M factor XIII', 3 units/ml hirudin, 1.5 mM CaCl$_2$, 4.4×10^{-5} M iodo[1-^{14}C]acetamide, NaCl to maintain $\mu = 0.15$ (●), and either 9.2×10^{-6} M IMCO fibrinogen (▲) or 4.2×10^{-5} M (■), and 2.1×10^{-5} M or H12DSK (○). Ordinate shows protein-bound isotope.

ing the fibrin-to-fibrin aggregation sites. A second domain is located at tne cross-linking sites in the COOH-terminal portions of the γ-chains and the mid-sections of the α-chains.[15, 19] The third domain resides in the midsection of the Aα chains;[7] it is already exposed in the intact fibrinogen molecule and functions by promoting the dissociation of the FXIII' hydrolytically modified zymogen ensemble.

FIGURE 7 gives an outline of the regulatory aspects discussed, within the framework of the clotting of fibrinogen and subsequent cross-linking of fibrin in plasma.

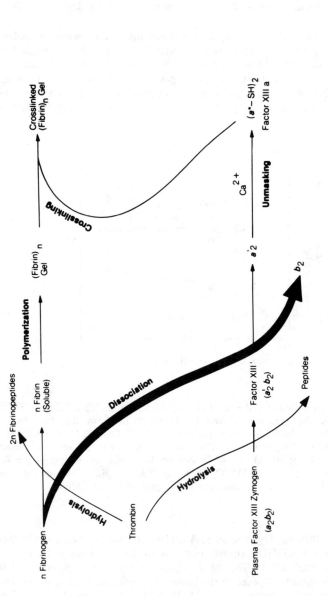

FIGURE 7. Role of fibrinogen in the dissociation of thrombin-activated plasma factor XIII. In this proposed seequence both thrombin and fibrinogen have dual functions. By hydrolyzing the same peptide bond in both fibrinogen and factor XIII zymogen, thrombin controls the rate of production of the substrate (i.e. fibrin) and also the generation of the cross-linking enzyme that catalyzes the formation of stabilizing γ-glutaminyl-ϵ-lysyl bonds in the substrate. Fibrinogen is not only the precursor of the essential substrate (fibrin) but also acts as a modulator by promoting the dissociation of thrombin-modified factor XIII ($a_2'b_2$; Factor XIII').

REFERENCES

1. LORAND, L. 1972. Fibrinoligase: The fibrin-stabilizing factor system of blood plasma. Ann. N.Y. Acad. Sci. **202:** 6–30.
2. LORAND, L. 1977. Haemorrhagic disorders of fibrin-stabilization. *In* Haemostasis: Biochemistry, Physiology and Pathology. D. Ogston & B. Bennett, Eds.: 405–423. John Wiley & Sons. New York.
3. LORAND, L., M. S. LOSOWSKY & K. J. A. MILOSZEWSKI. 1980. Human factor XIII: Fibrin stabilizing factor. Progress in Hemostasis and Thrombosis, Vol. 5. T. H. Spaet, Ed.: 245–290. Grune and Stratton, Inc. New York.
4. CURTIS, C. G., P. STENBERG, C-H.J. CHOU, A. GRAY, K. L. BROWN & L. LORAND. 1973. Titration and subunit localisation of active center cysteine in fibrinoligase (Thrombin-activated fibrin stabilizing factor). Biochem. Biophys. Res. Commun. **52:** 51–56.
5. CURTIS, C. G., K. L. BROWN, R. B. CREDO, R. A. DOMANIK, A. GRAY, P. STENBERG & L. LORAND. 1974. Calcium-dependent unmasking of active center cysteine during activation of fibrin stabilizing factor. Biochemistry **13:** 3774–3780.
6. LORAND, L., A. J. GRAY, K. BROWN, R. B. CREDO, C. G. CURTIS, R. A. DOMANIK & P. STENBERG. 1974. Dissociation of the subunit structure of fibrin stabilizing factor during activation of the zymogen. Biochem. Biophys. Res. Commun. **56:**(4) 914–922.
7. CREDO, R. B., C. G. CURTIS & L. LORAND. 1981. α-Chain domain of fibrinogen controls generation of fibrinoligase (Coagulation factor XIIIa). Calcium ion regulatory aspects. Biochemistry **20:** 3770–3778.
8. LORAND, L., C-H.J. CHOU & I. SIMPSON. 1972. Thiolester substrates for transamidating enzymes: Studies on fibrinoligase. Proc. Natl. Acad. Sci. USA **69:** 2645–2648.
9. CURTIS, C. G., P. STENBERG, K. L. BROWN, A. BARON, K. CHEN, A. GRAY, I. SIMPSON & L. LORAND. 1974. Kinetics of transamidating enzymes. Production of thiol in the reactions of thiol esters with fibrinoligase. Biochemistry **13:** 3257–3262.
10. STENBERG, P., C. G. CURTIS, D. WING, Y. S. TONG, R. B. CREDO, A. GRAY & L. LORAND. 1975. Amide formation in the enzymic reactions of thiol esters with amines. Biochem. J. **147:** 155–163.
11. CREDO, R. B., C. G. CURTIS & L. LORAND. 1978. Ca^{2+}-related regulatory function of fibrinogen. Proc. Natl. Acad. Sci. USA **75:** 4234–4237.
11a. LORAND, L., R. B. CREDO & T. J. JANUS. 1981. Factor XIII (Fibrin Stabilizing Factor). Methods Enzymol. **80:** 333–341.
11b. JANUS, T. J., R. B. CREDO, C. G. CURTIS, L. HAGGROTH & L. LORAND. 1981. Novel mode of activating human factor XIII zymogen without the hydrolytic cleavage of the *a* subunits. Fed. Proc. **40:** 1585 (Abst. 4959).
12. SCHWARTZ, M. L., S. V. PIZZO, R. L. HILL & P. A. McKEE. 1971. The subunit structures of human plasma and platelet factor XIII (Fibrin-stabilizing factor). J. Biol. Chem. **246:** 5851–5854.
13. SCHWARTZ, M. L., S. V. PIZZO, R. L. HILL & P. A. McKEE. 1973. Human factor XIII from plasma and platelets. Molecular weights, subunit structures, proteolytic activation and cross-linking of fibrinogen and fibrin. J. Biol. Chem. **248:** 1395–1407.
14. MOSESSON, M. W., D. K. GALANAKIS & J. S. FINLAYSON. 1974. Comparison of human plasma fibrinogen subfractions and early plasmic fibrinogen derivatives. J. Biol. Chem. **249:** 4656–4664.
15. FRETTO, L. J., E. W. FERGUSSON, H. M. STEINMAN & P. A. McKEE. 1978. Localisation of the α-chain cross-link acceptor sites of human fibrin. J. Biol. Chem. **253:** 2184–2195.

16. LORAND, L. 1952. Fibrino-peptide. Biochem. J. **52:** 200–203.
17. BETTELHEIM, F. R. & K. BAILEY. 1952. Products of action of thrombin on fibrinogen. Biochim. Biophys. Acta **9:** 578–579.
18. BLOMBACK, B. & A. VESTERMARK. 1958. Isolation of fibrino-peptides by chromatography. Ark. Kemi **12:** 173–182.
19. CHEN, R. & R. F. DOOLITTLE. 1971. γ-γ Cross-linking sites in human and bovine fibrin. Biochemistry **10:** 4486–4491.

DISCUSSION OF THE PAPER

H. ERICKSON (*Duke University Medical Center, Durham, NC*): Did you measure the affinity constant for factor XIII.

C. G. CURTIS: No.

ERICKSON: Do you know how the association may depend on the fibrinogen concentration? Have you varied the fibrinogen concentration?

CURTIS: Yes, it is dependent on the fibrinogen concentration and appears to reach its peak within the range of normal fibrinogen concentration.

C. SORIA (*Hôpital Lariboisière, Paris, France*): Have you performed the same assay using placental factor XIII, which contains only A subunits?

CURTIS: No.

"FIBRINOGEN OKLAHOMA": A STUDY OF INTERACTIONS

J. S. Finlayson

Division of Blood and Blood Products
Bureaus of Drugs and Biologics
Food and Drug Administration
Bethesda, Maryland 20205

The hemorrhagic disorder known as Fibrinogen Oklahoma [1] was initially thought to result from a defect in fibrin formation [2] or Factor XIIIa activity [3]— conclusions that were consistent with the behavior of the clot during certain laboratory tests. [2,3] My interest in this disorder began when Dr. James Hampton provided some partially purified fibrinogen (Blombäck [4] fraction I-O) from affected family members.

Chromatography of these fibrinogen samples on DEAE cellulose [5] yielded a normal elution pattern, including a normal profile of clottable protein and Factor XIII activity. [5] Furthermore, no electrophoretic differences from the normal fibrinogen pattern were observed, regardless of whether immunoelectrophoresis [6] was performed on unfractionated plasma from the propositus or sodium dodecyl sulfate (SDS) polyacrylamide gel electrophoresis [7] was carried out on fibrinogen isolated by glycine [8] or sodium sulfite [9] precipitation. Similarly, when the propositus's plasma was clotted in the presence of calcium, the resulting fibrin exhibited normal cross-linking as assessed by gel electrophoresis [10] or by determining the number of ϵ-amino lysyl cross-links by cyanoethylation according to Pisano *et al.*[11]

Fibrinopeptide release [12] (by thrombin) from the propositus's fibrinogen was compared with that from fibrinogen isolated [8] from pooled normal plasma; no difference was observed (FIG. 1). These fibrinogens were then used to prepare fibrin monomers,[13] and fibrin monomer aggregation was investigated.[12] Again, no difference was apparent (FIG. 2). These results were consistent with the findings that the thrombin clotting time of the propositus's fibrinogen was normal and that it did not delay the clotting of normal fibrinogen.

To explore the possibility that the abnormality might arise from increased susceptibility to proteolysis, fibrinogen was isolated [8] from several samples of the propositus's plasma and from several normal plasma pools, clotted with thrombin in the presence of dilute human plasmin (final concentrations 0.014– 0.069 CTA unit/ml) and incubated at 37°C. The lysis times for all clots fell in the same range. More significantly, Dr. Laurence Sherman found that autologous and homologous fibrinogen showed similar metabolic turnover times in the propositus (102 and 108 hours, respectively) and that these times were virtually the same as the mean value obtained for normal individuals (102 hours).

The distribution of total plasma fibrinogen into "low solubility" (i.e., conventionally prepared) and "high solubility" (analogous to Mosesson [14] fraction I-8) fractions was investigated by determining the proportion that remained soluble in 8% (v/v) ethanol at −2°C.[15] No consistent differences between the propositus's and normal plasma were found, either in my laboratory or that of

577

0077–8923/83/0408–0577 $01.75/0 © 1983, NYAS

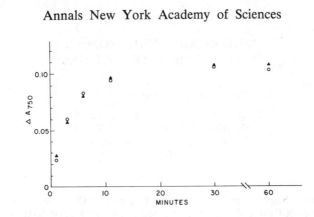

FIGURE 1. Fibrinopeptide release from fibrinogen after treatment with thrombin. Fibrinogen (93% clottable) was isolated by glycine precipitation;[8] 0.5-ml portions of a fibrinogen solution were incubated with thrombin for the time indicated (incubation mixture: 4.3 mg clottable protein/ml, 0.9 U.S. unit thrombin/ml). Peptide was determined by a modified Folin-Lowry method.[12] *Solid triangles:* fibrinogen from propositus's plasma; *open circles:* fibrinogen from pooled normal plasma.

Dr. Michael Mosesson. This result appeared to agree with Sherman's findings that the propositus neither exhibited abnormally high plasma levels of fraction I-8 nor produced excessive fraction I-8 during the metabolism of exogenous fibrinogen (see above).

A potential clue to the nature of this disorder was obtained by clotting the propositus's plasma in the presence of ethylenediaminetetraacetate (EDTA), winding the clot on a glass rod, allowing the wound clot to incubate in its

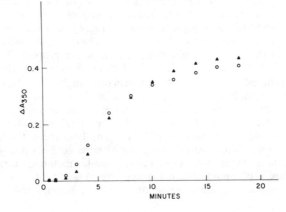

FIGURE 2. Fibrin monomer aggregation. Fibrinogen isolated by glycine precipitation[8] was clotted with thrombin, and the fibrin was harvested and dissolved in 0.02M acetic acid.[13] Fibrin monomer aggregation was initiated by adjusting the pH to 6.8 (final ionic strength 0.12, final protein concentration 0.24 mg/ml) as described previously.[12] *Solid triangles:* propositus's fibrin; *open circles:* normal fibrin.

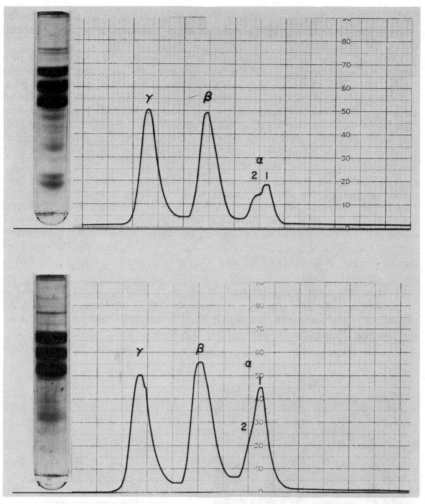

FIGURE 3. SDS polyacrylamide gel electrophoresis of fibrin from propositus's plasma. Portions of plasma were clotted with thrombin in the presence of 8 mM EDTA; the clots were wound on glass rods, allowed to incubate in their own serum at 37°C for 2 hours, rinsed, solubilized, reduced, and electrophoresed on 10% gels for 5 hours at 6 mA/gel.[10] The incubation mixture used to prepare the fibrin depicted in the lower half of the figure contained, in addition to EDTA, 1000 kallikrein inactivator units aprotinin/ml. The gels photographed were loaded with ~40 μg protein; those subjected to densitometry, ~12 μg protein. As shown, the anode is at the bottom of each gel and at the left of the tracings. The two very anodal bands seen in the upper gel are α remants[16] $\alpha/_{11}$ and $\alpha/_{12}$.

own serum at 37°C, and examining the resulting fibrin by gel electrophoresis.[10] After 2 hours, bands corresponding to α chain remnants [16] $\alpha/_{11}$ and $\alpha/_{12}$ were readily apparent, and densitometry indicated that they had been formed at the expense of the α chain (FIG. 3, *top*). When aprotinin was present in the incubation mixture, these bands were not seen and the α chain was essentially intact (FIG. 3, *bottom*), indicating that these remnants were being formed during the incubation.

Although these remnants also arose when normal plasma was treated in a similar manner,[10] they appeared much earlier when the propositus's plasma was clotted (FIG. 4). However, they were not formed when the propositus's purified fibrinogen was clotted and incubated under the same conditions. Wound clots were therefore prepared (in the presence of EDTA) from a series of samples of the propositus's plasma and that of normal donors. After clotting, they were allowed to incubate in their own sera at 37°C and observed periodically. In all cases, the time required for complete lysis of the propositus's clots was considerably less than that for normals.

Euglobulin clot lysis times were then determined.[17] In all cases, the lysis times determined for the propositus were markedly shorter than normal (2.5–7 hours versus 21–30 hours). Although these times could be lengthened somewhat by adding purified fibrinogen (either before or after precipitation of the euglobulin fraction) so as to equalize the fibrin content of the propositus's and the control clots, the difference persisted. The foregoing results were confirmed by Dr. Fletcher Taylor, who found that the propositus exhibited a shortened dilute-blood clot lysis time.[18]

It therefore seemed possible that this bleeding disorder might be the result of inadequate inhibitory activity. Accordingly, Dr. Peter Harpel analyzed the propositus's plasma for α_2-macroglobulin. Somewhat disappointingly, both the immunoelectrophoretic pattern and the plasma concentration of the α_2-macroglobulin (determined immunologically) were normal, a finding subse-

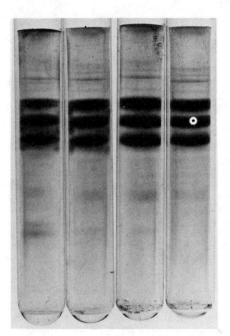

FIGURE 4. SDS polyacrylamide gel electrophoresis of fibrin from propositus's and pooled normal plasma. Clots were prepared and electrophoresed as summarized in the legend to FIGURE 3. The incubation mixtures contained, respectively (from left): propositus's plasma, normal plasma, propositus's plasma and aprotinin, normal plasma and aprotinin. Anode is at bottom.

quently confirmed in my laboratory. Furthermore, this inhibitor appeared to be functional.

With the description of α_2-proteinase inhibitor [19] and the report that its absence resulted in a bleeding diathesis,[20] it became of interest to reexamine the propositus's plasma. Laurell "rocket" immunoanalyses [21] carried out in association with Dr. Franco Carmassi have consistently shown that samples of the propositus's plasma have reduced levels of α_2-proteinase inhibitor. The concentrations found (50–75% of normal), however, are within the range reported to support normal hemostasis,[22] hence the functional significance of these measurements is not clear. Nevertheless, they are consistent with the observation of bleeding after trauma [3] and appear worthy of further investigation.

ACKNOWLEDGMENTS

At least as important as the interactions between the fibrinogen and the other proteins described above have been the cooperative interactions between the scientists that made this study possible. The author is grateful to all of those named in the text. In addition, special thanks are due to Lynne A. Reamer for skillful and devoted assistance and to the original investigator, Dr. James W. Hampton, whose continued interest and cooperation have allowed the study to reach its present stage.

REFERENCES

1. HAMPTON, J. W. & R. O. MORTON. 1970. Fibrinogen Oklahoma—recharacterization of a familial bleeding diathesis. Abstracts, XIII. International Congress of Hematology, p. 313.
2. HAMPTON, J. W., D. M. MAKEY & J. W. WILSON. 1962. Atoka diathesis: A familial disorder involving defective fibrin generation. Thromb. Diath. Haemorrh. **7:** 201.
3. HAMPTON, J. W., R. M. BIRD & D. M. HAMMARSTEN. 1965. Defective fibrinase activity in two brothers. J. Lab. Clin. Med. **65:** 469–474.
4. BLOMBÄCK, B. & M. BLOMBÄCK. 1956. Purification of human and bovine fibrinogen. Ark. Kemi **10:** 415–443.
5. FINLAYSON, J. S. 1968. Chromatographic purification of fibrinogen. In Fibrinogen. K. Laki, Ed.: 39–59. Marcel Dekker. New York.
6. SCHEIDEGGER, J. J. 1955. Une micro-méthode de l'immuno-électrophorèse. Int. Arch. Allergy **7:** 103–110.
7. WEBER, K. & M. OSBORN. 1969. The reliability of molecular weight determinations by dodecyl sulfate-polyacrylamide gel electrophoresis. J. Biol. Chem. **244:** 4406–4412.
8. KAZAL, L. A., S. AMSEL, O. P. MILLER & L. M. TOCANTINS. 1963. The preparation and some properties of fibrinogen precipitated from human plasma by glycine. Proc. Soc. Exp. Biol. Med. **113:** 989–994.
9. HUSEBY, R. M. & N. U. BANG. 1971. Fibrinogen. In Thrombosis and Bleeding Disorders. Theory and Methods. N. U. Bang, F. K. Beller, E. Deutsch & E. F. Mammen, Eds.: 222–247. Academic Press. New York.
10. FINLAYSON, J. S. & M. W. MOSESSON. 1973. Crosslinking of α chain remnants in human fibrin. Thromb. Res. **2:** 467–478.

11. PISANO, J. J., J. S. FINLAYSON & M. P. PEYTON. 1969. Chemical and enzymic detection of protein cross-links. Measurement of ε-(γ-glutamyl)lysine in fibrin polymerized by factor XIII. Biochemistry 8: 871–876.
12. GRALNICK, H. R., H. M. GIVELBER, J. R. SHAINOFF & J. S. FINLAYSON. 1971. Fibrinogen Bethesda: A congenital dysfibrinogenemia with delayed fibrinopeptide release. J. Clin. Invest. 50: 1819–1830.
13. BELITSER, V. A., T. V. VARETSKAJA & G. V. MALNEVA. 1968. Fibrinogen-fibrin interaction. Biochim. Biophys. Acta 154: 367–375.
14. MOSESSON, M. W. & S. SHERRY. 1966. The preparation and properties of human fibrinogen of relatively high solubility. Biochemistry 5: 2829–2835.
15. GRALNICK, H. R., H. M. GIVELBER & J. S. FINLAYSON. 1973. A new congenital abnormality of human fibrinogen: Fibrinogen Bethesda II. Thromb. Diath. Haemorrh. 29: 562–571.
16. MOSESSON, M. W., J. S. FINLAYSON, R. A. UMFLEET & D. GALANAKIS. 1972. Human fibrinogen heterogeneities. I. Structural and related studies of plasma fibrinogens which are high solubility catabolic intermediates. J. Biol. Chem. 247: 5210–5219.
17. KWAAN, H. C. 1972. Disorders of fibrinolysis. Med. Clin. N. Amer. 56: 163–176.
18. TAYLOR, F. B., JR., U. R. NILSSON, R. H. CREECH, E. T. CARROLL & J. G. BEISSWENGER. 1973. A new approach to the study of hypercoagulability. A discussion of the mechanism and application of a coagulolysis assay which measures both coagulative and fibrinolytic activities. Thromb. Diath. Haemorrh. Suppl. 54: 223–235.
19. MOROI, M. & N. AOKI. 1976. Isolation and characterization of α_2-plasmin inhibitor from human plasma. A novel proteinase inhibitor which inhibits activator-induced clot lysis. J. Biol. Chem. 251: 5956–5965.
20. KOIE, K., T. KAMIYA, K. OGATA, J. TAKAMATSU & M. KOHAKURA. 1978. α_2-Plasmin-inhibitor deficiency (Miyasato disease). Lancet 2: 1334–1336.
21. LAURELL, C.-B. 1966. Quantitative estimation of proteins by electrophoresis in agarose gel containing antibodies. Anal. Biochem. 15: 45–52.
22. AOKI, N., H. SAITO, T. KAMIYA, K. KOIE, Y. SAKATA & M. KOBAKURA. 1979. Congenital deficiency of α_2-plasmin inhibitor associated with severe hemorrhagic tendency. J. Clin. Invest. 63: 877–884.

SPECIFICITY OF FIBRONECTIN–FIBRIN
CROSS-LINKING *

Deane F. Mosher and Ruth B. Johnson

Department of Medicine
University of Wisconsin
Madison, Wisconsin 53706

INTRODUCTION

The concentration of fibronectin in serum is 20–50% less than the concentration in plasma [1]; this difference is greater if clotting is carried out at 0–4° C.[2] Loss of fibronectin into the clot at 22–37° C is due to factor XIII_a-catalyzed covalent cross-linking [by ϵ-(γ-glutamyl) lysine linkages] between fibronectin and the α chain of fibrin.[3, 4] Loss of fibronectin into the clot at lower temperatures is due to both noncovalent binding and covalent cross-linking.[2] The only other plasma protein known to be incorporated into the clot in a factor XIII_a-dependent manner is α_2-plasmin inhibitor.[5, 6] Assuming that the concentration of fibrinogen in plasma is 2400 $\mu g/ml$, of which 100% is incorporated into the clot; that the concentration of fibronectin in plasma is 320 $\mu g/ml$, of which 35% is incorporated into the clot; and that the concentration of α_2-plasmin inhibitor in plasma is 69 $\mu g/ml$, of which 24% is incorporated into the clot; the mass of the clot would be 94.9% fibrin, 4.4% fibronectin, and 0.7% α_2-plasmin inhibitor.

Fibronectin is necessary for cryoprecipitation of fibrinogen-fibrin complexes, even when the complexes are saturated with fibrin.[7] Fibronectin probably acts as a nucleus, because the ratio of fibronectin : fibrinogen : fibrin in the precipitated complexes is approximately 0.05 : 0.8 : 0.2. Fibronectin is also necessary for the formation of a precipitate in heparinized plasma at 2° C.[8] Although the heparin-precipitable fraction of normal plasma contains about 65% fibrinogen and 35% fibronectin, the precipitate can be formed from plasma that lacks fibrinogen. In a purified system, the amount of precipitation depends upon fibronectin concentration, heparin concentration, pH, ionic strength, and calcium ion concentration. For a given set of conditions, the amount of precipitation is increased if fibrinogen is also present. Optimal precipitation occurs when fibronectin and heparin are present in a ratio of 3 : 1. Fibrinogen or fibrin with intact α chains participate best in the cryoprecipitation or cold heparin precipitation reactions; thus, fibrinogen and fibrin molecules that lack the COOH-terminal region of the α chain are excluded from these precipitates.[7, 8]

Fibronectin that has been cross-linked to fibrin alters the properties of the clot. Fine clots (i.e., clots consisting of fine fibrils with few branch points, formed at high ionic strength and pH) to which fibronectin is cross-linked at 22° C, have half the elastic modulus of cross-linked fine clots that lack fibronectin, whereas coarse clots (i.e., clots consisting of coarse fibrils with many

* This work was supported by National Institutes of Health Grant HL 21644 and was performed while DFM was an Established Investigator of the American Heart Association and its Wisconsin Affiliate.

branch points, formed at low pH and ionic strength) to which fibronectin is cross-linked at 22° C have twice the elastic modulus of cross-linked coarse clots that lack fibronectin.[9] Cross-linking of fibronectin to fibrin profoundly enhances the attachment and spreading of cells on a fibrin-coated substratum.[10] Thus, covalent attachment of fibronectin to fibrin may be important for adhesion and migration of fibroblasts, endothelial cells, and monocytes into a wound.

In this paper, we present further information about the specificity of fibronectin-fibrin cross-linking.

METHODS

Human fibronectin, "peak I" fibrinogen, and factor XIII were purified as described previously.[3, 11, 12] One preparation of fibrinogen had degraded Aα chains presumably because of cleavage by a bacterial protease. Lysyl residues of fibronectin and fibrinogen were modified by incubation with approximately 0.3 M ethyl acetimidate.[13] Fibronectin was iodinated by a chloramine-T technique.[12] Human thrombin was a gift from Dr. John Fenton, II. TPCK-treated trypsin and soybean trypsin inhibitor were purchased from Worthington and Sigma, respectively. Monoclonal antifibronectin antibody I to the 31 kd fragment of early tryptic digests [14] was a gift from Drs. Dennis Smith and Leo Furcht; monoclonal antifibronectin antibody II was generated in our laboratory by Ms. Roxann Hanning.

Samples were analyzed on sodium dodecyl sulfate-polyacrylamide slab gels [15] after denaturation in 2% sodium dodecyl sulfate with or without 2% 2-mercaptoethanol. Densitometry was done on a SL-504-XL densitometer (Biomed Instruments). Electroblotting [16] was performed with an apparatus purchased from BioRad Laboratories.

Fibrinogen, 10 mg/ml in 0.1 sodium bicarbonate, 0.5 M sodium chloride, pH 8.3, was coupled to cyanogen bromide–activated agarose (Pharmacia) as recommended by the manufacturer. A 0.8 × 7.2 cm column was packed, and the support was converted to a fibrin-agarose by passing thrombin, 2 U/ml, through the column. Affinity chromatography on fibrin-agarose was performed at 4° C. The column was equilibrated and washed in Tris-buffered saline (0.01 M Tris, 0.15 sodium chloride, pH 7.4). Bound proteins were eluted with 1 M sodium bromide, 0.02 M sodium acetate, pH 5.0. The flow rate was usually 10 ml/hr.

Experimental details of cross-linking and digestion conditions are given in figure legends.

RESULTS AND DISCUSSION

Fibronectin and fibrin both have factor XIII$_a$-susceptible glutaminyl residues and thus have the potential of contributing glutaminyl residues, lysyl residues, or both to ε-(γ-glutamyl) lysine cross-links. In order to learn which protein contributes lysyl residues, we modified lysyl residues with ethyl acetimidate following the lead of Fuller and Doolittle [17] and studied the cross-linking of underived or amidinated fibronectin to underived or amidinated fibrin. Cross-linking was performed at 0° C to optimize fibronectin-fibrin cross-linking. We found that amidinated fibronectin could be cross-linked to underived fibrin, that underived fibronectin could *not* be cross-linked to amidinated fibrin, and that amidinated fibronectin could not be cross-linked to amidinated fibrin. The lysyl residues of amidinated fibrinogen were 99% modified, and amidinated fibrin formed a firm clot but did not cross-link to itself. Only 84% of lysyl residues of amidinated fibronectin were modified, but amidinated fibronectin in 10 mM dithiothreitol cross-linked to itself much less well than underived fibronectin cross-linked to itself.[13] Thus, we conclude that fibronectin-fibrin

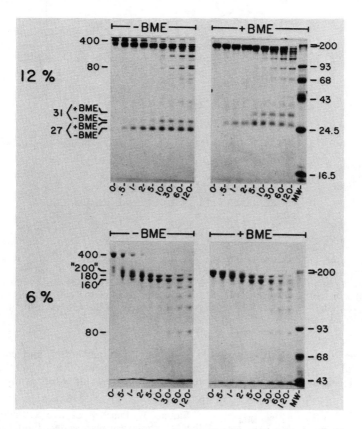

FIGURE 1. Fragmentation pattern during light trypsinization of fibronectin. Fibronectin, 2 mg/ml, was digested with trypsin, 1 μg/ml, at 37° C. At 0, 0.5, 1, 2, 5, 10, 30, 60, and 120 minutes, aliquots were removed, mixed with 5 μg/ml soybean trypsin inhibitor, then denatured with 2% sodium dodecyl sulfate, with and without 2% 2-mercaptoethanol. Samples were run on 6% and 12% polyacrylamide slab gels. Molecular weight standards in this and subsequent figures were: fibronectin, 200 kd; phosphorylase a, 93 kd; bovine serum albumin, 68 kd; ovalbumin, 43 kd; chymotrypsinogen, 24.5 kd; and, hemoglobin, 16.5 kd.

cross-linking involves lysyl residues in fibrin and factor XIII$_a$-reactive glutaminyl residues in fibronectin.

In previous studies, we and others have identified four fragments and hypothesized the existence of a fifth fragment in early tryptic digests of plasma fibronectin (FIGS. 1, 3, and 4): a basic 27 kd NH$_2$-terminal fragment that

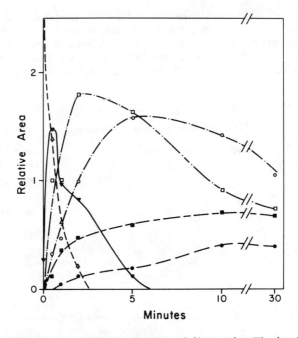

FIGURE 2. Time course of trypsin digestion of fibronectin. The bands in FIGURE 1 were scanned by densitometry and their relative areas were plotted vs. time of trypsin digestion. The areas of the 27 kd (■) and 31 kd (●) fragments are the averages of the bands on the two 12% gels. The areas of the 400 kd (△), 200 kd (▼), 180 kd (□), and 160 kd (○) fragments are from the 6% gel without reduction by 2-mercaptoethanol.

contains a factor XIII_a-susceptible glutaminyl residue,[12, 18, 19] and mediates cross-linking of fibronectin to collagen [12] and Staphylococci; [20] 160 kd and 180 kd gelatin-binding fragments that contain a free sulfhydryl group in an

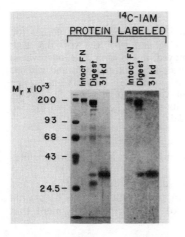

FIGURE 3. Labeling of free sulfhydryls in lightly trypsinized fibronectin. Fibronectin, 6 mg/ml, was digested for 10 min at 37° C with trypsin, 1 µg/ml. Digestion was terminated by adding soybean trypsin inhibitor, 5 µg/ml. Intact fibronectin (FN) 6 mg/ml; the digest; and isolated 31 kd fragment,[14] 1.5 mg/ml; were individually mixed with an equal volume of 8 M guanidine, and exposed free sulfhydryls were labeled with a 20-fold molar excess of [1-14C]-iodoacetamide (New England Nuclear), specific activity of 19.1 mCi/m mol, for 2 hr in the dark. Guanidine and unincorporated label were removed by extensive dialysis against Tris-buffered saline, and the proteins were analyzed after denaturation and reduction in sodium dodecyl sulfate on 10% polyacrylamide gels. Protein staining is shown on the left, and autoradiography is shown on the right.

80 kd trypsin-resistant region;[14] a 31 kd fragment that contains a second free sulfhydryl group;[14] and a COOH-terminal fragment that is the site of the interchain disulfide bonds.[21] The 27 kd and 31 kd fragments migrated further on

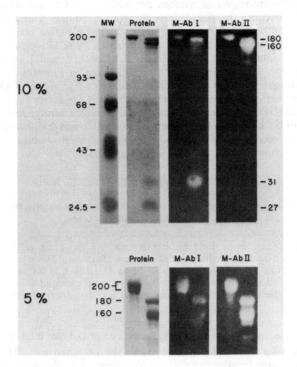

FIGURE 4. Relationships among A and B subunits of fibronectin and the 180 kd and 160 kd fragments of early tryptic digests. Intact fibronectin and fibronectin digested for 10 min at 37° C with trypsin, 1 μg/ml, were analyzed by electrophoresis in 2% sodium dodecyl sulfate on 10% and 5% polyacrylamide slab gels. The protein was transferred to nitrocellulose paper (Schleicher and Schuell) by electroblotting.[16] One section of paper was stained with 0.1% naphthol blue black in 45% methanol and 10% acetic acid to visualize protein; the section was destained in 45% methanol and 10% acetic acid. Replicate sections of paper were soaked in Tris-buffered saline containing 3% bovine albumin for 1 hr at 37° C, rinsed in saline, and soaked overnight in Tris-buffered saline containing 3% albumin, 10% fetal calf serum, and either 50 μg/ml purified mouse monoclonal anti-human fibronectin antibody I (M-Ab I), 0.2% ascites fluid containing monoclonal mouse anti-human fibronectin antibody II (M-Ab II), or 0.2% control ascites fluid (not shown). The sections were washed in Tris-buffered saline and soaked for 1 hr in Tris-buffered saline containing 3% albumin, 10% fetal calf serum, and 1% fluorescein isothiocyanate conjugated rabbit anti-mouse IgG (Miles-Yeda). The sections were washed with Tris-buffered saline, and fluorescent bands were photographed.

sodium dodecyl sulfate-polyacrylamide gels in their unreduced state than in their reduced state.[14] This is in accord with sequence studies that demonstrate five repeats of a homologous, disulfide-looped "finger" in the NH_2-terminal region and three repeats of the "finger" in the COOH-terminal region.[21]

It has been suggested [22, 23] that the larger of the two large fragments in early tryptic fragments of hamster fibronectin, *i.e.*, the 180 kd fragments, arises only from the larger of two closely spaced bands that can be seen in lightly loaded polyacrylamide gels of reduced fibronectin subunits, the so-called A and B subunits.[24] Because the size difference between human A and B subunits is at most 5 kd and because V8 protease and cyanogen bromide maps of separated A and B human subunits are identical,[24] it seems more likely that the 180 kd fragment is an intermediate in the digestion of both A and B subunits to the 160 kd fragment. Evidence for this notion is given in FIGURES 2 and 4. The first event during trypsinization of fibronectin is cleavage at the COOH-terminus to yield monomeric fragments very close in size to the intact reduced subunit. The 27 kd and 180 kd fragments are produced concurrently, followed by the appearance of 31 kd and 160 kd fragments (FIGS. 1 and 2). Monoclonal antibody I, which recognizes 31 kd and 180 kd fragments but not the 160 kd fragment, recognizes both A and B subunits (FIG. 4).

TABLE 1

BINDING OF FIBRONECTIN AND FIBRONECTIN FRAGMENTS TO
FIBRIN-AGAROSE AFFINITY COLUMN

Samples of fibronectin or fibronectin fragments, 0.2 mg–1 mg total protein, were applied to a fibrin-agarose column as described in METHODS. The trypsin digestion is described in FIGURE 3, and the 160–180 kd[12] and 31 kd and 27 kd[14] fibronectin fragments were purified as described earlier. Bound (B) and unbound (U) fractions were analyzed after denaturation and reduction in sodium dodecyl sulfate on polyacrylamide gels.

Fibrinogen or Fragments That Bound (B) or Did Not Bind (U)

Sample Applied	200 kd (Intact)	180 kd	160 kd	31 kd	27 kd
Intact fibronectin	B				
Tryptic digest		B	U	U	B
160–180 kd fragments		B	U		
31 kd fragment				U	
27 kd fragment					U

The 180 kd and 27 kd fragments of early tryptic digests of human fibronectin bound to fibrin-agarose at 4° C, whereas 160 kd and 31 kd fragments did not (TABLE 1). The 180 kd fragment bound from a mixture of 180 kd and 160 kd fragments. Purified 31 kd fragment (which probably represents the difference between the 180 kd and 160 kd fragments, as discussed above) did not bind. Isolated 27 kd fragment, however, also did not bind. Hörmann and Seidl [25] observed that 70 kd and 30 kd fragments of human fibronectin bound to fibrin-agarose after the gelatin-binding 70 kd catheptic D fragment was partially digested with plasmin. In similar experiments with hamster fibronectin, Sekiguchi *et al.*[22, 23] found binding to fibrin-agarose of 200 kd and 32 kd fragments of fibronectin (probably homologous to the human 180 kd and 27–30 kd fragments) and of 24 kd and 21 kd fragments of thermolysin-treated fibronectin. Thus, the binding to fibrin-agarose may be a multivalent reaction that requires interactions of fibronectin fragments with one another and with fibrin.

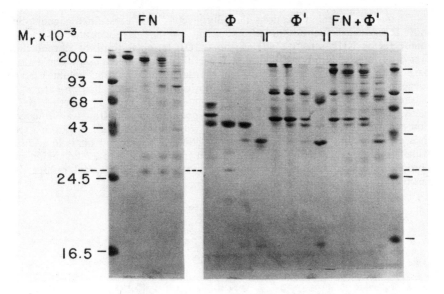

FIGURE 5. Fragmentation patterns of fibronectin, fibrinogen, cross-linked fibrin, and fibronectin cross-linked to fibrin. Solutions of fibronectin (FN, 650 μg/ml) and/or fibrinogen (φ, 730 μg/ml), in Tris-buffered saline containing 10 mM calcium ion, were untreated or were clotted (to form φ') and cross-linked for 2 hr at 4° C with thrombin, 1 U/ml, and factor XIII, 26 μg/ml. Solutions and clots were then digested with trypsin, 9 μg/ml. In each set, digestions were terminated at 0, 2, 10, and 60 min by dilution into electrophoresis buffer containing 2% 2-mercaptoethanol, and samples were analyzed by electrophoresis on 12% polyacrylamide slab gels. The dashed line points to the 27 kd fragment, which is present in digests of soluble fibronectin and absent in digests of fibronectin cross-linked to fibrin.

The 27 kd fragment could not be demonstrated in trypsinates of fibronectin cross-linked to fibrin at 0° C (FIG. 5). Labeled 27 kd fragment was cross-linked to bands of 86 kd and >180 kd when lightly trypsinzied ^{125}I-fibronectin was cross-linked to fibrin at 0° C (FIG. 6). The cross-linked complexes were susceptible to further trypsinization (FIG. 6, *track t15*). At 37° C, fibronectin was cross-linked more completely to fibrin with intact α chains than to fibrin with degraded α chains (FIG. 7). At 0° C, however, there was efficient cross-linking of fibronectin to fibrin with degraded α chains (not shown).

These experiments indicate that the 27 kd fragment is the major site of fibronectin-fibrin cross-linking. Such a conclusion is in accord with previous experiments which indicate that the 27 kd fragment is the site of fibronectin-collagen cross-linking [12] and that a collagen fragment inhibits fibronectin-fibrin cross-linking.[11] The major site for fibronectin cross-linking in fibrin appears to be the COOH-terminal portion of the α chain, which is extremely susceptible to proteolysis.[26] It is likely, therefore, that fibronectin is rapidly released from the clot when cross-linked fibrin is degraded by plasmin. The same Aα chain site probably participates in the interactions that cause fibronectin and fibrinogen/fibrin to precipitate in the cold.[7, 8]

There are several indications that fibronectin-fibrin cross-linking could be more complicated than outlined above. Richter *et al.*[27] have identified a second, minor factor XIII_a-sensitive site in the COOH-terminal third of fibronectin. Our experiments were not designed to test critically whether this site could also participate in cross-linking. Francis and Marder (ref. 28 and this volume) have suggested that a small subfraction of fibrinogen, isolated by ion-exchange chromatography, may be responsible for interactions with fibronectin. It is likely that our "peak I" fibrinogen was poor in the subfraction. Finally, fibronectin was able to cross-link to fibrin with degraded α chains, and there must be secondary sites in fibrin for fibronectin cross-linking. This is in accord with fluorescence polarization studies which indicate that the 27 kd region of fibronectin, which contains the factor XIII_a-susceptible glutaminyl residue, is flexible.[29]

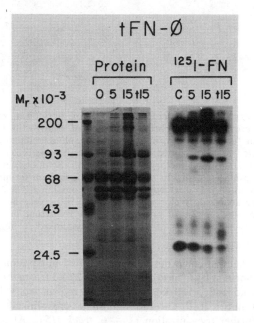

FIGURE 6. Factor XIII_a-mediated cross-linking of lightly trypsinized ^{125}I-fibronectin to fibrin. Labeled fibronectin in Tris-buffered saline was digested for 5 min at 37° C with trypsin, 1 μg/ml; digestion was terminated with soybean trypsin inhibitor, 7 μg/ml. The mixture was cooled to 0° C, and the following were added: fibrinogen, 1 mg/ml; factor XIII, 29 μg/ml; 10 mM calcium ion; and thrombin, 1 U/ml. At 0, 5, and 15 min after addition of thrombin, samples were diluted in electrophoresis buffer containing 2% 2-mercaptoethanol and analyzed by electrophoresis on 10% polyacrylamide gels followed by autoradiography. In addition, a replicate of the sample with 15 min incubation (t15) was subjected to further digestion for 10 min at 22° C with trypsin, 18 μg/ml, prior to denaturation and electrophoretic analysis. In control experiments, it was found that soybean trypsin inhibitor, 7 μg/ml, completely inhibited the fibronectin-degrading activity of trypsin, 1 μg/ml, when preincubated with trypsin, and that more extensive trypsinization (18 μg/ml for 10 min at 22° C) of ^{125}I-fibronectin caused formation of a labeled 85 kd fragment similar to that seen in the t15 track (also see FIG. 1).

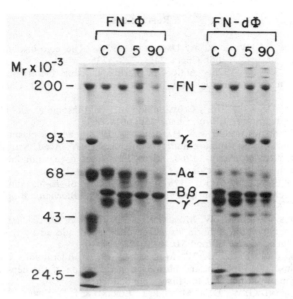

FIGURE 7. Cross-linking of fibronectin to fibrin made from intact fibrinogen (ϕ) or fibrinogen with degraded Aα chains (dϕ). Clotting and cross-linking of the following proteins in Tris-buffered saline containing 10 mM calcium was performed at 37° C: fibronectin, 370 μg/ml; fibrinogen, 1200 μg/ml; factor XIII, 29 μg/ml; and thrombin, 1 U/ml. At 0, 5, and 90 min, samples were mixed with electrophoresis buffer containing 2% 2-mercaptoethanol and analyzed on 8% polyacrylamide slab gels. The control incubation (C) lacked factor XIII and thrombin.

SUMMARY

Our experiments indicate that (1) non-covalent binding of fibronectin to fibrin is mediated by sites in a 27 kd NH$_2$-terminal region and a 31 kd COOH-terminal region of fibronectin; (2) the 31 kd region is probably present in both chains of the fibronectin dimer; and (3) covalent (factor XIII$_a$-mediated) cross-linking of fibronectin and fibrin is between a glutaminyl residue in the 27 kd region of fibronectin and a lysyl residue in the COOH-terminal two-thirds of the fibrin α chain.

ACKNOWLEDGMENTS

We thank Mr. Peter Schad for his help with the experiments shown in FIGURES 5, 6 and 7, Drs. Dennis Smith and Leo Furcht for the gift of monoclonal antifibronectin antibody I, and Ms. Roxann Hanning for preparing monoclonal antifibronectin antibody II.

REFERENCES

1. MOSESSON, M. W. & R. A. UMFLEET. 1970. The cold-insoluble globulin of human plasma. J. Biol. Chem. 245: 5728–5736.
2. MOSHER, D. F. 1976. Action of fibrin-stabilizing factor on cold-insoluble globulin and α_2-macroglobulin in clotting plasma. J. Biol. Chem. 251: 1639–1645.
3. MOSHER, D. F. 1975. Cross-linking of cold-insoluble globulin by fibrin-stabilizing factor. J. Biol. Chem. 250: 6614–6621.
4. IWANAGA, S., K. SUZUKI & S. HASHIMOTO. 1978. Bovine plasma cold-insoluble globulin: Gross structure and function. Ann. N.Y. Acad. Sci. 312: 56–73.
5. SAKATA, Y. & N. ACKI. 1980. Cross-linking of α_2-plasmin inhibitor to fibrin by fibrin-stabilizing factor. J. Clin. Invest. 65: 290–297.
6. TAMAKA, T. & N. ACKI. 1981. Cross-linking of α_2-plasmin inhibitor and fibronectin to fibrin by fibrin-stabilizing-factor. Biochem. Biophys. Acta 661: 280–286.
7. STATHAKIS, N. E., M. W. MOSESSON, A. B. CHEN & D. K. GALANAKIS. 1978. Cryoprecipitation of fibrin-fibrinogen complexes induced by the cold-insoluble globulin of plasma. Blood 51: 1211–1222.
8. STATHAKIS, N. E. & M. W. MOSESSON. 1977. Interactions among heparin, cold-insoluble globulin, and fibrinogen in formation of the heparin-precipitable fraction of plasma. J. Clin. Invest. 60: 855–865.
9. KAMYKOWSKI, G. W., D. F. MOSHER, L. LORAND & J. D. FERRY. 1981. Modification of shear modulus and creep compliance of fibrin clots by fibronectin. Biophys. Chem. 13: 25–28.
10. GRINNELL, F., M. FELD & D. MINTER. 1980. Fibroblast adhesion to fibrinogen and fibrin substrata: Requirement for cold-insoluble globulin (plasma fibronectin). Cell 19: 517–525.
11. MOSHER, D. F., P. E. SCHAD & H. K. KLEINMAN. 1979. Cross-linking of fibronectin to collagen by blood coagulation factor XIII$_a$. J. Clin. Invest. 64: 781–787.
12. MOSHER, D. F., P. E. SCHAD & J. M. VANN. 1980. Cross-linking of collagen and fibronectin by factor XIII$_a$: Localization of participating glutaminyl residues to a tryptic fragment of fibronectin. J. Biol. Chem. 255: 1181–1188.
13. MOSHER, D. F., P. E. SCHAD & H. K. KLEINMAN. 1979. Inhibition of blood coagulation factor XIII$_a$-mediated cross-linking between fibronectin and collagen by polyamines. J. Supramol. Struct. 11: 227–235.
14. SMITH, D. E., D. F. MOSHER, R. B. JOHNSON & L. T. FURCHT. 1982. Immunological identification of two sulfhydryl-containing fragments of human plasma fibronectin. J. Biol. Chem. 257: 5831–5838.
15. AMES, G. F.-L. 1974. Resolution of bacterial proteins by polyacrylamide gel electrophoresis on slabs. Membrane, soluble, and periplasmic fractions. J. Biol. Chem. 249: 634–644.
16. TOWBIN, H., T. STAEHELIN & J. GORDON. 1979. Electrophoretic transfer of proteins from polyacrylamide gels to nitrocellulose sheets: Procedure and some applications. Proc. Natl. Acad. Sci. USA 76: 4350–4354.
17. FULLER, G. M. & R. F. DOOLITTLE. 1966. The formation of cross-linked fibrins. Evidence for the involvement of lysine ε-amino groups. Biochem. Biophys. Res. Commun. 25: 694–700.
18. JILEK, F. & H. HÖRMANN. 1977. Cold-insoluble globulin: Cyanogen bromide and plasminolysis fragments containing a label introduced by transamidation. Hoppe-Seylers Z. Physiol. Chem. 358: 1165–1168.
19. MCDONAGH, R. P., J. MCDONAGH, T. E. PETERSEN, H. C. THØGERSEN, K. SKORSTENGAARD, L. SOTTRUP-JENSEN, S. MAGNUSSON, A. DELL & H. R. MORRIS. 1981. Amino acid sequence of the factor XIII$_a$ acceptor site in bovine plasma fibronectin. FEBS Lett. 127: 174–178.

20. MOSHER, D. F. & R. A. PROCTOR. 1980. Binding and factor XIII$_a$-mediated cross-linking of a 27 kilodalton fragment of fibronectin to *Staphylococcus aureus.* Science **209:** 927–929.
21. PETERSEN, T. E., H. C. THØGERSEN, K. SKORSTENGAARD, K. VIBE-PEDERSEN, P. SAHL, L. SOTTRUP-JENSEN & S. MAGNUSSON. 1983. Partial primary structure of bovine plasma fibronectin. Three different types of internal homology. Proc. Natl. Acad. Sci. USA **80:** 137–141.
22. SEKIGUCHI, K. & S. HAKOMORI. 1980. Identification of two fibrin-binding domains in plasma fibronectin and unequal distribution of these domains in two different subunits: A preliminary note. Biochem. Biophys. Res. Commun. **97:** 709–715.
23. SEKIGUCHI, K., M. FUKUDA & S. HAKOMORI. 1981. Domain structure of hamster plasma fibronectin. J. Biol. Chem. **256:** 6452–6462.
24. KURKINEN, M., T. VARTIO & A. VAHERI. 1980. Polypeptides of human plasma fibronectin are similar but not identical. Biochem. Biophys. Acta **624:** 490–498.
25. HÖRMANN, H. & M. SEIDL. 1980. Affinity chromatography on immobilized fibrin monomer, III. The fibrin affinity center of fibronectin. Hoppe-Seylers Z. Physiol. Chem. **361:** 1449–1452.
26. MARDER, V. J., C. W. FRANCIS & R. F. DOOLITTLE. 1982. Fibrinogen structure and physiology. *In* Hemostasis and Thrombosis. R. W. Colman, J. Hirsh, V. J. Marder & E. W. Salzman, Eds.: 145–163. J. B. Lippincott Company. Philadelphia, PA.
27. RICHTER, H., M. SEIDL & H. HÖRMANN. 1981. Location of heparin-binding sites of fibronectin. Detection of a hitherto unrecognized transamidase sensitive site. Hoppe-Seylers Z. Physiol. Chem. **362:** 399–408.
28. FRANCIS, C. W. & V. J. MARDER. 1982. Heterogeneity of normal fibrinogen reflecting two distinct high molecular weight variant γ chains. Clin. Res. **30:** 559A.
29. WILLIAMS, E. C. & D. F. MOSHER. 1982. Fluorescence polarization studies of dansylcadaverine-fibronectin and its interaction with a collagen fragment. Fed. Proc. **41:** 1438.

DISCUSSION OF THE PAPER

UNIDENTIFIED SPEAKER: Dr. Mosher, have you examined any other body fluids for degradation products? Synovial, or lung lavage fluids are examples of fluids in which there is a good chance that there may be such degradation.

D. F. MOSHER: No, we have not, although we are quite eager to. I should say that we have looked at saline insoluble extracts of whole tissue, and under these conditions, we can see a whole range of polymeric fibronectins in tissue both with and without reduction. This suggests that, although the circulating material is intact dimer, once it gets into the tissues themselves it forms both disulfide-bonded and perhaps transglutaminase-catalyzed multimers.

M. W. MOSESSON (*Mt. Sinai Medical Center, Milwaukee, WI*): Steve Carsons at Suny-Downstate Medical Center has studied synovial fluid fibronectin in osteoarthritis, rheumatoid arthritis, septic arthritis, and other conditions. Much of the fibronectin in these fluids, particularly that from inflammatory conditions, appears to be partially degraded.

T. EDGINGTON (*Research Institute of Scripps Clinic, La Jolla, CA*): Maybe I just want to try to put some words in your mouth. You can reject them if you wish. Do you think that the binding and cross-linking of fibronectin to the fibrin could play any role as far as cellular fixation in tissue sites?

MOSHER: Yes, I think that that is very likely.

J. MCDONAGH (*University of North Carolina at Chapel Hill*): I take it from your model that you think that the two chains of fibrinectin are identical.

MOSHER: Well, I think they are very similar, and I am not convinced that the difference between them has been identified.

N. U. BANG (*Lilly Laboratory of Clinical Research, Indianapolis, IN*): Dr. Mosher, you reported that fibronectin levels in patients with DIC is sharply reduced; I do not know how quantitative your blots are, but it certainly looked as though all these patients had about normal fibronectin levels. Is that correct?

MOSHER: I do not think that the blots can be used quantitatively. We have done quantitative measurements. Five of the samples had less than our normal level, which means they had about half of the mean normal level. One of those five samples was the channel that had the degraded fibronectin in it; the primary fibrinolysis patient did not have a low fibronectin level.

L. SHERMAN (*Washington University School of Medicine, St. Louis, MO*): With regard to the DIC issue, in a paper that is in press we have shown experimentally in acute DIC that there is a concomitant drop in fibronectin along with fibrinogen levels. This drop is dependent apparently on intact α chains, does not occur upon defibrination with oncrod, and does not occur if an animal is defibrinated previously with ancrod and then treated with thrombin. Thus, it appears related to a direct interaction of fibrin with fibronectin *in vivo*. We cannot agree with what you have seen in your patient samples, since we've been unable to demonstrate any circulating cross-linked fibronectin fibrin complexes.

MOSHER: I think that is a very important observation.

[Note added in proof: Two recent papers (Hayashi, M. & K. M. Yamada. 1983. Domain structure of the carboxyl-terminal half of human plasma fibronectin. J. Biol. Chem. **258**:3332–3340; Sekiguchi, K. & S. Hakomori. 1983. Domain structure of human plasma fibronectin. Differences and similarities between human and hamster fibronectins. J. Biol. Chem. **258**:3967–3973) suggest that the 180 kd fragment lacks a site of early trypsin cleavage which is present in the precursor of the 160 kd and 31 kd fragments.]

ENHANCEMENT OF THE STREPTOKINASE-INDUCED ACTIVATION OF HUMAN PLASMINOGEN BY HUMAN FIBRINOGEN AND HUMAN FIBRINOGEN FRAGMENT D_1*

Francis J. Castellino, Dudley K. Strickland, Joseph P. Morris,
James Smith, and Bakshy Chibber

Department of Chemistry
University of Notre Dame
Notre Dame, Indiana 46556

INTRODUCTION

The important role of fibrinogen and fibrin in activation of plasminogen is becoming widely appreciated. One distinction between plasminogen activators is their ability, or lack thereof, to interact with fibrin and, concomitantly, to have their activation capabilities toward fibrin-bound plasminogen greatly stimulated in the presence of fibrin. Activators related to urokinase do not interact with fibrin and are therefore believed to be responsible for activation of soluble fibrinogen. However, certain tissue activators strongly adsorb to fibrin and are presumed to function in activation of plasminogen bound to the clot.[1] The kinetic role of fibrin in stimulation of plasminogen activation is dependent on the activator employed. Wallen et al.,[2] found that the K_m value for Lys_{77}-plasminogen (plasminogen lacking the NH_2-terminal 76 residues) toward uterine tissue plasminogen activator is greatly decreased in the presence of fibrin, whereas, the k_{cat} is minimally affected. Binder and Sprag[3] found that when native Glu_1-plasminogen was employed, the k_{cat} for activation by tissue activator was greatly increased by fibrin and the K_m was essentially unaffected. Utilizing an activator purified from a human melanoma cell culture, Hoylaerts et al.,[4] observed that fibrin stimulated Glu_1-plasminogen activation by significantly decreasing the K_m for plasminogen and stimulated Lys_{77}-plasminogen activation, also chiefly through an effect on the K_m of plasminogen for the activator. The presence of fibrinogen also led to enhanced activation of both Glu_1- and Lys_{77}-plasminogen by this activator, through effects on both the K_m and k_{cat} of the activation.

Regarding activation of human plasminogen by streptokinase, it has been shown that a plasma protein appears to potentiate the activation.[5-7] This plasma protein has been identified with fibrinogen[8, 9] and degradation fragments of fibrinogen.[9, 10] Kinetic studies of the effect of fibrinogen and its degradation products on this activation and the role of the plasminogen lysine binding sites, presumed to be important in plasmin(ogen)-fibrin(ogen) interactions, in this process are simplified with the judicious utilization of various molecular forms of plasminogen lacking some or all of these lysine binding sites.

* This research was supported by Grant HL–13423 from the National Institutes of Health.

595

STRUCTURAL FEATURES OF HUMAN PLASMINOGEN

Human plasminogen is a single chain plasma glycoprotein of molecular weight approximately 93,000,[11] containing 790 amino acids of known sequence.[12] Based upon limited sequence homology of certain regions of plasminogen and prothrombin, it has been proposed that the NH_2-terminal 560 residues of human plasminogen can be represented by five kringle structures.[12] This region of the molecule, which represents the latent heavy chain of human plasmin,[13] contains all of the plasminogen carbohydrate,[14] as well as the lysine binding sites of plasminogen.[15] Upon limited digestion of plasminogen with elastase, several important regions of the plasminogen molecule can be isolated.[12, 16] Among these are three lower molecular weight forms of plasminogen that possess selectively degraded NH_2-terminal regions. Lys_{77}-plasminogn has been isolated and extensively employed in past studies and represents a form of native human plasminogen (Glu_1-plasminogen) containing the amino acid sequence Lys_{77}-Asn_{790}. Lys_{77}-plasminogen contains all of the kringle regions and lysine binding sites as in Glu_1-plasminogen. Val_{354}-plasminogen is a degraded form of native plasminogen and consists of amino acid residues Val_{354}-Asn_{790}. This plasminogen lacks the first three kringle regions of Glu_1-plasminogen and Lys_{77}-plasminogen and concomitantly lacks the tight lysine binding site present in plasminogen molecules containing kringles 1–3.[18] Val_{354}-plasminogen that does retain some lysine binding properties has been isolated from a limited elastolytic digestion of Glu_1-plasminogen.[16] Val_{442}-plasminogen contains amino acid residues Val_{442}-Asn_{790} and has also been purified from elastolytic digestions of native plasminogen. This form of the molecule is believed to possess only very weak lysine binding ability and lacks the first four kringle regions of plasminogen.[12] Val_{442}-plasminogen has been employed as a model of plasminogen that does not contain the lysine binding sites, a concept not entirely accurate.[19]

ACTIVATION OF HUMAN PLASMINOGEN BY STREPTOKINASE

Activation of human plasminogen by all known activators proceeds by the activator-catalyzed cleavage of the Arg_{560}-Val_{561} bond in human plasminogen.[13] The generated plasmin possesses the ability to catalyze cleavage of the Lys_{76}-Lys_{77} peptide bond from either plasmin, yielding the final plasmin, or from remaining Glu_1-plasminogen, yielding Lys_{77}-plasminogen.[11]

Since plasminogen activators must be capable of cleaving the Arg_{560}-Val_{561} peptide bond, it is an obvious conclusion that they must possess proteolytic activity. In one case, however, the bacterial plasminogen activator, streptokinase, does not possess proteolytic, esterolytic, or amidolytic activity, and a great deal of attention has been directed toward the mechanism of plasminogen activation by this activator.

On the basis of significant studies from several different laboratories (for a recent review see ref. 20), it now appears clear that at least two general steps are required for activation of human plasminogen (HPg) by streptokinase (SK). Briefly, in the first stage, molecular events believed to be important for formation of the enzyme responsible for plasminogen activation are described.

$$SK + HPg \rightleftharpoons SK \cdot HPg \rightarrow SK \cdot HPg' \rightarrow SK^* \cdot HPm$$
$$\downarrow \uparrow$$
$$SK + HPm$$

(1)

Here, SK and HPg interact to form an equimolar complex (SK·HPg), which undergoes rearrangement to form a modified complex, SK·HPg′, containing an active site in the plasminogen moiety. This complex is of limited stability and, intramolecularly, catalyzes cleavage of the Arg_{560}-Val_{561} peptide bond, in plasminogen, within the complex, in order to form the streptokinase plasmin complex (SK*·HPm), which also contains a proteolytically modified streptokinase moiety (SK*). This latter complex can also form from SK and free HPm. Whereas, plasmin alone cannot catalyze activation of plasminogen, the SK*·HPm complex (and presumably, the SK·HPg′ complex) is an effective enzyme for this conversion. Therefore, the second step of the activation proceeds as follows:

$$HPg \xrightarrow[\text{SK*·HPm}]{\text{SK·HPg′}} HPm \qquad (2)$$

where SK*·HPm (or SK·HPg′) functions as an enzyme in the conversion of HPg to HPm.

EFFECT OF FIBRINOGEN AND FIBRINOGEN FRAGMENT D ON THE STREPTOKINASE-CATALYZED ACTIVATION OF HUMAN PLASMINOGEN

We have recently shown that the presence of fibrinogen enhances the rate of human plasminogen activation approximately fourfold by catalytic levels of streptokinase, and that this property of fibrinogen appears to reside in the fibrinogen fragment D_1 domain.[10] Since the stimulatory effect of fibrinogen and fragment D_1 could be displayed at either step (1) or step (2), or both, above, we have designed kinetic experiments to attempt to isolate the site of stimulatory action of fibrinogen and fragment D. We first wished to determine whether the activity of the preformed SK*·Pm activator toward plasminogen activation is stimulated by fibrinogen or fragment D_1. Steady state kinetic data on the

TABLE 1

STEADY-STATE KINETIC PARAMETERS FOR ACTIVATION OF HUMAN PLASMINOGEN BY THE PREFORMED STREPTOKINASE-PLASMIN ACTIVATOR COMPLEX IN THE ABSENCE AND PRESENCE OF FIBRINOGEN FRAGMENT D_1*

Substrate	K_m, app (μM)	k_{cat} (min^{-1})	k_{cat}/K_m, app
Glu_1-Pg†	0.61 ± 0.15	27.8 ± 1.5	45.7
Glu_1-Pg + F-D_1‡	0.90 ± 0.2	36.8 ± 1.5	40.8
Lys_{77}-Pg	0.16 ± 0.05	68.7 ± 2.5	429
Lys_{77}-Pg + F-D_1	0.17 ± 0.05	71.2 ± 2.5	414
Val_{442}-Pg	0.71 ± 0.20	14.2 ± 1.0	20
Val_{442}-Pg + F-D_1	0.33 ± 0.10	15.6 ± 1.0	47.3

* Taken from reference 20.
† For Glu_1-Pg and Lys_{77}-Pg, the activator was the equimolar SK*·Lys_{77}-Pm complex. In the case of Val_{442}-Pg, the activator was the SK*·Val_{442}-Pm complex. The experimental details are given in reference 20.
‡ The concentration of fragment D_1 (F-D_1) was 3.0 μM, when present.

activation of various plasminogens by SK*·Pm in the presence of fragment D_1 are listed in TABLE 1. In the case of Glu_1-plasminogen and Lys_{77}-plasminogen, there is very little effect of fragment D_1 on the steady state activation parameters. A twofold decrease in the K_m for activation of Val_{442}-plasminogen is noted in the presence of fragment D_1. It, therefore, appears that very little influence of fragment D_1 (as well as fibrinogen) is exerted on the activity of the preformed activator. We next turned our attention to defining whether fibrinogen and fragment D_1 were capable of influencing events in step (1), above, concerned with the rate of formation of the plasminogen activator complex, since, by elimination, it did appear that this would be the central locus of their effects. While we were not able to separate all steps in the reactions within activator formation, we were able to evaluate the role of fibrinogen and fragment D_1 in the rate of formation of the SK·HPg' complex, viz,

$$SK + HPg \rightleftharpoons SK \cdot HPg \rightarrow SK \cdot HPg' \tag{3}$$

This was accomplished by adding an equimolar level of SK to HPg, in the presence of the active site titrant 3',6'-bis[(4-guanidinobenzoyl)oxy]-5-[N'-(4-carboxyphenyl)thioureido] spiro[isobenzofuran-1(3H), 9'-[9H]xanthen]-3-one (FDE). This reagent was synthesized by a procedure published by Mangel et al.[21] and the kinetic parameters obtained for its reaction with plasmin suggested that it was a rapid active site titrant with very low acyl-enzyme turnover properties. We found that when SK and HPg were mixed in its presence, the activator was trapped at the SK·HPg' step, and that the rate of formation of SK·HPg' could be continuously monitored, fluorometrically, since the released product p-[(p-guanidinobenzoyl)fluorescyl-6-thioureido] benzoic acid, exhibits fluorescence properties.[21] Because a fluorescence assay was employed, very low concentrations of SK and HPg could be used, thereby assuring conditions wherein reaction rates were dependent upon SK and HPg levels. By varying the concentration of fibrinogen and fragment D_1, at constant levels of SK and HPg, k_{obs} values and K_m values for fibrinogen and fragment D_1 for stimulation of reaction 3, above, were obtained for each of the three plasminogens used in this study. The results are listed in TABLE 2.

TABLE 2

EFFECT OF FIBRINOGEN AND FRAGMENT D_1 ON THE RATE OF FORMATION OF THE ACTIVE SITE IN THE STREPTOKINASE-PLASMINOGEN EQUIMOLAR COMPLEX *

Plasminogen Employed	$k_{obs} \times 10^3$ (sec^{-1})	K_m (μM)
Glu_1-Pg	9.0 ± 1.5	—
Glu_1-Pg + fragment D_1	44.1 ± 3.0	0.67 ± 0.1
Glu_1-Pg + fibrinogen	67.2 ± 4.0	0.25 ± 0.05
Lys_{77}-Pg	23.5 ± 2.0	—
Lys_{77}-Pg + fragment D_1	73.9 ± 4.5	0.50 ± 0.1
Lys_{77}-Pg + fibrinogen	59.0 ± 3.0	0.07 ± 0.01
Val_{442}-Pg	15.3 ± 2.0	—
Val_{442}-Pg + fragment D_1	102.2 ± 7.0	0.27 ± 0.1
Val_{442}-Pg + fibrinogen	134.3 ± 7.5	0.30 ± 0.15

* Taken from reference 20, which also lists the experimental details.

TABLE 3

EFFECT OF ε-ACA ON THE RATE OF FORMATION OF THE ACTIVE SITE IN THE
STREPTOKINASE·GLU$_1$-PLASMINOGEN COMPLEX IN THE ABSENCE AND
PRESENCE OF FIBRINOGEN FRAGMENT D$_1$*

[ε-ACA] (mM)	[Fragment D$_1$] (μM)	$k_{obs} \times 10^3$ (sec^{-1})
0	0	9.0 ± 1.5
0	2.0	44.0 ± 3.0
0.01	2.0	42.1 ± 3.0
0.05	2.0	42.6 ± 3.0
0.1	2.0	47.0 ± 3.5
1.0	2.0	63.2 ± 4.0
10.0	2.0	52.0 ± 3.5
25.0	2.0	54.0 ± 3.5

* The experimental details are as in TABLE 2 (described in ref. 20).

From these data, it is clear that the presence of saturating levels of both fibrinogen and fragment D$_1$ increase the k_{obs} for active site formation in the SK·Glu$_1$-Pg′complex by five- to sevenfold when compared to the rate obtained in the absence of these stimulators. Also, the K_m for fibrinogen is slightly lower than that for fragment D$_1$, suggesting that fibrinogen produces its effect at lower concentrations than fragment D$_1$.

Considering the data in TABLE 2 for active site formation in the SK·Lys$_{77}$-Pg′equimolar complex, in the absence of fibrinogen or fragment D$_1$, the k_{obs} for the rate of active site formation in the SK·Lys$_{77}$-Pg′equimolar complex is approximately twofold greater than that for the SK·Lys$_{77}$-Pg′equimolar complex. This value is increased approximately 2.5–3.5-fold at saturating levels of fibrinogen or fragment D$_1$. Again, from the comparative values of the K_m for each effector molecule, it appears as though fibrinogen stimulates at lower concentration than fragment D$_1$.

Regarding active site formation in the SK·Val$_{442}$-Pg′complex, large increases (seven- to eightfold) in k_{obs} are observed in the presence of fibrinogen and fragment D$_1$. The K_m values are very similar for each molecule, showing that their stimulatory effects are produced at virtually identical concentrations.

A notable conclusion to be drawn from the data of TABLE 2 concerns the fact that fibrinogen and fragment D$_1$ stimulate active site generation in the SK·Val$_{442}$-Pg′complex. Since Val$_{442}$-Pg does not possess the tight lysine binding site believed to be of importance to plasminogen-fibrin(ogen) interactions,[22] we conclude that this site does not mediate interactions necessary for stimulation by fibrinogen of the rate of active site formation in the SK-plasminogen complex. Of course, it is possible that a lysine binding site necessary for this interaction is produced in the SK·Val$_{442}$-Pg′complex or that an important lysine binding site is indeed present in Val$_{442}$-Pg.

In order to directly test the point as to whether the plasminogen lysine binding sites were important in stimulation of active site formation, by fibrinogen or fragment D$_1$, in the SK·Glu$_1$-Pg′equimolar complex, the effect of various levels of ε-amino caproic acid (ε-ACA) on the rate of formation of the active site in the equimolar SK·Glu$_1$-Pg′ complex was assessed. The results are listed in TABLE 3. As the ε-ACA level is increased from 0.01 mM to 25 mM, in the presence of 2 μM fragment D_1, the k_{obs} for the reaction does not decrease, in fact a slight stimulation is noted at high ϵ-ACA levels. Since concentrations of 25 mM ϵ-ACA are sufficient to saturate the strong lysine binding site on Glu_1-Pg [23] and to interact to a significant degree with all weak lysine binding sites on Glu_1-Pg,[23] we conclude that these sites on Glu_1-Pg are not important to the stimulatory effects of fibrinogen and fragment D_1 on the streptokinase-induced activation of human plasminogen.

REFERENCES

1. THORSEN, S., P. GLAS-GREENWALT & T. ASTRUP. 1972. Thromb. Diath. Haemorrh. **28:** 65–74.
2. WALLEN, P. & M. RANBY. 1980. Abstracts of the 28th Annual Colloquium in Protides and Biological Fluids. Abstract 67.
3. BINDER, B. R. & J. SPRAG. 1980. In Protides of the Biological Fluids. H. Peeters, Ed. Vol. **28:** 391–394. Pergamon Press. Oxford.
4. HOYLAERTS, M., D. C. RIJKEN, H. R. LIJNEN & D. COLLEN. 1982. J. Biol. Chem. **257:** 2912–2919.
5. TAKADA, A., Y. TAKADA & J. L. AMBRUS. 1970. J. Biol. Chem. **245:** 6389–6396.
6. TAKADA, A., Y. TAKADA & J. L. AMBRUS. 1972. Biochim. Biophys. Acta **263:** 610–618.
7. CHESTERMAN, C. M., S. A. CEDERHOLM-WILLIAMS, M. J. ALLINGTON & A. A. SHARP. 1977. Thromb. Res. **10:** 421–426.
8. CAMIOLO, S. M., G. MARKUS, J. L. EVERS & G. H. HOBIKA. 1980. Thromb. Res. **17:** 697–706.
9. VIOLAND, B. N., A. CASTELLINO & F. J. CASTELLINO. 1980. Thromb. Res. **19:** 705–710.
10. STRICKLAND, D. K., J. P. MORRIS & F. J. CASTELLINO. 1982. Biochemistry **21:** 721–728.
11. VIOLAND, B. N. & F. J. CASTELLINO. 1976. J. Biol. Chem. **251:** 3906–3912.
12. SOTTRUP-JENSEN, L., H. CLAEYS, M. ZAJDEL, T. E. PETERSEN & S. MAGNUSSON. 1978. In Progress in Chemical Fibrinolysis and Thrombolysis. J. F. Davidson, R. M. Rowen, M. M. Samama & P. C. Desnoyers, Eds. Vol. **3:** 191–209. Raven Press. New York.
13. ROBBINS, K. C., L. SUMMARIA, B. HSIEH & R. J. SHAH. 1967. J. Biol. Chem. **242:** 2333–2342.
14. HAYES, M. L., R. K. BRETTHAUER & F. J. CASTELLINO. 1975. Arch. Biochem. Biophys. **171:** 651–655.
15. RICKLI, E. E. & W. I. OTAVSKY. 1975. Eur. J. Biochem. **59:** 441–447.
16. POWELL, J. R. & F. J. CASTELLINO. 1981. Biochem. Biophys. Res. Commun. **102:** 46–52.
17. SUMMARIA, L., B. HSIEH & K. C. ROBBINS. 1967. J. Biol. Chem. **242:** 4279–4283.
18. LERCH, P. G., E. E. RICKLI, W. LERGIER & D. GILLESSEN. 1980. Eur. J. Biochem. **107:** 7–13.
19. CASTELLINO, F. J., V. A. PLOPLIS, J. R. POWELL & D. K. STRICKLAND. 1981. J. Biol. Chem. **256:** 4478–4482.
20. CASTELLINO, F. J. 1981. Chem. Rev. **81:** 431–446.
21. MANGEL, W. F., D. C. LIVINGSTON, J. R. BROCKELHURST, H. Y. LIU, G. A. PELTZ, J. F. CANNON, S. P. LEYTUS, J. A. WEHRYLY, B. L. SALTER & J. L. MOSHER. 1980. Cold Spring Harbor Symp. Quant. Biol. **44:** 669–680.
22. THORSEN, S. 1975. Biochem. Biophys. Acta **393:** 55–65.
23. MARKUS, G., J. L. DE PASQUALE & F. C. WISSLER. 1978. J. Biol. Chem. **253:** 727–732.

DISCUSSION OF THE PAPER

R. HANTGAN (*University of North Carolina School of Medicine, Chapel Hill*): I should like to point out an alternative interpretation of your differential scanning calorimetry experiment, which is as a thermodynamic experiment and not an indicator of confirmation. It may be that the presence of calcium stabilizes the native conformation of fragment D, thereby preventing it from being thermodenatured, rather than that it changes the conformation. The experiment does not necessarily speak to the conformation of fragment D.

F. J. CASTELLINO: That certainly could be true: it is either the one or the other.

FIBRIN, FIBRONECTIN, AND MACROPHAGES

Celso Bianco

Lindsley F. Kimball Research Institute
The New York Blood Center
New York, New York 10021

INTRODUCTION

Mononuclear phagocytes play a fundamental role in inflammation and in immune responses. Classically macrophages have been considered scavengers, in charge of uptake of particulates, bacteria, and cellular debris at inflammatory sites. More recently, extensive study of its cell biology and biochemistry has led to a much broader understanding of macrophage participation in biological processes.

The present review focuses on the interaction between mononuclear phagocytes, fibrin, and fibronectin.

The full expression of effector functions by macrophages follows a well-defined series of events. Mononuclear phagocytes arise from stem cells in the bone marrow, where they undergo differentiation and intense cell division. Mature monocytes leave the bone marrow and reach the circulation where they have a half-life of about 72 hours. Normally, blood monocytes migrate into tissues and differentiate into tissue macrophages, many of them recognized by their characteristic morphology and properties. Examples are alveolar macrophages, peritoneal macrophages, Kupffer cells, and osteoclasts. This subject has been extensively reviewed by Van Furth.[1] In the course of an inflammatory process, monocytes leave the vascular bed, migrate towards the site of injury, localize at that site, differentiate, and acquire the ability to perform a number of effector functions including enhanced phagocytosis, microbial killing and secretion of proteases. Plasma proteins clearly direct many of these processes. For instance, the migration of monocytes towards the site of injury seems to be facilitated by peptides generated during activation of the coagulation and of the complement systems that result in alteration of vascular permeability. C5a and fibrinopeptides are chemotactic attractants for monocytes, stimulating and orienting cell migration. Plasmin and Bb (a cleavage fragment of Factor B of the complement system) are spreading inducers that promote cell localization and adherence.[2] At the inflammatory site, monocytes differentiate into macrophages with enhanced phagocytic capacity, and show markedly increased secretion of neutral proteases such as plasminogen activators, collagenase, and elastase (reviewed in ref. 3).

The ability of fibronectin to bind to substances likely to be present at sites of injury, and the presence of plasma membrane receptors for fibronectin on monocytes may also determine monocyte localization at a site of injury. Furthermore, interaction of fibronectin with monocyte receptors seems to induce enhancement of phagocytic capability and secretion of neutral proteases. The following pages summarize evidence which support these statements.

602

0077–8923/83/0408–0602 $01.75/0 © 1983, NYAS

PLASMA FIBRONECTIN

Plasma fibronectin or Cold Insoluble Globulin (CIg) is present in human plasma in substantial amounts: 300 ± 100 $\mu g/ml$.[4] The majority of the molecules found in circulation are dimers with a molecular weight of 450,000 composed of two disulfide linked chains.[5] The two chains are identical except for an extra peptide at the COOH-terminus of one of the chains. Structurally and immunologically related forms are found on fibroblast surfaces, basement membranes, matrices, and extravascular fluids (reviewed in refs. 6–9). Despite some differences among the several forms of fibronectin, they share a number of common properties including binding sites for collagen,[10] fibrin and fibrinogen,[11] heparin,[12] and fibroblasts.[8] Interestingly, CIg binds with much higher avidity to denatured collagen than to native collagen,[13, 14] making gelatin, a product of mild hydrolysis of collagen, an excellent experimental substrate. The avidity of CIg for fibrin is also higher than that for fibrinogen.[12] Thus, two substances likely to be present at sites of injury are effective substrates for fibronectin.

FIBRONECTIN AND MACROPHAGES

Blumenstock *et al.* have demonstrated immunological and biochemical identity between CIg and a serum protein which they termed α2-opsonic protein.[15] This protein had previously been shown to promote uptake of a gelatin containing lipid emulsion by rat liver slices, suggesting an interaction with macrophages.[15, 16] These results were later confirmed using gelatin coated latex beads.[17–19] More recently Czop *et al.* indicated that a proteolytic fragment of fibronectin enhanced the uptake of rabbit erythrocytes and other particles capable of activating the alternative pathway of complement.[20]

RECEPTORS FOR FIBRONECTIN ON MONOCYTES

We have published a series of results that strongly suggest that human peripheral blood monocytes have a plasma membrane receptor for fibronectin.[21] Monocytes are adherent cells. In the presence of serum proteins, they firmly attach to glass and plastic surfaces, while other blood cells tend not to adhere. This property has been widely used for the preparation of enriched monocyte cultures. Coating of plastic or glass surfaces with collagen or gelatin prevents leukocyte adherence. However, the addition of fibronectin to these surfaces promotes monocyte adherence, in a concentration-dependent manner. The maximum number of monocytes was attained when fibronectin concentration reached 40 $\mu g/cm^2$ of substrate surface and the incubation between cells and surface was allowed to proceed for 30 minutes to 1 hour at room temperature. Interestingly, only monocytes bind to these fibronectin containing surfaces. Erythrocytes, neutrophils, and lymphocytes do not bind. The exclusiveness of monocyte attachment was confirmed by morphology, staining for nonspecific esterase and phagocytosis of IgG-coated erythrocytes. Occasionally, when a large number of platelets was present in the initial leukocyte preparations, binding to a gelatin surface was independent of the addition of fibronectin. The binding of monocytes to the fibronectin coated surfaces is also Mg^{++} dependent. A minimum of 1mEq/L is required for binding to occur, and a plateau is reached at 2 mEq/L. EDTA is highly effective in releasing monocytes from the surface. This is a practical procedure for preparation of pure suspensions of monocytes.

Fibronectin was shown to be the major plasma protein promoting adherence of monocytes to collagen in depletion-reconstitution experiments. Depletion of CIg from serum abolished binding, and addition of purified CIg fully reconstituted the activity. Preincubation experiments also indicated that the interaction of fibronectin with monocytes required prior interaction of fibronectin with its substrates. Soluble fibronectin did not bind to monocytes in detectable amounts. Possibly binding requires a multiplicity of receptor-ligand interactions in a cooperative manner, or conformational changes of the molecule of fibronectin prior to interaction with the cellular receptor.

FIBRONECTIN AND PHAGOCYTOSIS

We performed a number of experiments in which latex beads coated with gelatin and fibronectin, and erythrocytes coated with gelatin and fibronectin were offered to human monocytes in culture. These experiments showed that fibronectin promotes attachment of these particles to the monocyte surface. However, under the conditions in which the experiments were carried out, no ingestion was observed. It is our belief that fibronectin, by itself, is unable to promote particle ingestion. It can however, because of its ability to promote particle attachment, increase the rate of uptake of particles which would be ingested independently of fibronectin, including latex beads. Similar synergistic effects between a mediator of attachment and a mediator of phagocytosis have been observed with C3 and IgG in macrophages.[22] It should be noted that in our experiments heparin or 2-mercaptoethanol were not required for particle attachment.

Fibronectin-substrate complexes seem to play other roles in monocyte function. We have observed that the attachment of monocytes to fibronectin-coated gelatin, fibronectin-coated collagen and fibronectin-coated fibrin leads to increased expression of receptors for Fc of IgG and for C3. This enhancement was observed in experiments in which monocytes plated on plastic surfaces in the absence of CIg were compared to monocytes plated on fibronectin-coated surfaces. Many more erythrocytes coated with IgG (EIgG) or with IgM and complement (EIgMC) bound to monocytes plated on fibronectin than to monocytes plated on plastic. The increase in binding of EIgG reached 20-fold. This enhanced receptor activity was retained after the cells were released from the surface by EDTA treatment and subsequently reattached to a plastic surface in the absence of fibronectin.

MONOCYTE ATTACHMENT TO FIBRIN

Several lines of evidence indicate that monocytes recognize fibrin monomers and fibrin clots in the absence of fibronectin.[23-25] Thus, it was of interest to verify the consequences of fibronectin binding to fibrin in this interaction. Plastic culture plates with 16mm wells were coated with fibrin by two different methods. (1) They received 250 μl of 1 mg/ml fibrinogen peak I, were dried at 40° C, exposed to 1.5 units of thrombin for 15 minutes at 37° C and then washed with

PBS. (2) Fibrin monomers were prepared in solution by adding 1 unit of thrombin/mg fibrinogen, followed by dissolution of the clot with NaBr 0.1M in 0.07M sodium acetate buffer pH 5. After dialysis, the material was diluted in 0.3M Tris, 0.001M EDTA buffer pH 7.4 and added to the plates (250 μg/well). The plates were gently dried and washed with PBS prior to use. Purified plasma fibronectin was added to the plates, incubated for 30 minutes at room temperature and then mononuclear cells added for a further 30 minute period of incubation as previously described.[21] FIGURE 1 shows that in the absence of fibronectin, 2,000 monocytes/cm² were bound to the surface. However, at 80 μg of fibronectin/cm², over 22,000 monocytes bound to the same surface area. Thus, fibronectin enhanced monocyte binding by at least 10-fold. In the several experiments performed, the enhancement was more pronounced when fibrin monomers were used. However, this difference has to be interpreted with caution. Monocytes can bind to the underlying plastic, to fibrin and to fibronectin. In the experiments using method (1), high binding of monocytes

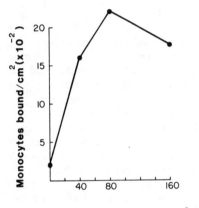

FIGURE 1. Attachment of human peripheral blood monocytes to a plastic surface pre-coated with fibrin monomers and with increasing amounts of fibronectin (CIg). The preparation of the surface is described in the text. The cells were allowed to adhere for a period of 30 min at room temperature. The number of monocytes bound was determined by phase microscopy using a calibrated grid.

occurred in the absence of fibronectin. Further experiments are required to determine the fraction of activity resulting from each component of the system when this procedure is utilized.

OTHER FUNCTIONAL CONSEQUENCES OF THE INTERACTION OF MONOCYTES WITH FIBRONECTIN-COATED SURFACES

As previously mentioned, binding of monocytes to fibronectin-coated surfaces leads to an enhancement of expression of receptors for Fc of IgG and for C3. Another important consequence of the interaction is the ability of monocytes bound to these surfaces to secrete neutral proteases. Monocytes and macrophages can be induced by a number of stimuli to secrete increased amounts of plasminogen activator, an elastase, a collagenase, and several proteinases. The importance of these enzymes at an inflammatory site is obvious. The secretion of neutral proteases by monocytes plated on plastic was compared to that of monocytes plated on a fibronectin-collagen surface. Collagen was

TABLE 1

PLASMINOGEN ACTIVATOR ACTIVITY OF SUPERNATANTS OF HUMAN PERIPHERAL BLOOD
MONOCYTES CULTIVATED IN THE PRESENCE AND IN THE ABSENCE OF A
CIg-GELATIN SURFACE *

Surface	^{125}I Released/hr/10^6 Cells
Collagen + CIg (50 μg/cm^2)	3,640
Plastic, no CIg added	414
Ratio	8.8

* Values obtained 24 hours after phagocytic load (latex).

chosen because of its greater stability under culture conditions. Mononuclear
cells separated as described [21] were added to 24-well, plastic culture plates in
which some of the wells had been precoated with acid-soluble collagen from
bovine tendon. The cells were suspended in medium containing 10% human
serum depleted of CIg by affinity chromatography on a gelatin-coated absorbent. The nonadherent cells were removed by washing, and the wells replenished with culture medium containing no serum. The cultures then received a
phagocytic load of 1.1μ latex beads (0.1% in volume). The uningested beads
were removed and the cells incubated for the stated periods of time (TABLES 1
and 2). The supernatants were removed and assayed. The elastase assay was
performed exactly as described by Banda and Werb using ^{3}H-elastin as a substrate. The assay for plasminogen activator was performed as described by
Gordon using ^{125}I-fibrin as a substrate and dog plasminogen purified by affinity
chromatography on lysine-Sepharose.[26] The results indicate that after a period
on a fibronectin-containing substrate, monocytes secrete appreciable amounts of
elastase and plasminogen activator.

TABLE 2

SECRETION OF EDTA-INHIBITABLE ELASTOLYTIC ACTIVITY BY HUMAN PERIPHERAL
BLOOD MONOCYTES CULTIVATED IN THE PRESENCE AND IN THE ABSENCE OF A
CIg-COLLAGEN

Surface	Phagocytic Stimulus (latex)	Units/hr/10^6 Cells * 7.5 hr	20 hr
Collagen + CIg (30 μg/cm^2)	no	1.95	5.90
	yes	4.69	23.45
Plastic, no CIg added	no	1.15	3.35
	yes	2.25	4.50

* Amount of enzyme that degrades 1 μg of elastase in one hour.

Conclusions

Plasma proteins are usually present at inflammatory sites. This fact, together with the observations discussed above, allows the construction of a possible scenario involving monocytes, fibronectin, and fibrin in inflammatory processes. Previously described phenomena suggest that chemotactic factors direct blood monocytes to the site of injury. There monocytes will encounter surfaces containing denatured collagen and fibrin-containing clots, both coated with fibronectin. Monocytes would tend to attach to these surfaces through the fibronectin receptors. This attachment would lead to differentiation of the cell, resulting in the expression of a larger number of receptors for Fc of IgG and for C3b. These cells would ingest opsonized particles more effectively. The interaction with the fibronectin-containing surface would also constitute the first step required for the secretion of neutral proteases. The phagocytic load would trigger the secretory event, as demonstrated in other systems.[28]

In essence, we propose that the interaction of fibronectin with the fibronectin receptor promotes the differentiation of the monocyte into an inflammatory macrophage, able to clear debris, dispose of contaminating infectious agents, degrade the damaged matrix, and destroy the blood clot in order to allow for tissue reconstruction. Recent observations indicate that a fibronectin produced by macrophages acts as a chemoattractant for fibroblasts.[30]

The observation that, among blood leukocytes, only monocytes carry a receptor for fibronectin also provides a plausible explanation for the different behavior of neutrophils and monocytes in acute inflammatory responses. During the first few hours following injury both neutrophils and monocytes migrate to the site. The predominance of neutrophils results from their higher number and speedier locomotion. Both cell types are retained at the site as a result of activation of the alternative pathway of complement fixation (Bb) and of the contact phase of blood coagulation.[29] During later phases of inflammation, macrophages predominate, probably because the generation of chemo-attractants and spreading factors has ceased, and fibronectin becomes the major protein responsible for monocyte retention.

Acknowledgments

The present review incorporates work developed in collaboration with Michael P. Bevilacqua, Michael W. Mosesson, David L. Amrani, Peter B. Milburn, and Martha A. Reichert. The manuscript was prepared by Kathleen Reichert. Their contribution is greatly appreciated. The help of Barbara Hosein in revising the manuscript is also acknowledged.

References

1. VAN FURTH, R., M. C. MARTINA, J. A. RAEBURN, T. L. VAN ZWET, R. CROFTON & A. B. VAN OUD ALBLAS. 1980. Characteristics, origin and kinetics of human and murine mononuclear phagocytes. *In* Mononuclear Phagocytes Functional Aspects, R. Van Furth, Ed. Vol. 1: 279–298. Martinus Nijhoff Publishers. The Hague, the Netherlands.
2. GOTZE, O., C. BIANCO & Z. A. COHN. 1979. The induction of macrophage spreading by factor B of the properdin system. J. Exp. Med. 149: 372–386.

3. COHN, Z. A. 1978. The activation of mononuclear phagocytes. J. Immunol. **121:** 813–816.
4. MOSESSON, M. W. & R. A. UMFLEET. 1970. The cold-insoluble globulin of plasma. I. Purification, primary characterization, and relationship to fibrinogen and other cold-insoluble fraction components. J. Biol. Chem. **245:** 5728–5736.
5. MOSESSON, M. W., A. B. CHEN & R. M. HUSEBY. 1975. The cold-insoluble globulin of human plasma: studies of its essential structural features. Biochem. Biophys. Acta **386:** 509–524.
6. MOSESSON, M. W. & D. AMRANI. 1980. The structure and biologic activities of plasma fibronectin. Blood **56:** 145–158.
7. MOSHER, D. F. 1980. Fibronectin. *In* Progress in Hemostasis and Thrombosis. T. H. Spaet, Ed. Grune and Stratton Publishers. **5:** 111–151.
8. VAHERI, A. & D. F. MOSHER. 1978. High molecular weight, cell surface associated glycoprotein (fibronectin) lost in malignant transformation. Biochem. Biophys. Acta **516:** 1–25.
9. YAMADA, K. M. & K. OLDEN. 1978. Fibronectins—adhesive glycoproteins of cell surface and blood. Nature (Lond.) **275:** 179–184.
10. ENGVALL, E. & E. RUOSLAHTI. 1977. Binding of soluble form of fibroblast surface protein, fibronectin, to collagen. Int. J. Cancer. **20:** 1–5.
11. RUOSLAHTI, E. & A. VAHERI. 1975. Interaction of soluble fibroblasts surface antigen with fibrinogen and fibrin. Identity with cold insoluble globulin of human plasma. J. Exp. Med. **141:** 497–501.
12. STATHAKIS, N. E., M. W. MOSESSON, A. B. CHEN & D. K. GALANAKIS. 1978. Cryoprecipitation of fibrin-fibrinogen complexes induced by the cold-insoluble globulin of plasma. Blood **51:** 1211–1222.
13. JILEK, F. & H. HORMANN. 1978. Cold-insoluble globulin (fibronectin): Affinity to soluble collagen of various types. Hoppe-Seylers Z. Physiol. Chem. **359:** 247–250.
14. ENGVALL, E., E. RUOSLAHTI & E. J. MILLER. 1978. Affinity of fibronectin to collagens of different genetic types and to fibrinogen. J. Exp. Med. **147:** 1584–1595.
15. BLUMENSTOCK, F. A., T. M. SABA, P. B. WEBER & R. LAFFIN. 1978. Biochemical and immunological characterization of human opsonic alpha–2–SB glycoprotein: Its identity with cold-insoluble globulin. J. Biol. Chem. **253:** 4287–4291.
16. BLUMENSTOCK, F. A., T. M. SABA, P. B. WEBER & E. CHO. 1976. Purification and biochemical characterization of a macrophage stimulating alpha–2–globulin opsonic protein. J. Reticuloendothel. Soc. **19:** 157–172.
17. MOLNAR, J., F. B. GELDER, M. Z. LAI, G. E. SIEFRING, R. B. CREDO & L. LORAND. 1979. Purification of opsonically active human and rat cold-insoluble globulin (plasma fibronectin). Biochemistry **18:** 3909–3916.
18. DORAN, J. E., A. R. MANSBERGER & A. C. REESE. 1980. Cold insoluble globulin-enhanced phagocytosis of gelatinized targets by macrophage monolayers: A model system. J. Reticuloendothel. Soc. **27:** 471–483.
19. GUDEWICS, P. W., J. MOLNAR, M. Z. LAI, D. W. BEEZHOLD, G. E. SIEFRING, R. B. CREDO & L. LORAND. 1980. Fibronectin-mediated uptake of gelatin-coated latex particles by peritoneal macrophages. J. Cell. Biol. **87:** 427–433.
20. CZOP, J. K., J. L. KADISH & K. F. AUSTEN. 1981. Augmentation of human monocyte opsonin-independent phagocytosis by fragments of human plasma fibronectin. Proc. Natl. Acad. Sci. USA **78:** 3649–3653.
21. BEVILACQUA, M. P., D. AMRANI, M. W. MOSESSON & C. BIANCO. 1981. Receptors for cold-insoluble globulin (plasma fibronectin) on human monocytes. J. Exp. Med. **153:** 42–60.
22. MANTOVANI, B., M. RABINOVITCH & V. NUSSENZWEIG. 1972. Phagocytosis of immune-complexes by macrophages. J. Exp. Med. **135:** 780–792.

23. COLVIN, R. B. & H. R. DVORAK. 1975. Fibrinogen fibrin on the surface of macrophages. Detection, distribution, binding requirements and possible role in macrophage adherence phenomena. J. Exp. Med. **142:** 1377–1390.
24. SHERMAN, L. A. & J. LEE. 1977. Specific binding of soluble fibrin to macrophages. J. Exp. Med. **145:** 76–85.
25. GONDA, S. R. & J. R. SHAINOFF. 1982. Adsorptive endocytosis of fibrin monomer by macrophages: evidence of a receptor for the amino terminus of the fibrin β chain. Proc. Natl. Acad. Sci. USA **79:** 4565–4569.
26. BANDA, M. J., H. F. DOVEY & Z. WERB. 1981. Elastinolytic enzymes. *In* Methods for Studying Mononuclear Phagocytes. D. O. Adams, P. J. Edelson & H. Koren, Eds.: 603–618. Academic Press. New York.
27. GORDON, S., Z. WERB, & Z. A. COHN. 1976. Methods for the detection of macrophage secretory enzymes. *In* In Vitro Methods in Cell Mediated Immunity. B. R. Bloom & J. R. David, Eds.: 341–352. Academic Press. New York.
28. GORDON, S., J. C. UNKELESS & Z. A. COHN. 1974. Induction of macrophage plasminogen activator by endotoxin and phagocytosis. Evidence for a two stage process. J. Exp. Med. **140:** 995–1010.
29. BIANCO, C., O. GOTZE & Z. A. COHN. 1979. Regulation of macrophage migration by products of the complement system. Proc. Natl. Acad. Sci. USA **76:** 888–891.
30. POSTLETHWAITE, A. E., J. KESKI-OJA, G. BALIAN & A. KANG. 1981. Induction of fibroblast chemotaxis by fibronectin. Localization of the chemotactic region to a 140,000-molecular weight non-gelatin-binding fragment. J. Exp. Med. **153:** 494–499.

Discussion of this paper appears on p. 634.

BINDING OF SOLUBLE FIBRIN TO MACROPHAGES

Laurence A. Sherman

Division of Laboratory Medicine
Departments of Pathology and Medicine
Washington University School of Medicine
St. Louis, Missouri

INTRODUCTION

The role of the reticuloendothelial system in removing fibrin from the circulation had until recently been considered to be phagocytosis of particulate fibrin, on the basis of experiments showing reticuloendothelial system (RES) clearance of injected particulate fibrin.[1, 2] In reality, substantial uncertainty existed about the true *in vivo* process because of a lack of data about the *in vivo* physical and biochemical characteristics of fibrin in the blood prior to incorporation into thrombi, RES clearance, or fibrinolysis. Of note is that circulating microparticulate fibrin has not been found in any pathological states. Alternate forms of circulating *in vivo* fibrin needed to be identified, and presumably also a mechanism of reticuloendothelial system uptake of fibrin in experimental disseminated intravascular coagulation (DIC) other than phagocytosis of particulate material. A number of reports have described a soluble fibrin-like material in blood from patients with thrombosis and other diseases. This material is initially soluble in the blood but can be precipitated in the cold, and was frequently called cryofibrinogen.[3, 4] The development of biochemical data concerning such soluble forms of fibrin was key for subsequent concepts of the function of phagocytic cells in fibrin clearance.

SOLUBLE FIBRIN

The phenomenon that soluble fibrin could be found in plasma from patients with thromboses or thrombosis-prone diseases has been noted in numerous studies utilizing a variety of methods of demonstrating the fibrin. In addition to cryoprecipitation, heparin is used to augment precipitate formation.[5] Ethanol [6] and protamine [7] also precipitate similar material from the plasma of such patients. The constituents of cryoprecipitable material were shown biochemically to be both fibrin and fibrinogen, by Shainoff and Page [8] and similarly for ethanol precipitates by Kierulf [6] and in our laboratory.[9] In later investigations other proteins such as fibronectin, F VIII and F XIII were found in cryoprecipitates made by varying techniques. Using purified proteins Wegrzynowicz et al.[10] demonstrated that increasing amounts of fibrinogen could enhance fibrin solubility *in vitro*. This suggested that a soluble complex could be formed between fibrinogen and fibrin, which prevented clot formation. At the same time Fletcher et al.[11] showed that after exposure of plasma to thrombin, fibrinogen antigen emerged earlier on gel chromatography, presumably now combined with itself or other proteins. A similar pattern was found on studying plasma from patients with thromboses and/or intravascular coagulation. This higher molecular weight

610

0077–8923/83/0408–0610 $01.75/0 © 1983, NYAS

peak was thought to be a complex of fibrin and fibrinogen. However, Bang *et al.* demonstrated a similar shift in fibronectin elution with defibrination, suggesting fibrin:fibronectin interaction.[12] All of these studies pointed towards the potential existence of soluble fibrinogen:fibrin complexes existing *in vivo*, possibly including degradation products or fibronectin as well.

Several laboratories investigated the clearance of complexes of fibrinogen and fibrin formed *in vitro* and injected into animals or injection of soluble fibrin alone.[13-16] A number of techniques for fibrin formation and complex stabilization were utilized with some variation in results. In some instances the complexes contained fibrin(ogen) degradation products as well. In other studies not cited, the conditions were such that the dissolved fibrin probably came out of solution as soon as it was exposed to blood pH and ionic strength. Consensus can be found that soluble fibrinogen:fibrin complexes and fibrin have a rapid complete blood clearance with a blood half-life of 1–13 hours, depending on the method of calculation. Secondly, these complexes remain soluble in the blood. Only a few reports have examined the solubility and dissociation questions. In an attempt to reproduce the high molecular weight complexes found in plasma by Fletcher *et al.*, we subjected fibrin/fibrinogen complexes to gel chromatography and found two major peaks. That portion which emerged in an earlier peak remained as a soluble complex *in vivo*, as assessed by gel chromatography of plasma samples after injection, and similar blood $t_{1/2}$ life of fibrin and fibrinogen. The later peak of unassociated fibrinogen and fibrin did not form a complex *in vivo*.[13] Thirdly, clearance is generally regarded as being effected by the reticuloendothelial systems, and delayed clearance of complexes occurs with antecedent reticuloendothelial system blockade. The dissociability of noncrosslinked complexes is in question and may in part relate to whether des A fibrin or des AB fibrin is utilized.

As noted, a potential source of conflicting results is the method of preparation. In FIGURE 1 is shown the clearance of two types of noncrosslinked complexes. The first technique employed formation of fibrin in 0.09MKCl, 0.05MNaBr phosphate buffer pH 6.3 and redissolving the clot in 0.27MTrisHCl, 0.03MTris acetate buffer pH 5.3 ("NaBr fibrin"). The fibrin was mixed with fibrinogen in Tris NaCl pH 7.4, and the complex separated on a Sepharose 4B column in 0.3M Tris, 0.001M EDTA, 0.15M TAME pH 7.4.[13] The second form of preparing fibrin utilized 4M urea 0.01M Tris pH 7.0 to dissolve the fibrin ("urea fibrin") and the same Sepharose column and buffers for separating fibrin:fibrinogen complexes from nonassociated molecules. It can be seen that the "urea-fibrin" complex disassociated from fibrinogen *in vivo* whereas the "NaBr fibrin" did not.

In prior clearance studies [9] we studied the *in vivo* clearance in homologous animals of cryoprofibrin which is a fibrin:fibrinogen complex presumably formed *in vivo* and precipitated out of plasma by ethanol in the cold. The clearance of this material was a complex curve with half lives of 3 and 13 hours for the two phases. Like "urea fibrin" but unlike "NaBr fibrin" dissociation occurred, as assessed crudely by ethanol fractionation techniques. Muller-Berghaus [14] has noted dissociation after injecting radio-labeled "urea fibrin" or fibrinogen. A difficulty in the these studies is the effect of urea. Urea fibrinogen itself has a shorter half life of 45–50 hours vs. 60–65 hours for fibrinogen prepared without urea. In all these studies the presence or absence of potentially important trace contaminants such as fibronectin in each preparation may differ.

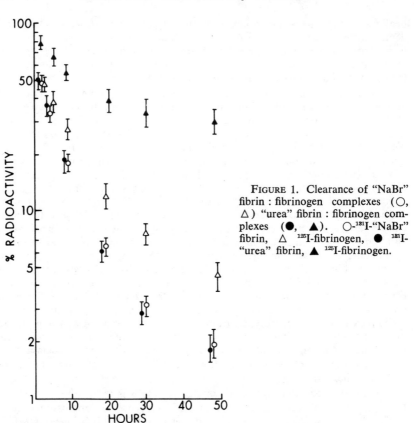

FIGURE 1. Clearance of "NaBr" fibrin : fibrinogen complexes (○, △) "urea" fibrin : fibrinogen complexes (●, ▲). ○-^{131}I-"NaBr" fibrin, △ ^{125}I-fibrinogen, ● ^{131}I-"urea" fibrin, ▲ ^{125}I-fibrinogen.

Of further complexity is the question of complex formation with fibronectin and fibrin(ogen) degradation products. Fibrin:fibrin or fibrin:fibrinogen are at least in part crosslinked on the basis of finding γ-γ dimers in plasma samples. In our laboratory we have been unable to demonstrate fibrin/fibrinonectin covalent binding *in vivo* using comparable techniques to defibrination experiments wherein γ-γ dimers can be found,[17] but noncovalent binding probably occurs.

These findings that fibrin could exist *in vivo* in a soluble form in a variety of complexes with fibrinogen, fibrinogen degradation products, and fibronectin, led to several studies of the uptake of fibrin by phagocytic cells. Colvin and Dvorak [18] demonstrated by immunofluorescent techniques fibrinogen related antigen on the surface of macrophages. Two immunofluorescent patterns, speckled and fibrillar, were thought to be fibrinogen and fibrin, respectively. In our laboratories [19] we examined the uptake of radiolabeled soluble fibrin by guinea pig macrophages in tissue culture. Initial experiments indicated uptake of both fibrin and fibrinogen. To exclude pinocytosis or related processes metabolic inhibitors were employed. Elicited peritoneal cells were harvested 72 hours after i.p. injection of Marcol 52. Adherent cells were incubated overnight in RPMI 1640, washed and resuspended in Hank's solution. The same results were found with a tenfold greater amount of fibrin bound vs. fibrinogen. Early fibrin, but not fibrinogen, degradation products were also bound. Cell membrane binding was relatively rapid (<1 hour) and essentially irreversible. Fibrin binding was saturable and Scatchard analysis yielded values of 3.6 μg of fibrin bound per 10^6 cells (FIG. 2). Removal of cell bound radioactivity by trypsin or plasmin confirmed that the cell fibrin association was the result of membrane binding. Similar removal was obtained by Colvin and Dvorak. Both laboratories also found Ca^{++} but not Mg^{++} to be necessary for binding. In later investigations Bang et al. found 2.3 μg of fibrin/degradation product complexes bound per 10^7 rabbit alveolar macrophages.[20]

Definition of the alteration in fibrin(ogen) as the result of thrombin action that led to greatly increased binding has had several approaches. One possible site of attachment would be the COOH end of the α-chain of fibrin which is readily accessible to surfaces or other molecules. In fibrin/fibrin crosslinking by FXIII α-chain crosslinking can produce large polymers. Similarly, the COOH end of the α-chain contains the site crosslinked by F XIII to fibronectin, albeit at a slower rate. High solubility low molecular weight fibrinogen fraction I-8 has intact NH_2-terminals, but is missing most of the COOH ends of the α-chains, and does not bind fibronectin. I-8 bears great similarity to early plasmin-derived fibrinogen species, which retain all of their ability to clot but have impaired

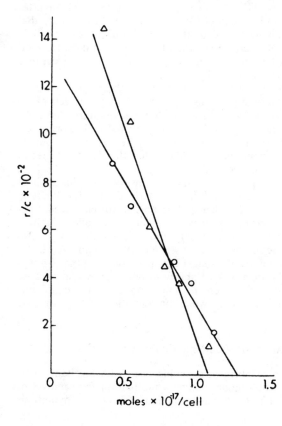

FIGURE 2. Scatchard analysis of moles ^{125}I-f bound with increasing concentrations. ^{125}I-f/F complex. Symbols ($\bigcirc$, $\triangle$) represent two separate sets of experiments. r/c, bound per unbound $\times 10^2$. (Reproduced from Sherman & Lee [19] with permission from the *Journal of Experimental Medicine*.)

TABLE 1 *

| Fibrin | Relative Macrophage Binding | |
	With Inhibitors	Without Inhibitors
I–4 (intact α)	100 (100)	100 (100)
I–8 (–COOHα)	88 (106)	86 (104)

* Results are relative amounts of fibrin bound with the amount of I–4 bound expressed as 100% The figures in parentheses are relative amounts on a molar basis assuming molecular weights of 325,000 for I–4 and 269,000 for I–8. Culture and conditions and inhibitors were the same as in reference 19.

polymerization. Fraction I-8 was compared to higher molecular weight fraction I-4 (intact α-chains) in their abilities to bind to macrophages. In metabolically inhibited cells 86% as much I-8 bound as I-4, and 88% as much without inhibitors (TABLE 1). Because I-8 has a lower molecular weight virtually the same number of molecules of I-8 were bound as I-4. This indicated the COOH end of the α-chain was not the site of attachment. Gonda and Shainoff [21] have found the synthetic polypeptide Gly-Pro-Arg, inhibited uptake of soluble fibrin by rabbit peritoneal macrophages, with use of metabolic inhibitors. This tripeptide corresponds to the new NH_2-terminal sequence of the α-chain after thrombin cleavage of fibrinopeptide A. Put together these studies indicate that the NH_2-terminal end and not the COOH end, of the fibrin α-chain is responsible for macrophage binding. It is of interest that when Esnouf and Marshall [22] used ancrod, which extensively degrades the α-chain, to produce fibrin, the clearance of fibrin in dogs was not affected by prior RES blockade.

The nature of the binding site on the macrophage membrane has been approached in several studies. Sherman and Lee [19] found that trypsinized metabolically inhibited macrophages bound fibrin as effectively as did untreated cells, over a one-hour period. Of interest was that with untreated cells fibronectin and its plasmin-derived fragments *reduced* fibrin binding by 20–45%, probably as a result of subtle complex function. It is likely that the fibrin:fibronectin macrophage interaction may differ depending on whether the fibrin is in a soluble or particulate form. It appeared that trypsin-sensitive proteins such as surface fibrinogen or fibronectin were not likely intermediaries for binding. Conversely, Jilek and Herman [23] found that adding fibronectin enhanced trypsinized macrophage uptake of fibrin, using cells without metabolic inhibitors incubated for a much longer time (overnight). Gonda and Shainoff [20] also observed fibrin uptake by macrophages in the absence of fibronectin. Against a simple involvement of fibronectin are the data (TABLE 1) showing that I-8 was bound by macrophages in spite of a lack of ability to bind fibronectin because of the missing COOH-terminal end of α-chain. The variation in the conditions of these several studies will require future work for resolution of the disparate results.

In Vivo SIGNIFICANCE OF SOLUBLE FIBRIN CLEARANCE

Teleologically, stabilization of fibrin in a soluble form with resultant clearance by macrophage binding would appear to be a mechanism for removal of fibrin from the circulation prior to thrombus formation. To test this hypothesis a quantitative approach was taken to assess RES capacity.[24] Radioiodinated soluble fibrin was injected in varying amounts to groups of animals. After one hour the animals were sacrificed and the organs analyzed for the presence of microthrombi. At lower doses, radioactivity was confined to intracellular pools. At 3–6 mg/kg fibrin injected microthrombi were found in kidney, liver, and spleen (TABLE 2). Radioautography confirmed concentrations of radioactivity around macrophages (FIG. 3). The amount of fibrin corresponds to immediate conversion to fibrin of 2–3% of the circulating fibrinogen pool. Muller-Berghaus et al.[13] noted microthrombi only in the spleen when 1 mg/kg of fibrin was injected. These several reports demonstrate a discrete but limited acute capacity of the RES to clear fibrin from the blood prior to thrombus formation. The requisite antecedent is maintenance of fibrin in a soluble form. The latter potentially involves fibrinogen, fibrinogen degradation products and fibronectin. Other variables besides concentration can influence the solubility of fibrin. For example, cationic proteins found in platelets and leukocytes can precipitate fibrin and conflicting data exist concerning a protective effect of neutropenia against fibrin deposition in experimental endotoxin-induced intravascular coagulation.[25, 26] The relative in vivo significance of soluble fibrin binding by macrophages vs microparticulate or other uptake is difficult to ascertain because of the preparation issues noted above. It is clear both in vitro and in vivo differences exist.

SOLUBLE FIBRIN AND MACROPHAGE FIBRINOLYTIC ENZYMES

Phagocytic cells synthesize and secrete a variety of proteinases some of which can degrade fibrin(ogen). In general little or no activity is found in resting cells, but increased amounts of proteinases are found in elicited (stimulated) cells.[27] Eliciting agents such as intraperitoneal oil or thioglycolate have generally been employed. We have studied the fibrinolytic enzymes from both mouse and guinea pig eluted macrophages.[28] With guinea pigs, peritoneal macrophages were harvested 72 hours after intraperitoneal injection of Marcol 52. Mouse peritoneal macrophages were harvested 3 days after intraperitoneal thioglycolate injection. Initially, a ^{125}I-labeled casein assay was used to detect proteolysis in the presence or absence of plasminogen. The media from incubated macrophages were tested and unexpected plasminogen independent proteolysis was found, contrary to prior reports.[27] A different assay was then used measuring ^{125}I-labeled fibrin solubilization in a fluid phase assay. The latter substrate was chosen as more specific and comparable to the in vivo setting. Again, proteolysis

TABLE 2

	Dose (mg/kg)		
	1	3	6
Trasylol	none	K,S	K,S,Li,Lu,H
Control	none	K	K,S,Li,Lu,H

* K(Kidney), S(Spleen), Li(Liver), Lu(Lung), H(Heart). Presence of the symbol denotes insoluble fibrin (thrombi) in organs of a majority of animals injected with that dose. Methods of experiments and assay systems are the same as in reference 24.

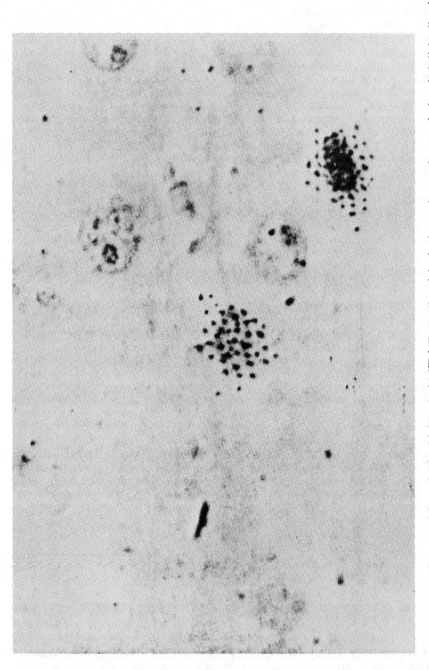

FIGURE 3. Autoradiograph of liver of animal injected with ^{125}I-f/F. Radioactivity is clustered around two reticuloendothelial cells with little radioactivity near the larger hepatocytes. Magnification ×2000. (Reproduced from Sherman et al.[24] with permission from the British Journal of Haematology.)

TABLE 3 *

Peritoneal Macrophages	% Fibrin Lysins		% Total Fibrinolytic Activity	
	Plasminogen Present	Plasminogen Absent	PA	NPDF
Mouse †	68 ± 8	44 ± 6	35	65
Guinea Pig ‡	55 ± 10	42 ± 9	24	76

* PA = plasminogen activator.   NPDF = non-plasminogen-dependent fibrinolytic activity. Conditions of eliciting guinea pig cells are in reference 19, those of mouse cells and the lysis assay are in reference 28.
† Mean ± S.D. of 5 experiments.
‡ Mean ± S.D. of 3 experiments.

occurred with and without plasminogen, but it was greater with. Plasminogen activator represents the increase (difference) when plasminogen was added (TABLE 3). Reactions were carried out in Tris H-Cl pH 8.0. Limited observations suggested greater non-plasmin fibrinolysis at neutral pH's. To pursue the biological relevance of these observations, several potential physiological stimulants were investigated using resting mouse macrophages in cell culture without metabolic inhibitors.[28] Plasmin, and plasmin-α2 macroglobulin complexes failed to stimulate release of fibrinolytic enzymes. In contrast small amounts (>5 μg/ml) of soluble fibrin in the media caused appreciable increases in both plasminogen activator and non-plasmin fibrinolytic activity over a 72-hour period (FIG. 4). (The soluble fibrin was a 1:3 fibrin:fibrinogen complex prepared using NaBr.) Incubation of soluble fibrin with cells was in serum free αMEM with lactalbumin hydrolysate, for a 72-hour period. As with the other studies most of the fibrinolytic activity was not plasminogen dependent.

Analysis of cleavage fragments of non-plasmin dependent fibrinolysis showed small polypeptides ~10,000 MW. No evidence was found for the classical X, Y, D, and E plasmin-derived products. Because plasmin-independent fibrinolysis can degrade the D and E fragments, the lack of D and E does not exclude the presence of plasminogen activator when mixtures of enzymes are present. In other studies Bang et al.[20] have identified unique fibrinolysis products, probably due to cathepsin D, using rabbit alveolar macrophages after BCG injection

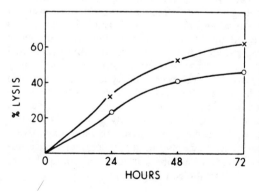

FIGURE 4. Development of proteolytic activity in serial samples of medium from f/F stimulated macrophages. (○) NPDF; (×) proteolysis in the presence of plasminogen (PA and NPDF). 60 μg of f/F were used to stimulate resident cells in α-MEM containing 0.05% LAH. (Reproduced from Sherman et al.[28] with permission from the Journal of the Reticuloendothelial Society.)

of the animals. The acid pH requirements found in that study indicate a lack of identity with the enzymes described here. Virtually no plasminogen activator was found, but the low pH used in the fibrinolysis studies may have prevented this activity from being observed. Banda and Werb [29] have shown macrophage elastase can degrade fibrinogen and Plow and coworkers have noted other neutral protease, fibrinolytic activity from leukocytes.[30, 31] These various enzymes have other effects on hemostasis including elastase degradation of fibronectin and F VIII proteolysis by neutral proteases from leukocytes.[32] It is also of interest that we have found soluble fibrin to be bound to metabolically inhibited granulocytes in greater amounts than bound to macrophages.[19] The response to soluble fibrin with release of fibrinolytic enzymes indicates a means of increasing the limited reticuloendothelial system capacity to clear fibrin from the blood. As with other stimulants of macrophage activity, increases in macrophage fibrinolytic activity have a lag period, at least *in vitro.* Thus these increases in fibrinolysis may be of little significance in acute DIC. Conversely data exist that ordinary thrombus formation may extend over several days, which would allow enhancement of cellular fibrinolysis.

DISCUSSION

Substantial added data are needed concerning soluble fibrin binding to macrophages, particularly concerning regeneration of binding sites and better biochemical understanding of the *in vivo* constituents of soluble fibrin complexes. The variations in amounts of various released enzymes are in part functions of different assay and cell systems. Specificity of induction due to different eliciting agents may also exist. Morland and Morland [33] have shown relative differences in individual lysosomal enzyme activities during induction of macrophage enzymes, by different eliciting agents. Disparate results between lysosomal (cathepsin D) and non-lysosomal enzymes (plasminogen activators) might also be expected. The conditions that we have employed using soluble fibrin were comparable to those which occur *in vivo* in a number of disease states. The required duration of exposure and the possible effect of fibrinogen derivatives other than fibrin require further study. The lack of binding we have noted for late fibrinogen degradation products [19] would suggest a relatively specific mechanism. Other hemostatic moieties cleared in the reticuloendothelial system such as antithrombin III-thrombin complexes may be active as well.

The focus of this paper has been on the interaction of soluble fibrin with phagocytic cells, chiefly macrophages. The influence clinically of unrelated stimuli to macrophages, on fibrin binding and levels of fibrinolytic enzymes is unknown. It also should be stressed that moieties in the blood may influence both the formation and clearance of soluble fibrin. The relationship of increased solubilization of fibrin to increasing fibrinogen concentration was previously noted. Increasing fibronectin lessens the development of hypofibrinogenemia in experimental intravascular coagulation, an effect possibly unrelated to the afore-noted potential effect of fibronectin on macrophage fibrin uptake. There is a variety of other known and posited factors which humerally and cellularly alter reticuloendothelial system effects on hemostasis and these are more extensively reviewed elsewhere,[34] but the data and concepts are still incomplete. Accordingly, a variety of complex *in vivo* interrelationships will need elaboration before we understand the dynamics of fibrin metabolism in patients.

REFERENCES

1. LEWIS, J. R. & I. L. F. SZETO. 1965. Clearance of infused fibrin transport. Fed. Proc. **24:** 840.
2. GANS, H. & J. R. LOWMAN. 1967. The uptake of fibrin and fibrin-degradation products by the isolated perfused rat liver. Blood **29:** 526.
3. GLUECK, H. I. & L. G. HERRMANN. 1964. Cold-precipitable fibrinogen, cryofibrinogen. Arch. Intern. Med. **113:** 748–757.
4. McKEE, P. A., J. M. KALBFLEISCH & R. M. BIRD. 1963. Incidence and significance of cryofibrinogenemia. J. Lab. Clin. Med. **61:** 203.
5. SMITH, R. T. & R. W. VON KORFF. 1957. Heparin precipitable fraction of human plasma. I. Isolation and characterization of fraction. J. Clin. Invest. **36:** 596.
6. KIERULF, P. 1973. Studies on soluble fibrin in plasma. II. N-terminal analysis of a modified fraction I (Cohn) from patient plasmas. Scand. J. Clin. Lab. Invest. **31:** 37.
7. LIPINSKA, B. & K. WOROWSKI. 1968. Detection of soluble fibrin monomer complexes in blood by means of protamine sulfate test. Thromb. Diath. Haemostas. **20:** 44.
8. SHAINOFF, J. R. & I. H. PAGE. 1962. Significance of cryoprofibrin in fibrinogen-fibrin conversion. J. Exp. Med. **116:** 687.
9. SHERMAN, L. A. 1972. Fibrinogen turnover: Demonstration of multiple pathways of catabolism. J. Lab. Clin. Med. **79:** 710.
10. WEGRZYNOWICZ, A., M. KOPEC & Z. S. LATALLO. 1971. Formation of soluble fibrin complexes and some factors affecting their solubility. Scand. J. Haematol. **13:** 49.
11. FLETCHER, A. P., H. ALKJAERSIG, J. O'BRIEN, et al. 1970. Blood hyper-coagulability and thrombosis. Trans. Assoc. Am. Physicians **83:** 159.
12. BANG, N. U., M. S. HANSEN, G. F. SMITH, et al. 1973. Properties of soluble fibrin polymers encountered in thrombotic states. In Present Status of Thrombosis. R. Losito, Ed.: 75. F. K. Schattauer Verlag. New York.
13. SHERMAN, L. A., S. HARWIG & J. LEE. 1975. In vitro formation and biological fate of fibrinogen:fibrin complexes. J. Lab. Clin. Med. **86:** 100.
14. MULLER-BERGHAUS, G., I. MAHN, G. KOVEKER & F.-D. MAUL. 1976. In vivo behavior of homologous urea-soluble ^{131}I-fibrinogen in rabbits: The effect of fibrinolysis inhibition. Brit. J. Haematol. **33:** 61.
15. CHANG, M. L. & N. U. BANG. 1977. Biological behavior of higher molecular weight products of fibrinolysis. J. Lab. Clin. Med. **90:** 216.
16. VON HUGO, R., R. HAFER, B. STEIN, et al. 1977. Incorporation of ^{125}I-fibrinogen in circulating soluble fibrin monomer complexes during hyper-coagulability. Thrombos. Res. **10:** 703.
17. SHERMAN, L. A. & J. LEE. 1982. Fibronectin: Blood turnovers in normal animals and during intravascular coagulation. Blood **80:** 558.
18. COLVIN, R. B. & H. F. DVORAK. 1975. Fibrinogen/fibrin on the surface of macrophages: Detection, distribution, binding requirements, and possible role in macrophage adherence phenomena. J. Exp. Med. **142:** 1377.
19. SHERMAN, L. A. & J. LEE. 1977. Specific binding of soluble fibrin to macrophages. J. Exp. Med. **145:** 76.
20. BANG, N. U., M. L. CHANG, L. E. MATTLER et al. 1981. Monocyte/macrophage-mediated catabolism of fibrinogen and fibrin. Ann. N.Y. Acad. Sci. **370:** 568.
21. GONDA, S. R. & J. R. SHAINOFF. 1982. Absorptive endocytosis of fibrin monomer by macrophages; Evidence of a receptor for the amino terminus of the fibrin α-chain. Proc. Natl. Acad. Sci. USA **79:** 4565.

22. Esnouf, M. P. & R. Marshall. 1968. The effect of blockade of the reticuloendothelial system and of hypotension on the response of dogs to *Ancitrodon rhodostoma* venom. Clin. Sci. **35:** 261.
23. Jilik, F. & H. Hormann. 1978. Fibronectin (cold-insoluble globulin), V[1] mediation of fibrin-monomer binding to macrophages. *In* Hoppe-Seyler's Physiol. Chem. **358:** 1603.
24. Sherman, L. A., J. Lee & A. Jacobson. 1977. Quantitation of the reticuloendothelial system clearance of soluble fibrin. Brit. J. Haematol. **37:** 231.
25. Muller-Berghaus, G., T. Eckhardt & W. Kramer. 1974. The role of leukocytes and platelets in the precipitation of fibrin *in vivo:* Mechanisms of the generation of microclots from soluble fibrin. Thrombos. Res. **4:** 895.
26. Lipinski, B., A. Nowak & V. Gurewich. 1974. The organ distribution of ^{125}I-fibrin in the generalized Shwartzman reaction and its relation to leucocytes. Brit. J. Haematol. **28:** 221.
27. Unkeless, J. C., S. Gordon & E. Reich. 1974. Secretion of plasminogen activator by stimulated macrophages. J. Exp. Med. **139:** 834.
28. Sherman, L. A., J. Lee & C. Stewart. 1981. Release of fibrinolytic enzymes from macrophages in response to soluble fibrin. J. Reticuloend. Soc. **30:** 317.
29. Banda, M. J. & Z. Werb. 1980. The role of macrophage elastase in the proteolysis of fibrinogen, plasminogen, and fibronectin. Fed. Proc. **39:** 1756. (abstract).
30. Plow, E. F. & T. S. Edgington. 1975. An alternative pathway for fibrinolysis. I. The cleavage of fibrinogen by leukocyte proteases at physiologic pH. J. Clin. Invest. **56:** 30.
31. Plow, E. F. 1982. Leukocyte elastase release during blood coagulation. A potential mechanism for activation of the alternative fibrinolytic pathway. J. Clin. Invest. **59:** 564.
32. Kopec, M., K. Bykowska, K. Lopaciuk, *et al.* 1980. Effects of neutral proteases from human leukocytes on structure and biological properties of human Factor VIII. Thromb. Haemostas. **43:** 211.
33. Morland, B. & J. Morland. 1978. Selective induction of lysosomal enzyme activities in mouse peritoneal macrophages. J. Reticuloendothel. Soc. **23:** 469.
34. Snedeker, P. W., J. E. Kaplan & T. M. Saba. 1978. Effect of traumatic shock and alteration of reticuloendothelial function on the vascular clearance of soluble fibrin. Physiologist **21:** 113.

Discussion of this paper appears on page 634.

FIBRINOGEN–FIBRIN INTERACTIONS WITH FIBROBLASTS AND MACROPHAGES *

Robert B. Colvin †· ‡

† *Immunopathology Unit, Department of Pathology*
Massachusetts General Hospital
and Harvard Medical School
Boston, Massachusetts 02114

A common feature of inflammatory responses *in vivo* is the extra-vascular accumulation of a fibrin-fibronectin matrix. Well-documented examples include healing skin and corneal wounds,[1-4] cell-mediated immune reaction (both delayed type hypersensitivity and graft rejection),[5-7] crescents in glomerulonephritis,[8] the stroma of tumors,[9, 10] and the neovascularization response.[11]

In all instances cell migration, activation, differentiation and proliferation occur, involvnig fibroblasts, macrophages, epithelium and other cells. We have hypothesized that fibrin and fibronectin may serve as the essential provisional matrix that promotes some of these cellular events.[2-4] In initial studies to test this hypothesis, we have examined the surface interaction of fibroblasts and macrophages with fibrinogen, particularly with regard to fibronectin.

FIBROBLASTS

Glass adherent human WI-38 fibroblasts grown in culture bind soluble human fibrinogen in a delicate fibrillar pattern that closely follows cell processes, as judged by immunofluorescence (FIG. 1).[12] Controls incubated in serum show no detectable fibrinogen. This binding occurs at 4 to 37° C, is calcium, magnesium, and factor XIII independent and is destroyed by brief trypsinization.

We believe that most or all of the binding is dependent on fibronectin. The binding pattern closely resembles the pattern of fibronectin on these cells (FIG. 2). All cells tested that had a surface fibronectin matrix also bound fibrinogen, while those without a fibronectin matrix did not. Furthermore, SV40-transformed WI-38 cells, which had a sparse, granular pattern of fibronectin, had an identical sparse granular pattern of fibrinogen binding (FIG. 3).

The nature of the bound material was examined using ^{125}I-labeled human fibrinogen. On reduced SDS acrylamide gels, the bound fibrinogen extracted from the cell monolayer had substantially the same pattern as the starting material (FIG. 4), except for a small peak probably corresponding to γ-chain dimers. We conclude that, under the conditions of the assay, the bound material was fibrinogen in native, or largely native, form.

* Supported in part by Public Grant Numbers CA-20822, CA-15889, CA-16881 and HL-17419, awarded by the National Cancer Institute and National Heart, Lung and Blood Institute, DHHS.

† These studies were done in collaboration with Richard A.F. Clark, Harold F. Dvorak, Phyllis I. Gardner, M. Elizabeth Hammond, Richard Kradin, Joan M. Lanigan, Michael W. Mosesson, Richard O. Roblin, and Evie L. Verderber.

‡ Address reprint requests to: Robert B. Colvin, M.D., Immunopathology Unit, Department of Pathology, Cox 5, Massachusetts General Hospital, Boston, MA 02114.

0077-8923/83/0408-0621 $01.75/0 © 1983, NYAS

The specificity of the binding was assessed in competition experiments (TABLE 1). Both cold fibrinogen and gelatin inhibited the uptake of labeled fibrinogen. I-9, a fibrinogen catabolite that lacks the region of the Aα chain that binds fibronectin, showed significantly less binding than intact fibrinogen. Fibrin monomer under the same conditions bound in greater amounts. As in the experiments of Sherman and Lee [13] with macrophages, the possible contribution of self-association of the fibrin prevents definitive conclusions regarding relative avidity for fibrin and fibrinogen. However, these data are consistent with the studies above that suggest a slight enrichment of cross-linked fibrin/ogen in the material eluted from cells incubated in ^{125}I-labeled fibrinogen (FIG. 4).

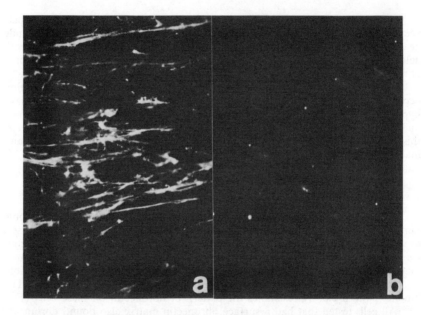

FIGURE 1. Unfixed glass adherent human fibroblasts (WI-38) stained with fluorescein anti-fibrinogen after incubation at 22° C for 30 minutes in 1.5 mg/ml of human fibrinogen (a) or 10% human serum (b). Although no intrinsic fibrinogen is detected on the cells incubated in serum, binding of fibrinogen can be detected after a short exposure to exogenous fibrinogen. A similar pattern was seen in the absence of calcium, magnesium and in factor XIII-deficient plasma. Immunofluorescence micrograph, × 510.

Others have shown that fibroblasts actively interact with fibrin, as judged by their ability to mediate clot retraction,[14] and that this property is lost on viral transformation.[15] The adherence of fibroblasts to cross-linked fibrin and fibronectin also has been well documented.[16] Finally, soluble fibronectin (or fragments) can serve as a chemotactic agent for fibroblasts.[17, 18] Thus the fibrin-fibronectin provisional matrix is a likely candidate for influencing fibroblast migration and localization in inflammatory reactions. The observations that tumor cells (including malignant fibroblasts) have diminished capacity to bind fibrinogen and mediate clot retraction is of some interest.[15] Whether this affects their anchorage in tissue or malignant behavior remains speculative.

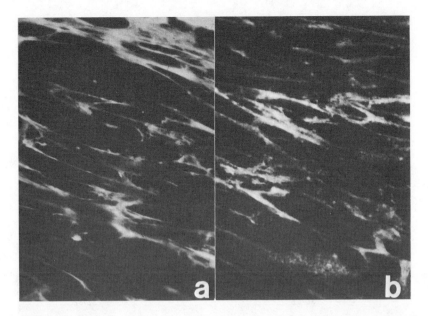

FIGURE 2. WI-38 fibroblasts stained with antifibronectin (a) and antifibrinogen (b), the latter after incubation in 50% plasma. Both fibronectin and fibrinogen were distributed as fibrils that closely follow cell processes in a pattern that is very similar. Immunofluorescence micrograph, × 510.

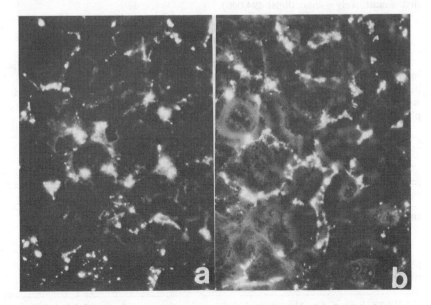

FIGURE 3. SV-40 transformed WI-38 fibroblasts stained for fibronectin (a) and fibrinogen (b) as in FIGURE 2. The components are in a similar granular pattern, but quite different from that in untransformed cells (FIG. 2). × 510.

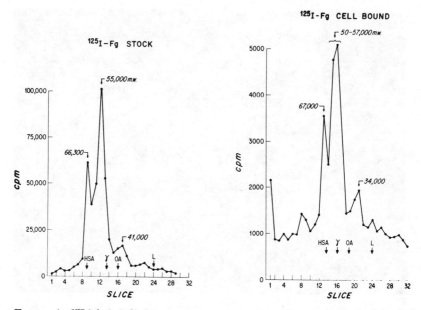

FIGURE 4. ^{125}I-labeled fibrinogen stock that was added to cells (*left panel*) and ^{125}I-labeled cell-bound material recovered from adherent WI-38 cells with 1M NaOH (*right panel*) analyzed after reduction on SDS 10% acrylamide disc gels and sliced at 1 mm intervals (12). The 3 peaks corresponding to the Aα, Bβ and γ chains are seen in the stock. The cell-bound material has in addition a small peak at 98,000 (slice 8) that is most likely γ chain dimer (94,000).

MACROPHAGES

Macrophages, ubiquitous cells in chronic inflammation, have many potential interactions with the clotting system, including production of procoagulants,[19, 20] plasminogen activator,[22] α$_2$-macroglobulin [23] and fibronectin.[18, 24, 25] It is not surprising, therefore, that fibrin is often associated with the surfaces of macrophages in inflammatory reactions *in vivo*, as illustrated in FIGURE 5.

TABLE 1

^{125}I-LABELED-FIBRINOGEN BINDING TO FIBROBLASTS *

Form of Fibrinogen	Inhibitor	% Binding
Intact	None	100 ± 31
	HSA (1.5 mg/ml)	95 ± 29
	Fibrinogen (1.5 mg/ml)	25 ± 7 †
	Gelatin (1.8 mg/ml)	47 ± 4 †
I-9	None	58 ± 10†
Fibrin monomer	None	792 ± 71†

* WI-38 human fibroblast monolayers incubated with 2.9 μg ^{125}I labeled human fibrinogen plus inhibitor for 30 at 22° C. 100% control was 38–99 ng/μg Lowry protein. Trypsin released 78 ± 1% of the counts. Data are from ref. 12.
  † p < 0.05 vs. control.

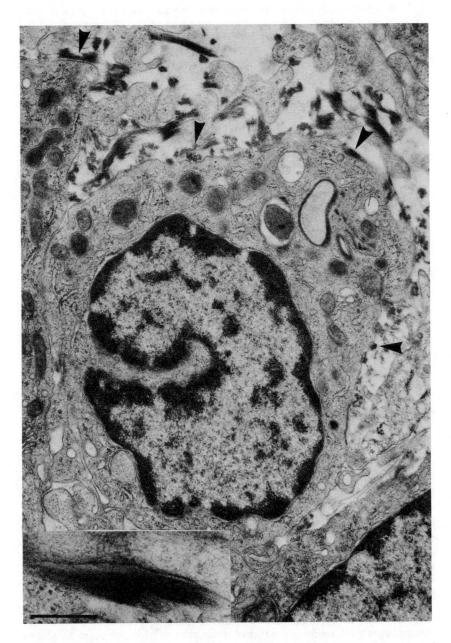

FIGURE 5. An electron micrograph of three adjacent macrophages in a human renal biopsy (acute pyelonephritis). Dark fibrillar bands of fibrin are present in the extracellular matrix and in close apposition to the cell membrane (*arrow heads*) over much of the cell surface. Insert shows characteristic fibrin periodicity. Bar corresponds to 300 nm in the insert and 1000 nm in the larger figure.

Nelson and collaborators some years ago observed that anticoagulants inhibit the adherence of macrophages to the mesothelium of the peritoneal cavity.[26-28] This adherence was induced by antigen challenge in guinea pigs primed for delayed hypersensitivity and was termed the macrophage disappearance reaction. Although thrombin could also induce macrophage adherence,[29] the mechanism, particularly the role of the clotting system, was obscure.

This prompted our search for surface bound fibrin and fibronectin on guinea pig macrophages that might mediate such adherence.[30, 31] Both resident and exudate macrophages have surface fibrin or fibrinogen that is readily demonstrable by immunofluorescence.[30] Two patterns are seen: a network, believed to be fibrin, and a speckled pattern thought to be fibrinogen (FIG. 6, TABLE 2). This conclusion was based on the lack of the network pattern in animals pre-

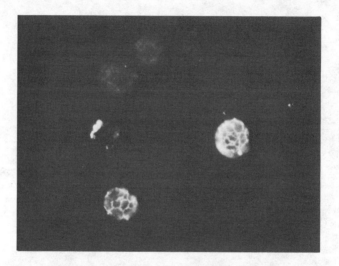

FIGURE 6. Immunofluorescence micrograph of a suspension of guinea pig resident peritoneal cells stained with fluorescein conjugated antifibrinogen. A bright net pattern is seen on two cells and a fainter, granular pattern on two, one of which has a brightly staining segmental patch. $\times$ 510.

treated with warfarin (TABLE 2). Hopper and associates have recently confirmed this view by the identification of γ dimers in extracts of macrophages displaying nets.[32] Heparin in the lavage fluid does not prevent the nets,[30] so they probably are formed *in vivo* by a low-grade activation of the clotting system. In support of this, the speckled pattern, but not the nets, could be restored by incubation in heparinized plasma.[30]

Curiously, not all macrophages displayed surface fibrin or fibrinogen. Within the peritoneal cavity about 25–50% of the cells were positive, but alveolar macrophages were uniformly negative, even after *in vitro* exposure to plasma. Whether this heterogeneity is due to the state of activation or defines stable subpopulations remains to be established.

We then attempted to relate these findings directly to the macrophage disappearance reaction (MDR). A typical MDR was elicited in animals sensitized with complete Freund's adjuvant (CFA) and challenged i.p. with 10 μg purified protein derivative (PPD) from mycobacteria (TABLE 3). Free peritoneal macrophages were reduced to 17% of the number recovered from sensitized controls challenged with saline. Preliminary experiments showed that macrophages isolated from the peritoneal cavity of animals undergoing an MDR included fibrin/ogen positive cells. To measure cell uptake and clearance of fibrinogen, we included ^{125}I-labeled fibrinogen in the antigen challenge solution injected i.p. Cell-associated ^{125}I-labeled fibrin/ogen was calculated as that fraction of injected ^{125}I-labeled fibrinogen which was recovered in the washed cell pellet per 10^8 macrophages counted. Sensitized animals challenged with PPD had a 2.4-fold increase in the [124]I-labeled fibrin/ogen uptake per macrophage compared to sensitized controls given saline (p < 0.05, TABLE 3). From the specific activity and the amount injected, the number of ^{125}I-labeled fibrin/ogen molecules/macrophage could be estimated (TABLE 3). The increase from 18,500 to 44,000 molecules/macrophage does not, however, include cell-

TABLE 2

CELL SURFACE FIBRIN/FIBRINOGEN ON NORMAL PERITONEAL MONONUCLEAR CELLS *

| | (Anticoagulants) | | % Cells with Surface Fibrin/Fibrinogen | | |
	Warfarin	Heparin	Total	Speckles	Nets
Experiment 1	—	—	45.1 ± 7.6	17.7 ± 5.3	27.4 + 6.2
	+	+	31.7 ± 12.8	31.7 ± 12.8	0.0 ± 0.0
Experiment 2	—	—	34.2 ± 6.5	8.8 ± 1.6	25.4 ± 6.4
	—	+	51.2 ± 4.8	18.0 + 3.0	25.7 ± 8.3

* Cells were harvested from the peritoneal cavity of normal guinea pigs in buffered saline solutions with calcium and magnesium that contained no anticoagulants or 10 U/ml heparin as indicated. Some animals were pretreated with warfarin. Cells were washed and stained in the same buffer with fluoresceinated anti-fibrinogen. The mean % of mononuclear cells exhibiting surface fluorescence ± SE is given for 4–6 animals. A total of 200–400 cells were counted in each sample. Data are from ref. 30.

bound autologous (cold) fibrinogen and thus underestimates the total number of molecules/cell.

Because previous studies had shown a marked retention of ^{125}I-labeled fibrin in the cutaneous tuberculin reaction,[5] the clearance of ^{125}I-labeled fibrinogen from the peritoneal cavity during the MDR was measured. No effect on clearance was detected in the MDR (t½ 2.3–3.1 hours), as might be expected if a generalized activation of the clotting system in the peritoneal cavity had converted a substantial fraction of the ^{125}I-labeled fibrinogen to insoluble fibrin. The form of the ^{125}I-labeled fibrinogen (soluble fibrin, FDP, etc.) was not determined.

Various controls were performed to determine the specificity of the observed effects. Non-sensitized animals challenged with PPD showed a slight, but nonsignificant, decrease (to 71%) in the number of free macrophages when compared to normal animals challenged with saline. No increase in ^{125}I-labeled fibrin/ogen uptake was found. ^{131}I-HSA recovered in the cell pellet measured

TABLE 3

125I-LABELED-FIBRINOGEN UPTAKE IN THE MACROPHAGE DISAPPEARANCE REACTION *

CFA Sensitized	PPD Challenge	Free Macrophages $\times 10^{-6}$	Cell Bound †		Average Cell Uptake of 125I-labeled Fibrinogen (Molecules/Macrophages)	Clearance from Peritoneal Fluid (t ½ hr)	
			125I-Fib	131I-Alb		125I-Fib	131I-Alb
+	+	2.7 ± 0.6	9.1 ± 2.7	0.33	44,000	3.1 ± 0.5	4.0 ± 0.7
+	0	15.7 ± 5.6	3.8 ± 1.1	0.25	18,000	3.2 ± 0.4	4.0 ± 0.4
0	+	5.8 ± 1.1	3.1 ± 0.7	0.30	15,000	2.7 ± 0.3	3.4 ± 0.5
0	0	8.2 ± 1.9	2.8 ± 0.9	0.16	13,600	2.3 ± 0.3	2.7 ± 0.4

* Sensitized (CFA) and control (unimmunized) animals with a 4 day glycogen-induced peritoneal exudate were challenged i.p. with 10 μg PPD in saline or with saline alone, containing 125I labeled human fibrinogen (28.6 μg, 0.125 μCi/μg) and 131I labeled human albumin (HSA). Three to four hours later the free peritoneal cells were harvested by lavage with 100 ml cold Hanks balanced salt solution with 10 U/ml heparin. Cells were washed three times and the radioactivity of the final cell pellet and of an aliquot of the initial lavage fluid was measured in a γ-counter. A small portion (<1%) of the cells were taken from the first wash for performing cell counts. Values represent the means from two experiments with 8–9 animals in each group.

† Percentage injected dose recovered in washed cell pellet/10⁸ macrophages.

simultaneously were 3–10% that of ^{125}I-labeled fibrinogen, consistent with the lack of cell-surface albumin detectable by immunofluorescence; this presumably is nonspecific and represents residual peritoneal fluid not washed from the cells and pinocytotic uptake. No differences were found in ^{131}I-HSA uptake or clearance in the various groups.

Thus, challenge with specific antigen decreases the numbers of free peritoneal macrophages at four hours, and specifically increases their uptake of ^{125}I-labeled fibrinogen, although levels of ^{125}I-labeled fibrinogen (or degradation products) in soluble form in the peritoneal cavity was not measurably affected. This suggests that the activation of the clotting system is not generalized but may be limited to the macrophages. Hopper found a similar increase in macrophage fibrinogen uptake and net formation at 42 hours (in the form of cross-linked fibrin) and demonstrated these cells had a diminished capacity to migrate.[32] Fibrin therefore could serve to immobilize macrophages in inflammatory sites; whether it activates the cells, or does more than simply cause aggregation, remains an interesting speculation.

We postulated that fibronectin might serve as a "receptor" for fibrin/fibrinogen.[30] We subsequently found that a population of peritoneal macrophages did indeed display surface fibronectin when stained with $F(ab')_2$ polyclonal or monoclonal antifibronectin antibody (ref. 31 and Colvin et al., submitted for publication). The pattern was granular; nets were not seen (FIG. 7). This has been confirmed by others.[33] The fibronectin was indistinguishable from the plasma form as judged by surface labeling and reduced SDS gel analysis. Just as in the fibrin/fibrinogen staining there were population differences: macrophages with surface fibronectin were most numerous and stained brightest in peritoneal exudates, whereas resident macrophages and alveolar macrophages had less fibronectin. This was not a permanent state, because during pulmonary injury induced with hyperoxia, the alveolar macrophages display a progressive increase in surface fibronectin, suggesting a change in their surface properties as a consequence of the inflammatory process.[34]

The relationship between fibrinogen and fibronectin on the surface was studied in co-capping experiments. Macrophages incubated with fluorescein-conjugated $F(ab')_2$ antifibronectin antibodies capped their surface fibronectin within 10–20 minutes at 37° C followed by ingestion. When such capped cells were fixed and stained with rhodamine-conjugated anti-fibrinogen, the caps were co-distributed. In contrast, surface cytophilic IgG on the macrophages did not co-cap with fibronectin. Although these data do not indicate which, if either, component is serving as a receptor, they do demonstrate their close association. Others have shown that fibronectin can restore fibrin binding to trypsinized macrophages,[35] and there is good evidence that monocytes and macrophages can interact directly with fibronectin, at least in an insoluble form [36–41] and serve as an opsonin.[39–40] A monoclonal antibody that blocks the binding of human monocytes to fibronectin has recently been reported.[42] The present data are consistent with the hypothesis that the fibronectin binds to the macrophage, which in turn binds to fibrinogen or fibrin.

The capping of fibronectin and fibrinogen on macrophages contrasts sharply with the inability of fibroblast fibronectin to show conventional capping and the lack of capping of certain other macrophage surface components such as IgG and histocompatibility antigens.[43] The data indicate that macrophage fibronectin is membrane associated and is connected to the appropriate contractile apparatus in the cell. We therefore examined the metabolic requirements for this process. Capping of macrophage surface fibronectin was abolished at 4°, by 2 µg/ml cytochalasin B, and by 1 mg/ml 2-deoxyglucose. Azide and cyanide, even at high concentrations (30 and 20mM, respectively) were ineffective. These data are in marked contrast to those for B lymphocyte immunoglobulin capping,[44] and suggest that fibronectin movement on the macrophage surface requires anerobic glycosis as an energy source as well as intact microfilaments.

The consequences of interaction of macrophages with fibronectin are not yet completely delineated. Increased "activity" of membrane Fc and C3

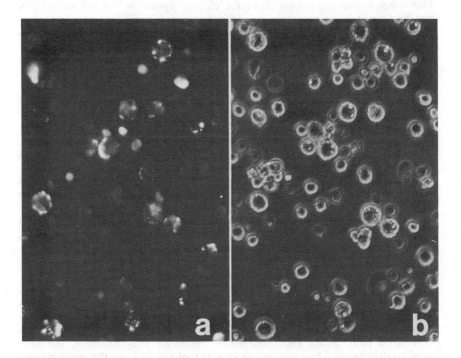

FIGURE 7. Immunofluorescence micrography (a) of oil-induced guinea pig peritoneal exudate cells stained in suspension with a murine monoclonal antibody to guinea pig plasma fibronectin (2D3). Delicate granular surface staining is seen over about half of the macrophages, several of which contain oil droplets by phase microscopy (b). × 320.

receptors occurs,[36] and increased spreading[31] and increased activity of a macrophage-derived growth factor[45] have been described. Exogenous fibronectin also is needed for maximal response of macrophages in the antigen induced in migration inhibition (Colvin et al., unpublished data). Isolated plasma fibronectin may not be an adequate substitute for fibronectin extracted from macrophages in this response.[33] It is not known whether this inadequacy is due to contaminating co-factors, to artifactual alterations from the purification procedure, or to inherent structural differences between plasma and macrophage fibronectins.

Epithelial Cells

Recent evidence from healing skin and corneal wounds indicates that the epithelium in both instances migrates on a substrate of fibrin and fibronectin.[2, 3] This occurs before more permanent basal lamina components, namely laminin and type IV collagen, are regenerated. While there is little evidence that epithelial cells adhere to fibronectin and fibrin *in vitro* (reviewed in ref. 45), our studies indicate that such interaction is likely to occur *in vivo*. Understanding of the role of fibrin and fibronectin in these settings may lead to new therapeutic approaches in wound healing.

Conclusion

It is evident that numerous and intimate interactions occur between cells and the clotting system. Elucidation of their biological significance is the next important research goal, because of the clear indications that fibrinogen and other clotting components have important functions beyond their traditional roles in hemostasis.

References

1. GRINNELL, F., R. E. BILLINGHAM & L. BURGESS. 1981. Distribution of fibronectin during wound healing in vivo. J. Invest. Dermatol. **76:** 181.
2. FUJIKAWA, L. S., C. S. FOSTER, T. J. HARRIST, J. M. LANIGAN & R. B. COLVIN. 1981. Fibronectin in healing rabbit corneal wounds. Lab. Invest. **45:** 120.
3. CLARK, R. A. F., J. M. LANIGAN, P. DELLAPELLE, E. MANSEAU, H. F. DVORAK & R. B. COLVIN. 1982. Fibronectin and fibrin provide a provisional matrix for epidermal cell migration during wound re-epithelization. J. Invest. Dermatol. **79:** 264.
4. CLARK, R. A. F., J. H. QUINN, H. J. WINN, J. M. LANIGAN, P. DELLAPELLE & R. B. COLVIN. 1982. Fibronectin is produced by blood vessels in response to injury. J. Exp. Med. **156:** 646.
5. COLVIN, R. B. & H. F. DVORAK. 1975. Role of the clotting system in cell-mediated hypersensitivity. II. Kinetics of fibrinogen/fibrin accumulation and vascular permeability changes in tuberculin and cutaneous basophil hypersensitivity reactions. J. Immunol. **114:** 377.
6. COLVIN, R. B., M. W. MOSESSON & H. F. DVORAK. 1979. Delayed-type hypersensitivity skin reactions in congenital afibrinogenemia lack fibrin deposition and induration. J. Clin. Invest. **63:** 1302.
7. CLARK, R. A. F., H. F. DVORAK & R. B. COLVIN. 1981. Fibronectin in delayed-type hypersensitivity skin reactions: Association with vessel permeability and endothelial cell activation. J. Immunol. **126:** 787.
8. PETTERSSON, E. E. & R. B. COLVIN. 1978. Cold-insoluble globulin (fibronectin, LETS protein) in normal and diseased human glomeruli: Papain-sensitive attachment to normal glomeruli and deposition in crescents. Clin. Immunol. Immunopathol. **11:** 425.
9. DVORAK, H. F., A. M. DVORAK, E. J. MANSEAU, L. WIBERG & W. H. CHURCHILL. 1979. Fibrin gel investment associated with line 1 and line 10 solid tumor growth, angiogenesis, and fibroplasia in guinea pigs. Role of cellular immunity myofibroblasts, microvascular damage and infarction in line 1 tumor regression. J. Natl. Cancer Inst. **62:** 1459.

10. DVORAK, H. F., G. R. DICKERSIN, A. M. DVORAK, E. J. MANSEAU & K. PYNE. 1981. Human breast carcinoma: Fibrin deposits and desmoplasia. Inflammatory cell type and distribution. Microvasculature and infarction. J. Natl. Cancer Inst. **67:** 335.

11. GOLUB, B. M., C. S. FOSTER & R. B. COLVIN. 1982. Fibronectin, laminin, type IV collagen, Factor VIII, and fibrin: Sequential analysis during experimental corneal neovascularization. Invest. Ophthal. Vis. Sci. **22**(Suppl): 27 (Abst).

12. COLVIN, R. B., P. I. GARDNER, R. O. ROBLIN, E. L. VERDERBER, J. M. LANIGAN & M. W. MOSESSON. 1979. Cell surface fibrinogen-fibrin receptors on cultured human fibroblasts: Association with fibronectin (cold-insoluble globulin, LETS protein) and loss in SV40 transformed cells. Lab. Invest. **11:** 464.

13. SHERMAN, L. A. & J. LEE. 1977. Specific binding of soluble fibrin to macrophages. J. Exp. Med. **145:** 76.

14. NIEWIAROWSKI, S., E. REGOECZI & J. F. MUSTARD. 1972. Adhesion of fibroblasts to polymerizing fibrin and retraction of fibrin produced by fibroblasts. Proc. Soc. Exp. Biol. Med. **140:** 199.

15. AZZARONE, B., C. CURATOLO, G. CARLONI, M. B. DONATI, L. MORASCA & A. MACIEIRA-COELHO. 1981. Fibrin clot retraction in normal and transformed avian fibroblasts. J. Natl. Cancer Inst. **67:** 89.

16. GRINNELL, F., M. FELD & D. MINTER. 1980. Fibroblast adhesion to fibrinogen and fibrin substrata: Requirement for cold-insoluble globulin (plasma fibronectin). Cell **19:** 517.

17. POSTLETHWAITE, A. E., J. KESKI-OJA, G. BALIAN & A. H. KANG. 1981. Induction of fibroblast chemotaxis by fibronectin. J. Exp. Med. **153:** 494.

18. TSUKAMOTO, Y., W. E. HELSEL & S. M. WAHL. 1981. Macrophage production of fibronectin, a chemoattractant for fibroblasts. J. Imunol. **127:** 673.

19. RICKLES, F. R., J. H. HARDRIN, F. A. PITLICK, L. W. HOYER & M. E. CONRAD. 1973. Tissue factor activity in lymphocyte cultures from normal individuals and patients with hemophilia A. J. Clin. Invest. **52:** 1427.

20. LEVY, G. A., & T. S. EDGINGTON. 1980. Lymphocyte cooperation is required for amplification of macrophage procoagulant activity. J. Exp. Med. **151:** 1232.

21. GECZY, C. L. & K. E. HOPPER. 1981. A mechanism of migration inhibition in delayed-type hypersensitivity reactions. II. Lymphokines promote procoagulant activity of macrophages in vitro. J. Immunol. **126:** 1059.

22. GORDON, S., J. C. UNKELESS & Z. A. COHN. 1974. Induction of macrophage plasminogen activator by endotoxin stimulation and phagocytosis. Evidence for a two-stage process. J. Exp. Med. **140:** 185.

23. HOVI, T., D. MOSHER & A. VAHERI. 1977. Cultured human monocyte synthesize and secrete α_2-macroglobulin. J. Exp. Med. **145:** 1580.

24. JOHANSSON, S., K. RUBIN, M. HOOK, T. AHLGREN & R. SELJELID. 1979. In vivo biosynthesis of cold-insoluble globulin (fibronectin) by mouse peritoneal macrophages. FEBS Letters **105:** 313.

25. ALITALO, K., T. HOVI & A. VAHERI. 1980. Fibronectin is produced by human monocytes. J. Exp. Med. **151:** 602.

26. NELSON, D. S. & S. V. BOYDEN. 1963. The loss of macrophages from peritoneal exudates following the injection of antigens into guinea pigs with delayed-type hypersensitivity. Immunology **6:** 264.

27. NELSON, D. S. 1965. The effects of anticoagulants and other drugs on cellular and cutaneous reactions to antigen in guinea pigs with delayed-type hypersensitivity. Immunology **9:** 219.

28. NELSON, D. S. & R. J. NORTH. 1965. The fate of peritoneal macrophages after the injection of antigen into guinea pigs with delayed-type hypersensitivity. Lab. Invest. **14:** 89.

29. JOKAY, I., & E. KARCZAG. 1973. Thrombin-induced macrophage disappearance reaction in mice. Experientia (Basel) **29:** 334.
30. COLVIN, R. B. & H. F. DVORAK. 1975. Fibrinogen/fibrin on the surface of macrophages: detection, distribution, binding requirements and possible role in macrophage adherence phenomena. J. Exp. Med. **142:** 1377.
31. COLVIN, R. B., J. LANIGAN, R. A. F. CLARK, T. H. EBERT, E. VERDERBER & M. E. HAMMOND. 1979. Macrophage fibronectin (cold-insoluble globulin, LETS protein). Fed. Proc. **38:** 1408.
32. HOPPER, K. E., C. L. GECZY & W. A. DAVIES. 1981. A mechanism of migration inhibition in delayed-type hypersensitivity reactions. I. Fibrin deposition on the surface of elicited peritoneal macrophages *in vivo*. J. Immunol. **126:** 1052.
33. REMOLD, H. G., J. E. SHAW & J. R. DAVID. 1981. A macrophage surface component related to fibronectin is involved in the response to migration inhibitory factor. Cell Immunol. **58:** 175.
34. KRADIN, R., Y. ZHU, C. HALES & R. COLVIN. 1983. Alveolar macrophages localize fibronectin on their surface following exposure to hyperoxia. Amer. Rev. Resp. Dis. (Abst). In press.
35. JILEK, F. & H. HORMAN. 1978. Fibronectin (cold insoluble globulin). V. Mediators of fibrin monomer binding to macrophages. Hoppe Seyler's Z. Physiol. Chem. **359:** 1603.
36. BEVILACQUA, M. P., D. AMRANI, M. W. MOSESSON & C. BIANCO. 1981. Receptors for cold-insoluble globulin (plasma fibronectin) on human monocytes. J. Exp. Med. **153:** 42.
37. DORAN, J. E., A. R. MANSBERGER & A. C. REESE. 1980. Cold insoluble globulin-enhanced phagocytosis of gelatinized targets by macrophage monolayer: A model system. J. Reticuloendothel. Soc. **28:** 471.
38. GUDEWICZ, P. W., J. MOLNAV, M. Z. LAI, D. W. BEEZHOLD, G. E. SIEFRIUG, R. B. CREDE & L. LORAND. 1980. Fibronectin mediated uptake of gelatin-coated particles by peritoneal macrophages. J. Cell Biol. **87:** 427.
39. VAN DE WATER, L., S. SCHOEDER, E. B. CRENSHAW & R. O. HYNES. 1981. Phagocytosis of gelatin-latex particles by a murine macrophage line is dependent on fibronectin and heparin. J. Cell Biol. **90:** 32.
40. VILLIGER, B., D. G. KELLEY, W. ENGLEMAN, C. KUHN & J. A. MCDONALD. 1981. Human alveolar macrophage fibronectin: Synthesis, secretion and ultra-structural localization during gelatin-coated latex particle binding. J. Cell Biol. **90:** 711.
41. BLUMENSTOCK, F. A., T. M. SABA, E. ROCCARIO, E. CHO & J. E. KAPLAN. 1981. Opsonic fibronectin after trauma and particle injection determined by a peritoneal macrophage monolayer assay. J. Reticuloendothel. Soc. **30:** 61.
42. ROURKE, F. J., L. MILKS, M. REICHERF, M. W. MOSESSON & C. BIANCO. 1982. Monocyte receptors for fibronectin: Analysis by monoclonal antibodies. Fed. Proc. **41:** 380.
43. SCHREINER, G. F. & E. R. UNANUE. 1976. Membrane and cytoplasmic changes in B lymphocytes induced by ligand-surface immunoglobulin interaction. Adv. Immunol. **24:** 38.
44. MARTIN, B. M., M. A. GIMBRONE, G. R. MAJEAU, E. K. UNANUE & R. S. COTRAN. 1981. Monocyte/macrophage derived growth factor production: Modulation by cold-insoluble globulin and extracellular matrix. Circulation **64:** 214.
45. COLVIN, R. B. 1983. Roles of fibronectin in wound healing. *In* Fibronectin. D. F. Mosher, Ed. Academic Press. New York.

Discussion of this paper follows on page 634.

FIBRIN, FIBRONECTIN, AND MACROPHAGES

Celso Bianco

BINDING OF SOLUBLE FIBRIN TO MACROPHAGES

Laurence A. Sherman

FIBRINOGEN-FIBRIN INTERACTIONS WITH FIBROBLASTS AND MACROPHAGES

Robert B. Colvin

———————◆———————

DISCUSSION OF THE PAPERS

L. J. WANGH (*Brandeis University, Waltham, MA*): I would like to ask Dr. Ritchie whether, in view of what Dr. Bianco told us about the importance of the surface for activation of macrophages and their binding in the activation, the surface has an effect on the production of HSF?

D. RITCHIE (*University of Texas Medical Branch, Galveston*): One of the slides that I showed on the production of HSF by macrophages demonstrated that after about 3 hours in a plastic tissue culture dish the cells attach flatten out and production of HSF increases quite dramatically. Whether or not this is due to cell surface interaction is debatable.

S. NIEWIAROWSKI (*Temple University Health Sciences Center, Philadelphia, PA*): I would like to ask Dr. Sherman, who showed that soluble fibrin binds better to macrophages as compared to fibrinogen, whether changes in the structure are the cause of this or whether it is due to the fact that fibrin is present as oligomers, and thus binds more because of multiple binding sites?

L. A. SHERMAN: We have looked at monomeric fibrin and we still find increased binding; we do not find a 15- to 20-fold increase but rather 6- or 7-fold.

NIEWIAROWSKI: Is it possible to accept that macrophages and fibroblasts contain surface fibrinogen receptors for fibrinogen which bind fibrinogen in a saturated and reversible manner?

SHERMAN: In studies that we have done on other cells we find some fibrinogen binding but with macrophages it binds more. Furthermore, we do not find any large amounts of fibrin binding to cells like fibroblasts or epithelial cells.

R. B. COLVIN: In our studies with fibroblasts, fibrin monomers that were prepared by the method of Sherman, bound in greater amounts. However, we were hesitant to conclude that this represented greater avidity because of the point you raised of self-association. The binding seemed to be directly related to the cellular fibronectin because cells without fibronectin did not bind fibrinogen, nor did fibrinogen fraction I-9 bind very well. This fragment lacks the ability to bind to fibronectin in solution.

E. PLOW (*Research Institute of Scripps Clinic, La Jolla, CA*): The association or the co-association of fibronectin and fibrinogen or fibrin binding and trying to relate that which one serves as the binding protein for the other may be a little premature. For example, on the platelet you have receptors for both

634

fibrin and fibronectin when they are thrombin-stimulated. You even have a disease, thrombasthenia, where neither protein binds. Yet we have reasonably good evidence that (1) fibronectin is not serving as the fibrinogen receptor and (2) fibrinogen or fibrin is not the fibronectin receptor on the platelet.

COLVIN: I certainly agree and I hope I made that clear in the talk. I believe that Dr. Bianco finds that fibronectin does not bind in soluble form very well to monocytes/macrophages. We find the same thing with guinea pig macrophages. Yet his data clearly indicate that in solid phase, fibronectin binds to these cells. There are at least two possible explanations. One is that it is a low avidity receptor and requires multiple binding sites displayed best in the solid phase, alternatively, and perhaps more interestingly, there may be some sort of conformational change in the fibronectin consequent to its binding to a particle. The latter might be analogous to the situation with CIq in serum, which is activated by conformational changes in IgG, it might be a homeostatic mechanism that keeps us from opsinizing ourselves to death.

M. P. BEVILACQUA (*Brigham and Women's Hospital, Boston, MA*): I would like to ask Dr. Colvin if he could discuss the possibility that the fibronectin he was looking at was synthesized by the cells?

COLVIN: Of course, that is a possibility. We have looked at the nature of the fibronectin by using surface labeling, and it appears identical to the plasma form of fibronectin. I think it is possible that it is made by the macrophages; but, in fact, if you look at macrophages cultured for a period of time, in the absence of exogenous fibronectin, you see very little fibronectin on the surface. Thus, it appears that something else is required to display the fibronectin on the surface.

J. HAWIGER (*Vanderbilt University School of Medicine, Nashville, TN*): In terms of receptors, we have noticed that it is possible to reverse fibrinogen binding in the presence of large amounts of unlabeled material. However, we have never been able to reverse fibrin binding to any significant extent.

UNIDENTIFIED SPEAKER: I wonder whether the state of activation of macrophages has anything to do with the availability of their receptors? Platelets, by analogy, do not bind fibrinogen when they are not activated and I wonder whether macrophages may undergo the same process of receptor exposure and regulation?

C. BIANCO: I was thinking about this same thing during the presentations on platelets. To my knowledge, this has not been described for monocytes or macrophages. What has been very well described is that we can recover macrophages at different stages of differentiation expressing different activities. If we recover macrophages that have been activated or induced by endotoxin or by thioglycollate, they will bind better to those surfaces and behave as if there were an increase in the number of fibronectin receptors.

J. ESTRADA (*Houston, TX*): Dr. Colvin, regarding the fibroblasts, you said that just the monomers of fibrin would bind and that the fibronectin was an essential step for this binding. Have you studied any fibrin binding under any other circumstances, as in tumors?

COLVIN: We have studied this to a limited extent using tissue culture lines and tumor lines, most of which lack fibronectin. Those that lacked it also did not bind fibrinogen. Those cells that had very sparse amounts of fibronectin such as SV-40 transformed human fibroblasts also had very sparse fibrin binding and the pattern was the same.

CONCLUDING REMARKS

Russell F. Doolittle

Department of Chemistry
University of California, San Diego
La Jolla, California 92037

I feel that this conference on the molecular biology of fibrinogen can be summarized by saying that the molecular part is well along and the biology is just beginning. Studies that have established the sequence of the fibrinogen molecule and its overall covalent structure, as well as many fundamental physical chemistry studies, are essentially complete at this point. They have been enormously important for many aspects of our understanding. Many other aspects depend greatly on this accomplishment; it was necessary in order to assign precise locations of events, for example.

Beyond the chemical work, the electron microscopy of fibrinogen and fibrin and the x-ray crystallography studies have revealed tremendous detail which a few years ago we probably could not have expected.

There is little doubt that fibrinogen is a linear array of tethered domains. Three of these are firmly established, two of which exist as duplets. The data presented by Drs. Cohen and Weisel suggest that there may be as many as seven discrete domains; Dr. Erickson has shown that there may be a resolvable mass corresponding to the elusive α chain COOH-terminals.

The degradation products of the molecule have been well characterized by a number of investigators, and these have provided us with a tremendous base upon which to design other experiments. The major heterogeneities of the molecule have been outlined, although their functional meaning remains mysterious.

Some of the cellular work, such as the identification of receptors on platelets, has been well defined, and we can expect that in the near future we will be hearing the specific details of the interaction between these receptors and the fibrinogen molecule.

The mechanics of fibrin polymerization have received considerable attention at this conference. The various steps in the process are now being singled out. Peptides, such as Gly-Pro-Arg-Pro, are able to arrest the process at defined points. Definitive results can soon be expected, also, in areas concerning the kinetics and the mechanism of fibrinopeptide release, as well as all the subsequent associative events. It is exciting to learn about the work that is being done on the complementary binding sites in the assembly process. This is something which certainly will require much work, however, before we can feel that we understand what happens.

We also hope that the x-ray crystallographic work that Dr. Cohen has been carrying out will soon give us even greater details about the fine structure of fibrinogen and the assembled fibrin clot. Exciting work on the interaction of fibrinogen with other proteins, including factor XIII, and plasminogen, and the interaction between streptokinase and plasminogen mediated in the presence of fibrinogen are certainly areas that should bring new information. In this regard I would like to point out that the study of abnormal fibrinogens has been one that is beginning to bear fruit; I would like to single out the studies reported by Dr. Henschen in which structure-function correlations are revealed, and the reports by the Sorias, relating to a fibrinogen that does not bind effectively to plasminogen and is associated with severe thromboembolic disease.

These are important leads. The description of the γ and the γ' chains is very interesting, although we still have no idea what function these may serve. Similarly, we know the detailed structure of the carbohydrate on the fibrinogen molecule, but we really do not have any genuine insights into its role. Studies relating to the regulation of fibrinogen synthesis, and the expression of the fibrinogen genes, particularly the work reported by Drs. Chung and Crabtree, is as overwhelming as it is impressive. It is enough to make an amino acid sequencer like myself want to retire from the trade, or better, to take up DNA sequencing. In the end, DNA sequencing may show us the details of how this remarkable molecule evolved.

ROLE OF MICROTUBULES IN FIBRINOGEN
SECRETION BY RAT LIVER CELLS

M. Maurice † and G. Feldmann

† *Unité de Recherches de Physiopathologie Hépatique*
INSERM U 24, Hôpital Beaujon
92118 Clichy Cedex France

Laboratoire d'Histologie, Embryologie, Cytogénétique
Faculté de Médecine Xavier-Bichat
75018 Paris, France

It is now well established that hepatocytes are the liver cells that produce fibrinogen.[1] As is the case for every secretory protein,[2] the polypeptide chains of fibrinogen, after their synthesis by the ribosomes of the rough endoplasmic reticulum, pass into the lumen of this organelle, then into the smooth endoplasmic reticulum, Golgi apparatus and secretory vesicles before reaching the plasma membrane of the hepatocyte.[3] Numerous and very different biochemical and cellular events occur during this process. Among them, the role of microtubules is still debated. Microtubules are elongated organelles, with an approximate diameter of 25 nm and a length of several microns.[4] The wall of the tubules is mainly composed of a specific protein, tubulin, of 110 000 daltons.[4] In liver cells, tubulin exists in two main forms, one called free tubulin and the other polymerized tubulin, the latter expressed as microtubules. The two forms are in equilibrium and their percentages in the normal rat liver is 60 and 40%, respectively, as recently reported.[5] Microtubules are randomly distributed throughout the hepatocyte cytoplasm; they are often clustered around the Golgi apparatus.[6] Up to now, no clear demonstration of morphological links between microtubules and endoplasmic reticulum, Golgi apparatus or secretory vesicles have been reported, although microtubules are sometimes very close to these organelles. Some electron micrographs also suggest that microtubules could reach the plasma membrane of the hepatocytes.[6] In 1973, Le Marchand and coworkers [7] observed that the perfusion of mouse livers with colchicine induced an inhibition of lipoprotein secretion without changing their hepatic synthesis. They proposed that the accumulation of lipoproteins in hepatocytes was due to the disruption of microtubules by colchicine. This disruption was explained by specific binding of the drug to free tubulin, impeding the formation of polymerized tubulin.[4] Two years later, we observed that *in vivo* administration of colchicine to rats induced an accumulation of fibrinogen in the hepatocytes [8]; the use of peroxidase-labeled antibodies against rat fibrinogen allowed us to observe that in this condition the protein accumulated in large amounts in every organelle engaged in the secretory process of fibrinogen. The role of microtubules was also suggested for other plasma proteins poduced by the liver,[9-11] with however some discrepant results [11] possibly owing to the fact that, colchicine in high doses [11, 12] can also induce other cellular effects such as necrosis of hepatocytes.

We have examined the role of microtubules in this investigation by studying a situation in which the secretion of fibrinogen is greatly enhanced, the acute inflammatory reaction. This reaction was induced by a single subcutaneous injection of turpentine to adult rats. In addition to the study of the different forms of tubulin with a [³H]colchicine binding assay,[5] we used ultrastructural morphometric methods [6] to measure the volume density of microtubules in the different cytoplasmic areas involved in fibrinogen secretion, the endoplasmic reticulum area, Golgi area, and sinusoidal area, the last area being the part of the hepatocyte located just beneath the plasma membrane. We found that 24 hours after the beginning of the inflammatory reaction, fibrinogen plasma concentration increased by 280%. At that time, liver polymerized tubulin rose by 58% compared to control rats whereas free tubulin remained unchanged. In the sinusoidal area the volume density of microtubules increased by 48% compared to control rats; it increased also around the Golgi apparatus because of the hypertrophy of this organelle that occurred during the reaction; no change, however, was observed in the endoplasmic reticulum area. The above observations demonstrate that when fibrinogen secretion by the liver cells is enhanced, there is a change in the equilibrium in the different tubulin pools with additional microtubules formed in the hepatocytes. These results bring new arguments favoring the view that microtubules play a role in fibrinogen secretion.

REFERENCES

1. FELDMANN, G. 1979. *In* Progress in Liver Diseases, Vol. 6. H. Popper & F. Schaffner, Eds.: 23. Grune and Stratton. New York.
2. PALADE, G. 1975. Science **189:** 347.
3. COURTOY, P., C. LOMBART, G. FELDMANN, N. MOGUILEVSKY & E. ROGIER. 1981. Lab. Invest. **44:** 105.
4. DUSTIN, P. 1978. Microtubules. Springer-Verlag. Berlin.
5. MAURICE, M., G. FELDMANN, B. BELLON & P. DRUET. 1980. Biochem. Biophys. Res. Commun. **97:** 355.
6. MAURICE, M. & G. FELDMANN. 1982. Exp. Mol. Pathol. **36:** 193.
7. LE MARCHAND, Y., A. SINGH, F. ASSIMACOPOULOS-JEANNET, L. ORCI, C. ROUILLER & B. JEANRENAUD. 1973. J. Biol. Chem. **248:** 6862.
8. FELDMANN, G., M. MAURICE, C. SAPIN & J. P. BENHAMOU. 1975. J. Cell Biol. **67:** 237.
9. LE MARCHAND, Y., C. PATZELT, F. ASSIMACOPOULOS-JEANNET, E. G. LOTEN & B. JEANRENAUD. 1974. J. Clin. Invest. **53:** 1512.
10. REDMAN, C. M., D. BANERJEE, K. HOWELL & G. E. PALADE. 1975. J. Cell Biol. **66:** 42.
11. REDMAN, C. M., D. BANERJEE, C. MANNING, C. Y. HUANG & K. GREEN. 1978. J. Cell Biol. **77:** 400.
12. DUBIN, M., M. MAURICE, G. FELDMANN & S. ERLINGER. 1980. Gastroenterology **79:** 646.

HUMAN FETAL FIBRINOGEN: ITS CHARACTERISTICS OF DELAYED FIBRIN FORMATION, HIGH SIALIC ACID AND AP PEPTIDE CONTENT ARE MORE MARKED IN PRE-TERM THAN IN TERM SAMPLES

Dennis K. Galanakis,[†] Jose Martinez,[‡] Cahir McDevitt, and Frederick Miller

Department of Pathology
State University of New York
Health Science Center,
Stony Brook, New York 11794

‡ *Cardeza Foundation for Hematologic Research*
Department of Medicine
Jefferson Medical College of Thomas Jefferson University
Philadelphia, Pennsylvania 19107

Human fetal fibrinogen, when compared with the adult form, displays prolonged clotting times [1-4] and delayed fibrin aggregation [4,5] and forms a more transparent,[4-6] electronmicroscopically distinct fibrin clot.[7,8] The present studies compare samples from (full) term and pre-term infants, describe its high fibrinopeptide AP and sialic acid content, and examine the role of its increased sialic acid on its fibrin formation characteristics. Fibrinogen, fraction I-2 ($\geq$97% clottable) was obtained from pooled adult, full-term (FT), and pre-term (P, 28–35 weeks gestation) infants at birth.[4] Sialic acid measurements,[9] performed on 6 P, 11 FT, and 13 adult samples, disclosed mean values (range) of 9.0(7.8–11.3), 7.9(5.6–14.0), 6.6(5.4–7.8) residues per mole of P, FT, and adult fibrinogen, respectively. Similarly, thrombin times [4] of single adult, FT, and P preparations were 31 ± 1, 52 ± 1, and 74 ± 3 seconds, respectively, indicating that P is more representative (than FT) of the fetal form of fibrinogen. Efforts to localize the increased sialic acid were made by measurements of reduced ^{3}H-labeled [10] samples and by direct assays [9] of isolated S-carboxymethyl chains.[11] Both assays disclosed similar Bβ:γ chain ratios ($\sim$1.5) in fetal and adult samples. Although some of the fetal Bβ chain preparations displayed somewhat higher sialic acid content than their adult counterparts, there were minor but appreciable sialic acid losses during the isolation [11] procedure, limiting the interpretation of these measurements; nevertheless, the results suggested that both fetal Bβ and γ chains contained increased sialic acid.

In order to assess the role of sialic acid on its functional behavior, fetal fibrinogen was exposed to *Vibrio cholerae* (Schwarz/Mann, Orangeburg, NY) neuraminidase,[12] and this reduced its sialic acid content to (a range of 4.5–6.3 res./mole, similar to) that of adult fibrinogen (see above). This treatment (in three preparations tested, and termed asialo-fibrinogen) corrected the P fibrinogen thrombin [4] clotting times (from 22 ± 1 seconds) to those of adult samples (15 ± 1 seconds), but the increased transparency of the asialo-fetal fibrin clots

* This work was supported in part by United States Public Health grant HL 27196.
† To whom to address correspondence.

(i.e. its decreased absorbance at 350 nm) remained uncorrected (not shown); this suggested that the distinctly limited fibril size [7, 8] in fetal fibrin is either not attributable to its increased sialic acid, particularly since adult asialo-fibrin clots do not display increased transparency,[13] or that other sialic acid residue(s), remaining on the molecule accounted for this effect. Also consistent with this conclusion were results of mixing experiments in which adult fibrin formed a more transparent clot in the presence of equal amounts of asialo-P fibrinogen (FIG. 1). Similar results were obtained with mixtures of adult and fetal fibrin, in agreement with studies on bovine fetal fibrin.[5] Thus, this inhibitory effect (on adult fibrin) was not dependent on prior cleavage of fetal fibrinopeptides.

Fetal fibrinopeptides were examined by use of a high performance liquid chromatographic (HPLC) procedure recently described by Kehl *et al.*[14] Comparison (FIG. 2) of single samples disclosed high AP peptide content in FT and P fibrinogen. The AP:A + AY ratios were 0.6, 1.3, and 1.9, for adult, FT, and P samples, respectively, consistent with the reported high phosphorus content of fetal fibrinogen [1] (which does not account for its prolonged clotting times [15]). The increased AP peptide in fetal fibrinogen as reflected in the AP:A + AY ratios remained relatively constant during the time course of

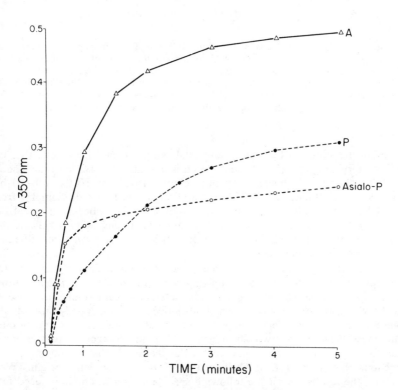

FIGURE 1. Adult fibrin aggregation in the presence of P or asialo-P fibrinogen. Fibrin in 0.25% acetic acide was diluted with approximately 30 volumes of 0.135M NaCl, 0.025M Tris-PO₄ pH 7, containing fibrinogen, 0.5 mg/ml. Final fibrin concentration was 0.5 mg/ml.

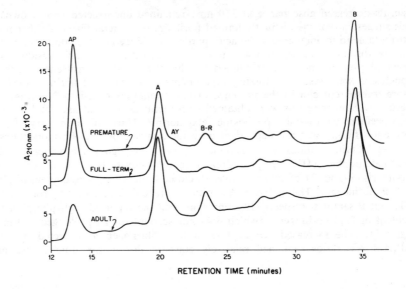

FIGURE 2. HPLC comparison of adult, FT and P fibrinopeptides released by incubating 5 mg of fibrinogen and 2 U of human thrombin/ml of buffer (0.15 M ammonium acetate, pH 8.5) for 2 hours, at room temperature. Approximately ⅓ of the amounts released from each preparation were applied to a (Whatman ODS–2) column; the separation procedure was that described by Kehl *et al.*,[13] using a Varian model 5020 chromatograph equipped with model 9176 strip chart recorder.

thrombin digestion (in both adult and fetal samples), and it suggests that the dephosphorylating capacity of fetal plasma is diminished, relative to that shown in adult plasma.[17] The higher AP peptide content (in P than FT) was invariably demonstrable in numerous (>10) preparations of each tested. Samples from each HPLC peak subjected to amino acid analysis (using a Beckman model 121 analyzer) disclosed no difference in the amino acid compositions between the respective fetal and adult fibrinopeptides, suggesting there was no difference in their respective amino acid sequences.

The foregoing results demonstrated that P is more characteristic of the fetal form than is FT fibrinogen, implying a mixture of adult and fetal fibrinogen in full-term blood. Although its characteristics seem similar to those of the liver disease-associated abnormal fibrinogen [12, 16] the persistence of clot transparency in asialo-fetal samples suggests structural (i.e., carbohydrate) differences between the two fibrinogens. Also, the foregoing analyses imply that the structure(s) (i.e., oligosaccharide chain(s)) that accounts for the functional behavior of fetal fibrinogen is of post-translational origin. What is more, the results permit the following speculation on thrombin-induced fetal fibrinogen/fibrin conversion. The oligosaccharide side chain(s) that contains the excess sialic acid residues and by steric or other hindrance effect limits lateral association between fibrin monomers, is likely to account for the distinctly finer febril network [7, 8] in fetal fibrin.

ACKNOWLEDGMENTS

We are grateful to Drs. R. J. Bachvaroff and F. Rappaport for use of their Chem. Research model 2000 chromatograph, to Mr. Barry Sunray for technical and to Mrs. Susan Volkes for secretarial assistance, and to Audio-Visual Services for illustrations and photography.

REFERENCES

1. WITT, I. & H. MULLER. 1970. Biochim. Biophys. Acta **221:** 402–404.
2. WITT, I., H. MULLER & W. KUNZER. 1969. Thromb. Diath. Haemorrh. **22:** 101–109.
3. MILLS, D. A. & S. KARPATKIN. 1972. Biochim. Biophys. Acta **285:** 398–403.
4. GALANAKIS, D. K. & M. W. MOSESSON. 1976. Blood **48:** 109–117.
5. GUILLIN, M. C. & D. MENACHÈ. 1973. Thromb. Res. **3:** 117–135.
6. BURSTEIN, M., S. LEWI & P. WALTER. 1954. Sang **25:** 102–107.
7. TESCH, R., R. TROLP & I. WITT. 1979. Thromb. Res. **16:** 239–243.
8. MÜLLER, M., W. BURCHARD & I. WITT. 1981. Thromb. Res. **24:** 339–346.
9. WARREN, L. 1959. J. Biol. Chem. **234:** 1971–1975.
10. MARTINEZ, J., K. MacDONALD & J. E. PALASCAK. 1983. Blood. In press.
11. GALANAKIS, D. K., M. W. MOSESSON & N. E. STATHAKIS. 1978. J. Lab. Clin. Med. **92:** 376–386.
12. MARTINEZ, J., J. E. PALASCAK & D. KWASNICK. 1978. J. Clin. Invest. **61:** 535–538.
13. MARTINEZ, J., J. PALASCAK & C. PETERS. 1977. J. Lab. Clin. Med. **89:** 367–377.
14. KEHL M., F. LOTTSPEICH & A. HENSCHEN. 1981. Hoppe-Seyler's Z. Physiol. Chem. **362:** 1661–1664.
15. WITT, I. & K. HASLER. 1972. Biochim. Biophys. Acta **271:** 357–362.
16. GRALNICK, H. R., H. GIVELBER & E. ABRAMS. 1978. N. Engl. J. Med. **299:** 221–225.
17. SHAINOFF, J. R., P. G. DYMENT, G. C. HOFFMAN & F. M. BUMPUS. 1972. Circulation 46(Suppl. II): 52 (abstract).

FIBRINOGEN LOUISVILLE: AN Aα16 ARG→HIS DEFECT THAT FORMS NO HYBRID MOLECULES IN HETEROZYGOUS INDIVIDUALS AND INHIBITS AGGREGATION OF NORMAL FIBRIN MONOMERS *

Dennis K. Galanakis,† Agnes Henschen,‡ Marie Keeling,§
Maria Kehl,‡ Rita Dismore,§ and Ellinor I. Peerschke

† Department of Pathology
SUNY Health Science Center
Stony Brook, New York 11794

‡ Max-Planck Institute for Biochemistry
Munich, Federal Republic of Germany

§ University of Louisville
Norton Children's Hospital
Louisville, Kentucky 40202

Fibrinogen-fibrin conversion is a complex process that remains incompletely understood. Useful information relating to this process has been obtained by studies of inherited abnormalities of fibrinogen of which the structural defect has been identified.[1-3] The present studies describe the third reported Aα16 arginine → histidine (Arg → His) heterozygous substitution defect, and detail certain unreported characteristics, which include inhibition of aggregation of normal fibrin monomers and the absence of hybrid molecules. The propositus was a 54-year-old male who had a life-long history of easy bruising and had bled excessively, requiring transfusion, during gastrectomy; affected family members had no history of undue bleeding or thrombotic disorder. The plasma abnormality was prolonged thrombin time, low coagulable and normal immuno-assayable fibrinogen in six of nine family members tested (FIG. 1). Propositus (fraction I-2) fibrinogen displayed prolonged thrombin times, and in analyses,[4] in which clot formation was monitored turbidimetrically (350 nm), thrombin-induced clot formation was delayed, while fibrin (prepared by use of thrombin) displayed normal fibrin aggregation. Propositus fibrinogen was 97% and 40% clottable with human thrombin (a gift from D. J. Fenton) and reptilase (Abbott Laboratories, No. Chicago, IL), respectively, suggesting an abnormal peptide A. These characteristics are in general agreement with those reported for fibrinogens Petoskey [5, 6] and Manchester.[7]

A high performance liquid chromatographic (HPLC) procedure [8] was employed to examine fibrinopeptides released by thrombin from propositus fibrinogen. This procedure, permits separation and measurement of peptides AP, A, AY, des-Arg B, and B. The presence of the normal and an abnormal retention peak for each of the peptide A forms was shown (FIG. 2). Samples from the abnormal (*, FIG. 2) HPLC peaks, subjected to amino acid analysis (Biotronik analyzer, W. Germany) showed that the normal peptide A lacked

* This work was supported in part by United States Public Health grant HL 27196, and by a grant from Deutche Forschungsgemeinschaft.
† To whom to address correspondence.

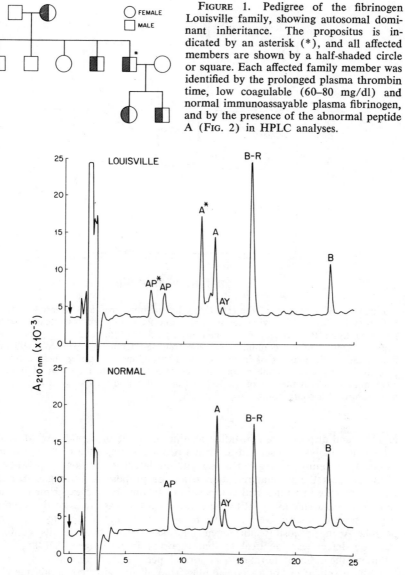

FIGURE 1. Pedigree of the fibrinogen Louisville family, showing autosomal dominant inheritance. The propositus is indicated by an asterisk (*), and all affected members are shown by a half-shaded circle or square. Each affected family member was identified by the prolonged plasma thrombin time, low coagulable (60–80 mg/dl) and normal immunoassayable plasma fibrinogen, and by the presence of the abnormal peptide A (FIG. 2) in HPLC analyses.

FIGURE 2. High pressure liquid chromatographic retention profile [8] of peptides A and B released from fibrinogen, (Louisville or normal, 5 mg/ml) in buffer (0.15 M ammonium acetate, pH 8.5) by bovine thrombin (2 U/ml) at room temperature. The amounts of A and B peptides applied to the column were those released from approximately 400 μg of fibrinogen. B-R:des-arginine peptide B. The various fibrinopeptides were identified by their amino acid sequence using Edman degradation and sequenator analyses.[9] An asterisk(*) indicates abnormal peptides, and the vertical arrows denote injection of sample. A linear 6–14% acetonitrile gradient, pH 6, was employed using a reverse phase column and equipment described elsewhere.[8]

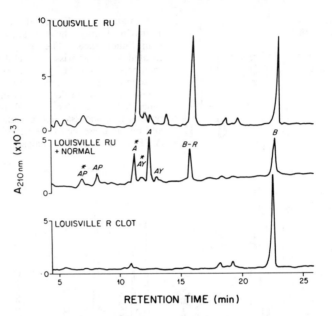

FIGURE 3. Comparison of the HPLC profiles of fibrinopeptides released by thrombin from fibrinogen Louisville reptilase-clottable(RC) and reptilase-unclottable(RU) fractions using the procedure outlined in legend to FIGURE 2. The amounts of peptides applied to the column were those released from approximately 600 μg of fibrinogen at 120 minutes of incubation with thrombin; (i.e. under these conditions, described elsewhere,[8] peptide B was released at a higher rate than that of Louisville A(*), not shown). A mixture of the RU and normal peptides (middle chromatograph) is also shown for comparison.

arginine and displayed one histidine residue per mole of peptide A. Moreover, sequenator analysis [9] disclosed a normal peptide A sequence up to position 15 (valine) indicating an Arg → His substitution defect in position 16, (TABLE 1). Material from the normal retention peaks of propositus fibrinogen displayed no abnormality in amino acid composition. In time course experiments, monitored by measurements of HPLC retention peaks, release rates of the abnormal peptides were much lower, relative to a normal control and to the normal peptide A in propositus fibrinogen (not shown). These results confirmed that the defect was identical to that reported in fibrinogen Petoskey.[5, 6] In addition, the ratio of normal to abnormal peptide A (approximately 1) was virtually identical in fibrinogen samples from all five affected family members. Also, the ratio of AP to A + AY peptides was identical in all affected and unaffected family members.

Studies were performed on the abnormal (reptilase unclottable:RU) form of propositus fibrinogen (which was isolated by removal of the reptilase-clottable (RC) fraction). RU samples when exposed to thrombin [8] released only abnormal peptide A and normal peptide B (FIG. 3). By contrast, no peptide A remained on RC samples. These results are consistent with the absence of hybrid molecules in both RC and RU samples, and in general

TABLE 1

AMINO ACID SEQUENCE OF LOUISVILLE PEPTIDE A *

	1	2	3	4	5	6	7	8	9	10	11	12	13	14	15	16
Louisville A	Ala	Asp	Ser	Gly	Glu	Gly	Asp	Phe	Leu	Ala	Glu	Gly	Gly	Gly	Val	His
	→	→	→	→	→	→	→	→	→	→	→	→	→	→	→	(→)
Normal A	Ala	Asp	Ser	Gly	Glu	Gly	Asp	Phe	Leu	Ala	Glu	Gly	Gly	Gly	Val	Arg
	→	→	→	→	→	→	→	→	→	→	→	→	→	→	→	(→)

* Amino acid sequence of Louisville (A*) and normal peptide A, determined [9] on samples obtained from the respective HPLC peaks shown in FIGURE 2. Sequenator analyses were carried out to position 15 (Val). The residue in position 16 was determined by (the absence of Arg and the presence of one His residue/mole peptide A in) analyses of amino acid content.

agreement with the biphasic fibrin aggregation obtained when fibrinogen Petoskey was treated with reptilase and then with thrombin.[5] Additional support for this interpretation was provided by limited NH_2-terminal sequence analyses [9] showing no Gly-Pro-Arg sequence (i.e., no fibrin) in RU in con-

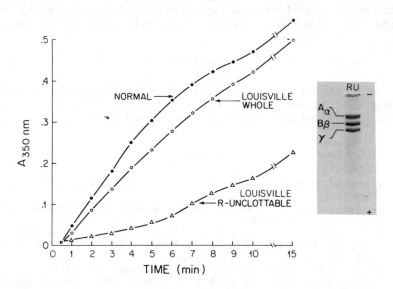

FIGURE 4. Fibrin aggregation in the presence of normal I–2 (●———●) and Louisville I–2 (○———○) or RU (△———△) fibrinogen fractions. Fibrin: fibrinogen ratio was 1:2.5. Fibrin in 0.25% acetic acid was added to 20–30 volumes of 0.05 M Tris HCl, .135 M NaCl, pH 7.4, containing fibrinogen and CaCl (20 mM). Final fibrin concentration was 200 μg/ml. The polyacrylamide gel (on the right) is that of the RU sample following its reduction and electrophoresis,[4] showing no appreciable loss of intact α chains. RU fraction was prepared by synerizing and removing the repeatedly washed clot (RC) formed by adding 25 μl reptilase (Abbott Laboratories, N. Chicago, IL.) to 5 mg fibrinogen (1–2) in 1 ml of buffer (0.15 M NaCl-.01 M Tris HCl, pH 7.4) containing 500 units Kunitz pancreatic trypsin inhibitor (Trasylol, FBA Pharmaceuticals, New York, NY).

trast to RC samples. Trace amounts of A peptide NH_2-terminal sequences were found in some RC samples (consistent with their HPLC profile, FIG. 3) and suggesting that not all normal peptide A had been washed from the clot. These results provided evidence for the conclusion that heterodimeric fibrinogen molecules (having one normal and one [Louisville] abnormal Aα chain) do not form during the subunit assembly of fibrinogen.

In other functional analyses, thrombin-prepared propositus (I-2) fibrin [4] displayed normal fibrin aggregation and the possible effect of the RU fibrinogen fraction on the aggregation of normal fibrin (prepared by use of thrombin, and termed T-fibrin) was examined (FIG. 4). A concentration-dependent inhibition of aggregation of normal fibrin was observed, relative to the normal fibrinogen control. This inhibition by the RU fraction, was appreciable at 1.5:1 RU:fibrin ratio and became more marked with higher ratios, whereas only modest inhibition could be shown with the parent (propositus I-2 fibrinogen, used in the same ratio. Not shown, was the much less pronounced RU inhibition on normal fibrin prepared by use of reptilase (i.e. R-fibrin). That is, a much higher RU:normal (R-fibrin) ratio was required to exert a similar extend of aggregation inhibition shown with the 2.5:1 RU:T-fibrin ratio. Moreover, the presence of $CaCl_2$ (20 mM) in the buffer did not correct this inhibitory effect of RU fibrinogen on normal fibrin. In platelet aggregation experiments,[10] Louisville I-2 as well as its RU fraction displayed normal support of ADP-induced aggregation of (aspirin-treated) gel-sieved platelets.

ACKNOWLEDGMENTS

We are grateful to Mr. Barry Sunray, Miss Christina Jonsson, and Miss Barbel Heinz for technical and to Mrs. Susan Volkes for secretarial assistance, and to Audio-Visual Services for illustrations and photography.

REFERENCES

1. BLOMBÄCK, M., B. BLOMBÄCK, E. F. MAMMEN & A. S. PRASAD. 1968. Nature **218:** 134–137.
2. FINLAYSON, J. W., L. A. REAMER, M. W. MOSESSON & D. MENACHE. 1980. Thromb. Res. **17:** 577–579.
3. SOUTHAN, C., A. HENSCHEN & F. LOTTSPEICH. 1982. In Fibrinogen–Recent Biochemical and Medical Aspects. A. Henschen, H. Graeff & F. Lottspeich, Eds.: 153–166. Walter de Gruyter. Berlin.
4. GALANAKIS, D. K. & M. W. MOSESSON. 1976. Blood **48:** 109–118.
5. HIGGINS, D. L., J. A. PENNER & J. A. SHAFER. 1981. Thromb. Res. **23:** 491–504.
6. HIGGINS, D. L. & J. A. SHAFER. 1981. J. Biol. Chem. **256:** 12013–12017.
7. HENSCHEN, A., et al. 1982. Fibrinogen Manchester. Submitted for publication.
8. KEHL, M., F. LOTTSPEICH & A. HENSCHEN. 1981. Hoppe-Seyler's Z. Physiol. Chem. **362:** 1661–1664.
9. EDMAN, P. & A. HENSCHEN. 1975. In Protein Sequence Determinations. S. B. Needleman, Ed. 2nd edit.: 232–279. Springer-Verlag. Berlin.
10. PEERSCHKE, E. I., M. B. ZUCKER, R. A. GRANT, J. J. EGAN & M. M. JOHNSON. 1980. Blood **55:** 841–847.

XENOPUS FIBRINOGEN: CHARACTERIZATION OF SUBUNITS, POST-TRANSLATIONAL MODIFICATIONS, *IN VITRO* TRANSLATION, AND GLUCOCORTICOID-REGULATED SYNTHESIS *

Lené J. Holland, John W. Weisel, and Lawrence J. Wangh

Department of Biology
Brandeis University
Waltham, Massachusetts 02254

Xenopus fibrinogen has a unique structure that distinguishes it from all known mammalian fibrinogens. Frog plasma fibrinogen was compared with human and bovine plasma fibrinogens. The subunits of each protein were identified by digestion with three enzymes: Bβ subunits with Venzyme, Aα subunits with batroxobin, both Aα and Bβ subunits with thrombin. This analysis revealed that *Xenopus* plasma fibrinogen is resolved into four polypeptides: the Bβ subunit (63K daltons), the Aα subunit (two forms: 59K and 55K daltons), and the γ subunit (52K daltons). The chains of *Xenopus* fibrin are: α (57K and 53K daltons), β (56.5K daltons), and γ (52K daltons). The subunits of radioactive fibrinogen secreted by cultured liver cells comigrate with those of plasma fibrinogen. Only the 59K dalton form of the Aα subunit is present in secreted fibrinogen, indicating that the 55K dalton form in the plasma arises by removal of a COOH-terminal peptide. The frog A and B fibrinopeptides are calculated to be 2K and 6.5 K daltons, respectively.

Thus, frog fibrinogen differs from both human and bovine fibrinogens in two respects. First, the *Xenopus* Bβ subunit has a higher molecular weight than the Aα subunit. In contrast, in both mammals the Aα subunit is larger than the Bβ subunit. Second, the *Xenopus* B fibrinopeptide has a calculated molecular weight of 6.5K daltons. The A and B fibrinopeptides of human and bovine fibrinogens are approximately 2K daltons.

Fibrinogen molecules that lack carbohydrate groups are secreted by *Xenopus* liver cells cultured in the presence of tunicamycin. Comparison of these molecules to intact fibrinogen molecules revealed that the Aα subunit is not glycosylated, the Bβ subunit is glycosylated at two sites, one of which is within the B fibrinopeptide, and the γ subunit is glycosylated at one site. The nonglycosylated B fibrinopeptide is 4K daltons while the A fibrinopeptide is 2K daltons. Thus, the unmodified B fibrinopeptide has approximately twice the number of amino acids as the A. This conclusion is in agreement with amino acid composition data for frog fibrinopeptides.[1] The large number of amino acids in the *Xenopus* B fibrinopeptide and its modification by glycosylation are features that are also found in the B fibrinopeptide of the lamprey.[2]

Each subunit of fibrinogen arises from a precursor polypeptide. These precursors were identified by thrombin and batroxobin digestion of *in vitro* translation products synthesized from total *Xenopus* liver mRNA. In contrast to the subunits of secreted fibrinogen, pre-Aα has a higher molecular

* This research was supported by National Science Foundation grants PCM–77–21578 and PCM–79–23483 to L.J.W. and by National Institutes of Health grant AM–17346 to C. Cohen.

weight than pre-Bβ. Secreted fibrinogen, nonglycosylated fibrinogen, *in vitro* translated precursors, and the thrombin generated polypeptides of each of these forms of fibrinogen were analyzed in a single SDS gel. This analysis demonstrated that: (1) the precursor of each fibrinogen polypeptide includes a signal peptide that is removed during the process of protein secretion; (2) the Bβ subunit is modified by an additional (unidentified) group located within the β portion of the polypeptide.

Fibrinogen synthesis in frogs, as in other animals, increases in response to physiological stress. Purified *Xenopus* liver parenchymal cells maintained in a defined, serum free, medium for several weeks cease the production of fibrinogen. Fibrinogen synthesis is sustained or reinduced when dexamethasone, a synthetic glucocorticoid, is added to the medium. We are using this cell culture system to investigate the molecular mechanisms of glucocorticoid regulation of fibrinogen synthesis.

REFERENCES

1. GLADNER, J. A. 1968. *In* Fibrinogen. K. Laki, Ed. Marcel Dekker. New York.
2. COTTRELL, B. A. & R. F. DOOLITTLE. 1976. Biochem. Biophys. Acta **453:** 426–438.

THE COMPLETE AMINO ACID SEQUENCE
OF COAGULOGEN ISOLATED FROM
LIMULUS POLYPHEMUS AMEBOCYTES

Toshiyuki Miyata, Masuyo Hiranaga, Mariko Umezu,
and Sadaaki Iwanaga

Department of Biology, Faculty of Science
Kyushu University
33, Fukuoka-812, Japan

A clottable protein, coagulogen, isolated from the amebocyte lysate of *Limulus polyphemus* consisted of a single basic polypeptide chain [1-4] and it formed a gel by the action of a clotting enzyme partially purified from the lysate (FIG. 1). The gelation involved limited proteolysis at the position of the Arg-18-Lys-19 and the Arg-46-Gly-47 located in the NH$_2$-terminal portion of *Limulus* coagulogen, liberating peptide C from its inner portion.[5] The resulting gel protein consisted of two chains of A and B, bridged by two disulfide linkages. The structural studies of the B chain (129 residues) and the previously established amino acid sequences of the A chain (18 residues) and peptide C (28 residues) made it now possible to establish the whole sequence of *Limulus* coagulogen. The coagulogen contained a total of 175 amino acid residues with a NH$_2$-terminal glycine and COOH-terminal serine.[6] The molecular weight was calculated to be 19,674. It contained 16 half-cystines in disulfide linkages and the 5 residues were located in clusters at the COOH-terminal 14-residue region (positions 162-175), as shown in FIG. 2.

The overall amino acid sequence of *Limulus* coagulogen was very close to the previously established coagulogen of *Tachypleus*,[7] having 70% sequence homology. The 16 half-cystines of these coagulogens were in the same linear position, suggesting a very similar conformation. Moreover, the COOH-terminal tripeptide regions of A chain and peptide C, Leu-Gly-Arg and Ser-Gly-Arg, which seems to interact with a clotting enzyme to liberate peptide C, were completely conserved in both coagulogens (FIG. 3).

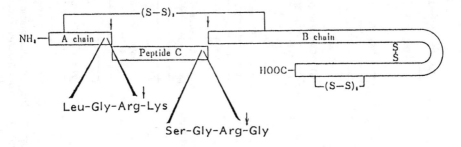

FIGURE 1. Alignment of the peptide segments in *Tachypleus* coagulogen. The arrows indicate the sites cleaved by a clotting enzyme.

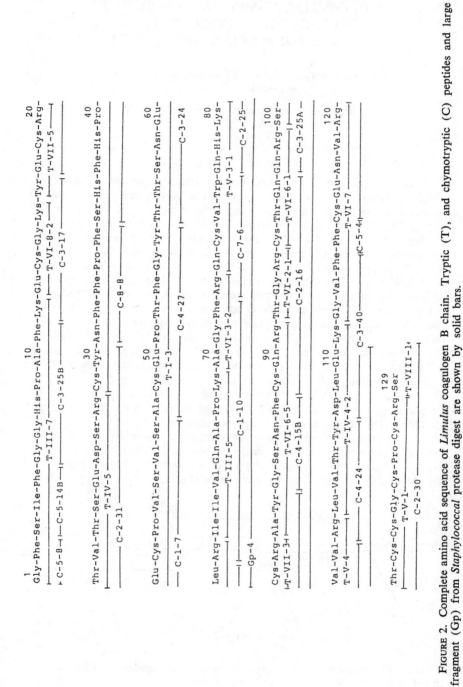

FIGURE 2. Complete amino acid sequence of *Limulus* coagulogen B chain. Tryptic (T), and chymotryptic (C) peptides and large fragment (Gp) from *Staphylococcal* protease digest are shown by solid bars.

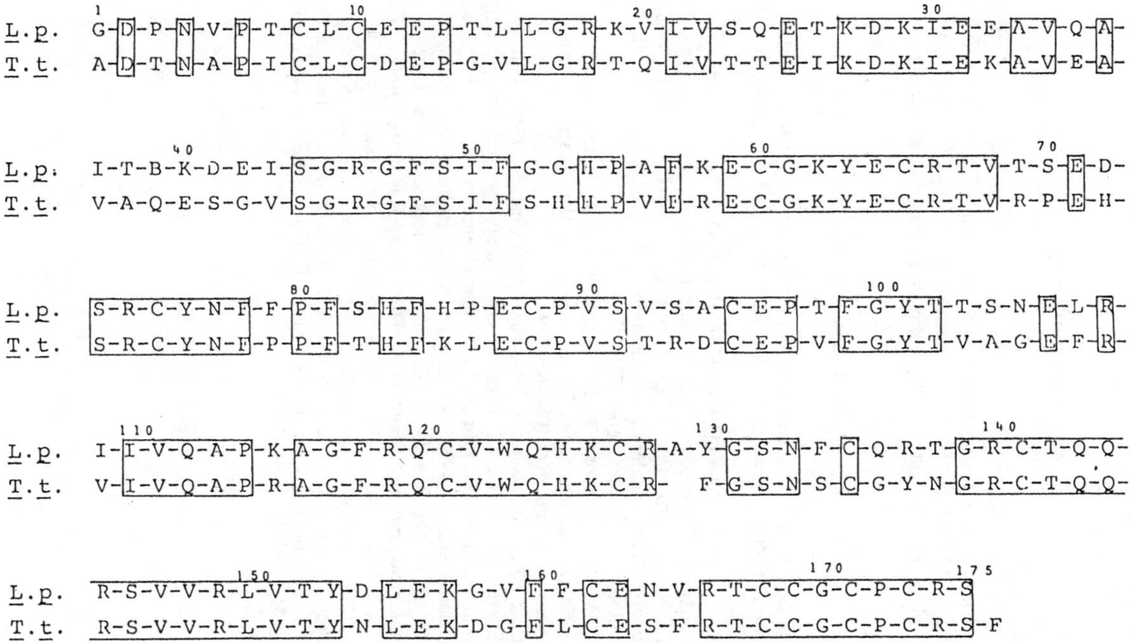

```
    1                                                          30
L.p. G-D-P-N-V-P-T-C-L-C-E-E-P-T-L-L-G-R-K-V-I-V-S-Q-E-T-K-D-K-I-E-E-A-V-Q-A
T.t. A-D-T-N-A-P-I-C-L-C-D-E-P-G-V-L-G-R-T-Q-I-V-T-T-E-I-K-D-K-I-E-K-A-V-E-A

     40                          50                 60                    70
L.p. I-T-B-K-D-E-I-S-G-R-G-F-S-I-F-G-G-H-P-A-F-K-E-C-G-K-Y-E-C-R-T-V-T-S-E-D
T.t. V-A-Q-E-S-G-V-S-G-R-G-F-S-I-F-S-H-H-P-V-E-R-E-C-G-K-Y-E-C-R-T-V-R-P-E-H

     80                 90                100
L.p. S-R-C-Y-N-F-F-P-F-S-H-F-H-P-E-C-P-V-S-V-S-A-C-E-P-T-F-G-Y-T-S-N-E-L-R
T.t. S-R-C-Y-N-F-P-P-F-T-H-F-K-L-E-C-P-V-S-T-R-D-C-E-P-V-F-G-Y-I-V-A-G-E-F-R

     110                120               130                140
L.p. I-T-V-Q-A-P-K-A-G-F-R-Q-C-V-W-Q-H-K-C-R-A-Y-G-S-N-F-C-Q-R-T-C-R-C-T-Q-Q
T.t. V-I-V-Q-A-P-R-A-G-F-R-Q-C-V-W-Q-H-K-C-R-F-G-S-N-S-C-G-Y-N-G-R-C-T-Q-Q

     150                 160                170             175
L.p. R-S-V-V-R-L-V-T-Y-D-L-E-K-G-V-F-F-C-E-N-V-R-T-C-C-G-C-P-C-R-S
T.t. R-S-V-V-R-L-V-T-Y-N-L-E-K-D-G-F-L-C-E-S-F-R-T-C-C-G-C-P-C-R-S-F
```

FIGURE 3. Amino acid sequence homologies between *Limulus* and *Tachypleus*[7] coagulogens. Chemically identical residues in the sequences are framed.

It should be noted that in coagulogen molecule the α-helix region, which is predicted by the method of Chou and Fasman,[8] is found mainly in the peptide C segment and that the β-sheet and reverse turn regions are distributed in the B chain segment. The functional sites which participate in gel formation remain to be defined.

ACKNOWLEDGMENTS

The authors wish to express their thanks to Drs. K. Sekiguchi, F. Shishikura and M. Niwa for their kind assistance in the bleeding of *Limulus polyphemus*. They also thank Misses S. Ezaki-Hashimoto and K. Usui for amino acid analyses and Miss Y. Kitazaki for her expert secretarial assistance.

REFERENCES

1. SOLUM, N. O. 1973. The coagulogen of *Limulus polyphemus* hemocytes. A comparison of the clotted and non-clotted forms of the molecule. Thromb. Res. **2:** 55–70.
2. GAFFIN, S. L. 1976. The clotting of the lysed white cells of *Limulus* induced by endotoxin I. Preparation and characterization of clot-forming proteins.
3. TAI, J. Y., R. C. SEID, JR., R. D. HUHN & T. Y. LIU. 1977. Studies on *Limulus* amebocyte lysate II. Purification of the coagulogen and the mechanism of clotting. J. Biol. Chem. **252:** 4773–4776.
4. MOSESSON, M. W., C. WOLFENSTEIN-TODEL, J. LEVNI & O. BERTRAND. 1979. Characterization of amebocyte coagulogen from the horseshoe crab (*Limulus polyphemus*). Thromb. Res. **14:** 765–779.
5. NAKAMURA, S., F. SHISHIKURA, S. IWANAGA, T. TAKAGI, K. TAKAHASHI, M. NIWA & K. SEKIGUCHI. 1980. Horseshoe crab coagulogens: Their structure and gelation mechanism. *In* Frontiers in Protein Chemistry. T. Y. Liu, G. Mamiya & K. Yasunobu, Eds.: 495–514. Elsevier North Holland, Inc. New York.
6. MIYATA, T., M. UMEZU & S. IWANAGA. 1981. The amino acid sequence studies on *Limulus polyphemus* coagulogen. Thrombos. Haemostas. **46**(1): 47.
7. TAKAGI, T., Y. HOKAMA, T. MORITA, S. IWANAGA, S. NAKAMURA & M. NIWA. 1979. Amino acid sequence studies on horseshoe crab (*Tachypleus tridentatus*) coagulogen and the mechanism of gel formation. *In* Biomedical Applications of the Horseshoe Crab (Limulidae). E. Cohen, Ed.: 169–184. Alan R. Liss, Inc. New York.

CARBOHYDRATE COMPOSITION OF NORMAL FIBRINOGEN COMPARED TO THE ABNORMAL FIBRINOGEN OF LIVER DISEASE *

J. Martinez, P. M. Keane, and P. B. Gilman

Cardeza Foundation for Hematologic Research
Thomas Jefferson University Hospital
Philadelphia, Pennsylvania 19107

The dysfibrinogenemia of liver disease is functionally characterized by impaired fibrin monomer polymerization, which is reflected in the prolongation of its thrombin time.[1-3] We have previously demonstrated that this abnormal fibrinogen has an increased sialic acid content, and that the increased sialic acid correlates with the prolongation of the thrombin time.[4] The prolongation of the thrombin time normalized after enzymatic removal of the sialic acid, suggesting that the excess of sialic acid was responsible for the functional defect.[4]

To further characterize the carbohdrate abnormality, we measured the galactose content of fibrinogen from both normal persons and those with liver disease. Galactose content was quantitated by the galactose oxidase method.[5] Results are shown in TABLE 1. All abnormal fibrinogens tested showed an increase of galactose content that was similar to the increase of sialic acid.

The subsequent sugar, *N*-acetylglucosamine, was measured in normal and abnormal fibrinogen after acid hydrolysis. The glucosamine was separated on a Dowex column and measured colorimetrically.[6] Results are shown in TABLE 2.

In both normal and abnormal fibrinogens, mannose was quantitated after acid hydrolysis. Free mannose was separated on a Dowex column and measured

TABLE 1

SIALIC ACID-GALACTOSE CONTENT OF FIBRINOGEN

	Sialic Acid *	Galactose *	Sialic Acid/Galactose Ratio
Normal †	6.0 ± 0.3	10.6 ± 0.5	0.57
Patient 1	7.6	14.9	0.51
Patient 2	7.4	13.3	0.55
Patient 3	7.5	13.5	0.55
Patient 4	9.6	16.9	0.58
Patient 5	8.0	15.5	0.51

* Residues per molecule of fibrinogen.
† Mean ± 1 SD of 5 normal fibrinogens.

* This work was supported in part by National Institutes of Health Research Grant HL–20092.

TABLE 2

SIALIC ACID—N-ACETYLGLUCOSAMINE CONTENT OF FIBRINOGEN

	Sialic Acid *	N-Acetylglucosamine *	Sialic Acid/N-Acetylglucosamine Ratio
Normal ‡	6.0 ± 0.3	18.3 ± 1.3	0.33
Patient 1	8.0	26.0	0.31
Patient 2	7.6	23.6	0.32
Patient 3	8.1	27.6	0.29
Patient 4	7.4	21.2	0.35

* Residues per molecule of fibrinogen.
‡ Mean ± 1 SD of 5 normal fibrinogens.

enzymatically.[7] Normal and abnormal fibrinogens had almost identical contents of mannose, as seen in TABLE 3.

The affinity of glycopeptides from normal and abnormal fibrinogen for Concanavalin-A was studied after Pronase digestion. Sialic acid was labeled with B³H by the method of Van Lenten and Ashwell.[8] The labeled fibrinogens were separately digested with Pronase and the glycopeptide fraction isolated in a Biogel P-6 column. The labeled glycopeptides were then chromatographed on a Con-A column.[9] Over 90% of the glycopeptide derived from normal fibrinogen were bound to the column and were eluted with 0.01M α-methyl-D-glucoside. In contrast, approximately 27% of the glycopeptide derived from the abnormal fibrinogen did not bind to the Con-A column.

The sugar composition of the abnormal fibrinogen when compared with that of the normal protein reveals an increase of sialic acid, galactose and N-acetylglucosamine. In contrast, both proteins contain similar amounts of mannose.

Chromatography of the glycopeptides on a Con-A Sepharose column shows decrease binding of the glycopeptides derived from fibrinogen of patients with liver disease, suggesting increased substitution of the α-mannose.

These studies—sugar composition and glycopeptide affinity for Con-A—suggest that in liver disease, while assembly of the fibrinogen core oligosaccharides is normal, there is an increased branching of the peripheral sugars distal to mannose.

TABLE 3

SIALIC ACID-MANNOSE CONTENT OF FIBRINOGEN

	Sialic Acid *	Mannose *	Sialic Acid/Mannose Ratio
Normal	6 ± 0.5	12.5 (12.0–12.9)	0.48
Patient 1	7.9	12.7	0.62
Patient 2	8.0	12.8	0.62
Patient 3	9.8	13.3	0.73

* Residues per molecule of fibrinogen.

REFERENCES

1. PALASCAK, J. E. & J. MARTINEZ. 1977. Dysfibrinogenemia associated with liver disease. J. Clin. Invest. **60:** 89–95.
2. GREEN, G., J. M. THOMSON, I. W. DYMOCK & L. POLLER. 1976. Abnormal fibrin polymerization in liver disease. Br. J. Haematol. **34:** 427–439.
3. GRALNICK, H. R., H. GIVELBER & E. ABRAMS. 1978. Dysfibrinogenemia associated with hepatoma. Increased carbohydrate content of the fibrinogen molecule. N. Engl. J. Med. **299:** 221–225.
4. MARTINEZ, J., J. E. PALASCAK & D. KWASNIAK. 1978. Abnormal sialic acid content of the dysfibrinogenemia associated with liver disease. J. Clin. Invest. **61:** 535–538.
5. FORD, J. D. & J. C. HAWORTH. 1964. The estimation of galactose in plasma using galactose oxidase. Clin. Chem. **10:** 1002–1006.
6. ELSON, L. A. & W. T. J. MORGAN. 1933. CCXLVIII. A colorimetric method for the determination of glucosamine and chondrosamine. Biochem. J. **27:** 1824–1828.
7. FINCH, P. R., R. YUEN, H. SCHACHTER & M. A. MOSCARELLO. 1969. Enzymic methods for the micro assay of D-mannose, D-glucose, D-galactose, and L-fucose from acid hydrolyzates of glycoproteins. Anal. Biochem. **31:** 296–305.
8. VAN LENTEN, L. & G. ASHWELL. 1971. Studies on the chemical and enzymatic modification of glycoproteins. J. Biol. Chem. **246:** 1889–1894.
9. KRUSIUS, T. & E. RUOSLAHTI. 1982. Carbohydrate structure of the Concanavalin-A molecular variants of α-fetoprotein. J. Biol. Chem. **257:** 3453–3457.

HIGH VOLTAGE ELECTRON MICROSCOPY
OF FIBRIN FILM

Michael F. Müller,* Hans Ris,† and John D. Ferry*

*Departments of Chemistry * and Zoology †
University of Wisconsin
Madison, Wisconsin 53706*

The initial work on fibrin films [1] goes back to 1947, when it was learned that fibrin films could be formed by gentle pressing of coarse fibrin clots. These films, in which the fibrous elements are largely oriented in the plane of the film, are rubbery and can be stretched more than 100% with substantial recovery after release of the stress. The molecular origin of extensibility and recovery of fibrin has remained in doubt, although it is unlikely that rubberlike elasticity could be involved with structures that are so massive on a molecular scale. The usual coarse fibrin films (obtained from clots formed at $pH = 6.3$, $\mu = 0.15$) with and without ligation have been extensively studied during recent years.[2, 3] Measurements of x-ray scattering, birefringence, and stress relaxation on stretched films have led to some conclusions about the molecular mechanism of deformation. Very recently, we also succeeded in preparing fine films by unilateral shrinkage of fine fibrin clots (at $pH = 8.5$, $\mu = 0.45$); these have very different properties. For the present study, we adopted the method of high voltage electron microscopy (HVEM) to compare the structures of (1) fine and coarse films, (2) ligated and unligated films, (3) unstretched films and those stretched uniaxially 45%.

Experimental Procedures

Strips of fibrin film (unstretched and stretched 45%) were fixed with glutaraldehyde and osmium tetroxide, stained with uranyl acetate and lead citrate, dehydrated, and imbedded in epon-araldite. Sections 2500 Å thick were prepared both parallel and perpendicular to the fibrin plane. Stereoscopic electron micrographs were obtained at the Madison HVEM facility.

Results

Analyses of the micrographs under the different conditions led to the following conclusions.

(1) Films from coarse clots show typical fibrillar structures with a high degree of lateral aggregation.

(2) Films from coarse unligated clots (formed with 1.5 mM ethylenediamine-tetraacetic acid) show significantly thicker bundles than those from coarse ligated clots (formed with 3 mM Ca^{++}).

(3) The fiber thickness in ligated coarse films is ca. 1500 Å and thus much higher than measured with classical techniques by polymerizing fibrinogen on a copper grid (ca. 800 Å).

(4) The fibrin bundles are twisted and a helical structure seems likely.

(5) Fibrin bundles are wrapped around each other and form thick fibrin cables.

(6) Incomplete lateral contact leads to anastomosis (branching) of the fibrin strands.

(7) Films from coarse and fine clots show remarkable differences. In contrast to the coarse structures, fine films show very little lateral aggregation and only thin (twisted) fibrin threads can be seen.

(8) Stretching of the films leads to an extensive and unexpected high degree of orientation.

REFERENCES

1. FERRY, J. D. & P. R. MORRISON. 1947. J. Am. Chem. Soc. **69:** 400.
2. ROSKA, F. R. & J. D. FERRY. 1982. Biopolymers **21:** 1811.
3. ROSKA, F. R., J. D. FERRY, J. S. LIN & J. W. ANDEREGG. 1982. Biopolymers **21:** 1833.

CONTROL OF FIBRINOGEN SYNTHESIS BY GLUCOCORTICOIDS IN THE ACUTE PHASE RESPONSE *

J. Pindyck, G. Beuving,† K. M. Hertzberg,
T. J. Liang, D. Amrani,‡ and G. Grieninger

Lindsley F. Kimball Research Institute
The New York Blood Center
New York, New York 10021

† *Spelderholt State Institute for Poultry Research*
Beekbergen, the Netherlands

The response of an animal to cellular injury and physiological stress (e.g. pregnancy) includes elevation in the circulating levels of glucocorticoid hormones and of a group of plasma proteins known as acute phase reactants (for review see ref. 1 and 2). The relationship between the rise in glucocorticoids and the increase in synthesis of acute phase proteins was explored both *in vivo,* in experimentally stressed chickens, and *in vitro,* in hormone-free hepatocyte cultures [3] prepared from chick embryos.

We have demonstrated that development of an inflammatory response in the chicken with turpentine (0.5 ml/kg, subcutaneously) results, as in mammals, in increased levels of fibrinogen. In addition, the α_1-globulin "M," which has been shown in our laboratory to be an adult-type protein, i.e., inducible in embryonic liver cell culture, is a member of this category of reactants. A twofold increase in fibrinogen levels, and an almost fivefold increase in "M" protein levels within three days of injury identify these as major acute phase protein in the chicken. Other proteins show more modest increases, whereas albumin levels fall precipitously during the same time period, rising to normal levels as the other protein levels return to base line by the tenth day.

The levels of corticosterone (the natural glucocorticoid hormone in the chicken) have been monitored, using a competitive protein-binding assay,[4] and show a notable correlation with the rise in the plasma levels of fibrinogen and the "M" protein. Highest hormone levels occur approximately 24 hours after injury, just when the plasma levels of these two proteins are increasing most rapidly.

Examination of the effect of dexamethasone on embryonic chicken hepatocytes cultured in hormone-free medium has revealed that fibrinogen and α_1-globulin "M" are extremely responsive to this hormone, each showing a two-to-threefold increase in synthesis. The corticosteroid effect, which was selective with respect to stimulating only a small percent of the total proteins, did not include the suppression of albumin synthesis observed in the injured animal; albumin synthesis was not affected.

* Supported by Grants HL–28444 and HL–09011 from the National Institutes of Health. G. G. is an Established Investigator of the American Heart Association
‡ Present Address: Mt. Sinai Medical Center, Milwaukee, Wisconsin 53233.

These findings provide insight into the role played by glucocorticoid hormones in regulating the production of acute phase proteins and indicate that hormone-free cultured cells give an opportunity to dissect the complex events observed during inflammation in the whole animal into its component parts.

REFERENCES

1. KOJ, A. 1974. *In* Structure and Function of Plasma Proteins. A. C. Allison, Ed. 1: 73–131. Plenum Press. New York.
2. KUSHNER, I. 1982. Ann. N.Y. Acad. Sci. 389: 39–48.
3. LIANG, T. J. & G. GRIENINGER. 1981. Proc. Natl. Acad. Sci. USA 78: 6972–6976.
4. BEUVING, G. & G. M. A. VONDER. 1977. J. Reprod. Fert. 51: 169–173.

PREFERENTIAL RESISTANCE OF FIBRINOGEN SYNTHESIS TO INHIBITION OF POLYPEPTIDE INITIATION *

P. Plant, O. Martini, G. Koch,† and G. Grieninger

The Lindsley F. Kimball Research Institute
The New York Blood Center
New York, New York 10021

† Section of Molecular Biology
Institute of Physiological Chemistry
University of Hamburg
Hamburg, Federal Republic of Germany

Regulation of hepatocellular biosynthesis of plasma proteins shows a high degree of selectivity, whether the mechanisms involved operate at a transcriptional or post-transcriptional level.[1] While selective gene transcription may explain the former, the mechanism(s) conferring post-transcriptional selectivity of stimulation are poorly understood. Using chick embryo hepatocyte cultures, we have shown that agents such as serum dramatically stimulate fibrinogen and albumin synthesis while only slightly affecting transferrin synthesis.[2] Preliminary findings in our laboratory indicate that these increases in the production of particular plasma proteins are not the result of increases in their respective mRNA levels. To test whether properties inherent in the specific mRNAs play a role in this control, we compared the efficiencies with which the mRNAs for fibrinogen, albumin, and transferrin are translated. Relative translational efficiencies were determined in intact cells by a procedure established in Koch's laboratory, which is based on the principle that differences in the ability of particular mRNAs to initiate polypeptide synthesis can be revealed by lowering the overall rate of initiation of protein synthesis. This is accomplished by raising the tonicity of the medium of cultured cells: "hypertonic initiation block." [3]

The tonicity of the culture medium of embryonic chick hepatocytes was increased by the addition of 130 mM excess NaCl, inhibiting the initiation of protein synthesis by 85%. Differences between the synthesis of hepatic proteins under the hypertonic and isotonic conditions were examined by SDS-polyacrylamide gel electrophoresis of extracts of cells that had been labeled for 20 minutes with ^{35}S-methionine. It was observed that incorporation of label into albumin constituted a larger percentage of total incorporation into hepatic proteins under hypertonic conditions than under isotonic conditions; the opposite was found for incorporation into transferrin. Synthesis of fibrinogen as well as albumin and transferrin were quantified by specific immunoprecipitation, separation on SDS-polyacrylamide gels and measurement of radioactivity in the respective polypeptides. It was found that the synthesis of fibrinogen and albumin was five-times more resistant to inhibition of polypeptide initiation

* Supported by Grants HL–28444 and HL–09011 from the National Institutes of Health. P. P. was supported by an Institutional Grant NRSA 5 T32 HL–07331 from the National Institutes of Health. G.G. is an Established Investigator of the American Heart Association.

than the synthesis of transferrin. Taken together, these findings indicate that the mRNAs for fibrinogen and albumin possess higher initiation efficiencies than transferrin mRNA or the mRNAs for most other hepatic proteins. Further support for this conclusion comes from cell-free translation experiments. Poly(A)-containing RNA was prepared from hepatocyte monolayers and translated in wheat germ lysates.[4] Fibrinogen, albumin and transferrin were immunoprecipitated from the translation products and quantitated as above. Translation of fibrinogen and albumin mRNA remained proportional to the amount of added RNA over a considerably wider range than does transferrin mRNA. The differences in translational efficiencies of these mRNAs detected both in culture and in the cell-free system may provide the basis for the differential effects of certain agents on synthesis of these plasma proteins.

REFERENCES

1. LIANG, T. J. & G. GRIENINGER. 1981. Proc. Natl. Acad. Sci. USA **78:** 6972–6976.
2. PLANT, P. W., T. J. LIANG, J. PINDYCK & G. GRIENINGER. 1981. Biochem. Biophys. Acta **655:** 407–412.
3. NUSS, D. L. & G. KOCH. 1976. J. Mol. Biol. **102:** 601–612.
4. PLANT, P., R. DEELEY & G. GRIENINGER. Manuscript submitted.

THROMBOEMBOLIC DISEASE ASSOCIATED WITH DEFECTIVE LYS-PLASMINOGEN BINDING RELATED TO AN ABNORMAL FIBRIN

C. Soria, J. Soria, and J. Caen

Laboratoire d'Hématologie
U 150 INSERM
Hôpital Lariboisière
75010 Paris, France

We have observed a family suffering from severe thromboembolic disease, which affected several family members and led to early death in two cases.

Five members (from three generations) of this family have a dysfibrinogenemia that was characterized by abnormal monomer polymerization. Release of fibrinopeptides A and B by thrombin and clot stabilization induced by activated factor XIII were normal.

We studied the dissolution of patient and control fibrin clots using a post-occlusion euglobulin solution. Significantly less degradation occurred in these patients' fibrin than occurred in control fibrin. It is unlikely that this decreased clot lysis resulted from an abnormality in the plasmin cleavage sites on the fibrinogen molecule, since the patients' fibrinogen incubated with plasmin was degraded similarly to control samples, as demonstrated by PAGE-SDS. However, we observed that Lys-plasminogen bound to the patients' clots in smaller amounts than it bound to clots prepared from healthy donors. This defective binding of Lys-plasminogen may explain the abnormal clot lysis found in this dysfibrinogenemia. It was observed in all patient samples tested, but was not found in a family member who does not suffer from dysfibrinogenemia.

We postulate that this dysfibrinogenemia induces an abnormal clot structure resulting in defective plasminogen binding. The defective clot lysis probably relates directly to the observed decreased binding of Lys-plasminogen to the patients' clot. Thus, we have described a deficiency in thrombolysis that may explain the recurrent thrombotic episodes observed in this family. Until now, defective thrombolysis has only been related to a decrease in fibrinolytic plasminogen activator or to an excess of inhibitor. This case would represent a new approach to understanding thromboembolic disease.

MONOCLONAL ANTIBODIES THAT REACT PREFERENTIALLY WITH FIBRINOGEN DEGRADATION PRODUCTS OR WITH CROSS-LINKED FIBRIN SPLIT PRODUCTS

J. Soria,* C. Soria,† C. Boucheix,‡ M. Mirshahi,* J. Y. Perrot,†
A. Bernadou,* M. Samama,* and C. Rosenfeld ‡

Hotel Dieu
75004 Paris, France

† *Hôpital Lariboisière*
U 150 INSERM
75010 Paris, France

‡ *Hôpital Paul Brousse*
U 253 INSERM
94800 Villejuif, France

In order to prepare antibodies that allow the determination and the differentiation of fibrinogen degradation products and fibrin degradation products, monoclonal antibodies were produced by hybridization between NS_1 myeloma cells and spleen cells from BALB/c mice hyperimmunized with early stage purified fibrin degradation products (FbDP).

We tested the supernatants from hybridoma cultures using two methods:

(1) In the screening test, the ability of the antibody to bind to undegraded fibrinogen, fibrinogen degradation products (isolated fragments D or E), or to fibrin split products immobilized on polystyrene wells was analyzed.

(2) In the competitive binding assay, the ability of fibrinogen, fibrinogen degradation products (fragment D or fragment E) or fibrin degradation products to inhibit the binding of selected monoclonal antibodies to the fragment D or the fragment E polystyrene coated wells was tested.

The antibodies may be classified into two groups:

(1) Antibodies against fibrinogen revealed epitopes localized on fragment D or on fragment E. One of the monoclonal anti-fragment E antibodies was of special interest because it revealed an epitope that was expressed in the same manner in fibrinogen and in fragment E, but which was very poorly expressed in cross-linked fibrin degradation products. We concluded that the fibrinogen-E domain determinant recognized by this monoclonal antibody is sequestered in cross-linked fibrin split products. The presence of this monoclonal antibody was probably due to a small quantity of free fragment E in the preparation of FbDP used for immunization.

(2) Anti-D neo-antibodies revealed epitopes not accessible in native fibrinogen, but which were exposed only on fragment D degradation product or on fibrin split products.

In the competitive inhibition binding assay, one of these anti-D neo-antibodies reacted preferentially with the cross-linked fibrin split products, showing very good discrimination between fragment D and fibrin split products-fragment D. The results indicate a difference in the reactivity of the epitope revealed by this antibody, depending on the change in its accessibility in cross-linked fibrin split products or in isolated fragment D.

The purpose of this study is to develop immunoenzymological assays. These monoclonal antibodies may be used in this field because they do not react with epitopes present in the other plasma proteins, as was shown by using the plasma of an afibrinogenemic patient.

REGULATION OF FIBRINOGEN RECEPTOR ON HUMAN PLATELETS BY CHANGES IN PLATELET CYCLIC AMP LEVELS *

Stanley E. Graber and Jack Hawiger

Departments of Medicine and Pathology
Vanderbilt University
Nashville, Tennessee 37203

and

Veterans Administration Medical Center
Nashville, Tennessee 37203

The membrane receptor for fibrinogen plays an essential role in aggregation of human platelets. In unstimulated platelets, this receptor is not available, but it becomes accessible to fibrinogen when platelets are incubated with aggregating agents such as ADP, epinephrine or thrombin.[1-3] Since PGI_2, a potent activator of platelet adenylate cyclase,[4] prevents exposure of the fibrinogen receptor in stimulated platelets,[1] we investigated the relationship between platelet cAMP levels and fibrinogen receptor status in thrombin treated platelets.

Binding of ^{125}I-fibrinogen to human platelets separated from plasma proteins was done in a steady state system at room temperature without stirring; 10^8 platelets treated with thrombin (0.05 U) followed by hirudin were used. Platelet bound ^{125}I-fibrinogen was separated from free ligand by rapid centrifugation through a mixture of dibutylphthalate and apiezon oil C.[1] Parallel samples were treated with trichloroacetic acid and assayed for cAMP by radioimmunoassay.[5]

A progressive elevation of platelet cAMP levels (1.4 to 16 pmol/sample) in response to increasing doses of PGI_2 (0.625 to 16 nM) corresponded to the degree of inhibition of fibrinogen binding induced by thrombin stimulation. Similar results were obtained with the diterpene forskolin, a recently described unique agonist of adenylate cyclase.[6, 7] This substance (0.1 to 10 μM) also produced a dose dependent increase in cAMP levels (1.94 to 11.3 pmol/sample) and progressive inhibition of fibrinogen binding (15 to 100%). Moreover, with both agonists the inhibition of fibrinogen binding was sustained for up to 2 hours and was accompanied by a large persistent increase in cAMP. Exogenous dibutyryl-cAMP (10^{-6} to 10^{-3} M) also resulted in a dose-related inhibition of fibrinogen binding (20 to 100%). In contrast, when platelets were treated with the adenylate cyclase inhibitor, 9-(tetrahydro-2-furyl) adenine,[8, 9] the effects of PGI_2 on cAMP levels and fibrinogen binding were markedly attenuated, i.e., cAMP levels decreased from 10.4 to 3.09 pmol/sample, while platelet bound fibrinogen increased from 3484 to 45,803 cpm.

These results, when taken together, provide convincing evidence that inhibition of fibrinogen binding to human platelets produced by PGI_2 is linked to its effect on cAMP levels. Furthermore, the observation that elevation of

* Supported by the National Institutes of Health and the Veterans Administration.

platelet cAMP by multiple mechanisms (forskolin treatment, exogenous dibutyryl-cAMP and PGI_2 treatment) always resulted in inhibition of fibrinogen binding suggests that an increase in platelet cAMP from any cause prevents exposure of the fibrinogen receptor. To our knowledge, this is the first clear demonstration that elevation of platelet cAMP levels modulates the activity of a specific receptor on the platelet surface. This effect on the fibrinogen receptor could be mediated by an action of cAMP on the receptor itself or, alternatively, could be a secondary phenomenon related to an action of cAMP on a more basic process such as movement of platelet cytoskeletal proteins. Despite the uncertainty of the mechanism, these experiments clearly demonstrate that cAMP modulates the availability of the fibrinogen receptor and, thus, further characterize the inhibitory action of cAMP on platelet function.

REFERENCES

1. HAWIGER, J., S. PARKINSON & S. TIMMONS. 1980. Nature 283: 5743.
2. MARGUERIE, G. A., E. F. PLOW & T. S. EDGINGTON. 1979. J. Biol. Chem. 254: 5357.
3. BENNETT, J. S. & G. VILAIRE. 1979. J. Clin. Invest. 64: 1393.
4. GORMAN, R. R., S. BUNTING & O. V. MILLER. 1977. Prostaglandins 13: 377.
5. BROOKER, G., J. F. HARPER, W. L. TERASAKI & R. D. MOYLAN. 1979. Adv. Cyclic Nucl. Res. 10: 1.
6. SEAMON, K. B., W. PADGETT & J. W. DALY. 1981. Proc. Natl. Acad. Sci. USA 78: 3363.
7. SEAMON, K. B. & J. W. DALY. 1981. J. Cyclic Nucl. Res. 7: 201.
8. WEINRYB, I. & M. MICHEL. 1974. Biochim. Biophys. Acta 334: 218.
9. HASLAM, R. J., M. M. L. DAVIDSON, J. E. B. FOX & J. A. LYNHAM. 1978. Thrombos. Haemostas. 40: 232.

EFFECTS OF EDTA, GLY-PRO-ARG-PRO, AND AMINO ACID REPLACEMENT ON THE THROMBIN-CATALYZED RELEASE OF FIBRINOPEPTIDES

S. D. Lewis, D. L. Higgins, and J. A. Shafer

Department of Biological Chemistry
The University of Michigan Medical School
Ann Arbor, Michigan 48109

Thrombin catalyzes release of fibrinopeptide A (FPA) and fibrinopeptide B (FPB) from the $A\alpha$- and $B\beta$-chains of fibrinogen in the conversion of fibrinogen to fibrin. It is generally believed that release of FPA occurs prior to the release of FPB. It has been proposed, however, that release of FPB can occur prior to release of FPA.[1] Evidence is presented here indicating that normally very little (<5%) of the FPB is released before FPA. Also steady state kinetic parameters for the release of FPA and FPB were evaluated using high perform- ance liquid chromatography to determine FPA and FPB.[2]

Our observations of the thrombin catalyzed release of fibrinopeptides from fibrinogen from patients with fibrinogen-Petoskey support the view that FPB release occurs primarily after release of FPA. Fibrinogen-Petoskey contains equal amounts of normal $A\alpha$-chains and abnormal $A\alpha$-chains, with a His replacement for Arg-$A\alpha$16.[2] The presence of a histidyl residue instead of an arginyl residue at the scissile bond results in a dramatic decrease in the rate (150-fold at pH 7.4) of thrombin catalyzed release of FPA from the abnormal $A\alpha$-chains. The FPA released from the abnormal chains has a COOH-terminal His and can be distinguished chromatographically from normal FPA. The retarded release of FPA-Petoskey caused by the Arg $\rightarrow$ His replacement in 50% of the $A\alpha$-chains was accompanied by a delayed release of 50% of the FPB, suggesting that release of FPB prior to release of FPA must be a very slow process. If it were not, the delayed release of FPB should not have cor- responded to the delayed release of FPA-Petoskey.

In addition to the release of FPA, the association of desA-fibrinogen mole- cules may be required to realize substantial rates of release of FPB, since EDTA and Gly-Pro-Arg-Pro, inhibitors of fibrin polymerization, inhibited the release of FPB, but did not inhibit the release of FPA.

The steady state kinetic parameters k_{cat} and K_m were determined from fits of the kinetic data to

$$[A\alpha]e/V = [A\alpha]/k_{cat} + K_m/k_{cat} \qquad (1)$$

where V is the velocity of FPA release, e is the thrombin concentration and $[A\alpha]$ is the concentration of $A\alpha$-chains. For the release of FPA at pH 7.4, 37° C, $k_{cat} = 84(\pm4)$ s^{-1}, $K_m = 7.2(\pm0.9)$ μM.

Methods were developed for direct determination of the specificity constant k_{cat}/K_m for the thrombin catalyzed release of FPA and FPB from fibrinogen.

When $[A\alpha] \ll K_{mA}$ and $[B\beta] \ll K_{mB}$, the Michaelis-Menten equation for the release of FPA becomes

$$-d[A\alpha]/dt = k_{catA}[A\alpha]/K_{mA} \tag{2}$$

Integration yields

$$\ln([A\alpha]/[A\alpha]_o) = -k_{cat}et/K_{mA} \tag{3}$$

The dependence of $\ln([A\alpha]/[A\alpha]_o)$ on time yielded values for k_{cat}/K_{mA} that corresponded to the quotient of the separately determined values of k_{cat} and K_m. Since release of FPB occurs after FPA, the appearance of FPB follows consecutive first order kinetics according to the equation

$$[FPB]/[FPB]_f = 1 + (k_2\exp(-k_1t) - k_1\exp(-k_2t))/(k_1 - k_2) \tag{4}$$

where $k_1 = k_{cat}e/K_{mA}$, $k_2 = k_{catB}e/K_{mB}$, [FPB] is the concentration of released FPB at time t, and $[FPB]_f$ is the final concentration of FPB. Values of $12(\pm 2)$ $\mu M^{-1}s^{-1}$ and $4.2(0.3)$ $\mu M^{-1}s^{-1}$ for k_{cat}/K_{mA} and k_{cat}/K_{mB}, respectively were obtained from fits of the kinetic data to Equations 3 and 4.

REFERENCES

1. MARTINELLI, R. A. & H. A. SCHERAGA. 1980. Biochemistry 19: 2343–2350.
2. HIGGINS, D. L. & J. A. SHAFER. 1981. J. Biol. Chem. 256: 12013–12017.

Index of Contributors

(Italicized page numbers refer to comments made in discussions.)